Asset Pricing and Portfolio Choice Theory

FINANCIAL MANAGEMENT ASSOCIATION

Survey and Synthesis Series

Asset Management: A Systematic Approach to Factor Investing
Andrew Ang

Asset Pricing and Portfolio Choice Theory
Kerry E. Back

Asset Pricing and Portfolio Choice Theory, 2nd Edition
Kerry E. Back

Beyond Greed and Fear: Understanding Behavioral Finance and the Psychology of Investing
Hersh Shefrin

Beyond the Random Walk: A Guide to Stock Market Anomalies and Low-Risk Investing
Vijay Singal

Consumer Credit and the American Economy
Thomas A. Durkin, Gregory Elliehausen, Michael E. Staten, and Todd J. Zywicki

Cracking the Emerging Markets Enigma
G. Andrew Karolyi

Debt Management: A Practitioner's Guide
John D. Finnerty and Douglas R. Emery

Dividend Policy: Its Impact on Firm Value
Ronald C. Lease, Kose John, Avner Kalay, Uri Loewenstein, and Oded H. Sarig

Efficient Asset Management: A Practical Guide to Stock Portfolio Optimization and Asset Allocation, 2nd Edition
Richard O. Michaud and Robert O. Michaud

Exchange-Traded Funds and the New Dynamics of Investing
Ananth N. Madhavan

Last Rights: Liquidating a Company
Dr. Ben S. Branch, Hugh M. Ray, Robin Russell

Managing Pension and Retirement Plans: A Guide for Employers, Administrators, and Other Fiduciaries
August J. Baker, Dennis E. Logue, and Jack S. Rader

Managing Pension Plans: A Comprehensive Guide to Improving Plan Performance
Dennis E. Logue and Jack S. Rader

Mortgage Valuation Models: Embedded Options, Risk, and Uncertainty
Andrew Davidson and Alex Levin

Real Estate Investment Trusts: Structure, Performance, and Investment Opportunities
Su Han Chan, John Erickson, and Ko Wang

Real Options: Managing Strategic Investment in an Uncertain World
Martha Amram and Nalin Kulatilaka

Real Options in Theory and Practice
Graeme Guthrie

Slapped by the Invisible Hand: The Panic of 2007
Gary B. Gorton

Survey Research in Corporate Finance: Bridging the Gap between Theory and Practice
H. Kent Baker, J. Clay Singleton, and E. Theodore Veit

The Financial Crisis of Our Time
Robert W. Kolb

The Search for Value: Measuring the Company's Cost of Capital
Michael C. Ehrhardt

Too Much Is Not Enough: Incentives in Executive Compensation
Robert W. Kolb

Trading and Exchanges: Market Microstructure for Practitioners
Larry Harris

Truth in Lending: Theory, History, and a Way Forward
Thomas A. Durkin and Gregory Elliehausen

Value Based Management with Corporate Social Responsibility, 2nd Edition
John D. Martin, J. William Petty, and James S. Wallace

Valuing the Closely Held Firm
Michael S. Long and Thomas A. Bryant

Working Capital Management
Lorenzo Preve and Virginia Sarria-Allende

Asset Pricing and Portfolio Choice Theory

SECOND EDITION

Kerry E. Back

OXFORD
UNIVERSITY PRESS

Oxford University Press is a department of the University of Oxford. It furthers the University's objective of excellence in research, scholarship, and education by publishing worldwide. Oxford is a registered trade mark of Oxford University Press in the UK and in certain other countries.

Published in the United States of America by Oxford University Press
198 Madison Avenue, New York, NY 10016, United States of America.

© Oxford University Press 2017

All rights reserved. No part of this publication may be reproduced, stored in a retrieval system, or transmitted, in any form or by any means, without the prior permission in writing of Oxford University Press, or as expressly permitted by law, by license or under terms agreed with the appropriate reproduction rights organization. Inquiries concerning reproduction outside the scope of the above should be sent to the Rights Department, Oxford University Press, at the address above

You must not circulate this work in any other form
and you must impose this same condition on any acquirer.

Library of Congress Cataloging-in-Publication Data
Names: Back, K. (Kerry), author.
Title: Asset pricing and portfolio choice theory / Kerry E. Back.
Description: 2nd edition. | Oxford ; New York : Oxford University Press, [2017] |
Series: Financial Management Association survey and synthesis
series | Includes bibliographical references and index.
Identifiers: LCCN 2016007558 (print) | LCCN 2016014435 (ebook) |
ISBN 9780190241148 (alk. paper) | ISBN 9780190241155 (Updf) | ISBN 9780190241162 (Epub)
Subjects: LCSH: Capital assets pricing model. | Portfolio management.
Classification: LCC HG4636 .B33 2016 (print) | LCC HG4636 (ebook) |
DDC 332.63/2042—dc23
LC record available at http://lccn.loc.gov/2016007558

To Diana, my angel.

CONTENTS

Preface to the First Edition xv
Preface to the Second Edition xvi

Asset Pricing and Portfolio Puzzles xvii

PART ONE Single-Period Models

1. Utility and Risk Aversion 3
    1.1. Utility Functions and Risk Aversion 4
    1.2. Certainty Equivalents and Second-Order Risk Aversion 8
    1.3. Linear Risk Tolerance 11
    1.4. Utility and Wealth Moments 16
    1.5. Risk Aversion for Increments to Random Wealth 17
    1.6. Notes and References 19

2. Portfolio Choice 27
    2.1. First-Order Condition 29
    2.2. Single Risky Asset 32
    2.3. Multiple Risky Assets 35
    2.4. CARA-Normal Model 38
    2.5. Mean-Variance Preferences 41
    2.6. Linear Risk Tolerance and Wealth Expansion Paths 43
    2.7. Beginning-of-Period Consumption 47
    2.8. Notes and References 48

3. Stochastic Discount Factors 52
    3.1. Basic Relationships Regarding SDFs 53
    3.2. Arbitrage, the Law of One Price, and Existence of SDFs 56
    3.3. Complete Markets and Uniqueness of the SDF 59
    3.4. Risk-Neutral Probabilities 61
    3.5. Orthogonal Projections of SDFs onto the Asset Span 62
    3.6. Hansen-Jagannathan Bounds 67
    3.7. Hedging and Optimal Portfolios with Quadratic Utility 70

3.8. Hilbert Spaces and Gram-Schmidt Orthogonalization 72
3.9. Notes and References 75

4. Equilibrium and Efficiency 79
   4.1. Pareto Optima 80
   4.2. Competitive Equilibria 83
   4.3. Complete Markets 84
   4.4. Aggregation and Efficiency with Linear Risk Tolerance 86
   4.5. Beginning-of-Period Consumption 93
   4.6. Notes and References 95

5. Mean-Variance Analysis 99
   5.1. Graphical Analysis 100
   5.2. Mean-Variance Frontier of Risky Assets 101
   5.3. Mean-Variance Frontier with a Risk-Free Asset 106
   5.4. Orthogonal Projections and Frontier Returns 111
   5.5. Frontier Returns and Stochastic Discount Factors 117
   5.6. Separating Distributions 118
   5.7. Notes and References 122

6. Factor Models 127
   6.1. Capital Asset Pricing Model 128
   6.2. General Factor Models 135
   6.3. Jensen's Alpha and Performance Evaluation 142
   6.4. Statistical Factors 145
   6.5. Arbitrage Pricing Theory 147
   6.6. Empirical Performance of Popular Models 150
   6.7. Notes and References 155

7. Representative Investors 162
   7.1. Pareto Optimality Implies a Representative Investor 163
   7.2. Linear Risk Tolerance 165
   7.3. Consumption-Based Asset Pricing 167
   7.4. Coskewness-Cokurtosis Pricing Model 171
   7.5. Rubinstein Option Pricing Model 172
   7.6. Notes and References 175

PART TWO Dynamic Models

8. Dynamic Securities Markets 183
   8.1. Portfolio Choice Model 184

CONTENTS ix

    8.2. Stochastic Discount Factor Processes 187
    8.3. Arbitrage and the Law of One Price 192
    8.4. Complete Markets 192
    8.5. Bubbles, Transversality Conditions, and Ponzi Schemes 195
    8.6. Inflation and Foreign Exchange 198
    8.7. Notes and References 198

9. Dynamic Portfolio Choice 202
    9.1. Euler Equation 202
    9.2. Static Approach in Complete Markets 205
    9.3. Orthogonal Projections for Quadratic Utility 206
    9.4. Introduction to Dynamic Programming 208
    9.5. Dynamic Programming for Portfolio Choice 212
    9.6. CRRA Utility with IID Returns 219
    9.7. Notes and References 227

10. Dynamic Asset Pricing 233
    10.1. CAPM, CCAPM, and ICAPM 234
    10.2. Testing Conditional Models 246
    10.3. Competitive Equilibria 247
    10.4. Gordon Model and Representative Investors 249
    10.5. Campbell-Shiller Linearization 251
    10.6. Risk-Neutral Probabilities 254
    10.7. Notes and References 256

11. Explaining Puzzles 260
    11.1. External Habits 260
    11.2. Rare Disasters 266
    11.3. Epstein-Zin-Weil Utility 268
    11.4. Long-Run Risks 276
    11.5. Uninsurable Labor Income Risk 279
    11.6. Notes and References 283

12. Brownian Motion and Stochastic Calculus 289
    12.1. Brownian Motion 290
    12.2. Itô Integral and Itô Processes 292
    12.3. Martingale Representation 298
    12.4. Itô's Formula 299
    12.5. Geometric Brownian Motion 303
    12.6. Covariation of Itô Processes and General Itô's Formula 305

12.7. Conditional Variances and Covariances 308
12.8. Transformations of Models 309
12.9. Notes and References 311

13. Continuous-Time Markets 318
    13.1. Asset Price Dynamics 318
    13.2. Intertemporal Budget Constraint 322
    13.3. Stochastic Discount Factor Processes 323
    13.4. Valuation via SDF Processes 330
    13.5. Complete Markets 333
    13.6. Markovian Model 335
    13.7. Real and Nominal SDFs and Interest Rates 336
    13.8. Notes and References 337

14. Continuous-Time Portfolio Choice and Pricing 342
    14.1. Euler Equation 343
    14.2. Representative Investor Pricing 343
    14.3. Static Approach to Portfolio Choice 344
    14.4. Introduction to Dynamic Programming 349
    14.5. Markovian Portfolio Choice 352
    14.6. CCAPM, ICAPM, and CAPM 357
    14.7. Notes and References 360

15. Continuous-Time Topics 367
    15.1. Fundamental Partial Differential Equation 367
    15.2. Fundamental PDE and Optimal Portfolio 369
    15.3. Risk-Neutral Probabilities 370
    15.4. Jump Risks 374
    15.5. Internal Habits 380
    15.6. Verification Theorem 387
    15.7. Notes and References 390

PART THREE Derivative Securities

16. Option Pricing 401
    16.1. Uses of Options and Put-Call Parity 403
    16.2. "No Arbitrage" Assumptions 406
    16.3. Changing Probabilities 407
    16.4. Black-Scholes Formula 409
    16.5. Fundamental Partial Differential Equation 413

16.6. Delta Hedging and Greeks 415
16.7. American Options and Smooth Pasting 419
16.8. Dividends 423
16.9. Notes and References 424

17. Forwards, Futures, and More Option Pricing 432
    17.1. Forward Measures 432
    17.2. Forwards and Futures 433
    17.3. Margrabe, Black, and Merton Formulas 437
    17.4. Implied and Local Volatilities 443
    17.5. Stochastic Volatility 445
    17.6. Notes and References 449

18. Term Structure Models 458
    18.1. Forward Rates 459
    18.2. Factor Models and the Fundamental PDE 460
    18.3. Affine Models 461
    18.4. Quadratic Models 469
    18.5. Expectations Hypotheses 469
    18.6. Fitting the Yield Curve and HJM Models 474
    18.7. Notes and References 477

19. Perpetual Options and the Leland Model 485
    19.1. Perpetual Options 486
    19.2. More Time-Independent Derivatives 492
    19.3. Perpetual Debt with Endogenous Default 494
    19.4. Optimal Static Capital Structure 498
    19.5. Optimal Dynamic Capital Structure 500
    19.6. Finite Maturity Debt 505
    19.7. Notes and References 509

20. Real Options and q Theory 513
    20.1. An Indivisible Investment Project 515
    20.2. q Theory 518
    20.3. Irreversible Investment as a Series of Real Options 524
    20.4. Dynamic Programming for Irreversible Investment 530
    20.5. Irreversible Investment and Perfect Competition 535
    20.6. Berk-Green-Naik Model 541
    20.7. Notes and References 546

PART FOUR Beliefs, Information, and Preferences

21. Heterogeneous Beliefs 553
    21.1. State-Dependent Utility Formulation 554
    21.2. Aggregation in Single-Period Markets 555
    21.3. Aggregation in Dynamic Markets 558
    21.4. Short Sales Constraints and Overpricing 562
    21.5. Speculative Trade and Bubbles 564
    21.6. Notes and References 565

22. Rational Expectations Equilibria 569
    22.1. No-Trade Theorem 570
    22.2. Normal-Normal Updating 573
    22.3. Fully Revealing Equilibria 577
    22.4. Grossman-Stiglitz Model 578
    22.5. Hellwig Model 583
    22.6. Notes and References 586

23. Learning 591
    23.1. Estimating an Unknown Drift 592
    23.2. Portfolio Choice with an Unknown Expected Return 594
    23.3. More Filtering Theory 597
    23.4. Learning Expected Consumption Growth 603
    23.5. A Regime-Switching Model 605
    23.6. Notes and References 608

24. Information, Strategic Trading, and Liquidity 613
    24.1. Glosten-Milgrom Model 614
    24.2. Kyle Model 616
    24.3. Glosten Model of Limit Order Markets 620
    24.4. Auctions 624
    24.5. Continuous-Time Kyle Model 632
    24.6. Notes and References 642

25. Alternative Preferences 651
    25.1. Experimental Paradoxes 652
    25.2. Betweenness Preferences 658
    25.3. Rank-Dependent Preferences 663
    25.4. First-Order Risk Aversion 665
    25.5. Ambiguity Aversion 666
    25.6. Notes and References 673

Appendices

A. Some Probability and Stochastic Process Theory 679
    A.1. Random Variables 679
    A.2. Probabilities 680
    A.3. Distribution Functions and Densities 681
    A.4. Expectations 681
    A.5. Convergence of Expectations 682
    A.6. Interchange of Differentiation and Expectation 683
    A.7. Random Vectors 684
    A.8. Conditioning 685
    A.9. Independence 686
    A.10. Equivalent Probability Measures 687
    A.11. Filtrations, Martingales, and Stopping Times 688
    A.12. Martingales under Equivalent Measures 688
    A.13. Local Martingales 689
    A.14. The Usual Conditions 690

Bibliography 691
Index 715

PREFACE TO THE FIRST EDITION

This book is intended as a textbook for the introductory finance Ph.D. course in asset pricing theory, or for a two-semester sequence of such courses. It includes the "classical" results for single-period, discrete-time, and continuous-time models, and a small part on other topics. A first-semester course on single-period and discrete-time models could be based on Chapters 1–11 and most of the topics (Chapters 18–22). A second semester introducing continuous-time models could be based on Chapters 12–17, the remainder of the topics chapters, and other readings.

In order to make the book accessible and useful to students having a variety of abilities and interests, I have tried to limit the mathematical sophistication required to read the main text, while including detailed calculations and proofs as appendices to many of the sections. Also, each chapter concludes with a "notes and references" section, and many of these briefly introduce additional concepts and results.

The exercises are an important part of the book. Some introduce topics not covered in the text, some provide results that are needed in later chapters, and some request details of calculations or proofs that were omitted from the text. I have broken complex exercises into multiple parts, in order to provide road maps for the students. A solutions manual is available for adopting instructors.

I have attempted to give credit to original sources in the "notes and references" sections. Rubinstein (2006) was a useful resource in this regard. Certainly, I will have omitted some important references, for which I apologize.

This book has been used in draft form at several universities during the past few years. I thank Shmuel Baruch, David Chapman, Mike Gallmeyer, Philipp Illeditsch, Jessica Wachter, and Guofu Zhou for helpful advice.

Kerry Back
Rice University
August 2009

PREFACE TO THE SECOND EDITION

I am grateful for the warm reception the first edition of this book received, and I am grateful for this opportunity to make some revisions and additions. Each chapter of the book has been extensively revised in this new edition. There has also been some reorganization of sections and chapters. The new material in this edition includes

- Pricing and preferences for skewness and kurtosis
- Jensen's alpha and performance evaluation
- Campbell-Shiller linearization
- Rare disasters
- Long-run risks
- Jump processes
- Leland capital structure model
- q theory
- Learning (filtering)
- Glosten-Milgrom model
- Glosten model of limit-order markets
- Auctions

I am grateful to Lorenzo Garlappi and his students for compiling a detailed errata for the first edition, and I am grateful to Shmuel Baruch and Jonathan Berk for helpful comments on this edition.

Kerry Back
Rice University
November 2015

ASSET PRICING AND PORTFOLIO PUZZLES

A theory in financial economics is a set of hypotheses about how people behave. The implications of the hypotheses are derived by mathematical reasoning and compared to empirical evidence to test the theory. As in other scientific disciplines, financial theories are often found to be in conflict with factual evidence, prompting revisions and elaborations of the theories. These conflicts are called "puzzles" or "anomalies." Usually, there are many revisions proposed in response to any given puzzle. The true resolution of the puzzle may be an amalgam of various revisions or it may be some factor not yet appreciated. Sorting out these issues leads to active debates within the finance literature. These unresolved issues and the debates they spawn are what make our profession lively, though they may sometimes cause us to be the two-handed economists that Harry Truman lamented.[1]

Before beginning the study of asset pricing and portfolio choice theory, it is useful to gain some understanding of some of the more important puzzles. Asset pricing puzzles can be broadly classified as cross-sectional or aggregate. Both types are described below. Portfolio choice theory is also confronted with puzzles, some of which are described at the end of this chapter. There are statistical issues concerning some of these empirical "facts," and different results have been obtained in some cases using different methods. Also, particularly with the cross-sectional anomalies, there are concerns with data mining (if 100 possible anomalies are examined, we should expect to find 5 that are significant at the 5% level even if none are true anomalies). Nevertheless, the issues listed below have occupied substantial parts of both the theoretical and the empirical literature on asset pricing and portfolio choice.

---

1. Truman is supposed to have said, "Give me a one-handed economist! All my economists say 'on the one hand ... on the other ...'"

Some of the puzzles listed below are not discussed further in this book. The primary goal of this book is to teach the basic theory with which these empirical findings are in conflict. To also describe in detail the debates about the conflicts would require surveying most of the literature on portfolio choice and asset pricing theory, which is obviously out of the question. However, an attempt is made in this book to describe some of the key approaches to resolving some of the key puzzles.

Cross-Sectional Anomalies

The basic tenet of asset pricing theory is that securities that have higher risks should have higher average returns to compensate investors for bearing their risks. A difference in the expected returns of two securities that have similar risks is an anomaly. If such a difference were to exist, investors should demand more of the security with the high expected return and less of the security with the low expected return, causing the price of the former to rise and the latter to fall until they have the same expected return. When this does not happen, it is a puzzle. Apparently, we are not measuring risks correctly, or investors are not optimizing or are optimizing something other than what we think they should be (a supposition that leads to the area of behavioral finance, which is discussed in only the last chapter of this book).

There are many cross-sectional anomalies regarding stock returns. Harvey, Liu, and Zhu (2014) catalogue over 200 firm-level characteristics that have been shown in academic papers to have correlations with stock returns that are not explained by standard measures of risk. Some of the anomalies pertain to multiple asset classes—stocks, bonds, and currencies. Asness, Moskowitz, and Pedersen (2013) document the pervasiveness of the following effects across asset classes:

- Value: assets that are cheaper—in the sense of having a lower price compared to earnings per share or book value per share or in the sense of having suffered long-term price declines—have higher returns that are not explained by standard measures of risk.
- Momentum: assets that have been rising in price relative to other similar assets continue to outperform in the future in a manner that is not explained by standard measures of risk.

- short-term interest rates are high (Fama and Schwert, 1977),
- the previous three- to five-year return was high (Fama and French, 1988b),
- the dividend-price ratio is low (Fama and French, 1988a),
- market risk is high (Campbell, 1987).

There are also predictability puzzles regarding bonds and currencies called forward premium anomalies or failures of the expectations hypotheses. In a nutshell, long-term bonds do better when the term structure of interest rates is steeply upward sloping, meaning long-term bond prices are low (Fama, 1984b), and investments in foreign currencies do better when foreign interest rates are higher (Fama, 1984a). These are also value effects: bonds do well when they are cheap relative to the coupons they pay (compared to short-term interest rates), and foreign currencies do well when they are cheap relative to the interest rates they pay (compared to domestic interest rates).

Portfolio Puzzles

Investors, especially individual investors, seem to violate some of the recommendations from standard portfolio choice models. Some of the key violations are listed below. Another issue within portfolio choice theory is trying to reconcile it with the advice given by practicing investment advisers ("put 100-minus-your-age percent of your wealth in stocks" is an example of advice that is commonly given).

- home bias: investors invest excessively in assets of their home country and local area (French and Poterba, 1991; Coval and Moskowitz, 1999),
- the disposition effect: investors sell winners too early and losers too late (Shefrin and Statman, 1985),
- underdiversification: investors do not adequately diversify (Barber and Odean, 2000),
- excessive trading: investors trade too much, generating excessive transactions costs (Barber and Odean, 2000).

PART ONE

Single-Period Models

1

Utility and Risk Aversion

CONTENTS

1.1 Utility Functions and Risk Aversion	4
1.2 Certainty Equivalents and Second-Order Risk Aversion	8
1.3 Linear Risk Tolerance	11
1.4 Utility and Wealth Moments	16
1.5 Risk Aversion for Increments to Random Wealth	17
1.6 Notes and References	19

The first part of this book addresses the decision problem of an investor in a single-period framework. We suppose the investor makes certain decisions at the beginning of the period (how much of her wealth to spend and how to invest what she does not spend), and the generally random investment returns and any other income (such as labor income) determine her wealth at the end of the period. The investor has preferences for spending at the beginning of the period and wealth at the end of the period, and these preferences, in conjunction with the available investment opportunities, determine her choices.

Preferences are for real wealth, meaning wealth adjusted for inflation. We can also say that preferences are for consumption, measured in terms of some standard basket of goods. The allocation of resources across goods is a topic in microeconomics, not finance. Following the common practice in finance, we will abstract from that issue and assume there is only a single consumption good. We will typically use it as the numeraire (meaning the unit in which prices are measured, so the price of the consumption good is always 1). However, on occasion, it will be convenient to treat consumption and wealth as if they were denominated in dollars. You are free to substitute any other currency or to translate "dollar" as "unit of the consumption good."

Models with more than one time period are studied later in the book. Introducing multiple periods typically introduces state dependence in the investor's preferences for end-of-period wealth. Specifically, the investor cares about the investment opportunities available at the end of the period as well as her wealth, because it is the combination of wealth and investment opportunities that determines the possibilities for future wealth (and consumption). The correlation of wealth with changes in investment opportunities is usually an important consideration in multiperiod settings; however, in the single-period model it is assumed that preferences depend only on the (marginal) probability distribution of wealth.

It is assumed in most of the book that each investor satisfies certain axioms of rationality, which imply that the investor's choices are those that maximize the expected value of a utility function. Specifically, letting c_0 denote beginning-of-period consumption and $\tilde{c}_1$ denote end-of-period consumption (which equals end-of-period wealth), assume there is a function v such that the investor maximizes the expected value of $v(c_0, \tilde{c}_1)$.[1] In many parts of the book—in particular, in this chapter and the next—the probabilities with respect to which expected values are computed can be subjective probabilities; that is, we do not need to assume the investor knows the true probabilities of the outcomes implied by her choices.

In this chapter and in many other places in the book, we simplify our model and focus on end-of-period wealth. We can do this by assuming c_0 is optimally chosen and by considering the derived utility function $w \mapsto v(c^*, w)$ where c^* denotes the optimal beginning-of-period consumption. Denoting this function of w by $u(w)$ and denoting expectation by E, assume the investor chooses her investments to maximize the expected utility $\mathsf{E}[u(\tilde{w})]$.

1.1 UTILITY FUNCTIONS AND RISK AVERSION

A utility function u is said to represent preferences over wealth gambles if $\mathsf{E}[u(\tilde{w}_1)] \geq \mathsf{E}[u(\tilde{w}_2)]$ for any random $\tilde{w}_1$ and $\tilde{w}_2$ such that $\tilde{w}_1$ is at least as preferred as $\tilde{w}_2$. When making decisions under certainty, the utility function representing preferences is unique only up to monotone transforms. However, for decisions under uncertainty, utility functions are unique up to monotone affine transforms: If two utility functions u and f represent the same preferences over all wealth gambles, then u must be a monotone affine transform of f; that is, there exist a constant a and a constant $b > 0$ such that $u(w) = a + bf(w)$ for every w.

---

1. Throughout Part I of this book, a tilde ˜ is used to denote a random variable.

Concavity and Risk Aversion

An investor is said to be (weakly) risk averse if

$$u(w) \geq E[u(\tilde{w})] \qquad (1.1)$$

for any random $\tilde{w}$ with mean w. An equivalent definition is that a risk-averse investor would prefer to avoid a fair bet, meaning that if $\tilde{\varepsilon}$ is a zero-mean random variable and w is a constant, then

$$u(w) \geq E[u(w+\tilde{\varepsilon})]. \qquad (1.2)$$

These inequalities are known as Jensen's inequality. They are equivalent to concavity of the utility function.[2] Concavity is preserved by monotone affine transforms (though not by general monotone transforms), so, for given preferences, either all utility functions representing the preferences are concave or none are.

Strict concavity is equivalent to strict risk aversion, meaning strict preference for a sure thing over a gamble with the same mean (strict inequality in (1.1) and (1.2), unless $\tilde{w} = w$ or $\tilde{\varepsilon} = 0$ with probability 1). For a differentiable function u, concavity is equivalent to nonincreasing marginal utility ($u'(w_1) \leq u'(w_0)$ if $w_1 > w_0$), and strict concavity is equivalent to decreasing marginal utility ($u'(w_1) < u'(w_0)$ if $w_1 > w_0$). For a twice differentiable function u, concavity is equivalent to $u''(w) \leq 0$ for all w, and strict concavity is implied by $u''(w) < 0$ for all w.

Figures 1.1 and 1.2 illustrate the relationship between concavity and risk aversion. Figure 1.1 considers a simple two-outcome gamble and uses the definition of concavity to deduce that an investor with concave utility would prefer the mean of the gamble. Figure 1.2 sketches the proof of Jensen's inequality, using the fact that a tangent line to the graph of a concave function lies above the graph of the function.

Coefficients of Risk Aversion

The coefficient of absolute risk aversion at a wealth level w is defined as

$$\alpha(w) = -\frac{u''(w)}{u'(w)},$$

where the primes denote derivatives. The second derivative of the utility function measures its concavity; dividing by the first derivative eliminates

---

2. A function u is concave if for any w_1 and w_2 and any $\lambda \in [0,1]$, $u(\lambda w_1 + [1-\lambda]w_2) \geq \lambda u(w_1) + [1-\lambda]u(w_2)$. The function u is strictly concave if the inequality is strict for any $\lambda \in (0,1)$.

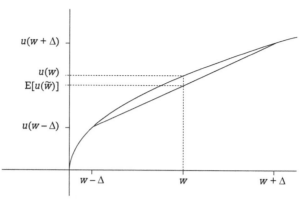

Figure 1.1 Example of concavity and risk aversion. The random variable $\tilde{w}$ takes the values $w - \Delta$ and $w + \Delta$ with equal probabilities. The expected utility of $\tilde{w}$ is the height of the line segment connecting the points $(w - \Delta, u(w - \Delta))$ and $(w + \Delta, u(w + \Delta))$ at its midpoint w. When the utility function is concave, the graph of the function lies above the line segment, so $u(w) > \mathsf{E}[u(\tilde{w})]$.

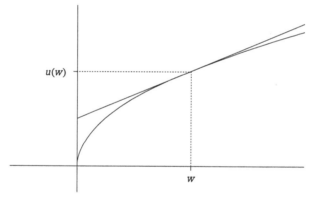

Figure 1.2 Proof of Jensen's inequality. When u is concave, there is a number b such that the line $y = u(x) + b \times (x - w)$ is tangent to the graph of u at the point $(x, y) = (w, u(w))$ and lies above the graph at all other points. Thus, in all states of the world, $u(w) + b \times (\tilde{w} - w) \geq u(\tilde{w})$. Take w as the mean of $\tilde{w}$ and take expectations of both sides to obtain Jensen's inequality: $u(w) \geq \mathsf{E}[u(\tilde{w})]$.

the dependence on the arbitrary scaling of the utility—that is, the coefficient of absolute risk aversion is unaffected by a monotone affine transform of the utility function. Hence, it depends on the preferences, not on the particular utility function chosen to represent the preferences. Note that $\alpha(w) \geq 0$ for any risk-averse investor, because concavity implies $u'' \leq 0$.

Utility and Risk Aversion

Clearly, a high value of α indicates a high curvature of the utility function; moreover, as is explained in the next section, this implies high aversion to risk.

The coefficient of relative risk aversion is defined as

$$\rho(w) = w\alpha(w) = -\frac{wu''(w)}{u'(w)}.$$

The coefficient of risk tolerance is defined as

$$\tau(w) = \frac{1}{\alpha(w)} = -\frac{u'(w)}{u''(w)}.$$

Section 1.2 describes the sense in which $\alpha(w)$ and $\rho(w)$ measure risk aversion. We will also see why $\alpha(w)$ is called the coefficient of *absolute* risk aversion and why $\rho(w)$ is called the coefficient of *relative* risk aversion.

Aggregate Absolute Risk Aversion

Because risk can be shared among investors, the aggregate risk tolerance in the economy is frequently important. If there are H investors with coefficients of absolute risk aversion α_h and coefficients of risk tolerance $\tau_h = 1/\alpha_h$, then the aggregate risk tolerance is defined as $\tau = \sum_{h=1}^{H} \tau_h$. The aggregate absolute risk aversion is defined to be the reciprocal of the aggregate risk tolerance:

$$\alpha = \frac{1}{\sum_{h=1}^{H} 1/\alpha_h}.$$

This is equal to the harmonic mean of the absolute risk aversion coefficients divided by H.[3]

Signs of Higher Derivatives

Section 1.3 explains why it is reasonable to assume investors have decreasing absolute risk aversion (DARA), meaning that $\alpha'(w) < 0$. By differentiating α, it is easily seen that this implies $u''' > 0$. Thus, whenever we need to know the sign of the third derivative of the utility function, it is reasonable to assume that it is positive. This is the same as the marginal utility being a convex function. In combination with DARA, another reasonable condition called decreasing absolute prudence (Section 1.6) implies that the fourth derivative of the utility

---

3. The harmonic mean of numbers $x_1,\ldots,x_n$ is the reciprocal of the average reciprocal: $1/\left[(1/n)\sum_{i=1}^{n} 1/x_i\right]$.

function is negative. Thus, it is reasonable to assume that the first four (and perhaps more) derivatives of the utility function alternate in sign.

1.2 CERTAINTY EQUIVALENTS AND SECOND-ORDER RISK AVERSION

Let w be the mean of a random $\tilde{w}$, so $\tilde{w} = w + \tilde{\varepsilon}$, where $\mathsf{E}[\tilde{\varepsilon}] = 0$. A constant x is said to be the certainty equivalent of $\tilde{w}$ for an individual with utility function u if

$$u(x) = \mathsf{E}[u(w + \tilde{\varepsilon})].$$

A constant π is said to be the risk premium of $\tilde{w}$ if $w - \pi$ is the certainty equivalent, that is,

$$(w - \pi) = \mathsf{E}[u(w + \tilde{\varepsilon})]. \tag{1.3}$$

In other words, starting at wealth w, π is the largest amount the individual would pay to avoid the gamble $\tilde{\varepsilon}$.[4] This concept is illustrated in Figure 1.3.

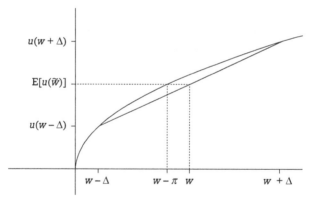

Figure 1.3 Certainty equivalent and risk premium. The random variable $\tilde{w}$ takes the values $w - \Delta$ and $w + \Delta$ with equal probabilities. The expected utility of $\tilde{w}$ is the height of the line segment connecting the points $(w - \Delta, u(w - \Delta))$ and $(w + \Delta, u(w + \Delta))$ at its midpoint w. The certainty equivalent of $\tilde{w}$ is $w - \pi$ shown on the horizontal axis, and the amount a person with utility function u would pay to obtain the mean instead of the gamble is π, the distance between the two vertical dotted lines.

---

4. This use of the word "premium" is from insurance. The term "risk premium" is used in a different way in most of the book, meaning the extra expected return an investor earns from holding a risky asset.

Approximate Risk Premia for Small Gambles

We will show that at the end of the section that, for small gambles,

$$\pi \approx \frac{1}{2}\sigma^2 \alpha(w), \quad (1.4)$$

where σ^2 is the variance of $\tilde{\varepsilon}$. Thus, the amount an investor would pay to avoid the gamble is approximately proportional to her coefficient of absolute risk aversion. The meaning of the approximation is that $\pi/\sigma^2 \to \alpha(w)/2$ if we consider a sequence of gambles with variances $\sigma^2 \to 0$. The approximation is derived at the end of this section.

The distinction between absolute and relative risk aversion can be seen by contrasting (1.4) with the following: Let $\pi = vw$ be the risk premium of $w + w\tilde{\varepsilon}$, where w is a constant and $\tilde{\varepsilon}$ is a zero-mean random variable with variance σ^2. Then,

$$v \approx \frac{1}{2}\sigma^2 \rho(w). \quad (1.5)$$

Thus, the proportion v of initial wealth w that an investor would pay to avoid a gamble equal to the proportion $\tilde{\varepsilon}$ of initial wealth depends on her relative risk aversion and the variance of $\tilde{\varepsilon}$. The result (1.5) follows immediately from (1.4): Let $w\tilde{\varepsilon}$ be the gamble we considered when discussing absolute risk aversion; then, the variance of the gamble is $w^2\sigma^2$; thus,

$$\pi \approx \frac{1}{2}w^2\sigma^2\alpha(w) \quad \Leftrightarrow \quad v \approx \frac{1}{2}w\sigma^2\alpha(w) = \frac{1}{2}\sigma^2\rho(w).$$

To make (1.4) more concrete, consider flipping a coin for \$1. In other words, take $\tilde{\varepsilon} = \pm 1$ with equal probabilities. The standard deviation of this gamble is 1, so the variance is 1 also. Condition (1.4) says that an investor would pay approximately \$1 $\times \alpha(w)/2$ to avoid the gamble. If she would pay 10 cents to avoid it, then $\alpha(w) \approx 0.2$.

To make (1.5) more concrete, let w be your wealth and consider flipping a coin where you win 10% of w if the coin comes up heads and lose 10% of w if it comes up tails. This is a large gamble, so the approximation in (1.5) may not be very good. Nevertheless, it can help us interpret (1.5). The standard deviation of the random variable $\tilde{\varepsilon}$ defined as $\tilde{\varepsilon} = \pm 0.1$ with equal probabilities is 0.1, and its variance is $0.01 = 1\%$. According to (1.5), an investor would pay approximately

$$\frac{1}{2}\rho(w) \times 1\%$$

of her wealth to avoid the gamble. If she would pay exactly 2% of her wealth to avoid this 10% gamble, then (1.5) says that $\rho(w) \approx 4$.

Second-Order Risk Aversion

We interpreted (1.4) as stating that the risk premium is proportional to absolute risk aversion. Obviously, we can also say that the risk premium is proportional to the variance of the gamble. This is called second-order risk aversion. Second-order risk aversion is a consequence of expected utility maximization. There are different theories of preferences, some of which are discussed in Chapter 25, that imply first-order risk aversion, meaning that the risk premium is proportional to the standard deviation of the gamble. First-order risk aversion entails much larger risk premia for small gambles, because the standard deviation divided by the variance goes to infinity as the variance goes to zero (just as $x/x^2 \to \infty$ as $x \to 0$).

To derive (1.4) at any given w, take $\tilde{\varepsilon}$ to be a zero-mean random variable with unit variance and consider the gamble $\sigma\tilde{\varepsilon}$ for a constant $\sigma > 0$. Define the risk premium $\pi(\sigma)$ by

$$u(w - \pi(\sigma)) = \mathsf{E}[u(w + \sigma\tilde{\varepsilon})]. \tag{1.6}$$

Obviously, $\pi(0) = 0$. Thus, a second-order Taylor series expansion of $\pi(\sigma)$ around $\sigma = 0$ gives[5]

$$\pi(\sigma) \approx \pi'(0)\sigma + \frac{1}{2}\pi''(0)\sigma^2. \tag{1.7}$$

We will show that expected utility maximization implies that $\pi'(0) = 0$, so the risk premium is approximately proportional to the variance. Furthermore, we will show that, in the constant of proportionality, $\pi''(0)$ is the coefficient of absolute risk aversion at w.

Differentiate both sides of (1.6) with respect to σ to obtain

$$-u'(w - \pi(\sigma))\pi'(\sigma) = \mathsf{E}[u'(w + \sigma\tilde{\varepsilon})\tilde{\varepsilon}]. \tag{1.8}$$

Evaluating this at $\sigma = 0$ yields

$$-u'(w)\pi'(0) = \mathsf{E}[u'(w)\tilde{\varepsilon}] = u'(w)\mathsf{E}[\tilde{\varepsilon}] = 0,$$

since the mean of $\tilde{\varepsilon}$ is zero by assumption. Thus, $\pi'(0) = 0$ as claimed before. To derive (1.4) from (1.7), it now suffices to show that $\pi''(0)$ is the coefficient of absolute risk aversion at w. To deduce this, differentiate (1.8) again. This yields

$$-u'(w - \pi(\sigma))\pi''(\sigma) + u''(w - \pi(\sigma))[\pi'(\sigma)]^2 = \mathsf{E}[u''(w + \sigma\tilde{\varepsilon})\tilde{\varepsilon}^2]. \tag{1.9}$$

---

5. We simply assume that π is sufficiently differentiable to justify this expansion. We also assume that we can interchange differentiation and expectation in the following argument. Both assumptions can be avoided at the expense of a slightly lengthier argument.

Evaluate (1.9) at $\sigma = 0$, use the fact that $\pi'(0) = 0$, and use the fact that $\tilde{\varepsilon}$ has zero mean and unit variance to obtain

$$-u'(w)\pi''(0) = \mathsf{E}[u''(w)\tilde{\varepsilon}^2] = u''(w).$$

Thus, $\pi''(0) = -u''(w)/u'(w)$.

1.3 LINEAR RISK TOLERANCE

A large amount of financial research is based on a special class of utility functions. This is the class of utility functions having linear risk tolerance (LRT), meaning that the risk tolerance at wealth w is

$$\tau(w) = A + Bw \qquad (1.10)$$

for some constants A and B.[6] The parameter B is called the cautiousness parameter. We also say that these utility functions have hyperbolic absolute risk aversion (HARA), due to the fact that the graph of the function

$$\alpha(w) = \frac{1}{A + Bw}.$$

is a hyperbola. Note that any LRT utility function with a positive cautiousness parameter has increasing risk tolerance and therefore decreasing absolute risk aversion.

This class of functions contains two important subclasses: the class of utility functions having constant absolute risk aversion (CARA) and the class of utility functions having constant relative risk aversion (CRRA)

Constant Absolute Risk Aversion

Constant absolute risk aversion means that absolute risk aversion is the same at every wealth level. Thus, risk tolerance is $\tau(w) = A = 1/\alpha$, where α is absolute risk aversion. It is left as an exercise (Exercise 1.12) to demonstrate that every CARA utility function is a monotone affine transform of the utility function

$$u(w) = -e^{-\alpha w},$$

where α is a constant and equal to the absolute risk aversion. This is called negative exponential utility.

---

6. Generally, in this book, a distinction is made between linear and affine functions, a linear function being of the form Bw and an affine function including a constant (intercept): $A + Bw$. However, we make an exception in the term "linear risk tolerance," which is firmly entrenched in the literature.

CARA utility is characterized by an absence of wealth effects. This absence applies to the risk premium discussed in the previous section and also to portfolio choice. For the risk premium, note that

$$u(w - \pi) = -e^{-\alpha w} e^{\alpha \pi},$$

and

$$u(w + \tilde{\varepsilon}) = -e^{-\alpha w} e^{-\alpha \tilde{\varepsilon}},$$

so

$$u(w - \pi) = \mathsf{E}[u(w + \tilde{\varepsilon})] \quad \Leftrightarrow \quad e^{\alpha \pi} = \mathsf{E}\left[e^{-\alpha \tilde{\varepsilon}}\right],$$

implying

$$\pi = \frac{1}{\alpha} \log \mathsf{E}\left[e^{-\alpha \tilde{\varepsilon}}\right], \tag{1.11}$$

which is independent of w. Thus, an individual with CARA utility will pay the same to avoid a fair gamble no matter what her initial wealth might be. This seems somewhat unreasonable, as is discussed further below.

If the gamble $\tilde{\varepsilon}$ is normally distributed, then the risk premium (1.11) can be calculated more explicitly. We use the fact, which has many applications in finance, that if $\tilde{x}$ is normally distributed with mean μ and variance σ^2, then[7]

$$\mathsf{E}\left[e^{\tilde{x}}\right] = e^{\mu + \frac{1}{2}\sigma^2}. \tag{1.12}$$

In the case at hand, $\tilde{x} = -\alpha \tilde{\varepsilon}$, which has mean zero and variance $\alpha^2 \sigma^2$. Thus,

$$\mathsf{E}\left[e^{-\alpha \tilde{\varepsilon}}\right] = e^{\frac{1}{2}\alpha^2 \sigma^2},$$

and (1.11) implies

$$\pi = \frac{1}{2}\alpha \sigma^2. \tag{1.13}$$

This shows that the approximate formula (1.4) is exact when absolute risk aversion is constant and the gamble is normally distributed.

---

7. It is useful to compare (1.12) to Jensen's inequality. Jensen's inequality states that if f is a concave function and $\tilde{x}$ is a random variable (not necessarily normal) with mean μ, then $\mathsf{E}[f(\tilde{x})] \leq f(\mu)$. On the other hand, if f is convex (meaning that $-f$ is concave), then we have the opposite inequality: $\mathsf{E}[f(\tilde{x})] \geq f(\mu)$. The exponential function is convex, so Jensen's inequality tells us that

$$\mathsf{E}\left[e^{\tilde{x}}\right] \geq e^{\mu}.$$

So, we can ask, By how much must we scale up the right-hand side to make it equal the left-hand side? The formula (1.12) says that we must multiply the right-hand side by the number $e^{\sigma^2/2}$ (which is larger than 1) when $\tilde{x}$ is normally distributed. Formula (1.12) is encountered in statistics as the moment-generating function of the normal distribution.

Consider flipping a fair coin for $1,000. Formula (1.11) says that the amount an individual with CARA utility would pay to avoid the gamble is the same whether she starts with wealth of $1,000 or wealth of $1,000,000,000. One might think that in the latter case the gamble would seem much more trivial, and, since it is a fair gamble, the individual would pay very little to avoid it. On the other hand, she might pay a significant amount to avoid gambling all of her wealth. If so—that is, if an individual would pay less with an initial wealth of $1,000,000,000 than with an initial wealth of $1,000 to avoid a given gamble—then she has DARA utility.

Constant Relative Risk Aversion

Constant relative aversion means that relative risk aversion is the same at all wealth levels. Let ρ denote the relative risk aversion. Then, absolute risk aversion is ρ/w, so risk tolerance is $\tau(w) = Bw = w/\rho$. Note that any CRRA utility function (with positive risk aversion) is a DARA utility function.

Any monotone CRRA utility function is a monotone affine transform of one of the following functions (Exercise 1.12): (i) $u(w) = \log w$, where $\log$ is the natural logarithm, (ii) $u(w)$ equals a positive power, less than one, of w, or (iii) $u(w)$ equals minus a negative power of w. The last two cases (power utility) can be consolidated by writing

$$u(w) = \frac{1}{\gamma}w^\gamma$$

where $\gamma < 1$ and $\gamma \neq 0$. A slightly more convenient formulation, which we will adopt, is to write

$$u(w) = \frac{1}{1-\rho}w^{1-\rho} \qquad (1.14)$$

where $\rho = 1 - \gamma$ is a positive constant different from 1. We can easily check that ρ is the coefficient of relative risk aversion of the utility function (1.14). Logarithmic utility has constant relative risk aversion equal to 1, and an investor with power utility (1.14) is said to be more risk averse than a log-utility investor if $\rho > 1$ and to be less risk averse than a log-utility investor if $\rho < 1$. The three cases $\rho > 1$, $\rho = 1$, and $\rho < 1$ are illustrated in Figure 1.4.

The fraction of wealth an individual with CRRA utility would pay to avoid a gamble that is proportional to initial wealth is independent of the individual's wealth. To see this, let $\tilde{\varepsilon}$ be a zero-mean gamble. An individual will pay πw to avoid the gamble $\tilde{\varepsilon}w$ if

$$u((1-\pi)w) = \mathsf{E}[u((1+\tilde{\varepsilon})w)]. \qquad (1.15)$$

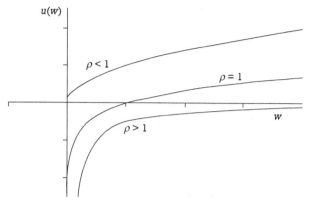

Figure 1.4 Utility functions with constant relative risk aversion. Three CRRA utility functions are shown. The lowest function has relative risk averion $\rho > 1$, the middle function has relative risk aversion $\rho = 1$ (log utility), and the highest function has relative risk aversion $\rho < 1$.

We can confirm (Exercise 1.5) that π is independent of w for CRRA utility by using the facts that $\log(xy) = \log x + \log y$ and $(xy)^\gamma = x^\gamma y^\gamma$.

Logarithmic utility is a limiting case of power utility obtained by taking $\rho \to 1$, in the sense that a monotone affine transform of power utility converges to the natural logarithm function as $\rho \to 1$. Specifically,

$$\frac{1}{1-\rho} w^{1-\rho} - \frac{1}{1-\rho} \to \log w$$

as $\rho \to 1$ for each $w > 0$ (by l'Hôpital's rule).

Shifted CRRA Utility

There are five types of utility functions in the LRT class (Exercise 1.12). We have seen three: negative exponential, logarithmic, and power. The remaining two are

Shifted Logarithmic For some constant ζ and every $w > \zeta$,

$$u(w) = \log(w - \zeta).$$

where $\log$ is the natural logarithm function. The risk tolerance is

$$\tau(w) = w - \zeta.$$

Shifted Power For a constant ζ and a constant ρ with $\rho \neq 0$ and $\rho \neq 1$ and for w such that $(w-\zeta)/\rho > 0$, and including $w = \zeta$ if $\rho < 1$,

$$u(w) = \frac{\rho}{1-\rho}\left(\frac{w-\zeta}{\rho}\right)^{1-\rho}.$$

The risk tolerance is

$$\tau(w) = \frac{w-\zeta}{\rho}.$$

Obviously, the shifted log utility function includes logarithmic utility as a special case ($\zeta = 0$). Also, the shifted power utility function includes power utility as a special case (when $\rho > 0$, the additional factor ρ^ρ in the definition of shifted power utility is irrelevant). For the shifted utility functions with $\rho > 0$, we can interpret the constant ζ as a subsistence level of consumption and interpret the utility as the utility of consumption in excess of the subsistence level. This interpretation probably makes more sense when $\zeta > 0$, but we do not require $\zeta > 0$ to use the utility functions. For the shifted power utility function with $\rho < 0$, ζ is a satiation (bliss) level of wealth: For $w > \zeta$ either the utility function is not defined (in the real numbers) or it is decreasing in wealth, as in the quadratic case discussed below.

There are three different cases for the shifted power utility function, the first two of which parallel the cases for power utility.

(i) $\rho > 1$. The utility is proportional to $-(w-\zeta)^{-\eta}$, where $\eta = \rho - 1 > 0$. It is defined for $w > \zeta$ and is monotone increasing up to zero as $w \to \infty$.
(ii) $0 < \rho < 1$. The utility is proportional to $(w-\zeta)^\eta$, where $\eta = 1 - \rho \in (0,1)$. It is zero at ζ and is monotone increasing up to infinity as $w \to \infty$.
(iii) $\rho < 0$. The utility is proportional to $-(\zeta - w)^\eta$, where $\eta = 1 - \rho > 1$. It is defined for $w \leq \zeta$ and is monotone increasing up to zero as $w \uparrow \zeta$.

Quadratic Utility

A special case of category (iii) of the shifted power utility function is $\rho = -1$, in which case the utility is

$$-\frac{1}{2}(w-\zeta)^2 = -\frac{1}{2}\zeta^2 + \zeta w - \frac{1}{2}w^2.$$

This is the case of quadratic utility, which has a special importance in finance theory, because it implies mean-variance preferences. Specifically, the investor's expected utility is, ignoring the additive constant $-\zeta^2/2$,

$$\zeta \mathsf{E}[\tilde{w}] - \frac{1}{2}\mathsf{E}[\tilde{w}^2] = \zeta \mathsf{E}[\tilde{w}] - \frac{1}{2}\mathsf{E}[\tilde{w}]^2 - \frac{1}{2}\mathrm{var}(\tilde{w}),$$

where $\mathrm{var}(\tilde{w})$ denotes the variance of $\tilde{w}$. Thus, preferences over gambles depend only on their means and variances when an investor has quadratic utility. Quadratic utility is defined for $w > \zeta$, but it is decreasing in wealth for $w > \zeta$.

If the parameter ζ is larger than any wealth level the investor could achieve, then it is not a problem that the quadratic utility is decreasing for wealth above ζ. A greater problem with quadratic utility is that it has increasing absolute risk aversion, even for $w < \zeta$. This property of increasing absolute risk aversion (decreasing risk tolerance) is shared by every shifted power utility function with $\rho < 0$. As will be seen in Chapter 2, this implies that risky assets are inferior goods for an investor with quadratic utility: such an investor would invest less in risky assets if her wealth were higher. This is a very unattractive assumption.

1.4 UTILITY AND WEALTH MOMENTS

We can often represent the expected utility of a gamble as an infinite series in the central moments of the gamble's distribution. Let w denote the mean of a gamble $\tilde{w}$, and suppose the utility function has a convergent Taylor series expansion around w. This means that

$$u(\tilde{w}) = u(w) + u'(w)(\tilde{w}-w) + \frac{1}{2}u''(w)(\tilde{w}-w)^2 + \cdots + \frac{1}{n!}u^{(n)}(w)(\tilde{w}-w)^n + \cdots$$

in each state of the world, where $u^{(n)}$ denotes the nth derivative of the utility function (for convenience, we are using both this notation and primes to denote derivatives here). Assuming we can interchange expectation with the limit in the infinite series, we obtain

$$\mathsf{E}[u(\tilde{w})] = u(w) + \frac{1}{2}u''(w)\mathsf{E}[(\tilde{w}-w)^2] + \sum_{n=3}^{\infty}\frac{1}{n!}u^{(n)}(w)\mathsf{E}[(\tilde{w}-w)^n]. \quad (1.16)$$

Risk aversion implies that $u''(w) < 0$, so, in view of the formula (1.16), it is reasonable to say that a risk-averse investor dislikes variance. As mentioned in Section 1.1, the third derivative of a DARA utility function is positive. Therefore, the coefficient on the third central moment $\mathsf{E}[(\tilde{w}-w)^3]$ is positive in (1.16). The third central moment divided by the standard deviation cubed is called skewness. Therefore, it is reasonable to say that an investor with DARA utility prefers positive skewness to negative skewness. As mentioned in Section 1.1, the combination of DARA utility and decreasing absolute prudence implies that the fourth derivative of the utility function is negative. The fourth central moment divided by the variance squared is called kurtosis (sometimes 3 is subtracted, in which case normal distributions have zero kurtosis). So, it is reasonable to say that decreasing absolute prudence implies a dislike of kurtosis.

It is tempting to truncate the infinite sum in (1.16) and to argue that we can approximate expected utility by a function of a finite number of central moments. However, this is dangerous to do. For most utility functions, and absent any restrictions on the distributions of gambles being considered, any function of a finite number (say n) of central moments will fail to represent preferences, in the sense that there exist two gambles with all of the first n moments indicating one preference, whereas the actual expected utilities imply the opposite preference. See Section 1.6 and Exercise 1.4 for further information on this point. Of course, expected quadratic utility depends only on the first two central moments, as discussed in the previous section, and, if only normal distributions are being considered, then expected utility again depends on only the first two central moments (see the end-of-chapter notes for further discussion of this point).

1.5 RISK AVERSION FOR INCREMENTS TO RANDOM WEALTH

A risk-averse person is a person who prefers w to $\tilde{w}$ whenever w is the mean of $\tilde{w}$. Frequently, we are interested in analyzing an increment to wealth rather than total wealth. It is useful to know when a person who starts with a random wealth $\tilde{x}$ prefers $\tilde{x} + w$ to $\tilde{x} + \tilde{w}$. This is not always true, even assuming risk aversion, because $\tilde{w}$ may hedge some of the risk in $\tilde{x}$; hence, it is possible for $\tilde{x} + \tilde{w}$ to be less risky than $\tilde{x} + w$ and to be preferred by a risk-averse investor to $\tilde{x} + w$.

To rule out the possibility that $\tilde{w}$ hedges $\tilde{x}$, we want to assume $\tilde{w}$ is unrelated to $\tilde{x}$. There are three different ways of formalizing the concept of "unrelated."

(1) Two random variables $\tilde{w}$ and $\tilde{x}$ are said to be independent if

$$\text{prob}(\tilde{x} \leq a \text{ and } \tilde{w} \leq b) = \text{prob}(\tilde{x} \leq a) \times \text{prob}(\tilde{w} \leq b)$$

for all a and b.

(2) $\tilde{w}$ is said to be mean independent of $\tilde{x}$ if

$$\mathsf{E}[\tilde{w} \mid \tilde{x}] = \mathsf{E}[\tilde{w}].$$

(3) $\tilde{w}$ and $\tilde{x}$ are uncorrelated if $\text{cov}(\tilde{x}, \tilde{w}) = 0$.

In (2), the random variable $E[\tilde{w} \mid \tilde{x}]$ is called the conditional expectation of $\tilde{w}$ given $\tilde{x}$. It is defined in Appendix A. Usually, it depends on the realization of $\tilde{x}$ (is a function of $\tilde{x}$) and hence is a random variable. It is the probability-weighted average value of $\tilde{w}$, with the probabilities being conditional on knowledge of $\tilde{x}$. Observing $\tilde{x}$ will generally lead to updating of the probabilities of various events, and this produces the dependence of $E[\tilde{w} \mid \tilde{x}]$ on $\tilde{x}$. However, when $\tilde{w}$ is mean independent of $\tilde{x}$, then, by definition, the conditional expectation $E[\tilde{w} \mid \tilde{x}]$ is the same for all values of $\tilde{x}$ and equal to $E[\tilde{w}]$.

Some important facts about conditional expectations are

(a) The law of iterated expectations states that the expectation of the conditional expectation is just the unconditional expectation; that is,

$$E[E[\tilde{w} \mid \tilde{x}]] = E[\tilde{w}].$$

(b) If $\tilde{y}$ depends only on $\tilde{x}$ in the sense that $\tilde{y} = g(\tilde{x})$ for some function g, then $E[\tilde{y}\tilde{w} \mid \tilde{x}] = \tilde{y}E[\tilde{w} \mid \tilde{x}]$. The interpretation is that if $\tilde{x}$ is known, then $\tilde{y}$ is known, so it is like a constant, pulling out of the expectation.

(c) Jensen's inequality applies to conditional expectations. Recall that Jensen's inequality states that

$$E[u(\tilde{y})] \leq u(E[\tilde{y}])$$

for any concave function u. This generalizes to conditional expectations as

$$E[u(\tilde{y}) \mid \tilde{x}] \leq u(E[\tilde{y} \mid \tilde{x}]).$$

Calling u a utility function, the left-hand side is the conditional expected utility and the right-hand side is the utility of the conditional expectation.

(d) Independence implies mean independence: If $\tilde{w}$ and $\tilde{x}$ are independent, then $E[\tilde{w} \mid \tilde{x}] = E[\tilde{w}]$. The interpretation is that knowing $\tilde{x}$ tells you nothing about the average value of $\tilde{w}$ when $\tilde{x}$ is independent of $\tilde{w}$.

(e) Mean independence implies uncorrelated (Exercise 1.6).

A risk-averse person prefers $\tilde{x} + w$ to $\tilde{x} + \tilde{w}$ whenever $\tilde{w}$ is mean independent of $\tilde{x}$, where, as before, w denotes the mean of $\tilde{w}$. In view of (d) above, this implies that $\tilde{x} + w$ is also preferred to $\tilde{x} + \tilde{w}$ when $\tilde{x}$ and $\tilde{w}$ are independent. To establish the result, suppose that a utility function u is concave and $\tilde{w}$ is mean independent of $\tilde{x}$. Then,

$$E[\tilde{x} + \tilde{w} \mid \tilde{x}] = \tilde{x} + w.$$

By Jensen's inequality for conditional expectations,

$$E[u(\tilde{x} + \tilde{w}) \mid \tilde{x}] \leq u(E[\tilde{x} + \tilde{w} \mid \tilde{x}]) = u(\tilde{x} + w).$$

Taking expectations and applying the law of iterated expectations on the left-hand side yields

$$\mathsf{E}[u(\tilde{x}+\tilde{w})] \leq \mathsf{E}[u(\tilde{x}+w)]. \tag{1.17}$$

Thus, $\tilde{x}+w$ is preferred to $\tilde{x}+\tilde{w}$. Some related results are described in the next section.

There are examples in which $\tilde{x}$ and $\tilde{w}$ are uncorrelated, but a risk-averse investor prefers $\tilde{x}+\tilde{w}$ to $\tilde{x}+w$. Thus, the stronger condition of mean independence is needed to ensure that every risk-averse investor prefers $\tilde{x}+w$ to $\tilde{x}+\tilde{w}$. This is in spite of the fact that $\tilde{x}+w$ mean-variance dominates $\tilde{x}+\tilde{w}$ whenever $\tilde{w}$ and $\tilde{x}$ are uncorrelated—to see the mean-variance dominance, calculate the variance of $\tilde{x}+\tilde{w}$ as

$$\operatorname{var}(\tilde{x}+\tilde{w}) = \operatorname{var}(\tilde{x}) + \operatorname{var}(\tilde{w}) + 2\operatorname{cov}(\tilde{x},\tilde{w})$$
$$= \operatorname{var}(\tilde{x}) + \operatorname{var}(\tilde{w}) > \operatorname{var}(\tilde{x}) = \operatorname{var}(\tilde{x}+w).$$

1.6 NOTES AND REFERENCES

Whether probabilities can ever be regarded as objective is a point of contention. The classic reference on this issue is Savage (1954), who argues for the personalistic (subjective) point of view. This view motivates Bayesian statistics, and the objective view underlies the frequentist approach to statistics.

Axioms of rationality implying expected utility maximization were first presented by von Neumann and Morgenstern (1947), assuming objective probabilities. The formulation of von Neumann and Morgenstern masks a critical axiom, which has come to be known as the independence axiom. In Herstein and Milnor (1953), which is a fairly definitive formulation and extension of the von Neumann–Morgenstern result, this axiom takes the form: If gamble A is preferred to gamble B, and C is any other gamble, then the compound lottery consisting of a one-half chance of A and a one-half chance of C should be preferred the compound lottery consisting of a one-half chance of B and a one-half chance of C. This axiom is consistently violated in some experimental settings, as is discussed in Chapter 25.

Savage (1954) extends the von Neumann-Morgenstern result to the setting of subjective probabilities. Naturally, this also requires a version of the independence axiom (Savage's sure thing principle).

Arrow (1971) argues that the utility function should be bounded on the set of possible outcomes (boundedness follows from his monotone continuity axiom). Note that all of the LRT utility functions (on their maximal domains) are either unbounded above or unbounded below or both. Based in part on the argument that a utility function should be bounded, Arrow (1965) suggests that utility functions should have increasing relative risk aversion. Brocas, Carrillo, Giga, and Zapatero (2015) provide experimental evidence that most people have increasing relative risk aversion. There is a particular subclass of the LRT class that are DARA utility functions with increasing relative risk aversion (Exercise 1.10).

An unbounded utility function is somewhat problematic conceptually. For example, if the utility function is unbounded above, then there exists a gamble (a generalized St. Petersburg paradox—see Exercise 1.9) with infinite expected utility, meaning that it would be preferred to any constant wealth, no matter how large. A common response in support of unbounded utility functions is that an expected utility of $+\infty$ can be achieved only via an unbounded gamble (if u is concave, then $\mathsf{E}[u(\tilde{w})] = \infty$ implies $\mathsf{E}[\tilde{w}] = \infty$) and such gambles are surely not in anyone's choice set.

Bounded utility functions (defined on unbounded domains) are also somewhat paradoxical. If the utility function is bounded above, then there exist w and x such that, for all y, the person prefers w to a gamble in which she obtains $w - x$ and $w + y$ with equal probabilities. Thus, the possibility of losing x is so unattractive that no amount of possible gain can compensate for it. This is explored in the context of constant relative risk aversion $\rho > 1$ in Exercise 1.2(b).

The concepts of absolute and relative risk aversion are due to Arrow (1965) and Pratt (1964, 1976) and are often called the Arrow-Pratt measures of risk aversion. Arrow (1965) relates the risk-aversion coefficients to portfolio choice, as is discussed in the next chapter. Pratt (1964) shows that (up to monotone affine transforms) the only CARA utility function is the negative exponential function and the only CRRA utility functions are the log and power functions. Mossin (1968) appears to be the first to characterize the LRT utility functions.

Pratt (1964) derives the result in Section 1.2 that the risk premium for a small gamble is approximately proportional to the variance of the gamble. This result means that expected utility maximizers are approximately risk neutral with respect to small gambles. Aversion to small and large gambles in the context of CRRA utility is considered in Exercise 1.2. Exercise 1.2(a) is based on an observation made by Kandel and Stambaugh (1991). Roughly speaking, a "reasonable" aversion to small bets seems to imply an "unreasonable" aversion to large bets, when relative risk aversion is constant. Rabin (2000) derives a result of this form that applies to any expected utility preferences.

Pratt (1964) also shows that nonincreasing absolute risk aversion is equivalent to requiring smaller risk premia at higher wealth levels, in the sense that $\pi_1 \leq \pi_0$ whenever π_0 is the premium for a zero-mean gamble at initial wealth w and π_1 is the premium for the same gamble at initial wealth $w+a$ for $a>0$.[8] Dybvig and Lippman (1983) show that this is equivalent to the following: If an individual will accept a gamble (having necessarily a positive expected value if the individual is risk averse) at any wealth level, then she will also accept the gamble at any higher wealth level.

These implications for risk premia or choices at different wealth levels also apply to the risk premia or choices of different individuals: Assuming the absolute risk aversion of person 1 at each wealth level is at least as large as that of person 2 and both start at the same initial wealth, then the risk premium required by person 1 for any zero mean gamble is at least as large as that required by person 2, and if person 1 will accept a particular gamble, then person 2 will also. Pratt (1964) shows that person 1 being more risk averse in the sense of having a (weakly) higher absolute risk aversion at each wealth level is equivalent to the utility function of person 1 being a concave transformation of that of person 2: $u_1(w) = f(u_2(w))$ for a concave function f. In this sense, "more risk averse" is equivalent to the utility function being "more concave."

Ross (1981) defines a stronger concept of nonincreasing risk aversion, involving the premia for gambles when uncertainty is unavoidable. Let $\tilde{w}$ and $\tilde{\varepsilon}$ be any gambles such that $\tilde{\varepsilon}$ has a zero mean and is mean independent of $\tilde{w}$ ($\mathsf{E}[\tilde{\varepsilon} \mid \tilde{w}] = 0$). Let $a > 0$ be a constant. Let π_0 and π_1 be the risk premia for $\tilde{\varepsilon}$ when initial wealth is the random amounts $\tilde{w}$ and $\tilde{w}+a$ respectively, meaning

$$\mathsf{E}[u(\tilde{w}-\pi_0)] = \mathsf{E}[u(\tilde{w}+\tilde{\varepsilon})] \quad \text{and} \quad \mathsf{E}[u(\tilde{w}+a-\pi_1)] = \mathsf{E}[u(\tilde{w}+a+\tilde{\varepsilon})].$$

Then, an individual exhibits nonincreasing risk aversion in Ross' sense if $\pi_1 \leq \pi_0$. Machina (1982) proposes an even stronger concept of nonincreasing risk aversion, requiring $\pi_1 \leq \pi_0$ whenever a is a positive random variable. He shows, surprisingly, that this is inconsistent with expected utility maximization. Epstein (1985) proposes a yet stronger concept, suggesting we should have $\pi_1 \leq \pi_0$ if we replace $\tilde{w}+a$ by any gamble that is weakly preferred to $\tilde{w}$. He shows, under some technical conditions, that this implies mean-variance preferences.

Pratt and Zeckhauser (1987) consider yet another concept that is stronger than nonincreasing risk aversion: They define preferences to exhibit proper risk aversion if, whenever each of two independent gambles is independent of initial wealth and individually undesirable, then the sum of the gambles is undesirable. Assuming expected utility maximization, this means that for any gambles $\tilde{w}$, $\tilde{x}$, and $\tilde{y}$ which are mutually independent, if $\mathsf{E}[u(\tilde{w})] \geq \mathsf{E}[u(\tilde{w}+\tilde{x})]$ and $\mathsf{E}[u(\tilde{w})] \geq \mathsf{E}[u(\tilde{w}+\tilde{y})]$, then $\mathsf{E}[u(\tilde{w})] \geq \mathsf{E}[u(\tilde{w}+\tilde{x}+\tilde{y})]$. The interpretation is that adding

---

8. This is certainly suggested by the characterization of risk premia in Section 1.2; however, the result in Section 1.2 is only an approximate result for small gambles. To go from the local result of Section 1.2 to global results, we have to integrate the risk aversion coefficient, as in Exercise 1.12.

background risk in the form of $\tilde{y}$ cannot make the unattractive risk $\tilde{x}$ attractive. In terms of risk premia, proper risk aversion is equivalent to either of the following: (i) adding background risk $\tilde{y}$ to $\tilde{w}$ (weakly) increases the risk premium of $\tilde{x}$, or (ii) the risk premium of $\tilde{x} + \tilde{y}$ is at least as large as the sum of the separate risk premia of $\tilde{x}$ and $\tilde{y}$. Pratt and Zeckhauser show that CARA and CRRA utilities are proper in this sense.

Kimball (1990) defines $-u'''(w)/u''(w)$ to be the absolute prudence of a utility function u at wealth w and $-wu'''(w)/u''(w)$ to be relative prudence. There are many parallels between prudence and risk aversion (prudence is the risk aversion of the marginal utility function). Kimball relates prudence to precautionary premia, paralleling the relation of risk aversion to risk premia. One of Kimball's results is that decreasing absolute prudence implies that the precautionary premium decreases as initial wealth is increased (see Exercise 2.9 for the definition of "precautionary premium"). Kimball (1993) introduces a strengthening of proper risk aversion (which he calls standard risk aversion) and shows that it is equivalent to the combination of decreasing absolute risk aversion and decreasing absolute prudence. It is easy to see that the combination of decreasing absolute risk aversion and decreasing absolute prudence implies a positive fourth derivative of the utility function. Haas (2007) shows that investors with decreasing absolute prudence dislike kurtosis, defining kurtosis in terms of the weight in the tails of the distribution (rather than in terms of the fourth central moment) and eschewing the Taylor series expansion argument.

The observation that the third derivative of a DARA utility function is positive, implying a preference for positive skewness, is made by Arditti (1967). Brockett and Kahane (1992) show that if the derivatives of the utility function alternate in sign, then, for each finite n and any prespecified ordering of the first n central moments of two random variables $\tilde{w}_1$ and $\tilde{w}_2$, there exist $\tilde{w}_1$ and $\tilde{w}_2$ that satisfy the prespecified ordering and are such that $E[u(\tilde{w}_1)] > E[u(\tilde{w}_2)]$. For example, we can require that $\tilde{w}_2$ have a higher mean, lower variance, higher skewness, and lower kurtosis and still have $\tilde{w}_1$ preferred to $\tilde{w}_2$. Exercise 1.4 presents an example from Brockett and Kahane (1992).

Some pairs of gambles are ranked the same way by all investors with monotone preferences or by all monotone risk-averse investors. Let F denote the cumulative distribution function of a random variable $\tilde{x}$ and G the cumulative distribution function of a random variable $\tilde{y}$. Then, $\tilde{x}$ is said to first-order stochastically dominate $\tilde{y}$ if $F(a) \leq G(a)$ for every constant a. This means that $\tilde{x}$ has more mass in the upper tail than $\tilde{y}$ at whatever level a we choose to define the tail. First-order stochastic dominance is equivalent to $E[u(\tilde{x})] \geq E[u(\tilde{y})]$ for every monotone function u (Quirk and Saposnik, 1962). The random variable $\tilde{x}$ is said to second-order stochastically dominate $\tilde{y}$ if

$$\int_{-\infty}^{b} F(a)\,da \leq \int_{-\infty}^{b} G(a)\,da$$

Utility and Risk Aversion

for each b. This is equivalent to either of the following: (i) $\mathsf{E}[u(\tilde{x})] \geq \mathsf{E}[u(\tilde{y})]$ for every monotone concave function u (Hadar and Russell, 1969), or (ii) the distribution of $\tilde{y}$ equals that of $\tilde{x} + \tilde{z} + \tilde{\varepsilon}$ where $\tilde{z}$ is a nonpositive random variable, and $\mathsf{E}[\tilde{\varepsilon} \mid \tilde{x} + \tilde{z}] = 0$ (Strassen, 1965). That (ii) implies (i) is the subject of Exercise 1.11.

Rothschild and Stiglitz (1970) give related results, establishing the equivalence of the following: (i) $\mathsf{E}[u(\tilde{x})] \geq \mathsf{E}[u(\tilde{y})]$ for every concave—not necessarily monotone—function u, (ii) the distribution of $\tilde{y}$ equals the distribution of adding a mean-preserving spread to $\tilde{x}$, and (iii) the distribution of $\tilde{y}$ equals the distribution of $\tilde{x} + \tilde{\varepsilon}$ where $\mathsf{E}[\tilde{\varepsilon} \mid \tilde{x}] = 0$. That (iii) implies (i) is shown in Section 1.5. For more on these equivalences, see Leshno, Levy, and Spector (1997) and Machina and Pratt (1997).

EXERCISES

1.1. Calculate the risk tolerance of each of the LRT utility functions (negative exponential, log, power, shifted log, and shifted power) to verify the formulas for risk tolerance given in Section 1.3.

1.2. Consider a person with constant relative risk aversion ρ who has wealth w.
(a) Suppose she faces a gamble in which she wins or loses some amount x with equal probabilities. Derive a formula for the amount π that she would pay to avoid the gamble; that is, find π satisfying

$$u(w - \pi) = \frac{1}{2}u(w - x) + \frac{1}{2}u(w + x)$$

when u is log or power utility. Do not use the approximation (1.5).
(b) Suppose instead that she is offered a gamble in which she loses x or wins y with equal probabilities. Find the maximum possible loss x at which she would accept the gamble; that is, find x satisfying

$$u(w) = \frac{1}{2}u(w - x) + \frac{1}{2}u(w + y)$$

when u is log or power utility. Do not use the approximation (1.5).
(c) Suppose the person has wealth of $100,000 and faces a gamble as in Part (a). Use the answer in Part (a) to calculate the amount she would pay to avoid the gamble, for various values of ρ (say, between 0.5 and 40), and for $x = \$100$, $x = \$1,000$, $x = \$10,000$, and $x = \$25,000$. For large gambles, do large values of ρ seem reasonable? What about small gambles?

(d) Suppose $\rho > 1$, and the person is offered a gamble as in Part (b). Show that she will reject the gamble no matter how large y is if

$$\frac{x}{w} \geq 1 - 0.5^{1/(\rho-1)} \quad \Leftrightarrow \quad \rho \geq \frac{\log(0.5) + \log(1 - x/w)}{\log(1 - x/w)}.$$

For example, with wealth of \$100,000, the person would reject a gamble in which she loses \$10,000 or wins 1 trillion dollars with equal probabilities when ρ satisfies this inequality for $x/w = 0.1$. What values of ρ (if any) seem reasonable?

1.3. This exercise is a very simple version of a model of the bid-ask spread presented by Stoll (1978). Consider an individual with constant absolute risk aversion α. Assume $\tilde{w}$ and $\tilde{x}$ are joint normally distributed with means μ_w and μ_x, variances σ_w^2 and σ_x^2, and correlation coefficient ρ.

(a) Compute the maximum amount the individual would pay to obtain $\tilde{w}$ when starting with $\tilde{x}$; that is, compute BID satisfying

$$E[u(\tilde{x})] = E[u(\tilde{x} + \tilde{w} - \text{BID})].$$

(b) Compute the minimum amount the individual would require to accept the payoff $-\tilde{w}$ when starting with $\tilde{x}$; that is, compute ASK satisfying

$$E[u(\tilde{x})] = E[u(\tilde{x} - \tilde{w} + \text{ASK})].$$

1.4. Calculate the mean, variance, and skewness of the following two random variables:

$$\tilde{w}_1 = \begin{cases} 2.45 & \text{with probability } 0.5141, \\ 7.49 & \text{with probability } 0.4859, \end{cases}$$

$$\tilde{w}_2 = \begin{cases} 0 & \text{with probability } 0.12096, \\ 4.947 & \text{with probability } 0.750085, \\ 10 & \text{with probability } 0.128955. \end{cases}$$

You should see that $\tilde{w}_2$ has a higher mean, lower variance, and higher skewness than $\tilde{w}_1$. Show that, nevertheless, $\tilde{w}_1$ is preferred to $\tilde{w}_2$ by a CARA investor with absolute risk aversion equal to 1, by a CRRA investor with relative risk aversion equal to $1/2$, and by an investor with shifted log utility $\log(1 + w)$.

1.5. Consider a person with constant relative risk aversion ρ.

(a) Verify that the fraction of wealth she will pay to avoid a gamble that is proportional to wealth is independent of initial wealth (that is, show that π defined in (1.15) is independent of w for logarithmic and power utility).

(b) Consider a gamble $\tilde{\varepsilon}$. Assume $1+\tilde{\varepsilon}$ is lognormally distributed; specifically, assume $1+\tilde{\varepsilon} = e^{\tilde{z}}$, where $\tilde{z}$ is normally distributed with variance σ^2 and mean $-\sigma^2/2$. By the rule for means of exponentials of normals, $\mathsf{E}[\tilde{\varepsilon}] = 0$. Show that π defined in (1.15) equals

$$1 - e^{-\rho\sigma^2/2}.$$

Note: This is consistent with the approximation (1.5), because a first-order Taylor series expansion of the exponential function e^x around $x = 0$ shows that $e^x \approx 1 + x$ when $|x|$ is small.

1.6. Use the law of iterated expectations to show that if $\mathsf{E}[\tilde{\varepsilon} \mid \tilde{y}] = 0$, then $\mathrm{cov}(\tilde{y}, \tilde{\varepsilon}) = 0$ (thus, mean independence implies uncorrelated).

1.7. Let $\tilde{y} = e^{\tilde{x}}$, where $\tilde{x}$ is normally distributed with mean μ and variance σ^2. Show that

$$\frac{\mathrm{stdev}(\tilde{y})}{\mathsf{E}[\tilde{y}]} = \sqrt{e^{\sigma^2} - 1}.$$

1.8. The notation and concepts in this exercise are from Appendix A. Suppose there are three possible states of the world which are equally likely, so $\Omega = \{\omega_1, \omega_2, \omega_3\}$ with $\mathbb{P}(\{\omega_1\}) = \mathbb{P}(\{\omega_2\}) = \mathbb{P}(\{\omega_3\}) = 1/3$. Let $\mathcal{G}$ be the collection of all subsets of Ω:

$$\mathcal{G} = \{\emptyset, \{\omega_1\}, \{\omega_2\}, \{\omega_3\}, \{\omega_1, \omega_2\}, \{\omega_1, \omega_3\}, \{\omega_2, \omega_3\}, \Omega\}.$$

Let $\tilde{x}$ and $\tilde{y}$ be random variables, and set $a_i = \tilde{x}(\omega_i)$ for $i = 1, 2, 3$. Assume no two of the a_i are the same. Suppose $\tilde{y}(\omega_1) = b_1$ and $\tilde{y}(\omega_2) = \tilde{y}(\omega_3) = b_2 \ne b_1$.

(a) What is $\mathrm{prob}(\tilde{x} = a_j \mid \tilde{y} = b_i)$ for $i = 1, 2$ and $j = 1, 2, 3$?
(b) What is $\mathsf{E}[\tilde{x} \mid \tilde{y} = b_i]$ for $i = 1, 2$?
(c) What is the σ–field generated by $\tilde{y}$?

1.9. Suppose an investor has log utility $u(w) = \log w$ for each $w > 0$.

(a) Construct a gamble $\tilde{w}$ such that $\mathsf{E}[u(\tilde{w})] = \infty$. Verify that $\mathsf{E}[\tilde{w}] = \infty$.
(b) Construct a gamble $\tilde{w}$ such that $\tilde{w} > 0$ in each state of the world and $\mathsf{E}[u(\tilde{w})] = -\infty$.
(c) Given a constant wealth w, construct a gamble $\tilde{\varepsilon}$ with $w + \tilde{\varepsilon} > 0$ in each state of the world, $\mathsf{E}[\tilde{\varepsilon}] = 0$ and $\mathsf{E}[u(w+\tilde{\varepsilon})] = -\infty$.

1.10. Which LRT utility functions are DARA utility functions with increasing relative risk aversion, for some parameter values? Which of these utility functions are monotone increasing and bounded on the domain $w \geq 0$?

1.11. Show that condition (ii) in the discussion of second-order stochastic dominance in the end-of-chapter notes implies condition (i); that is, assume $\tilde{y} = \tilde{x} + \tilde{z} + \tilde{\varepsilon}$ where $\tilde{z}$ is a nonpositive random variable and $\mathsf{E}[\tilde{\varepsilon} \mid \tilde{x} + \tilde{z}] = 0$ and show that $\mathsf{E}[u(\tilde{x})] \geq \mathsf{E}[u(\tilde{y})]$ for every monotone concave function u.

1.12. Show that any monotone LRT utility function is a monotone affine transform of one of the five utility functions: negative exponential, log, power, shifted log, or shifted power. Hint: Consider first the special cases (i) risk tolerance $= A$ and (ii) risk tolerance $= Bw$. In case (i) use the fact that

$$\frac{u''(w)}{u'(w)} = \frac{d\log u'(w)}{dw}$$

and in case (ii) use the fact that

$$\frac{wu''(w)}{u'(w)} = \frac{d\log u'(w)}{d\log w}$$

to derive formulas for $\log u'(w)$ and hence $u'(w)$ and hence $u(w)$. For the case $A \neq 0$ and $B \neq 0$, define

$$v(w) = u\left(\frac{w - A}{B}\right),$$

show that the risk tolerance of v is Bw, apply the results from case (ii) to v, and then derive the form of u.

1.13. Show that risk neutrality $[u(w) = w$ for all $w]$ can be regarded as a limiting case of negative exponential utility as $\alpha \to 0$ by showing that there are monotone affine transforms of negative exponential utility that converges to w as $\alpha \to 0$. Hint: Take an exact first-order Taylor series expansion of negative exponential utility, expanding in α around $\alpha = 0$. Writing the expansion as $c_0 + c_1 \alpha$, show that

$$\frac{-e^{-\alpha w} - c_0}{\alpha} \to w$$

as $\alpha \to 0$.

Portfolio Choice

CONTENTS

2.1 First-Order Condition	29
2.2 Single Risky Asset	32
2.3 Multiple Risky Assets	35
2.4 CARA-Normal Model	38
2.5 Mean-Variance Preferences	41
2.6 Linear Risk Tolerance and Wealth Expansion Paths	43
2.7 Beginning-of-Period Consumption	47
2.8 Notes and References	48

This chapter describes the optimal portfolio of an investor who can invest in a given set of assets at the beginning of a period of time. Her portfolio and the values of shares at the end of the period determine her wealth. She maximizes the expected utility of end-of-period wealth. Let n denote the number of assets. Let $\tilde{x}_i$ denote the payoff (per share) of asset i and $p_i \geq 0$ the price (per share) of asset i, for $i = 1, \ldots, n$. If $p_i > 0$, then the return of asset i is defined as

$$\tilde{R}_i = \frac{\tilde{x}_i}{p_i}.$$

For each unit of the consumption good invested, the investor obtains $\tilde{R}_i$. The term "return" is used in this book more generally for the payoff of a portfolio with a unit price. The rate of return is defined as[1]

$$\tilde{r}_i = \tilde{R}_i - 1 = \frac{\tilde{x}_i - p_i}{p_i}.$$

---

1. What we are calling the return is often called the gross return, and you will often encounter the term "return" being used for what we are calling the rate of return. In this book, gross returns appear more frequently than rates of return, so we use the shorter name "returns" for them.

If there is a risk-free asset, then R_f denotes its return. The risk premium of a risky asset is defined to be $\mathsf{E}[\tilde{R}_i] - R_f$. This extra average return is an investor's compensation for bearing the risk of the asset. Explaining why different assets have different risk premia is the main goal of asset pricing theory.

Except for Section 2.7, this chapter addresses the optimal investment problem, assuming consumption at the beginning of the period is already determined.[2] Let w_0 denote the amount invested at the beginning of the period, and let θ_i denote the number of shares the investor chooses to hold of asset i. The investor may have some possibly random income $\tilde{y}$ at the end of the period which she consumes in addition to the end-of-period portfolio value. There are many possible sources of such income, but the largest source is salary and wages. For simplicity, we will consistently call end-of-period income "labor income." The investor's choice problem is

$$\max_{(\theta_1,\ldots,\theta_n)} \mathsf{E}[u(\tilde{w})] \quad \text{subject to} \quad \sum_{i=1}^n \theta_i p_i = w_0 \quad \text{and} \quad \tilde{w} = \tilde{y} + \sum_{i=1}^n \theta_i \tilde{x}_i. \tag{2.1}$$

In (2.1), we have represented a portfolio in terms of the number of shares held of each asset. Alternatively, we can represent it in terms of the amount $\phi_i = \theta_i p_i$ of the consumption good invested in each asset. Assuming the asset prices are positive, the choice problem is

$$\max_{(\phi_1,\ldots,\phi_n)} \mathsf{E}[u(\tilde{w})] \quad \text{subject to} \quad \sum_{i=1}^n \phi_i = w_0 \quad \text{and} \quad \tilde{w} = \tilde{y} + \sum_{i=1}^n \phi_i \tilde{R}_i. \tag{2.1'}$$

Another way to represent a portfolio that is often convenient is in terms of the fraction $\pi_i = \phi_i / w_0$ of initial wealth invested in each asset. Assuming the asset prices are positive, the choice problem is

$$\max_{(\pi_1,\ldots,\pi_n)} \mathsf{E}[u(\tilde{w})] \quad \text{subject to} \quad \sum_{i=1}^n \pi_i = 1 \quad \text{and} \quad \tilde{w} = \tilde{y} + w_0 \sum_{i=1}^n \pi_i \tilde{R}_i. \tag{2.1''}$$

Note that $\theta_i < 0$ is allowed in problem (2.1)—likewise, $\phi_i < 0$ is allowed in (2.1') and $\pi_i < 0$ allowed in (2.1''). This means that we are assuming investors can borrow assets and sell them, which is called short selling or, more simply, shorting (a positive position in an asset is called a long position). If an asset has a positive payoff and an investor has shorted it, then the term $\theta_i \tilde{x}_i$ contributes negatively to end-of-period wealth. This reflects the fact that the

---

2. We can nevertheless assume, for concreteness, that the consumption good is the numeraire for the beginning-of-period asset prices.

investor must pay the value of the asset to the person from whom it was borrowed (equivalently, the investor must buy the asset at its end-of-period value and return it to the person from whom it was borrowed). Short selling a risk-free asset means borrowing the consumption good. In actual markets, there are generally restrictions on borrowing and short selling; in particular, there are typically margin requirements. These restrictions are ignored in (2.1)–(2.1″). There is an implicit constraint in (2.1)–(2.1″) that $\tilde{w}$ be in the domain of definition of the utility function with probability 1. For example, for log or power utility, we must have $\tilde{w} \geq 0$ (or $\tilde{w} > 0$) with probability 1. This solvency constraint indirectly constrains borrowing and short selling. For other utility functions, for example CARA utility, there is not even a solvency constraint in (2.1)–(2.1″). Solvency constraints, short sales prohibitions, and the like can be added at the expense of introducing Kuhn-Tucker conditions into the first-order conditions. Of course, the theory that we will develop ignoring those constraints also applies when such constraints exist but do not bind at the optimum.

2.1 FIRST-ORDER CONDITION

Most asset pricing theories are based on the first-order condition for the portfolio choice problem (2.1). A rigorous derivation of the first-order condition is given at the end of this section. Glossing over technical issues for the moment, the first-order condition can be derived as follows. Substituting in the second constraint in (2.1), the Lagrangean for the choice problem (2.1) is

$$\mathsf{E}\left[u\left(\tilde{y} + \sum_{i=1}^n \theta_i \tilde{x}_i\right)\right] - \gamma\left(\sum_{i=1}^n \theta_i p_i - w_0\right),$$

where γ is the Lagrange multiplier. The first-order condition is that the partial derivatives of the Lagrangean with respect to the θ_i are zero at the optimum. Assume that differentiation and expectation can be interchanged:

$$\frac{\partial}{\partial \theta_i}\mathsf{E}[u(\cdot)] = \mathsf{E}\left[\frac{\partial}{\partial \theta_i}u(\cdot)\right]. \tag{2.2}$$

Then, at the optimum, we have the first-order conditions

$$(\forall i) \qquad \mathsf{E}\left[u'(\tilde{w})\tilde{x}_i\right] - \gamma p_i = 0, \tag{2.3}$$

where

$$\tilde{w} = \tilde{y} + \sum_{i=1}^n \theta_i \tilde{x}_i$$

is the optimal end-of-period wealth (θ denoting the optimal portfolio).

There are several different ways to write (2.3), each of which is important in asset pricing theory. First, we can rearrange (2.3) as

$$(\forall i) \qquad \mathsf{E}\left[\frac{u'(\tilde{w})}{\gamma}\tilde{x}_i\right] = p_i. \qquad (2.4\mathrm{a})$$

If $p_i \neq 0$, then we can also write (2.4a) in terms of the return of asset i as

$$\mathsf{E}\left[\frac{u'(\tilde{w})}{\gamma}\tilde{R}_i\right] = 1. \qquad (2.4\mathrm{b})$$

If $p_j \neq 0$ also, then (2.4b) implies

$$\mathsf{E}\left[u'(\tilde{w})\left(\tilde{R}_i - \tilde{R}_j\right)\right] = 0. \qquad (2.4\mathrm{c})$$

Finally, if there is a risk-free asset, then we can take $\tilde{R}_j = R_f$ in (2.4c), leading to

$$\mathsf{E}\left[u'(\tilde{w})\left(\tilde{R}_i - R_f\right)\right] = 0. \qquad (2.4\mathrm{d})$$

The random variable $\tilde{R}_i - \tilde{R}_j$ in (2.4c) is the payoff of the portfolio consisting of a unit of the consumption good invested in asset i and an equal short position in asset j. The payoff of a zero-cost portfolio such as this is called an excess return. Moreover, two random variables $\tilde{y}$ and $\tilde{z}$ satisfying $\mathsf{E}[\tilde{y}\tilde{z}] = 0$ are called orthogonal. So, the first-order condition (2.4c) is expressed thus: Marginal utility evaluated at the optimal portfolio is orthogonal to each excess return.

The simple intuition for (2.4c) is that the expectation

$$\mathsf{E}\left[u'(\tilde{w})\left(\tilde{R}_i - \tilde{R}_j\right)\right] \qquad (2.5)$$

is the marginal value of adding the zero-cost portfolio to the optimal portfolio. To see this, consider investing an additional Δ units of the consumption good in asset i and investing the same amount less in (or shorting) asset j. Then, the end-of-period wealth would be $\tilde{w} + \Delta(\tilde{R}_i - \tilde{R}_j)$ instead of being $\tilde{w}$. The change in utility per unit of the consumption good is

$$\frac{\mathsf{E}[u(\tilde{w} + \Delta(\tilde{R}_i - \tilde{R}_j)) - u(\tilde{w})]}{\Delta} \approx \frac{\mathsf{E}[u'(\tilde{w})\Delta(\tilde{R}_i - \tilde{R}_j)]}{\Delta} = \mathsf{E}\left[u'(\tilde{w})\left(\tilde{R}_i - \tilde{R}_j\right)\right],$$

using a first-order Taylor series expansion around $\tilde{w}$. If this were positive, then adding a little of the zero-cost portfolio to the optimal portfolio θ would yield a portfolio even better than the optimal portfolio, which is of course impossible. If the expectation (2.5) were negative, then reducing the holding of the zero-cost portfolio by a little (that is, adding a little of the portfolio with payoff $\tilde{R}_j - \tilde{R}_i$) would lead to an improvement in utility. Because it is impossible to improve upon the optimum, the expectation (2.5) must be zero at the optimal wealth $\tilde{w}$.

The key assumption needed to derive the first-order condition (see (2.6) and (2.7) below) is that it is actually feasible to add to and subtract from the optimal portfolio a little of the zero-cost portfolio with payoff $\tilde{R}_i - \tilde{R}_j$. A simple example in which it is infeasible to do this and the first-order condition (2.4) fails is when there is a risk-free asset, the utility function is only defined for nonnegative wealth (as with CRRA utility), the risky asset returns are normally distributed, there is no labor income $\tilde{y}$, and the expected return of some asset i is different from the risk-free return. In this example, regardless of the expected returns of the risky assets, the only feasible portfolio is to invest all wealth in the risk-free asset. In other words, the constraint that wealth be in the domain of the utility function is binding at the optimum. Investing all wealth in the risk-free asset produces a constant optimal wealth w, and $E[u'(w)(\tilde{R}_i - R_f)] = u'(w)E[\tilde{R}_i - R_f] \neq 0$.

Thus, there are cases in which the first-order condition (2.4) does not hold. However, those cases are generally ignored in this book. Unless otherwise noted, we simply assume that the first-order condition holds.

We will prove (2.4c) when $p_i > 0$ and $p_j > 0$. Let θ denote the optimal portfolio, so

$$\tilde{w} = \tilde{y} + \sum_{i=1}^{n} \theta_i \tilde{x}_i$$

is the optimal wealth. Suppose the utility function is defined for all $w > \underline{w}$, where $\underline{w}$ is some constant, possibly equal to $-\infty$. Assume the utility function is concave and differentiable.

Assume there exists $\epsilon > 0$ such that

$$\tilde{w}(\omega) + \delta(\tilde{R}_i(\omega) - \tilde{R}_j(\omega)) > \underline{w} \qquad (2.6)$$

in all states of the world ω and all δ such that $|\delta| \leq \epsilon$. Assume further that

$$E[u(\tilde{w} + \delta(\tilde{R}_i - \tilde{R}_j))] > -\infty \qquad (2.7)$$

for all δ such that $|\delta| \leq \epsilon$. The optimality of $\tilde{w}$ implies

$$E\left[\frac{u(\tilde{w} + \delta(\tilde{R}_i - \tilde{R}_j)) - u(\tilde{w})}{\delta}\right] \leq 0 \qquad (2.8)$$

for all $\delta > 0$.

We will use the following property of any concave function u: For any $w > \underline{w}$ and any real a,

$$\frac{u(w + \delta a) - u(w)}{\delta} \uparrow \quad \text{as } \delta \downarrow, \qquad (2.9)$$

taking $\delta > 0$ and sufficiently small that $w + \delta a > \underline{w}$. To prove (2.9), consider $0 < \delta_2 < \delta_1$. Define $\lambda = \delta_2/\delta_1$. Apply the definition of concavity in footnote 2 of Chapter 1 with $w_2 = w$ and $w_1 = w + \delta_1 a$, noting that

$$\lambda w_1 + (1 - \lambda) w_2 = w_2 + \lambda(w_1 - w_2) = w + \delta_2 a.$$

This yields

$$u(w+\delta_2 a) \geq \frac{\delta_2}{\delta_1} u(w+\delta_1 a) + \left(1 - \frac{\delta_2}{\delta_1}\right) u(w),$$

so

$$\frac{u(w+\delta_2 a) - u(w)}{\delta_2} \geq \frac{u(w+\delta_1 a) - u(w)}{\delta_1},$$

as claimed.

Apply (2.9) in each state of the world with $w = \tilde{w}(\omega)$ and $a = \tilde{R}_j(\omega) - \tilde{R}_i(\omega)$. This shows that the expression inside the expectation in (2.8) is monotonically increasing as δ decreases. We can write this expression as

$$\frac{u(\tilde{w} + \delta(\tilde{R}_i - \tilde{R}_j)) - u(\tilde{w})}{\delta(\tilde{R}_i - \tilde{R}_j)} (\tilde{R}_i - \tilde{R}_j),$$

showing that it converges to

$$u'(\tilde{w})(\tilde{R}_i - \tilde{R}_j)$$

in each state of the world as $\delta \downarrow 0$. The monotone convergence theorem (Appendix A.5) in conjunction with (2.8) therefore yields

$$\mathsf{E}[u'(\tilde{w})(\tilde{R}_i - \tilde{R}_j)] \leq 0.$$

Repeating the argument with i and j reversed yields

$$\mathsf{E}[u'(\tilde{w})(\tilde{R}_j - \tilde{R}_i)] \leq 0,$$

so (2.4c) holds.

2.2 SINGLE RISKY ASSET

This section studies portfolio choice with a risk-free asset, a single risky asset with return $\tilde{R}$, and no labor income ($\tilde{y} = 0$). Let μ denote the mean and σ^2 the variance of $\tilde{R}$. The investor chooses an amount ϕ to invest in the risky asset, leaving $w_0 - \phi$ to invest in the risk-free asset. This leads to wealth

$$\begin{aligned}\tilde{w} &= \phi \tilde{R} + (w_0 - \phi) R_f \\ &= w_0 R_f + \phi(\tilde{R} - R_f). \end{aligned} \quad (2.10)$$

The first-order condition is

$$\mathsf{E}\left[u'(\tilde{w})(\tilde{R} - R_f)\right] = 0. \quad (2.11)$$

Investment Is Positive If the Risk Premium Is Positive

If the risk premium is nonzero and the investor has strictly monotone utility, then it cannot be optimal for her to invest 100% of her wealth in the risk-free asset. If it were, then $\tilde{w}$ would be nonrandom, which means that $u'(\tilde{w})$ could be taken out of the expectation in (2.11), leading to

$$u'(\tilde{w})\mathsf{E}[\tilde{R} - R_f],$$

which is nonzero by assumption. Therefore, putting 100% of wealth in the risk-free asset contradicts the first-order condition (2.11).

In fact, if $\mu > R_f$, then it is optimal to invest a positive amount in the risky asset, and if $\mu < R_f$, then it is optimal to short the risky asset. We can deduce this from the second-order risk aversion property (1.4) discussed in Chapter 1. The essence of the argument is that the risk premium provided by the market is first order in the trade size whereas the risk premium required is second order, so the premium provided is larger than the premium required whenever the investment is small.

We will only analyze the case $\mu > R_f$, because the other case is symmetric. We will show that

$$\mathsf{E}[u(\tilde{w})] > u(w_0 R_f),$$

where $\tilde{w}$ is the end-of-period wealth given in (2.10) and ϕ is positive and sufficiently small, implying that some investment in the risky asset is better than none. Write the end-of-period wealth (2.10) as

$$\tilde{w} = w_0 R_f + \phi(\mu - R_f) + \phi(\tilde{R} - \mu)$$

$$\stackrel{\text{def}}{=} w + \tilde{\varepsilon},$$

where $w = w_0 R_f + \phi(\mu - R_f)$ and $\tilde{\varepsilon} = \phi(\tilde{R} - \mu)$. From the second-order risk aversion property (1.4), it follows that

$$\mathsf{E}[u(\tilde{w})] = \mathsf{E}[u(w + \tilde{\varepsilon})] = u(w - \pi),$$

where[3]

$$\pi \approx \frac{1}{2}\phi^2 \sigma^2 \alpha(w).$$

For sufficiently small ϕ,

$$\phi(\mu - R_f) > \frac{1}{2}\phi^2 \sigma^2 \alpha(w) \approx \pi,$$

---

3. We are now using σ^2 to denote the variance of $\tilde{R}$, so the variance of $\tilde{\varepsilon}$ is $\phi^2 \sigma^2$.

so $\phi(\mu - R_f) > \pi$, meaning that the risk premium provided by the market is greater than the risk premium required. This implies that

$$w - \pi > w - \phi(\mu - R_f) = w_0 R_f,$$

so

$$\mathsf{E}[u(\tilde{w})] = u(w - \pi) > u(w_0 R_f)$$

for sufficiently small ϕ.

DARA Implies a Risky Asset Is a Normal Good

If an investor has DARA utility (which, as noted before, includes CRRA utility), then the optimal investment in the risky asset is larger when initial wealth is larger. In economics, if consumers purchase more of a good when their incomes rise, the good is called normal. If consumers purchase less when their incomes rise, the good is called inferior. Thus, the risky asset is a normal good for investors with DARA utility. It seems very unlikely that a risky asset would be an inferior good, so this is another justification for assuming DARA utility.

Assume the asset has a positive risk premium, so the optimal investment is positive. As is standard practice in economics, we derive this comparative statics result (the dependence of investment on initial wealth) by differentiating the first-order condition. Here, we differentiate (2.11), with the random wealth $\tilde{w}$ being defined in (2.10) and assuming the optimal investment ϕ is a continuously differentiable function of w_0. Because the first-order condition holds for all w_0, the derivative of (2.11) with respect to w_0 must be zero; thus, using the formula (2.10) to compute the derivative, we have

$$\mathsf{E}\left[(\tilde{R} - R_f)u''(\tilde{w})\left\{R_f + (\tilde{R} - R_f)\frac{d\phi}{dw_0}\right\}\right] = 0.$$

Therefore,

$$R_f \mathsf{E}\left[(\tilde{R} - R_f)u''(\tilde{w})\right] + \mathsf{E}\left[(\tilde{R} - R_f)^2 u''(\tilde{w})\right]\frac{d\phi}{dw_0} = 0,$$

implying

$$\frac{d\phi}{dw_0} = \frac{-R_f \mathsf{E}\left[(\tilde{R} - R_f)u''(\tilde{w})\right]}{\mathsf{E}\left[(\tilde{R} - R_f)^2 u''(\tilde{w})\right]}. \quad (2.12)$$

The denominator in (2.12) is negative, due to risk aversion. Our claim is that the numerator is also negative, leading to $d\phi/dw_0 > 0$. This is obviously equivalent to

$$\mathsf{E}\left[(\tilde{R} - R_f)u''(\tilde{w})\right] > 0, \quad (2.13)$$

which is established below.

It may be surprising that (2.13) can be true. The second derivative is negative due to risk aversion and we are assuming the risk premium $\mathsf{E}[\tilde{R} - R_f]$ is positive, so one might think the expectation in (2.13) should be negative. The expectation can be written as

$$\mathsf{E}[\tilde{R} - R_f]\mathsf{E}[u''(\tilde{w})] + \mathrm{cov}(\tilde{R} - R_f, u''(\tilde{w})).$$

As just explained, the first term is negative. Thus, for (2.13) to be true, the covariance must be positive (and sufficiently large). Keeping in mind that $u'' < 0$, a positive covariance means that large values of $\tilde{R} - R_f$ must correspond to values of $u''(\tilde{w})$ that are small in absolute value, and small or negative values of $\tilde{R} - R_f$ must correspond to values of $u''(\tilde{w})$ that are large in absolute value. In other words, there must be less concavity ($u''(\tilde{w})$ closer to zero) when the return $\tilde{R}$, and hence the wealth $\tilde{w}$, is larger. This is precisely what we are assuming—decreasing absolute risk aversion.

To prove (2.13), define $w_f = w_0 R_f$ (the wealth level when $\tilde{R} = R_f$) and substitute

$$u''(\tilde{w}) = -\alpha(\tilde{w})u'(\tilde{w})$$
$$= -\alpha(w_f)u'(\tilde{w}) + [\alpha(w_f) - \alpha(\tilde{w})]u'(\tilde{w})$$

in the left-hand side of (2.13) to obtain

$$-\alpha(w_f)\mathsf{E}\left[(\tilde{R} - R_f)u'(\tilde{w})\right] + \mathsf{E}\left[[\alpha(w_f) - \alpha(\tilde{w})](\tilde{R} - R_f)u'(\tilde{w})\right].$$

The first term in this expression is zero, due to the first-order condition (2.11). The second term is positive because

$$[\alpha(w_f) - \alpha(\tilde{w})](\tilde{R} - R_f)$$

is everywhere positive, due to $\tilde{w}$ being greater than w_f whenever $\tilde{R} > R_f$ and the assumption that absolute risk aversion is a decreasing function of wealth. Therefore, (2.13) holds.

2.3 MULTIPLE RISKY ASSETS

This section introduces matrix algebra notation for the portfolio choice model and explains the portfolio variance formula and the benefits of diversification. A specific example is solved in the next section.

Matrix Notation

Suppose there is a risk-free asset with return R_f and n risky assets with returns $\tilde{R}_i$. Let $\tilde{\mathbf{R}}$ denote the n–dimensional column vector with $\tilde{R}_i$ as its ith element, let μ denote the vector of expected returns (the n–dimensional column vector with ith element $\mathsf{E}[\tilde{R}_i]$), and let ι denote an n–dimensional column vector of 1's. Let ϕ_f denote the investment in the risk-free asset, let ϕ_i denote the investment in risky asset i, and let ϕ denote the n–dimensional column vector with ϕ_i as its ith element.

The budget constraint of the investor is

$$\phi_f + \sum_{i=1}^n \phi_i = w_0,$$

where w_0 is the given initial wealth. This can also be written as

$$\phi_f = w_0 - \iota'\phi,$$

where $'$ denotes the transpose operator. The end-of-period wealth is

$$\phi_f R_f + \sum_{i=1}^n \phi_i \tilde{R}_i = \phi_f R_f + \phi'\tilde{\mathbf{R}} = w_0 R_f + \phi'(\tilde{\mathbf{R}} - R_f \iota),$$

and the expected end-of-period wealth is

$$w_0 R_f + \phi'(\mu - R_f \iota) = w_0[R_f + \pi'(\mu - R_f \iota)], \tag{2.14}$$

where $\pi_i = \phi_i/w_0$ is the fraction of wealth invested in the ith risky asset.

Covariance Matrix and Portfolio Variance

Let Σ denote the covariance matrix of the risky asset returns. The (i,j)th element of Σ is $\operatorname{cov}(\tilde{R}_i, \tilde{R}_j)$. The diagonal elements are variances. In matrix notation,

$$\Sigma = \mathsf{E}[(\tilde{\mathbf{R}} - \mu)(\tilde{\mathbf{R}} - \mu)'].$$

The variance of end-of-period wealth produced by a portfolio ϕ is

$$\operatorname{var}\left(\sum_{i=1}^n \phi_i \tilde{R}_i\right) = \phi'\Sigma\phi. \tag{2.15}$$

Portfolio Choice

To see this, note that, because the square of a scalar equals the scalar multiplied by its transpose, the variance is

$$\begin{aligned}\mathsf{E}\big[(\phi'(\tilde{\mathbf{R}}-\mu))^2\big] &= \mathsf{E}\big[\phi'(\tilde{\mathbf{R}}-\mu)(\tilde{\mathbf{R}}-\mu)'\phi\big] \\ &= \phi'\mathsf{E}[(\tilde{\mathbf{R}}-\mu)(\tilde{\mathbf{R}}-\mu)']\phi \\ &= \phi'\Sigma\phi.\end{aligned}$$

We can also write the variance of end-of-period wealth as

$$w_0^2 \pi'\Sigma\pi, \qquad (2.16)$$

where $\pi_i = \phi_i/w_0$ for $i = 1, \ldots, n$.

If there are only two risky assets, then

$$\pi'\Sigma\pi = \pi_1^2 \operatorname{var}(\tilde{R}_1) + \pi_2^2 \operatorname{var}(\tilde{R}_n) + 2\pi_1\pi_2 \operatorname{cov}(\tilde{R}_1, \tilde{R}_2).$$

In general,

$$\pi'\Sigma\pi = \sum_{i=1}^{n} \pi_i^2 \operatorname{var}(\tilde{R}_i) + 2\sum_{i=1}^{n}\sum_{j=i+1}^{n} \pi_i\pi_j \operatorname{cov}(\tilde{R}_i, \tilde{R}_j).$$

The number of covariance terms is $n(n-1)/2$. This grows large quite rapidly. For example, with 1,000 stocks, there are 499,500 covariance terms. Practical portfolio choice models either aggregate assets into a relatively small number of categories (asset classes) or use a statistical factor model (Section 6.4) to reduce the number of covariances that must be estimated.

Because variances $\phi'\Sigma\phi$ must be nonnegative, any covariance matrix is positive semidefinite. If a symmetric positive semidefinite matrix is nonsingular, then it is positive definite (see, for example, Pemberton and Rau, 2011). We can ensure nonsingularity and positive definiteness of the covariance matrix Σ by eliminating redundant assets, so we will generally assume that it is nonsingular and positive definite.[4]

---

4. If Σ is singular, then there is a nonzero vector ϕ such that $\Sigma\phi = 0$. Of course, this implies $\phi'\Sigma\phi = 0$, so in this circumstance the portfolio ϕ of risky assets is risk free. We can scale ϕ such that $\iota'\phi = 1$, meaning that ϕ represents a unit-cost portfolio. Hence, in the absence of arbitrage opportunities, we must have $\phi'\tilde{\mathbf{R}} = R_f$, showing that the risk-free asset is redundant. We can also rearrange the equation $\phi'\tilde{\mathbf{R}} = R_f$ to see, for any i such that $\phi_i \neq 0$, that the return of asset i is equal to the return of a portfolio of the other risky assets and the risk-free asset; thus, there is a redundant risky asset. If it were eliminated, the opportunities available to investors would be unchanged.

Principle of Diversification

For a very simple example, assume the n assets are mutually uncorrelated (the covariance terms are all zero) and they all have the same variance σ^2. Then, $\pi'\Sigma\pi = \sigma^2 \sum_{i=1}^n \pi_i^2$. The portfolio in this example that has minimum risk, among portfolios that are fully invested in risky assets (meaning the π_i sum to one), is the equally weighted portfolio: $\pi_i = 1/n$ for each i. This is an illustration of the general principle that diversification reduces risks.

In this example, portfolio risk can be made negligible through diversification if there is a large number of assets. Setting $\pi_i = 1/n$ for each i, we have

$$\pi'\Sigma\pi = \sigma^2 \sum_{i=1}^n \frac{1}{n^2} = \frac{\sigma^2}{n} \to 0$$

as $n \to \infty$. The unrealistic feature of this example is that the assets are uncorrelated. Generally, we find positive correlations between assets; for example, all stocks tend to go up when the market goes up. Thus, risk cannot generally be eliminated by diversification. However, this example shows that the risk coming from the term

$$\sum_{i=1}^n \pi_i^2 \, \text{var}(\tilde{R}_i)$$

can generally be made small by diversification. This leaves the risk from the covariances as the important risk in a portfolio.

2.4 CARA-NORMAL MODEL

This section considers an investor with CARA utility and normally distributed asset returns. This is an especially tractable model. Neither CARA utility nor normally distributed returns is an attractive hypothesis, but the model provides a simple illustration of portfolio choice in a single-period model. Assume there is a risk-free asset and no labor income $\tilde{y}$.

A Single Risky Asset

Suppose there is a single risky asset. Given an amount ϕ invested in the risky asset, the realized utility of the investor is

$$-\exp(-\alpha \tilde{w}) = -\exp\left(-\alpha[w_0 R_f + \phi(\tilde{R} - R_f)]\right).$$

The random variable
$$-\alpha[w_0 R_f + \phi(\tilde{R} - R_f)]$$
is normally distributed with mean
$$-\alpha w_0 R_f - \alpha\phi(\mu - R_f)$$
and variance $\alpha^2\phi^2\sigma^2$. Therefore, using the fact that the expectation of the exponential of a normally distributed random variable is the exponential of the mean plus half of the variance, the expected utility is

$$\mathsf{E}[-\exp(-\alpha\tilde{w})] = -\exp\left(-\alpha\left[w_0 R_f + \phi(\mu - R_f) - \frac{1}{2}\alpha\phi^2\sigma^2\right]\right). \quad (2.17)$$

Certainty Equivalent

Equation (2.17) states that

$$w_0 R_f + \phi(\mu - R_f) - \frac{1}{2}\alpha\phi^2\sigma^2 \quad (2.18)$$

is the certainty equivalent of the random wealth $\tilde{w} = w_0 R_f + \phi(\tilde{R} - R_f)$. Maximizing expected utility is equivalent to maximizing the utility of the certainty equivalent, which is equivalent to maximizing the certainty equivalent itself. In summary, the optimal portfolio ϕ is the portfolio that maximizes (2.18). Note that (2.18) depends only on mean and variance (and the parameters α and w_0). This is a general property of normal distributions (Section 2.5).

Optimal Portfolio and Absence of Wealth Effects

Differentiating the certainty equivalent (2.18) with respect to ϕ and setting the derivative equal to zero yields

$$\phi = \frac{\mu - R_f}{\alpha\sigma^2}. \quad (2.19)$$

Thus, the optimal amount ϕ to invest is an increasing function of the risk premium $\mu - R_f$, a decreasing function of the variance σ^2, and a decreasing function of the investor's absolute risk aversion α. Note that $\phi > 0$ when the risk premium is positive, as shown more generally before. Also, note that ϕ does not depend on the initial wealth w_0. This is another illustration of the absence of wealth effects discussed in Section 1.3. An investor with CARA utility would invest the same amount in the risky asset whether her initial wealth were $1,000 or $1,000,000,000. Obviously, this depends on the assumption that the investor can

Optimal Fraction of Wealth to Invest

The optimal fraction of initial wealth to invest in the risky asset is $\pi = \phi/w_0$, where ϕ is given by (2.19). Thus,

$$\pi = \frac{\mu - R_f}{(\alpha w_0)\sigma^2}. \qquad (2.20)$$

This formula—or the multivariate version of it—is often used in a normative way to advise investors. The parameters $\mu - R_f$ and σ can be estimated from historical data, but α is a matter of individual preference. It is simpler (and more common) to think of αw_0 as being the risk aversion parameter of the investor's preferences. To give some guidance on what values of αw_0 are reasonable, we can do calculations of the following sort. If it is optimal for an investor to put 60% of her wealth in the risky asset when the risk premium $\mu - R_f$ is 8% and the standard deviation is 20%, then $\alpha w_0 = 3.33$. As another illustration, Maginn, Tuttle, McLeavey, and Pinto (2007, p. 241) advise that a value of αw_0 of "1 to 2 represents a relatively low degree of risk aversion" and a value of "6 to 8 represents a high degree of risk aversion."

Certainty Equivalent with Multiple Risky Assets

Suppose there are n risky assets with returns $\tilde{R}_i$ that are joint normally distributed. As is the case with a single normally distributed risky asset, maximizing expected CARA utility with multiple normally distributed assets is equivalent to solving a mean-variance problem: Choose ϕ to maximize

$$(\mu - R_f \iota)'\phi - \frac{1}{2}\alpha\phi'\Sigma\phi. \qquad (2.21)$$

In the practice of investment advising, the objective function (2.21) is usually expressed in terms of the fractions of wealth invested. Set $\pi = \phi/w_0$. Substitute in (2.21) and cancel a w_0 factor to obtain

$$(\mu - R_f \iota)'\pi - \frac{1}{2}(\alpha w_0)\pi'\Sigma\pi.$$

Portfolio Choice

The first term is the portfolio risk premium, and the second term is the portfolio variance penalized by the factor $\alpha w_0/2$. As discussed above, αw_0 is usually regarded as the preference parameter to be input into this objective.

Optimal Portfolio with Multiple Risky Assets

Differentiate (2.21) with respect to ϕ and equate the derivative to zero to obtain

$$\mu - R_f \iota - \alpha \Sigma \phi = 0$$

with solution

$$\phi = \frac{1}{\alpha} \Sigma^{-1}(\mu - R_f \iota). \tag{2.22}$$

This is a straightforward generalization of the formula (2.19) for the optimal portfolio of a CARA investor with a single normally distributed asset. We will see in Chapter 5 that the portfolio $\Sigma^{-1}(\mu - R_f \iota)$ has a special significance in mean-variance analysis even when asset returns are not normally distributed.

The formula (2.22) implies the single-asset formula (2.19) if the return of the asset is independent of all other asset returns. For such an asset i, (2.22) implies

$$\phi_i = \frac{\mu_i - R_f}{\alpha \sigma_i^2}, \tag{2.23}$$

where σ_i^2 is the variance of $\tilde{R}_i$. In general, (2.22) states that the demand for each asset i depends on the entire vector of risk premia and the covariances between asset i and the other assets.

2.5 MEAN-VARIANCE PREFERENCES

We have seen two examples of mean-variance preferences: normal returns with CARA utility in the previous section, and quadratic utility in Section 1.3. This section addresses the general question: Under what conditions do investors rank portfolios based on the means and variances of their payoffs?

Assume there is no labor income, so the end-of-period wealth of an investor is the payoff of the investor's portfolio. Regardless of the distribution of the asset payoffs, an investor with quadratic utility $u(w) = -\frac{1}{2}(w-\zeta)^2$ chooses portfolios based on mean and variance to maximize

$$\zeta E[\tilde{w}] - \frac{1}{2} E[\tilde{w}]^2 - \frac{1}{2}\text{var}(\tilde{w}),$$

as discussed in Section 1.3. An alternative question is— For what payoff distributions will all investors, regardless of their utility functions, rank portfolios based on mean and variance?

Let $\tilde{x}_1, \ldots, \tilde{x}_n$ denote the payoffs of the risky assets, and let $\tilde{x}$ denote the column vector with $\tilde{x}_i$ as its ith component. A sufficient condition for portfolios to be ranked based on mean and variance is that $\tilde{x}$ have a multivariate normal distribution. In this circumstance, each portfolio payoff is a linear combination of joint normally distributed variables and therefore has a normal distribution. Moreover, a normal distribution is entirely characterized by its mean and variance. Thus, if $\tilde{x}$ is multivariate normal and if portfolios θ and ψ have payoffs with the same mean and variance, then the payoffs of θ and ψ have the exact same distribution, and all investors must be indifferent between them. Only mean and variance can matter if payoffs are normally distributed.

In general, all investors rank portfolios based on mean and variance if and only if the distributions of portfolio payoffs are completely characterized by their means and variances, as with normal distributions. If there is a risk-free asset, a necessary and sufficient condition for this to be the case is that $\tilde{x}$ have an elliptical distribution. If $\tilde{x}$ has a density function, then $\tilde{x}$ is said to be elliptically distributed if there is a positive definite matrix Σ and a vector μ such that the density function is constant on each set

$$\{x \mid (x - \mu)' \Sigma^{-1} (x - \mu) = a\}$$

for $a > 0$.[5] These sets are ellipses. If the $\tilde{x}_i$ have finite variances, then μ is the vector of means and Σ is the covariance matrix of $\tilde{x}$. The class of elliptical distributions includes distributions that are bounded (and hence can satisfy limited liability) and distributions with "fat tails" (and hence may match empirical returns better than normal distributions do). It also includes distributions that do not have finite means and variances, in which case μ is interpreted as a location parameter and Σ as a scale parameter, and investors have location-scale preferences.

If an investor has random labor income, then she will care about how her portfolio hedges or exacerbates the risk of the income. For example, an investor with quadratic utility and labor income $\tilde{y}$ will choose a portfolio with return $\tilde{R}$ that maximizes

---

5. For general random vectors (not necessarily having density functions), the definition of being elliptical is as follows. A matrix C is orthogonal if $CC' = C'C = I$ where $'$ denotes the transpose operator and I is the identity matrix. A random vector $\tilde{z}$ is spherically distributed if the distribution of $\tilde{z}$ is the same as the distribution of $C\tilde{z}$ for every orthogonal matrix C. A random vector $\tilde{x}$ is elliptically distributed if there is a nonsingular matrix A and vector μ such that $\tilde{z}$ defined as $\tilde{z} = A(\tilde{x} - \mu)$ is spherically distributed.

$$\zeta \mathsf{E}[w_0 \tilde{R} + \tilde{y}] - \frac{1}{2}\mathsf{E}[w_0 \tilde{R} + \tilde{y}]^2 - \frac{1}{2}\mathrm{var}(w_0 \tilde{R} + \tilde{y}).$$

The variance here equals

$$w_0^2 \,\mathrm{var}(\tilde{R}) + \mathrm{var}(\tilde{y}) + 2w_0 \,\mathrm{cov}(\tilde{R}, \tilde{y}).$$

Therefore, the covariance between the portfolio return and the labor income affects the investor's expected utility, implying that the investor cares about more than just the mean and variance of the portfolio return. A portfolio that has a negative covariance with the labor income reduces the investor's risk. For example, an investor may want to short sell stocks in the industry in which she works. An example with normal returns and CARA utility is given in Exercise 2.3. The quadratic utility example is revisited in Section 3.7.

2.6 LINEAR RISK TOLERANCE AND WEALTH EXPANSION PATHS

The previous section establishes that the amounts a CARA investor invests in normally distributed assets are independent of her initial wealth. This section explains how optimal investments depend on initial wealth in the more general case of LRT utility and risky assets with general returns (not necessarily normally distributed). Suppose that there is a risk-free asset and that the utility function has risk tolerance $\tau(w) = A + Bw$. Assume there is no labor income ($\tilde{y} = 0$).

Parallel Wealth Expansion Paths

Let ϕ_i denote the optimal investment in risky asset i. It is shown at the end of the section that

$$\phi_i = \xi_i A + \xi_i B R_f w_0, \qquad (2.24)$$

where ξ_i is a number that does not depend on A or on w_0. It does depend on B and on the distribution of the asset returns. Thus, the optimal investment is an affine (constant plus linear) function of initial wealth w_0. The function $w_0 \mapsto \phi_i$ is called a wealth expansion path (or income expansion path or Engel curve).

If two investors have LRT utility functions with the same cautiousness parameter B, then the coefficient $\xi_i B R_f$ on initial wealth w_0 in (2.24) is the same for both investors. We say that their wealth expansion paths are parallel (because the graphs of the functions $w_0 \mapsto \phi_i$ are parallel lines). This plays an important role in Section 4.4, when see that a competitive equilibrium is Pareto optimal if the economy is populated by investors with LRT utility functions and the same cautiousness parameter. Moreover, we will see that, because wealth expansion paths are parallel, competitive equilibrium prices in such an economy do not depend on how wealth is distributed among investors.

Two-Fund Separation

Let ϕ denote the total investment in risky assets, so $w_0 - \phi$ is the amount invested in the risk-free asset. The ratio ϕ_i/ϕ is the fraction of the investor's total investment in risky assets that is allocated to asset i. We will show that this ratio is also independent of A and w_0. Thus, if all investors have LRT utility functions with the same cautiousness parameter, then the ratio ϕ_i/ϕ must be the same for all investors and therefore must equal the market capitalization (price times shares outstanding) of asset i divided by the market capitalization of all risky assets. Thus, all investors hold the market portfolio of risky assets. This is an example of two-fund separation, which means that all investors allocate their wealth across two funds, in this case the risk-free asset and the market portfolio of risky assets.

CARA and CRRA Utility

Special cases of (2.24) are

$$\text{CARA utility:} \quad B = 0, \text{ so } \phi_i = \xi_i A, \tag{2.24a}$$

$$\text{CRRA utility:} \quad A = 0, \text{ so } \frac{\phi_i}{w_0} = \xi_i B R_f. \tag{2.24b}$$

The case of CARA utility is of course the case considered in the previous section, and, as in the previous section, (2.24a) shows that the optimal amount to invest in each risky asset is independent of initial wealth. However, we allow here for nonnormal return distributions. For CRRA utility, (2.24b) states that the optimal fraction of initial wealth to invest in each risky asset is independent of initial wealth.

It is slightly more convenient here to let n denote the number of risky assets, so there are $n+1$ assets, including the risk-free asset, which we index as asset 0. Then, $\phi = \sum_{i=1}^{n} \phi_i$. Given the optimal investment ϕ_i, define ξ_i so that (2.24) holds; that is, set

$$\xi_i = \frac{\phi_i}{A + B R_f w_0}. \tag{2.25}$$

We want to show that ξ_i does not depend on A or on w_0. Set $\xi = \sum_{i=1}^n \xi_i$. If ξ_i does not depend on A or on w_0, then neither does ξ. Moreover, $\phi_i/\phi = \xi_i/\xi$. Therefore, showing that ξ_i does not depend on A or on w_0 suffices to show that ϕ_i/ϕ does not depend on A or on w_0.

The wealth achieved by the investor is

$$\tilde{w} = \left(w_0 - \sum_{i=1}^n \phi_i\right) R_f + \sum_{i=1}^n \phi_i \tilde{R}_i$$

$$= w_0 R_f + \sum_{i=1}^n \phi_i(\tilde{R}_i - R_f). \qquad (2.26)$$

For negative exponential utility, we can write the expected utility as

$$-e^{-\alpha w_0 R_f} \mathsf{E}\left[\exp\left(-\alpha \sum_{i=1}^n \phi_i(\tilde{R}_i - R_f)\right)\right].$$

To maximize this expected utility is equivalent to maximizing

$$-\mathsf{E}\left[\exp\left(-\alpha \sum_{i=1}^n \phi_i(\tilde{R}_i - R_f)\right)\right].$$

Substituting $\xi_i = \phi_i/A = \alpha\phi_i$, the optimization problem is to maximize

$$-\mathsf{E}\left[\exp\left(-\sum_{i=1}^n \xi_i(\tilde{R}_i - R_f)\right)\right],$$

and this does not depend on w_0 or A; hence, ξ_i is independent of w_0 and A.

For CRRA utility, define $\pi_i = \phi_i/w_0 = BR_f \xi_i$, and write the wealth (2.26) as

$$\tilde{w} = w_0\left[R_f + \sum_{i=1}^n \pi_i(\tilde{R}_i - R_f)\right]. \qquad (2.27)$$

For logarithmic utility, the expected utility equals

$$\log w_0 + \log\left[R_f + \sum_{i=1}^n \pi_i(\tilde{R}_i - R_f)\right],$$

and maximizing this is equivalent to maximizing

$$\log\left[R_f + \sum_{i=1}^n \pi_i(\tilde{R}_i - R_f)\right].$$

This optimization problem does not depend on w_0, so the optimal π_i and hence ξ_i do not depend on w_0. For power utility, the expected utility equals

$$w_0^{1-\rho} \frac{1}{1-\rho} \left[R_f + \sum_{i=1}^n \pi_i(\tilde{R}_i - R_f) \right]^{1-\rho}.$$

Maximizing this is equivalent to maximizing the same thing without the constant factor $w_0^{1-\rho}$, an optimization problem that does not depend on w_0, so we conclude that ξ_i is independent of w_0 for power utility also.

Now consider shifted logarithmic and shifted power utility, recalling that the risk tolerance is

$$\tau(w) = \frac{w - \zeta}{\rho},$$

so $A = -\zeta/\rho$ and $B = 1/\rho$, with $\rho = 1$ for log utility. It is convenient to solve these portfolio choice problems in two steps: First, invest ζ/R_f in the risk-free asset, and then invest $w_0 - \zeta/R_f$ optimally in the risk-free and risky assets. This is without loss of generality, because the amount invested in the risk-free asset in the first step can be disinvested in the second step if this is optimal. The first investment produces ζ, so the total wealth achieved is

$$\tilde{w} = \zeta + \left(w_0 - \frac{\zeta}{R_f}\right)\tilde{R},$$

where $\tilde{R}$ denotes the return on the second investment. This can be written as

$$\tilde{w} = \zeta + \left(w_0 - \frac{\zeta}{R_f}\right)\left[R_f + \sum_{i=1}^n \pi_i(\tilde{R}_i - R_f)\right],$$

where we define

$$\pi_i = \frac{\phi_i}{w_0 - \zeta/R_f} = BR_f\xi_i. \tag{2.28}$$

This implies that the utility achieved is, for shifted log,

$$\log\left[\left(w_0 - \frac{\zeta}{R_f}\right)\left[R_f + \sum_{i=1}^n \pi_i(\tilde{R}_i - R_f)\right]\right]$$

and, for shifted power,

$$\frac{\rho}{1-\rho}\left[\frac{1}{\rho}\left(w_0 - \frac{\zeta}{R_f}\right)\left[R_f + \sum_{i=1}^n \pi_i(\tilde{R}_i - R_f)\right]\right]^{1-\rho}.$$

In either case, the logic of the previous paragraph leads to the conclusion that the optimal π_i and hence ξ_i are independent of w_0 and A.

2.7 BEGINNING-OF-PERIOD CONSUMPTION

Consider now the problem of choosing consumption optimally at the beginning of the period in addition to choosing an optimal portfolio. Call the beginning of the period date 0 and the end of the period date 1. Now let w_0 denote the beginning-of-period wealth before consuming. This includes the value of any shares held plus any date–0 endowment. Letting $v(c_0, c_1)$ denote the utility function, the choice problem is

$$\max \; \mathsf{E}[v(c_0, \tilde{c}_1)] \text{ subject to } c_0 + \sum_{i=1}^n \theta_i p_i = w_0 \text{ and } \tilde{c}_1 = \tilde{y} + \sum_{i=1}^n \theta_i \tilde{x}_i. \quad (2.29)$$

Substituting in the second constraint, the Lagrangean for this problem is

$$\mathsf{E}\left[v\left(c_0, \tilde{y} + \sum_{i=1}^n \theta_i \tilde{x}_i\right)\right] - \gamma \left(c_0 + \sum_{i=1}^n \theta_i p_i - w_0\right),$$

and the first-order conditions are

$$\mathsf{E}\left[\frac{\partial}{\partial c_0} v(c_0, \tilde{c}_1)\right] = \gamma, \quad (2.30\text{a})$$

$$(\forall i) \quad \mathsf{E}\left[\frac{\partial}{\partial c_1} v(c_0, \tilde{c}_1) \tilde{x}_i\right] = \gamma p_i. \quad (2.30\text{b})$$

The system (2.30) implies the following:

$$(\forall i) \quad \mathsf{E}\left[\frac{\partial}{\partial c_1} v(c_0, \tilde{c}_1) \tilde{x}_i\right] = p_i \mathsf{E}\left[\frac{\partial}{\partial c_0} v(c_0, \tilde{c}_1)\right]. \quad (2.31)$$

As before, this is a necessary condition for optimality provided it is feasible, starting from the optimal portfolio, to add a little or subtract a little of each asset i.

The first-order condition (2.31) simplifies if utility is time additive, meaning that $v(c_0, c_1) = u_0(c_0) + u_1(c_0)$ for functions u_0 and u_1. In this case, the first-order condition is

$$(\forall i) \quad \mathsf{E}[u_1'(\tilde{c}_1) \tilde{x}_i] = p_i u_0'(c_0). \quad (2.32)$$

This equation is called the Euler equation. Usually, it is assumed that $u_1 = u$ and $u_2 = \delta u$ for some function u and constant $\delta \in (0,1)$. Time-additive utility leads to strong results. For example, it produces the Consumption-based Capital Asset Pricing Model (CCAPM) of Breeden (1979); see Chapters 10 and 14. However, it is also a strong assumption. In particular, it links the way an investor trades off consumption at different dates with the investor's tolerance for risk, which should probably be distinct aspects of preferences. A precise statement of the link for CRRA utility is that the elasticity of intertemporal substitution (EIS) equals the reciprocal of the coefficient of relative risk aversion (Exercise 2.6).

2.8 NOTES AND REFERENCES

Arrow (1965) is the source of the results in Section 2.2 that (i) an investor in a market with a risk-free and a single risky asset invests a positive amount in the risky asset if its risk premium is positive and (ii) the investment is increasing in initial wealth if absolute risk aversion is decreasing. An application of result (i) to insurance markets is as follows. Suppose that an uninsured individual has final wealth $w - \tilde{\varepsilon}$. Suppose the individual can buy insurance at a cost y per unit, meaning that, if an amount x of insurance is chosen, then the final wealth is $w - \tilde{\varepsilon} - xy + x\tilde{\varepsilon}$. We can write this as $w - y + (x-1)(\tilde{\varepsilon} - y)$. If the insurance is actually unfair, meaning that $y > \mathsf{E}[\tilde{\varepsilon}]$, then it is optimal to choose less than full insurance, because in that circumstance the choice problem is equivalent to a portfolio choice problem in which the risky asset has a negative risk premium ($\mathsf{E}[\tilde{\varepsilon}] - y < 0$), implying a short position ($x - 1 < 0$) is optimal.

The CARA-normal model is a special case of mean-variance optimization, studied by Markowitz (1952, 1959) and addressed further in Chapter 5. The fact that investors have mean-variance preferences when returns are elliptically distributed is shown by Owen and Rabinovitch (1983) and Chamberlain (1983a). Owen and Rabinovitch (1983) give several examples of elliptical distributions. Chamberlain (1983a) also gives necessary and sufficient conditions for mean-variance preferences in the absence of a risk-free asset.

Mossin (1968) establishes the result (2.24) that optimal investments are affine in wealth when an investor has LRT utility. Cass and Stiglitz (1970) establish the two-fund separation result of Section 2.6 when investors have LRT utility functions with the same cautiousness parameter. They also show that this condition is a necessary condition on preferences for two-fund separation to hold with a risk-free asset and for all distributions of risky asset returns. They give other conditions on preferences that are necessary and sufficient for two-fund separation in complete markets and in markets without a risk-free asset.

Exercises 2.8–2.9 illustrate the concepts of precautionary savings and precautionary premia. More on those topics can be found in Kimball (1990).

EXERCISES

2.1. Suppose there is a risk-free asset with return $R_f = 1.05$ and three risky assets each of which has an expected return equal to 1.10. Suppose the covariance matrix of the risky asset returns is

$$\Sigma = \begin{pmatrix} 0.09 & 0.06 & 0 \\ 0.06 & 0.09 & 0 \\ 0 & 0 & 0.09 \end{pmatrix}.$$

Portfolio Choice

Suppose the returns are normally distributed. What is the optimal fraction of wealth to invest in each of the risky assets for a CARA investor with $\alpha w_0 = 2$? Why is the optimal investment higher for the third asset than for the other two?

2.2. Suppose there is a risk-free asset and n risky assets with payoffs $\tilde{x}_i$ and prices p_i. Assume the vector $\tilde{x} = (\tilde{x}_1 \cdots \tilde{x}_n)'$ is normally distributed with mean μ_x and nonsingular covariance matrix Σ_x. Let $p = (p_1 \cdots p_n)'$. Suppose there is consumption at date 0 and consider an investor with initial wealth w_0 and CARA utility at date 1:
$$u_1(c) = -e^{-\alpha c}.$$
Let θ_i denote the number of shares the investor considers holding of asset i and set $\theta = (\theta_1 \cdots \theta_n)'$. The investor chooses consumption c_0 at date 0 and a portfolio θ, producing wealth $(w_0 - c_0 - \theta'p)R_f + \theta'\tilde{x}$ at date 1.

(a) Show that the optimal vector of share holdings is
$$\theta = \frac{1}{\alpha}\Sigma_x^{-1}(\mu_x - R_f p). \qquad (2.33)$$

(b) Suppose all of the asset prices are positive, so we can define returns $\tilde{x}_i/p_i$. Explain why (2.33) implies (2.22). Note: This is another illustration of the absence of wealth effects. Neither date-0 wealth nor date-0 consumption affects the optimal portfolio for a CARA investor.

2.3. Suppose there is a risk-free asset with return R_f and a risky asset with return $\tilde{R}$. Consider an investor who maximizes expected end-of-period utility of wealth and who has CARA utility and invests w_0. Suppose the investor has labor income $\tilde{y}$ at the end of the period, so her end-of-period wealth is $\phi_f R_f + \phi \tilde{R} + \tilde{y}$, where ϕ_f denotes the investment in the risk-free asset and ϕ the investment in the risky asset.

(a) Suppose $\tilde{y}$ and $\tilde{R}$ are independent. Show that the optimal ϕ is the same as if there were labor income. Hint: Use the law of iterated expectations as in Section 1.5 and the fact that if $\tilde{v}$ and $\tilde{x}$ are independent random variables, then $E[\tilde{v}\tilde{x}] = E[\tilde{v}]E[\tilde{x}]$.

(b) Define $b = \mathrm{cov}(\tilde{y},\tilde{R})/\mathrm{var}(\tilde{R})$, $a = (E[\tilde{y}] - bE[\tilde{R}])/R_f$, and $\tilde{\varepsilon} = \tilde{y} - aR_f - b\tilde{R}$. Show that $\tilde{y} = aR_f + b\tilde{R} + \tilde{\varepsilon}$ and that $\tilde{\varepsilon}$ has a zero mean and is uncorrelated with $\tilde{R}$. Note: This is an example of an orthogonal projection, which is discussed in more generality in Section 3.5.

(c) Suppose $\tilde{y}$ and $\tilde{R}$ have a joint normal distribution. Using the result of the previous part, show that the optimal ϕ is $\phi^* - b$, where ϕ^* denotes the optimal investment in the risky asset when there is no labor income.

2.4. Consider a CARA investor with n risky assets having normally distributed returns, as studied in Section 2.4, but suppose there is no risk-free asset, so the budget constraint is $\iota'\phi = w_0$. Show that the optimal portfolio is

$$\phi = \frac{1}{\alpha}\Sigma^{-1}\mu + \left(\frac{\alpha w_0 - \iota'\Sigma^{-1}\mu}{\alpha \iota'\Sigma^{-1}\iota}\right)\Sigma^{-1}\iota.$$

Note: As will be seen in Section 5.2, the two vectors $\Sigma^{-1}\mu$ and $\Sigma^{-1}\iota$ play an important role in mean-variance analysis even without the CARA-normal assumption.

2.5. Suppose there is a risk-free asset and n risky assets. Consider an investor with quadratic utility

$$\zeta E[\tilde{w}] - \frac{1}{2}E[\tilde{w}]^2 - \frac{1}{2}\text{var}(\tilde{w})$$

and no labor income. Show that the optimal portfolio for the investor is

$$\phi = \frac{1}{1+\kappa^2}(\zeta - w_0 R_f)\Sigma^{-1}(\mu - R_f\iota),$$

where

$$\kappa^2 = (\mu - R_f\iota)'\Sigma^{-1}(\mu - R_f\iota).$$

Hint: In the first-order conditions, define $\gamma = (\mu - R_f\iota)'\phi$, solve for ϕ in terms of γ, and then compute γ. Note: We will see in Chapter 5 that κ is the maximum Sharpe ratio of any portfolio.

2.6. Consider a utility function $v(c_0, c_1)$. The marginal rate of substitution (MRS) is defined to be the negative of the slope of an indifference curve and is equal to

$$\text{MRS}(c_0, c_1) = \frac{\partial v(c_0, c_1)/\partial c_0}{\partial v(c_0, c_1)/\partial c_1}.$$

The elasticity of intertemporal substitution (EIS) is defined as

$$\frac{d\log(c_1/c_0)}{d\log \text{MRS}(c_0, c_1)},$$

where the MRS is varied holding utility constant. Show that, if

$$v(c_0, c_1) = \frac{1}{1-\rho}c_0^{1-\rho} + \frac{\delta}{1-\rho}c_1^{1-\rho},$$

then the EIS is $1/\rho$.

2.7. Consider the portfolio choice problem with only a risk-free asset and with consumption at both the beginning and end of the period. Assume the investor

Portfolio Choice

has time-additive power utility, so she solves

$$\max \quad \frac{1}{1-\rho}c_0^{1-\rho} + \delta\frac{1}{1-\rho}c_1^{1-\rho} \quad \text{subject to} \quad c_0 + \frac{1}{R_f}c_1 = w_0.$$

As shown in Exercise 2.6, the investor's EIS is $1/\rho$.

(a) Show that the optimal consumption-to-wealth ratio c_0/w_0 is a decreasing function of R_f if the EIS is greater than 1 and an increasing function of R_f if the EIS is less than 1. Note: The effect of changing R_f is commonly broken down into an income effect and a substitution effect. This shows that the substitution effect dominates when the EIS is high and the income effect dominates when the EIS is low.

(b) For given c_0 and $\tilde{c}_1$, show that the solution of the investor's optimization problem implies that R_f must be lower when the EIS is higher.

2.8. Consider the portfolio choice problem with only a risk-free asset and with consumption at both the beginning and end of the period. Suppose the investor has time-additive utility with $u_0 = u$ and $u_1 = \delta u$ for a common function u and discount factor δ. Suppose the investor has labor income $\tilde{y}$ at the end of the period, so she chooses c_0 to maximize

$$u(c_0) + \delta E[u((w_0 - c_0)R_f + \tilde{y})].$$

Suppose the investor has convex marginal utility ($u''' > 0$) and suppose that $E[\tilde{y}] = 0$. Show that the optimal c_0 is smaller than if $\tilde{y} = 0$. Note: This illustrates the concept of precautionary savings—the risk imposed by $\tilde{y}$ results in higher savings $w_0 - c_0$.

2.9. Letting c_0^* denote optimal consumption in the previous exercise, define the precautionary premium π by

$$u'((w_0 - \pi - c_0^*)R_f) = E[u'((w_0 - c_0^*)R_f + \tilde{y})].$$

(a) Show that c_0^* would be the optimal consumption of the investor if she had no labor income and had initial wealth $w_0 - \pi$.
(b) Assume the investor has CARA utility. Show that the precautionary premium is independent of initial wealth (again, no wealth effects with CARA utility).

3

Stochastic Discount Factors

CONTENTS

3.1	Basic Relationships Regarding SDFs	53
3.2	Arbitrage, the Law of One Price, and Existence of SDFs	56
3.3	Complete Markets and Uniqueness of the SDF	59
3.4	Risk-Neutral Probabilities	61
3.5	Orthogonal Projections of SDFs onto the Asset Span	62
3.6	Hansen-Jagannathan Bounds	67
3.7	Hedging and Optimal Portfolios with Quadratic Utility	70
3.8	Hilbert Spaces and Gram-Schmidt Orthogonalization	72
3.9	Notes and References	75

Security prices are determined by discounting future payoffs. We discount because cash or consumption is worth less in the future than in the present, and we discount because of risk. If investors were risk neutral, then they would discount only because of the time value of money. The price of an asset with payoff $\tilde{x}$ would be computed as $p = \mathsf{E}[\tilde{x}]/R_f$, assuming a risk-free asset with return R_f exists. We call this "discounting at the risk-free rate." However, risk-averse investors discount both because of the time value of money and because of risk. A stochastic discount factor encapsulates both discounts in a single entity.

As in the previous chapter, let $\tilde{x}_i$ denote the payoff of asset i, let p_i denote its price, and let $\tilde{R}_i$ denote the return $\tilde{x}_i/p_i$. A stochastic discount factor (SDF) is any random variable $\tilde{m}$ such that

$$(\forall i) \qquad p_i = \mathsf{E}[\tilde{m}\tilde{x}_i]. \qquad (3.1)$$

This definition is of fundamental importance in asset pricing theory. In fact, an asset pricing theory is simply a set of hypotheses that implies some particular form for $\tilde{m}$.

Sections (3.2) and (3.3) describe conditions related to the existence and uniqueness of SDFs. Section 3.4 describes a method of pricing built on SDFs and equivalent to SDFs called risk-neutral pricing. The remainder of the chapter presents results relating to the unique SDF that is spanned by the asset payoffs. This is the orthogonal projection of any SDF onto the span of the assets. Orthogonal projection is an important concept, and it appears throughout this book in various contexts.

3.1 BASIC RELATIONSHIPS REGARDING SDFs

Arrow Securities and State Prices

If there are only finitely many states of the world, say $\omega_1, \ldots, \omega_k$, then (3.1) can be written as

$$(\forall i) \qquad p_i = \sum_{j=1}^{k} \tilde{m}(\omega_j)\tilde{x}_i(\omega_j)\,\mathrm{prob}_j, \qquad (3.2)$$

where prob_j denotes the probability of the jth state. Consider a security that pays one unit of the consumption good in a particular state ω_j and zero in all other states (called an Arrow security in recognition of the seminal work of Arrow, 1953). Let q_j (rather than p) denote the price of the Arrow security for state ω_j. The price of an Arrow security is called a state price. Applying (3.2) to the Arrow security yields $q_j = \tilde{m}(\omega_j)\,\mathrm{prob}_j$, implying $\tilde{m}(\omega_j) = q_j/\mathrm{prob}_j$. Thus, the value of the SDF in a particular state of the world is the ratio of the corresponding state price to the probability of the state. If there are infinitely many states of the world, then we can interpret $\tilde{m}$ similarly, though a little more care is obviously needed because individual states will generally have zero probabilities.[1]

Because $\tilde{m}$ specifies the price of a unit of the consumption good in each state per unit probability, it is also called a state price density. Another name for $\tilde{m}$ is "pricing kernel." The multiplicity of names (there are even others besides these) is one indicator of the importance of the concept.

Definition in Terms of Returns and Orthogonality to Excess Returns

If each p_i is positive, then the definition (3.1) of an SDF is equivalent to

$$(\forall i) \qquad \mathsf{E}\big[\tilde{m}\tilde{R}_i\big] = 1. \qquad (3.3)$$

---

1. Technically, $\tilde{m}$ is the Radon-Nikodym derivative of the set function that assigns prices to events (sets of states) relative to the probabilities of events (Appendix A.10).

This implies

$$(\forall i,j) \qquad E[\tilde{m}(\tilde{R}_i - \tilde{R}_j)] = 0. \tag{3.4}$$

Thus, an SDF is orthogonal to excess returns. Moreover, if $\tilde{m}$ is an SDF, then (3.1)–(3.4) hold for portfolios as well as individual assets. To see this, consider a portfolio consisting of θ_i shares of asset i for each i. Multiply both sides of (3.1) by θ_i and add over i to obtain

$$E[\tilde{m}\tilde{x}] = p, \tag{3.5}$$

where $\tilde{x} = \sum_{i=1}^n \theta_i \tilde{x}_i$ is the payoff of the portfolio and $p = \sum_{i=1}^n \theta_i p_i$ is the price of the portfolio. If a portfolio has a positive price (cost), then

$$E[\tilde{m}\tilde{R}] = 1, \tag{3.6}$$

where $\tilde{R} = \tilde{x}/p$ is the return of the portfolio. Frequently, we work exclusively with returns—instead of prices and payoffs—and we define an SDF as a random variable $\tilde{m}$ such that $E[\tilde{m}\tilde{R}] = 1$ for all returns $\tilde{R}$.

Marginal Utility Defines an SDF

The first-order condition (2.4a) states that

$$u'(\tilde{w}) = \gamma \tilde{m} \tag{3.7}$$

for an SDF $\tilde{m}$ and constant γ. Due to concavity of utility, marginal utility is a decreasing function of wealth. Hence, (3.7) implies that wealth is a decreasing function of $\tilde{m}$. This is intuitive: Investors consume less in states that are more expensive.

The parameter γ in (3.7) is the marginal utility or expected marginal utility of beginning-of-period consumption. To see this, consider the model of Section 2.7 in which beginning-of-period consumption is optimally chosen. Denoting terminal wealth by $\tilde{c}_1$ and beginning-of-period consumption by c_0 as in Section 2.7, the first-order condition (2.31) yields

$$\frac{\partial v(c_0, \tilde{c}_1)}{\partial c_1} = E\left[\frac{\partial v(c_0, \tilde{c}_1)}{\partial c_0}\right] \tilde{m}, \tag{3.8}$$

which implies that γ in (3.7) equals $E[\partial v(c_0, \tilde{c}_1)/\partial c_0]$. This simplifies if we assume time-additive utility, meaning that there are functions u_0 and u_1 such that $v(c_0, c_1) = u_0(c_0) + u_1(c_1)$. In this circumstance, (3.8) implies

$$\tilde{m} = \frac{u_1'(\tilde{c}_1)}{u_0'(c_0)}. \tag{3.9}$$

Thus, the investor's marginal rate of substitution (MRS) between date–0 consumption and date–1 consumption is an SDF. Comparing (3.9) to (3.7) with $u = u_1$ and $\tilde{w} = \tilde{c}_1$, we see that $\gamma = u_0'(c_0)$.

A leading special case is when the functions u_0 and u_1 are the same except for a discounting of future utility u_1. If there is a function u and discount factor $0 < \delta < 1$ such that $u_0 = u$ and $u_1 = \delta u$, then the SDF $\tilde{m}$ in (3.8) is

$$\tilde{m} = \frac{\delta u'(\tilde{c}_1)}{u'(c_0)}. \qquad (3.10)$$

A Formula for Risk Premia

The introduction to Chapter 2 states that asset pricing theory is concerned with explaining the risk premia of different assets. The introduction to this chapter states that asset pricing theory is about SDFs. It is important to understand that these two statements are consistent. Use the fact that the covariance of any two random variables is the expectation of their product minus the product of their expectations to write (3.6) as

$$1 = \mathrm{cov}(\tilde{m}, \tilde{R}) + \mathsf{E}[\tilde{m}]\mathsf{E}[\tilde{R}]. \qquad (3.11)$$

Suppose there is a risk-free asset. Then, (3.6) with $\tilde{R} = R_f$ implies $\mathsf{E}[\tilde{m}] = 1/R_f$. Substituting this in (3.11) and rearranging gives the following formula for the risk premium of any asset or portfolio with return $\tilde{R}$:

$$\mathsf{E}[\tilde{R}] - R_f = -R_f \mathrm{cov}(\tilde{m}, \tilde{R}). \qquad (3.12)$$

This shows that risk premia are determined by covariances with any SDF. An asset pricing theory is a theory that specifies a particular form for $\tilde{m}$ and therefore implies via (3.12) that risk premia are determined by covariances with some set of variables. We will see this in Chapter 6 for the single-period model, in Chapter 10 for a discrete-time model, and in Chapter 14 for continuous time.

Formulas for Prices

Parallel to formula (3.12) for risk premia is a formula for prices in terms of expected future cash flows adjusted for covariances with an SDF and discounted at the risk-free rate. For any asset with end-of-period payoff $\tilde{x}$ and price p and any SDF $\tilde{m}$, we have

$$p = \mathsf{E}[\tilde{m}\tilde{x}] = \mathsf{E}[\tilde{m}]\mathsf{E}[\tilde{x}] + \mathrm{cov}(\tilde{m}, \tilde{x}) = \frac{\mathsf{E}[\tilde{x}] + R_f \mathrm{cov}(\tilde{m}, \tilde{x})}{R_f}, \qquad (3.13)$$

using $E[\tilde{m}] = 1/R_f$ for the last equality. For example, if the payoff is uncorrelated with the SDF, then the price is the expected payoff discounted at the risk-free rate. Most assets' payoffs are negatively correlated with an SDF, so the risk adjustment $R_f \text{cov}(\tilde{m}, \tilde{x})$ is a deduction from the expected payoff. To see why the correlation is usually negative, consider the marginal utility formula (3.9). Marginal utility is high when consumption is low, so SDFs are usually high when asset payoffs are low. We will study this more carefully in the context of equilibrium models later.

Finally, we can also represent prices as expected future cash flows discounted at a risk-adjusted rate. Rearrange (3.12) with $\tilde{R} = \tilde{x}/p$ as

$$p = \frac{E[\tilde{x}]}{R_f - R_f \text{cov}(\tilde{m}, \tilde{R})}. \tag{3.14}$$

This formula is widely used in practice, for example, with the risk adjustment based on the capital asset pricing model (CAPM)—see Chapter 6.

3.2 ARBITRAGE, THE LAW OF ONE PRICE, AND EXISTENCE OF SDFs

The remainder of this chapter analyzes the existence and structure of SDFs without reference to the first-order condition (3.7). This section describes two important concepts regarding securities markets and the relations of the concepts to the existence of SDFs. A summary of the results is as follows:

Existence of Strictly Positive SDF	$\Leftrightarrow$	No Arbitrage Opportunities
	$\Rightarrow$	Law of One Price
	$\Leftrightarrow$	Existence of SDF

Let n denote the number of assets, including the risk-free asset if it exists. Set $p = (p_1 \cdots p_n)'$, and interpret a portfolio θ as a column vector. A random variable $\tilde{x}$ is said to be a marketed payoff if it is the payoff of a portfolio, meaning that $\tilde{x} = \sum_{i=1}^{n} \theta_i \tilde{x}_i$ for some $\theta \in \mathbb{R}^n$. The "law of one price" is said to hold if each marketed payoff has a unique cost. This means that if there are two portfolios producing the same payoff, then they have the same cost. Mathematically,

$$(\forall \theta, \hat{\theta} \in \mathbb{R}^n) \quad \sum_{i=1}^{n} \theta_i \tilde{x}_i = \sum_{i=1}^{n} \hat{\theta}_i \tilde{x}_i \quad \Rightarrow \quad p'\theta = p'\hat{\theta}.$$

When we write equality of two random variables (as for the portfolio payoffs here), we always mean that they are equal with probability 1.

Stochastic Discount Factors

An arbitrage opportunity is defined to be a portfolio θ satisfying

(i) $p'\theta \leq 0$,
(ii) $\sum_{i=1}^{n} \theta_i \tilde{x}_i \geq 0$ with probability 1, and
(iii) either $p'\theta < 0$ or $\sum_{i=1}^{n} \theta_i \tilde{x}_i > 0$ with positive probability (or both).

Thus, an arbitrage opportunity is a portfolio that requires no investment at date 0, has a nonnegative value at date 1, and either produces income at date 0 or has a positive value with positive probability at date 1. If there is an arbitrage opportunity, then no investor with a strictly monotone utility function can have an optimal portfolio, because she will want to exploit the arbitrage opportunity at infinite scale.

A very mild assumption regarding a securities market is that no arbitrage opportunities exist. An even weaker assumption is that the law of one price holds. It is a weaker assumption, because a failure of the law of one price is an arbitrage opportunity. Specifically, if θ and $\hat{\theta}$ produce the same payoff, but θ is cheaper than $\hat{\theta}$, then buying θ and short selling $\hat{\theta}$ is an arbitrage opportunity. Consequently, if there are no arbitrage opportunities, then the law of one price must hold.

The Law of One Price and SDFs

If the law of one price holds, then there is an SDF. Thus, the price $p'\theta$ of any payoff $\sum_{i=1}^{n} \theta_i \tilde{x}_i$ can be computed as $\mathsf{E}[\tilde{m} \sum_{i=1}^{n} \theta_i \tilde{x}_i]$ for some $\tilde{m}$. In fact, the law of one price is equivalent to the existence of an SDF. This is true with infinitely many states of the world (limiting attention to payoffs with finite variances), but it is easier to see when there are only finitely many states.

For the remainder of this section, suppose there are k possible states of the world. We make no assumptions regarding the number of states versus the number of assets, so k can be smaller than, equal to, or larger than n. Denote the payoff of asset i in state j as x_{ij}. A state-price vector is defined to be a vector $(q_1 \cdots q_k)$ satisfying

$$\begin{pmatrix} x_{11} & \cdots & x_{1k} \\ \vdots & \vdots & \vdots \\ x_{n1} & \cdots & x_{nk} \end{pmatrix} \begin{pmatrix} q_1 \\ \vdots \\ q_k \end{pmatrix} = \begin{pmatrix} p_1 \\ \vdots \\ p_n \end{pmatrix}. \qquad (3.15)$$

This equation means that the price p_i of each asset i equals the sum of the payoffs of asset i in the various states of the world multiplied by the state price of each state. As discussed in the introduction to this chapter, a state price is the price of the Arrow security that pays 1 unit of the consumption good in that state and 0 in all other states. Denote the matrix in (3.15) by X and denote the vectors by q

and p, so we can write (3.15) as

$$Xq = p. \tag{3.16}$$

As remarked in Section 3.1, given a state price vector q, we can compute an SDF as $m_j = q_j / \operatorname{prob}_j$, where prob_j denotes the probability of state j. With this construction, we have, for each asset i,

$$p_i = \sum_{j=1}^{k} x_{ij} q_j = \sum_{j=1}^{k} x_{ij} m_j \operatorname{prob}_j \stackrel{\text{def}}{=} \mathsf{E}[\tilde{x}_i \tilde{m}].$$

Thus, an SDF exists if and only if a state price vector exists.

In this finite-state world, the law of one price can be expressed as

$$(\forall \theta, \hat{\theta}) \quad X'\theta = X'\hat{\theta} \;\Rightarrow\; p'\theta = p'\hat{\theta}. \tag{3.17}$$

This is equivalent to $X'\theta = 0 \Rightarrow p'\theta = 0$. This means that p is orthogonal to each vector θ that is orthogonal to each of the columns of X. This is true if and only if p is a linear combination of the columns of X, which is the meaning of (3.16).[2] Thus, the law of one price is equivalent to the existence of an SDF.

Arbitrage and SDFs

If there are no arbitrage opportunities, then there must be a strictly positive SDF. It is natural that state prices (prices of Arrow securities) should be positive, but positivity is not implied by the law of one price alone. We will establish the relation between absence of arbitrage opportunities and the existence of a strictly positive SDF assuming finitely many states of the world. It is also true if there are infinitely many states of the world; however, the proof with infinitely many states is more difficult and will not be given here.[3]

---

2. In more detail, the set of vectors Xq for $q \in \mathbb{R}^k$ is the linear subspace of $\mathbb{R}^n$ spanned by the columns of X. The set of vectors θ such that $X'\theta = 0$ is the orthogonal complement of the subspace. The condition that $p'\theta = 0$ whenever $X'\theta = 0$ means that p is in the orthogonal complement of the orthogonal complement. However, the orthogonal complement of the orthogonal complement of a subspace of $\mathbb{R}^n$ is the subspace itself. So, p is in the subspace (that is, $p = Xq$ for some q) if and only if p is in the orthogonal complement of the orthogonal complement of the subspace (that is, $p'\theta = 0$ whenever $X'\theta = 0$).

3. It is in fact true that there is a bounded strictly positive SDF $\tilde{m}$, implying that the expectation $\mathsf{E}[\tilde{m}\tilde{x}]$ exists whenever $\mathsf{E}[\tilde{x}]$ exists. The most constructive proof with infinitely many states relies on the fact that a CARA investor has an optimal portfolio in any market in which there are no arbitrage opportunities. This is a nontrivial fact (and also will not be proven here), but given the existence of an optimal portfolio, the CARA investor's first-order condition implies that an SDF can be constructed as a constant times marginal utility.

Stochastic Discount Factors

The proof that absence of arbitrage opportunities implies the existence of a strictly positive SDF relies on some version of the separating hyperplane theorem. In the following, we invoke the version known as Tucker's Complementarity Theorem (Rockafellar, 1970, Theorem 22.7). The converse is also true: Existence of a strictly positive SDF implies absence of arbitrage opportunities. The converse is easy to show and is left as an exercise.

We will show that absence of arbitrage opportunities implies existence of a strictly positive SDF in the finite-state model. Consider the $(k+1) \times n$ matrix Y defined as

$$Y = \begin{pmatrix} -p' \\ X' \end{pmatrix}.$$

Set $M = \{Y\theta \mid \theta \in \mathbb{R}^n\}$. The assumption of no arbitrage opportunities means that the zero vector is the only nonnegative vector in M. It follows from Tucker's Complementarity Theorem that there exists a strictly positive vector v that is orthogonal to M, meaning that $v'Y\theta = 0$ for all $\theta \in \mathbb{R}^n$. Define $q_j = v_{j+1}/v_1$ for $j = 1, \ldots, k$. The equality $v'Y\theta = 0$ can be written as

$$-v_1 p' \theta + (v_2 \cdots v_{k+1}) X' \theta = 0,$$

which implies that

$$q'X'\theta = p'\theta.$$

Because this holds for all $\theta \in \mathbb{R}^n$, $p = Xq$.

3.3 COMPLETE MARKETS AND UNIQUENESS OF THE SDF

The securities market is said to be complete if, for any $\tilde{w}$, there exists a portfolio θ such that

$$\sum_{i=1}^n \theta_i \tilde{x}_i = \tilde{w}. \tag{3.18}$$

Thus, any desired distribution of wealth across states of the world can be achieved by choosing the appropriate portfolio (if cost is not a constraint).

It should be apparent that true completeness is a rare thing. For example, if there are infinitely many states of the world, then (3.18) is an infinite number of equalities, which we are supposed to satisfy by choosing a finite-dimensional vector $\theta \in \mathbb{R}^n$. This is impossible. Note that there must be infinitely many states if we want the security payoffs $\tilde{x}_i$ to be normally distributed, or to be lognormally distributed, or to have any other continuous distribution. Thus, single-period markets with finitely many continuously distributed assets are not complete.

On the other hand, if significant gains are possible by improving risk sharing, then we would expect assets to be created to enable those gains to be realized. Also, as is shown later, dynamic trading can dramatically increase the "span" of securities markets. The real impediments to achieving at least approximately complete markets are moral hazard and adverse selection. For example, there are very limited opportunities for obtaining insurance against employment risk, due to moral hazard.

In any case, completeness is a useful benchmark against which to compare actual security markets. As remarked above, to have complete markets in a single-period model with a finite number of securities, there must be only finitely many possible states of the world. In the remainder of this section, consider the finite-state model introduced in Section 3.2. In that model, the definition of market completeness is equivalent to: For each $w \in \mathbb{R}^k$, there exists $\theta \in \mathbb{R}^n$ such that

$$X'\theta = w, \tag{3.19}$$

where X is the $n \times k$ matrix of asset-state payoffs x_{ij} introduced in (3.15). This system of equations has a solution for each $w \in \mathbb{R}^k$ if and only if X has rank k. Thus, in particular, market completeness implies $n \geq k$; that is, there must be at least as many securities as states of the world.

If the market is complete and the law of one price holds, then there is a unique solution q to the equation $Xq = p$, and hence a unique SDF. We know that there is some solution because the law of one price holds. To see that the solution must be unique, premultiply both sides of the equation $Xq = p$ by X' and then invert the nonsingular matrix $X'X$ to obtain[4]

$$q = (X'X)^{-1}X'p. \tag{3.20}$$

Thus, $(X'X)^{-1}X'p$ is the only possible solution of the equation $Xq = p$ when the market is complete.

---

4. The matrix $X'X$ is nonsingular because X has rank k, due to the assumption of completeness. If the market is complete and there are exactly as many assets as states of the world, then X is square and nonsingular, so X^{-1} exists. Therefore, $Xq = p$ implies $q = X^{-1}p$. This is consistent with (3.20), because when X is nonsingular we have

$$(X'X)^{-1}X'p = X^{-1}(X')^{-1}X'p = X^{-1}p.$$

3.4 RISK-NEUTRAL PROBABILITIES

Another way to represent prices is via a risk-neutral probability. To distinguish between a risk-neutral probability and the probability under which we have been taking expectations, it is common to call the latter (whether objective or subjective) the physical probability.[5] A risk-neutral probability is defined from the physical probability and a strictly positive SDF. Assume there is a strictly positive SDF $\tilde{m}$, and, for each event A, let 1_A denote the indicator function of A, that is, for each state of the world ω, $1_A(\omega) = 1$ if $\omega \in A$ and $1_A(\omega) = 0$ if $\omega \notin A$.

Suppose first that there is a risk-free asset. For each event A, define

$$\mathbb{Q}(A) = R_f \mathsf{E}[\tilde{m} 1_A]. \qquad (3.21)$$

Then, $\mathbb{Q}$ is the risk-neutral probability associated to $\tilde{m}$. In a finite-state world, (3.21) means that the risk-neutral probability of each state is the product of R_f, the physical probability of the state, and the value of $\tilde{m}$ in the state (it is also the product of R_f with the state price). In general, $\mathbb{Q}$ is a probability: $\mathbb{Q}(A) \geq 0$, $\mathbb{Q}(\Omega) = 1$, where Ω is the set of all states of the world, and if $A_1, A_2, \ldots$ is a sequence of disjoint events, then $\mathbb{Q}(\cup A_i) = \sum \mathbb{Q}(A_i)$. As with any probability, there is an expectation operator associated with $\mathbb{Q}$. Denote it by E^*. The definition of $\mathbb{Q}$ can be restated as

$$\mathsf{E}^*[1_A] = R_f \mathsf{E}[\tilde{m} 1_A],$$

because the expectation of an indicator function is the probability of the event. More generally, the definition of $\mathbb{Q}$ implies that

$$\mathsf{E}^*[\tilde{x}] = R_f \mathsf{E}[\tilde{m}\tilde{x}] \qquad (3.22)$$

for every $\tilde{x}$ for which the expectation $\mathsf{E}[\tilde{m}\tilde{x}]$ exists. Because the price of any payoff $\tilde{x}$ is $\mathsf{E}[\tilde{m}\tilde{x}]$, equation (3.22) implies that the price of any payoff $\tilde{x}$ is

$$\frac{1}{R_f} \mathsf{E}^*[\tilde{x}]. \qquad (3.23)$$

Thus, we can compute the price by taking the expectation relative to the probability $\mathbb{Q}$ and then discounting at the risk-free rate. As remarked before, prices would equal expected payoffs discounted at the risk-free rate if investors were risk neutral, so the probability $\mathbb{Q}$ is called a risk-neutral probability.

---

5. The terms "physical probability measure" and "physical measure" are also often used. A risk-neutral probability is also called an "equivalent martingale measure." See the end-of-chapter notes. In general, what we are calling a "probability" is more precisely called a "probability measure." The shorter term "probability" could be confused with a real number (the probability of an event). It should generally be clear from context whether we mean a real number or a function mapping events into real numbers, but when confusion seems likely, we refer to the function as a probability measure.

Let $\tilde{R}$ denote the return $\tilde{x}/p$, where $p = \mathsf{E}[\tilde{m}\tilde{x}]$. Equation (3.22) can be rearranged as

$$\mathsf{E}^*[\tilde{R}] = R_f. \quad (3.24)$$

Thus, the expected return of each asset is the risk-free return when expectations are taken relative to a risk-neutral probability.

If there is no risk-free asset, then risk-neutral probabilities are defined in the same way by substituting $1/\mathsf{E}[\tilde{m}]$ for R_f in (3.21). In a complete market, there is a unique SDF and hence a unique risk-neutral probability. In an incomplete market, different SDFs $\tilde{m}$ define different risk-neutral probabilities $\mathbb{Q}$.

3.5 ORTHOGONAL PROJECTIONS OF SDFs ONTO THE ASSET SPAN

We will see that any finite-variance SDF $\tilde{m}$ is equal to $\tilde{m}_p + \tilde{\varepsilon}$ where $\tilde{m}_p$ is the unique SDF spanned by the assets and $\tilde{\varepsilon}$ is orthogonal to the assets. By "spanned by the assets," it is meant that $\tilde{m}_p$ is the payoff of some portfolio.[6] By "orthogonal to the assets," it is meant that $\mathsf{E}[\tilde{\varepsilon}\tilde{x}_i] = 0$ for each asset i. The SDF $\tilde{m}_p$ is called the orthogonal projection of $\tilde{m}$ onto the span of the assets.

As with any payoff, the cost of the payoff $\tilde{m}_p$ is the expected value of its product with an SDF. Because $\tilde{m}_p$ is both a payoff and an SDF, its cost is $\mathsf{E}[\tilde{m}_p^2]$. Dividing the payoff $\tilde{m}_p$ by its cost produces the return $\tilde{R}_p = \tilde{m}_p/\mathsf{E}[\tilde{m}_p^2]$. This return is used at the end of the section in a discussion of hedging and portfolio optimization for a quadratic utility investor with random labor income. It appears again in Chapters 5 and 7 in the characterization of the mean-variance frontier (it is an inefficient frontier return).

Projections and Regressions

To explain orthogonal projections, it may be useful to first consider ordinary least-squares estimates of linear regression coefficients. This should be a familiar topic, and it is helpful to understand that an orthogonal projection (on a finite-dimensional space) is the same thing as a linear regression. The usual multivariate linear regression model is written as

$$y = X\hat{\beta} + \varepsilon,$$

---

6. The span of the assets is the set of linear combinations of the asset payoffs. These are the random variables that equal the payoff of some portfolio. Thus, "span of the assets" is a synonym for "set of marketed payoffs."

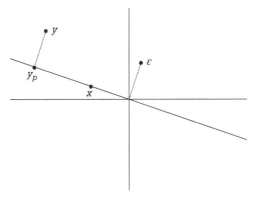

Figure 3.1 Orthogonal projection. y_p is the orthogonal projection of y onto the line spanned by x. Specifically, $y = y_p + \varepsilon$, where y_p is spanned by x (that is, $y_p = bx$ for some constant b) and ε is orthogonal to x (that is, the inner product of ε and x is zero).

where y is a $T \times 1$ vector of observations of the dependent variable, X is a $T \times K$ matrix of observations on K independent variables (one of which may be a constant), $\hat{\beta}$ denotes the $K \times 1$ vector of estimated regression coefficients, and ε is the $T \times 1$ vector of residuals. The vector $X\hat{\beta}$ is the vector of predicted values of the dependent variable, given the observations of the independent variables. Denote it by y_p ("p" for "predicted" or "projected"). The fact that y_p is of the form $X\hat{\beta}$ for some $\hat{\beta}$ means that y_p is a linear combination of the columns of the X matrix (is spanned by the columns of X).

The vector $\hat{\beta}$ is chosen to minimize the sum of squared errors $(y - y_p)'(y - y_p)$, which is equivalent to choosing y_p as the closest point to y in the span of the columns of X. This is also equivalent to choosing y_p so that the error $\varepsilon = y - y_p$ is orthogonal to the columns of X. In other words, $\hat{\beta}$ is defined by the equations $X'(y - X\hat{\beta}) = 0$. Assuming $X'X$ is invertible,[7] this is equivalent to

$$\hat{\beta} = (X'X)^{-1}X'y \quad \text{and} \quad y_p = X(X'X)^{-1}X'y. \qquad (3.25)$$

The formula for the orthogonal projection $\tilde{m}_p$ is analogous to that for y_p. An example of an orthogonal projection is shown in Figure 3.1.

---

7. The matrix $X'X$ is invertible if the columns of X are linearly independent (that is, if there is no multicollinearity). If $X'X$ is not invertible, then the projection y_p is still uniquely defined, but the vector $\hat{\beta}$ is not unique. See Section 3.8 for a general discussion of projections.

Projecting SDFs

Assume all asset payoffs $\tilde{x}_i$ have finite variances and the law of one price holds. Then, there is an SDF $\tilde{m}$ with a finite variance. We will describe its projection $\tilde{m}_p$ onto the span of the assets. If all of the asset prices are positive, then we can equivalently project onto the span of the returns, because the span of the asset returns equals the span of the asset payoffs.

Let $\tilde{X}$ denote the column vector of dimension n that has $\tilde{x}_i$ as its ith element. If there is a risk-free asset, then include its payoff as an element of $\tilde{X}$. For $\tilde{m}_p$ to be in the span of the asset payoffs means that $\tilde{m}_p = \tilde{X}'\theta$ for some $\theta \in \mathbb{R}^n$.[8] We are defining orthogonality in terms of the probability-weighted inner product, that is, the expectation. Thus, the condition that the residual $\tilde{m} - \tilde{m}_p$ be orthogonal to the assets is[9]

$$\mathsf{E}[\tilde{X}(\tilde{m} - \tilde{X}'\theta)] = 0.$$

This can be solved as[10]

$$\mathsf{E}[\tilde{X}\tilde{m}] = \mathsf{E}[\tilde{X}\tilde{X}']\theta \quad \Rightarrow \quad \theta = \mathsf{E}[\tilde{X}\tilde{X}']^{-1}\mathsf{E}[\tilde{X}\tilde{m}] \quad (3.26a)$$

$$\Rightarrow \quad \tilde{m}_p = \mathsf{E}[\tilde{X}\tilde{m}]'\mathsf{E}[\tilde{X}\tilde{X}']^{-1}\tilde{X}. \quad (3.26b)$$

Here we have assumed that the matrix $\mathsf{E}[\tilde{X}\tilde{X}']$ is invertible. If it is not invertible, then there are multiple portfolios θ satisfying $\tilde{X}'\theta = \tilde{m}_p$, but the projection $\tilde{m}_p$ is still uniquely defined (Section 3.8).

Note that $\mathsf{E}[\tilde{X}\tilde{m}]$ in (3.26b) is the n-dimensional column vector with $\mathsf{E}[\tilde{x}_i\tilde{m}]$ as its ith element. By the definition of an SDF, this ith element is p_i. Thus, $\mathsf{E}[\tilde{X}\tilde{m}] = p$, and the formula (3.26b) is equivalent to

$$\tilde{m}_p = p'\mathsf{E}[\tilde{X}\tilde{X}']^{-1}\tilde{X}. \quad (3.27)$$

Because the SDF $\tilde{m}$ with which we started does not appear in (3.27), this shows that the projection $\tilde{m}_p$ is unique—the same for every SDF $\tilde{m}$—as claimed at the beginning of the section. Figure 3.2 illustrates the projection $\tilde{m}_p$ in a simple example.

It turns out that we rarely use the formula (3.27). It usually suffices just to know that there is a unique SDF $\tilde{m}_p$ that is the payoff of a portfolio. However, we will frequently use the formula given in the next subsection for projecting onto the

---

8. Notice that the independent variables form the columns of the X matrix in the linear regression, but the assets form the rows of the vector $\tilde{X}$. Thus, $X\beta$ is replaced by $\tilde{X}'\theta$, and the positions of the transposes are reversed in going from (3.25) to (3.26).

9. Notice that $\tilde{X}(\tilde{m} - \tilde{X}'\theta)$ denotes multiplication of the column vector $\tilde{X}$ by the scalar $\tilde{m} - \tilde{X}'\theta$.

10. In (3.26b), and in subsequent formulas, $\tilde{m}_p$ is written as $\theta'\tilde{X}$ instead of $\tilde{X}'\theta$. Of course, these are the same, because the transpose of a scalar equals itself.

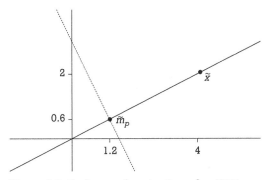

Figure 3.2 Orthogonal projection of an SDF.
There are two states of the world, and they are equally likely. There is no risk-free asset. There is a single asset, and it pays 4 in the first state and 2 in the second. That is, $\tilde{x} = (x_1, x_2) = (4, 2)$. The price of the asset is $p = 3$. An SDF is any vector (m_1, m_2) satisfying the equation $(1/2)x_1 m_1 + (1/2)x_2 m_2 = p$, which is equivalent to $2m_1 + m_2 = 3$. The span of the assets is the set $\{(y_1, y_2) \mid (y_1, y_2) = \theta(x_1, x_2) \text{ for some } \theta\}$ and is shown as the solid line. The set of SDFs is the dotted line. The intersection of the set of SDFs with the span of the assets is the SDF $\tilde{m}_p = (6/5, 3/5)$.

span of a constant and a set of random variables. It appears in a variety of contexts later in the book.

Projections That Include a Constant

We present here a general formula for projections that we will frequently use. The first application is for SDFs in the next section, but there are many more applications later. Let $\tilde{Y}$ denote an n–dimensional vector of random variables, and let $\tilde{X}$ denote a k–dimensional vector of random variables, and assume all of the random variables have finite variances. Define[11]

$$\mathrm{Cov}(\tilde{Y}, \tilde{X}) = \mathsf{E}[(\tilde{Y} - \overline{Y})(\tilde{X} - \overline{X})'],$$

---

11. Notice that the order of the vectors $\tilde{X}$ and $\tilde{Y}$ matters here. Interchanging them is equivalent to transposing: $\mathrm{Cov}(\tilde{X}, \tilde{Y}) = \mathrm{Cov}(\tilde{Y}, \tilde{X})'$.

where $\overline{Y} = \mathsf{E}[\tilde{Y}]$ and $\overline{X} = \mathsf{E}[\tilde{X}]$. This is an $n \times k$ matrix. Use $\mathrm{Cov}(\tilde{X})$ as shorthand for $\mathrm{Cov}(\tilde{X},\tilde{X})$. This is the covariance matrix of the k random variables that constitute the vector $\tilde{X}$. Assume that $\mathrm{Cov}(\tilde{X})$ is nonsingular.

We are interested in projecting each of the random variables that constitute $\tilde{Y}$ onto the span of a constant and the random variables that constitute $\tilde{X}$. It is convenient for future applications to do all of the projections at the same time and to express the result in matrix notation. However, the result would be the same if we projected each element of $\tilde{Y}$ separately and then stacked the individual projections.

Stacking the constants into an n–dimensional vector A, the projection of $\tilde{Y}$ is $A + B\tilde{X}$, where B is an $n \times k$ matrix and

$$\tilde{Y} = A + B\tilde{X} + \tilde{\varepsilon}, \tag{3.28}$$

with $\mathsf{E}[\tilde{\varepsilon}] = 0$ and $\mathsf{E}[\tilde{\varepsilon}\tilde{X}'] = 0$. The latter equality is for an $n \times k$ matrix, and it means that $\mathsf{E}[\tilde{\varepsilon}_i \tilde{X}_j] = 0$ for all $i = 1, \ldots, n$ and all $j = 1, \ldots, k$. Define

$$B = \mathrm{Cov}(\tilde{Y}, \tilde{X})\, \mathrm{Cov}(\tilde{X})^{-1}, \tag{3.29}$$

$$A = \overline{Y} - B\overline{X}, \tag{3.30}$$

$$\tilde{\varepsilon} = \tilde{Y} - \overline{Y} - B(\tilde{X} - \overline{X}). \tag{3.31}$$

Then, (3.28) holds. To show that $A + B\tilde{X}$ is the projection, we need to show that $\mathsf{E}[\tilde{\varepsilon}] = 0$ and $\mathsf{E}[\tilde{\varepsilon}\tilde{X}'] = 0$. The former follows directly from taking expectations in (3.31). For the latter, use the fact that $\mathsf{E}[\tilde{\varepsilon}] = 0$ to obtain

$$\mathsf{E}[\tilde{\varepsilon}\tilde{X}'] = \mathsf{E}[\tilde{\varepsilon}(\tilde{X} - \overline{X})'].$$

Substitute from (3.31) to obtain

$$\mathsf{E}[\tilde{\varepsilon}(\tilde{X} - \overline{X})'] = \mathsf{E}[(\tilde{Y} - \overline{Y})(\tilde{X} - \overline{X})'] - B\mathsf{E}[(\tilde{X} - \overline{X})(\tilde{X} - \overline{X})']$$
$$= \mathrm{Cov}(\tilde{Y}, \tilde{X}) - B\,\mathrm{Cov}(\tilde{X})$$
$$= \mathrm{Cov}(\tilde{Y}, \tilde{X}) - \mathrm{Cov}(\tilde{Y}, \tilde{X}) = 0.$$

This completes the verification that $\tilde{\varepsilon}$ has a zero mean and is orthogonal to $\tilde{X}$. Thus,

$$\overline{Y} + \mathrm{Cov}(\tilde{Y}, \tilde{X})\, \mathrm{Cov}(\tilde{X})^{-1}(\tilde{X} - \overline{X}) \tag{3.32}$$

is the orthogonal projection of $\tilde{Y}$ onto the span of a constant and $\tilde{X}$. This includes the case in which $\tilde{Y}$ is univariate ($n = 1$). In that case, $\mathrm{Cov}(\tilde{Y}, \tilde{X})$ is a k–dimensional row vector. When $\tilde{Y}$ and $\tilde{X}$ are both univariate, then using lowercase letters to denote the random variables the orthogonal projection of $\tilde{y}$ onto the span of a constant and $\tilde{x}$ is

$$\bar{y} + \frac{\mathrm{cov}(\tilde{y}, \tilde{x})}{\mathrm{var}(\tilde{x})}(\tilde{x} - \bar{x}). \tag{3.33}$$

Each element of the column vector $B\tilde{X}$ in the projection of $\tilde{Y}$ is the linear combination of the $\tilde{X}_j$ that has maximum correlation with the corresponding element of $\tilde{Y}$. To see this, for each $i = 1, \ldots, n$, let b_i denote the ith row of B oriented as a column vector. Consider any other k–dimensional column vector b. The covariance of $\tilde{y}_i$ with $b'\tilde{X}$ is

$$\mathrm{cov}(\tilde{y}_i, b'\tilde{X}) = \mathrm{cov}(b'_i\tilde{X}, b'\tilde{X}) + \mathrm{cov}(\tilde{\varepsilon}_i, b'\tilde{X}) = \mathrm{cov}(b'_i\tilde{X}, b'\tilde{X})$$

because $\mathsf{E}[\tilde{\varepsilon}] = 0$ and $\mathsf{E}[\tilde{\varepsilon}\tilde{X}'] = 0$. This implies that

$$\mathrm{corr}(\tilde{y}_i, b'\tilde{X}) = \frac{\mathrm{cov}(\tilde{y}_i, b'\tilde{X})}{\mathrm{stdev}(\tilde{y}_i)\,\mathrm{stdev}(b'\tilde{X})}$$

$$= \frac{\mathrm{cov}(b'_i\tilde{X}, b'\tilde{X})}{\mathrm{stdev}(\tilde{y}_i)\,\mathrm{stdev}(b'\tilde{X})}$$

$$= \frac{\mathrm{corr}(b'_i\tilde{X}, b'\tilde{X})\,\mathrm{stdev}(b'_i\tilde{X})}{\mathrm{stdev}(\tilde{y}_i)}.$$

Thus, the correlation of $\tilde{y}_i$ with $b'\tilde{X}$ depends on b only via the correlation of $b'_i\tilde{X}$ with $b'\tilde{X}$. This is maximized at $b = b_i$.

3.6 HANSEN-JAGANNATHAN BOUNDS

This section derives lower bounds on the standard deviations of SDFs due to Hansen and Jagannathan (1991). The Hansen-Jagannathan bounds have real economic significance. As discussed previously, an asset pricing model is a specification of an SDF $\tilde{m}$. A model can be rejected by the Hansen-Jagannathan bound if $\tilde{m}$ is not sufficiently variable. An illustration of the economic significance of the Hansen-Jagannathan bound with a risk-free asset is given in Exercise 7.2.

Continue to assume that asset payoffs have finite variances and the law of one price holds. Therefore, an SDF exists.

Hansen-Jagannathan Bound with a Risk-Free Asset

Assume there is a risk-free asset. Then, (3.12) holds for any SDF $\tilde{m}$, which we repeat here:

$$\mathsf{E}[\tilde{R}] - R_f = -R_f \mathrm{cov}(\tilde{m}, \tilde{R}).$$

Letting $\mathrm{corr}(\tilde{m}, \tilde{R})$ denote the correlation of $\tilde{m}$ with $\tilde{R}$, we can write this as

$$\mathrm{corr}(\tilde{m}, \tilde{R}) \times \mathrm{stdev}(\tilde{m}) = -\frac{\mathsf{E}[\tilde{R}] - R_f}{R_f \, \mathrm{stdev}(\tilde{R})}.$$

Because the correlation is between -1 and 1, this implies

$$\text{stdev}(\tilde{m}) \geq \frac{|E[\tilde{R}] - R_f|}{R_f \, \text{stdev}(\tilde{R})}. \tag{3.34}$$

Recalling that $1/R_f = E[\tilde{m}]$, we can rewrite this as

$$\frac{\text{stdev}(\tilde{m})}{E[\tilde{m}]} \geq \frac{|E[\tilde{R}] - R_f|}{\text{stdev}(\tilde{R})}. \tag{3.35}$$

The Sharpe ratio of a risky asset with return $\tilde{R}$ is the ratio of the risk premium $E[\tilde{R}] - R_f$ to the risk $\text{stdev}(\tilde{R})$. Thus, the ratio on the right-hand side of (3.35) is the absolute value of the Sharpe ratio of the return $\tilde{R}$. Hence, the ratio of the standard deviation of any SDF to its mean must be at least as large as the maximum absolute Sharpe ratio of all returns. This is one version of the Hansen-Jagannathan (1991) bounds.

Minimum Standard Deviation SDF

The SDF with the minimum standard deviation is the unique SDF $\tilde{m}_p$ in the span of the assets. Furthermore, for this SDF, $\text{stdev}(\tilde{m})/E[\tilde{m}]$ equals the Sharpe ratio of some portfolio; in other words, the inequality in (3.35) is an equality for the SDF $\tilde{m}_p$ and the return $\tilde{R}$ with the maximum Sharpe ratio.

We leave the calculation of the return $\tilde{R}$ with the maximum Sharpe ratio for Chapter 5, but we will show here that $\tilde{m}_p$ has the minimum standard deviation. This is an easy calculation. For any SDF $\tilde{m}$, we have $\tilde{m} = \tilde{m}_p + \tilde{\varepsilon}$, where $\tilde{\varepsilon}$ is orthogonal to the assets. The orthogonality implies $\tilde{\varepsilon}$ is orthogonal to $\tilde{m}_p$; furthermore, given the existence of a risk-free asset, the orthogonality implies $E[\tilde{\varepsilon}] = 0$. Therefore,

$$\text{cov}(\tilde{m}_p, \tilde{\varepsilon}) = E[\tilde{m}_p \tilde{\varepsilon}] - E[\tilde{m}_p]E[\tilde{\varepsilon}] = 0.$$

This implies

$$\text{var}(\tilde{m}) = \text{var}(\tilde{m}_p) + \text{var}(\tilde{\varepsilon}) \geq \text{var}(\tilde{m}_p),$$

so the variance of $\tilde{m}$ is minimized by taking $\tilde{\varepsilon} = 0$, that is, $\tilde{m} = \tilde{m}_p$.

Hansen-Jagannathan Bound with No Risk-Free Asset

Now, suppose that there is no risk-free asset. For any SDF $\tilde{m}$ and any return $\tilde{R}$, (3.11) holds, namely,

$$1 = \text{cov}(\tilde{m}, \tilde{R}) + E[\tilde{m}]E[\tilde{R}].$$

Therefore,

$$\text{corr}(\tilde{m}, \tilde{R}) \times \text{stdev}(\tilde{m}) \times \text{stdev}(\tilde{R}) = 1 - \mathsf{E}[\tilde{m}]\mathsf{E}[\tilde{R}],$$

implying

$$\text{stdev}(\tilde{m}) \geq \frac{|\mathsf{E}[\tilde{m}]\mathsf{E}[\tilde{R}] - 1|}{\text{stdev}(\tilde{R})}. \qquad (3.36)$$

The maximum of the right-hand side of (3.36) over all returns $\tilde{R}$ defines a lower bound on the standard deviation of any SDF, with the lower bound depending on the mean $\mathsf{E}[\tilde{m}]$.

Projecting SDFs onto a Constant and the Asset Span

To refine the bound (3.36), we consider another projection of SDFs. This is identical to the previous projection if there is a risk-free asset. Stack the payoffs of the risky assets in a column vector $\tilde{X}_0$. Let Σ_x denote the covariance matrix of $\tilde{X}_0$ and assume it is nonsingular. Let μ_x denote the mean of $\tilde{X}_0$, and let ν denote the mean of $\tilde{m}$. From (3.32), we have that

$$\tilde{m}_{vp} \stackrel{\text{def}}{=} \nu + \text{Cov}(\tilde{m}, \tilde{X}_0) \Sigma_x^{-1} (\tilde{X}_0 - \mu_x) \qquad (3.37)$$

is the orthogonal projection of $\tilde{m}$ onto the span of a constant and the risky assets. Furthermore,

$$\text{Cov}(\tilde{m}, \tilde{X}_0) = \mathsf{E}[(\tilde{m} - \nu)(\tilde{X}_0 - \mu_x)'] = \mathsf{E}[\tilde{m}\tilde{X}_0]' - \nu\mu_x' = (p_0 - \nu\mu_x)',$$

where $p_0 = \mathsf{E}[\tilde{m}\tilde{X}_0]$ is the vector of prices of the risky assets. Therefore,

$$\tilde{m}_{vp} = \nu + (p_0 - \nu\mu_x)' \Sigma_x^{-1} (\tilde{X}_0 - \mu_x). \qquad (3.38)$$

When a risk-free asset exists, the mean of an SDF is $1/R_f$, so $\nu = 1/R_f$. In that case, the span of a constant and the risky assets is the span of all of the assets, so it must be that $\tilde{m}_{vp} = \tilde{m}_p$. Thus, (3.38) is simply a new formula for the projection $\tilde{m}_p$ when there is a risk-free asset. If there is no risk-free asset, then $\tilde{m}_{vp}$ is an SDF for any value of ν. We can see this from

$$\begin{aligned}
\mathsf{E}[\tilde{m}_{vp}\tilde{X}_0] &= \mathsf{E}\left[\{\nu + (p_0 - \nu\mu_x)' \Sigma_x^{-1}(\tilde{X}_0 - \mu_x)\}\tilde{X}_0\right] \\
&= \nu\mu_x + (p_0 - \nu\mu_x) \Sigma_x^{-1} \Sigma_x \\
&= \nu\mu_x + p_0 - \nu\mu_x = p_0.
\end{aligned}$$

If all of the asset prices are positive, then we can define the return of each asset, and the span of the returns is the same as the span of the asset payoffs. So, the projection of an SDF $\tilde{m}$ onto the span of a constant and the asset payoffs can also be written as

$$\tilde{m}_{vp} = v + \text{Cov}(\tilde{m}, \tilde{\mathbf{R}}) \Sigma^{-1}(\tilde{\mathbf{R}} - \mu), \tag{3.39}$$

where $\tilde{\mathbf{R}}$ denotes the vector of risky asset returns, Σ is the covariance matrix of the returns, and $\mu = \mathsf{E}[\tilde{\mathbf{R}}]$.

Hansen-Jagannathan Bound with No Risk-Free Asset Revisited

Suppose there is no risk-free asset, and for a given real number v, consider the SDFs $\tilde{m}$ with $\mathsf{E}[\tilde{m}] = v$. The minimum variance SDF with this mean is the orthogonal projection $\tilde{m}_{vp}$ defined in (3.38). This can be seen by the same reasoning as in the case with a risk-free asset. Let $\tilde{m}$ be any SDF with $\mathsf{E}[\tilde{m}] = v$. Then, $\tilde{m} = \tilde{m}_{vp} + \tilde{\varepsilon}$, where (i) $\tilde{m}_{vp}$ is spanned by the assets and a constant, and (ii) $\tilde{\varepsilon}$ is orthogonal to the assets and a constant. The orthogonality implies $\mathsf{E}[\tilde{m}_{vp}\tilde{\varepsilon}] = 0$ and $\mathsf{E}[\tilde{\varepsilon}] = 0$. Thus, by the same calculations as before, we conclude that

$$\text{var}(\tilde{m}) = \text{var}(\tilde{m}_{vp}) + \text{var}(\tilde{\varepsilon}) \geq \text{var}(\tilde{m}_{vp}).$$

Hence, (3.36) implies that, for each real v,

$$\text{stdev}(\tilde{m}) \geq \text{stdev}(\tilde{m}_{vp}) \geq \max_{\tilde{R}} \frac{|v\mathsf{E}[\tilde{R}] - 1|}{\text{stdev}(\tilde{R})}.$$

3.7 HEDGING AND OPTIMAL PORTFOLIOS WITH QUADRATIC UTILITY

This section characterizes the optimal portfolio of a quadratic utility investor who has labor income $\tilde{y}$. The investor hedges the risk of her labor income by selling the projection of the labor income onto the span of the assets. In other words, she sells the portfolio for which the payoff best approximates her labor income. For example, a worker may want to short sell stocks of companies in the industry in which she works. The result that the investor hedges labor income $\tilde{y}$ by selling its projection also holds for CARA utility with normal returns (Exercise 2.3). Here we do not assume normal returns.

We use orthogonal projections to derive the characterization. Let the utility function be $u(w) = -(w - \zeta)^2/2$. Suppose the investor has labor income $\tilde{y}$. The investor chooses a marketed payoff $\tilde{x}$ subject to the budget constraint to maximize $\mathsf{E}[u(\tilde{x} + \tilde{y})]$. As before, the first-order condition is that $u'(\tilde{x} + \tilde{y}) = \gamma \tilde{m}$ for a constant γ (the Lagrange multiplier) and an SDF $\tilde{m}$. Because $u'(\tilde{x} + \tilde{y}) = \zeta -$

$\tilde{x} - \tilde{y}$, the first-order condition implies

$$\tilde{x} = \zeta - \tilde{y} - \gamma \tilde{m}. \tag{3.40}$$

Because $\tilde{x}$ is a marketed payoff, it equals its own projection onto the span of the assets. Therefore, it equals the projection of the right-hand side of (3.40) onto the span of the assets. Projections are linear (the projection of the sum of two random variables is the sum of the projections), so we have

$$\tilde{x} = \zeta_p - \tilde{y}_p - \gamma \tilde{m}_p, \tag{3.41}$$

where the subscript p denotes the projection in each case. The three terms on the right-hand side are as follow:

- ζ_p is the projection of the constant ζ onto the span of the assets. If there is a risk-free asset, then ζ is in the span, and $\zeta_p = \zeta$.
- $\tilde{y}_p$ is the best approximation to $\tilde{y}$ in the asset span. The investor hedges her labor income by selling the best approximation.
- The investor shorts the asset (or portfolio) that has payoff $\tilde{m}_p$ and corresponding return $\tilde{R}_p = \tilde{m}_p / \mathsf{E}[\tilde{m}_p^2]$.

Adding these pieces might give the impression that the investor invests in the risk-free asset (if one exists) and short sells risky assets. However, this is incorrect, because the portfolio with return $\tilde{R}_p$ can include a long position in the risk-free asset, so shorting it may partially or completely offset the long position in the risk-free asset created by the first piece of the portfolio. In fact, we will see in Section 5.4 that a return is mean-variance efficient if and only it is short the portfolio with return $\tilde{R}_p$ and long the risk-free asset. In the presence of a risk-free asset, the three parts listed above can be described thus: short the best approximation to labor income and invest the proceeds plus initial wealth in a portfolio on the mean-variance frontier (see Exercise 5.2 for the distinction between frontier and efficient portfolios for quadratic utility investors).

To obtain a more precise formula for the extent to which the investor shorts the asset with return $\tilde{R}_p$, use the budget constraint to calculate the Lagrange multiplier γ in (3.41). We have

$$w_0 = \mathsf{E}[\tilde{m}_p \tilde{x}] = \mathsf{E}[\tilde{m}_p \zeta_p] - \mathsf{E}[\tilde{m}_p \tilde{y}_p] - \gamma \mathsf{E}[\tilde{m}_p^2],$$

implying

$$\gamma = \frac{\mathsf{E}[\tilde{m}_p \zeta_p] - \mathsf{E}[\tilde{m}_p \tilde{y}_p] - w_0}{\mathsf{E}[\tilde{m}_p^2]}$$

and

$$\tilde{x} = \zeta_p - \tilde{y}_p - \left(\mathsf{E}[\tilde{m}_p \zeta_p] - \mathsf{E}[\tilde{m}_p \tilde{y}_p] - w_0\right) \tilde{R}_p. \tag{3.42}$$

Thus,
$$\mathsf{E}[\tilde{m}_p \zeta_p] - \mathsf{E}[\tilde{m}_p \tilde{y}_p] - w_0 \tag{3.43}$$

is the amount (number of units of the consumption good) by which the investor shorts the asset with return $\tilde{R}_p$. We can also write (3.43) as

$$\zeta \mathsf{E}[\tilde{m}_p] - \mathsf{E}[\tilde{m}_p \tilde{y}] - w_0,$$

because the residuals in the projections of ζ and $\tilde{y}$ are orthogonal to $\tilde{m}_p$.

3.8 HILBERT SPACES AND GRAM-SCHMIDT ORTHOGONALIZATION

The results on the law of one price, arbitrage, and existence of SDFs presented earlier in the chapter in a finite-state model also apply to models with payoffs having finite variances. The analysis of such models is facilitated by extending some linear algebra notions in $\mathbb{R}^n$ to the space of finite-variance random variables. This section provides a quick description of the main ideas.

The concept of Gram-Schmidt orthogonalization is quite useful and is discussed further in the end-of-chapter notes. The other results presented in this section, while also useful, are used in this book only in a few proofs and exercises (for example, Exercise 3.9). Hence, they can be skipped if those proofs and exercises are to be skipped.

It is conventional to denote the space of finite-variance random variables by $\mathcal{L}^2$, the $\mathcal{L}$ being a reference to Henri Lebesgue, who is responsible for the general definition of integrals and expectations.[12] Addition and scalar multiplication are defined in $\mathcal{L}^2$ in the obvious way: Given $\tilde{x}$ and $\tilde{y}$ in $\mathcal{L}^2$, the sum is the random variable $\tilde{z}$ defined by $\tilde{z}(\omega) = \tilde{x}(\omega) + \tilde{y}(\omega)$ in each state of the world ω, and given a constant a, $a\tilde{x}$ is the random variable $\tilde{z}$ defined by $\tilde{z}(\omega) = a\tilde{x}(\omega)$ in each state of the world ω. Sums and scalar multiples of finite-variance random variables also have finite variances.

The square root of the second moment of a random variable $\tilde{x} \in \mathcal{L}^2$ is defined to be its norm, which is denoted by $\|\tilde{x}\|$; that is, $\|\tilde{x}\|^2 = \mathsf{E}[\tilde{x}^2]$. The inner product of any two random variables $\tilde{x}, \tilde{y} \in \mathcal{L}^2$ is denoted by $\langle \tilde{x}, \tilde{y} \rangle$ and defined to be the expectation of their product: $\langle \tilde{x}, \tilde{y} \rangle = \mathsf{E}[\tilde{x}\tilde{y}]$. If $\tilde{x}, \tilde{y} \in \mathcal{L}^2$ have zero means, then obviously $\|\tilde{x}\| = \text{stdev}(\tilde{x})$ and $\langle \tilde{x}, \tilde{y} \rangle = \text{cov}(\tilde{x}, \tilde{y})$. The space $\mathcal{L}^2$ equipped with the norm and inner product has many of the properties of $\mathbb{R}^n$ (with norm

---

12. Actually, we regard any random variables $\tilde{x}$ and $\tilde{y}$ that are equal with probability 1 as being equivalent, and the space $\mathcal{L}^2$ is the space of equivalence classes. However, this issue is unimportant for our purposes, and we will speak of the elements of $\mathcal{L}^2$ as random variables.

Stochastic Discount Factors

$\sqrt{\sum_{i=1}^{n} x_i^2}$ and inner product $\sum_{i=1}^{n} x_i y_i$). The space $\mathcal{L}^2$ is an example of a Hilbert space. All of the properties of $\mathcal{L}^2$ stated below are true of any Hilbert space, including $\mathbb{R}^n$.

For $\tilde{x}, \tilde{y} \in \mathcal{L}^2$, define

$$\text{proj}_{\tilde{y}}\tilde{x} = \frac{\langle \tilde{x}, \tilde{y}\rangle}{\langle \tilde{y}, \tilde{y}\rangle} \tilde{y}.$$

This is the orthogonal projection of $\tilde{x}$ onto the line $\{a\tilde{y} \mid a \in \mathbb{R}\} \subset \mathcal{L}^2$. The residual $\tilde{\varepsilon} \stackrel{\text{def}}{=} \tilde{x} - \text{proj}_{\tilde{y}}\tilde{x}$ is orthogonal to $\tilde{y}$, in the sense that $\langle \tilde{\varepsilon}, \tilde{y}\rangle = 0$. Let $\tilde{x}_1, \ldots, \tilde{x}_k \in \mathcal{L}^2$ be linearly independent (none is a linear combination of the others). The Gram-Schmidt orthogonalization of $\tilde{x}_1, \ldots, \tilde{x}_k$ is $\tilde{z}_1, \ldots, \tilde{z}_k$ defined as follows:

$$\begin{aligned}
\tilde{y}_1 &= \tilde{x}_1, & \tilde{z}_1 &= \frac{\tilde{y}_1}{\|\tilde{y}_1\|}, \\
\tilde{y}_2 &= \tilde{x}_2 - \text{proj}_{\tilde{y}_1}\tilde{x}_2, & \tilde{z}_2 &= \frac{\tilde{y}_2}{\|\tilde{y}_2\|}, \\
\tilde{y}_3 &= \tilde{x}_3 - \text{proj}_{\tilde{y}_1}\tilde{x}_3 - \text{proj}_{\tilde{y}_2}\tilde{x}_3, & \tilde{z}_3 &= \frac{\tilde{y}_3}{\|\tilde{y}_3\|}, \\
&\vdots & &\vdots \\
\tilde{y}_k &= \tilde{x}_k - \sum_{i=1}^{k-1} \text{proj}_{\tilde{y}_i}\tilde{x}_k, & \tilde{z}_k &= \frac{\tilde{y}_k}{\|\tilde{y}_k\|}.
\end{aligned}$$

The $\tilde{z}_i$ span the same subspace as do the $\tilde{x}_i$ (see below for definitions) and have the property that $\|\tilde{z}_i\| = 1$ for each i and $\langle \tilde{z}_i, \tilde{z}_j \rangle = 0$ for $i \ne j$. They are called orthonormal.

A set $M \subset \mathcal{L}^2$ is called a subspace if it is closed under scalar multiplication and addition, meaning that (i) if $\tilde{x} \in M$ and a is a constant, then $a\tilde{x} \in M$, and (ii) if $\tilde{x} \in M$ and $\tilde{y} \in M$, then $\tilde{x} + \tilde{y} \in M$.

A sequence $\tilde{x}_1, \tilde{x}_2, \ldots$, in $\mathcal{L}^2$ is said to converge to $\tilde{x} \in \mathcal{L}^2$ if $\|\tilde{x}_n - \tilde{x}\| \to 0$. A set $D \subset \mathcal{L}^2$ is called closed if for any sequence $\tilde{x}_1, \tilde{x}_2, \ldots$ in D converging to $\tilde{x} \in \mathcal{L}^2$, we have $\tilde{x} \in D$.

If M is a closed linear subspace of $\mathcal{L}^2$, then for every $\tilde{x} \in \mathcal{L}^2$ there exists a unique (up to null events) $\tilde{x}_p \in M$ such that the residual $\tilde{x} - \tilde{x}_p$ is orthogonal to M, meaning that $\langle \tilde{x} - \tilde{x}_p, \tilde{y}\rangle = 0$ for every $\tilde{y} \in M$. This $\tilde{x}_p$ is also the unique closest point in M to $\tilde{x}$, meaning that it minimizes $\|\tilde{x} - \tilde{x}'_p\|$ over $\tilde{x}'_p \in M$. It is called the orthogonal projection of $\tilde{x}$ on M.

The linear span of a set $\{\tilde{x}_1, \ldots, \tilde{x}_n\} \subset \mathcal{L}^2$ is defined to be the set of all random variables

$$\sum_{i=1}^{n} a_i \tilde{x}_i$$

where the a_i are constants. The linear span is a subspace of $\mathcal{L}^2$. The linear span M of a finite set $\{\tilde{x}_1, \ldots, \tilde{x}_n\}$ is a closed linear subspace, so any $\tilde{x} \in \mathcal{L}^2$ has a unique orthogonal projection on M. Let $k \le n$ denote the maximum number of linearly independent elements of M (k is called the dimension of M) and construct the Gram-Schmidt orthogonalization $\tilde{z}_1,\ldots,\tilde{z}_k$ of k such elements. The unique orthogonal projection of any $\tilde{x} \in \mathcal{L}^2$ on M equals $\sum_{i=1}^k \operatorname{proj}_{\tilde{z}_i} \tilde{x}$.

Let D be any subset of $\mathcal{L}^2$. A function $f : D \to \mathbb{R}$ is continuous if for any sequence $\tilde{x}_1, \tilde{x}_2, \ldots$ in D converging to $\tilde{x} \in D$, we have $f[\tilde{x}_n] \to f[\tilde{x}]$.[13]

Let M be a subspace of $\mathcal{L}^2$. A function $f : M \to \mathbb{R}$ is linear if $f[a\tilde{x}] = af[\tilde{x}]$ and $f[\tilde{x}+\tilde{y}] = f[\tilde{x}] + f[\tilde{y}]$ for all constants a and $\tilde{x},\tilde{y} \in M$.

If M is a closed linear subspace of $\mathcal{L}^2$ and $f : M \to \mathbb{R}$ is a continuous linear function, then there exists a unique $\tilde{m} \in M$ such that $f[\tilde{x}] = \langle \tilde{x}, \tilde{m} \rangle$ for every $\tilde{x} \in M$. This representation of f as $\tilde{x} \mapsto \langle \tilde{x}, \tilde{m} \rangle$ is called the Riesz representation of f.

If M is the linear span of a finite set $\{\tilde{x}_1,\ldots,\tilde{x}_n\}$, then every linear function $f : M \to \mathbb{R}$ is continuous and hence has a Riesz representation as $f[\tilde{x}] = \langle \tilde{x}, \tilde{m} \rangle$ for a unique $\tilde{m} \in M$.

For all $\tilde{x}, \tilde{y} \in \mathcal{L}^2$, the Cauchy-Schwartz inequality holds:

$$|\langle \tilde{x},\tilde{y}\rangle| \le \|\tilde{x}\| \cdot \|\tilde{y}\|.$$

This is equivalent to the statement that the correlation of any two random variables must be between -1 and $+1$. To see this, note that the correlation is unaffected by subtracting constants, so suppose $\tilde{x}$ and $\tilde{y}$ are two finite-variance random variables with zero means. The Cauchy-Schwartz inequality states that the absolute value of the covariance of $\tilde{x}$ and $\tilde{y}$ is less than or equal to the product of their standard deviations.

The Cauchy-Schwartz inequality implies

$$\|\tilde{x}\| \ge \left| \frac{\langle \tilde{x},\tilde{y}\rangle}{\|\tilde{y}\|} \right| = \left| \left\langle \tilde{x}, \frac{\tilde{y}}{\|\tilde{y}\|} \right\rangle \right|,$$

which implies further that

$$\|\tilde{x}\| \ge |\langle \tilde{x},\tilde{z}\rangle|$$

for every $\tilde{z}$ with $\|\tilde{z}\| = 1$. Because

$$\|\tilde{x}\| = \left| \left\langle \tilde{x}, \frac{\tilde{x}}{\|\tilde{x}\|} \right\rangle \right|,$$

we actually have

$$\|\tilde{x}\| = \max_{\|\tilde{z}\|=1} |\langle \tilde{x},\tilde{z}\rangle|.$$

---

13. The square-bracket notation is used here to distinguish a real-valued function (sometimes called a functional) of a random variable, such as the expectation $E[\tilde{x}]$ or cost $C[\tilde{x}]$ seen earlier, from random variables $\omega \mapsto g(\tilde{x}(\omega))$ where $g : \mathbb{R} \to \mathbb{R}$.

3.9 NOTES AND REFERENCES

The origin of the SDF concept is a bit murky, but all of the ideas described in the introduction to the chapter are clearly expressed in Rubinstein (1976), and many appear in Dreze (1970) and Beja (1971). All of this is built upon the fundamental state price concept due to Arrow (1953).

Dybvig and Ross (1989) term the equivalence of the following conditions the "fundamental theorem of asset pricing":

(i) Absence of arbitrage opportunities;
(ii) Existence of a strictly positive SDF;
(iii) Existence of an optimum for an investor with strictly monotone utility.

The fact (i) $\Rightarrow$ (iii) means that any arbitrage-free prices are equilibrium prices in some economy. Namely, take H investors who have an optimum and assume each is endowed with the optimal portfolio. This is an example of an autarkic (no trade) equilibrium.

Ross (1978b) shows that the absence of arbitrage opportunities implies the existence of a strictly positive SDF in a market with no arbitrage opportunities, when there are only finitely many states of the world. The existence of a strictly positive SDF in a market with infinitely many states of the world and no arbitrage opportunities is due to Dalang, Morton, and Willinger (1990). The existence of an optimal portfolio for a CARA investor in such a market and the consequent existence of a strictly positive SDF is due to Rogers (1994). For a survey of this topic, see Delbaen and Schachermayer (2006).

Chamberlain and Rothschild (1983) show that the law of one price implies the existence of an (not necessarily strictly positive) SDF when asset payoffs have finite variances. They also introduce the projection of SDFs onto the asset span.

The concept of a risk-neutral probability is introduced by Cox and Ross (1976a,b) and is developed systematically by Harrison and Kreps (1979), under the name "equivalent martingale measure." In Chapter 9, we will again use orthogonal projections to characterize the optimal portfolio of a quadratic utility investor, as in Section 3.5, but in a dynamic model, following Cochrane (2014). The Hansen-Jagannathan bounds are from Hansen and Jagannathan (1991). Hansen and Jagannathan also derive minimum variance bounds for strictly positive SDFs, a topic not covered in this chapter. Luttmer (1996) extends the bounds to economies with frictions.

Risk-neutral probabilities combine physical probabilities with an SDF. Risk-neutral probabilities can in principle be calculated from market option prices, given the presence of traded options at enough strike prices. These are forward-looking probabilities. There is no obvious way to calculate forward-looking physical probabilities. Therefore, it is natural to ask what we can learn about physical probabilities (and the SDF) from risk-neutral probabilities (that is, can we disentangle the product of physical probabilities and the SDF into the separate factors). Under different assumptions about how asset prices evolve, this problem has been tackled by Jackwerth and Rubinstein (1996), Dybvig and Rogers (1997), Jackwerth (2000), Ross (2015), and Borovička, Hansen, and Scheinkman (forthcoming). These results are called recovery theorems.

In the Gram-Schmidt orthogonalization in Section 3.8, $\tilde{y}_j$ is the residual from the orthogonal projection of $\tilde{x}_j$ on $\tilde{y}_1,\ldots,\tilde{y}_{j-1}$, for $j=1,\ldots,k$. If the $\tilde{x}_j$ have zero means, then

$$\tilde{x}_j = \sum_{i=1}^{j-1} \frac{\text{cov}(\tilde{x}_j,\tilde{y}_i)}{\text{var}(\tilde{y}_i)}\tilde{y}_i + \tilde{y}_j.$$

In general, the Gram-Schmidt orthogonalization can be represented as follows: Let Σ denote the matrix with $\mathsf{E}[\tilde{x}_i\tilde{x}_j]$ as its (i,j)th element (the covariance matrix if the $\tilde{x}_i$ have zero means). By linear independence, Σ is positive definite ($u'\Sigma u = \mathsf{E}[(\sum_{i=1}^{k} u_i\tilde{x}_i)^2] > 0$ if $u \neq 0$). For any symmetric positive-definite matrix Σ, there exists a unique lower triangular matrix L with positive diagonal elements, called the Cholesky decomposition of Σ, such that $LL' = \Sigma$. Set $\tilde{X} = (\tilde{x}_1\cdots\tilde{x}_k)'$, $\tilde{Y} = (\tilde{y}_1\cdots\tilde{y}_k)'$, and $\tilde{Z} = (\tilde{z}_1\cdots\tilde{z}_k)'$. The Gram-Schmidt orthogonalization of $\tilde{X}$ is $\tilde{Z}$ defined by

$$\tilde{Z} = L^{-1}\tilde{X}. \tag{3.44}$$

To see this, note that, in the definition of the Gram-Schmidt orthogonalization, $\tilde{X} = A\tilde{Y}$, where A is a lower triangular matrix. Specifically, A has 1's on its diagonal and $\mathsf{E}[\tilde{x}_i\tilde{y}_j]/\mathsf{E}[\tilde{y}_j^2]$ as the (i,j)th element below the diagonal. Also, $\tilde{Z} = D^{-1}\tilde{Y}$, where D is the diagonal matrix with $\sqrt{\mathsf{E}[\tilde{y}_i^2]}$ as its ith diagonal element. Moreover, $\mathsf{E}[\tilde{Z}\tilde{Z}'] = I$, because the $\tilde{z}_i$ are orthonormal. Setting $L = AD$ therefore yields (i) $\tilde{X} = A\tilde{Y} = AD\tilde{Z} = L\tilde{Z}$, (ii) $\Sigma = \mathsf{E}[\tilde{X}\tilde{X}'] = L\mathsf{E}[\tilde{Z}\tilde{Z}']L' = LL'$, and (iii) the ith diagonal element of L is $\sqrt{\mathsf{E}[\tilde{y}_i^2]} > 0$.

EXERCISES

3.1. Assume there are two possible states of the world: ω_1 and ω_2. There are two assets, a risk-free asset returning R_f in each state, and a risky asset with initial price equal to 1 and date–1 payoff $\tilde{x}$. Normalize the unit of measurement of the risk-free asset so that its price is 1 at date 0 and its payoff per share is R_f. Let $R_d = \tilde{x}(\omega_1)$ and $R_u = \tilde{x}(\omega_2)$. Assume without loss of generality that $R_u > R_d$.

(a) What inequalities between R_f, R_d, and R_u are equivalent to the absence of arbitrage opportunities?

(b) Assuming there are no arbitrage opportunities, compute the unique vector of state prices, and compute the unique risk-neutral probabilities of states ω_1 and ω_2.
(c) Suppose another asset is introduced into the market that pays $\max(\tilde{x} - K, 0)$ for some constant K. Compute the price at which this asset should trade, assuming there are no arbitrage opportunities.

3.2. Assume there are three possible states of the world: ω_1, ω_2, and ω_3. Assume there are two assets: a risk-free asset returning R_f in each state, and a risky asset with return R_1 in state ω_1, R_2 in state ω_2, and R_3 in state ω_3. Assume the probabilities are $1/4$ for state ω_1, $1/2$ for state ω_2, and $1/4$ for state ω_3. Assume $R_f = 1.0$, and $R_1 = 1.1$, $R_2 = 1.0$, and $R_3 = 0.9$.

(a) Prove that there are no arbitrage opportunities.
(b) Describe the one-dimensional family of state-price vectors (q_1, q_2, q_3).
(c) Describe the one-dimensional family of SDFs

$$\tilde{m} = (m_1, m_2, m_3),$$

where m_i denotes the value of the SDF in state ω_i. Verify that $m_1 = 4$, $m_2 = -2$, $m_3 = 4$ is an SDF.
(d) Compute the projection of SDFs onto the span of the risk-free and risky assets by applying the formula (3.33) for the projection of a random variable $\tilde{y}$ onto the span of a constant and a random variable $\tilde{x}$. Take $\tilde{y}$ to be the SDF $m_1 = 4$, $m_2 = -2$, $m_3 = 4$ and $\tilde{x}$ to be the risky asset return $R_1 = 1.1$, $R_2 = 1.0$, $R_3 = 0.9$.
(e) The projection in part (d) is by definition the payoff of some portfolio. What is the portfolio?

3.3. Assume there is a risk-free asset. Let $\tilde{R}$ denote the vector of risky asset returns, let μ denote the mean of $\tilde{R}$, and let Σ denote the covariance matrix of $\tilde{R}$. Let ι denote a vector of 1's. Derive the following formula for the SDF $\tilde{m}_p$ from the projection formula (3.32):

$$\tilde{m}_p = \frac{1}{R_f} + \left(\iota - \frac{1}{R_f}\mu\right)' \Sigma^{-1}(\tilde{R} - \mu). \tag{3.45}$$

3.4. Suppose two random vectors $\tilde{X}$ and $\tilde{Y}$ are joint normally distributed. Explain why the orthogonal projection (3.32) equals $E[\tilde{Y} \mid \tilde{X}]$.

3.5. Show that, if there is a strictly positive SDF, then there are no arbitrage opportunities.

3.6. Show by example that the law of one price can hold but there can still be arbitrage opportunities.

3.7. Suppose there is an SDF $\tilde{m}$ with the property that for every function g there exists a portfolio θ (depending on g) such that

$$\sum_{i=1}^{n} \theta_i \tilde{x}_i = g(\tilde{m}).$$

Consider an investor with no labor income $\tilde{y}$. Show that her optimal wealth is a function of $\tilde{m}$. Hint: For any feasible $\tilde{w}$, define $\tilde{w}^* = \mathsf{E}[\tilde{w} \mid \tilde{m}]$, and show that $\tilde{w}^*$ is both budget feasible and at least as preferred as $\tilde{w}$, using the result of Section 1.5. Note: The assumption in this exercise is a weak form of market completeness. The exercise is inspired by Chamberlain (1988).

3.8. Suppose there is a risk-free asset. Adopt the notation of Exercise 3.3, and assume the risky asset returns have a joint normal distribution. Show that the optimal portfolio of risky assets for an investor with no labor income is $\pi = \delta \Sigma^{-1}(\mu - R_f \iota)$ for some real number δ by applying the reasoning of Exercise 3.7 with $\tilde{m} = \tilde{m}_p$, using the formula (3.45) for $\tilde{m}_p$, and using the result of Exercise 3.4.

3.9. Assume the payoff of each asset has a finite variance and the law of one price holds. Apply facts stated in Section 3.8 to show that there is a unique SDF $\tilde{m}_p$ in the span of the asset payoffs. Show that the orthogonal projection of any other SDF onto the span of the asset payoffs equals $\tilde{m}_p$.

4

Equilibrium and Efficiency

CONTENTS

4.1 Pareto Optima	80
4.2 Competitive Equilibria	83
4.3 Complete Markets	84
4.4 Aggregation and Efficiency with Linear Risk Tolerance	86
4.5 Beginning-of-Period Consumption	93
4.6 Notes and References	95

This chapter presents the definitions of competitive equilibrium and Pareto optimum. Competitive equilibria in complete markets are Pareto optimal. Also, competitive equilibria are Pareto optimal if all investors have LRT utility functions with the same cautiousness parameter even if markets are incomplete (provided only that a risk-free asset exists). Pareto optimality is useful for asset pricing, because it implies the existence of a representative investor, as is explained in Chapter 7. This means that assets can be priced by an SDF formed from a marginal utility of market wealth.

Gorman aggregation means that equilibrium prices are independent of the initial distribution of wealth across investors. Gorman aggregation is possible for all asset payoff distributions if and only if investors have LRT utility functions with the same cautiousness parameter.

It is assumed except in Section 4.5 that beginning-of-period consumption is already determined, so the focus is on the investment problem. Section 4.5 shows that the results also hold when investors choose both beginning-of-period consumption and investments optimally. It is assumed throughout the chapter that all investors agree on the probabilities of the different possible states of the world.

4.1 PARETO OPTIMA

Suppose there are H investors, indexed as $h = 1, \ldots, H$, with utility functions u_h. A social objective is to allocate the aggregate end-of-period wealth $\tilde{w}_m$ ("m" for "market") to investors in such a way that it is impossible to further increase the expected utility of any investor without reducing the expected utility of another. An allocation with this property is called Pareto optimal, in recognition of the economist Vilfredo Pareto. As is discussed further below, Pareto optimality in a securities market is an issue of efficient risk sharing.

Formally, an allocation $(\tilde{w}_1, \ldots, \tilde{w}_H)$ is defined to be Pareto optimal if (i) it is feasible—that is, $\sum_{h=1}^{H} \tilde{w}_h = \tilde{w}_m$—and (ii) there does not exist any other feasible allocation $(\tilde{w}'_1, \ldots, \tilde{w}'_H)$ such that

$$E[u_h(\tilde{w}'_h)] \geq E[u_h(\tilde{w}_h)]$$

for all h, with

$$E[u_h(\tilde{w}'_h)] > E[u_h(\tilde{w}_h)]$$

for some h. For the sake of brevity, the term "allocation" will mean feasible allocation in the remainder of the chapter.

EXAMPLE

A simple example of an allocation that does not involve efficient risk sharing and hence is not Pareto optimal is as follows. Suppose there are two risk-averse investors and two possible states of the world, with $\tilde{w}_m$ being the same in both states, say, $\tilde{w}_m = 6$, and with the two states being equally likely. The allocation

$$\tilde{w}_1 = \begin{cases} 2 & \text{in state 1} \\ 4 & \text{in state 2} \end{cases}$$

$$\tilde{w}_2 = \begin{cases} 4 & \text{in state 1} \\ 2 & \text{in state 2} \end{cases}$$

is not Pareto optimal, because both investors would prefer to receive 3 in each state, which is also a feasible allocation.

Later in this section, we will see that the allocation in this example cannot be Pareto optimal even if the states have different probabilities—in which case it is not necessarily true that each investor would prefer to receive 3 in each state. The reason it cannot be Pareto optimal is that the aggregate wealth is constant across the two states (6 in each state), and, as we will see, in any Pareto-optimal allocation each investor's wealth must be constant across sets of states in which market wealth is constant. At a Pareto optimum, there must be perfect insurance against everything except the uncertainty of aggregate wealth, and insurance is imperfect in the above example.

Social Planner's Problem

It is a standard result from microeconomics that a Pareto optimum maximizes a weighted average of utility functions. We can prove this as follows: Consider an allocation $(\tilde{w}'_1, \ldots, \tilde{w}'_H)$. Define

$$\bar{u}_h = E[u_h(\tilde{w}'_h)]$$

for each h. If the allocation is Pareto optimal, then it must solve

$$\max_{\tilde{w}_2, \ldots, \tilde{w}_H} E\left[u_1\left(\tilde{w}_m - \sum_{h=2}^{H} \tilde{w}_h\right)\right] \quad \text{subject to} \quad E[u_h(\tilde{w}_h)] \geq \bar{u}_h \text{ for } h = 2, \ldots, H. \tag{4.1}$$

The Lagrangean for this problem is

$$E\left[u_1\left(\tilde{w}_m - \sum_{h=2}^{H} \tilde{w}_h\right)\right] + \sum_{h=2}^{H} \lambda_h E[u_h(\tilde{w}_h)] - \sum_{h=2}^{H} \lambda_h \bar{u}_h.$$

Because of concavity, the optimum for (4.1) maximizes the Lagrangean. Taking $\lambda_1 = 1$ shows that a Pareto-optimal allocation must solve

$$\max \sum_{h=1}^{H} \lambda_h E[u_h(\tilde{w}_h)] \quad \text{subject to} \quad \sum_{h=1}^{H} \tilde{w}_h = \tilde{w}_m \tag{4.2}$$

for some $(\lambda_1, \ldots, \lambda_H)$. The problem (4.2) is called the social planner's problem. As usual, the equality $\sum_{h=1}^{H} \tilde{w}_h = \tilde{w}_m$ means equality in each state of the world (or with probability 1), so the constraint in (4.2) is really a system of constraints, one for each state of the world.

There are no constraints in the social planner's problem that operate across multiple states of the world, so, to achieve the maximum value of the objective function, it suffices to maximize separately in each state of the world. In other words, the social planner's problem is equivalent to

$$\max \sum_{h=1}^{H} \lambda_h u_h(w_h) \quad \text{subject to} \quad \sum_{h=1}^{H} w_h = \tilde{w}_m(\omega) \tag{4.3}$$

in each state of the world ω. The Lagrangean for (4.3) in state ω is

$$\sum_{h=1}^{H} \lambda_h u_h(w_h) - \tilde{\eta}(\omega) \left(\sum_{h=1}^{H} w_h - \tilde{w}_m(\omega) \right),$$

with $\tilde{\eta}(\omega)$ being the Lagrange multiplier, and the first-order condition that is solved by an interior Pareto-optimal allocation $(\tilde{w}_1, \ldots, \tilde{w}_H)$ is

$$(\forall h, \omega) \qquad \lambda_h u_h'(\tilde{w}_h(\omega)) = \tilde{\eta}(\omega). \qquad (4.4)$$

Sharing Rules

If an allocation $(\tilde{w}_1, \ldots, \tilde{w}_H)$ is Pareto optimal, then each individual must be allocated higher wealth in states in which market wealth is higher. This says nothing about which individuals get higher wealth than others, only that all individuals must share in market prosperity, and all must suffer (relatively speaking) when market wealth is low. As we will see, this is a simple consequence of the first-order condition (4.4) and risk aversion.

For any two investors j and h, the first-order condition (4.4) implies

$$\lambda_j u_j'(\tilde{w}_j(\omega)) = \lambda_h u_h'(\tilde{w}_h(\omega)). \qquad (4.5)$$

Consider two different states ω_1 and ω_2. Divide (4.5) in state ω_1 by (4.5) in state ω_2 to obtain

$$\frac{u_j'(\tilde{w}_j(\omega_1))}{u_j'(\tilde{w}_j(\omega_2))} = \frac{u_h'(\tilde{w}_h(\omega_1))}{u_h'(\tilde{w}_h(\omega_2))}. \qquad (4.6)$$

This is the familiar result from microeconomics that marginal rates of substitution must be equalized across individuals at a Pareto optimum. Here, wealth (consumption) in different states of the world plays the role of different commodities in the usual consumer choice problem.

Assuming strict risk aversion (strictly diminishing marginal utilities), the equality (4.6) of marginal rates of substitutions produces the following chain of implications:

$$\tilde{w}_j(\omega_1) > \tilde{w}_j(\omega_2) \Rightarrow \frac{u_j'(\tilde{w}_j(\omega_1))}{u_j'(\tilde{w}_j(\omega_2))} < 1$$
$$\Rightarrow \frac{u_h'(\tilde{w}_h(\omega_1))}{u_h'(\tilde{w}_h(\omega_2))} < 1$$
$$\Rightarrow \tilde{w}_h(\omega_1) > \tilde{w}_h(\omega_2).$$

Because this is true for every pair of investors, it follows that at a Pareto optimum all investors must have higher wealth in states in which market wealth is higher.

Thus, each investor's wealth is related in a one-to-one fashion with market wealth; that is, letting f_h denote the one-to-one relationship for investor h, we have $\tilde{w}_h(\omega) = f_h(\tilde{w}_m(\omega))$ in each state of the world ω. The functions f_h are called sharing rules.

A particular consequence of these sharing rules is that if market wealth is the same in two different states of the world, then each investor's wealth in a Pareto-optimal allocation must be constant across the two states of the world. This shows that the example in Section 4.1 is inconsistent with Pareto optimality when investors are strictly risk averse.

If investors have LRT utility functions with the same cautiousness parameter, then the sharing rules f_h must be affine (linear plus a constant) functions. Specifically, if an allocation $(\tilde{w}_1, \ldots, \tilde{w}_H)$ is Pareto optimal, then there exist constants a_h and $b_h > 0$ for each h such that

$$\tilde{w}_h(\omega) = a_h + b_h \tilde{w}_m(\omega) \tag{4.7}$$

for each ω. Thus, the wealths of different investors move together and do so in a linear way. This is shown in Section 4.4.

4.2 COMPETITIVE EQUILIBRIA

A competitive equilibrium is characterized by two conditions: (i) markets clear, and (ii) each agent optimizes, taking prices as given. We take production decisions as given and model the economy as a pure exchange economy. Thus, part (ii) means that each investor chooses an optimal portfolio.

To define a competitive equilibrium formally, let n denote the number of assets, including the risk-free asset if it exists. Let $\overline{\theta}_{hi}$ denote the number of shares of asset i owned by investor h before trade at date 0, for $i = 1, \ldots, n$ and $h = 1, \ldots, H$. Set $\overline{\theta}_h = (\overline{\theta}_{h1} \cdots \overline{\theta}_{hn})'$. The value of the shares, which of course depends on the asset prices, is the investor's wealth at date 0.[1] Assume investor h invests her date–0 wealth in a portfolio $\theta_h = (\theta_{h1}, \ldots, \theta_{hn})$ of the n assets. We can allow investors to have labor income at date 1 which they consume in addition to their portfolio values.[2] Let $\tilde{y}_h$ denote the labor income of investor h at date 1.

---

1. If we included consumption at date 0, we would allow investors to have other wealth at date 0, which we would call consumption-good endowments, and/or allow for the assets to have paid dividends in the consumption good before trade at date 0. Because in equilibrium all of the assets must be held by investors, total consumption of all investors at date 0 must equal total consumption-good endowments plus dividends. This variation of the model is discussed in Section 4.5.

2. As before, we use the term "labor income" to encompass all end-of-period endowments of investors.

Let $\tilde{x}_i$ denote the payoff of asset i. Let θ_m denote the market portfolio, meaning that $\theta_m = (\theta_{m1} \cdots \theta_{mn})'$ where $\theta_{mi} = \sum_{h=1}^{H} \overline{\theta}_{hi}$, which is the total supply of asset i. A competitive equilibrium is a set of prices $(p_1, \ldots, p_n)$ and a set of portfolios $(\theta_1, \ldots, \theta_H)$ such that markets clear,[3] that is,

$$\sum_{h=1}^{H} \theta_h = \theta_m, \quad (4.8)$$

and such that each investor's portfolio is optimal; that is, for each h, θ_h solves

$$\max \; \mathsf{E}[u_h(\tilde{w}_h)] \quad (4.9)$$

subject to $\sum_{i=1}^{n} \theta_{hi} p_i = \sum_{i=1}^{n} \overline{\theta}_{hi} p_i$ and $\tilde{w}_h = \tilde{y}_h + \sum_{i=1}^{n} \theta_{hi} \tilde{x}_i.$

4.3 COMPLETE MARKETS

In the complete-markets model of Section 3.3, standard results about competitive equilibria in pure exchange economies apply. This includes the First Welfare Theorem, which states that competitive equilibrium allocations are Pareto optimal.

Assume there are only finitely many states of the world, the market is complete, and the law of one price holds. Adopt the notation of Section 3.2 and define

$$U_h(w) = \sum_{j=1}^{k} \mathrm{prob}_j u_h(w_j)$$

for any $w \in \mathbb{R}^k$, where k is the number of states of the world. Thus, $U_h(w)$ is the expected utility of the random wealth defined by w. Let $y_h = (\tilde{y}_h(\omega_1) \cdots \tilde{y}_h(\omega_k))' \in \mathbb{R}^k$ denote the date–1 labor income of investor h. Then, (4.9) can be expressed as:

$$\max_{\theta} \; U_h(X'\theta + y_h) \quad \text{subject to} \quad p'\theta = p'\overline{\theta}_h. \quad (4.10)$$

Let q denote the unique vector of state prices, so $Xq = p$. Due to completeness, for any $w \in \mathbb{R}^k$, there exists a portfolio θ such that $X'\theta + y_h = w$. Moreover, the

---

3. As is standard in microeconomics, equilibrium prices can be scaled by a positive constant and remain equilibrium prices because the set of budget-feasible choices for each investor (in this model, the set of feasible portfolios θ_h) is unaffected by the scaling—technically, budget constraints are homogeneous of degree zero in prices. In microeconomics, it is common to resolve this indeterminacy by requiring the price vector to lie in the unit simplex. As mentioned in Chapter 1, it is customary to resolve it in finance by setting the price of the consumption good equal to 1 (making the consumption good the numeraire). Choosing a different numeraire scales all of the returns $\tilde{x}_i/p_i$ by the same factor.

cost of this portfolio is $p'\theta = q'X'\theta = q'(w - y_h)$. Hence, (4.10) can be expressed as:

$$\max_{w} U_h(w) \quad \text{subject to} \quad q'w = q'y_h + p'\bar{\theta}_h. \tag{4.11}$$

Set $\bar{w}_h = y_h + X'\bar{\theta}_h$. This is the end-of-period wealth of the investor in the absence of trade. We have $q'\bar{w}_h = q'y_h + q'X'\bar{\theta}_h = q'y_h + p'\bar{\theta}_h$. Therefore, (4.10) can be expressed as:

$$\max_{w} U_h(w) \quad \text{subject to} \quad q'w = q'\bar{w}_h. \tag{4.12}$$

This is a standard consumer choice problem from microeconomics, in which wealth in different states plays the role of consumption of different goods and the prices are the state prices q. The economy in which investors have endowments $\bar{w}_h \in \mathbb{R}^k$ and solve the consumer choice problem (4.12), with the price vector q being determined by market clearing, is called an Arrow-Debreu economy, in recognition of Arrow and Debreu (1954) and other work of those authors. There is a one-to-one relationship between equilibria of the Arrow-Debreu economy and equilibria of the securities market, with the price vector $p \in \mathbb{R}^n$ in the securities market and the price vector $q \in \mathbb{R}^k$ in the Arrow-Debreu economy being related by $p = Xq$. Market wealth in this economy is

$$w_m \stackrel{\text{def}}{=} \sum_{h=1}^{H} y_h + X'\theta_m = \sum_{h=1}^{H} \bar{w}_h.$$

The following are true:

(a) An equilibrium of the Arrow-Debreu economy is an equilibrium of the securities market. More precisely, suppose $(q, w_1^*, \ldots, w_H^*)$ is such that w_h^* solves (4.12) for each h and markets clear:

$$\sum_{h=1}^{H} w_h^* = w_m. \tag{4.13}$$

Then, there exists a portfolio θ_h^* for each investor solving (4.9) such that $w_h^* = X'\theta_h^* + y_h$ for each h and the securities market clears, that is, (4.8) holds.

(b) An equilibrium of the securities market is an Arrow-Debreu equilibrium. More precisely, suppose $(p, \theta_1^*, \ldots, \theta_H^*)$ is such that θ_h^* solves (4.9) for each h and (4.8) holds. Then, $w_h^* = X'\theta_h^* + y_h$ solves (4.12) for each h and (4.13) holds.

To prove (a), let θ_h^* be any solution to $w_h^* = X'\theta_h^* + y_h$ for $h < H$ and set $\theta_H^* = \theta_m - \sum_{h<H} \theta_h^*$. Obviously, (4.8) holds. Also, we have

$$X'\theta_H^* = X'\theta_m - \sum_{h<H} X'\theta_h^*$$

$$= \sum_{h=1}^{H}(\overline{w}_h - y_h) - \sum_{h<H}(w_h^* - y_h)$$

$$= w_H^* - y_H,$$

using the definition $\overline{w}_h = X'\overline{\theta}_h - y_h$ for the second equality and (4.13) for the third. The optimality of the portfolios follows from the equivalence of the choice problems.

To verify that (4.13) holds in (b), note that

$$\sum_{h=1}^{H} w_h^* = \sum_{h=1}^{H} X'\theta_h^* + \sum_{h=1}^{H} y_h$$

$$= X'\theta_m + \sum_{h=1}^{H} y_h$$

$$= w_m,$$

using $w_h^* = X'\theta_h^* + y_h$ for the first equality and (4.8) for the second. Again, the optimality of the w_h^* follows from the equivalence of the choice problems.

4.4 AGGREGATION AND EFFICIENCY WITH LINEAR RISK TOLERANCE

Competitive equilibria are Pareto optimal in complete markets. They are also Pareto optimal when all investors have LRT utility functions with the same cautiousness parameter, even if markets are incomplete. Assume investor h has linear risk tolerance $\tau_h(w) = A_h + Bw$ and the cautiousness parameter B is the same for all investors. Except for the first subsection, which characterizes Pareto optima, it is assumed in this section that there is no labor income and there is a risk-free asset. For convenience, let n denote the number of risky assets, there being $n + 1$ assets including the risk-free asset. Index the risk-free asset as asset 0.

To simplify, but still address the most important cases, assume that utility functions do not have a bliss level; that is, exclude the shifted power case with

$\rho < 0$.[4] The shifted power utility function will be written as

$$\frac{1}{1-\rho}(w-\zeta)^{1-\rho},$$

and log and power utility are special cases of shifted log and shifted power utility, respectively. Thus, we are assuming one of the following conditions holds:

(a) Each investor has CARA utility, with possibly different absolute risk aversion coefficients α_h.
(b) Each investor has shifted CRRA utility with the same coefficient $\rho > 0$ ($\rho = 1$ meaning shifted log and $\rho \neq 1$ meaning shifted power) and with possibly different (and possibly zero) shifts ζ_h.

SHARING RULES WITH LINEAR RISK TOLERANCE

This subsection shows that any Pareto-optimal allocation involves an affine sharing rule as in (4.7). Moreover, $\sum_{h=1}^{H} a_h = 0$ and $\sum_{h=1}^{H} b_h = 1$, where a_h and b_h are the coefficients in (4.7). The converse—that affine sharing rules produce Pareto-optimal allocations—is left as an exercise (Exercises 4.5 and 4.6). Given a Pareto optimum $(\tilde{w}_1, \ldots, \tilde{w}_H)$, let $\lambda_1, \ldots, \lambda_H$ be weights such that the Pareto optimum solves the social planner's problem (4.2). We can show:

(a) If each investor h has CARA utility with absolute risk aversion coefficient α_h, then (4.7) holds with

$$a_h = \tau_h \left[\log(\lambda_h \alpha_h) - \sum_{j=1}^{H} \frac{\tau_j}{\tau} \log(\lambda_j \alpha_j) \right] \quad \text{and} \quad b_h = \frac{\tau_h}{\tau}, \quad (4.14)$$

where $\tau_h = 1/\alpha_h$ is the coefficient of risk tolerance, and $\tau = \sum_{j=1}^{H} \tau_j$.

(b) If each investor h has shifted CRRA utility with the same coefficient $\rho > 0$, then, setting $\zeta = \sum_{h=1}^{H} \zeta_h$, (4.7) holds with

$$a_h = \zeta_h - b_h \zeta \quad \text{and} \quad b_h = \frac{\lambda_h^{1/\rho}}{\sum_{j=1}^{H} \lambda_j^{1/\rho}}. \quad (4.15)$$

Note that the two cases are somewhat different, because the weights λ_h in the social planner's problem affect only the intercepts a_h in the CARA case,

---

4. To prove the First Welfare Theorem, we need to assume that investors are not satiated at the equilibrium allocation, so this requires some additional hypotheses when $\rho < 0$.

whereas for shifted CRRA utility, an investor with a higher weight λ_h has a higher coefficient b_h, that is, an allocation $\tilde{w}_h$ with a greater sensitivity to market wealth $\tilde{w}_m$. The result for the CARA case is another manifestation of the absence of wealth effects. In a Pareto-optimal competitive equilibrium, the weight the social planner places on an individual investor depends on, among other things, the investor's wealth. Wealthier investors have higher weights, other things equal. With CARA utility, weights and hence wealths do not affect equilibrium/optimal exposures of investors to market wealth risk.

Note that (4.15) implies the sharing rule $\tilde{w}_h = a_h + b_h \tilde{w}_m$ in the shifted CRRA case can be written in the perhaps more transparent form:

$$\tilde{w}_h - \zeta_h = b_h(\tilde{w}_m - \zeta). \tag{4.16}$$

The CARA case will be proven. The shifted CRRA case, which is similar, is left as an exercise.

We need to solve the social planner's problem (4.3) in each state of the world. Specializing the first-order condition (4.4) to the case of CARA utility, it becomes

$$(\forall h) \qquad \lambda_h \alpha_h e^{-\alpha_h \tilde{w}_h} = \tilde{\eta}.$$

We need to find $\tilde{\eta}$, which we can do by (i) solving for $\tilde{w}_h$:

$$\begin{aligned}\tilde{w}_h &= -\frac{1}{\alpha_h} \log \tilde{\eta} + \frac{\log(\lambda_h \alpha_h)}{\alpha_h} \\ &= -\tau_h \log \tilde{\eta} + \tau_h \log(\lambda_h \alpha_h),\end{aligned} \tag{4.17}$$

(ii) adding over investors to obtain

$$\tilde{w}_m = -\tau \log \tilde{\eta} + \sum_{\ell=1}^{H} \tau_\ell \log(\lambda_\ell \alpha_\ell),$$

and then (iii) solving for $\tilde{\eta}$ as

$$-\log \tilde{\eta} = \frac{1}{\tau} \tilde{w}_m - \frac{1}{\tau} \sum_{\ell=1}^{H} \tau_\ell \log(\lambda_\ell \alpha_\ell).$$

Substitute this back into (4.17) to obtain

$$\tilde{w}_h = \frac{\tau_h}{\tau} \tilde{w}_m - \frac{\tau_h}{\tau} \sum_{\ell=1}^{H} \tau_\ell \log(\lambda_\ell \alpha_\ell) + \tau_h \log(\lambda_h \alpha_h).$$

This establishes the affine sharing rule (4.14).

Gorman Aggregation

A price vector $(p_0, p_1, \ldots, p_n)$ is called an equilibrium price vector if there exist portfolios $\theta_1, \ldots, \theta_H$ such that the prices and portfolios form an equilibrium. Consider equilibrium price vectors in which $p_i \neq 0$ for each i, so we can apply the portfolio choice results of Section 2.6, which are expressed in terms of returns. We will show that the set of such equilibrium price vectors does not depend on the initial wealth distribution. In microeconomics, this property is called Gorman aggregation, and it relies on wealth expansion paths being linear and parallel, as shown for this model in (2.24).

Walras's law implies that the market for the risk-free asset clears if the markets for the other n assets clear,[5] so markets clear if and only if

$$(\forall i = 1, \ldots, n) \quad \sum_{h=1}^{H} \phi_{hi} = p_i \theta_{mi}, \qquad (4.18)$$

where $\phi_{hi} = p_i \theta_{hi}$ denotes the investment in asset i by investor h.

For convenience, we repeat (2.24) here:

$$\phi_{hi} = \xi_i A_h + \xi_i B R_f w_{h0}, \qquad (4.19)$$

Recall that it is shown in Section 2.6 that ξ_i is independent of A_h and w_{h0}. In our current model, investors differ only with regard to A_h and w_{h0}, so ξ_i is the same for each investor h. Consequently, the aggregate investment in risky asset i is

$$\sum_{h=1}^{H} \phi_{hi} = \xi_i A + \xi_i B R_f w_{m0}, \qquad (4.20)$$

where $A = \sum_{h=1}^{H} A_h$ and where $w_{m0} = \sum_{j=0}^{n} p_j \overline{\theta}_j$ is aggregate initial wealth.

---

5. Walras's law states in general that any single market, in which the price is nonzero, must clear if all other markets clear, and it is a consequence of budget equations. A proof in our context is as follows. We take the risk-free asset to have a positive payoff, and in equilibrium its price must be nonzero if investors have strictly monotone utilities. Summing the budget constraints of the H investors gives

$$p_0 \sum_{h=1}^{H} \theta_{h0} + \sum_{i=1}^{n} p_i \sum_{h=1}^{H} \theta_{hi} = p_0 \sum_{h=1}^{H} \overline{\theta}_{h0} + \sum_{i=1}^{n} p_i \sum_{h=1}^{H} \overline{\theta}_{hi}$$

$$= p_0 \theta_{m0} + \sum_{i=1}^{n} p_i \theta_{mi},$$

where asset 0 is the risk-free asset. Market clearing for the other n assets implies

$$\sum_{i=1}^{n} p_i \sum_{h=1}^{H} \theta_{hi} = \sum_{i=1}^{n} p_i \theta_{mi}.$$

Therefore, $p_0 \sum_{h=1}^{H} \theta_{h0} = p_0 \theta_{m0}$ which implies that the market for the risk-free asset clears.

Combining (4.18) and (4.20), markets clear if and only if

$$\xi_i A + \xi_i B R_f \sum_{j=0}^{n} p_j \bar{\theta}_j = p_i \theta_{mi} \qquad (4.21)$$

for each i. The coefficients ξ_i depend on the returns and therefore on the prices. Hence, (4.21) is not an explicit formula for the equilibrium prices. However, because the ξ_i are the same for each investor and do not depend on investors' initial wealths, this characterization of equilibrium shows that equilibrium prices do not depend on the initial wealth distribution across investors.[6]

Implementing Affine Sharing Rules

When markets are incomplete, there are allocations that cannot be achieved via security trading. However, when investors have LRT utility functions with the same cautiousness parameter, Pareto-optimal sharing rules are affine, as shown earlier in this section. This implies, as will be shown here, that any Pareto-optimal allocation can be implemented in the securities market. This property is sometimes described as the market being effectively complete.

As previously, take asset 0 to be risk free. We need the risk-free asset to generate the intercepts a_h in the sharing rules.[7] We can assume without loss of generality that asset 0 is in zero net supply ($\theta_{m0} = 0$). This is without loss of generality because we can also take one of the other assets to be risk free, if there is actually a positive supply of the risk-free asset.

To implement the Pareto-optimal allocation, include an investment of a_h/R_f in asset 0 in the portfolio of investor h. As noted before, $\sum_{h=1}^{H} a_h = 0$, so the total investment in asset 0 of all investors is zero, equaling the supply. For each asset $i = 1, \ldots, n$, set the number of shares held by investor h to be $\theta_{hi} = b_h \theta_{mi}$. Because $\sum_{h=1}^{H} b_h = 1$, the total number of shares held by investors of asset i is θ_{mi}. Thus, the proposed portfolios are feasible, that is, the market clears. Because we have taken asset 0 to be in zero net supply, market wealth is

$$\tilde{w}_m = \sum_{i=1}^{n} \theta_{mi} \tilde{x}_i,$$

---

6. It would be more precise to say that relative prices p_i/p_j are independent of the initial wealth distribution, because the absolute prices p_i have one degree of indeterminacy (footnote 3).

7. Note that the intercepts a_h are zero for log and power utility (shifted CRRA utility with zero shifts). Consequently, the result that any Pareto-optimal allocation can be achieved via security trading holds in those cases even when there is no risk-free asset.

and we have for each investor h that

$$\tilde{w}_h = a_h + \sum_{i=1}^{n} b_h \theta_{mi} \tilde{x}_i = a_h + b_h \tilde{w}_m.$$

Thus, these portfolios implement the affine sharing rules.

Note that the relative investment in any two assets $i, j \in \{1, \ldots, n\}$ for any investor h is

$$\frac{b_h \theta_{mi} p_i}{b_h \theta_{mj} p_j} = \frac{\theta_{mi} p_i}{\theta_{mj} p_j}.$$

This means that the relative investment is equal to the relative market capitalizations of the two assets (market capitalization is price times shares outstanding). Thus, we say that each investor holds a combination of the risk-free asset and the market portfolio of risky assets. This is called two-fund separation. Section 2.6 shows, using the result on parallel Engel curves repeated above as (4.19), that each investor chooses the market portfolio in a competitive market. Thus, Pareto-optimal portfolios and equilibrium portfolios seem to coincide, both equaling the market portfolio. This is shown more explicitly in the next subsection.

First Welfare Theorem with Linear Risk Tolerance

This subsection shows that any competitive equilibrium in this economy is Pareto optimal. The key fact is the effective completeness of markets established in the previous subsection. We will show that any allocation that is Pareto dominated is Pareto dominated by a Pareto optimum. Therefore, if a competitive equilibrium were not Pareto optimal, it would be Pareto dominated by an allocation that can be implemented in the securities market. But this dominant allocation could not be budget feasible for each investor; because, if it were, it would have been chosen instead of the supposed competitive equilibrium allocation. Adding budget constraints across investors shows that the Pareto-dominant allocation is not feasible, which is a contradiction. The remainder of this section provides the details of the proof.

We argue by contradiction. Consider a competitive equilibrium allocation

$$(\tilde{w}_1, \ldots, \tilde{w}_H)$$

and suppose there is a feasible Pareto-superior allocation. Without loss of generality, suppose that the first investor's expected utility can be feasibly increased without reducing the expected utility of the other investors. Let

$\bar{u}_h = E[u_h(\tilde{w}_h)]$ for $h > 1$. We now define a Pareto optimum that increases the expected utility of the first investor without changing the expected utilities of other investors.

(a) If each investor h has CARA utility with some absolute risk aversion coefficient α_h, define a_h for $h > 1$ by

$$E[u_h(a_h + b_h \tilde{w}_m)] = \bar{u}_h,$$

where $b_h = \tau_h/\tau$ and $\tau = \sum_{h=1}^{H} \tau_h$. Set $a_1 = -\sum_{h=2}^{h} a_h$ and

$$(\forall h) \qquad \tilde{w}'_h = a_h + b_h \tilde{w}_m.$$

(b) If each investor h has shifted CRRA utility with the same coefficient $\rho > 0$ ($\rho = 1$ meaning shifted log and $\rho \neq 1$ meaning shifted power) and some shift ζ_h, define b_h for $h > 1$ by

$$E[u_h(\zeta_h + b_h(\tilde{w}_m - \zeta))] = \bar{u}_h,$$

where $\zeta = \sum_{h=1}^{H} \zeta_h$. Set $b_1 = 1 - \sum_{h=2}^{H} b_h$ and

$$(\forall h) \qquad \tilde{w}'_h = \zeta_h + b_h(\tilde{w}_m - \zeta).$$

In case (a), each random wealth $\tilde{w}'_h$ is feasible for each investor and $\sum_{h=1}^{H} \tilde{w}'_h = \tilde{w}_m$. The allocation $(\tilde{w}'_1, \ldots, \tilde{w}'_H)$ is Pareto optimal (Exercise 4.5) and can be implemented in the securities market. Because we assumed the first investor's utility could be increased without decreasing the utility of other investors and because $(\tilde{w}'_1, \ldots, \tilde{w}'_H)$ is a Pareto optimum that does not change the expected utility of investors $2, \ldots, H$, we must have $E[u_1(\tilde{w}'_1)] > E[u_1(\tilde{w}_1)]$. From here, the proof of the contradiction follows the same lines as the usual proof of the First Welfare Theorem: Because of strictly monotone utilities, the random wealth $\tilde{w}'_h$ must cost at least as much as $\tilde{w}_h$ for each h and cost strictly more for investor $h = 1$. Adding the investor's budget constraints shows that $\tilde{w}_m = \sum_{h=1}^{H} \tilde{w}'_h$ costs more than $\tilde{w}_m = \sum_{h=1}^{H} \tilde{w}_h$, which is a contradiction.

In case (b), if $\sum_{h=2}^{H} b_h < 1$, then $\tilde{w}'_1$ is feasible for investor 1, and the same reasoning as in the previous paragraph leads to a contradiction—see Exercise 4.6 for the fact that the allocation in (b) is Pareto optimal. In the next paragraph, we show that $\sum_{h=2}^{H} b_h \geq 1$ also produces a contradiction.

Define $\tilde{w}^*_m = \tilde{w}_m - \zeta_1$. Because $\tilde{w}_1 \geq \zeta_1$, we have

$$\tilde{w}^*_m \geq \tilde{w}_m - \tilde{w}_1 = \sum_{h=2}^{H} \tilde{w}_h.$$

Note that for $h > 1$

$$\tilde{w}'_h = \zeta_h + b_h(\tilde{w}_m - \zeta) = \zeta_h + b_h\left(\tilde{w}^*_m - \sum_{h=2}^{H}\zeta_h\right).$$

If $\sum_{h=2}^{H} b_h = 1$, then the allocation $(w'_2, \ldots, \tilde{w}'_H)$ is a Pareto-optimal allocation of the wealth $\tilde{w}^*_m$ among investors $2, \ldots, H$. However, it leaves only $\tilde{w}_m - \tilde{w}^*_m = \zeta_1$ for investor 1, which is either infeasible or the worst possible level of wealth for investor 1. Hence, it is impossible to give each investor $h > 1$ the expected utility $E[u_h(\tilde{w}'_h)] = E[u_h(\tilde{w}_h)]$ while increasing the expected utility of the first investor above $E[u_1(\tilde{w}_1)]$, contradicting our maintained hypothesis. If $\sum_{h=2}^{H} b_h > 1$, then the allocation $(w'_2, \ldots, \tilde{w}'_H)$ actually dominates (for investors $2, \ldots, H$) the Pareto-optimal allocation

$$\tilde{w}''_h = \zeta_h + \frac{b_h}{\sum_{h=2}^{H} b_h}\left(\tilde{w}^*_m - \sum_{h=2}^{H}\zeta_h\right)$$

of the wealth $\tilde{w}^*_m$ among investors $2, \ldots, H$. Hence, the same reasoning leads to a contradiction.

4.5 BEGINNING-OF-PERIOD CONSUMPTION

Including beginning-of-period consumption does not materially change any of the results of this chapter. As mentioned before, in this model with no production and only one consumption good, Pareto optimality is about efficient risk sharing. Whether Pareto optimality can be achieved in competitive markets depends on the nature of asset markets and on investors' risk tolerances regarding date–1 consumption. Of course, the allocation of consumption at date 0 is also relevant for Pareto optimality, but any allocation of date–0 consumption can be achieved by trading the consumption good against assets at date 0, so competitive equilibria are Pareto optimal under the same circumstances described earlier. Here are a few details to support this claim.

Suppose investors have endowments y_{h0} at date 0 and choose consumption at date 0 as well as asset investments. If there are k states of the world at date 1 and the market is complete, then there is an equivalence between the securities market and an Arrow-Debreu economy with $k+1$ goods, namely, consumption at date 0 and consumption in each of the k states at date 1. We can normalize an equilibrium price vector in the Arrow-Debreu market by taking the price of good 0 (consumption at date 0) to equal 1. The other k prices (q_j for $j = 1, \ldots, k$) are then the state prices for the k states; that is, q_j is the date–0 price of one unit of wealth (consumption) at date 1 in state j, as discussed in Section 4.3.

The Arrow-Debreu economy is equivalent to the securities market in which the price of asset i is given by $p_i = \sum_{j=1}^{k} q_j x_{ij}$. The Pareto optimality of a competitive equilibrium follows from the First Welfare Theorem, just as it does with consumption only at date 1.

A Pareto-optimal allocation of date–0 consumption and date–1 consumption must maximize

$$\mathsf{E}\left[\sum_{h=1}^{H} \lambda_h v_h(c_{h0}, \tilde{c}_{h1})\right] \qquad (4.22)$$

for positive weights λ_h subject to the economy's resource constraints, where v_h denotes the utility function of investor h. Let $(c_{10}^*, \ldots, c_{H0}^*)$ be the allocation of date–0 consumption at a Pareto optimum and define

$$u_h(c_1) = v_h(c_{h0}^*, c_1)$$

for each h. The allocation of date–1 consumption at the Pareto optimum must maximize

$$\mathsf{E}\left[\sum_{h=1}^{H} \lambda_h u_h(\tilde{c}_{h1})\right]$$

subject to the economy's resource constraints. Therefore, all of the results of Section 4.1 regarding the social planner's problem, the first-order condition, and Pareto-optimal sharing rules apply to the economy with date–0 consumption.

To apply the results on LRT utility with date–0 consumption, we want u_h defined above to have linear risk tolerance $\tau_h(w) = A_h + Bw$ at any allocation of date–0 consumption that is part of a Pareto optimum. To ensure this, assume each investor has time-additive utility $v_h(c_0, c_1) = u_{h0}(c_0) + u_{h1}(c_1)$, and assume the utility functions u_{h1} have linear risk tolerance $\tau_h(w) = A_h + Bw$. With time-additive utility, an allocation is Pareto optimal if and only if there are positive weights λ_h such that the date–0 allocation maximizes

$$\sum_{h=1}^{H} \lambda_h u_{h0}(c_{h0}), \qquad (4.23a)$$

and the date–1 allocation maximizes

$$\mathsf{E}\left[\sum_{h=1}^{H} \lambda_h u_{h1}(\tilde{c}_{h1})\right], \qquad (4.23b)$$

subject to the economy's resource constraints. It follows from the results in Section 4.4 that a Pareto optimum involves an affine sharing rule for date–1 consumption, relative asset prices p_i/p_j are independent of the initial wealth distribution, any affine sharing rule for date–1 consumption can be implemented in the securities market, and a competitive equilibrium allocation maximizes

(4.23b), for some positive weights λ_h, subject to the date–1 resource constraint. This last fact implies that a competitive equilibrium is Pareto optimal (Exercise 4.8).

4.6 NOTES AND REFERENCES

Arrow (1953) shows that the welfare theorems of Arrow (1951) and Debreu (1954) apply to complete securities markets. The term "Gorman aggregation" refers to Gorman (1953). The two-fund separation result of Section 4.4, also discussed in Section 2.6, is due to Cass and Stiglitz (1970). Rubinstein (1974) shows that Gorman aggregation is possible under the conditions of Section 4.4 and also in some circumstances when investors have heterogeneous beliefs. The relation of aggregation to affine sharing rules is studied in a more general setting by Wilson (1969). DeMarzo and Skiadas (1998) extend Rubinstein's results to markets with asymmetric information.

Ross (1976b) shows that call and put options on traded assets can increase the span of traded assets, to the extent that any random variable that depends only on asset payoffs is the payoff of some portfolio (whether such a market is complete depends on whether labor income is spanned by the asset payoffs). This is addressed in Exercise 4.10, in the context of a single risky asset, interpreted as the market portfolio. In this case, as Breeden and Litzenberger (1978) show, the state prices are the prices of option portfolios called butterfly spreads.

There are only a few cases other than those considered in Section 4.4 in which an equilibrium can be analytically computed with heterogeneous investors. One case in which quite a bit can be said is when there are two investors (or two classes of investors) having CRRA utility with one being twice as risk averse as the other. This case, which is studied by Wang (1996) in a continuous-time model, is addressed in Exercise 4.3.

EXERCISES

4.1. Suppose there are n risky assets with normally distributed payoffs $\tilde{x}_i$. Assume all investors have CARA utility and no labor income $\tilde{y}_h$. Define α to be the aggregate absolute risk aversion as in Section 1.1. Assume there is a risk-free asset in zero net supply. Let $\bar{\theta} = (\bar{\theta}_1 \cdots \bar{\theta}_n)'$ denote the vector of supplies of the n risky assets. Let μ denote the mean and Σ the covariance matrix of the vector $\tilde{X} = (\tilde{x}_1 \cdots \tilde{x}_n)'$ of asset payoffs. Assume Σ is nonsingular. Suppose the utility functions of investor h are

$$u_0(c) = -e^{-\alpha_h c} \quad \text{and} \quad u_1(c) = -\delta_h e^{-\alpha_h c}.$$

Let $\bar{c}_0$ denote the aggregate endowment $\sum_{h=1}^{H} y_{h0}$ at date 0.

(a) Use the result of Exercise 2.2 on the optimal demands for the risky assets to show that the equilibrium price vector is

$$p \stackrel{\text{def}}{=} \frac{1}{R_f}(\mu - \alpha \Sigma \bar{\theta}). \tag{4.24}$$

(b) Interpret the risk adjustment vector $\alpha \Sigma \bar{\theta}$ in (4.24), explaining in economic terms why a large element of this vector implies an asset has a low price relative to its expected payoff.

(c) Assume δ_h is the same for all h (denote the common value by δ). Use the market-clearing condition for the date–0 consumption good to deduce that the equilibrium risk-free return is

$$R_f \stackrel{\text{def}}{=} \frac{1}{\delta} \exp\left(\alpha \left(\bar{\theta}' \mu - \bar{c}_0\right) - \frac{1}{2}\alpha^2 \bar{\theta}' \Sigma \bar{\theta}\right). \tag{4.25}$$

(d) Explain in economic terms why the risk-free return (4.25) is higher when $\bar{\theta}' \mu$ is higher and lower when δ, $\bar{c}_0$, or $\bar{\theta}' \Sigma \bar{\theta}$ is higher.

(e) Assume different investors may have different discount factors δ_h. Set $\tau_h = 1/\alpha_h$ and $\tau = \sum_{h=1}^{H} \tau_h$ (which is aggregate risk tolerance). Show that the equilibrium risk-free return is (4.25) when we define

$$\delta = \prod_{h=1}^{H} \delta_h^{\tau_h/\tau} \quad \Leftrightarrow \quad \log \delta = \sum_{h=1}^{H} \frac{\tau_h}{\tau} \log \delta_h.$$

4.2. Adopt the assumptions of the previous exercise, but assume there is no risk-free asset and there is consumption only at date 1. Show that the vector

$$p = \gamma \left(\mu - \alpha \Sigma \bar{\theta}\right)$$

is an equilibrium price vector for any $\gamma > 0$. Note: When $\gamma < 0$, this is also an equilibrium price vector, but each investor has a negative marginal value of wealth. In this model, investors are forced to hold assets because there is no date–0 consumption. When $\gamma < 0$, they are forced to invest in undesirable assets and would be better off if they had less wealth. Including consumption at date 0 or changing the budget constraint to $p'\theta \leq p'\bar{\theta}_h$ instead of $p'\theta = p'\bar{\theta}_h$ (in other words, allowing free disposal of wealth) eliminates the equilibria with $\gamma < 0$.

4.3. Suppose there are two investors, the first having constant relative risk aversion $\rho > 0$ and the second having constant relative risk aversion 2ρ.

Equilibrium and Efficiency

(a) Show that the Pareto-optimal sharing rules are

$$\tilde{w}_1 = \tilde{w}_m + \eta - \sqrt{\eta^2 + 2\eta\tilde{w}_m}, \quad \text{and} \quad \tilde{w}_2 = \sqrt{\eta^2 + 2\eta\tilde{w}_m} - \eta,$$

for $\eta > 0$. Hint: Use the first-order condition and the quadratic formula. Because η is arbitrary in $(0, \infty)$, there are many equivalent ways to write the sharing rules.

(b) Suppose the market is complete and satisfies the law of one price. Show that the SDF in a competitive equilibrium is

$$\tilde{m} = \gamma \left(\sqrt{\eta^2 + 2\eta\tilde{w}_m} - \eta \right)^{-2\rho}$$

for positive constants γ and η.

4.4. Suppose each investor h has a concave utility function, and suppose an allocation $(\tilde{w}_1, \ldots, \tilde{w}_m)$ of market wealth $\tilde{w}_m$ satisfies the first-order condition

$$u'_h(\tilde{w}_h) = \gamma_h \tilde{m}$$

for each investor h, where $\tilde{m}$ is an SDF and is the same for each investor. Show that the allocation solves the social planner's problem (4.2) with weights $\lambda_h = 1/\gamma_h$. Note: The first-order condition holds with the SDF being the same for each investor in a competitive equilibrium of a complete market, because there is a unique SDF in a complete market. Recall that γ_h in the first-order condition is the marginal value of beginning-of-period wealth (Section 3.1). Thus, the weights in the social planner's problem can be taken to be the reciprocals of the marginal values of wealth. Other things equal, investors with high wealth have low marginal values of wealth and hence have high weights in the social planner's problem.

4.5. Suppose all investors have CARA utility. Consider an allocation

$$\tilde{w}_h = a_h + b_h \tilde{w}_m$$

where $b_h = \tau_h/\tau$ and $\sum_{h=1}^{H} a_h = 0$. Show that the allocation is Pareto optimal. Hint: Show that it solves the social planner's problem with weights λ_h defined as $\lambda_h = \tau_h e^{a_h/\tau_h}$.

4.6. Suppose all investors have shifted CRRA utility with the same coefficient $\rho > 0$. Suppose $\tilde{w}_m > \zeta$. Consider an allocation

$$\tilde{w}_h = \zeta_h + b_h(\tilde{w}_m - \zeta)$$

where $\sum_{h=1}^{H} b_h = 1$. Show that the allocation is Pareto optimal. Hint: Show that it solves the social planner's problem with weights λ_h defined as $\lambda_h = b_h^\rho$.

4.7. Show that if each investor has shifted CRRA utility with the same coefficient $\rho > 0$ and shift ζ_h, then, as asserted in Section 4.4, any Pareto-optimal allocation involves an affine sharing rule.

4.8. Consider an economy with date–0 consumption as in Section 4.5. Assume the investors have time-additive utility and the date–1 allocation solves the social planner's problem (4.1). Using the first-order condition (3.9), show that the equilibrium allocation is Pareto optimal. Hint: Using the first-order condition (4.4) with $\tilde{\eta} = \tilde{\eta}_1$, show that

$$(\forall h) \qquad \lambda_h u'_{h0}(c_{h0}) = R_f \mathsf{E}[\tilde{\eta}_1].$$

4.9. Consider a model with date–0 endowments y_{h0} and date–0 consumption c_{h0}. Suppose all investors have log utility, a common discount factor δ, and no date–1 labor income. Show that, in a competitive equilibrium, the date–0 value of the market portfolio is $\delta \sum_{h=1}^{H} c_{h0}$.

4.10. Suppose the payoff of the market portfolio $\tilde{w}_m$ has k possible values. Denote these possible values by $a_1 < \cdots < a_k$. For convenience, suppose $a_i - a_{i-1}$ is the same number Δ for each i. Suppose there is a risk-free asset with payoff equal to 1. Suppose there are $k - 1$ call options on the market portfolio, with the exercise price of the ith option being a_i. The payoff of the ith option is $\max(0, \tilde{w}_m - a_i)$.

(a) Show for each $i = 1, \ldots, k - 2$ that a portfolio that is long one unit of option i and short one unit of option $i + 1$ pays Δ if $\tilde{w}_m \geq a_{i+1}$ and 0 otherwise. (This portfolio of options is a bull spread.)

(b) Consider the following k portfolios. Show that the payoff of portfolio i is 1 when $\tilde{w}_m = a_i$ and 0 otherwise. (Thus, these are Arrow securities for the events on which $\tilde{w}_m$ is constant.)

  (i) $i = 1$: long one unit of the risk-free asset, short $1/\Delta$ units of option 1, and long $1/\Delta$ units of option 2. (This portfolio of options is a short bull spread.)

  (ii) $1 < i < k - 1$: long $1/\Delta$ units of option $i - 1$, short $2/\Delta$ units of option i, and long $1/\Delta$ units of option $i + 1$. (These portfolios are butterfly spreads.)

  (iii) $i = k - 1$: long $1/\Delta$ units of option $k - 2$ and short $2/\Delta$ units of option $k - 1$.

  (iv) $i = k$: long $1/\Delta$ units of option $k - 1$.

(c) Given any function f, define $\tilde{z} = f(\tilde{w}_m)$. Show that there is a portfolio of the risk-free asset and the call options with payoff equal to $\tilde{z}$.

Mean-Variance Analysis

CONTENTS

5.1	Graphical Analysis	100
5.2	Mean-Variance Frontier of Risky Assets	101
5.3	Mean-Variance Frontier with a Risk-Free Asset	106
5.4	Orthogonal Projections and Frontier Returns	111
5.5	Frontier Returns and Stochastic Discount Factors	117
5.6	Separating Distributions	118
5.7	Notes and References	122

This chapter describes the portfolios that are on the mean-variance frontier, meaning that their returns have minimum variance among all portfolios with the same expected return. The study of these portfolios can be motivated by the assumption that investors have mean-variance preferences, but understanding the mean-variance frontier has importance beyond that special case. The broader importance of mean-variance analysis for asset pricing is discussed in Sections 5.3 and 5.5 and Chapter 6.

Section 5.6 departs somewhat from the main theme of the chapter, presenting a condition on the joint distribution of returns sufficient to imply that investors' optimal portfolios are combinations of two mutual funds (that is, two-fund separation holds). This complements the result of Section 2.6, discussed again in Section 4.4, that two-fund separation holds if investors have linear risk tolerance $\tau(w) = A + Bw$ with the same cautiousness parameter B. The result is included in this chapter because the condition implies that investors' optimal portfolios are on the mean-variance frontier.

The following notation will be used: There are n risky assets, $\tilde{R}_i$ is the return of asset i, $\tilde{\mathbf{R}}$ is the n–dimensional column vector with $\tilde{R}_i$ as its ith element, μ is the column vector of expected returns (the n–vector with ith element $\mathsf{E}[\tilde{R}_i]$), Σ

is the covariance matrix (the $n \times n$ matrix with $\text{cov}(\tilde{R}_i, \tilde{R}_j)$ as its (i,j)th element), and ι is an n–dimensional column vector of 1's. In some parts of the chapter, the existence of a risk-free asset with return $R_f > 0$ is assumed.

Portfolios will be defined as the fractions of wealth invested in the risky assets. Denote a portfolio as a column vector π. If the portfolio is fully invested in the risky assets, then the components π_i must sum to 1, which can be represented as $\iota'\pi = 1$, where ' denotes the transpose operator. The return of a portfolio fully invested in the risky assets is a weighted average of the risky asset returns, namely, $\pi'\tilde{\mathbf{R}}$. It has mean $\pi'\mu$ and, as explained in Section 2.4, its variance is $\pi'\Sigma\pi$.

Assume the covariance matrix Σ is nonsingular, which is equivalent to no portfolio of the risky assets being risk free. A covariance matrix is positive semidefinite ($\pi'\Sigma\pi$ is the variance of a portfolio and cannot be negative), and a nonsingular symmetric positive-semidefinite matrix is positive definite. Therefore, both Σ and Σ^{-1} are positive definite (see, for example, Pemberton and Rau, 2011).

If the risky assets all have the same expected return, then all portfolios of risky assets have the same expected return, and the only way to trade off mean and variance is by varying the proportion invested in the risk-free asset (if it is assumed to exist). This case is not very interesting, so assume that at least two of the risky assets have different expected returns. This means that μ is not a scalar multiple of ι.

5.1 GRAPHICAL ANALYSIS

It is conventional to depict the mean-variance frontier in a two-dimensional plot of risk (standard deviation) and expected return. To introduce this, we start with plotting portfolios obtained by combining two assets.

Suppose there are two risky assets ($n = 2$) and consider a portfolio π (with $\iota'\pi = 1$). The expected portfolio return is

$$\pi'\mu = \pi_1\mu_1 + \pi_2\mu_2.$$

The standard deviation of the portfolio return is

$$\sqrt{\pi'\Sigma\pi} = \sqrt{\pi_1^2 \operatorname{var}(\tilde{R}_1) + \pi_2^2 \operatorname{var}(\tilde{R}_2) + 2\pi_1\pi_2 \operatorname{cov}(\tilde{R}_1\tilde{R}_2)}.$$

Figure 5.1 illustrates the expected returns and risks of various portfolios π.

We are also interested in combining a risky asset (or a portfolio of risky assets) with a risk-free asset. Letting $\tilde{R}$ denote the return of the risky asset, a portfolio of the risky and risk-free asset has return $\pi\tilde{R} + (1-\pi)R_f$, where the fraction of wealth π invested in the risky asset can be any real number. Letting μ denote the

Mean-Variance Analysis

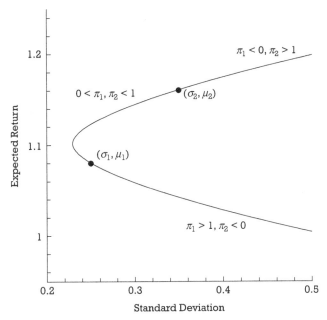

Figure 5.1 Expected returns and risks of portfolios of two risky assets. The standard deviations σ_i and expected returns μ_i of two assets $i = 1, 2$ are plotted. The hyperbola is the locus of (standard deviation, expected return) pairs of portfolios of the two assets. In this example, $\mu_2 > \mu_1$. Expected returns above μ_2 are achieved by shorting asset 1 ($\pi_1 < 0$) and investing more than 100% in asset 2. Expected returns below μ_1 are achieved by shorting asset 2 ($\pi_2 < 0$) and investing more than 100% in asset 1. Expected returns between μ_1 and μ_2 are achieved with long positions in both assets.

expected return of the risky asset, the portfolio has expected return $\pi\mu + (1-\pi)R_f$ and standard deviation $|\pi| \cdot \text{stdev}(\tilde{R})$. Figure 5.2 illustrates the risk and expected returns that can be achieved by varying the weight π on the risky asset.

5.2 MEAN-VARIANCE FRONTIER OF RISKY ASSETS

We want to describe the mean-variance frontier of risky assets, so we ignore the risk-free asset and constrain portfolios to satisfy $\iota'\pi = 1$. To compute the mean-variance frontier, we can solve the problem

$$\min \quad \frac{1}{2}\pi'\Sigma\pi \quad \text{subject to} \quad \mu'\pi = \mu_{\text{targ}} \quad \text{and} \quad \iota'\pi = 1. \tag{5.1}$$

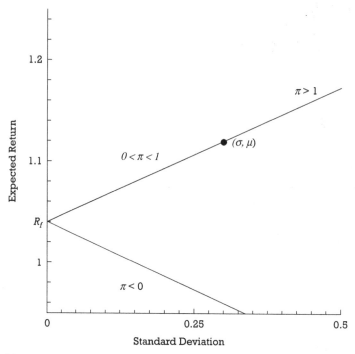

Figure 5.2 Expected returns and risks of portfolios of a risky and risk-free asset. The standard deviation and expected return of the risky asset are depicted at (σ, μ). The plot shows all of the (standard deviation, expected return) pairs of portfolios of the risky and risk-free assets. Expected returns above μ are achieved by levering (shorting the risk-free asset), in which case the weight on the risky asset is $\pi > 1$. Expected returns below R_f are achieved by shorting the risky asset, meaning $\pi < 0$. Expected returns between R_f and μ are achieved by being long both assets, meaning $0 < \pi < 1$.

Here, μ_{targ} is the target expected return for the portfolio. A portfolio solving this problem for some μ_{targ} is said to be on the mean-variance frontier. The mean-variance frontier is traced out by varying μ_{targ}. The factor $1/2$ is included here only for convenience—obviously, minimizing one-half the variance is equivalent to minimizing the variance.

Global Minimum Variance Portfolio

Before analyzing the problem (5.1), it is convenient to first solve the problem without the constraint $\mu'\pi = \mu_{\text{targ}}$. With no constraint on the portfolio expected return, the problem is to find the minimum risk portfolio. The portfolio of risky assets with minimum risk is called the global minimum variance (GMV) portfolio.

The Lagrangean for the optimization problem

$$\min \tfrac{1}{2}\pi'\Sigma\pi \quad \text{subject to} \quad \iota'\pi = 1 \tag{5.2}$$

is

$$\tfrac{1}{2}\pi'\Sigma\pi - \gamma(\iota'\pi - 1),$$

where γ is the Lagrange multiplier, and the first-order condition is

$$\Sigma\pi = \gamma\iota \quad \Leftrightarrow \quad \pi = \gamma\Sigma^{-1}\iota.$$

We can compute the Lagrange multiplier γ by imposing the constraint $\iota'\pi = 1$. This produces

$$\pi = \frac{1}{\iota'\Sigma^{-1}\iota}\Sigma^{-1}\iota. \tag{5.3}$$

The portfolio (5.3) is the GMV portfolio. Denote it by π_{gmv}.

Mean-Variance Frontier

Now, consider the optimization problem (5.1). The Lagrangean is

$$\tfrac{1}{2}\pi'\Sigma\pi - \delta(\mu'\pi - \mu_{\text{targ}}) - \gamma(\iota'\pi - 1),$$

where δ and γ are the Lagrange multipliers, and the first-order condition is

$$\Sigma\pi = \delta\mu + \gamma\iota. \tag{5.4}$$

Together with the constraints, the first-order condition is necessary and sufficient for a solution. Solving the first-order condition gives

$$\pi = \delta\Sigma^{-1}\mu + \gamma\Sigma^{-1}\iota. \tag{5.5}$$

This means that π is a linear combination of the two vectors $\Sigma^{-1}\mu$ and $\Sigma^{-1}\iota$. The weights δ and γ in the linear combination are determined by the constraints in (5.1). It is convenient to define the following constants:

$$A = \mu'\Sigma^{-1}\mu, \quad B = \mu'\Sigma^{-1}\iota, \quad C = \iota'\Sigma^{-1}\iota. \tag{5.6}$$

The assumption that μ is not a scalar multiple of ι implies[1] that $AC > B^2$. With this notation, applying the constraints in (5.1) to π defined in (5.5) yields

$$\mu_{\text{targ}} = \delta A + \gamma B, \tag{5.7a}$$

$$1 = \delta B + \gamma C. \tag{5.7b}$$

We can solve this system for δ and γ. We distinguish two cases. The first is rather trivial, but we address it for completeness. Note that the expected return of the GMV portfolio is $\mu' \pi_{\text{gmv}} = B/C$.

B = 0 In this case, $\mu' \pi_{\text{gmv}} = 0$, which means that the expected return of the GMV portfolio is zero (and the expected *rate of return* is -100%). In this case, the system (5.7) is solved by $\delta = \mu_{\text{targ}}/A$, and $\gamma = 1/C$. Note that $(1/C)\Sigma^{-1}\iota = \pi_{\text{gmv}}$. The frontier portfolio in this case is

$$\pi = \frac{\mu_{\text{targ}}}{A}\Sigma^{-1}\mu + \pi_{\text{gmv}}. \tag{5.8}$$

B ≠ 0 In this case, define

$$\pi_{\text{mu}} = \frac{1}{B}\Sigma^{-1}\mu.$$

This is a fully invested portfolio of the risky assets; that is, it satisfies $\iota' \pi_{\text{mu}} = 1$. We can write the formula (5.5) for the frontier portfolio as

$$\pi = \delta B \pi_{\text{mu}} + \gamma C \pi_{\text{gmv}}.$$

The constraint (5.7b) states that $\gamma C = 1 - \delta B$, so, defining $\lambda = \delta B$, we can write the portfolio as

$$\pi = \lambda \pi_{\text{mu}} + (1-\lambda)\pi_{\text{gmv}}. \tag{5.9}$$

Now, we can compute λ in terms of μ_{targ} from

$$\mu_{\text{targ}} = \mu'\pi = \lambda \mu' \pi_{\text{mu}} + (1-\lambda)\mu' \pi_{\text{gmv}}$$

$$= \lambda \frac{A}{B} + (1-\lambda)\frac{B}{C} \quad \Rightarrow \quad \lambda = \frac{BC\mu_{\text{targ}} - B^2}{AC - B^2}. \tag{5.10}$$

---

1. This is a special case of the Cauchy-Schwarz inequality (Section 3.8), using $x'\Sigma^{-1}y$ as the inner product of vectors x and y. A proof is as follows. Define $\theta = \mu - (B/C)\iota$. Then, $\iota'\Sigma^{-1}\theta = 0$. Hence,

$$A = \mu'\Sigma^{-1}\mu = \left(\frac{B}{C}\iota + \theta\right)' \Sigma^{-1}\left(\frac{B}{C}\iota + \theta\right) = \frac{B^2}{C} + \theta'\Sigma^{-1}\theta > \frac{B^2}{C},$$

implying $AC > B^2$. The strict inequality in the above line follows from the fact that μ is not a scalar multiple of ι, which implies that $\theta \neq 0$, and from the positive definiteness of Σ^{-1}.

Two-Fund Spanning

Any two portfolios π_a and π_b on the mean-variance frontier span the entire frontier in the sense that, if π_c is a third portfolio on the frontier, then

$$\pi_c = \lambda \pi_a + (1-\lambda)\pi_b$$

for some λ. This is shown for $\pi_a = \pi_{mu}$ and $\pi_b = \pi_{gmv}$ in (5.9) when $B \neq 0$, but it holds more generally. Again, we consider the two possible cases $B = 0$ and $B \neq 0$. Let π_a, π_b, and π_c denote frontier portfolios with $\pi_a \neq \pi_b$.

B = 0 From (5.8), we see that

$$\pi_i = \frac{\mu'\pi_i}{A}\Sigma^{-1}\mu + \pi_{gmv}$$

for $i \in \{a,b,c\}$. Any two distinct frontier portfolios have different expected returns, so $\mu'\pi_a \neq \mu'\pi_b$. It follows that

$$\lambda \pi_a + (1-\lambda)\pi_b = \frac{\lambda \mu'\pi_a + (1-\lambda)\mu'\pi_b}{A}\Sigma^{-1}\mu + \pi_{gmv} = \pi_c$$

if

$$\lambda \mu'\pi_a + (1-\lambda)\mu'\pi_b = \mu'\pi_c \quad \Leftrightarrow \quad \lambda = \frac{\mu'\pi_c - \mu'\pi_b}{\mu'\pi_a - \mu'\pi_b}.$$

B ≠ 0 From (5.8), we see that there exist λ_a, λ_b, and λ_c such that $\pi_i = \lambda_i \pi_{mu} + (1-\lambda_i)\pi_{gmv}$ for $i \in \{a,b,c\}$. Because $\pi_a \neq \pi_b$, it must be that $\lambda_a \neq \lambda_b$. It follows that

$$\lambda \pi_a + (1-\lambda)\pi_b = [\lambda \lambda_a + (1-\lambda)\lambda_b]\pi_{mu}$$
$$+ [\lambda(1-\lambda_a) + (1-\lambda)(1-\lambda_b)]\pi_{gmv} = \pi_c$$

if

$$\lambda = \frac{\lambda_c - \lambda_b}{\lambda_a - \lambda_b}.$$

The Mean-Standard Deviation Trade-Off

The variance of any frontier portfolio π is $\pi'\Sigma\pi$. Using the formula (5.8) when $B = 0$ and the formulas (5.9) and (5.10) when $B \neq 0$ and some straightforward algebra, we can compute that, regardless of whether $B = 0$ or $B \neq 0$,

$$\pi'\Sigma\pi = \frac{A - 2B\mu_{targ} + C\mu_{targ}^2}{AC - B^2}. \tag{5.11}$$

Thus, the variance is a quadratic function of the mean μ_{targ}. The set of (standard deviation, mean) pairs for frontier portfolios is the set of pairs

$$\left(\sqrt{\frac{A - 2B\mu_{\text{targ}} + C\mu_{\text{targ}}^2}{AC - B^2}}, \mu_{\text{targ}} \right)$$

obtained by varying μ_{targ} from $-\infty$ to ∞. This set is a hyperbola. It has the same general shape as that traced out by portfolios of two risky assets in Figure 5.1—not surprisingly, since the frontier is spanned by two portfolios, as just observed.

If a frontier portfolio has an expected return less than the expected return B/C of the GMV portfolio, then we say that it is on the inefficient part of the mean-variance frontier. It is inefficient in the sense that moving to the GMV portfolio increases expected return and reduces variance, improving both. On the other hand, there is a real trade-off between mean and variance for expected returns above B/C. Frontier portfolios with expected return above B/C are said to be mean-variance efficient or to be on the efficient part of the mean-variance frontier. They are efficient in the sense that it is impossible to increase expected return without increasing risk, and it is impossible to reduce risk without reducing expected return.

5.3 MEAN-VARIANCE FRONTIER WITH A RISK-FREE ASSET

Suppose now that there is a risk-free asset with return R_f. Continue to let π denote a portfolio of the risky assets, μ the vector of expected returns of the risky assets, and Σ the covariance matrix of the risky assets. The fraction of wealth invested in the risk-free asset is $1 - \iota'\pi$. The variance of a portfolio return is still $\pi'\Sigma\pi$, but the expected return is

$$(1 - \iota'\pi)R_f + \mu'\pi = R_f + (\mu - R_f\iota)'\pi.$$

Thus, the risk premium is $(\mu - R_f\iota)'\pi$.

Frontier Portfolios

The minimum variance problem is

$$\min \; \frac{1}{2}\pi'\Sigma\pi \quad \text{subject to} \quad R_f + (\mu - R_f\iota)'\pi = \mu_{\text{targ}}. \tag{5.12}$$

The first-order condition is $\Sigma \pi = \delta(\mu - R_f \iota)$ for some Lagrange multiplier δ. This implies

$$\pi = \delta \Sigma^{-1}(\mu - R_f \iota). \tag{5.13}$$

This equation and the constraint imply

$$\delta = \frac{\mu_{\text{targ}} - R_f}{(\mu - R_f \iota)' \Sigma^{-1}(\mu - R_f \iota)}. \tag{5.14}$$

Therefore, the frontier portfolios are

$$\pi = \frac{\mu_{\text{targ}} - R_f}{(\mu - R_f \iota)' \Sigma^{-1}(\mu - R_f \iota)} \Sigma^{-1}(\mu - R_f \iota). \tag{5.15}$$

Capital Market Line

In (standard deviation, mean) space, the frontier consists of two rays (forming a cone) emanating from $(0, R_f)$. The upper part of the cone is the efficient part of the frontier and is called the capital market line. The capital market line consists of the frontier portfolios with expected returns above R_f. From formula (5.15), we see that these portfolios—the efficient portfolios—are all positive scalar multiples ($\mu_{\text{targ}} - R_f > 0$) of the vector $\Sigma^{-1}(\mu - R_f \iota)$. The lower part of the cone is the inefficient part of the frontier. The inefficient frontier portfolios are negative scalar multiples of $\Sigma^{-1}(\mu - R_f \iota)$. The mean-variance frontier is illustrated in Figures 5.3 and 5.4 below.

Maximum Sharpe Ratio

The slope of the capital market line is the maximum possible Sharpe ratio. To calculate the maximum Sharpe ratio, note that the standard deviation of a frontier portfolio is

$$\sqrt{\pi' \Sigma \pi} = \frac{|\mu_{\text{targ}} - R_f|}{\sqrt{(\mu - R_f \iota)' \Sigma^{-1}(\mu - R_f \iota)}}.$$

Therefore, the Sharpe ratio (ratio of risk premium to standard deviation) of a frontier portfolio is

$$\pm \sqrt{(\mu - R_f \iota)' \Sigma^{-1}(\mu - R_f \iota)}. \tag{5.16}$$

The maximum possible Sharpe ratio is the positive square root in (5.16).

Tangency Portfolio

The frontier (cone) with a risk-free asset is generally tangent to the frontier (hyperbola) formed from the risky assets only. The portfolio of risky assets at the tangency point is called the tangency portfolio. It belongs to both frontiers. The tangency portfolio exists whenever the expected return of the GMV portfolio is different from the risk-free return. This means that $B/C \neq R_f$; equivalently, $\iota' \Sigma^{-1} (\mu - R_f \iota) \neq 0$. When this is true, we can define

$$\pi_{\text{tang}} = \frac{1}{\iota' \Sigma^{-1} (\mu - R_f \iota)} \Sigma^{-1} (\mu - R_f \iota). \qquad (5.17)$$

Note that $\iota' \pi_{\text{tang}} = 1$, so π_{tang} is a portfolio fully invested in the risky assets. The portfolio π_{tang} is of the form (5.15) with

$$\frac{\mu_{\text{targ}} - R_f}{(\mu - R_f \iota)' \Sigma^{-1} (\mu - R_f \iota)} = \frac{1}{\iota' \Sigma^{-1} (\mu - R_f \iota)}. \qquad (5.18)$$

Therefore, it is on the mean-variance frontier defined by the risky and risk-free assets. Any solution to the optimization problem (5.12) that is feasible for the optimization problem (5.1)—that is, it satisfies $\iota' \pi = 1$—must also solve the optimization problem (5.1). Therefore, π_{tang} must be on both frontiers.

From (5.18), we can see that the tangency portfolio has a positive risk premium (that is, $\mu_{\text{targ}} - R_f > 0$, where μ_{targ} is the mean of the tangency portfolio return) and consequently is on the efficient part of the frontier if and only if

$$\iota' \Sigma^{-1} (\mu - R_f \iota) > 0.$$

This is equivalent to $B/C > R_f$. Thus, if the expected return of the GMV portfolio is larger than the risk-free return, then the tangency portfolio is mean-variance efficient among all portfolios of the risky and risk-free assets (it plots on the upper part of the cone). This is illustrated in Figure 5.3. The opposite case $B/C < R_f$ is shown in Figure 5.4. If $B/C = R_f$, then there is no tangency portfolio. Graphically, the hyperbola is contained strictly within the cone. The demonstration of that is left as an exercise.

Two-Fund Spanning

Any two frontier portfolios span the entire frontier. Thus, we have two-fund spanning again. To see this, consider any three frontier portfolios. They are of the form (5.13) for constants δ_1, δ_2, and δ_3 (the δ's depend on the portfolio expected returns via (5.14)). We can always write $\delta_3 = \lambda \delta_1 + (1 - \lambda) \delta_2$ by setting

Mean-Variance Analysis

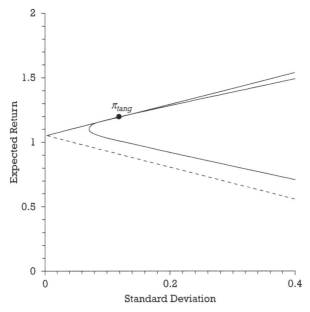

Figure 5.3 Mean-variance frontier with $B/C > R_f$. The hyperbola is the locus of (standard deviation, mean) pairs corresponding to frontier portfolios of risky assets. In this example, the efficient part of the frontier is the tangent line to the hyperbola emanating from $(0, R_f)$, and the dashed line is the inefficient part of the frontier. The standard deviation and mean of the tangency portfolio return is the tangency point, labeled π_{tang}. In this example, the expected return of the GMV portfolio is 1.10 (an expected rate of return of 10%), and the risk-free return R_f is 1.05 (a risk-free rate of 5%). Thus, $B/C = 1.10 > 1.05 = R_f$.

$\lambda = (\delta_3 - \delta_2)/(\delta_1 - \delta_2)$. Therefore, the first two portfolios span the third. This is true even if $\delta_3 = 0$, so any two frontier portfolios span the risk-free asset.

If the expected return of the GMV portfolio is different from R_f, then the risk-free asset and tangency portfolio span the mean-variance frontier. Except for the tangency portfolio itself, all portfolios proportional to the tangency portfolio satisfy either $\iota'\pi < 1$ or $\iota'\pi > 1$. In the former case, the portfolio involves a positive (long) position in the risk-free asset, and in the latter a negative (short) position in the risk-free asset. The tangency portfolio is special only because it consists entirely of risky assets; thus, the two funds—tangency portfolio and risk-free asset—consist of different assets.

If the tangency portfolio is inefficient, then all efficient portfolios consist of a long position in the risk-free asset and a short position in the tangency portfolio. If all investors hold mean-variance efficient portfolios, then the tangency portfolio

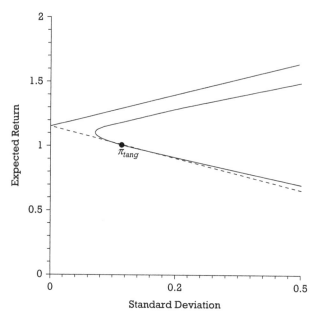

Figure 5.4 Mean-variance frontier with $B/C < R_f$. The hyperbola is the locus of (standard deviation, mean) pairs corresponding to frontier portfolios of risky assets. In this example, the efficient part of the frontier is the solid line emanating from $(0, R_f)$, and the inefficient part is the dashed line that is tangent to the hyperbola. The standard deviation and mean of the tangency portfolio return is at the tangency point, labeled π_{tang}. In this example, the expected return of the GMV portfolio is 1.10 (an expected rate of return of 10%), and the risk-free return R_f is 1.15 (meaning a risk-free rate of 15%). Thus, $B/C = 1.10 < 1.15 = R_f$.

must be efficient, because investors in aggregate must be long risky assets. As explained before, efficiency of the tangency portfolio is equivalent to the expected return of the GMV portfolio being above the risk-free return, as shown in Figure 5.2.

Covariances and Risk Premia

Chapter 6 discusses formulas for asset risk premia in terms of covariances or betas of asset returns with respect to pricing factors. It is useful to preview those results here by studying the first-order condition for the portfolio choice problem (5.12) a bit further. Recall that the first-order condition is $\Sigma \pi = \delta(\mu - R_f \iota)$.

The vector $\Sigma\pi$ is the vector of covariances of the portfolio π with the risky assets.[2] Thus, the first-order condition is that the vector of covariances should be proportional (with proportionality factor δ) to the vector of risk premia. To eliminate δ, multiply both sides of the first-order condition by π' to obtain $\pi'\Sigma\pi = \delta\pi'(\mu - R_f\iota)$. Solve for δ and substitute into the first-order condition to obtain

$$\mu - R_f\iota = \frac{1}{\delta}\Sigma\pi = \frac{\pi'(\mu - R_f\iota)}{\pi'\Sigma\pi}\Sigma\pi. \qquad (5.19)$$

The scalar $\pi'(\mu - R_f\iota)$ is the risk premium of the portfolio π. The scalar $\pi'\Sigma\pi$ is the variance of the portfolio π. Thus, the vector $(\pi'\Sigma\pi)^{-1}\Sigma\pi$ is the vector of betas of the asset returns with respect to the portfolio π.[3] We conclude that the first-order condition for mean-variance optimization in the presence of a risk-free asset is equivalent to the statement that asset risk premia equal the products of asset betas with the portfolio risk premium.

5.4 ORTHOGONAL PROJECTIONS AND FRONTIER RETURNS

The mean-variance frontier is analyzed via a different approach in this section. This analysis is equivalent to the previous analysis in a single-period model with a finite number of assets, which is what we are studying in this chapter. There are some results that are more easily obtained by this approach than by the calculus approach presented earlier. However, the real power of this different approach is in markets with infinitely many assets (an idealization considered in the study of the Arbitrage Pricing Theory, to be discussed in the next chapter) and in markets with dynamic trading. All that is needed is the existence of an SDF (not necessarily strictly positive) and the existence of orthogonal projections on the asset (or return) space. The analysis in this section applies to markets with and without a risk-free asset (and to the risky assets only by excluding the risk-free asset if it exists). The return $\tilde{R}_p$ and excess return $\tilde{e}_p$ described below

---

2. By this, we mean that
$$\Sigma\pi = \mathsf{E}\big[(\tilde{\mathbf{R}} - \mu)(\pi'\tilde{\mathbf{R}} - \pi'\mu)\big].$$
To see that this equality is true, just replace the term in the second set of parentheses with its transpose (which can be done because it is a scalar) and then move the constant π outside the expectation.

3. The betas are the covariances divided by the variance. They are also the coefficients in the orthogonal projections of the returns on the portfolio return produced by π. This is discussed further in Chapter 6.

vary depending on the market (and whether the risk-free asset is included or excluded).

Assume the asset returns have finite variances and the law of one price holds, so, as discussed in Section 3.2, there exists at least one SDF. As before, let $\tilde{m}_p$ denote the unique orthogonal projection of any SDF onto the span of the assets. Being in the span of the assets means that it is the payoff of some portfolio. Because $\tilde{m}_p$ is both an SDF and a payoff, its cost is $\mathsf{E}[\tilde{m}_p^2]$. As in Section 3.5, define the return $\tilde{R}_p = \tilde{m}_p/\mathsf{E}[\tilde{m}_p^2]$.

Representing the Expectation Operator on the Space of Excess Returns

The set of excess returns is the set of random variables (i) $\pi'\tilde{\mathbf{R}}$ with $\iota'\pi = 0$ if there is no risk-free asset, and (ii) $\pi'(\tilde{\mathbf{R}} - R_f \iota)$ if there is a risk-free asset. In either case, excess returns are the payoffs of zero-cost portfolios. For example, going long asset i and short an equal amount of asset j produces the excess return $a(\tilde{R}_i - \tilde{R}_j)$, where a denotes the amount invested long (= size of the short position). The set of excess returns is a finite-dimensional linear subspace of the space of finite-second-moment random variables; thus, as discussed in Section 3.8, orthogonal projections onto this set are well defined.

Let $\tilde{e}_p$ denote the projection of the random variable that is identically equal to 1 onto the space of excess returns. This means that $\tilde{e}_p$ is an excess return and $1 = \tilde{e}_p + \tilde{\xi}$, where $\tilde{\xi}$ is orthogonal to all excess returns. Thus, for each excess return $\tilde{x}$,

$$\mathsf{E}[\tilde{e}_p \tilde{x}] = \mathsf{E}\left[\left(1 - \tilde{\xi}\right)\tilde{x}\right] = \mathsf{E}[\tilde{x}].$$

Because $\mathsf{E}[\tilde{e}_p \tilde{x}] = \mathsf{E}[\tilde{x}]$, we say that $\tilde{e}_p$ represents the expectation operator on the space of excess returns.[4]

Return Decomposition and Frontier Returns

We will see that any return $\tilde{R}$ equals $\tilde{R}_p + b\tilde{e}_p + \tilde{\varepsilon}$, for some constant b and some excess return $\tilde{\varepsilon}$ that has zero mean and is uncorrelated with $\tilde{R}_p$ and $\tilde{e}_p$. Moreover, the frontier portfolios are the portfolios with returns $\tilde{R}_p + b\tilde{e}_p$ for some b.

To derive these facts, let $\tilde{R}$ be a return, and let $a\tilde{R}_p + b\tilde{e}_p$ be its orthogonal projection onto the span of $\tilde{R}_p$ and $\tilde{e}_p$. Then,

$$\tilde{R} = a\tilde{R}_p + b\tilde{e}_p + \tilde{\varepsilon},$$

---

4. $\tilde{e}_p$ is the Riesz representation of the linear function $\tilde{x} \mapsto \mathsf{E}[\tilde{x}]$ on the space of excess returns (Section 3.8).

Mean-Variance Analysis

where $\tilde{\varepsilon}$ is orthogonal to $\tilde{R}_p$ and $\tilde{e}_p$. We will use the following facts:

1. $\mathsf{E}[\tilde{R}_p\tilde{\varepsilon}] = 0$.
2. $\mathsf{E}[\tilde{e}_p\tilde{\varepsilon}] = 0$.
3. $\mathsf{E}[\tilde{m}_p\tilde{R}] = 1$.
4. $\mathsf{E}[\tilde{m}_p\tilde{R}_p] = 1$.
5. $\mathsf{E}[\tilde{m}_p\tilde{e}_p] = 0$.
6. $\mathsf{E}[\tilde{m}_p\tilde{\varepsilon}] = 0$.
7. $a = 1$ (so $\tilde{R} = \tilde{R}_p + b\tilde{e}_p + \tilde{\varepsilon}$).
8. $\mathsf{E}[\tilde{R}_p\tilde{e}_p] = 0$.
9. $\tilde{R}_p + b\tilde{e}_p$ is a return.
10. $\tilde{\varepsilon}$ is an excess return.
11. $\mathsf{E}[\tilde{\varepsilon}] = 0$.
12. $\mathrm{cov}(\tilde{R}_p, \tilde{\varepsilon}) = 0$.
13. $\mathrm{cov}(\tilde{e}_p, \tilde{\varepsilon}) = 0$.
14. $\mathsf{E}[\tilde{R}] = \mathsf{E}[\tilde{R}_p + b\tilde{e}_p]$.
15. $\mathrm{var}(\tilde{R}) = \mathrm{var}(\tilde{R}_p + b\tilde{e}_p) + \mathrm{var}(\tilde{\varepsilon})$.

The proofs of these facts are as follows:

1. $\tilde{\varepsilon}$ is orthogonal to $\tilde{R}_p$.
2. $\tilde{\varepsilon}$ is orthogonal to $\tilde{e}_p$.
3. $\tilde{R}$ is a return.
4. $\tilde{R}_p$ is a return.
5. $\tilde{e}_p$ is an excess return.
6. Fact 1 and the proportionality of $\tilde{R}_p$ to $\tilde{m}_p$.
7. Facts 3–6, which imply

$$1 = \mathsf{E}[\tilde{m}_p\tilde{R}] = a\mathsf{E}[\tilde{m}_p\tilde{R}_p] + b\mathsf{E}[\tilde{m}_p\tilde{e}_p] + \mathsf{E}[\tilde{m}_p\tilde{\varepsilon}] = a.$$

8. Fact 5 and the proportionality of $\tilde{R}_p$ to $\tilde{m}_p$.
9. $\tilde{R}_p$ is a return and $\tilde{e}_p$ is an excess return.
10. $\tilde{R}$ is a return, Fact 9, and Fact 7 (which implies $\tilde{\varepsilon} = \tilde{R} - \tilde{R}_p - b\tilde{e}_p$).
11. Facts 2 and 10 and the fact that $\tilde{e}_p$ represents the expectation operator on the space of excess returns, implying $0 = \mathsf{E}[\tilde{e}_p\tilde{\varepsilon}] = \mathsf{E}[\tilde{\varepsilon}]$.
12. Facts 1 and 11.
13. Facts 2 and 11.
14. Facts 7 and 11.
15. Facts 12 and 13.

From Facts 9, 14, and 15, we conclude that $\tilde{R}_p + b\tilde{e}_p$ is a return with the same mean as $\tilde{R}$ and a lower variance than $\tilde{R}$ (unless $\tilde{\varepsilon} = 0$). Thus, the frontier portfolios are the portfolios with returns $\tilde{R}_p + b\tilde{e}_p$.

Two-Fund Spanning

There is two-fund spanning: For any $b_1 \neq b_2$, the returns $\tilde{R}_p + b_1\tilde{e}_p$ and $\tilde{R}_p + b_2\tilde{e}_p$ span the mean-variance frontier. This is easy to see: For any b, we can write

$$\tilde{R}_p + b\tilde{e}_p = \lambda(\tilde{R}_p + b_1\tilde{e}_p) + (1-\lambda)(\tilde{R}_p + b_2\tilde{e}_p)$$

for

$$\lambda = \frac{b - b_2}{b_1 - b_2}.$$

Thus, any frontier return $\tilde{R}_p + b\tilde{e}_p$ is spanned by any two other frontier returns. For example, the frontier is spanned by $\tilde{R}_p$ and $\tilde{R}_p + b\tilde{e}_p$ for any $b \neq 0$.

Risk-Free Return Proxies

There are three frontier returns that serve as analogues of the risk-free return in various contexts. If there is a risk-free asset, then all of these returns are equal to the risk-free return. The returns are

Minimum Variance Return:

$$\tilde{R}_p + b_m\tilde{e}_p \quad \text{where} \quad b_m = \frac{E[\tilde{R}_p]}{1 - E[\tilde{e}_p]}, \tag{5.20}$$

Zero Beta Return:

$$\tilde{R}_p + b_z\tilde{e}_p \quad \text{where} \quad b_z = \frac{\text{var}[\tilde{R}_p]}{E[\tilde{R}_p]E[\tilde{e}_p]}, \tag{5.21}$$

Constant Mimicking Return:

$$\tilde{R}_p + b_c\tilde{e}_p \quad \text{where} \quad b_c = \frac{E[\tilde{R}_p^2]}{E[\tilde{R}_p]}. \tag{5.22}$$

If there is no risk-free asset, then $\tilde{R}_p + b_m\tilde{e}_p$ is the return of the GMV portfolio as described in Section 5.2. The return $\tilde{R}_p + b_z\tilde{e}_p$ is the frontier return that is uncorrelated with $\tilde{R}_p$. The return $\tilde{R}_p + b_c\tilde{e}_p$ is the projection of the random variable that is identically equal to b_c onto the space of returns. It has the property that

$$E[\tilde{R}(\tilde{R}_p + b_c\tilde{e}_p)] = b_c E[\tilde{R}] \tag{5.23}$$

for every return $\tilde{R}$. The verification of these facts is left for the exercises.

If there is a risk-free asset, then it is obvious that the risk-free return R_f has the minimum variance, is uncorrelated with $\tilde{R}_p$, and mimics a constant (that is, $E[R_f\tilde{R}] = R_f E[\tilde{R}]$ for every return $\tilde{R}$). Thus,

$$\tilde{R}_p + b_m\tilde{e}_p = \tilde{R}_p + b_z\tilde{e}_p = \tilde{R}_p + b_c\tilde{e}_p = R_f. \tag{5.24}$$

We can also show (Exercise 5.5) that

$$\tilde{R}_p + R_f\tilde{e}_p = R_f, \tag{5.25}$$

which means that $b_m = b_z = b_c = R_f$. Note that (5.25) provides a formula for the excess return $\tilde{e}_p$ when there is a risk-free asset: $\tilde{e}_p = (R_f - \tilde{R}_p)/R_f$.

Inefficiency of $\tilde{R}_p$

If there is a risk-free asset, then $\tilde{R}_p$ is on the inefficient part of the mean-variance frontier (plots on the lower part of the cone). If there is no risk-free asset and $E[\tilde{R}_p] > 0$—which is equivalent to the expected return of the GMV portfolio being positive—then $\tilde{R}_p$ is on the inefficient part of the frontier (plots on the lower part of the hyperbola). This is the usual circumstance (with limited liability, all returns are nonnegative). Certainly, $E[\tilde{R}_p] > 0$ if $\tilde{m}_p$ is strictly positive.

To establish the inefficiency of $\tilde{R}_p$, it suffices to show that $E[\tilde{R}_p]$ is less than the expected return of the minimum-variance portfolio, which is $E[\tilde{R}_p] + b_m E[\tilde{e}_p]$. In other words, it suffices to show that $b_m E[\tilde{e}_p] > 0$.

Because $\tilde{e}_p$ represents the expectation operator on the space of excess returns, the following are true:

16. $E[\tilde{e}_p^2] = E[\tilde{e}_p]$.
17. $\text{var}(\tilde{e}_p) = E[\tilde{e}_p](1 - E[\tilde{e}_p])$.

Thus,

$$b_m E[\tilde{e}_p] = \frac{E[\tilde{R}_p]E[\tilde{e}_p]}{1 - E[\tilde{e}_p]} = \frac{E[\tilde{R}_p]E[\tilde{e}_p]^2}{\text{var}(\tilde{e}_p)},$$

which has the same sign as $E[\tilde{R}_p]$. This also shows that $E[\tilde{R}_p + b_m\tilde{e}_p]$ has the same sign as $E[\tilde{R}_p]$.

Maximum Sharpe Ratio

We have already seen that $\tilde{R}_p$ and $\tilde{R}_p + b\tilde{e}_p$ span the mean-variance frontier for any $b \neq 0$, so, when there is a risk-free asset, $\tilde{R}_p$ and $\tilde{R}_p + b_m \tilde{e}_p = R_f$ span the frontier. Thus, any frontier return is of the form[5]

$$\lambda \tilde{R}_p + (1-\lambda) R_f = R_f + \lambda (\tilde{R}_p - R_f). \tag{5.26}$$

Hence, the risk premium of any frontier portfolio is $\lambda(\mathsf{E}[\tilde{R}_p] - R_f)$, and the standard deviation is $|\lambda| \operatorname{stdev}(\tilde{R}_p)$. The efficient frontier portfolios are of the form (5.26) with $\lambda < 0$ and have a Sharpe ratio equal to

$$\frac{R_f - \mathsf{E}[\tilde{R}_p]}{\operatorname{stdev}(\tilde{R}_p)}. \tag{5.27}$$

This is the maximum possible Sharpe ratio.

Equality in the HJ Bound with a Risk-Free Asset

The Hansen-Jagannathan bound, with a risk-free asset, is that the ratio of standard deviation to mean of any SDF is at least as large as the Sharpe ratio of any portfolio. The Hansen-Jagannathan bound is derived in Section 3.6, and it is shown there that the orthogonal projection $\tilde{m}_p$ defined in Section 3.5 has the minimum standard-deviation-to-mean ratio. Now, we will see that the Sharpe ratio (5.27) of the efficient portfolios equals $\operatorname{stdev}(\tilde{m}_p)/\mathsf{E}[\tilde{m}_p]$. Because $\tilde{R}_p$ is proportional to $\tilde{m}_p$, this is equivalent to

$$\frac{\operatorname{stdev}(\tilde{R}_p)}{\mathsf{E}[\tilde{R}_p]} = \frac{R_f - \mathsf{E}[\tilde{R}_p]}{\operatorname{stdev}(\tilde{R}_p)}. \tag{5.28}$$

Equation (5.28) is a consequence of the following facts:

18. $\mathsf{E}[\tilde{R}_p^2] = 1/\mathsf{E}[\tilde{m}_p^2]$.
19. $R_f \mathsf{E}[\tilde{R}_p] = 1/\mathsf{E}[\tilde{m}_p^2]$.
20. $\operatorname{var}(\tilde{R}_p) = R_f \mathsf{E}[\tilde{R}_p] - \mathsf{E}[\tilde{R}_p]^2$.

The proofs of these are as follows:

---

5. To see this more directly, observe that the frontier returns are

$$\tilde{R}_p + b\tilde{e}_p = \tilde{R}_p + b\frac{R_f - \tilde{R}_p}{R_f} = \lambda \tilde{R}_p + (1-\lambda) R_f,$$

where $\lambda = 1 - b/R_f$.

18. The definition $\tilde{R}_p = \tilde{m}_p/\mathsf{E}[\tilde{m}_p^2]$, which implies $\mathsf{E}[\tilde{R}_p^2] = \mathsf{E}[\tilde{m}_p\tilde{R}_p]/\mathsf{E}[\tilde{m}_p^2]$, and Fact 4.
19. The definition $\tilde{R}_p = \tilde{m}_p/\mathsf{E}[\tilde{m}_p^2]$, which implies $\mathsf{E}[\tilde{R}_pR_f] = \mathsf{E}[\tilde{m}_pR_f]/\mathsf{E}[\tilde{m}_p^2]$, and $\mathsf{E}[\tilde{m}_pR_f] = 1$.
20. Facts 18 and 19, which imply $\mathsf{E}[\tilde{R}_p^2] = R_f\mathsf{E}[\tilde{R}_p]$, and the definition of variance.

Fact 20 implies (5.28) directly.

5.5 FRONTIER RETURNS AND STOCHASTIC DISCOUNT FACTORS

The first-order condition for portfolio choice produces an SDF that is a function of the investor's wealth (namely, the SDF is proportional to the investor's marginal utility of wealth). This formula is useful, particularly for a representative investor (Chapter 7). However, it is frequently convenient to be able to construct SDFs from asset returns instead of wealth or consumption. Constructing SDFs that are affine functions of returns is intimately related to mean-variance analysis. Specifically, there is an SDF that is an affine function of a return if and only if the return is on the mean-variance frontier and not equal to the constant-mimicking return (so not equal to R_f if there is a risk-free asset). This is simplest to establish when there is a risk-free asset.

When there is a risk-free asset, an affine function of a return $\tilde{R}$ is a linear combination of the risk-free return and $\tilde{R}$ (the intercept in the affine function is some multiple of R_f); hence, it is spanned by the assets. Thus, an SDF $\tilde{m}$ is an affine function of a return $\tilde{R}$ if and only if $\tilde{m} = \tilde{m}_p$. Furthermore, $\tilde{m}_p$ is a linear combination of R_f and $\tilde{R}$ if and only if $\tilde{R}_p$ is a linear combination of R_f and $\tilde{R}$, and this is equivalent to $\tilde{R}$ being spanned by R_f and $\tilde{R}_p$, which is equivalent to mean-variance efficiency of $\tilde{R}$. The following establishes the result in the absence of a risk-free asset.

First, we will show that, for any frontier return $\tilde{R}_* = \tilde{R}_p + b\tilde{e}_p$ with $b \neq b_c$,

$$\tilde{x} = -\beta b + \beta \tilde{R}_*$$

is an SDF for some constant β. We can write $\tilde{x}$ as

$$\tilde{x} = \beta\tilde{R}_p + \beta b(\tilde{e}_p - 1).$$

Both $\tilde{R}_p$ and $\tilde{e}_p - 1$ are orthogonal to excess returns. Therefore, $\tilde{x}$ is orthogonal to excess returns, implying, for any return $\tilde{R}$,

$$\mathsf{E}[\tilde{x}\tilde{R}] = \mathsf{E}[\tilde{x}\tilde{R}_p] + \mathsf{E}[\tilde{x}(\tilde{R} - \tilde{R}_p)] = \mathsf{E}[\tilde{x}\tilde{R}_p].$$

Moreover, using $\mathsf{E}[\tilde{R}_p \tilde{e}_p] = 0$ (Fact 8 from Section 5.4), we have

$$\mathsf{E}[\tilde{x}\tilde{R}_p] = \beta \mathsf{E}[\tilde{R}_p^2] - \beta b \mathsf{E}[\tilde{R}_p].$$

Hence, setting

$$\beta = \frac{1}{\mathsf{E}[\tilde{R}_p^2] - b\mathsf{E}[\tilde{R}_p]},$$

we have, for every return $\tilde{R}$,

$$\mathsf{E}[\tilde{x}\tilde{R}] = \beta \mathsf{E}[\tilde{R}_p^2] - \beta b \mathsf{E}[\tilde{R}_p] = 1.$$

This shows that $\tilde{x}$ is an SDF.

Now, given any return $\tilde{R}_* = \tilde{R}_p + b\tilde{e}_p + \tilde{\varepsilon}$, suppose

$$\tilde{m} = \gamma + \beta \tilde{R}_*$$

is an SDF for some constants γ and β. We will show that $\tilde{\varepsilon} = 0$ and therefore $\tilde{R}_*$ is a frontier return; moreover, $b \neq b_c$. Using the definition of an SDF and the mutual orthogonality of $\tilde{R}_p$, $\tilde{e}_p$, and $\tilde{\varepsilon}$, we can calculate

$$1 = \mathsf{E}[\tilde{m}\tilde{R}_*] = \gamma \mathsf{E}[\tilde{R}_p] + \gamma b \mathsf{E}[\tilde{e}_p] + \beta \mathsf{E}[\tilde{R}_p^2] + \beta b^2 \mathsf{E}[\tilde{e}_p^2] + \beta \mathsf{E}[\tilde{\varepsilon}^2], \quad (5.29)$$

$$1 = \mathsf{E}[\tilde{m}\tilde{R}_p] = \gamma \mathsf{E}[\tilde{R}_p] + \beta \mathsf{E}[\tilde{R}_p^2], \quad (5.30)$$

$$0 = \mathsf{E}[\tilde{m}\tilde{e}_p] = \gamma \mathsf{E}[\tilde{e}_p] + \beta b \mathsf{E}[\tilde{e}_p^2]. \quad (5.31)$$

Subtracting (5.30) and b times (5.31) from (5.29) yields $\mathsf{E}[\tilde{\varepsilon}^2] = 0$. Thus, $\tilde{R}_*$ is a frontier return. Moreover, (5.31) and $\mathsf{E}[\tilde{e}_p^2] = \mathsf{E}[\tilde{e}_p]$ (Fact 16) implies $\gamma = -\beta b$, so $\tilde{m} = -\beta b + \beta \tilde{R}_*$. This implies $\mathsf{E}[\tilde{m}\tilde{R}_p] = 0$, contradicting (5.30), if $b = b_c$, so we conclude $b \neq b_c$.

5.6 SEPARATING DISTRIBUTIONS

Section 4.4 shows that, with a risk-free asset and no labor income, two-fund separation holds if all investors have linear risk tolerance $\tau(w) = A + Bw$ with the same cautiousness parameter B. This is independent of the distribution of risky asset payoffs. We observed in this chapter that the mean-variance frontier is spanned by any two frontier portfolios, so two-fund separation holds when all investors hold mean-variance efficient portfolios. Thus, if returns are elliptically distributed, then two-fund separation holds. This section asks more

Mean-Variance Analysis

generally: What distributions of risky asset payoffs imply two-fund separation for all investor preferences?

Assume there is a risk-free asset and no labor income. Index the risk-free asset as asset 0. For notational convenience, assume each investor has a zero endowment of asset 0 (which we can do by allowing one of the other assets to be risk free, if there are nonzero endowments of the risk-free asset). Let n denote the number of other assets, and assume there is a random variable $\tilde{v}$ such that the payoff $\tilde{x}_i$ of asset i, for $i = 1, \ldots, n$, satisfies

$$\tilde{x}_i = a_i + b_i \tilde{v} + \tilde{\varepsilon}_i \quad \text{with} \quad \mathsf{E}[\tilde{\varepsilon}_i | \tilde{v}] = 0 \tag{5.32a}$$

for constants a_i and b_i. As before, let θ_{mi} denote the aggregate number of shares outstanding of asset i, for $i = 1, \ldots, n$. Assume

$$\sum_{i=1}^{n} \theta_{mi} b_i \neq 0 \quad \text{and} \quad \sum_{i=1}^{n} \theta_{mi} \tilde{\varepsilon}_i = 0. \tag{5.32b}$$

Condition (5.32a) is an assumption about the residuals in the orthogonal projection of each $\tilde{x}_i$ on a constant and $\tilde{v}$; specifically, it is assumed that each residual is mean independent of $\tilde{v}$ rather than just being uncorrelated. An equivalent statement is that the expectation of $\tilde{x}_i$ conditional on $\tilde{v}$ is an affine function of $\tilde{v}$. Joint elliptical (and joint normal) random variables have this property. Note that we are not making any assumption about the correlation of $\tilde{\varepsilon}_i$ and $\tilde{\varepsilon}_j$ for $i \neq j$, as we do in a statistical factor model (Section 6.5).

Condition (5.32b) means that the market payoff has no residual risk but is not risk free (is exposed to the common factor), which is expressed as "the market portfolio is well diversified." From (5.32b), the payoff of the market portfolio is

$$\tilde{w}_m = a_m + b_m \tilde{v},$$

where

$$\tilde{w}_m = \sum_{i=1}^{n} \theta_{mi} \tilde{x}_i, \quad a_m = \sum_{i=1}^{n} \theta_{mi} a_i, \quad \text{and} \quad b_m = \sum_{i=1}^{n} \theta_{mi} b_i \neq 0.$$

Regarding investors, assume they have monotone preferences, are risk averse, and can feasibly hold the market portfolio, in the sense that, for each h,

$$\mathsf{E}\left[u_h\left(\delta_h \tilde{w}_m + a_h\right)\right] > -\infty, \tag{5.33}$$

for some $\delta_h > \sum_{i=1}^{n} \bar{\theta}_{hi} b_i / b_m$ and $a_h > 0$.

As remarked in Section 4.2, there is one degree of indeterminacy in equilibrium prices. Normalize the size of a share of the risk-free asset so that its payoff is one unit of the consumption good at date 1. Use this size share of the risk-free asset as the numeraire at date 0, so $R_f = 1$.[6] Define $p_i = a_i + b_i \lambda$ for a constant λ. We will see that these are equilibrium prices, for some constant λ.

Set $\tilde{z} = \tilde{v} - \lambda$. Then,

$$\tilde{x}_i = p_i + b_i \tilde{z} + \tilde{\varepsilon}_i,$$

implying

$$\tilde{R}_i = R_f + \beta_i \tilde{z} + \tilde{\xi}_i \quad \text{with} \quad E[\tilde{\xi}_i | \tilde{z}] = 0, \tag{5.34a}$$

where $\beta_i = b_i/p_i$ and $\tilde{\xi}_i = \tilde{\varepsilon}_i/p_i$. Note that the risk premium of asset i is $\beta_i E[\tilde{z}]$. Moreover, the market portfolio $\pi_{mi} = p_i \theta_{mi} / \sum_{j=1}^{n} p_j \theta_{mj}$ is well diversified in the sense that

$$\sum_{i=1}^{n} \pi_{mi} \beta_i \neq 0 \quad \text{and} \quad \sum_{i=1}^{n} \pi_{mi} \tilde{\xi}_i = 0. \tag{5.34b}$$

Condition (5.34a) and the existence of a well-diversified portfolio constitute a necessary and sufficient condition for two-fund separation when there is a risk-free asset.

The following facts about this model are established below:

(a) Two-fund separation holds.
(b) There exists a constant λ such that the p_i are equilibrium prices. Moreover, $\lambda \neq E[\tilde{v}]$, implying $E[\tilde{z}] \neq 0$.
(c) The optimal portfolio of each investor has the minimum variance among all portfolios with the same mean payoff (it is on the mean-variance frontier).
(d) The Capital Asset Pricing Model holds:

$$E[\tilde{R}_i] = R_f + \frac{\text{cov}(\tilde{R}_i, \tilde{R}_m)}{\text{var}(\tilde{R}_m)} (E[\tilde{R}_m] - R_f) \tag{5.35}$$

for each i, where $\tilde{R}_m$ denotes the market return.

The last fact anticipates Chapter 6.

---

6. The real risk-free return is still endogenous (that is, it depends on investors' preferences and the investment opportunities), because the price of the consumption good at date 0 in units of the risk-free asset is endogenous.

Mean-Variance Analysis

(a) Two-fund separation can be deduced from either (5.32) or (5.34). To avoid the implicit assumption of positive prices p_i in (5.34), we will use (5.32). If an investor has initial wealth w_0 and chooses to hold θ_i shares of asset i for $i = 1, \ldots, n$, then her end-of-period wealth is

$$\sum_{i=1}^n \theta_i \tilde{x}_i + \left(w_0 - \sum_{i=1}^n \theta_i p_i\right) R_f = w_0 + \sum_{i=1}^n \theta_i b_i (\tilde{v} - \lambda) + \sum_{i=1}^n \theta_i \tilde{\varepsilon}_i.$$

If $\theta_i = \delta \theta_{mi}$ for some δ, then the end-of-period wealth is

$$w_0 + \delta b_m (\tilde{v} - \lambda).$$

Given any $\theta_1, \ldots, \theta_n$ and setting

$$\delta = \frac{\sum_{i=1}^n \theta_i b_i}{b_m},$$

it follows from aversion to mean-independent noise (Section 1.5) that the portfolio $\delta \theta_m$ is preferred to $(\theta_1 \cdots \theta_n)'$. Thus, each investor chooses to hold a combination of the risk-free asset and the well-diversified portfolio.

(b) Each investor h holds a portfolio $\delta_h \theta_m$. The markets for assets $i = 1, \ldots, n$ clear if $\sum_{h=1}^H \delta_h = 1$, and then the market for asset 0 clears by Walras's law. The end-of-period wealth of investor h is

$$\tilde{w}_{h1} = w_{h0} + \delta_h b_m (\tilde{v} - \lambda) = w_{h0} + \delta_h (\tilde{w}_m - p_m),$$

where

$$p_m = \sum_{i=1}^n \theta_{mi} p_i = a_m + b_m \lambda.$$

If $p_m = \mathsf{E}[\tilde{w}_m]$, then risk aversion implies that $\delta_h = 0$ is optimal (thus, $p_m \neq \mathsf{E}[\tilde{w}_m]$ in equilibrium; equivalently, $\lambda \neq \mathsf{E}[\tilde{v}]$). On the other hand, as $p_m \downarrow -\infty$, we will show that it is eventually optimal to choose

$$\delta_h \geq \frac{\sum_{i=1}^n \overline{\theta}_{hi} b_i}{b_m},$$

implying $\sum_{h=1}^H \delta_h \geq 1$. By continuity, there is some p_m (that is, some λ) such that $\sum_{h=1}^H \delta_h = 1$. To see the nature of optimal demands as $p_m \downarrow -\infty$, substitute for w_{h0} and define

$$\kappa_h = \delta_h - \frac{\sum_{i=1}^n \overline{\theta}_{hi} b_i}{b_m}.$$

to write the end-of-period wealth as

$$\tilde{w}_{h1} = \sum_{i=1}^n \overline{\theta}_{hi}(a_i + b_i \tilde{v}) + \kappa_h(\tilde{w}_m - p_m).$$

For any fixed $\kappa_h > 0$, $\tilde{w}_{h1} \uparrow \infty$ as $p_m \downarrow -\infty$. Hence, by the monotone convergence theorem, expected utility converges to ∞ or the upper bound of the utility function. It follows that $\kappa_h > 0$ is superior to $\kappa_h \leq 0$ for sufficiently small p_m.

(c) The risk premium of each asset i is $\beta_i \mathsf{E}[\tilde{z}]$, so the expected return of a portfolio π is

$$R_f + \mathsf{E}[\tilde{z}] \sum_{i=1}^n \pi_i \beta_i.$$

Therefore, two portfolios with the same expected return must have the same $\sum_{i=1}^n \pi_i \beta_i$. Consider any δ and any portfolio that has the same expected return as $\delta \pi_m$. The return of the portfolio is

$$R_f + \tilde{z} \sum_{i=1}^n \pi_i \beta_i + \sum_{i=1}^n \pi_i \tilde{\xi}_i = R_f + \delta \tilde{z} \sum_{i=1}^n \pi_{mi} \beta_i + \sum_{i=1}^n \pi_i \tilde{\xi}_i = \tilde{R}_m + \sum_{i=1}^n \pi_i \tilde{\xi}_i.$$

Therefore, the variance of the portfolio return is

$$\mathrm{var}(\tilde{R}_m) + \sum_{i=1}^n \pi_i^2 \mathrm{var}(\tilde{\xi}_i) \geq \mathrm{var}(\tilde{R}_m).$$

(d) We have

$$\tilde{w}_m = a_m + b_m \lambda + b_m \tilde{z} = p_m + b_m \tilde{z}.$$

Also, $R_f = 1$. Therefore,

$$\tilde{R}_m = R_f + \beta_m \tilde{z},$$

where $\beta_m = b_m / p_m$. This implies $\mathrm{cov}(\tilde{R}_i, \tilde{R}_m) = \beta_i \beta_m \mathrm{var}(\tilde{z})$, $\mathrm{var}(\tilde{R}_m) = \beta_m^2 \mathrm{var}(\tilde{z})$, and $\mathsf{E}[\tilde{R}_m] - R_f = \beta_m \mathsf{E}[\tilde{z}]$, yielding

$$R_f + \frac{\mathrm{cov}(\tilde{R}_i, \tilde{R}_m)}{\mathrm{var}(\tilde{R}_m)} (\mathsf{E}[\tilde{R}_m] - R_f) = R_f + \beta_i \mathsf{E}[\tilde{z}] = \mathsf{E}[\tilde{R}_i].$$

5.7 NOTES AND REFERENCES

Mean-variance analysis is due to Markowitz (1952, 1959). Chamberlain and Rothschild (1983) show that the frontier is spanned by the projection $\tilde{m}_p$ and the payoff that represents the expectation operator on the space of payoffs. Hansen and Richard (1987) show that the frontier is spanned by the projections $\tilde{R}_p$ and $\tilde{e}_p$, as described in Section 5.4, and show that $\tilde{R}_p$ is the return with the minimum second moment (Exercise 5.4), implying the inefficiency of $\tilde{R}_p$.

Hansen and Jagannathan (1991) establish (5.28): When there is a risk-free asset, the maximum Sharpe ratio equals the minimum standard-deviation-to-mean ratio of any SDF. They also describe the returns achieving the upper bounds in (3.36). Those returns are affinely related to the corresponding SDFs, as in Section 5.5.

Ross (1978a) shows that condition (5.34a) and the existence of a well-diversified portfolio are necessary and sufficient for two-fund separation to hold for all investor preferences when there is a risk-free asset. He also gives necessary and sufficient conditions for k–fund separation, for any k, with and without a risk-free asset. Ross shows that normal distributions satisfy (5.34). Chamberlain (1983a) shows that elliptical distributions satisfy (5.34).

An important topic not considered in this chapter is the nature of the mean-variance frontier when there are restrictions on short sales. That issue is discussed by Ross (1977) and Dybvig (1984). Exercise 5.9 asks for a characterization of the mean-variance efficient portfolios when the rate at which an investor borrows exceeds the rate at which she can lend.

It is notoriously difficult to estimate the mean-variance frontier from sample data. There are two issues. One is that it is very difficult to estimate expected returns. The standard error (that is, the standard deviation in repeated samples) of a sample mean is $\sigma/\sqrt{T}$ where T is the number of observations and σ is the standard deviation of the random variable whose mean is being estimated. The annual standard deviation of a stock return is in the neighborhood of 30%. If we use 20 years of data, then the standard error is 6.7%. This means that a 95% confidence interval for the population mean is $\pm 13\%$, which is so wide that it is practically useless. For example, if the sample mean is 10%, then the 95% confidence interval is -3% to 23%. Of course, we could use more years of data, but then we have to be confident that the expected return has not changed over a longer time period. Furthermore, using more frequently sampled data helps not at all—it produces greater precision as the sampling frequency increases due to the increase in T, but it is greater precision of a higher frequency return, and when translated to annual returns, the precision is unchanged.[7] The second problem is that there are so many covariances to estimate: n variances and $n(n-1)/2$ covariances with n assets (for example, 499,500 covariances

---

7. This is exactly true for continuously compounded returns and approximately true for annually compounded returns. It is exactly true for continuously compounded returns, because continuously compounded returns are sums of continuously compounded returns—an annual return is the sum of monthly returns, etc.

with 1,000 assets). Common methods motivated by these issues include statistical factor models (Section 6.4), shrinkage (Ledoit and Wolf, 2004), and imposing constraints on portfolio weights in mean-variance optimization (Jagannathan and Ma, 2003). Nevertheless, the remaining difficulties are such that Cochrane (2014) states: "The results of $\Sigma^{-1}\mu$ are so sensitive to input assumptions that commercial optimizers and both regulatory and commercial risk management calculations add layers of ad hoc constraints, so many layers that the answers barely merit being called implementations of $\Sigma^{-1}\mu$. Much of the money management industry amounts to selling one or another attempted solution to estimating and computing $\Sigma^{-1}\mu$, at fees commensurate with the challenge of the problem."

EXERCISES

5.1. Suppose there are two risky assets with means $\mu_1 = 1.08$, $\mu_2 = 1.16$, standard deviations $\sigma_1 = 0.25$, $\sigma_2 = 0.35$, and correlation $\rho = 0.30$. Calculate the GMV portfolio and locate it on Figure 5.1.

5.2. Assume there is a risk-free asset. Consider an investor with quadratic utility $-(\tilde{w} - \zeta)^2/2$ and no labor income.

(a) Explain why the result of Exercise 2.5 implies that the investor will choose a portfolio on the mean-variance frontier.
(b) Under what circumstances will the investor choose a mean-variance efficient portfolio? Explain the economics of the condition you derive.
(c) Re-derive the answer to Part (b) using the orthogonal projection characterization of the quadratic utility investor's optimal portfolio presented in Section 3.5.

5.3. Suppose that the risk-free return is equal to the expected return of the GMV portfolio ($R_f = B/C$). Show that there is no tangency portfolio. Hint: Show there are no δ and λ satisfying

$$\delta \Sigma^{-1}(\mu - R_f \iota) = \lambda \pi_{\text{mu}} + (1-\lambda) \pi_{\text{gmv}}.$$

Recall that we are assuming μ is not a scalar multiple of ι.

5.4. Show that $\mathsf{E}[\tilde{R}^2] \geq \mathsf{E}[\tilde{R}_p^2]$ for every return $\tilde{R}$ (thus, $\tilde{R}_p$ is the minimum second-moment return). The returns having a given second moment a are the returns satisfying $\mathsf{E}[\tilde{R}^2] = a$, which is equivalent to

$$\text{var}(\tilde{R}) + \mathsf{E}[\tilde{R}]^2 = a;$$

Mean-Variance Analysis

thus, they plot on the circle $x^2 + y^2 = a$ in (standard deviation, mean) space. Use the fact that $\tilde{R}_p$ is the minimum second-moment return to illustrate graphically that $\tilde{R}_p$ must be on the inefficient part of the frontier, with and without a risk-free asset (assuming $E[\tilde{R}_p] > 0$ in the absence of a risk-free asset).

5.5. Write any return $\tilde{R}$ as $\tilde{R}_p + (\tilde{R} - \tilde{R}_p)$ and use the fact that $1 - \tilde{e}_p$ is orthogonal to excess returns—because $\tilde{e}_p$ represents the expectation operator on the space of excess returns—to show that

$$\tilde{x} \stackrel{\text{def}}{=} \frac{1}{E[\tilde{R}_p]} (1 - \tilde{e}_p)$$

is an SDF. When there is a risk-free asset, $\tilde{x}$, being spanned by a constant and an excess return, is in the span of the returns and hence must equal $\tilde{m}_p$. Use this fact to demonstrate (5.25).

5.6. Establish the properties claimed for the risk-free return proxies:

(a) Show that $\text{var}(\tilde{R}) \geq \text{var}(\tilde{R}_p + b_m \tilde{e}_p)$ for every return $\tilde{R}$.
(b) Show that $\text{cov}(\tilde{R}_p, \tilde{R}_p + b_z \tilde{e}_p) = 0$.
(c) Prove (5.23), showing that $\tilde{R}_p + b_c \tilde{e}_p$ represents the constant b_c times the expectation operator on the space of returns.

5.7. If all returns are joint normally distributed, then $\tilde{R}_p$, $\tilde{e}_p$, and $\tilde{\varepsilon}$ are joint normally distributed in the orthogonal decomposition $\tilde{R} = \tilde{R}_p + b\tilde{e}_p + \tilde{\varepsilon}$ of any return $\tilde{R}$ (because $\tilde{R}_p$ is a return and $\tilde{e}_p$ and $\tilde{\varepsilon}$ are excess returns). Assuming all returns are joint normally distributed, use the orthogonal decomposition to compute the optimal return for a CARA investor.

5.8. Assume there is a risk-free asset.

(a) Using the formula (3.45) for $\tilde{m}_p$, compute λ such that

$$\tilde{R}_p = \lambda \pi'_{\text{tang}} \mathbf{R} + (1 - \lambda) R_f.$$

(b) Show that λ in Part (a) is negative when $R_f < B/C$ and positive when $R_f > B/C$. Note: This shows that $\tilde{R}_p$ is on the inefficient part of the frontier, because the portfolio generating $\tilde{R}_p$ is short the tangency portfolio when the tangency portfolio is efficient and long the tangency portfolio when it is inefficient.

5.9. Consider the problem of choosing a portfolio π of risky assets, a proportion $\phi_b \geq 0$ of initial wealth to borrow, and a proportion $\phi_\ell \geq 0$ of initial wealth to lend to maximize the expected return $\pi'\mu + \phi_\ell R_\ell - \phi_b R_b$ subject to the constraints $(1/2)\pi'\Sigma\pi \leq k$ and $\iota'\pi + \phi_\ell - \phi_b = 1$. Assume $B/C > R_b > R_\ell$, where B and C

are defined in (5.6). Define
$$\pi_b = \frac{1}{\iota'\Sigma^{-1}(\mu - R_b\iota)}\Sigma^{-1}(\mu - R_b\iota),$$
$$\pi_\ell = \frac{1}{\iota'\Sigma^{-1}(\mu - R_\ell\iota)}\Sigma^{-1}(\mu - R_\ell\iota).$$

Using the Kuhn-Tucker conditions, show that the solution is either (i) $\pi = (1 - \phi_\ell)\pi_\ell$ for $0 \leq \phi_\ell \leq 1$, (ii) $\pi = \lambda\pi_\ell + (1 - \lambda)\pi_b$ for $0 \leq \lambda \leq 1$, or (iii) $\pi = (1 + \phi_b)\pi_b$ for $\phi_b \geq 0$.

6

Factor Models

CONTENTS

6.1	Capital Asset Pricing Model	128
6.2	General Factor Models	135
6.3	Jensen's Alpha and Performance Evaluation	142
6.4	Statistical Factors	145
6.5	Arbitrage Pricing Theory	147
6.6	Empirical Performance of Popular Models	150
6.7	Notes and References	155

This chapter discusses formulas for expected returns in terms of the covariances or betas of returns with some random variables. These models are at the heart of cross-sectional studies of asset pricing. Common models are the Capital Asset Pricing Model (CAPM) which asserts that the priced factor is the return on the market portfolio, and the Fama-French-Carhart model (Fama and French, 1993; Carhart, 1997), which uses the market (stock index) return and three excess returns as the factors. Those models are discussed in Sections 6.1 and 6.6.

Occasionally we will call a factor model a factor *pricing* model or a factor model *for risk premia* or something similar. The purpose is to distinguish the model from what in this book is called a statistical factor model but which is generally called simply a factor model. A statistical factor model explains the covariances of asset returns in terms of some factors. A factor pricing model explains risk premia in terms of factors. Section 2.3 shows that the risk of a diversified portfolio depends primarily on covariances. Thus, a statistical factor model explains diversified portfolio risks. It seems natural therefore that it should also explain risk premia. In other words, it should also be a factor pricing model. This is approximately true. The result "statistical factor model ⇒ factor pricing model" is called the arbitrage pricing theory and is discussed in Section 6.5.

We have already seen one example of a factor (pricing) model. Equation (3.12) shows that, if there is a risk-free asset, then risk premia are determined by covariances with any SDF. In the absence of a risk-free asset, the same reasoning leading to (3.12) yields

$$\mathsf{E}[\tilde{R}] = \frac{1}{\mathsf{E}[\tilde{m}]} - \frac{1}{\mathsf{E}[\tilde{m}]}\operatorname{cov}(\tilde{m}, \tilde{R}) \qquad (6.1)$$

for any return $\tilde{R}$ and any SDF $\tilde{m}$ with $\mathsf{E}[\tilde{m}] \neq 0$. According to (6.1), a return that has a zero covariance with $\tilde{m}$ has an expected return equal to $1/\mathsf{E}[\tilde{m}]$. This is called the zero-beta return. Denoting the zero-beta return by R_z, the factor model (6.1) can be written as

$$\mathsf{E}[\tilde{R}] - R_z = -R_z \operatorname{cov}(\tilde{m}, \tilde{R}). \qquad (6.2)$$

All factor models are special cases of this formula obtained by specifying a particular SDF.

6.1 CAPITAL ASSET PRICING MODEL

The CAPM states that there is a constant R_z such that each return $\tilde{R}$ satisfies

$$\mathsf{E}[\tilde{R}] - R_z = \frac{\operatorname{cov}(\tilde{R}, \tilde{R}_m)}{\operatorname{var}(\tilde{R}_m)}(\mathsf{E}[\tilde{R}_m] - R_z), \qquad (6.3)$$

where $\tilde{R}_m$ is the market return. The ratio $\operatorname{cov}(\tilde{R}, \tilde{R}_m)/\operatorname{var}(\tilde{R}_m)$ is called the market beta of the return $\tilde{R}$—it is the coefficient in the regression (orthogonal projection) of $\tilde{R}$ on a constant and $\tilde{R}_m$—see (3.33). Any return $\tilde{R}$ that has a zero market beta must have expected return equal to R_z, according to (6.3), so R_z is called the zero-beta return. If there is a risk-free asset, then R_z must be the risk-free return (because (6.3) also holds for the risk-free return when it exists). Using the zero-beta return as a proxy risk-free return, the CAPM states that the risk premium of an asset equals its market beta multiplied by the market risk premium. The market risk premium is called the price of risk. According to the model, the price of risk is the same for all assets, and different assets have different risk premia because they have different betas.

The CAPM formula looks the same when expressed in terms of rates of return instead of gross returns. Using a lowercase r to denote a rate of return (a gross return minus 1), and an overbar to denote an expected value, (6.3) implies

$$\bar{r} = r_z + \beta(\bar{r}_m - r_z), \qquad (6.4)$$

Factor Models

where

$$\beta = \frac{\text{cov}(\tilde{R}, \tilde{R}_m)}{\text{var}(\tilde{R}_m)} = \frac{\text{cov}(\tilde{r}, \tilde{r}_m)}{\text{var}(\tilde{r}_m)}.$$

The graph of the function $\beta \mapsto \bar{r}$ given by (6.4) is called the security market line.

Cost of Capital

The CAPM is widely used in corporate finance to determine required rates of return. We can substitute $\tilde{R} = \tilde{x}/p$ into (6.3) and rearrange to obtain

$$p = \frac{\mathsf{E}[\tilde{x}]}{R_z + \beta(\mathsf{E}[\tilde{R}_m] - R_z)} = \frac{\bar{x}}{1 + r_z + \beta(\bar{r}_m - r_z)}. \tag{6.5}$$

Thus, the price is the expected payoff discounted at the rate

$$r_z + \beta(\bar{r}_m - r_z). \tag{6.6}$$

Other assets or projects with the same market beta should be evaluated as discounted expected values with the same discount rate, if the CAPM is true. The rate (6.6) is called the required rate of return or the cost of capital. While the CAPM is the model that is most widely used in practice for determining costs of capital, the same transformation from the expected return formula (6.3) to the cost of capital formula (6.6) can be made in any factor model.

CAPM and Alphas

By projection, any return $\tilde{R}$ can be represented as

$$\tilde{R} = \alpha + \beta \tilde{R}_m + \tilde{\varepsilon},$$

where $\tilde{\varepsilon}$ has zero mean and is uncorrelated with $\tilde{R}_m$. This implies that

$$\mathsf{E}[\tilde{R}] = \alpha + \beta \mathsf{E}[\tilde{R}_m].$$

Therefore, the CAPM is equivalent to the statement that $\alpha = (1 - \beta)R_z$. Of course, the parameter R_z is the same for all assets, which gives this condition its bite.

If there is a risk-free asset, then it is convenient to do a different projection, producing a different intercept α. This projection uses excess returns:

$$\tilde{R} - R_f = \alpha + \beta(\tilde{R}_m - R_f) + \tilde{\varepsilon}.$$

The CAPM is equivalent to the statement that the intercept α in this projection is zero.

Market Portfolio and Market Return

The market return is the return of the market portfolio. The market portfolio is expressed in terms of fractions of wealth invested, where wealth is market wealth and the investment in each asset is the market capitalization (price times shares outstanding) of the asset. Let n denote the number of assets, including the risk-free asset if it exists, and, as usual, denote the beginning-of-period price of asset i by p_i, and denote the end-of-period payoff of asset i by $\tilde{x}_i$. Let θ_{mi} denote the total number of shares outstanding of asset i.

Suppose there is no labor income. Then, the market portfolio is π_m, where, for $i = 1, \ldots, n$,

$$\pi_{mi} = \frac{\theta_{mi} p_i}{\sum_{j=1}^{n} \theta_{mj} p_j}.$$

The market return is

$$\tilde{R}_m = \frac{\sum_{i=1}^{n} \theta_{mi} \tilde{x}_i}{\sum_{j=1}^{n} \theta_{mj} p_j} = \frac{\sum_{i=1}^{n} \theta_{mi} p_i \tilde{R}_i}{\sum_{j=1}^{n} \theta_{mj} p_j}$$

$$= \sum_{i=1}^{n} \frac{\theta_{mi} p_i}{\sum_{j=1}^{n} \theta_{mj} p_j} \tilde{R}_i$$

$$= \sum_{i=1}^{n} \pi_{mi} \tilde{R}_i.$$

If investors have labor incomes that are spanned by the asset payoffs, then the market portfolio and market return can be defined in a similar fashion. Being spanned by the asset payoffs means that the labor income $\tilde{y}_h$ of each investor h satisfies

$$\tilde{y}_h = \sum_{i=1}^{n} \theta_{hi} \tilde{x}_i$$

for some θ_h. This situation is equivalent to one in which each investor has no end-of-period income but is endowed with an additional θ_{hi} shares of asset i.[1] Call the additional shares "pseudo shares". By including the pseudo shares in the number of shares outstanding of each asset, the market portfolio and market return can be defined exactly as above. Of course, this market return is not directly observable, which makes it difficult to empirically test the model.

---

1. Recall that we are assuming there are no restrictions on short sales. Hence, when labor income is spanned by the asset payoffs, then it can be sold in advance by short selling the portfolio that has a payoff equal to the labor income. The assumption that labor income is spanned and can be sold in advance in this way is quite unrealistic.

The more realistic case of labor incomes that are not spanned by the asset payoffs is considered later in this section. The difficulty in that case is that unspanned random variables do not have unique prices (different SDFs imply different prices), so the value at date 0 of all market wealth including labor income is not defined. This makes it impossible to define a market return that includes labor income. Nevertheless, under some assumptions, a version of the CAPM can be established, in which the priced risk is the beta with respect to market wealth, including labor income instead of the beta with respect to a market return.

Mean-Variance Efficiency

The CAPM (6.3) is equivalent to the market portfolio being on the mean-variance frontier. It is easy to see one part of this equivalence (the CAPM implies the market return is on the mean-variance frontier) without referring to the algebra of Chapter 5. The CAPM (6.3) implies that, if a return $\tilde{R}$ has the same expected return as does $\tilde{R}_m$, then its beta is equal to 1, so

$$\operatorname{cov}(\tilde{R}, \tilde{R}_m) = \operatorname{var}(\tilde{R}_m).$$

Because a covariance can be no larger than the product of the standard deviations, this implies that

$$\operatorname{stdev}(\tilde{R}) \geq \operatorname{stdev}(\tilde{R}_m).$$

To deduce the reverse implication (the market being on the mean-variance frontier implies the CAPM), note that, in terms of the matrix notation of Chapter 5, the CAPM (6.3) is expressed as

$$\mu - R_z \iota = \frac{\pi_m'(\mu - R_z \iota)}{\pi_m' \Sigma \pi_m} \Sigma \pi_m, \tag{6.7}$$

where π_m is the market portfolio. Recall from Section 5.3 that $\Sigma \pi$ is the vector of covariances of the risky asset returns with the return of a portfolio π. Thus, $\Sigma \pi_m$ is the vector with ith element equal to $\operatorname{cov}(\tilde{R}_i, \tilde{R}_m)$. The ratio in (6.7) is the ratio of the market risk premium to the variance of the market return.

If there is a risk-free asset, then the CAPM (6.7) with $R_z = R_f$ is identical to condition (5.19), which characterizes a portfolio as being on the mean-variance frontier. In the absence of a risk-free asset, being on the mean-variance frontier implies the first-order condition (5.4), which we repeat here for convenience:

$$\Sigma \pi = \delta \mu + \gamma \iota. \tag{6.8}$$

In this condition, δ and γ are Lagrange multipliers. Assume for the moment that $\delta \neq 0$, and define $R_z = -\gamma/\delta$. Then, the first-order condition (6.8) implies

$$\Sigma \pi = \delta(\mu - R_z \iota) \qquad (6.9)$$

for some constants δ and R_z. We can determine δ by premultiplying (6.9) by π' to obtain

$$\pi' \Sigma \pi = \delta \pi'(\mu - R_z \iota),$$

which implies

$$\delta = \frac{\pi' \Sigma \pi}{\pi'(\mu - R_z \iota)}.$$

Thus, being on the mean-variance frontier in the absence of a risk-free asset (and with $\delta \neq 0$) implies

$$\Sigma \pi = \frac{\pi' \Sigma \pi}{\pi'(\mu - R_z \iota)} (\mu - R_z \iota)$$

for a constant R_z, which implies the CAPM (6.7).

The caveat $\delta \neq 0$ is a minor qualification. The portfolio satisfying the first-order condition (6.8) with $\delta = 0$ is the GMV portfolio. Thus, the CAPM holds in the absence of a risk-free asset if the market portfolio is on the mean-variance frontier and not equal to the GMV portfolio. A parallel but even more trivial qualification applies in the presence of a risk-free asset: The CAPM holds in the presence of a risk-free asset if the market portfolio is on the mean-variance frontier and not equal to the risk-free asset.

When Is the Market Portfolio on the Mean-Variance Frontier?

Assume that labor incomes are spanned by the asset payoffs and treat the labor incomes as pseudo shares as described earlier. In particular, include the pseudo shares in each investor's portfolio as a substitute for her labor income. Therefore, the market portfolio is well defined.

The market portfolio is on the mean-variance frontier if each investor's optimal portfolio is on the mean-variance frontier. This is because the market portfolio is a weighted average of the investors' portfolios (weighting by each investor's initial wealth) and because a weighted average of frontier portfolios is also a frontier portfolio, as shown in Chapter 5.

Each investor will choose an efficient portfolio on the mean-variance frontier if either (a) she has quadratic utility or (b) returns belong to the class of separating distributions. Thus, either quadratic utility or separating distributions is sufficient to imply the CAPM when labor incomes are spanned by the asset payoffs.[2] Recall that the class of separating distributions includes elliptical distributions and in particular normal distributions (Section 5.6).

The SDF in the CAPM

The CAPM (6.3) implies that there is an SDF that is an affine function of the market return. Specifically, set $\psi = (\mathsf{E}[\tilde{R}_m] - R_z)/\mathrm{var}(\tilde{R}_m)$ and define

$$\tilde{m} = \frac{1}{R_z}\left(1 + \psi \mathsf{E}[\tilde{R}_m] - \psi \tilde{R}_m\right). \tag{6.10}$$

The CAPM implies that $\tilde{m}$ is an SDF. We can verify this by computing, for any return $\tilde{R}$,

$$\mathsf{E}[\tilde{m}\tilde{R}] = \frac{1}{R_z}\left(\mathsf{E}[\tilde{R}] + \psi \mathsf{E}[\tilde{R}]\mathsf{E}[\tilde{R}_m] - \psi \mathsf{E}[\tilde{R}\tilde{R}_m]\right)$$
$$= \frac{1}{R_z}\left\{\mathsf{E}[\tilde{R}] - \psi \,\mathrm{cov}(\tilde{R}, \tilde{R}_m)\right\}$$
$$= 1,$$

using the CAPM (6.3) for the last equality.

Notice that $\tilde{m}$ in (6.10) can be negative for certain values of the market return. For example, under the reasonable assumptions $R_z > 0$ and $\psi > 0$, we have

$$\tilde{R}_m > \mathsf{E}[\tilde{R}_m] + \frac{1}{\psi} \quad \Rightarrow \quad \tilde{m} < 0.$$

If the CAPM is true, then there are three possibilities: (i) the market return $\tilde{R}_m$ is bounded in such a way that $\tilde{m} > 0$ with probability 1; (ii) markets are incomplete and (6.10) is zero or negative with positive probability, but there exists another SDF that is strictly positive; or (iii) there is an arbitrage opportunity. If the CAPM holds, markets are complete, and there are no arbitrage opportunities, then the market return must be bounded. To put this another way, it is unreasonable to assume that markets are complete, the CAPM applies to all assets, and the market

---

2. This is true ignoring the exceptional case that there is no risk-free asset and the market portfolio equals the GMV portfolio. That case can be ruled out by ensuring that investors all choose portfolios on the efficient part of the frontier not equal to the GMV portfolio. With quadratic utility, this requires that the bliss point be sufficiently far away—see Exercise 5.2 for the condition ensuring a quadratic utility investor chooses an efficient portfolio when a risk-free asset exists.

return is, for example, normally or lognormally distributed, because this implies the existence of arbitrage opportunities.[3]

The CAPM with Unspanned Endowments

If investors have labor incomes that are not spanned by the asset payoffs, then the CAPM holds in a modified form if (a) investors have quadratic utility or (b) labor incomes and asset payoffs are joint normally distributed. The modified version of the CAPM is that there exist constants R_z and λ such that

$$\mathsf{E}[\tilde{R}] = R_z + \lambda \frac{\mathrm{cov}(\tilde{R}, \tilde{w}_m)}{\mathrm{var}(\tilde{w}_m)} \qquad (6.11)$$

for each return $\tilde{R}$, where $\tilde{w}_m$ is end-of-period market wealth (the sum of asset payoffs and labor incomes). Again, R_z is called the zero-beta return and, by the same reasoning as before, it must equal the risk-free return if there is a risk-free asset. The proof of (6.11), which is given below, relies on the SDF factor model (6.1) and the fact that marginal utility defines an SDF. The linearity of marginal utility in wealth is used for quadratic utility, and Stein's lemma is used for normal distributions.

First, suppose that each investor h has quadratic utility $u_h(w) = \zeta_h w - \frac{1}{2}w^2$, and assume $\sum_h \zeta_h \neq \mathsf{E}[\tilde{w}_m]$. The first-order condition for portfolio choice is that marginal utility at the optimal wealth be proportional to an SDF, so

$$\zeta_h - \tilde{w}_h = \gamma_h \tilde{m}_h$$

for some constant γ_h and SDF $\tilde{m}_h$, where $\tilde{w}_h$ denotes the optimal end-of-period wealth of investor h. Adding across investors gives

$$\zeta - \tilde{w}_m = \gamma \tilde{m},$$

where

$$\zeta = \sum_{h=1}^{H} \zeta_h, \quad \gamma = \sum_{h=1}^{H} \gamma_h, \quad \tilde{m} = \frac{\sum_{h=1}^{H} \gamma_h \tilde{m}_h}{\sum_{h=1}^{H} \gamma_h}.$$

Moreover, $\tilde{m}$ is an SDF. The assumption $\sum_h \zeta_h \neq \mathsf{E}[\tilde{w}_m]$ implies $\mathsf{E}[\tilde{m}] \neq 0$. Substituting $\tilde{m} = (\zeta - \tilde{w}_m)/\gamma$ in (6.1) produces (6.11).

---

3. If the market is complete and there are only finitely many assets, then there must be only finitely many states of the world (Section 3.3), so every random variable is bounded. Hence, this discussion is meaningful primarily for markets with an infinite number of assets, for example, markets in which options are assumed to be traded at every strike price.

Now consider the joint normality hypothesis. Assume investors' utility functions are strictly monotone, concave, and twice continuously differentiable. Substitute $u'_h(\tilde{w}_h) = \gamma_h \tilde{m}_h$ in (6.1) to obtain

$$\mathsf{E}[\tilde{R}] = \frac{\gamma_h}{\mathsf{E}[u'_h(\tilde{w}_h)]} - \frac{1}{\mathsf{E}[u'_h(\tilde{w}_h)]} \operatorname{cov}(u'_h(\tilde{w}_h), \tilde{R}).$$

Stein's lemma (see the end-of-chapter notes) and the joint normality hypothesis imply

$$\operatorname{cov}(u'_h(\tilde{w}_h), \tilde{R}) = \mathsf{E}[u''_h(\tilde{w}_h)] \operatorname{cov}(\tilde{w}_h, \tilde{R}).$$

Therefore,

$$\mathsf{E}[\tilde{R}] = \alpha_h + \psi_h \operatorname{cov}(\tilde{w}_h, \tilde{R}), \qquad (6.12)$$

where

$$\alpha_h = \frac{\gamma_h}{\mathsf{E}[u'_h(\tilde{w}_h)]} \quad \text{and} \quad \psi_h = -\frac{\mathsf{E}[u''_h(\tilde{w}_h)]}{\mathsf{E}[u'_h(\tilde{w}_h)]}.$$

Divide both sides of (6.12) by ψ_h and add over investors to obtain

$$\left(\sum_{h=1}^H \frac{1}{\psi_h}\right) \mathsf{E}[\tilde{R}] = \sum_{h=1}^H \frac{\alpha_h}{\psi_h} + \operatorname{cov}(\tilde{w}_m, \tilde{R}).$$

This implies

$$\mathsf{E}[\tilde{R}] = R_z + \psi \operatorname{cov}(\tilde{w}_m, \tilde{R}), \qquad (6.13)$$

where

$$R_z = \sum_{h=1}^H \frac{\alpha_h}{\psi_h} \bigg/ \sum_{h=1}^H \frac{1}{\psi_h} \quad \text{and} \quad \psi = 1 \bigg/ \sum_{h=1}^H \frac{1}{\psi_h}.$$

Set $\lambda = \psi \operatorname{var}(\tilde{w}_m)$ in (6.13) to obtain (6.11).

6.2 GENERAL FACTOR MODELS

In the factor models most commonly used in finance, the factors are returns or excess returns. This includes the CAPM and the Fama-French-Carhart model. It is also possible for factors to be macroeconomic variables such as the percent change in gross domestic product or to be other financial variables such as the change in the short-term interest rate. In fact, it might seem more natural in general to use macroeconomic variables as priced risk factors than to use returns or excess returns. However, as we will see, it is without loss of generality to use returns or excess returns as factors, because we can always project other factors onto the span of the returns or the span of the excess returns and use the projections as factors.

A convenience of using returns or excess returns as factors is that the corresponding prices of risk are determined. The price of risk corresponding to a return factor is the return's risk premium (using the zero beta return as a proxy for the risk-free return if there is no risk-free asset). The price of risk corresponding to an excess return is the mean of the excess return. In contrast, for general factors, the prices of risk are not determined by theory.

Factor models are closely related to SDFs, as stated at the beginning of this chapter. Given any factor model with $R_z \neq 0$, an affine function of the factors is an SDF. Conversely, if an SDF is an affine function of random variables and the SDF has a nonzero mean, then there is a factor model with the random variables as factors. Thus, the search for factor models is equivalent to a search for SDFs and affine representations of SDFs.

Factor models are also closely related to mean-variance analysis. There is a factor model with a single return as the factor if and only if the return is on the mean-variance frontier (and not equal to the risk-free return if there is a risk-free asset or to the GMV return if there is no risk-free asset). All of the facts just stated will be established in the remainder of this section.

Definition of Single-Factor Models

We say that there is a single-factor model with factor $\tilde{f}$ if there exist constants R_z and λ such that, for each return $\tilde{R}$,

$$\mathsf{E}[\tilde{R}] - R_z = \lambda \frac{\mathrm{cov}(\tilde{f},\tilde{R})}{\mathrm{var}(\tilde{f})}. \qquad (6.14)$$

The ratio $\mathrm{cov}(\tilde{f},\tilde{R})/\mathrm{var}(\tilde{f})$ is the beta of $\tilde{R}$ with respect to $\tilde{f}$. As before, R_z is called the zero-beta return. If there is a risk-free asset, then taking $\tilde{R} = R_f$ in (6.14) shows that $R_z = R_f$.

The number λ in (6.14) is called the factor risk premium or price of risk. It defines the extra expected return an asset earns for each unit increase in its beta. If $\lambda > 0$, an investor gets compensated for holding extra risk in the form of a higher expected return when risk is measured by the beta with respect to $\tilde{f}$. In the CAPM, λ is the market risk premium.

When the Single Factor Is a Return or Excess Return

If the factor $\tilde{f}$ is a return $\tilde{R}_*$, then applying (6.14) with $\tilde{R} = \tilde{R}_*$ shows that the price of risk λ is the risk premium $\mathsf{E}[\tilde{R}_*] - R_z$ of the factor. Furthermore, the reasoning in Section 6.1 shows that the factor model is equivalent to the return $\tilde{R}_*$ being on the mean-variance frontier (excluding the risk-free return or GMV return).

If the factor $\tilde{f}$ is an excess return $\tilde{R}_1 - \tilde{R}_2$, then applying (6.14) with $\tilde{R} = \tilde{R}_1$ and applying it again with $\tilde{R} = \tilde{R}_2$ and then subtracting yields

$$E[\tilde{R}_1] - E[\tilde{R}_2] = \frac{\lambda}{\mathrm{var}(\tilde{f})}[\mathrm{cov}(\tilde{f},\tilde{R}_1) - \mathrm{cov}(\tilde{f},\tilde{R}_2)] = \frac{\lambda}{\mathrm{var}(\tilde{f})} \mathrm{var}(\tilde{f}) = \lambda,$$

so $E[\tilde{f}] = \lambda$. Thus, the price of risk is the mean of the factor. When the factor is an excess return and there is a risk-free asset, then, as with the CAPM, the factor model is equivalent to the α's being zero in the projections

$$\tilde{R} - R_f = \alpha + \beta \tilde{f} + \tilde{\varepsilon}.$$

Definition of Multifactor Models

Given random variables $\tilde{f}_1, \ldots, \tilde{f}_k$ and a return $\tilde{R}$, the orthogonal projection of $\tilde{R}$ on the span of the random variables and a constant is

$$E[\tilde{R}] + \beta'(\tilde{F} - E[\tilde{F}]),$$

where $\tilde{F} = (\tilde{f}_1 \cdots \tilde{f}_k)'$, and

$$\beta = \Sigma_F^{-1} \mathrm{Cov}(\tilde{F},\tilde{R}). \qquad (6.15)$$

This is shown in (3.32). In this equation, Σ_F is the (assumed to be invertible) covariance matrix of $\tilde{F}$, and $\mathrm{Cov}(\tilde{F},\tilde{R})$ denotes the column vector of dimension k with ith element $\mathrm{cov}(\tilde{f}_i,\tilde{R})$.[4] We call β the vector of multiple regression betas. We say that there is a multifactor model with factors $\tilde{f}_1, \ldots, \tilde{f}_k$ if there exist constants R_z and λ, with λ being a k–dimensional vector, such that, for each return $\tilde{R}$,

$$E[\tilde{R}] = R_z + \lambda'\beta. \qquad (6.16)$$

In this equation, β depends on the return $\tilde{R}$. However, λ is the same for all returns. Each λ_j in (6.16), for $j \in \{1, \ldots, k\}$, is called the risk premium or price of risk of factor j. The β_j's are measures of risk. The factor model (6.16) states that the risk premium of each asset (using the zero-beta return as a proxy risk-free return) is the sum of the asset's risks (its β_j's) multiplied by the prices of risk (the λ_j's).

---

4. To relate (6.15) to (3.32), note that $\beta' = \mathrm{Cov}(\tilde{F},\tilde{R})'\Sigma_F^{-1} = \mathrm{Cov}(\tilde{R},\tilde{F})\Sigma_F^{-1}$.

Prices of Risk in Multifactor Models

If an individual factor in a multifactor model is a return, then its price of risk is the risk premium of the return, just as in a single-factor model. Likewise, if an individual factor in a multifactor model is an excess return, then its price of risk is its mean, just as in a single-factor model. This is verified below. If all of the factors are excess returns and there is a risk-free asset, then, as with the CAPM, the multifactor model is equivalent to the α's being zero in the projections

$$\tilde{R} - R_f = \alpha + \beta'\tilde{F} + \tilde{\varepsilon}.$$

Consider a k–factor model and suppose for any $j \in \{1,\ldots,k\}$ that factor j is a return. Substitute $\tilde{R} = \tilde{f}_j$ in (6.16) to obtain

$$\mathsf{E}[\tilde{f}_j] = R_z + \lambda' \Sigma_F^{-1} \mathrm{Cov}(\tilde{F},\tilde{f}_j). \tag{6.17}$$

The vector $\mathrm{Cov}(\tilde{F},\tilde{f}_j)$ is the jth column of the matrix Σ_F. Therefore,

$$\Sigma_F^{-1} \mathrm{Cov}(\tilde{F},\tilde{f}_j)$$

is the jth column of the identity matrix, meaning that it has a 1 in the jth place and 0 elsewhere. Hence, $\lambda' \Sigma_F^{-1} \mathrm{Cov}(\tilde{F},\tilde{f}_j) = \lambda_j$, and (6.17) shows that $\lambda_j = \mathsf{E}[\tilde{f}_j] - R_z$.

Now suppose that factor j is an excess return. For any return $\tilde{R}$, $\tilde{R} + \tilde{f}_j$ is also a return, so

$$\mathsf{E}[\tilde{R}] = R_z + \lambda' \Sigma_F^{-1} \mathrm{Cov}(\tilde{F},\tilde{R}),$$

$$\mathsf{E}[\tilde{R} + \tilde{f}_j] = R_z + \lambda' \Sigma_F^{-1} \mathrm{Cov}(\tilde{F},\tilde{R}) + \lambda' \Sigma_F^{-1} \mathrm{Cov}(\tilde{F},\tilde{f}_j).$$

Subtract the first equation from the second to obtain

$$\mathsf{E}[\tilde{f}_j] = \lambda' \Sigma_F^{-1} \mathrm{Cov}(\tilde{F},\tilde{f}_j).$$

This means that $\mathsf{E}[\tilde{f}_j] = \lambda_j$.

Covariances as Measures of Risks

We can always write factor models in terms of covariances instead of betas: (6.14) is equivalent to

$$\mathsf{E}[\tilde{R}] = R_z + \psi \, \mathrm{cov}(\tilde{f},\tilde{R}), \tag{6.18}$$

where $\psi = \lambda / \mathrm{var}(\tilde{f})$, and (6.16) is equivalent to

$$\mathsf{E}[\tilde{R}] = R_z + \psi' \, \mathrm{Cov}(\tilde{F},\tilde{R}), \tag{6.19}$$

where $\psi = \Sigma_F^{-1}\lambda$. The term "price of risk" is often used (and is sometimes used in this book) for the coefficient ψ_j of the covariance $\mathrm{cov}(\tilde{F}_j, \tilde{R})$. Whether "price of risk" refers to the coefficient of the covariance in (6.19) or to the coefficient of the beta in (6.16) must be determined by context.

Factor Models Are Equivalent to SDFs

There is an equivalence between factor models and SDFs. Specifically, if there is a factor model with $R_z \neq 0$, then an affine function of the factors is an SDF. The construction of the affine function generalizes the CAPM construction (6.10). Conversely, if there exist numbers $a, b_1, \ldots, b_k$ and random variables $\tilde{f}_1, \ldots, \tilde{f}_k$ such that

$$\tilde{m} \stackrel{\text{def}}{=} a + b_1 \tilde{f}_1 + \cdots + b_k \tilde{f}_k$$

is an SDF with $\mathsf{E}[\tilde{m}] \neq 0$, then there is a factor model with $\tilde{f}_1, \ldots, \tilde{f}_k$ as the factors. Thus, factor models are essentially equivalent to affine representations of SDFs.

The discussion in Section 6.1 about the market return being bounded if the CAPM holds, the market is complete, and there are no arbitrage opportunities applies in any factor model. Specifically, the factors must be bounded if a factor model holds in a complete market; otherwise, there are arbitrage opportunities.

Let $\tilde{F} = (\tilde{f}_1, \ldots, \tilde{f}_k)'$ and suppose that $\tilde{m} = a + b'\tilde{F}$ is an SDF for some constant a and constant vector b. Assume $\mathsf{E}[\tilde{m}] \neq 0$. Then, (6.1) implies

$$\begin{aligned} \mathsf{E}[\tilde{R}] &= \frac{1}{\mathsf{E}[\tilde{m}]} - \frac{1}{\mathsf{E}[\tilde{m}]} \mathrm{cov}(b'\tilde{F}, \tilde{R}) \\ &= R_z + \psi' \mathrm{Cov}(\tilde{F}, \tilde{R}), \end{aligned}$$

where $R_z = 1/\mathsf{E}[\tilde{m}] \neq 0$ and $\psi = R_z b$. Therefore, there is a k–factor model with $\tilde{F}$ as the vector of factors.

To establish the converse, suppose there is a k–factor model (6.19) with $\tilde{F} = (\tilde{f}_1, \ldots, \tilde{f}_k)'$ as the vector of factors and $R_z \neq 0$. Define

$$\tilde{m} = \frac{1}{R_z} - \frac{1}{R_z} \psi'(\tilde{F} - \mathsf{E}[\tilde{F}]). \tag{6.20}$$

The calculation that (6.20) is an SDF is the same as the calculation that (6.10) is an SDF.

Factor Models Are Equivalent to Mean-Variance Efficiency

If a set of random variables spans a return on the mean-variance frontier—that is, if there is a frontier return that is a linear combination of the random variables—then there is a factor model with those variables as the factors (Exercise 6.1). This is true except for the trivial case that the frontier return is the risk-free return (or the GMV return if no risk-free asset exists). Conversely, if there is a factor pricing model with factors that are excess returns and there is a risk-free asset, then a linear combination of the factors plus the risk-free return is a return that is on the mean-variance frontier. To see that this is true, suppose there is a factor model

$$E[\tilde{R}] = R_f + \sum_{i=1}^{k} \lambda_i \operatorname{cov}(\tilde{R}, \tilde{f}_i)$$

for all returns $\tilde{R}$. Assume the factors $\tilde{f}_i$ are excess returns. Define the return

$$\tilde{R}_* = R_f + \sum_{i=1}^{k} \lambda_i \tilde{f}_i.$$

Then,

$$E[\tilde{R}] = R_f + \operatorname{cov}(\tilde{R}, \tilde{R}_*)$$

for all returns $\tilde{R}$. In particular,

$$E[\tilde{R}_*] = R_f + \operatorname{var}(\tilde{R}_*).$$

Thus,

$$\frac{E[\tilde{R}_*] - R_f}{\operatorname{var}(\tilde{R}_*)} = 1,$$

and

$$E[\tilde{R}] = R_f + \frac{\operatorname{cov}(\tilde{R}, \tilde{R}_*)}{\operatorname{var}(\tilde{R}_*)} (E[\tilde{R}_*] - R_f)$$

for all returns $\tilde{R}$. This implies that $\tilde{R}_*$ is on the mean-variance frontier, just as the CAPM implies that the market return is on the mean-variance frontier.

If the factors $\tilde{f}_i$ are not excess returns, then they can be replaced by their projections on the space of excess returns (see below). So, if there is a risk-free asset, then any factor model implies that a linear combination of the projections of the factors on the space of excess returns plus the risk-free asset is a return on the mean-variance frontier.

Orthogonalizing Factors

We can always take factors to have zero means and unit variances and to be mutually uncorrelated by using $\tilde{G}$ instead of $\tilde{F}$ as the vector of factors, where

$$\tilde{G} = L^{-1}(\tilde{F} - \mathsf{E}[\tilde{F}]),$$

and L is the Cholesky decomposition of Σ_F (Section 3.9). The vector $\tilde{G}$ is the Gram-Schmidt orthogonalization of $\tilde{F} - \mathsf{E}[\tilde{F}]$. From (6.16), we have

$$\mathsf{E}[\tilde{R}] = R_z + \psi' \operatorname{Cov}(\tilde{G}, \tilde{R}), \qquad (6.21)$$

where $\psi = L^{-1}\lambda$.

Projecting Factors onto Returns and Excess Returns

We can always replace any or all of the factors in a factor model by returns or excess returns or a combination thereof. For each factor, the return (or excess return) that can be used as a substitute for the factor is the one that has maximum correlation with the factor, and it is obtained by orthogonal projection of the factor onto the space of returns (or the space of excess returns) and a constant. It is called the factor-mimicking return (or factor-mimicking excess return). The following shows that the factor model still holds after such a substitution.

Consider a k–factor model, and let

$$\tilde{x} = \gamma + \beta \tilde{R}_*$$

denote the orthogonal projection of a factor $\tilde{f}_j$ onto the span of a constant and the returns. For any return $\tilde{R}$, we have

$$\operatorname{cov}(\tilde{R}, \tilde{f}_j) = \operatorname{cov}(\tilde{R}, \tilde{x}) = \beta \operatorname{cov}(\tilde{R}, \tilde{R}_*),$$

so we can substitute in (6.19) to obtain

$$\mathsf{E}[\tilde{R}] = R_z + \psi_j \beta \operatorname{cov}(\tilde{R}, \tilde{R}_*) + \sum_{i \neq j} \psi_i \operatorname{cov}(\tilde{R}, \tilde{f}_i)$$

for any return $\tilde{R}$. This shows that there is a k–factor model with $\tilde{R}_*$ replacing $\tilde{f}_j$. The fact that $\beta \tilde{R}_*$ has maximum correlation with $\tilde{f}_j$ among all random variables in the span of the returns is shown in Section 3.5. Of course, the correlation of $\tilde{R}_*$ with $\tilde{f}_j$ is the same as the correlation of $\beta \tilde{R}_*$ with $\tilde{f}_j$. Thus, $\tilde{R}_*$ is the return having maximum correlation with $\tilde{f}_j$.

Substituting excess returns for factors is very similar. Now let

$$\tilde{x} = \gamma + \beta \tilde{e}_*$$

denote the projection of a factor $\tilde{f}_j$ on a constant and the space of excess returns. The residual $\tilde{f}_j - \tilde{x}$ is orthogonal to a constant (has zero mean) and to each excess return, so it is uncorrelated with each excess return. This implies

$$\operatorname{cov}(\tilde{e}, \tilde{f}_j) = \operatorname{cov}(\tilde{e}, \tilde{x}) = \beta \operatorname{cov}(\tilde{e}, \tilde{e}_*)$$

for each excess return $\tilde{e}$. Choose an arbitrary return and call it $\tilde{R}_0$. Write any return $\tilde{R}$ as $\tilde{R} = \tilde{R}_0 + (\tilde{R} - \tilde{R}_0)$. Then, we have

$$\begin{aligned}\operatorname{cov}(\tilde{R}, \tilde{f}_j) &= \operatorname{cov}(\tilde{R}_0, \tilde{f}_j) + \operatorname{cov}(\tilde{R} - \tilde{R}_0, \tilde{f}_j) \\ &= \operatorname{cov}(\tilde{R}_0, \tilde{f}_j) + \beta \operatorname{cov}(\tilde{R} - \tilde{R}_0, \tilde{e}_*) \\ &= \operatorname{cov}(\tilde{R}_0, \tilde{f}_j) + \beta \operatorname{cov}(\tilde{R}, \tilde{e}_*) - \beta \operatorname{cov}(\tilde{R}_0, \tilde{e}_*) \\ &= \operatorname{cov}(\tilde{R}_0, \tilde{f}_j - \tilde{x}) + \beta \operatorname{cov}(\tilde{R}, \tilde{e}_*).\end{aligned}$$

Thus, we have a k–factor model: For all returns $\tilde{R}$,

$$\mathsf{E}[\tilde{R}] = \hat{R}_z + \hat{\psi}_j \operatorname{cov}(\tilde{R}, \tilde{e}_*) + \sum_{i \neq j} \psi_i \operatorname{cov}(\tilde{R}, \tilde{f}_i),$$

where we define $\hat{\psi}_j = \psi_j \beta$ and $\hat{R}_z = R_z + \psi_j \operatorname{cov}(\tilde{R}_0, \tilde{f}_j - \tilde{x})$.

6.3 JENSEN'S ALPHA AND PERFORMANCE EVALUATION

Assume there is a risk-free asset. The security market line is defined in Section 6.1 as the graph of the function $\beta \mapsto r_f + \beta(\bar{r}_m - r_f)$, where β denotes the market beta, that is, $\beta = \operatorname{cov}(\tilde{r}, \tilde{r}_m)/\operatorname{var}(\tilde{r}_m)$. If the CAPM holds, then each asset's market beta and expected rate of return plot on the security market line. In this section, we consider a single-factor model with a return $\tilde{R}_b$ as the factor, but we assume the model fails; that is, we assume there are some assets with expected returns above or below that predicted by the model. Call $\tilde{R}_b$ the benchmark return. We consider the interpretation of deviations from a security market line in which the market return is replaced by the benchmark return. We do not rule out the possibility that the benchmark is the market return: A special case of our analysis is when the benchmark is the market return and the CAPM fails.

As discussed previously in this chapter, a factor model with the excess return $\tilde{R}_b - R_f$ as the factor is equivalent to the α's being zero in the projections

$$\tilde{R} - R_f = \alpha + \beta(\tilde{R}_b - R_f) + \tilde{\varepsilon}. \tag{6.22}$$

A standard way of assessing the performance of an investment manager is to investigate whether the return $\tilde{R}$ the manager produces has a positive α in (6.22) relative to a given benchmark return $\tilde{R}_b$. If so, we say that the manager creates alpha and conclude that an investment in the manager may be warranted. This type of performance evaluation was first suggested by Jensen (1969), and the α in (6.22) is often called Jensen's alpha.

The justification for this method of performance evaluation lies in mean-variance analysis. If a return has a positive alpha, then combining the return with the benchmark return and the risk-free asset in correct proportions produces a return that mean-variance dominates the benchmark. This is shown below. Thus, an investor who has mean-variance preferences and whose default position is to hold the benchmark portfolio would benefit from investing some funds with a manager who produces a positive alpha.

A return has a positive alpha if and only if its Sharpe ratio exceeds a certain critical threshold. The Sharpe ratio is defined as risk premium divided by risk, with risk measured as the standard deviation; that is, the Sharpe ratio is $\mathsf{E}[\tilde{R} - R_f]/\operatorname{stdev}(\tilde{R})$. We show below that

$$\alpha > 0 \quad \Leftrightarrow \quad \frac{\mathsf{E}[\tilde{R} - R_f]}{\operatorname{stdev}(\tilde{R})} > \frac{\mathsf{E}[\tilde{R}_b - R_f]}{\operatorname{stdev}(\tilde{R}_b)} \times \operatorname{corr}(\tilde{R}, \tilde{R}_b). \qquad (6.23)$$

Thus, the critical threshold is the Sharpe ratio of the benchmark scaled down by the correlation between the return and the benchmark. This highlights the importance of correlation in determining whether adding some of a return will improve mean-variance efficiency. In the extreme case, a return uncorrelated with the benchmark only needs to have a positive Sharpe ratio (an expected return above the risk-free return) in order to improve mean-variance efficiency. This motivates so-called absolute return strategies, which attempt to achieve zero correlation with the benchmark. The only hurdle such strategies must surpass to improve mean-variance efficiency for an investor holding the benchmark is to beat the risk-free return on average.

It is important to recognize that a positive alpha for a return does not imply that the return mean-variance dominates the benchmark. All that can be said is that a marginal investment in the return and corresponding reduction in the benchmark investment will mean-variance dominate the benchmark. However, this is enough for performance evaluation. The usual issue is whether an investor should invest some funds in a manager, not whether the investor should move all of her funds to the manager.

The results stated above do not depend on the existence of a risk-free asset. They apply also when a risk-free asset does not exist by substituting the zero-beta return for the risk-free return. However, the proof is given below only for the case in which a risk-free asset exists.

First, we establish (6.23). We have

$$\alpha > 0 \quad \Leftrightarrow \quad \mathsf{E}[\tilde{R} - R_f] > \frac{\mathrm{cov}(\tilde{R}, \tilde{R}_b)}{\mathrm{var}(\tilde{R}_b)} \mathsf{E}[\tilde{R}_b - R_f]$$

$$\Leftrightarrow \quad \frac{\mathsf{E}[\tilde{R} - R_f]}{\mathrm{stdev}(\tilde{R})} > \frac{\mathrm{corr}(\tilde{R}, \tilde{R}_b)\,\mathrm{stdev}(\tilde{R}_b)}{\mathrm{var}(\tilde{R}_b)} \mathsf{E}[\tilde{R}_b - R_f]$$

$$= \frac{\mathsf{E}[\tilde{R}_b - R_f]}{\mathrm{stdev}(\tilde{R}_b)}\,\mathrm{corr}(\tilde{R}, \tilde{R}_b).$$

Now, we show that there is a return involving a positive investment in $\tilde{R}$ that mean-variance dominates $\tilde{R}_b$ when $\tilde{R}$ has a positive alpha with respect to $\tilde{R}_b$. Let

$$\tilde{R} = R_f + \alpha + \beta(\tilde{R}_b - R_f) + \tilde{\varepsilon},$$

where $\mathsf{E}[\tilde{\varepsilon}] = 0$ and $\mathrm{cov}(\tilde{R}_b, \tilde{\varepsilon}) = 0$. For any $\lambda > 0$ and $k > 0$, consider the return

$$\tilde{R}_{\lambda k} = \tilde{R}_b + \lambda[\tilde{R} - R_f - \beta(\tilde{R}_b - R_f)] - \lambda k(\tilde{R}_b - R_f).$$

The excess return in square braces is $\alpha + \tilde{\varepsilon}$, so we have

$$\tilde{R}_{\lambda k} = \tilde{R}_b + \lambda[\alpha + \tilde{\varepsilon}] - \lambda k(\tilde{R}_b - R_f).$$

Furthermore,

$$\mathsf{E}[\tilde{R}_{\lambda k}] = \mathsf{E}[\tilde{R}_b] + \lambda\alpha - \lambda k \mathsf{E}[\tilde{R}_b - R_f],$$
$$\mathrm{var}(\tilde{R}_{\lambda k}) = (1 - \lambda k)^2 \mathrm{var}(\tilde{R}_b) + \lambda^2 \mathrm{var}(\tilde{\varepsilon}).$$

Holding k fixed, differentiate with respect to λ and evaluate at $\lambda = 0$. We obtain

$$\left.\frac{\mathrm{d}\mathsf{E}[\tilde{R}_{\lambda k}]}{\mathrm{d}\lambda}\right|_{\lambda=0} = \alpha - k\mathsf{E}[\tilde{R}_b - R_f],$$

$$\left.\frac{\mathrm{d}\,\mathrm{var}(\tilde{R}_{\lambda k})}{\mathrm{d}\lambda}\right|_{\lambda=0} = -2k\,\mathrm{var}(\tilde{R}_b).$$

Obviously, at $\lambda = 0$, $\tilde{R}_{\lambda k} = \tilde{R}_b$. The derivatives tell us whether the mean and variance of $\tilde{R}_{\lambda k}$ are smaller or larger than that of $\tilde{R}_b$ for small $\lambda > 0$. If $\alpha > 0$, then for small $k > 0$, the derivative of the mean is positive and the derivative of the variance is negative. Therefore, for small $k > 0$ and small $\lambda > 0$, $\tilde{R}_{\lambda k}$ has both a larger mean and a smaller variance than $\tilde{R}_b$ when $\alpha > 0$.

6.4 STATISTICAL FACTORS

The previous sections of this chapter discuss factors that explain expected returns. They are called pricing factors. This section discusses factors that explain covariances. We will call them statistical factors. The next section shows that statistical factors are also pricing factors, at least approximately.

The simplest statistical factor model is called the market model. It assumes that the residuals from projections of returns on a constant and the market return are uncorrelated across assets. Thus, the covariance of any two asset returns is solely due to their common correlation with the market. The projections on a constant and the market return (plus residuals) are

$$\tilde{R}_i = \mu_i + \beta_i \tilde{R}_m + \tilde{\varepsilon}_i.$$

The assumption that the residuals are uncorrelated implies

$$\text{cov}(\tilde{R}_i, \tilde{R}_j) = \beta_i \beta_j \, \text{var}(\tilde{R}_m).$$

Thus, covariances are determined by betas. In this model, the residuals are called idiosyncratic risks or firm-specific risks, and the terms $\beta_i \tilde{R}_m$ are called the systematic risks of the assets.

The market model cannot be literally true, because, letting π_m denote the market portfolio (with $\sum_i \pi_{mi} = 1$), we have

$$\tilde{R}_m = \sum_i \pi_{mi} \tilde{R}_i = \sum_i \pi_{mi} \mu_i + \left(\sum_i \pi_{mi} \beta_i \right) \tilde{R}_m + \sum_i \pi_{mi} \tilde{\varepsilon}_i.$$

This is possible only if

$$\sum_i \pi_{mi} \mu_i = 0, \quad \sum_i \pi_{mi} \beta_i = 1, \quad \sum_i \pi_{mi} \tilde{\varepsilon}_i = 0.$$

However, the condition $\sum_i \pi_{mi} \tilde{\varepsilon}_i = 0$ implies a linear dependence among the residuals that is incompatible with their all being mutually uncorrelated.

If the market model were correct, then estimation of covariances would be much simpler than otherwise. For example, with 1,000 assets, there are 499,500 covariances (excluding the 1,000 variances) but only 1,000 betas. Even if the model is incorrect, it is possible that the model error introduced by assuming it is less than the statistical error that would be created by estimating so many distinct covariances independently. However, it is also likely that there are better statistical factor models. For example, the market model residuals of two oil companies are certainly correlated due to their common dependence on oil prices. Including additional factors is probably warranted.

In general, a statistical factor model consists of a set of random variables $\tilde{f}_1, \ldots, \tilde{f}_k$ such that the residuals in the projections

$$\tilde{R}_i = \mu_i + \sum_{j=1}^{k} \beta_{ij}(\tilde{f}_j - \mathsf{E}[\tilde{f}_j]) + \tilde{\varepsilon}_i \tag{6.24}$$

are uncorrelated across assets.[5] Again, the residuals are called idiosyncratic risks. Let n denote the number of assets, let B denote the $n \times k$ matrix of the β_{ij}, and stack the equations (6.24) for $i = 1, \ldots, n$. The stacked equations can be written as

$$\tilde{\mathbf{R}} = \mu + B(\tilde{F} - \mathsf{E}[\tilde{F}]) + \tilde{\varepsilon}, \tag{6.25}$$

where $\tilde{\mathbf{R}}$ is the vector of returns, $\mu = \mathsf{E}[\tilde{\mathbf{R}}]$, $\tilde{F}$ is the vector of factors, and $\tilde{\varepsilon}$ is the vector of residuals. Using the fact that the residuals in the projections are uncorrelated with the factors, we see that the covariance matrix of the vector of returns is $\Sigma = B\Sigma_F B' + D$, where Σ_F denotes the covariance matrix of $\tilde{F} = (\tilde{f}_1 \cdots \tilde{f}_k)'$, and where D is the diagonal matrix containing the residual variances. The condition that D is diagonal is the condition that defines this model as a statistical factor model. We can always do the projections (6.24) and derive the formula $B\Sigma_F B' + D$ for the covariance matrix of returns. However, if the various residuals were correlated, then D would have nonzero off-diagonal elements.

We can always use $\tilde{F} - \mathsf{E}[\tilde{F}]$ as the vector of statistical factors, in which case the factors would have zero means. By taking the Gram-Schmidt orthogonalization, we can also assume the factors to be uncorrelated and to have unit variances, in which case they are called orthonormal. When the factors are orthonormal, the covariance matrix of the returns is $\Sigma = BB' + D$.

Frequently, statistical factors are regarded as unobservable (latent), and we want to estimate them from return data. We might as well work with the orthornormal versions that have zero means. Previously, we have projected returns onto factors, but here it is useful to project the factors onto the returns. From (3.32), the projection of the zero-mean orthonormal factors onto the returns is given by

$$\mathrm{Cov}(\tilde{F}, \tilde{\mathbf{R}}) \Sigma^{-1} (\tilde{\mathbf{R}} - \mu).$$

From (6.25), we have $\mathrm{Cov}(\tilde{F}, \tilde{\mathbf{R}}) = \Sigma_F B' = B'$. Also, as noted previously, $\Sigma = BB' + D$. Therefore, the projection of the zero-mean orthonormal factors onto the returns is given by

$$B'(BB' + D)^{-1}(\tilde{\mathbf{R}} - \mu). \tag{6.26}$$

---

5. Stronger assumptions are sometimes made. For example, the idiosyncratic risks may be assumed to be independent of each other and independent of the factors, or they may be assumed to be mean independent of the factors.

We can use this projection to estimate the factors if we have estimates of B, D, and μ. The matrices B and D can be estimated by maximizing the likelihood that $BB' + D$ is the covariance matrix of the returns subject to the condition that D be diagonal.

6.5 ARBITRAGE PRICING THEORY

The Arbitrage Pricing Theory (APT) asserts that statistical factors must also be pricing factors. Unlike, for example, the CAPM, which is derived from equilibrium considerations (investor optimization and market clearing), the APT is derived from the statistical factor model and the absence of arbitrage opportunities. To be more precise, it is derived from the statistical factor model and the existence of an SDF. It *does not* depend on there being a strictly positive SDF.

The intuition behind the APT is that, in a diversified portfolio, the risk contributed by the idiosyncratic components of security returns should be negligible, due to a law of large numbers effect. Intuitively, investors should hold diversified portfolios and hence only be exposed to the systematic risk sources (statistical factors). It then seems sensible that investors would require risk premia as compensation only for holding systematic risks. Hence, the risk premium of each asset should depend only on the asset's exposure to the common risk sources and not on its idiosyncratic risk. Thus, an assumption about the correlations of assets implies a conclusion about the pricing of assets.

To gain an understanding of the APT, it is useful to consider first the very special case in which the idiosyncratic risks $\tilde{\varepsilon}_i$ are all zero (the returns are spanned by a constant and the factors). Assume the factor model (6.25) holds and assume $\tilde{\varepsilon}_i = 0$ for each i. For any SDF $\tilde{m}$ and asset i,

$$1 = \mathrm{E}[\tilde{m}\tilde{R}_i] = \mu_i \mathrm{E}[\tilde{m}] + \sum_{j=1}^{k} \beta_{ij} \mathrm{E}[\tilde{m}\tilde{f}_j]$$

$$= \left(\mu_i + \sum_{j=1}^{k} \beta_{ij} \mathrm{E}[\tilde{f}_j]\right) \mathrm{E}[\tilde{m}] + \sum_{j=1}^{k} \beta_{ij} \mathrm{cov}(\tilde{m},\tilde{f}_j)$$

$$= \mathrm{E}[\tilde{R}_i]\mathrm{E}[\tilde{m}] + \sum_{j=1}^{k} \beta_{ij} \mathrm{cov}(\tilde{m},\tilde{f}_j). \tag{6.27}$$

Assume $\mathrm{E}[\tilde{m}] \neq 0$ and rearrange to obtain

$$\mathrm{E}[\tilde{R}_i] = \frac{1}{\mathrm{E}[\tilde{m}]} - \frac{1}{\mathrm{E}[\tilde{m}]} \sum_{j=1}^{k} \beta_{ij} \mathrm{cov}(\tilde{m},\tilde{f}_j). \tag{6.28}$$

Thus, there is a factor model with $1/\mathsf{E}[\tilde{m}]$ as the zero-beta return and with

$$\lambda_j = -\frac{1}{\mathsf{E}[\tilde{m}]} \operatorname{cov}(\tilde{m}, \tilde{f}_j)$$

as the price of risk of factor j for each $j = 1, \ldots, k$.

Now consider the more interesting case in which $\tilde{\varepsilon}_i$ is nonzero. This adds the term $\mathsf{E}[\tilde{m}\tilde{\varepsilon}_i]$ to the right-hand side of (6.27). Recall that $\tilde{\varepsilon}_i$ has a zero mean. If it also has a price of zero, in the sense that $\mathsf{E}[\tilde{m}\tilde{\varepsilon}_i] = 0$, then we obtain (6.28) just as when $\tilde{\varepsilon}_i = 0$. In general, by following the algebra above, we can deduce that

$$\mathsf{E}[\tilde{R}_i] = \frac{1}{\mathsf{E}[\tilde{m}]} - \frac{1}{\mathsf{E}[\tilde{m}]} \sum_{j=1}^{k} \beta_i \operatorname{cov}(\tilde{m}, \tilde{f}_j) - \frac{\mathsf{E}[\tilde{m}\tilde{\varepsilon}_i]}{\mathsf{E}[\tilde{m}]}. \qquad (6.29)$$

Based on a comparison of (6.28) and (6.29), the term $-\mathsf{E}[\tilde{m}\tilde{\varepsilon}_i]/\mathsf{E}[\tilde{m}]$ is called the pricing error. Denote it by δ_i.

Why should the pricing errors or equivalently the prices $\mathsf{E}[\tilde{m}\tilde{\varepsilon}_i]$ be zero? As discussed above, the answer is that the $\tilde{\varepsilon}_i$ represent risks that can be diversified away, because they are uncorrelated with each other and with the factors. In a statistical factor model, the covariance matrix of returns is $B\Sigma_F B' + D$, and D is diagonal with diagonal terms $\operatorname{var}(\tilde{\varepsilon}_i)$, so the variance of a portfolio π is

$$\pi' B\Sigma_F B' \pi + \pi' D \pi = \pi' B\Sigma_F B' \pi + \sum_{i=1}^{n} \pi_i^2 \operatorname{var}(\tilde{\varepsilon}_i).$$

As discussed in Section 2.3, the term $\sum_{i=1}^{n} \pi_i^2 \operatorname{var}(\tilde{\varepsilon}_i)$ can be made negligible via diversification. Consider, for example, a portfolio that has $1/n$ of its value in each of the n assets. Then,

$$\sum_{i=1}^{n} \pi_i^2 \operatorname{var}(\tilde{\varepsilon}_i) = \frac{1}{n^2} \sum_{i=1}^{n} \operatorname{var}(\tilde{\varepsilon}_i) \leq \frac{1}{n} \times \max_{i=1,\ldots,n} \operatorname{var}(\tilde{\varepsilon}_i).$$

Thus, the total idiosyncratic risk is near zero when n is large and the $\tilde{\varepsilon}_i$ are bounded risks (say, $\operatorname{var}(\tilde{\varepsilon}_i) \leq \sigma^2$ for a constant σ and all i). It seems plausible in this circumstance that an asset's expected return should not depend on its idiosyncratic risk, meaning $\mathsf{E}[\tilde{m}\tilde{\varepsilon}_i] = 0$. Equivalently, it seems plausible that, if the $\tilde{\varepsilon}_i$ are unimportant in this sense, then there should be an SDF $\tilde{m}$ that depends only on (is an affine function of) the systematic risks $\tilde{f}_1, \ldots, \tilde{f}_k$.

There are two problems with the above argument. First, with only finitely many assets, the idiosyncratic risk of a diversified portfolio may be small, but it is not zero; thus, there could still be some small risk premia associated with the idiosyncratic risks of assets. Second, it may not be possible for all investors to hold well-diversified portfolios (portfolios with zero idiosyncratic risk) because

Factor Models

the market portfolio may not be well diversified—for example, the first asset may represent a large part of the total market.

As a result of these issues, the conclusion of the APT is only that if there is a "large" number of assets, then "most" of the pricing errors are "small," and therefore (6.28) is approximately true for most assets. Somewhat more formally, the APT is as follows: Consider an infinite sequence of assets with returns $\tilde{R}_1, \tilde{R}_2, \ldots$. Suppose there is an SDF $\tilde{m}$ with $\mathsf{E}[\tilde{m}] \neq 0$. Then, for any real number $\delta > 0$, there are only finitely many assets with pricing errors δ_i for which $|\delta_i| \geq \delta$. Any finite subset of assets is a "small" subset of an infinite set. This is the sense in which "most" assets have small pricing errors (smaller than any arbitrary $\delta > 0$).[6]

A proof of the APT is given below. The assumptions are stronger than necessary. This is discussed further in the end-of-chapter notes. The intuition for the APT in terms of the residual risks being diversifiable away and hence earning negligible risk premia really does not appear in the proof. The economics of the problem is embedded in the assumption that an SDF exists for the infinite sequence of returns.

Suppose there is a finite-variance random variable $\tilde{m}$ satisfying $\mathsf{E}[\tilde{m}] \neq 0$ and $\mathsf{E}[\tilde{m}\tilde{R}_i] = 1$ for all i. Assume $\mathrm{var}(\tilde{\varepsilon}_i) \leq \sigma^2$ for each i and $\mathrm{cov}(\tilde{\varepsilon}_i, \tilde{\varepsilon}_j) = 0$ for $i, j = 1, 2, \ldots$ and $i \neq j$.

Let ℓ^2 denote the space of sequences $x = (x_1, x_2, \ldots)$ such that $\sum_{i=1}^{\infty} x_i^2 < \infty$. Define the norm of $x \in \ell^2$ to be

$$\|x\| = \sqrt{\sum_{i=1}^{\infty} x_i^2} \qquad (6.30)$$

and the inner product of x and π in ℓ^2 to be

$$\langle x, \pi \rangle = \sum_{i=1}^{\infty} x_i \pi_i.$$

With these definitions, ℓ^2 is a Hilbert space, and (see Section 3.8)

$$\|x\| = \max_{\|\pi\|=1} |\langle x, \pi \rangle|. \qquad (6.31)$$

Fix for the moment an integer n, and let x denote the sequence given by $x_i = \mathsf{E}[\tilde{m}\tilde{\varepsilon}_i] = -\delta_i \mathsf{E}[\tilde{m}]$ for $i = 1, \ldots, n$ and $x_i = 0$ for $i > n$. The definition (6.30) gives us

---

6. The more precise conclusion of the APT is $\sum_{i=1}^{\infty} \delta_i^2 < \infty$. This implies the statement in the text, because if an infinite number of the pricing errors were larger than δ, then the sum of squared pricing errors would be infinite.

$$\sqrt{\sum_{i=1}^{n} \delta_i^2} = \frac{\|x\|}{|E[\tilde{m}]|}.$$

From (6.31), we have

$$\|x\| = \max_{\|\pi\|=1} |\langle x, \pi \rangle|$$

$$= \max_{\|\pi\|=1} \left| \sum_{i=1}^{n} \pi_i E[\tilde{m}\tilde{\varepsilon}_i] \right|$$

$$= \max_{\|\pi\|=1} \left| E\left[\tilde{m} \sum_{i=1}^{n} \pi_i \tilde{\varepsilon}_i \right] \right|$$

$$\leq \sqrt{E[\tilde{m}^2]} \max_{\|\pi\|=1} \sqrt{E\left[\left(\sum_{i=1}^{n} \pi_i \tilde{\varepsilon}_i\right)^2\right]}$$

$$= \sqrt{E[\tilde{m}^2]} \max_{\|\pi\|=1} \sqrt{\sum_{i=1}^{n} \pi_i^2 \operatorname{var}(\tilde{\varepsilon}_i)}$$

$$= \sqrt{E[\tilde{m}^2]} \max_{i} \sqrt{\operatorname{var}(\tilde{\varepsilon}_i)}$$

$$\leq \sigma \sqrt{E[\tilde{m}^2]}, \tag{6.32}$$

using the Cauchy-Schwartz inequality in $\mathcal{L}^2$ for the first inequality in the string above and the boundedness of the variances of the $\tilde{\varepsilon}_i$ for the second. Because this bound is independent of n, we conclude that $\sum_{i=1}^{\infty} \delta_i^2 < \infty$.

6.6 EMPIRICAL PERFORMANCE OF POPULAR MODELS

This section presents some basic empirical evidence regarding the CAPM and the Fama-French-Carhart model. The Fama-French-Carhart model uses three factors in addition to the market excess return, called SMB, HML, and UMD. SMB (small minus big) is the return of a portfolio that is long small-cap stocks and short large-cap stocks.[7] HML (high minus low) is the return of a portfolio that is long high book-to-market (value) stocks and short low book-to-market (growth) stocks. UMD (up minus down) is the return of a portfolio that is long stocks

---

7. Small-cap and large-cap are terms that refer to a firm's market capitalization (price times shares outstanding). When we refer to size, we mean market capitalization.

with high momentum and short stocks with low momentum (with momentum defined as the return over the previous year excluding the most recent month).

The evidence presented below concerns portfolios of stocks formed from sorts on size, book-to-market, and momentum. The Fama-French-Carhart model is designed to explain the returns of these portfolios. As shown below, it is somewhat successful in that regard. This is important, because value and momentum are strong effects that exist across multiple asset classes, as discussed in the introductory chapter. However, the model fails to explain the returns of portfolios sorted on numerous other characteristics. Such failures are called anomalies, as discussed in the introductory chapter. See Harvey, Liu, and Zhu (2014) for a catalogue of papers documenting failures of the model.[8]

All of the data used in constructing the tables below come from Kenneth French's data library.[9] The time period analyzed is 1970–2013, but similar results hold for different periods. The data are for returns of portfolios formed from sorting stocks on size (market capitalization) and another characteristic (book-to-market or momentum). The sorts are then intersected and stocks within each cell are value weighted to form portfolios. There are two reasons for sorting on size and another characteristic and intersecting the sorts rather than just sorting on the characteristic. First, the characteristic may be cross-sectionally correlated with size. For example, size and book-to-market are both defined in terms of market capitalization and are negatively correlated in the cross section of stocks, due to market capitalization being in the denominator of book-to-market. Thus, a sort on book-to-market will be contaminated by size: Low book-to-market stocks are likely to be large-cap stocks and high book-to-market stocks are likely to be small-cap stocks. Hence, a sort on book-to-market alone cannot clarify whether differences in returns are associated with differences in book-to-market or with differences in size. Second, small-cap stocks are costly to trade, so an anomaly that occurs only within small-cap stocks may not be exploitable at any reasonable scale. Such an anomaly is of less interest than an anomaly that occurs within mid-cap and large-cap stocks. Hence, it is useful to analyze an anomaly within each size category separately.

---

8. As mentioned in the introductory chapter, data mining is a concern in evaluating these rejections of the model. The point of Harvey, Liu, and Zhu (2014) is that, due to data mining, the hurdle for declaring a characteristic to be a determinant of expected returns should be increased over time as more characteristics are examined. They argue that most of the documented correlations between characteristics and average returns are probably spurious.

9. http://mba.tuck.dartmouth.edu/pages/faculty/ken.french/data_library.html.

In all cases below, we estimate the regression model

$$\tilde{R} - R_f = \alpha + \sum_{j=1}^{k} \beta_j \tilde{f}_j + \tilde{\varepsilon}$$

or the model

$$\tilde{R}_1 - \tilde{R}_2 = \alpha + \sum_{j=1}^{k} \beta_j \tilde{f}_j + \tilde{\varepsilon}$$

for a return $\tilde{R}$ or for two returns $\tilde{R}_1$ and $\tilde{R}_2$, and where either

- CAPM: $k = 1$ and $\tilde{f}_1 = \tilde{R}_m - R_f$, or
- Fama-French-Carhart: $k = 4$ and the factors are $\tilde{R}_m - R_f$, SMB, HML, and UMD.

In all cases, we are testing the null hypothesis $\alpha = 0$, which is implied by the factor model.

Portfolios Based on Sorts by Size and Book-to-Market Ratio

Early evidence against the CAPM was that it failed to explain the superior performance of small-cap stocks versus large-cap stocks and failed to explain the superior performance of value (high book-to-market or high earnings-to-price) versus growth (low book-to-market or low earnings-to-price) stocks. Table 6.1 shows that value stocks have significant CAPM alphas except in the largest size quintile. The pattern of value beating growth discussed in the introductory chapter for average excess returns also holds for CAPM alphas. Likewise, the pattern of small-cap stocks outperforming large-cap stocks, except in the growth quintile, also generally holds for CAPM alphas.

The SMB and HML factors were created to explain the performance of the size- and book-to-market-sorted portfolios. Table 6.2 shows that the Fama-French-Carhart alphas are generally much smaller than the CAPM alphas. An exception is the small-growth portfolio, for which the alpha is reduced by less than 20% and remains statistically significant. The Fama-French-Carhart t-statistics are below 2 except in the corners of the 5×5 sort (small growth, small value, and big growth). This corner effect reflects an interaction between size and value/growth. The value effect is much stronger among small-cap stocks than among large-cap stocks, and the size effect is much stronger among value stocks than among growth stocks. In fact, as noted before, the size effect reverses among growth stocks—large-cap growth stocks

Table 6-1. CAPM ALPHAS AND t-STATISTICS FOR SIZE- AND BOOK-TO-MARKET-SORTED PORTFOLIOS, 1970–2013.

	Alphas					t-Stats				
	Growth	2	3	4	Value	Growth	2	3	4	Value
Small	−0.62	0.12	0.25	0.47	0.56	−2.81	0.68	1.57	2.92	3.02
2	−0.29	0.11	0.35	0.42	0.42	−1.80	0.82	2.69	3.16	2.49
3	−0.21	0.20	0.27	0.35	0.56	−1.61	1.80	2.28	2.87	3.57
4	−0.04	0.04	0.20	0.34	0.33	−0.32	0.38	1.68	2.89	2.30
Big	−0.07	0.12	0.07	0.16	0.21	−0.85	1.57	0.68	1.39	1.49

Alphas Are in % per Month. t-Statistics Are Based on Newey-West Standard Errors.

Table 6-2. FAMA-FRENCH-CARHART ALPHAS AND t-STATISTICS FOR SIZE- AND BOOK-TO-MARKET-SORTED PORTFOLIOS, 1970–2013.

	Alphas					t-Stats				
	Growth	2	3	4	Value	Growth	2	3	4	Value
Small	−0.51	0.01	0.03	0.15	0.16	−4.63	0.14	0.44	2.29	2.22
2	−0.12	0.01	0.09	0.07	−0.06	−1.81	0.19	1.57	1.13	−0.84
3	−0.01	0.08	0.02	0.01	0.16	−0.15	1.04	0.22	0.18	1.55
4	0.13	−0.06	0.00	0.05	−0.04	1.85	−0.63	−0.03	0.62	−0.45
Big	0.16	0.06	−0.07	−0.09	−0.12	2.93	0.88	−0.89	−1.44	−1.15

Alphas Are in % per Month. t-Statistics Are Based on Newey-West Standard Errors.

outperform small-cap growth stocks in terms of average excess returns, CAPM alphas, and Fama-French-Carhart alphas.

Table 6.3 shows the alphas of portfolios that are long value stocks and short growth stocks (the extreme quintiles) within each size category. For the CAPM, all of the point estimates are positive (value beats growth), and they are statistically significant in the three smallest size quintiles. The alphas are generally smaller for the Fama-French-Carhart model, but they are also estimated more precisely, so the t-statistics do not fall as much as do the alphas. Interestingly, both the small-cap and large-cap alphas are significant in the Fama-French-Carhart model, but they are of opposite signs. Value underperforms growth among large-cap stocks relative to the predicted returns from the Fama-French-Carhart model. This is another manifestation of the interaction described earlier—the value effect is much stronger among small-cap stocks than among large-cap stocks.

Table 6-3. CAPM AND FAMA-FRENCH-CARHART ALPHAS AND t-STATISTICS FOR VALUE-MINUS-GROWTH PORTFOLIOS, 1970–2013.

	CAPM		FFC	
	Alpha	t-Stat	Alpha	t-Stat
Small	1.18	5.79	0.66	5.94
2	0.71	3.42	0.06	0.66
3	0.77	3.73	0.17	1.40
4	0.37	1.77	−0.17	−1.68
Big	0.28	1.49	−0.28	−2.35

Alphas Are in % per Month. t-Statistics Are Based on Newey-West Standard Errors.

Table 6-4. CAPM ALPHAS AND t-STATISTICS FOR SIZE- AND MOMENTUM-SORTED PORTFOLIOS, 1970–2013.

	Alphas					t-Stats				
	Down	2	3	4	Up	Down	2	3	4	Up
Small	−0.72	0.08	0.34	0.48	0.74	−3.19	0.50	2.25	3.11	3.58
2	−0.63	0.06	0.28	0.45	0.54	−3.22	0.41	2.25	3.68	3.44
3	−0.43	0.03	0.18	0.29	0.51	−2.30	0.22	1.59	2.84	3.69
4	−0.50	0.03	0.18	0.29	0.40	−2.59	0.21	1.73	3.36	3.22
Big	−0.51	0.04	−0.05	0.12	0.21	−2.68	0.31	−0.67	1.74	1.84

Alphas Are in % per Month. t-Statistics Are Based on Newey-West Standard Errors.

Portfolios Based on Sorts by Size and Momentum

As discussed in the introductory chapter, there is a momentum effect in average stock returns that is even stronger than value/growth. The momentum effect is not explained by the CAPM. Table 6.4 presents the CAPM alphas of size- and momentum-sorted portfolios. The CAPM alphas of the stocks with the lowest momentum are negative and statistically significant in every size category. The CAPM alphas of the stocks with the highest momentum are positive and statistically significant in every size category (for the largest-cap stocks, the alpha is significant at 10% but not at 5%). The absolute values of the alphas diminish as size increases but not to the same extent that value/growth alphas diminish as size increases.

The momentum effect is not explained by the CAPM; neither is it explained by the Fama-French three-factor model (the three factors are the market excess return, SMB, and HML). This is the motivation for the UMD factor included in the Fama-French-Carhart model. Table 6.5 presents the Fama-French-Carhart alphas of the size- and momentum-sorted portfolios. The model is somewhat successful in explaining the returns. Except for the smallest size category, the alphas are generally 2% per year or less.

Table 6-5. FAMA-FRENCH-CARHART ALPHAS AND t-STATISTICS FOR SIZE- AND MOMENTUM-SORTED PORTFOLIOS, 1970–2013.

	Alphas					t-Stats				
	Down	2	3	4	Up	Down	2	3	4	Up
Small	−0.38	−0.02	0.10	0.17	0.36	−2.69	−0.35	1.48	2.15	3.39
2	−0.19	0.06	0.07	0.15	0.17	−1.86	0.84	0.99	2.13	2.12
3	0.09	0.08	0.06	−0.01	0.14	0.86	0.99	0.70	−0.09	1.96
4	0.06	0.17	0.12	0.08	0.02	0.52	1.94	1.33	0.95	0.20
Big	0.10	0.34	−0.02	−0.04	−0.10	0.71	4.03	−0.24	−0.63	−1.35

Alphas Are in % per Month. t-Statistics Are Based on Newey-West Standard Errors.

Table 6-6. CAPM AND FAMA-FRENCH-CARHART ALPHAS AND t-STATISTICS FOR UP MINUS DOWN PORTFOLIOS, 1970–2013.

	CAPM		FFC	
	Alpha	t-Stat	Alpha	t-Stat
Small	1.46	6.74	0.74	5.89
2	1.16	5.28	0.36	2.95
3	0.94	4.00	0.05	0.46
4	0.90	3.56	−0.04	−0.34
Big	0.73	2.69	−0.20	−1.32

Alphas Are in % per Month. t-Statistics Are Based on Newey-West Standard Errors.

Table 6.6 presents the alphas of the portfolios that are long the highest-momentum stocks and short the lowest-momentum stocks within each size category. The CAPM alphas are statistically significant within every size category and are economically large—in the range of 10–15% per year. The Fama-French-Carhart alphas are substantially smaller and are significant only in the two smallest size categories.

6.7 NOTES AND REFERENCES

The CAPM is credited to Sharpe (1964) or to some combination of Sharpe, Treynor (1999), Lintner (1969), and Mossin (1966). Black (1972) extends the CAPM to markets without a risk-free asset and to markets in which a risk-free asset exists but borrowing is not possible. Brennan (1971) extends the CAPM to markets with different borrowing and lending rates (Exercise 6.5). Ross (1977) discusses the validity of the CAPM under various types of short-sales restrictions. Berk (1997) gives necessary and sufficient conditions for the CAPM that subsume the separating distributions assumption discussed in Section 5.6.

Jagannathan and Wang (1996) is a seminal attempt to test the CAPM with human capital counted as part of market wealth. The result known as Stein's lemma used in the proof of the CAPM in Section 6.1 when there is unspanned labor income and when asset returns and aggregate wealth are joint normally distributed is due to Stein (1973). A proof can also be found in the appendix of Rubinstein (1976).

The equivalence between a return being on the mean-variance frontier and factor pricing with that return as the factor is observed by Roll (1977) and Ross (1977). Dybvig and Ingersoll (1982) show that the CAPM is equivalent to there being an SDF that is an affine function of the market return and observe that, if markets are complete and the CAPM applies to all asset returns, then there are arbitrage opportunities if the market return is unbounded above.

As remarked in the text, performance evaluation using Jensen's alpha stems from Jensen (1969). Dybvig and Ross (1985a) show that a positive alpha implies that adding some of the asset to the benchmark will improve mean-variance efficiency, with or without a risk-free asset. A different calculation showing that adding some of an asset with a positive alpha to the benchmark improves the utility of mean-variance investors is suggested in Exercise 6.4. Dybvig and Ross (1985b) describe the pitfalls of using Jensen's alpha for performance evaluation when superior performance is due to superior information.

The market model is introduced by Sharpe (1963), who uses it to simplify estimation of the covariance matrix for mean-variance portfolio optimization, as discussed in Section 6.4. For the estimation of latent statistical factor models, see, for example, Anderson (2003).

The APT is due to Ross (1976a). Chamberlain and Rothschild (1983) show that the conclusion of the APT holds if the factor structure assumption is relaxed to allow weak correlation of the residuals, in the sense that the covariance matrix of the residuals is not required to be diagonal with bounded diagonal elements, but instead it is only assumed that the eigenvalues of the covariance matrix of the residuals are bounded as the number of assets goes to infinity. When this condition holds, we say that there is an approximate factor structure. The proof in Section 6.5 is due to Reisman (1988). We can see from the proof that the condition needed on the residuals is that

$$\max_{\|w\|=1} \mathsf{E}\left[\left(\sum_{i=1}^{n} w_i \tilde{\varepsilon}_i\right)^2\right] = \max_{\|w\|=1} w' \Sigma_{\varepsilon,n} w$$

be bounded independently of n, where $\Sigma_{\varepsilon,n}$ denotes the covariance matrix of the first n residuals. This condition is equivalent to the maximum eigenvalue of $\Sigma_{\varepsilon,n}$ being bounded independently of n.

The existence of an SDF for an infinite sequence of assets does not follow directly from the law of one price or even from the absence of arbitrage opportunities. Kreps (1981) and Chamberlain and Rothschild (1983) give sufficient conditions. The nature of these conditions is that there is no sequence of portfolios that converges, in some sense, to an arbitrage opportunity.

Shanken (1982) makes the following observations: (i) returns may satisfy a statistical factor model, but portfolios of the returns may not satisfy the same statistical factor model (thus, a statistical factor model is not robust to repackagings of the securities), (ii) returns may satisfy a statistical factor model, but portfolios may satisfy a different statistical factor model with even a different number of factors, and (iii) exact APT pricing in a statistical factor model satisfied by returns may be inconsistent with exact APT pricing in a statistical factor model satisfied by portfolios. An example from Shanken (1982) is presented in Exercise 6.7. Shanken concludes that the hypothesis "statistical factor model $\Rightarrow$ exact APT pricing" is untestable. Dybvig and Ross (1985c) argue that it is still reasonable to test the hypothesis "statistical factor model $\Rightarrow$ exact APT pricing," as an approximation to "statistical factor model $\Rightarrow$ approximate APT pricing," but only in circumstances where the approximation is good, which requires in particular that residual variances be small (see the bound (6.32)). They observe that the portfolios constructed by Shanken may have large residual variances.

Reisman (1992) proves the following: Suppose there is an infinite sequence of assets having an approximate factor structure with k factors $\tilde{g}_1, \ldots \tilde{g}_k$, and suppose there is an SDF for the infinite sequence of assets. Consider any other k random variables $\tilde{f}_1, \ldots, \tilde{f}_k$ having finite variance. Assume the $k \times k$ matrix $\mathsf{E}[\tilde{f}\tilde{g}']$ is nonsingular. Then, there is an approximate factor model with the $\tilde{f}_j$'s as the factors. This means that if k variables can be used to approximately price assets, then so can essentially any other k variables. Thus, "approximate factor model" is a much weaker concept than "exact factor model." See also Gilles and LeRoy (1991) and Shanken (1992). The result of Reisman (1992) forms part of the basis of a critique of asset pricing tests by Lewellen, Nagel, and Shanken (2007).

Exercise 6.6 presents a simple version of the equilibrium APT of Connor (1984). It is closely related to the theory of separating distributions due to Ross (1978a) and presented in Section 5.6. Another related result is due to Chamberlain (1983b): All of the pricing errors in the APT are zero if and only if there is a risky well-diversified portfolio on the mean-variance frontier (see Exercise 6.1 for the "if" part of this).

Bounds on the APT pricing errors are deduced by Grinblatt and Titman (1983) and Dybvig (1983), without assuming an infinite number of assets. Exercise 7.9 presents a simple version of Dybvig's result.

The original research on the size effect in stock returns appears in Banz (1981), and the original research on the value effect appears in Basu (1983). This research is summarized and extended in Fama and French (1992), a paper that became widely known as the "beta is dead" paper. The original research on the momentum effect appears in Jegadeesh and Titman (1993). The SMB and HML factors are proposed by Fama and French (1993, 1996). The UMD factor is proposed by Carhart (1997). Davis, Fama, and French (2000) remark on the failure of the Fama-French model to correctly price the small growth portfolio, as discussed in Section 6.6. Cochrane (2011, p. 1099) comments on the interaction between the size and value effects, noting that returns are better explained in cross-sectional regressions of returns on characteristics when a size × book-to-market characteristic is included. Recently proposed factor models extend the Fama-French model by including factors based on profitability and asset growth (Fama and French, 2015; Hou, Xue, and Zhang, 2015).

Exercise 6.8 presents a highly simplified version of the model of neglected assets due to Merton (1987). In that model, all investors have mean-variance preferences, but some investors do not consider investing in some assets, perhaps because they are unaware of the existence of the assets or because they do not have enough information about the distributions of the asset returns to consider investing in them. In this environment, the market portfolio is not mean-variance efficient, and the CAPM does not hold.

EXERCISES

6.1. Assume there exists a return $\tilde{R}_*$ that is on the mean-variance frontier and is an affine function of a vector $\tilde{F}$; that is, $\tilde{R}_* = a + b'\tilde{F}$. Assume either (i) there is a risk-free asset and $\tilde{R}_* \neq R_f$, or (ii) there is no risk-free asset and $\tilde{R}_*$ is different from the GMV return. Show that there is a factor model with factors $\tilde{F}$.

6.2. Assume returns are normally distributed, investors have CARA utility, and there is no labor income. Derive the CAPM from the portfolio formula (2.22), that is, from

$$\phi_h = \frac{1}{\alpha_h} \Sigma^{-1}(\mu - R_f \iota),$$

where α_h denotes the absolute risk aversion of investor h. Show that the price of risk is $\alpha w_0 \operatorname{var}(\tilde{R}_m)$, where α is the aggregate absolute risk aversion defined in Section 1.1 and $w_0 = \sum_{h=1}^{H} \phi_h$ is the market value of risky assets at date 0.

6.3. Assume there is a risk-free asset, and assume that a factor model holds in which each factor $\tilde{f}_1, \ldots, \tilde{f}_k$ is an excess return.

(a) Show that each return $\tilde{R}$ on the mean-variance frontier equals

$$R_f + \sum_{j=1}^{k} \beta_j \tilde{f}_j \qquad (6.33)$$

for some $\beta_1, \ldots, \beta_k$. In other words, show that the risk-free return and the factors span the mean-variance frontier.

(b) Show that a return of the form (6.33) is on the mean-variance frontier if and only if $\beta = (\beta_1 \cdots \beta_k)'$ satisfies

$$\beta = \delta \Sigma_F^{-1} \lambda$$

for some δ, where Σ_F is the (assumed to be nonsingular) covariance matrix of $\tilde{F} = (\tilde{f}_1 \cdots \tilde{f}_k)'$, and $\lambda = \mathsf{E}[\tilde{F}] \in \mathbb{R}^k$.

6.4. Suppose there is a risk-free asset and suppose Jensen's alpha in (6.22) is positive. Consider an investor with initial wealth w_0 who holds the benchmark portfolio and therefore has terminal wealth $w_0 \tilde{R}_b$. Assume $\mathsf{E}[u'(w_0 \tilde{R}_b)] > 0$. Consider the return

$$\tilde{R}_1 = \tilde{R} + (1-\beta)(\tilde{R}_b - R_f) = \tilde{R}_b + \alpha + \tilde{\varepsilon}.$$

Show that

$$\mathsf{E}[u'(w_0 \tilde{R}_b)(\tilde{R}_1 - \tilde{R}_b)] = \alpha \mathsf{E}[u'(w_0 \tilde{R}_b)] > 0 \qquad (6.34)$$

if utility is quadratic or if $\tilde{R}$ and $\tilde{R}_b$ are joint normally distributed. Note: Condition (6.34) implies that the expected utility of a convex combination $\lambda \tilde{R}_1 + (1-\lambda)\tilde{R}_b$ is greater than the expected utility of $\tilde{R}_b$ for sufficiently small $\lambda > 0$. Thus, this exercise shows that a positive Jensen's alpha implies that utility improvements are possible if utility is quadratic or returns are joint normal.

6.5. Suppose investors can borrow and lend at different rates. Let R_b denote the return on borrowing and R_ℓ the return on lending. Suppose $B/C > R_b > R_\ell$, where B and C are defined in (5.6). Suppose each investor chooses a mean-variance efficient portfolio, as described in Exercise 5.9. Show that the CAPM holds with $R_\ell \leq R_z \leq R_b$.

6.6. Assume the asset returns $\tilde{R}_i$ for $i = 1, \ldots, n$ satisfy

$$\tilde{R}_i = \mathsf{E}[\tilde{R}_i] + \mathrm{Cov}(\tilde{F}, \tilde{R}_i)' \Sigma_F^{-1} (\tilde{F} - \mathsf{E}[\tilde{F}]) + \tilde{\varepsilon}_i,$$

where each $\tilde{\varepsilon}_i$ is mean independent of the factors $\tilde{F}$, that is, $\mathsf{E}[\tilde{\varepsilon}_i | \tilde{F}] = 0$ (note it is not being assumed that $\mathrm{cov}(\tilde{\varepsilon}_i, \tilde{\varepsilon}_j) = 0$). Assume markets are complete and the market return is well diversified in the sense of having no idiosyncratic risk:

$$\tilde{R}_m = \mathsf{E}[\tilde{R}_m] + \mathrm{Cov}(\tilde{F}, \tilde{R}_m)' \Sigma_F^{-1} (\tilde{F} - \mathsf{E}[\tilde{F}]).$$

Show that there is a factor model with factors $\tilde{F}$. Hint: Pareto optimality implies sharing rules $\tilde{w}_h = f_h(\tilde{w}_m)$.

6.7. Suppose two assets satisfy a statistical factor model with a single factor:

$$\tilde{R}_1 = \mathsf{E}[\tilde{R}_1] + \tilde{f} + \tilde{\varepsilon}_1,$$
$$\tilde{R}_2 = \mathsf{E}[\tilde{R}_2] - \tilde{f} + \tilde{\varepsilon}_2,$$

where $\mathsf{E}[\tilde{f}] = \mathsf{E}[\tilde{\varepsilon}_1] = \mathsf{E}[\tilde{\varepsilon}_2] = 0$, $\mathrm{var}(\tilde{f}) = 1$, $\mathrm{cov}(\tilde{f}, \tilde{\varepsilon}_1) = \mathrm{cov}(\tilde{f}, \tilde{\varepsilon}_2) = 0$, and $\mathrm{cov}(\tilde{\varepsilon}_1, \tilde{\varepsilon}_2) = 0$. Assume $\mathrm{var}(\tilde{\varepsilon}_1) = \mathrm{var}(\tilde{\varepsilon}_2) = \sigma^2$. Define $\tilde{R}_1^* = \tilde{R}_1$ and $\tilde{R}_2^* = \pi \tilde{R}_1 + (1-\pi) \tilde{R}_2$ with $\pi = 1/(2+\sigma^2)$.

(a) Show that $\tilde{R}_1^*$ and $\tilde{R}_2^*$ do not satisfy a statistical factor model with the single factor $\tilde{f}$.

(b) Show that $\tilde{R}_1^*$ and $\tilde{R}_2^*$ satisfy a statistical factor model with zero factors, that is,

$$\tilde{R}_1^* = \mathsf{E}[\tilde{R}_1^*] + \tilde{\varepsilon}_1^*,$$
$$\tilde{R}_2^* = \mathsf{E}[\tilde{R}_2^*] + \tilde{\varepsilon}_2^*,$$

where $\mathsf{E}[\tilde{\varepsilon}_1^*] = \mathsf{E}[\tilde{\varepsilon}_2^*] = 0$ and $\mathrm{cov}(\tilde{\varepsilon}_1^*, \tilde{\varepsilon}_2^*) = 0$.

(c) Assume exact APT pricing with nonzero risk premium λ for the two assets in the single-factor model, that is, $\mathsf{E}[\tilde{R}_i] - R_f = \lambda \, \mathrm{cov}(\tilde{R}_i, \tilde{f})$ for $i = 1, 2$. Show that there cannot be exact APT pricing in the zero-factor model for $\tilde{R}_1^*$ and $\tilde{R}_2^*$.

6.8. Assume there are H investors with CARA utility and the same absolute risk aversion α. Assume there is a risk-free asset. Assume there are two risky assets with payoffs $\tilde{x}_i$ that are joint normally distributed with mean vector μ and nonsingular covariance matrix Σ. Assume H_U investors are unaware of the second asset and invest only in the risk-free asset and the first risky asset. If all investors invested in both assets ($H_U = 0$), then the equilibrium price vector would be

$$p^* = \frac{1}{R_f}\mu - \frac{\alpha}{HR_f}\Sigma\bar{\theta},$$

where $\bar{\theta}$ is the vector of supplies of the risky assets (Exercise 4.1). Assume $0 < H_U < H$, and set $H_I = H - H_U$.

(a) Show that the equilibrium price of the first asset is $p_1 = p_1^*$, and the equilibrium price of the second asset is

$$p_2 = p_2^* - \frac{\alpha}{HR_f}\left(\frac{H_U}{H_I}\right)\left(\mathrm{var}(\tilde{x}_2) - \frac{\mathrm{cov}(\tilde{x}_1, \tilde{x}_2)^2}{\mathrm{var}(\tilde{x}_1)}\right) < p_2^*.$$

Factor Models

(b) Show that there exist $A > 0$ and λ such that

$$E[\tilde{R}_1] = R_f + \lambda \cdot \frac{\text{cov}(\tilde{R}_1, \tilde{R}_m)}{\text{var}(\tilde{R}_m)}, \qquad (6.35a)$$

$$E[\tilde{R}_2] = A + R_f + \lambda \frac{\text{cov}(\tilde{R}_2, \tilde{R}_m)}{\text{var}(\tilde{R}_m)}, \qquad (6.35b)$$

$$\lambda = E[\tilde{R}_m] - R_f - A\pi_2, \qquad (6.35c)$$

where $\pi_2 = p_2 \bar{\theta}_2 / (p_1 \bar{\theta}_1 + p_2 \bar{\theta}_2)$ is the relative date–0 market capitalization of the second risky asset. (Note that λ is less than in the CAPM, and the second risky asset has a positive alpha, relative to λ.)

6.9. Suppose there is no risk-free asset and the minimum-variance return is different from the constant-mimicking return, that is, $b_m \neq b_c$. From Section 6.2, we know there is a factor model with the constant-mimicking return as the factor:

$$E[\tilde{R}] = R_z + \psi \, \text{cov}(\tilde{R}, \tilde{R}_p + b_c \tilde{e}_p) \qquad (6.36)$$

for every return $\tilde{R}$. From Section 6.2, we can conclude there is an SDF that is an affine function of the constant-mimicking return unless $R_z = 0$. However, the existence of an SDF that is an affine function of the constant-mimicking return would contradict the result of Section 5.5. So, it must be that $R_z = 0$ in (6.36). Calculate R_z to demonstrate this.

6.10. Suppose there is no risk-free asset and the minimum-variance return is different from the constant-mimicking return, that is, $b_m \neq b_c$. From Section 5.5, we know that there is an SDF that is an affine function of the minimum-variance return:

$$\tilde{m} = \gamma + \beta (\tilde{R}_p + b_m \tilde{e}_p) \qquad (6.37)$$

for some γ and β. From Section 6.2, we know that there is no factor model with the minimum-variance return as the factor. However, because there is an SDF that is an affine function of the minimum-variance return, we also know from Section 6.2 that there *is* a factor model with the minimum-variance return as the factor unless $E[\tilde{m}] = 0$. So it must be that $E[\tilde{m}] = 0$ for the SDF $\tilde{m}$ in (6.37). Calculate $E[\tilde{m}]$ to demonstrate this.

7

Representative Investors

CONTENTS

7.1	Pareto Optimality Implies a Representative Investor	163
7.2	Linear Risk Tolerance	165
7.3	Consumption-Based Asset Pricing	167
7.4	Coskewness-Cokurtosis Pricing Model	171
7.5	Rubinstein Option Pricing Model	172
7.6	Notes and References	175

We say that a competitive equilibrium of an economy admits a representative investor if the equilibrium prices are also equilibrium prices of an economy consisting of a single investor who owns all of the assets and endowments of the original economy. The wealth of this representative investor is the market wealth of the original economy, so her first-order condition is that her marginal utility evaluated at market wealth be proportional to an SDF. This is important because it produces an SDF that depends only on market wealth. The assumption that there is a representative investor is made frequently in finance to simplify valuation.

By way of comparison, the asset pricing formula

$$E[u'(\tilde{w})(\tilde{R}_i - \tilde{R}_j)] = 0 \qquad (7.1)$$

derived in Chapter 2 shows that the marginal utility of any investor is proportional to an SDF. However, this depends on the individual investor's wealth, which is generally unobservable. There are two basic routes to replacing individual variables with aggregate variables in asset pricing formulas: (i) if the formula is affine in the individual variable, we can add across investors, or (ii) under some circumstances, there is a representative investor for whom the relation holds, with the individual variable replaced by the aggregate. An example of (i) appears in the proof of the CAPM at the end of Section 6.1. This chapter is concerned with (ii).

The notation of Chapter 4 is used throughout this chapter. Assume there is consumption at date 0, because this is the assumption usually made in applications. The results are also valid with consumption only at date 1.

7.1 PARETO OPTIMALITY IMPLIES A REPRESENTATIVE INVESTOR

Assume each investor h has time-additive utility $u_{h0}(c_0) + u_{h1}(c_1)$, with the functions u_{h0} and u_{h1} being concave. The social planner's utility for aggregate consumption is the maximum weighted sum of individual utilities that can be achieved from the given aggregate consumption. Given positive weights λ_h, the social planner's utility functions are

$$u_0(c) \stackrel{\text{def}}{=} \max \left\{ \sum_{h=1}^{H} \lambda_h u_{h0}(c_h) \,\middle|\, \sum_{h=1}^{H} c_h = c \right\}, \quad (7.2a)$$

$$u_1(c) \stackrel{\text{def}}{=} \max \left\{ \sum_{h=1}^{H} \lambda_h u_{h1}(c_h) \,\middle|\, \sum_{h=1}^{H} c_h = c \right\}. \quad (7.2b)$$

Concavity of the individual utility functions u_{h0} and u_{h1} implies that the social planner's utility functions u_0 and u_1 are also concave (Exercise 7.7).

At a Pareto-optimal competitive equilibrium, the social planner's MRS is an SDF. This follows from the first-order conditions at the competitive equilibrium and from the envelope theorem. The details are provided below. At the equilibrium prices, the social planner—acting as a price-taking investor who initially owns all of the assets and endowments of the economy—would choose to hold all of the assets of the economy. That is, markets would clear. This follows from the first-order condition that the social planner's MRS be an SDF and from concavity. Thus, the equilibrium prices would also be equilibrium prices if the social planner were the only investor in the economy. Consequently, we call the social planner a representative investor.

Different Pareto optima typically imply different vectors $(\lambda_1, \ldots, \lambda_H)$ of weights in the social planning problem and therefore different representative investor utility functions. The weights for Pareto-optimal competitive equilibria typically depend on the distributions of endowments and asset holdings across investors, because changing the distributions changes the competitive equilibrium. Therefore, different distributions usually lead to different representative investors. However, as shown in the next section, if investors have LRT utility functions with the same cautiousness parameter and all have the same discount factor, then the utility function of the representative investor does not depend on the distributions of endowments and asset holdings. This is another facet of Gorman aggregation (Section 4.4).

Let c_0 and $\tilde{c}_1$ denote the total amounts of the consumption good available; that is,

$$c_0 = \sum_{h=1}^{H} y_{h0},$$

$$\tilde{c}_1 = \sum_{h=1}^{H} \tilde{y}_{h1} + \sum_{i=1}^{n} \overline{\theta}_i \tilde{x}_i.$$

We want to show that at a Pareto optimum the social planner's MRS is an SDF. Consider any Pareto optimum. As shown in Chapter 4, there exist weights $\lambda_1, \ldots, \lambda_H$ such that the date-0 allocation solves the optimization problem in (7.2a) for $c = c_0$ and the date-1 allocation solves the optimization problem in (7.2b) for $c = \tilde{c}_1$ in each state of the world. The first-order conditions for the maximization problems are

$$(\forall h) \quad \lambda_h u'_{h0}(c_{h0}) = \eta_0, \tag{7.3a}$$

$$(\forall h) \quad \lambda_h u'_{h1}(\tilde{c}_{h1}) = \tilde{\eta}_1. \tag{7.3b}$$

for Lagrange multipliers η_0 and $\tilde{\eta}_1$. By the envelope theorem,[1]

$$(\forall h) \quad \lambda_h u'_{h0}(c_{h0}) = u'_0(c_0), \tag{7.4a}$$

$$(\forall h) \quad \lambda_h u'_{h1}(\tilde{c}_{h1}) = u'_1(\tilde{c}_1). \tag{7.4b}$$

---

[1]. The interpretation of (7.4) is that the value for the social planner of a small amount of additional consumption can be computed by assuming the consumption is assigned to any individual investor, holding the consumption of other investors constant. A proof of the envelope theorem, as it applies in this context, is as follows: Consider (7.4a). Assuming the conditions of the implicit function theorem hold, the first-order condition for the optimization problem (7.2a) defines the c_h as continuously differentiable functions of c in a neighborhood of aggregate consumption c_0. Thus,

$$u'_0(c_0) = \sum_{h=1}^{H} \lambda_h u'_h(c_h(c_0)) \frac{dc_h}{dc}.$$

From the first-order condition (7.3a) and the fact that the derivatives dc_h/dc must sum to 1, we conclude that this equals η_0, and using the first-order condition (7.3a) again, we see that it equals $\lambda_h u'_{h0}(c_{h0})$.

At a competitive equilibrium, each investor holds an optimal portfolio, so the first-order condition (3.9) holds; namely,

$$\frac{u'_{h1}(\tilde{c}_{h1})}{u'_{h0}(c_{h0})}$$

is an SDF. Therefore, (7.4) implies that the social planner's MRS

$$\frac{u'_1(\tilde{c}_1)}{u'_0(c_0)} \tag{7.5}$$

is an SDF.

7.2 LINEAR RISK TOLERANCE

This section calculates the utility function of the representative investor when investors have time-additive LRT utility functions with the same cautiousness parameter. The utility function of the representative investor is also time additive and has linear risk tolerance with the same cautiousness parameter. Assume that each investor has the same utility for date–1 consumption as for date–0 consumption except for discounting the utility of date–1 consumption by a factor δ that is the same for all investors. Assume the utility function of each investor h has linear risk tolerance $\tau_h(c) = A_h + Bc$, and assume that all investors have the same cautiousness parameter B. As in Sections 4.4 and 4.5, assume $B \geq 0$. Assume also that there is a risk-free asset and that there are no date–1 endowments of the consumption good. Sections 4.4 and 4.5 show that a competitive equilibrium is Pareto optimal in this economy, so a representative investor exists. Furthermore, the following are true:

(A) Suppose each investor h has utility function

$$-e^{-\alpha_h c_0} - \delta e^{-\alpha_h c_1}$$

for some $\alpha_h > 0$. Then, up to a monotone affine transform, the utility function of the representative investor is

$$-e^{-\alpha c_0} - \delta e^{-\alpha c_1}, \tag{7.6}$$

where

$$\alpha = \frac{1}{\sum_{h=1}^{H} 1/\alpha_h}. \tag{7.7}$$

(B) Suppose each investor h has utility function

$$\log(c_0 - \zeta_h) + \delta \log(c_1 - \zeta_h)$$

for some ζ_h. Then, up to a monotone affine transform, the utility function of the representative investor is

$$\log(c_0 - \zeta) + \delta \log(c_1 - \zeta), \tag{7.8}$$

where $\zeta = \sum_{h=1}^{H} \zeta_h$.

(C) Suppose $\rho > 0$ and each investor h has utility function

$$\frac{1}{1-\rho}(c_0 - \zeta_h)^{1-\rho} + \frac{\delta}{1-\rho}(c_1 - \zeta_h)^{1-\rho}$$

for some ζ_h. Then, up to a monotone affine transform, the utility function of the representative investor is

$$\frac{1}{1-\rho}(c_0 - \zeta)^{1-\rho} + \frac{\delta}{1-\rho}(c_1 - \zeta)^{1-\rho}, \tag{7.9}$$

where $\zeta = \sum_{h=1}^{H} \zeta_h$.

The proof of (A) will be given. The proofs of (B) and (C) are very similar and are left for the exercises (Exercise 7.8).

Consider the problem

$$\max \; -\sum_{h=1}^{H} \lambda_h e^{-\alpha_h c_h} \tag{7.10}$$

subject to the resource constraint $\sum_{h=1}^{H} c_h = c$. This problem is analyzed in Section 4.1 (for a random aggregate consumption $\tilde{c}$ but solved state by state). The solution is

$$c_h = a_h + b_h c,$$

where a_h and b_h are defined in (4.14). In particular,

$$b_h = \frac{\alpha}{\alpha_h},$$

where α is the aggregate absolute risk aversion (7.7). Therefore, the maximum value in (7.10) is

$$-\sum_{h=1}^{H} \lambda_h e^{-\alpha_h(a_h + b_h c)} = -\sum_{h=1}^{H} \lambda_h e^{-\alpha_h a_h - \alpha c}$$

$$= -e^{-\alpha c} \sum_{h=1}^{H} \lambda_h e^{-\alpha_h a_h}.$$

Representative Investors

Apply this formula at date 0 and date 1. This shows that the utility function of the representative investor is

$$\max \left\{ -\sum_{h=1}^{H} \lambda_h e^{-\alpha_h c_{h0}} - \sum_{h=1}^{H} \lambda_h \delta_h e^{-\alpha_h c_{h1}} \,\bigg|\, \sum_{h=1}^{H} c_{h0} = c_0, \sum_{h=1}^{H} c_{h1} = c1 \right\}$$

$$= -\sum_{h=1}^{H} \lambda_h e^{-\alpha_h a_h} \left(e^{-\alpha c_0} + \delta e^{-\alpha c_1} \right). \tag{7.11}$$

The sum

$$\sum_{h=1}^{H} \lambda_h e^{-\alpha_h a_h}$$

is a positive constant (it does not depend on c_0 or c_1), so (7.11) is a monotone affine transform of (7.6) as claimed.

7.3 CONSUMPTION-BASED ASSET PRICING

When there is a representative investor, asset risk premia are determined by a single-factor model with the representative investor's MRS as the factor. This follows from (6.1) and the fact that (7.5) is an SDF. Set

$$R_z = \frac{u_0'(c_0)}{E[u_1'(\tilde{c}_1)]}. \tag{7.12}$$

Substitute (7.5) for $\tilde{m}$ in (6.1) to obtain

$$E[\tilde{R}] - R_z = -\frac{R_z}{u_0'(c_0)} \operatorname{cov}(\tilde{R}, u_1'(\tilde{c}_1)). \tag{7.13}$$

Thus, risk premia are determined by covariances (or betas) with respect to a function u_1' of aggregate consumption $\tilde{c}_1$.

This formula is important and intuitive. Note that the coefficient $-R_z/u_0'(c_0)$ in (7.13) is negative. Thus, the higher the covariance of an asset's return with marginal utility, the lower is the risk premium of the asset. Marginal utility is high when aggregate consumption is low, so assets having high covariances with marginal utility are assets that pay well when aggregate consumption is low. These are very desirable assets to hold for hedging. Thus, it is very reasonable that they should trade at high prices—equivalently, that they should earn low risk premia. Typically, even a high covariance of an asset return with marginal utility is a negative (but small in absolute value) covariance, because most assets pay better when aggregate consumption is high than when it is low. This implies a small but positive risk premium.

We can deduce the CAPM (6.11) from (7.13) by assuming the representative investor has quadratic utility or by assuming that returns are normally distributed. The calculations are the same as in Section 6.1, with the simplification that we do not need to add across investors. The price of risk λ in (6.11) takes a simple form when there is a representative investor (Exercises 6.2 and 7.1).

Any positive decreasing function can be a marginal utility function (just define the utility by integrating the marginal utility), so (7.13) produces a plethora of asset pricing models: For any positive decreasing function f, we can assume that there is a representative investor with $u_1' = f$ and thereby deduce the pricing formula

$$E[\tilde{R}] - R_z = \lambda \frac{\text{cov}(\tilde{R}, f(\tilde{c}_1))}{\text{var}(f(\tilde{c}_1))} \tag{7.14}$$

for some price of risk λ. We can test this for specific f or for f within some parametric class.

CRRA Utility and a Generalization of the CAPM

A very common application of (7.13) is to assume that all investors have CRRA utility with the same risk aversion and same discount factor, so there is a representative investor with time-additive CRRA utility as shown in Section 7.2. The SDF defined by the representative investor's MRS is

$$\tilde{m} = \delta \left(\frac{\tilde{c}_1}{c_0} \right)^{-\rho}, \tag{7.15}$$

where $\rho = 1$ for log utility. Assume there is a risk-free asset. Then, the consumption-based asset pricing formula (7.13) becomes

$$E[\tilde{R}] - R_f = -\delta R_f \, \text{cov}\left(\tilde{R}, (\tilde{c}_1/c_0)^{-\rho}\right). \tag{7.16}$$

Assume $\tilde{c}_1$ is spanned by the assets, so the market return $\tilde{R}_m$ can be defined as $\tilde{c}_1$ divided by the date–0 price of $\tilde{c}_1$. The date–0 price of $\tilde{c}_1$ is

$$E[\tilde{m}\tilde{c}_1] = \delta E[(\tilde{c}_1/c_0)^{-\rho} \tilde{c}_1] = \delta c_0 E[(\tilde{c}_1/c_0)^{1-\rho}].$$

Define $\nu = \delta E[(\tilde{c}_1/c_0)^{1-\rho}]$, so the date–0 price of $\tilde{c}_1$ is νc_0, and the market return is

$$\tilde{R}_m = \frac{\tilde{c}_1}{\nu c_0}. \tag{7.17}$$

Use this formula to substitute for $\tilde{c}_1/c_0$ in (7.16). Then, for each return $\tilde{R}$,

$$E[\tilde{R}] - R_f = \psi \, \text{cov}\left(\tilde{R}, \tilde{R}_m^{-\rho}\right), \tag{7.18}$$

Representative Investors

where
$$\psi = -\delta R_f v^{-\rho}. \tag{7.19}$$

Thus, when there is a representative investor with CRRA utility, risk premia are determined by a factor model with a power of the market return as the factor. By substituting $\tilde{R} = \tilde{R}_m$ in (7.18) and rearranging, we obtain an alternate formula for ψ:

$$\psi = \frac{\mathsf{E}[\tilde{R}_m] - R_f}{\mathrm{cov}(\tilde{R}_m, \tilde{R}_m^{-\rho})}. \tag{7.20}$$

Thus, the factor model can be written as

$$\mathsf{E}[\tilde{R}] - R_f = \frac{\mathsf{E}[\tilde{R}_m] - R_f}{\mathrm{cov}(\tilde{R}_m, \tilde{R}_m^{-\rho})} \, \mathrm{cov}\left(\tilde{R}, \tilde{R}_m^{-\rho}\right). \tag{7.21}$$

This generalizes the result that quadratic utility implies the CAPM: Recall that quadratic utility is shifted CRRA utility with $\rho = -1$, and note that substituting $\rho = -1$ in (7.21) produces the CAPM.

Finally, note that the risk-free return in this model is

$$R_f = \frac{1}{\mathsf{E}[\tilde{m}]} = \frac{1}{\delta \mathsf{E}[(\tilde{c}_1/c_0)^{-\rho}]}. \tag{7.22}$$

Combining this with (7.17) yields

$$\frac{\mathsf{E}[\tilde{R}_m]}{R_f} = \frac{\mathsf{E}[\tilde{c}_1]\mathsf{E}[\tilde{c}_1^{-\rho}]}{\mathsf{E}[\tilde{c}_1^{1-\rho}]}. \tag{7.23}$$

We will relate this result to the equity premium puzzle in the next subsection.

CRRA Utility, Lognormal Consumption, and Puzzles

We can obtain more explicit versions of the formulas (7.22) and (7.23) by assuming lognormal consumption growth, meaning that $\log(\tilde{c}_1/c_0) = \mu + \sigma \tilde{\xi}$ for constants μ and σ and a standard normal $\tilde{\xi}$. Using the formula for the mean of an exponential of a normal, (7.22) becomes

$$R_f = \frac{1}{\delta} e^{\rho\mu - \rho^2\sigma^2/2}. \tag{7.22'}$$

Also, (7.23) becomes

$$\frac{\mathsf{E}[\tilde{R}_m]}{R_f} = e^{\rho\sigma^2}. \tag{7.23'}$$

Finally, Exercise 1.7 implies that

$$\sigma^2 = \log\left(1 + \frac{\mathrm{var}(\tilde{c}_1/c_0)}{\mathsf{E}[\tilde{c}_1/c_0]^2}\right). \tag{7.24}$$

Take the period length to be one year, so $\tilde{c}_1/c_0$ is annual consumption growth. Mehra and Prescott (1985) report the mean and standard deviation of annual consumption growth over the period 1889–1978 as being 1.018 and 0.036, respectively. Substituting these numbers into (7.24) implies $\sigma^2 = 0.00125$. Over the same time period, they report the sample average annual market return and risk-free return of the U.S. stock market as being 1.0698 and 1.008. Substituting these and $\sigma^2 = 0.00125$ into (7.23′) yields $\rho = 47.6$. This is unreasonably high.[2] This is the equity premium puzzle: The variability in consumption growth is too small to justify the large historical equity premium, unless investors are unreasonably risk averse.

The other preference parameter δ can be deduced from the sample statistics and the calculated value of ρ by using (7.22′). From $\mathsf{E}[\tilde{c}_1/c_0] = e^{\mu+\sigma^2/2}$, we obtain

$$\mu = \log \mathsf{E}\left[\frac{\tilde{c}_1}{c_0}\right] - \frac{1}{2}\sigma^2 = 0.017.$$

With this substitution, (7.22′) implies $\delta = 0.55$, which is as unreasonable as $\rho = 47.6$.[3]

It is worthwhile to calculate the implications of reasonable values for the preference parameters ρ and δ. Mehra and Prescott (2003) suggest using $\delta = 0.99$ and $\rho = 10$. Given these preference parameters and the consumption growth statistics reported above, (7.22′) and (7.23′) imply $\mathsf{E}[R_m] = 1.141$ and $R_f = 1.127$. Thus, the risk-free rate should be 12.7% and the equity premium should be $14.1\% - 12.7\% = 1.4\%$. These numbers are far from the historical averages of 0.8% for the risk-free rate and 6.18% for the equity premium. Thus, the historical risk-free rate is too low and the equity premium too large, given the consumption growth statistics and reasonable values for the preference parameters in this model.

While the data reject this model, it still has some features that are intuitive and should survive in more general models. For example, (7.22′) implies that the continuously compounded interest rate is

$$\log R_f = -\log \delta + \rho\mu - \frac{1}{2}\rho^2\sigma^2. \tag{7.25}$$

The interest rate (7.25) is smaller when δ is larger, because a higher weight on future utility means investors are more inclined to save; a lower interest rate

---

2. A person with constant relative risk aversion of 47.6 will refuse a gamble in which she loses 1.5% of her wealth and wins y with equal probabilities, no matter how large y is (Exercise 1.2).

3. Starting from a constant nonrandom consumption path $c_1 = c_0$, $\delta = 0.55$ means that an investor would require a return of $1/\delta \approx 1.8$ units of consumption in one year to induce her to forgo a unit of consumption today. This is an unreasonable level of impatience.

offsets this inclination and enables the market for the risk-free asset to clear. The interest rate is higher when μ is larger, because high expected consumption growth makes investors inclined to borrow against future consumption; a higher interest rate reduces this inclination and clears the market for the risk-free asset. The interest rate is smaller when σ is higher, because extra risk in future consumption makes investors inclined to save more, which must be offset by a lower interest rate. Note that the effect of ρ on the interest rate is ambiguous. This is a result of ρ playing two roles in the consumption/portfolio choice problem: as the coefficient of risk aversion and as the reciprocal of the EIS.

7.4 COSKEWNESS-COKURTOSIS PRICING MODEL

We can approximate the representative investor's marginal utility via a Taylor series expansion and substitute the approximation into the pricing model (7.13). If we stop at the first order term, then we obtain an approximation of the CAPM (which is exact when utility is quadratic, because then there are no higher order terms). This is explored further in Chapter 10 in a dynamic model, where it is reasonable to take the time periods to be short, so changes in consumption are small and the approximation is better.[4] If we include the second order term, we obtain the coskewness pricing model, and if we include the second and third order terms, we obtain the coskewness-cokurtosis pricing model. Here, we explain the coskewness-cokurtosis model.

Take a Taylor series expansion of the representative investor's marginal utility around the mean $\bar{c}_1$ of aggregate consumption $\tilde{c}_1$. Keeping only the first three terms, we have

$$u_1'(\tilde{c}_1) \approx u_1'(\bar{c}_1) + u_1''(\bar{c}_1)(\tilde{c}_1 - \bar{c}_1) + \frac{1}{2}u'''(\bar{c}_1)(\tilde{c}_1 - \bar{c}_1)^2 + \frac{1}{6}u''''(\bar{c}_1)(\tilde{c}_1 - \bar{c}_1)^3. \tag{7.26}$$

Assume that $\tilde{c}_1$ is spanned by the assets, so we can define the market return as $\tilde{R}_m = \tilde{c}_1/\mathsf{E}[\tilde{m}\tilde{c}_1]$. Substitute the market return for $\tilde{c}_1$ in (7.26) and substitute (7.26) into (7.13) to obtain

$$\mathsf{E}[\tilde{R}] - R_z \approx \psi_1 \operatorname{cov}(\tilde{R}, \tilde{R}_m) + \psi_2 \operatorname{cov}(\tilde{R}, (\tilde{R}_m - \bar{R}_m)^2) + \psi_3 \operatorname{cov}(\tilde{R}, (\tilde{R}_m - \bar{R}_m)^3), \tag{7.27}$$

---

4. The model with the first-order term is called the Consumption-based Capital Asset Pricing Model, and it is exact in continuous time (Chapter 14).

where

$$\psi_1 = -\frac{R_z u_1''(\bar{c}_1) \mathrm{E}[\tilde{m}\tilde{c}_1]}{u_0'(c_0)},$$

$$\psi_2 = -\frac{R_z u_1'''(\bar{c}_1) \mathrm{E}[\tilde{m}\tilde{c}_1]^2}{u_0'(c_0)},$$

$$\psi_3 = -\frac{R_z u_1''''(\bar{c}_1) \mathrm{E}[\tilde{m}\tilde{c}_1]^3}{u_0'(c_0)}.$$

The formula (7.27) states that there is an approximate factor model in which the factors are $\tilde{R}_m$, $(\tilde{R}_m - \overline{R}_m)^2$, and $(\tilde{R}_m - \overline{R}_m)^3$. The terms $\mathrm{cov}(\tilde{R}, (\tilde{R}_m - \overline{R}_m)^2)$ and $\mathrm{cov}(\tilde{R}, (\tilde{R}_m - \overline{R}_m)^3)$ are called coskewness and cokurtosis, respectively. The model (7.27) is called the coskewness-cokurtosis pricing model.

If the representative investor has DARA utility and decreasing absolute prudence, then the third derivative of the utility function is positive and the fourth derivative is negative, as noted in Section 1.4. Therefore, $\lambda_1 > 0$, $\lambda_2 < 0$, and $\lambda_3 > 0$. Given these signs, the coskewness-cokurtosis pricing model asserts that assets with high risk premia are assets that have (i) high covariance with the market, (ii) low (meaning negative and large in absolute value) coskewness with the market, and/or (iii) high cokurtosis with the market. This model is a reasonable extension of the CAPM. It is much more reasonable to assume DARA utility and decreasing absolute prudence than it is to assume quadratic utility. However, as noted in Section 1.4, the approximation (7.26) may not be adequate.

7.5 RUBINSTEIN OPTION PRICING MODEL

This section presents another application of representative investor pricing. Assume again that all investors have CRRA utility with the same risk aversion and the same discount factor, so there is a representative investor with time-additive CRRA utility. Assume also that a risk-free asset is traded. Finally, assume lognormal consumption growth: $\log(\tilde{c}_1/c_0) = \mu + \sigma\tilde{\xi}$ for constants μ and σ and a standard normal $\tilde{\xi}$. Rubinstein (1976) shows that the Black-Scholes (1973) formula for the value of an option on the market portfolio is valid in this model. We will derive the formula in this section.

For convenience, set $\tilde{z} = \mu + \sigma\tilde{\xi}$. Under our assumptions,

$$\delta\left(\frac{\tilde{c}_1}{c_0}\right)^{-\rho} = \delta e^{-\rho\tilde{z}}$$

is an SDF, where as always $\rho = 1$ for log utility. The continuously compounded risk-free rate r is given by (7.25). The date–1 payoff of the market portfolio is $\tilde{c}_1$.

Denote the date–0 price of the market portfolio by S. We can compute S as

$$S = \delta E\left[\left(\frac{\tilde{c}_1}{c_0}\right)^{-\rho}\tilde{c}_1\right]$$

$$= \delta c_0 E\left[e^{(1-\rho)\tilde{z}}\right]$$

$$= \delta c_0 e^{(1-\rho)\mu+(1-\rho)^2\sigma^2/2}. \tag{7.28}$$

Consider a call option with exercise price K written on the payoff $\tilde{c}_1$ of the market portfolio.[5] The option pays

$$\max(\tilde{c}_1 - K, 0) \tag{7.29}$$

at date 1. Assume the option is introduced in zero net supply. Because of the Pareto optimality of the equilibrium, the option will not actually be traded.[6] However, we are interested in computing its equilibrium price (the price at which no one wants to buy or sell it).

We can write the payoff (7.29) of the option as

$$c_0 \max\left(\frac{\tilde{c}_1}{c_0} - \hat{K}, 0\right) = c_0 \max\left(e^{\tilde{z}} - \hat{K}, 0\right), \tag{7.30}$$

where $\hat{K} = K/c_0$. The price of the payoff (7.30) is

$$\delta c_0 E\left[e^{-\rho\tilde{z}}\max\left(e^{\tilde{z}} - \hat{K}, 0\right)\right]. \tag{7.31}$$

Calculation of (7.31) is straightforward and presented below. The result is that the price of the option is given by the Black-Scholes (1973) formula:

$$SN(d_1) - e^{-r}KN(d_2), \tag{7.32a}$$

where

$$d_1 = \frac{\log S - \log K + r + \sigma^2/2}{\sigma}, \tag{7.32b}$$

$$d_2 = d_1 - \sigma, \tag{7.32c}$$

and where N denotes the cumulative distribution function of a standard normal random variable.

---

5. Call options are defined in Chapter 16.

6. Trading the option is inconsistent with affine sharing rules: If any investor is long or short the option, then the investor's date–1 consumption will not be an affine function of aggregate consumption.

Let f denote the density function of the normally distributed variable $\tilde{z}$. Note that $\max\left(e^{\tilde{z}} - \hat{K}, 0\right) \neq 0$ if and only if $\tilde{z} > \log \hat{K}$. We can therefore write (7.31) as

$$\delta c_0 \int_{\log \hat{K}}^{\infty} e^{-\rho z} \left(e^z - \hat{K}\right) f(z)\, dz,$$

or alternatively as

$$\delta c_0 \int_{\log \hat{K}}^{\infty} e^{(1-\rho)z} f(z)\, dz - \delta K \int_{\log \hat{K}}^{\infty} e^{-\rho z} f(z)\, dz, \qquad (7.33)$$

recalling that $c_0 \hat{K} = K$ for the second term. We will show below, by standard calculus, that, for any a and b,

$$\int_a^{\infty} e^{bz} f(z)\, dz = \exp\left(b\mu + \frac{1}{2} b^2 \sigma^2\right) N\left(\frac{\mu + b\sigma^2 - a}{\sigma}\right). \qquad (7.34)$$

Substituting this in (7.33), the price of the option is

$$\delta c_0 \exp\left((1-\rho)\mu + \frac{1}{2}(1-\rho)^2 \sigma^2\right) N\left(\frac{\mu + (1-\rho)\sigma^2 - \log \hat{K}}{\sigma}\right)$$

$$- \delta K \exp\left(-\rho\mu + \frac{1}{2}\rho^2 \sigma^2\right) N\left(\frac{\mu - \rho\sigma^2 - \log \hat{K}}{\sigma}\right). \qquad (7.35)$$

Substitute the formula (7.25) for r, the formula (7.28) for S, and the formula $K = c_0 \hat{K}$ in (7.35) to obtain (7.32).

It remains to establish (7.34). The integral is

$$\frac{1}{\sqrt{2\pi \sigma^2}} \int_a^{\infty} \exp\left(bz - \frac{(z-\mu)^2}{2\sigma^2}\right) dz$$

$$= \exp\left(b\mu + \frac{1}{2} b^2 \sigma^2\right) \frac{1}{\sqrt{2\pi \sigma^2}} \int_a^{\infty} \exp\left(-\frac{(z - \mu - b\sigma^2)^2}{2\sigma^2}\right) dz$$

$$= \exp\left(b\mu + \frac{1}{2} b^2 \sigma^2\right) \operatorname{prob}(\tilde{w} \geq a),$$

where $\tilde{w}$ is a normally distributed variable with mean $\mu + b\sigma^2$ and variance σ^2. The probability in the last line equals

$$\operatorname{prob}\left(\frac{\tilde{w} - \mu - b\sigma^2}{\sigma} \geq \frac{a - \mu - b\sigma^2}{\sigma}\right) = 1 - N\left(\frac{a - \mu - b\sigma^2}{\sigma}\right)$$

$$= N\left(\frac{\mu + b\sigma^2 - a}{\sigma}\right).$$

This confirms (7.34).

7.6 NOTES AND REFERENCES

The results on a representative investor with LRT utility in Section 7.2 are due to Rubinstein (1974). The factor pricing model (7.21) is tested (and rejected) by Hansen and Singleton (1982, 1983, 1984). The coskewness-cokurtosis pricing model is due to Rubinstein (1973), Kraus and Litzenberger (1976), and Dittmar (2002).

Mehra and Prescott (1985) define the equity premium and risk-free rate puzzles. Those issues are discussed further in Chapters 10 and 11. The formulas (7.22′) and (7.23′) for the risk-free return and log equity premium based on lognormal consumption growth can be generalized. We need to evaluate $\mathsf{E}[\tilde{c}_1^\theta]$ for $\theta = -\rho$ and $\theta = 1 - \rho$. We can write these as

$$\mathsf{E}\left[e^{\theta \log \tilde{c}_1}\right].$$

This is the moment-generating function of $\log \tilde{c}_1$ evaluated at θ. Writing the moment-generating function as a power series in θ and omitting high-order terms, Martin (2013a) shows that the log risk-free return and log equity premium are approximately linear functions of the mean, variance, skewness, and kurtosis of log consumption growth.

The approach to option pricing presented in Section 7.5 is due to Rubinstein (1976). Rubinstein also shows that the Black-Scholes formula applies to options on other assets, provided the continuously compounded rate of return of the asset is joint normally distributed with the consumption growth rate. The extension to other assets requires only the evaluation of a bivariate integral rather than the univariate integral computed at the end of Section 7.5. The details are presented in the appendix of Rubinstein (1976).

Exercise 7.3 presents a lower bound on the market risk premium due to Martin (2015). Martin shows that the risk-neutral variance of the market return that appears in the bound can be calculated from equity index option prices. Exercise 7.9 presents a bound on the APT pricing errors, assuming a representative investor with CARA utility. This is due to Dybvig (1983). Actually, Dybvig does not assume CARA utility but instead assumes only that $u''' \geq 0$ and that the absolute risk aversion of the representative investor is bounded above. The other key assumption is that the residuals in the statistical factor model are bounded below, which would follow from limited liability.

EXERCISES

7.1. Assume there is a representative investor with quadratic utility $u(w) = -(\zeta - w)^2$. Assume $\mathsf{E}[\tilde{w}_m] \neq \zeta$. Show that λ in the CAPM (6.11) equals

$$\frac{\mathrm{var}(\tilde{w}_m)}{\mathsf{E}[\tau(\tilde{w}_m)]},$$

where $\tau(w)$ denotes the coefficient of risk tolerance of the representative investor at wealth level w. (Thus, the risk premium is higher when market wealth is riskier or when the representative investor is more risk averse.)

7.2. Assume there is a risk-free asset and a representative investor with power utility, so (7.15) is an SDF. Let $\tilde{z} = \log(\tilde{c}_1/c_0)$ and assume $\tilde{z}$ is normally distributed with mean μ and variance σ^2. Let κ denote the maximum Sharpe ratio of all portfolios.

(a) Use the Hansen-Jagannathan bound (3.35) to show that

$$\rho \geq \frac{\sqrt{\log(1+\kappa^2)}}{\sigma} \quad \Leftrightarrow \quad \kappa \leq \sqrt{e^{\rho^2\sigma^2} - 1}. \tag{7.36}$$

Hint: Apply the result of Exercise 1.7. Note that (7.36) implies risk aversion must be larger if consumption volatility is smaller or the maximum Sharpe ratio is larger. Also, using the approximation $\log(1+x) \approx x$, the lower bound on ρ in (7.36) is approximately κ/σ, and, using the approximation $e^x \approx 1+x$, the upper bound on κ is approximately $\rho\sigma$.

(b) Explain why the weak inequalities in (7.36) must be equalities when the market is complete.

(c) In the data analyzed by Mehra and Prescott (1985), the standard deviation of the market return is 16.54%. Derive a lower bound on the representative investor's risk aversion ρ by using this and the other Mehra-Prescott data in Section 7.3 in conjunction with (7.36). Note: You will find this bound is much more reasonable than the estimate of 47.6 presented in Section 7.3. This is because of two offsetting factors: Both the risk premium of the market and the volatility of the market are higher in the data than the model would predict, given reasonable values of δ and ρ.

(d) Use $\delta = 0.99$ and $\rho = 10$ and the Mehra-Prescott data on the mean and standard deviation of consumption growth to compute the standard deviation of the theoretical market return (7.17). Note: This is a very simple illustration of the excess volatility puzzle.

7.3. Assume there is a risk-free asset, and let $\tilde{m}$ be an SDF.

(a) Show that each return $\tilde{R}$ satisfies

$$\mathsf{E}[\tilde{R}] - R_f = \frac{\text{var}^*(\tilde{R})}{R_f} - \text{cov}(\tilde{m}\tilde{R}, \tilde{R}),$$

where var* denotes variance under the risk-neutral probability corresponding to $\tilde{m}$.

(b) Assume there is a representative investor with constant relative risk aversion ρ, so $\gamma u'(\tilde{R}_m) \stackrel{\text{def}}{=} \gamma \tilde{R}_m^{-\rho}$ is an SDF for some constant γ. Show that $u'(x)x$ is a decreasing function of x if and only if $\rho > 1$.

(c) If $f(x)$ is a decreasing function of x, then $\operatorname{cov}(f(\tilde{x}), \tilde{x}) < 0$ for any random variable $\tilde{x}$. Using this fact and the above results, explain why

$$\mathsf{E}[\tilde{R}_m] - R_f \geq \frac{\operatorname{var}^*(\tilde{R}_m)}{R_f}$$

when $\rho > 1$.

7.4. Assume there is a representative investor with constant relative risk aversion ρ. Assume there is a risk-free asset and the market is complete. Use the fact that $\tilde{R}_p$ and R_f span the mean-variance frontier to show that each mean-variance efficient return is of the form $a - b\tilde{R}_m^{-\rho}$ for $b > 0$.

7.5. Assume there is a representative investor with utility function u. The first-order condition

$$\mathsf{E}[u'(\tilde{R}_m)(\tilde{R}_1 - \tilde{R}_2)] = 0$$

must hold for all returns $\tilde{R}_1$ and $\tilde{R}_2$. Assume there is a risk-free asset. Consider any return $\tilde{R}$. By orthogonal projection, we have

$$\tilde{R} - R_f = \alpha + \beta(\tilde{R}_m - R_f) + \tilde{\varepsilon}$$

for some α and β, where $\mathsf{E}[\tilde{\varepsilon}] = \mathsf{E}[\tilde{R}_m \tilde{\varepsilon}] = 0$.

(a) Use the first-order condition in conjunction with the returns $\tilde{R}_m$ and

$$\tilde{R}_* \stackrel{\text{def}}{=} \tilde{R} + (1-\beta)(\tilde{R}_m - R_f) = \tilde{R}_m + \alpha + \tilde{\varepsilon}$$

to show that

$$\alpha = -\frac{\mathsf{E}[u'(\tilde{R}_m)\tilde{\varepsilon}]}{\mathsf{E}[u'(\tilde{R}_m)]}.$$

(b) Use the result of the previous part to derive the CAPM when there is a representative investor and the residual $\tilde{\varepsilon}$ of each asset return is mean independent of the market return (which is true, for example, when returns are joint elliptically distributed; Chu, 1973).

7.6. Assume in (7.16) that $\log \tilde{R}$ and $\log(\tilde{c}_1/c_0)$ are joint normally distributed. Specifically, let $\log \tilde{R} = \tilde{y}$ and $\log(\tilde{c}_1/c_0) = \tilde{z}$ with $\mathsf{E}[\tilde{y}] = \mu_y$, $\operatorname{var}(\tilde{y}) = \sigma_y^2$, $\mathsf{E}[\tilde{z}] = \mu$, $\operatorname{var}(\tilde{z}) = \sigma^2$, and $\operatorname{corr}(\tilde{y}, \tilde{z}) = \gamma$.

(a) Show that
$$\mu = -\log\delta + \rho\gamma\sigma\sigma_y + \rho\mu - \frac{1}{2}\rho^2\sigma^2 - \frac{1}{2}\sigma_y^2.$$

(b) Let $r = \log R_f$ denote the continuously compounded risk-free rate. Using (7.22′), show that
$$\mu = r + \rho\gamma\sigma\sigma_y - \frac{1}{2}\sigma_y^2. \tag{7.37}$$

Note: $\gamma\sigma\sigma_y$ is the covariance of the continuously compounded rate of return $\tilde{y}$ with the continuously compounded consumption growth rate $\tilde{z}$, so (7.37) has the usual form

$$\text{Expected Return} = \text{Risk-Free Return} + \psi \times \text{Covariance},$$

with $\psi = \rho$, except for the extra term $-\sigma_y^2/2$. The extra term, which involves the total and hence idiosyncratic risk of the return, is usually called a Jensen's inequality term, because it arises from the fact that $\mathsf{E}[e^{\tilde{y}}] = e^{\mu_y + \sigma_y^2/2} > e^{\mu_y}$.

7.7. Show that if u_{h0} and u_{h1} are concave for each h, then the social planner's utility functions u_0 and u_1 are concave.

7.8. Use the results on affine sharing rules in Section 4.4 to establish (7.8) and (7.9) in Section 7.2.

7.9. Suppose there is a risk-free asset in zero net supply and the risky asset returns have a statistical factor structure
$$\tilde{R}_i = a_i + b_i'\tilde{F} + \tilde{\varepsilon}_i,$$
where the $\tilde{\varepsilon}_i$ have zero means and are independent of each other and of $\tilde{F}$. Assume there is no labor income and there is a representative investor with CARA utility. Let α denote the risk aversion of the representative investor. Let π denote the vector of market weights. Denote initial market wealth by w_0 and end-of-period market wealth by $\tilde{w}_m = w_0\tilde{R}_m$. Let δ_i denote the APT pricing error defined in Section 6.5. Assume $\tilde{\varepsilon}_i \geq -\gamma$ with probability 1, for some constant γ. Via the following steps, show that
$$|\delta_i| \leq \frac{\alpha w_0 \pi_i \exp(\alpha\gamma w_0 \pi_i) \operatorname{var}(\tilde{\varepsilon}_i)}{R_f}.$$

(a) Show that
$$\delta_i = \frac{\mathsf{E}[\exp(-\alpha\tilde{w}_m)\tilde{\varepsilon}_i]}{R_f \mathsf{E}[\exp(-\alpha\tilde{w}_m)]}.$$

(b) Show that
$$\delta_i = \frac{\mathsf{E}[\exp(-\alpha w_0 \pi_i \tilde{\varepsilon}_i)\tilde{\varepsilon}_i]}{R_f \mathsf{E}[\exp(-\alpha w_0 \pi_i \tilde{\varepsilon}_i)]}.$$

Hint: Use independence and the fact that end-of-period market wealth is
$$\tilde{w}_m = w_0 \sum_{j \neq i} \pi_j \tilde{R}_j + w_0 \pi_i \tilde{R}_i$$
$$= w_0 \sum_{j \neq i} \pi_j \tilde{R}_j + w_0 \pi_i a_i + w_0 \pi_i b_i' \tilde{F} + w_0 \pi_i \tilde{\varepsilon}_i.$$

(c) Show that
$$\mathsf{E}[\exp(-\alpha w_0 \pi_i \tilde{\varepsilon}_i)] \geq 1.$$

Hint: Use Jensen's inequality.

(d) Show that
$$\left| \mathsf{E}[\exp(-\alpha w_0 \pi_i \tilde{\varepsilon}_i)\tilde{\varepsilon}_i] \right| \leq \alpha w_0 \pi_i \exp(\alpha \gamma w_0 \pi_i) \operatorname{var}(\tilde{\varepsilon}_i).$$

Hint: Use an exact first-order Taylor series expansion of the exponential function.

PART TWO

Dynamic Models

Dynamic Securities Markets

CONTENTS

8.1	Portfolio Choice Model	184
8.2	Stochastic Discount Factor Processes	187
8.3	Arbitrage and the Law of One Price	192
8.4	Complete Markets	192
8.5	Bubbles, Transversality Conditions, and Ponzi Schemes	195
8.6	Inflation and Foreign Exchange	198
8.7	Notes and References	198

This chapter introduces the dynamic model of securities markets that we study in this part of the book. It extends various concepts from the first part of the book to dynamic markets. We consider investors who choose consumption and portfolios at discrete dates that we denote as $t = 0, 1, 2, \ldots$. The unit in which time is measured is not important for our purposes, though it is convenient to assume that the dates are equally spaced. The economy is assumed to exist for an indefinite (that is, for an infinite) amount of time, though investors may have finite horizons.

A sequence of random variables is called a stochastic process, or, more briefly, a process. Stochastic processes will be denoted by capital Roman letters or lower-case Greek letters, using a time subscript to denote a variable at a particular time and the same letter without a time subscript to denote the entire sequence of variables. Tildes will no longer be used to denote random variables, because whether something is random depends on the vantage point in time.

Suppose there are n securities traded at each date. The cash flow (dividend) paid by security i at date t is denoted by D_{it}. Any dividends paid at date 0 have already been paid at the time the analysis begins. The price of security i at date t

is denoted by P_{it}. This is the ex-dividend price, that is, the price at which it trades after payment of its dividend at date t.

We can allow the set of securities to change over time. For example, when bonds mature, they are redeemed and disappear from the investment set. This extension is straightforward and is omitted only to simplify the notation. Despite the fact that the set of securities is unchanging in our formal model, we will sometimes make reference to a zero-coupon bond, which is a security that pays 1 unit of the consumption good at a particular date and 0 at all other dates. Such a bond is also called a pure discount bond or simply a discount bond.

Investors know the history of prices and dividends through date t, and perhaps the realizations of other random variables, before making consumption and portfolio decisions at date t. Their consumption and portfolio decisions can only depend on information available at the time they are made. We express this by saying that the consumption and portfolio processes must be adapted to the available information (Appendix A.11). We denote expectation conditional on date t information by E_t.

Assume each asset price is positive in each state of the world. The return of asset i from date t to date $t+1$ is

$$R_{i,t+1} = \frac{P_{i,t+1} + D_{i,t+1}}{P_{i,t}}.$$

Stack the returns as an n–dimensional column vector R_{t+1}.

One of the n assets may be risk free in the sense that $R_{i,t+1}$ is known when consumption and portfolio decisions are made at date t. In this case, the return will be written as $R_{f,t+1}$. However, the risk-free return may vary (randomly) over time. In other words, the risk-free return from date t to $t+1$ may not be known until date t, and, in particular, it need not be the same as the risk-free return from date $t-1$ to date t. If such an asset exists, then it is often called a money market account. The advantage of this terminology, relative to simply calling the asset risk free, is that it is less likely to lead to confusion between an asset being risk free for a single period or over a multiperiod horizon—the former is a money market account and the latter is a zero-coupon bond.

8.1 PORTFOLIO CHOICE MODEL

The investor has some initial wealth W_0. At any date t, the investor may be endowed with other income denoted by Y_t. As before, we call all such income labor income. Denote consumption at date t by C_t. Denote the value of the investor's portfolio at date t including Y_t but before consuming by W_t. We will call W_t the investor's wealth. However, it would be more precise to call W_t

financial wealth, because it does not include the value of future labor income $Y_{t+1}, Y_{t+2}, \ldots$.

Intertemporal Budget Constraint

After consuming, the investor invests $W_t - C_t$ in a portfolio π_t. The portfolio π_t is an n–dimensional column vector $\pi_t = (\pi_{t1} \cdots \pi_{tn})'$ specifying the fraction of the portfolio value that is invested in each asset i. It must satisfy $\iota'\pi_t = 1$. The return on the investor's portfolio between dates t and $t+1$ is $\pi_t' R_{t+1}$. Based on consumption and portfolio decisions and the returns of the assets, the investor's wealth evolves as

$$W_{t+1} = Y_{t+1} + (W_t - C_t)\pi_t' R_{t+1}. \tag{8.1}$$

Equation (8.1) is called the intertemporal budget constraint.

Time-Additive Utility

Suppose the investor has time-additive utility and a constant discount factor $0 < \delta < 1$. Let u denote the utility function in each period. The investor maximizes

$$\mathsf{E}\left[\sum_{t=0}^{\infty} \delta^t u(C_t)\right] \tag{8.2a}$$

if her horizon is infinite or

$$\mathsf{E}\left[\sum_{t=0}^{T} \delta^t u(C_t)\right] \tag{8.2b}$$

if she has a finite horizon T.[1] Occasionally, we will also analyze a finite-horizon model with an investor who seeks only to maximize the expected utility of terminal wealth $\mathsf{E}[u(W_T)]$.

---

1. An easy extension of the model is to replace (8.2b) with

$$\mathsf{E}\left[\sum_{t=0}^{T} \delta^t u(C_t) + U(W_T - C_T)\right],$$

where U is interpreted as the utility of providing a bequest.

Ponzi Schemes

In the infinite-horizon case, some constraint is needed to ensure that the investor does not continually borrow to finance current consumption and continually roll over the debt without ever repaying it. To do this is to "run a Ponzi scheme." Ponzi schemes are impossible if an investor can never consume more than her financial wealth in any period, that is, if $C_t \leq W_t$ for each t and in each state of the world. This is a "no borrowing" constraint, because it means the investor can never borrow against future labor income Y to finance current consumption. In reality, the ability to borrow against future income is indeed limited, and a no-borrowing constraint may be a reasonable approximation. Weaker constraints that also preclude Ponzi schemes are that $\lim_{t \to \infty} W_t \geq 0$ or condition (8.24) in Section 8.5.

Random Horizons

No investor literally has an infinite horizon, but neither does an investor know exactly the date T at which her horizon ends. The infinite-horizon model can accommodate a finite but random horizon. Suppose the investor dies after consuming at a random date τ. Suppose there exists $0 < \gamma < 1$ such that

$$\text{prob}(\tau \geq t) = \gamma^t$$

for each $t \geq 0$. This probability distribution has the property that the probability of dying at t conditional on having lived to t is

$$\frac{\gamma^t - \gamma^{t+1}}{\gamma^t} = 1 - \gamma.$$

Suppose the date of death is independent of the asset returns and labor income process, and the investor seeks to maximize

$$E\left[\sum_{t=0}^{\tau} \alpha^t u(C_t)\right] \tag{8.2c}$$

for a discount factor $0 < \alpha < 1$. This is equivalent to maximizing the infinite-horizon objective (8.2a) with discount factor $\delta = \alpha \gamma$. In this situation, there are two reasons the investor discounts the future, the first being the usual reason that consumption tomorrow is worth less than consumption today, and the second being that the investor may not survive until tomorrow. Each of these produces impatience, and the multiplication of the two discount factors α and γ results in a lower discount factor (greater impatience) than that associated with either reason alone.

To see that (8.2c) is equivalent to (8.2a) with $\delta = \alpha\gamma$, note that the expectation of

$$\sum_{t=0}^{\tau} \alpha^t u(C_t)$$

conditional on the return, income, portfolio, and consumption processes is just the expectation over τ, which is

$$\sum_{t=0}^{\infty} \text{prob}(\tau \geq t)\alpha^t u(C_t) = \sum_{t=0}^{\infty} \gamma^t \alpha^t u(C_t).$$

By iterated expectations, the unconditional expectation is therefore (8.2a) with $\delta = \alpha\gamma$.

8.2 STOCHASTIC DISCOUNT FACTOR PROCESSES

A sequence of random variables $M_1, M_2, \ldots$ is called an SDF process if M_t depends only on date–t information for each t and if, for each i and $t \geq 0$,

$$M_t P_{it} = \mathsf{E}_t \left[M_{t+1}(P_{i,t+1} + D_{i,t+1}) \right], \qquad (8.3)$$

where we take $M_0 = 1$. An equivalent statement of (8.3) is that

$$M_t = \mathsf{E}_t \left[M_{t+1} R_{i,t+1} \right]. \qquad (8.4)$$

By iterating on (8.3), we can see that, for any finite $T > t$,

$$M_t P_{it} = \sum_{s=t+1}^{T} \mathsf{E}_t \left[M_s D_{is} \right] + \mathsf{E}_t \left[M_T P_{iT} \right]. \qquad (8.5)$$

The simplest example of an SDF process is in a world without uncertainty in which there is an asset that has a constant return $R_f = 1 + r_f$. In that case, equation (8.4) becomes

$$\frac{M_{t+1}}{M_t} = \frac{1}{1+r_f}.$$

Starting with $M_0 = 1$, the solution of this is

$$M_t = \frac{1}{(1+r_f)^t}.$$

Single-Period SDF

For any date $t \geq 0$, call a random variable Z_{t+1} a single-period SDF for the period starting at t and ending at $t+1$ if

$$(\forall i) \qquad P_{it} = \mathsf{E}_t\left[Z_{t+1}(P_{i,t+1} + D_{i,t+1})\right]. \qquad (8.6)$$

If M is an SDF process and $M_t \neq 0$, then $Z_{t+1} = M_{t+1}/M_t$ is a single-period SDF. Conversely, given a sequence $Z_1, Z_2, \ldots$ of single-period SDFs, we can define an SDF process by compounding the single-period SDFs as[2]

$$M_t = \prod_{s=1}^{t} Z_s. \qquad (8.7)$$

In the world without uncertainty discussed above, the single-period SDF is $Z_t = (1+r_f)^{-1}$ each period, and $M_t = \prod_{s=1}^{t}(1+r_f)^{-1}$.

If Z_{t+1} is a single-period SDF and $\mathsf{E}_t[Z_{t+1}] \neq 0$, then the same algebra as in a single-period model shows

$$(\forall i) \qquad \mathsf{E}_t[R_{i,t+1}] = \frac{1}{\mathsf{E}_t[Z_{t+1}]} - \frac{1}{\mathsf{E}_t[Z_{t+1}]} \mathrm{cov}_t(R_{i,t+1}, Z_{t+1}). \qquad (8.8)$$

We call this a conditional factor model, because the expectation and covariance are conditional on date–t information. Also, the zero-beta return $1/\mathsf{E}_t[Z_{t+1}]$ depends on date–t information. Conditional factor models are described in Chapter 10.

Self-Financing Wealth Processes

Given a portfolio process π, the wealth process W defined by

$$W_{t+1} = (\pi_t' R_{t+1}) W_t, \qquad (8.9)$$

starting from any $W_0 \geq 0$, is called a self-financing wealth process. Notice that (8.9) is the intertemporal budget constraint (8.1) with $Y_{t+1} = C_t = 0$.

A special case of a self-financing wealth process is a dividend-reinvested asset price, defined as follows: Take π_t in (8.9) to be the ith basis vector (the n–dimensional column vector with a 1 in the ith place and 0 elsewhere), and take $W_0 = P_{i0}$. Denote the corresponding self-financing wealth process by

---

2. To see that (8.7) defines an SDF process, simply multiply both sides of (8.6) by M_t and note that $M_t Z_{t+1} = M_{t+1}$ when (8.7) holds.

S_i. Then, (8.9) states that
$$\frac{S_{i,t+1}}{S_{it}} = R_{i,t+1}.$$

Thus, the ratio of the dividend-reinvested prices is the total return (including dividends) on the asset. A more direct definition of a dividend-reinvested price, which justifies its name, is given below.

At any date τ, the dividend on one share of asset i can be used to purchase $D_{i\tau}/P_{i\tau}$ additional shares of asset i. If you do this at each date $\tau \geq 1$ starting with one share at date 0, then, at date t, you will own

$$\prod_{\tau=1}^{t}\left(1 + \frac{D_{i\tau}}{P_{i\tau}}\right) \tag{8.10}$$

shares, which are worth

$$S_{it} \stackrel{\text{def}}{=} P_{it} \prod_{\tau=1}^{t}\left(1 + \frac{D_{i\tau}}{P_{i\tau}}\right). \tag{8.11}$$

Thus, S_{it} is the value of the shares owned at date t of asset i if you start with a single share and reinvest all dividends. Note that

$$S_{i,t+1} \stackrel{\text{def}}{=} P_{i,t+1} \prod_{\tau=1}^{t+1}\left(1 + \frac{D_{i\tau}}{P_{i\tau}}\right)$$

$$= (P_{i,t+1} + D_{i,t+1}) \prod_{\tau=1}^{t}\left(1 + \frac{D_{i\tau}}{P_{i\tau}}\right)$$

$$= R_{i,t+1} S_{it}, \tag{8.12}$$

confirming that S_i is the dividend-reinvested asset price defined above.

The Martingale Property

A stochastic process X is said to be a martingale if

$$(\forall t) \quad X_t = \mathsf{E}_t[X_{t+1}].$$

Martingales are important in finance and also intuitive. If you play a fair game, for example, flipping a fair coin with a gain of $1 on HEADS and a loss of $1 on TAILS, then your wealth is a martingale: The expected gain is zero, so expected future wealth equals current wealth. Financial markets are generally not fair games (for example, the stock market goes up on average). However, *properly normalized* wealth processes are martingales. "Proper normalization" means multiplication by an SDF process. A caveat is that this statement applies only to wealth processes with no interim withdrawals or additions of wealth. This caveat is natural: If you play the fair coin-tossing game but spend $1 each time you win, then obviously your wealth will not be a martingale.

A wealth process with no interim withdrawals or additions of wealth is what we have termed a self-financing wealth process. For any self-financing wealth process W and any SDF process M,

$$M_t W_t = \mathsf{E}_t[M_{t+1} W_{t+1}] \tag{8.13}$$

for each $t \geq 0$. This is the sense in which properly normalized wealth processes are martingales. Equation (8.13) is established below. In fact, it is shown that a stochastic process M with $M_0 = 1$ is an SDF process if and only if MW is a martingale for each self-financing wealth process W.

To derive (8.13), consider a wealth process W satisfying the self-financing condition (8.9) in conjunction with a portfolio process π. Stack the equations (8.4) to obtain

$$M_t \iota = \mathsf{E}_t[M_{t+1} R_{t+1}].$$

Multiply both sides by π_t', using $\pi_t' \iota = 1$ on the left-hand side, to obtain

$$M_t = \mathsf{E}_t[M_{t+1}(\pi_t' R_{t+1})],$$

and then multiply both sides by W_t. Using (8.9) on the right-hand side, we see that (8.13) holds. Conversely, if (8.13) holds for every self-financing wealth process, then it holds for all dividend-reinvested asset prices, which implies the definition (8.4) of an SDF process.

Valuation over a Finite Horizon

If W is a self-financing wealth process and M is an SDF process, then

$$W_0 = \mathsf{E}[M_T W_T] \tag{8.14}$$

for any T. This can be seen by iterating on (8.13), using the law of iterated expectations, and recalling that we are taking $M_0 = 1$.[3] Equation (8.14) states

---

3. Namely, we have

$W_0 = \mathsf{E}[M_1 W_1] = \mathsf{E}[\mathsf{E}_1[M_2 W_2]] = \mathsf{E}[M_2 W_2] = \mathsf{E}[\mathsf{E}_2[M_3 W_3]] = \mathsf{E}[M_3 W_3] = \cdots = \mathsf{E}[M_T W_T].$

that the date–0 cost W_0 of achieving wealth W_T at date T can be computed by using M_T as an SDF for the period of time starting at date 0 and ending at date T. This obviously generalizes the concept of an SDF in a single-period model.

More generally, for any self-financing wealth process W and SDF process M and any dates $t < T$,

$$M_t W_t = \mathsf{E}_t[M_T W_T]. \tag{8.15}$$

This also follows directly by iterating on the martingale property (8.13).[4] If $M_t \neq 0$, then (8.15) implies

$$W_t = \mathsf{E}_t\left[\frac{M_T}{M_t} W_T\right], \tag{8.16}$$

which shows that the date t cost W_t of achieving wealth W_T at date T can be computed by using M_T/M_t as an SDF for the period of time beginning at date t and ending at date T.

For non-self-financing wealth processes, a variation of (8.15) is true. Suppose C, Y, and W satisfy the intertemporal budget constraint in conjunction with some portfolio process π. Let M be a strictly positive SDF process. Then, for each $T > t$,

$$W_t + \sum_{s=t+1}^{T} \mathsf{E}_t\left[\frac{M_s}{M_t} Y_s\right] = \sum_{s=t}^{T-1} \mathsf{E}_t\left[\frac{M_s}{M_t} C_s\right] + \mathsf{E}_t\left[\frac{M_T}{M_t} W_T\right]. \tag{8.17}$$

The left-hand side is wealth at t, including both financial wealth W_t and the date–t value of labor income $Y_{t+1}, \ldots, Y_T$. The right-hand side is the date–t cost of consumption $C_t, \ldots, C_{T-1}$ and terminal wealth W_T.

The proof of (8.17) is by induction. First, note that (8.17) is trivially true at $t = T$. Suppose it is true for $t = \tau+1, \tau+2, \ldots, T$ for some τ. The intertemporal budget constraint implies

$$\mathsf{E}_\tau[M_{\tau+1} W_{\tau+1}] = \mathsf{E}_\tau[M_{\tau+1} Y_{\tau+1}] + (W_\tau - C_\tau)\pi_\tau' \mathsf{E}_\tau[M_{\tau+1} R_{\tau+1}]$$
$$= \mathsf{E}_\tau[M_{\tau+1} Y_{\tau+1}] + M_\tau(W_\tau - C_\tau), \tag{8.18}$$

using the facts $\mathsf{E}_\tau[M_{\tau+1} R_{\tau+1}] = M_\tau \iota$ and $\pi_\tau' \iota = 1$ for the second equality. Equation (8.17) at $t = \tau+1$ states

$$M_{\tau+1} W_{\tau+1} = -\sum_{s=\tau+2}^{T} \mathsf{E}_{\tau+1}[M_s Y_s] + \sum_{s=\tau+1}^{T-1} \mathsf{E}_{\tau+1}[M_s C_s] + \mathsf{E}_{\tau+1}[M_T W_T].$$

---

4. It is a general property (or even the definition) of a martingale X that $X_t = \mathsf{E}_t[X_T]$ for any $t < T$.

Substitute this into (8.18) and use iterated expectations to obtain

$$-\sum_{s=\tau+2}^{T} \mathsf{E}_\tau[M_s Y_s] + \sum_{s=\tau+1}^{T-1} \mathsf{E}_\tau[M_s C_s] + \mathsf{E}_\tau[M_T W_T]$$
$$= \mathsf{E}_\tau[M_{\tau+1} Y_{\tau+1}] + M_\tau(W_\tau - C_\tau).$$

Divide by M_τ and rearrange to see that (8.17) holds for $t = \tau$. Therefore, it holds for all t.

8.3 ARBITRAGE AND THE LAW OF ONE PRICE

An arbitrage opportunity for a finite horizon T is a self-financing wealth process such that either

(i) $W_0 < 0$ and $W_T \geq 0$ with probability 1, or
(ii) $W_0 = 0$, $W_T \geq 0$ with probability 1, and $W_T > 0$ with positive probability.

If there are no arbitrage opportunities for the finite horizon T, then there is a strictly positive SDF process $M_1, \ldots, M_T$ (see the end-of-chapter notes). If there are no arbitrage opportunities for each finite horizon T, then there is an infinite-horizon strictly positive SDF process $M_1, M_2, \ldots$. Of course, strict positivity implies that the single-period SDFs M_{t+1}/M_t exist and are strictly positive.

A dynamic securities market is said to satisfy the law of one price if $W_0 = W_0^*$ whenever W and W^* are self-financing wealth processes that are equal at any date $t \geq 0$. If there are no arbitrage opportunities for each finite horizon T, then the law of one price holds. Also, if there is an SDF process (whether strictly positive or not), then the law of one price holds.

8.4 COMPLETE MARKETS

A random variable x is a marketed date–t payoff if there is a self-financing wealth process W such that $W_t = x$. A consumption process $C = (C_0, \ldots, C_T)$ is marketed if C_t is a marketed payoff for each $t = 0, \ldots, T$. The market is complete if, for each t, each random variable x that depends only on date–t information is a marketed date–t payoff. If the market is complete and satisfies the law of one price, then there can be at most one SDF process, because the condition $\mathsf{E}[M_t x] = \mathsf{E}[M_t^* x]$ for every x that depends on date–t information and the requirement that

M_t and M_t^* depend only on date–t information imply $M_t = M_t^*$.[5] If the market is complete and there are no arbitrage opportunities, then there is a unique SDF process, and it is strictly positive.

As in a single-period model, a discrete-time securities market (with a finite number of assets) can be complete only if the number of events that can be distinguished at any date t is finite. An example of how information might unfold in a dynamic market is presented in Figure 8.1. In this example, there are twelve disjoint events that can be distinguished at date 3, corresponding to the twelve date–3 nodes. Note that each node defines a unique path through the tree from date 0 to date 3. These events are called atoms of the date–3 information set, which means that they contain no nonempty proper subsets that are distinguishable at date 3. Label these events (nodes) as $A_1, \ldots, A_{12}$, starting from the top. There are five atoms in the date–2 information set. Starting from the top node at date 2, these five events are

$$A_1 \cup A_2, \quad A_3 \cup A_4 \cup A_5, \quad A_6 \cup A_7 \cup A_8, \quad A_9 \cup A_{10}, \quad A_{11} \cup A_{12}.$$

Likewise, there are two atoms in the information set at date 1:

$$\bigcup_{i=1}^{5} A_i \quad \text{and} \quad \bigcup_{i=6}^{12} A_i.$$

This is obviously a very simple example. However, it illustrates how information resolves when there is a finite number of distinguishable events at each date.

Suppose there are only finitely many events distinguishable at each date t. Let k_t denote the number of atoms in the date–t information set. In this circumstance, a random variable x depends only on date–t information if and only if x is constant on each atom of the date–t information set. Hence, each random variable that depends only on date–t information can be identified with a k_t–dimensional vector. Moreover, a consumption process $C = (C_0, \ldots, C_T)$ can be identified with a vector of dimension $k = \sum_{t=0}^{T} k_t$, where $k_0 = 1$. In a finite-horizon complete market, each basis vector e_j of $\mathbb{R}^k$ is a marketed consumption process.[6] In a finite-horizon complete market satisfying the law of one price, there is a unique cost q_j for each basis vector e_j. Any consumption process is a linear combination

---

5. As always, we use the convention here that equality between random variables means equality with probability 1. The random variables M_t and M_t^* could differ on a zero-probability event, but that would be immaterial.

6. The index j of e_j defines a date t and an atom of the date–t information set as follows: If $j = 1$, then $t = 0$, and e_j corresponds to the consumption process $C_0 = 1$, $C_t = 0$ for $t > 0$. If $j > 1$, let t be such that $\sum_{s=0}^{t-1} k_s < j \leq \sum_{s=0}^{t} k_s$. Then, e_j corresponds to the consumption process $C_t = 1$ on the atom $j - \sum_{s=0}^{t-1} k_s$ of the date–t information set, $C_t = 0$ otherwise, and $C_s = 0$ for all $s \neq t$.

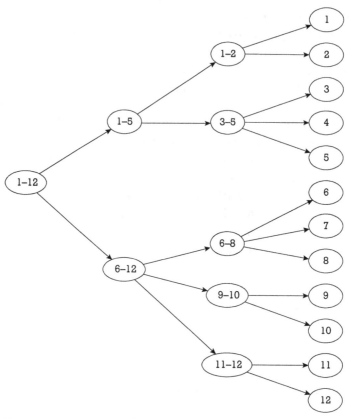

Figure 8.1 Dynamic resolution of uncertainty. In this example, there are four dates 0, 1, 2, and 3, starting with date 0 on the left and ending with date 3 on the right. There are 12 states of the world, corresponding to the date–3 nodes. The state of the world is learned over time. For example, by date 1 it is learned whether the state is one of the top five nodes (which can be reached from the top node at date 1) or one of the bottom seven (which can be reached from the bottom node at date 1).

$\sum_{j=1}^{k} c_j e_j$ of the basis vectors, and its unique cost is $\sum_{j=1}^{k} c_j q_j$. The q_j's can be interpreted as date-state prices.

In the example shown in Figure 8.1, a minimum of three assets is needed in order for the market to be complete. This is because there are at most three branches extending from any node. This generalizes the requirement that there be at least as many assets as states of the world in order to obtain completeness in a single-period model. The market in the example is complete with exactly three assets if none of the assets is redundant at any of the nodes from which three branches extend and if there are two nonredundant assets at each of the nodes from which two branches extend. A slightly simpler example in which there are only two branches extending from each node is considered in the exercises.

A very important feature of a dynamic securities market is that the number of assets needed to complete the market can be many fewer than the number of states of the world. If we consider the economy to end at date 3 in the example in Figure 8.1, then there are twelve states of the world, corresponding to the twelve date–3 nodes, but only three assets are needed to complete the market.

If investors were required to select portfolios at date 0 and hold them through date 3 in this example, then twelve assets would be needed in order for all date–3 payoffs to be marketed. In general, keeping the information structure fixed, the more frequently investors can trade, the smaller is the number of assets needed for markets to be complete. In the limit, with continuous trading, markets can be complete with a finite number of assets even when investors observe continuously distributed random variables. This is shown in Chapter 13.

8.5 BUBBLES, TRANSVERSALITY CONDITIONS, AND PONZI SCHEMES

Let M be a strictly positive SDF process. The question addressed in this section is under what conditions (8.5) and (8.17) are true for $T = \infty$ in the form[7]

$$P_t = \sum_{s=t+1}^{\infty} \mathsf{E}_t \left[\frac{M_s}{M_t} D_s \right] \qquad (8.19)$$

and

$$W_t + \sum_{s=t+1}^{\infty} \mathsf{E}_t \left[\frac{M_s}{M_t} Y_s \right] = \sum_{s=t}^{\infty} \mathsf{E} \left[\frac{M_s}{M_t} C_s \right]. \qquad (8.20)$$

We will call the right-hand side of (8.19) the fundamental value of the asset and the right-hand side of (8.20) the fundamental value of the consumption process. So, the question is whether prices equal fundamental values.

Bubbles

Consider equation (8.19). Suppose for simplicity that P and D are nonnegative. Denote the fundamental value by

$$F_t = \sum_{s=t+1}^{\infty} \mathsf{E}_t \left[\frac{M_s}{M_t} D_s \right],$$

---

7. We drop the i subscript from (8.5) for the sake of convenience.

and set $B_t = P_t - F_t$. Taking the limit as $T \to \infty$ in (8.5) yields

$$P_t = F_t + \lim_{T \to \infty} \mathsf{E}_t\left[\frac{M_T}{M_t} P_{iT}\right] \geq F_t. \tag{8.21}$$

Thus, under the nonnegativity assumption, the price must be at least as large as the fundamental value. In other words, $B_t \geq 0$.

If $B \neq 0$, then we call it a bubble in the price of the asset. The absence of arbitrage opportunities for finite horizons does not rule out bubbles. The simplest example of a bubble is as follows. Suppose there is no uncertainty and there is an asset with constant return $R_f = 1 + r_f$. As discussed in section 8.2, the unique SDF process is $M_t = (1 + r_f)^{-t}$. Consider an asset that never pays any dividends, so its fundamental value is zero, and assume that its initial price is $P_0 = 1$. Thus, the value of the bubble is 1 at date 0. To rule out arbitrage opportunities for all finite horizons, the asset must have return equal to R_f each period. Because the asset pays no dividends, the return is purely capital gain; thus, we must have $B_{t+1}/B_t = R_f$ for each t. Equivalently, $P_t = B_t = (1 + r_f)^t$. Therefore, the bubble grows at rate r_f.

To see the continually growing nature of bubbles in general, consider the definition $B_t = P_t - F_t$ written as

$$B_t = P_t - \sum_{s=t+1}^{\infty} \mathsf{E}_t\left[\frac{M_s}{M_t} D_s\right].$$

Multiply by M_t, do the same at $t+1$, subtract the two, and then take the conditional expectation at date t to obtain

$$\mathsf{E}_t[M_{t+1}B_{t+1}] - M_t B_t = \mathsf{E}_t[M_{t+1}P_{t+1}] - M_t P_t + \mathsf{E}_t[M_{t+1}D_{t+1}].$$

The definition (8.3) of an SDF process states that the right-hand side is zero, so

$$B_t = \mathsf{E}_t\left[\frac{M_{t+1}}{M_t} B_{t+1}\right].$$

This is the general version of the result $B_t = B_{t+1}/R_f$ in the certainty example in the previous paragraph. A bubble must grow enough so that its discounted future value—discounting both for risk and for the time value of money—equals its current value.

Transversality Condition and Bubbles

From (8.21), we see that there is no bubble in the price of the asset if and only if

$$\lim_{T \to \infty} \mathsf{E}_t[M_T P_{iT}] = 0. \tag{8.22}$$

Condition (8.22) is called a transversality condition. Assumptions such as the existence of a finite number of infinitely lived investors and the market being in equilibrium will ensure that the transversality condition holds, ruling out bubbles. See the end-of-chapter notes for further discussion.

Transversality Conditions for Wealth Processes and Ponzi Schemes

Now, assume C and Y in (8.20) are nonnegative. Take the limit in (8.17) as $T \to \infty$. Assume the infinite sums have finite values. Then, (8.20) holds if and only if

$$\lim_{T \to \infty} \mathsf{E}_t[M_T W_T] = 0. \tag{8.23}$$

This is also a transversality condition. If the infinite sums have finite values and

$$\lim_{T \to \infty} \mathsf{E}_t[M_T W_T] < 0,$$

then the right-hand side of (8.20) exceeds the left-hand side, meaning that the value of future consumption exceeds wealth, including the value of future labor income. This results from a Ponzi scheme, borrowing or shorting assets to finance consumption and never repaying. If an investor is allowed to do this, then the investor will not have an optimum. To preclude Ponzi schemes, it is sufficient to impose the constraint

$$\lim_{T \to \infty} \mathsf{E}_t[M_T W_T] \geq 0. \tag{8.24}$$

This constraint is generous in the sense of allowing borrowing against future labor income. A nonnegative wealth (no borrowing) constraint is probably more reasonable.

Optimization subject to the constraint (8.24) will result in the transversality condition (8.23) holding for the optimal wealth process W. Otherwise, we would have $\lim_{T \to \infty} \mathsf{E}_t[M_T W_T] > 0$, which is tantamount to "leaving money on the table."

Rather than imposing the constraint (8.24), we could rule out Ponzi schemes by, for example, constraining wealth processes to be bounded across dates and states of the world and assume $\lim_{T \to \infty} \mathsf{E}_t[M_T] = 0$. The condition $\lim_{T \to \infty} \mathsf{E}_t[M_T] = 0$ means that the price at date t of a zero-coupon bond maturing at T (a security that pays one unit of consumption at T and nothing at any other time) is near zero if T is sufficiently large. This is a very mild assumption. If W is bounded, then $\lim_{T \to \infty} \mathsf{E}_t[M_T] = 0$ implies the transversality condition (8.23).

A different transversality condition appears in Section 9.6. Condition (8.23) means that providing for wealth W_T sufficiently far into the future should constitute a negligible part of today's budget. The condition in Section 9.6 means that the value of wealth W_T sufficiently far into the future should constitute a negligible part of today's expected lifetime utility.

8.6 INFLATION AND FOREIGN EXCHANGE

To this point, we have considered asset prices denominated in units of the single consumption good; that is, they are "real" prices. However, it is easy to transform the analysis to nominal prices. As remarked in Section 8.2, a stochastic process M with $M_0 = 1$ is an SDF process if and only if MW is a martingale for each self-financing wealth process W (where W is in units of the consumption good). Let Z denote the price in dollars (or other currency) of the consumption good at date t. Then, the nominal value of the wealth process W is $W^* \stackrel{\text{def}}{=} ZW$. If M is an SDF process for real prices, then $M^* \stackrel{\text{def}}{=} Z_0 M/Z$ is an SDF process for nominal prices, because $M_0^* = 1$ and $M^*W^* = Z_0 MW^*/Z = Z_0 MW$ is a martingale. It is usually convenient to define $\hat{Z}_t = Z_t/Z_0$, which denotes the price of the consumption good "in date-0 dollars." Then, the nominal SDF process is $M^* = M/\hat{Z}$.

It is also easy to incorporate multiple currencies. For specificity, call one currency the domestic currency and the other the foreign currency. Let M^d be a nominal SDF process in the domestic currency. Let W^f denote the nominal value in the foreign currency of a self-financing wealth process. Let X denote the exchange rate, meaning the price of a unit of the foreign currency in units of the domestic currency. The price in the domestic currency of the asset with foreign currency value W^f is XW^f. Because M^d is a nominal SDF process for assets denominated in the domestic currency, $M^d XW^f$ is a martingale. Thus, $M^f \stackrel{\text{def}}{=} M^d X/X_0$ is a nominal SDF process for assets denominated in the foreign currency.

8.7 NOTES AND REFERENCES

The assumption that the discount factor for date $t+1$ utility, relative to date t utility, does not change as date $t+1$ approaches is important. Suppose, to the contrary, that at each t the investor seeks to maximize

$$u(c_t) + \beta \mathsf{E}_t \left[\sum_{s=t+1}^{\infty} \delta^{s-t} u(c_s) \right] \tag{8.25}$$

for some β. This means that at each date t the investor discounts date $t+1$ utility by $\beta\delta$ but discounts all subsequent utility, relative to the previous date, by δ. These are inconsistent preferences, in the sense that at t the investor anticipates maximizing

$$u(c_t) + \mathsf{E}_\tau\left[\sum_{s=\tau+1}^\infty \delta^{s-\tau}u(c_s)\right]$$

at all dates $\tau > t$ but instead when τ arrives maximizes

$$u(c_\tau) + \beta\mathsf{E}_\tau\left[\sum_{s=\tau+1}^\infty \delta^{s-\tau}u(c_s)\right].$$

The plans made at t for decisions at subsequent dates will not be optimal when those dates arrive. What is perhaps even more important is that the decisions made at t, which are optimal given the plans made at t for future actions, are not generally optimal in conjunction with the actions actually chosen at subsequent dates. This is called dynamic inconsistency. For example, if $\beta < 1$, the investor will be disinclined to save at each date t because of the discounting by β, anticipating incorrectly that she will be more inclined to save at subsequent dates. This is irrational but perhaps fairly common, in relation to saving and to other activities (dieting, quitting smoking, and so on). Utility of the form (8.25) is called hyperbolic discounting. See Strotz (1956), Phelps and Pollak (1986), and Rubinstein (2003).

If the investment opportunity set is constant over time and there is no labor income, then there is a sense in which the portfolio that is optimal for a single-period log-utility investor is asymptotically superior to any other portfolio (see Exercise 8.2). However, this does not imply that the log-optimal portfolio is optimal for an investor with non-log utility.

There is a large literature on bubbles. Tirole (1982) shows that there can be no bubbles in a fully dynamic rational expectations equilibrium with a finite number of infinitely lived risk-neutral investors who have common prior beliefs. LeRoy (2004) provides a survey and argues that rational bubbles are a reasonable explanation for apparently excessive market volatility. Chapter 21 presents a model of bubbles in markets with heterogeneous beliefs and short sales constraints due to Harrison and Kreps (1978).

EXERCISES

8.1. Suppose there is a risk-free asset with constant return R_f each period. Suppose there is a single risky asset with dividends given by

$$D_{t+1} = \begin{cases} \lambda_h D_t & \text{with probability } 1/2, \\ \lambda_\ell D_t & \text{with probability } 1/2, \end{cases}$$

where $\lambda_h > \lambda_\ell$ are constants, and $D_0 > 0$. Suppose the price of the risky asset satisfies $P_t = kD_t$ for a constant k. Suppose the information in the economy consists of the history of dividends, so the information structure can be represented by a tree as in Figure 8.1 with two branches emanating from each node (corresponding to the outcomes h and ℓ). For each date $t > 0$ and each path, let v_t denote the number of dates $s \leq t$ such that $D_s = \lambda_h D_{s-1}$, so $D_t = D_0 \lambda_h^{v_t} \lambda_\ell^{t-v_t}$. Recall that, for $0 \leq n \leq t$, the probability that $v_t = n$ is the binomial probability

$$2^{-t} \frac{t!}{n!(t-n)!}.$$

(a) State a condition implying that there are no arbitrage opportunities for each finite horizon T.
(b) Assuming the condition in part (a) holds, show that there is a unique single-period SDF from each date t to $t+1$, given by

$$Z_{t+1} = \begin{cases} z_h & \text{if } D_{t+1}/D_t = \lambda_h, \\ z_\ell & \text{if } D_{t+1}/D_t = \lambda_\ell, \end{cases}$$

for some constants z_h and z_ℓ. Calculate z_h and z_ℓ in terms of R_f, k, λ_h and λ_ℓ.
(c) Assuming the condition in part (a) holds, show that there is a unique SDF process M, and show that M_t depends on v_t and the parameters R_f, k, λ_h, and λ_ℓ.
(d) Consider $T < \infty$ and the random variable

$$x = \begin{cases} 1 & \text{if } D_{t+1} = \lambda_h D_t \text{ for each } t < T, \\ 0 & \text{if } D_{t+1} = \lambda_\ell D_t \text{ for any } t < T. \end{cases}$$

Calculate the self-financing wealth process that satisfies $W_T = x$.
(e) Suppose there is a representative investor with time-additive utility, constant relative risk aversion ρ, and discount factor δ. Assume the risk-free asset is in zero net supply. Calculate R_f and k in terms of λ_h, λ_ℓ, ρ, and δ.
(f) Given the formula for k in the previous part, what restriction on the parameters λ_h, λ_ℓ, ρ, and δ is needed to obtain $k > 0$? Show that this restriction is equivalent to

$$E\left[\sum_{t=1}^{\infty} \delta^t D_t^{1-\rho}\right] < \infty.$$

8.2. Suppose the return vectors $R_1, R_2, \ldots$ are independent and identically distributed. Let w be a positive constant. Assume $\max_\pi \mathsf{E}[\log(\pi' R_t)] > -\infty$ and let π^* be a solution to
$$\max_\pi \mathsf{E}[\log(\pi' R_t)].$$
Let W^* be the wealth process defined by the intertemporal budget constraint (8.1) with $\pi_t = \pi^*$ and $Y_t = C_t = 0$ for each t and $W_0^* = w$. Consider any other portfolio π for which
$$\mathsf{E}[\log(\pi' R_t))] < \max_\pi \mathsf{E}[\log(\pi' R_t)].$$
Let W be the wealth process defined by the intertemporal budget constraint (8.1) with $\pi_t = \pi$ and $Y_t = C_t = 0$ for each t and $W_0 = w$. Show that, with probability 1, there exists T (depending on the state of the world) such that
$$W_t^* > W_t$$
for all $t \geq T$. Hint: Apply the strong law of large numbers to $(1/T) \log W_T^*$ and to $(1/T) \log W_T$.

8.3. Consider any $T < \infty$, and suppose C_t is a marketed date–t payoff, for $t = 0, \ldots, T$. Show that there exists a wealth process W and portfolio process π such that C, W, and π satisfy
$$W_{t+1} = (W_t - C_t)\pi_t' R_{t+1} \tag{8.26}$$
for $t = 0, \ldots, T-1$, and $C_T = W_T$. Hint: Add up the wealth processes and take a weighted average of the portfolio processes associated with the individual payoffs. This result is applied in Section 9.2.

9

Dynamic Portfolio Choice

CONTENTS

9.1 Euler Equation	202
9.2 Static Approach in Complete Markets	205
9.3 Orthogonal Projections for Quadratic Utility	206
9.4 Introduction to Dynamic Programming	208
9.5 Dynamic Programming for Portfolio Choice	212
9.6 CRRA Utility with IID Returns	219
9.7 Notes and References	227

The first three sections of this chapter discuss topics in portfolio choice that are very much the same in dynamic markets as in single-period markets: the Euler equation, solving for optimal consumption in complete markets using the unique SDF process, and solving for optimal consumption for quadratic utility with orthogonal projections. The remainder of the chapter explains dynamic programming, which is a method for converting a dynamic optimization problem into a series of single-period problems. Section 9.4 explains dynamic programming in general, and Section 9.5 explains its application to portfolio choice. Key topics for portfolio choice are the envelope condition and hedging demands. Section 9.6 illustrates dynamic programming by solving for optimal portfolios with independent and identically distributed (IID) returns and CRRA utility.

9.1 EULER EQUATION

Consider a risk-averse investor with initial wealth W_0 and time-additive utility (8.2a) or (8.2b). The first-order condition for dynamic portfolio choice is called the Euler equation. It is:

Dynamic Portfolio Choice

$$(\forall i, t) \quad u'(C_t) = E_t\left[\delta u'(C_{t+1})R_{i,t+1}\right]. \tag{9.1}$$

This states that the investor is indifferent at the margin between consuming a bit more at t and investing a bit more in asset i to increase consumption at $t+1$. Assuming strictly monotone utility, the Euler equation is equivalent to the statement that the MRS

$$Z_{t+1} \stackrel{\text{def}}{=} \frac{\delta u'(C_{t+1})}{u'(C_t)}$$

is a single-period SDF. Compounding the single-period SDFs, we see that the Euler equation is equivalent to the statement that

$$M_t \stackrel{\text{def}}{=} \frac{\delta^t u'(C_t)}{u'(C_0)} \tag{9.2}$$

is an SDF process.

As in a single-period model, the first-order condition must hold at any optimum from which it is feasible to make small variations in the consumption and portfolio decisions. For each asset i, each date t and each event A observable at t, assume there is some $\epsilon > 0$ such that each of the following variations on the optimum produces finite expected utility:

(i) It is feasible for the investor to reduce consumption by ϵ at t when A occurs, to invest ϵ in asset i, and to consume the value of this additional investment at $t+1$.

(ii) It is feasible for the investor to increase consumption by ϵ at t when A occurs, to finance this consumption by investing ϵ less in (or shorting) asset i, and to restore wealth to its optimal level by consuming less at $t+1$.

Under this assumption, the first-order condition is derived below (by the same logic as in the single-period model).

It is also shown below that the first-order condition is sufficient for optimality when the horizon is finite. In the infinite-horizon case, the first-order condition plus the transversality condition are jointly sufficient for optimality, if a constraint is imposed to preclude Ponzi schemes.

Necessity of the Euler Equation To derive (9.1) at an optimum, first consider reducing consumption by ϵ at t when A occurs and investing ϵ in asset i, consuming the value of the investment at $t+1$. Consumption changes at t by $-\epsilon \, 1_A$ and at $t+1$ by $\epsilon \, 1_A R_{i,t+1}$. The resulting change in expected utility is

$$E\left[1_A \delta^t \{u(C_t - \epsilon) - u(C_t)\}\right] + E\left[1_A \delta^{t+1}\{u(C_{t+1} + \epsilon R_{i,t+1}) - u(C_{t+1})\}\right] \leq 0.$$

Letting $\epsilon \to 0$ (and using the monotone convergence theorem as in Section 2.1), we conclude that

$$\mathsf{E}\left[-1_A \delta^t u'(C_t)\right] + \mathsf{E}\left[1_A \delta^{t+1} u'(C_{t+1}) R_{i,t+1}\right] \leq 0.$$

Now, consider increasing consumption by ϵ at t when A occurs by investing less in asset i. Reasoning in the same way, we obtain

$$\mathsf{E}\left[1_A \delta^t u'(C_t)\right] + \mathsf{E}\left[-1_A \delta^{t+1} u'(C_{t+1}) R_{i,t+1}\right] \leq 0.$$

Therefore,

$$\mathsf{E}\left[1_A u'(C_t)\right] = \mathsf{E}\left[1_A \delta u'(C_{t+1}) R_{i,t+1}\right].$$

Because this is true for each event A observable at t, the Euler equation (9.1) follows from the definition of a conditional expectation (Appendix A.8).

Sufficiency of the Euler Equation Suppose u is concave. Suppose the Euler equation holds for a consumption process C that satisfies the intertemporal budget constraint (8.1) in conjunction with some wealth and portfolio processes. Let $(\hat{C}, \hat{W}, \hat{\pi})$ be any other solution of the intertemporal budget constraint (8.1). By concavity,

$$u(C_t) - u(\hat{C}_t) \geq u'(C_t)(C_t - \hat{C}_t). \tag{9.3}$$

Suppose the horizon is finite. Because there is no bequest motive, we can assume $C_T = W_T$ and $\hat{C}_T = \hat{W}_T$. By (8.17) and (9.2),

$$W_0 + \frac{1}{u'(C_0)} \sum_{t=1}^{T} \delta^t \mathsf{E}\left[u'(C_t) Y_t\right] = \frac{1}{u'(C_0)} \sum_{t=0}^{T} \delta^t \mathsf{E}\left[u'(C_t) C_t\right],$$

and

$$W_0 + \frac{1}{u'(C_0)} \sum_{t=1}^{T} \delta^t \mathsf{E}\left[u'(C_t) Y_t\right] = \frac{1}{u'(C_0)} \sum_{t=0}^{T} \delta^t \mathsf{E}\left[u'(C_t) \hat{C}_t\right].$$

Hence,

$$\sum_{t=0}^{T} \delta^t \mathsf{E}\left[u'(C_t)(C_t - \hat{C}_t)\right] = 0.$$

It follows from this and (9.3) that

$$\sum_{t=0}^{T} \delta^t \mathsf{E}\left[u(C_t) - u(\hat{C}_t)\right] \geq 0,$$

which shows that the solution of the Euler equation is optimal.

Consider the infinite-horizon case. Assume

$$\sum_{t=1}^{\infty} \mathsf{E}[u'(C_t)Y_t]$$

exists and is finite. Suppose the constraint (8.24) is imposed to preclude Ponzi schemes (with $M_t = \delta^t u'(C_t)/u'(C_0)$). Suppose the wealth process corresponding to the solution of the Euler equation satisfies the transversality condition

$$\lim_{T \to \infty} \mathsf{E}[\delta^T u'(C_T) W_T] = 0. \qquad (9.4)$$

Then, as explained in Section 8.5, we have

$$W_0 + \frac{1}{u'(C_0)} \sum_{t=0}^{\infty} \delta^t \mathsf{E}[u'(C_t)Y_t] = \frac{1}{u'(C_0)} \sum_{t=0}^{\infty} \delta^t \mathsf{E}[u'(C_t)C_t],$$

and

$$W_0 + \frac{1}{u'(C_0)} \sum_{t=0}^{\infty} \delta^t \mathsf{E}[u'(C_t)Y_t] \geq \frac{1}{u'(C_0)} \sum_{t=0}^{\infty} \delta^t \mathsf{E}\left[u'(C_t)\hat{C}_t\right].$$

Hence, (9.3) implies

$$\sum_{t=0}^{\infty} \delta^t \mathsf{E}\left[u(C_t) - u(\hat{C}_t)\right] \geq 0.$$

9.2 STATIC APPROACH IN COMPLETE MARKETS

In a complete market, we can collapse the intertemporal budget constraints (one constraint for each date and state of the world) into a single budget constraint. We can then solve for optimal consumption in the same way that we can solve for it in a complete single-period market. This approach to solving for optimal consumption in a dynamic model is called the static approach.

Assume the market is complete and there are no arbitrage opportunities. Then, as discussed in Sections 8.3 and 8.4, there is a unique SDF process M, and M is strictly positive. Consider an investor with a finite horizon T (the analysis can be extended to an infinite horizon at the expense of dealing with Ponzi schemes and transversality conditions). The investor chooses consumption and portfolio processes, and her wealth process is then determined by the intertemporal budget constraint. As before, assume there is no bequest motive, so $C_T = W_T$. From (8.17), the consumption process must satisfy

$$W_0 + \sum_{t=1}^{T} \mathsf{E}[M_t Y_t] = \sum_{t=0}^{T} \mathsf{E}[M_t C_t]. \tag{9.5}$$

Furthermore, by Exercise 8.3, for any consumption process satisfying (9.5), there exists a portfolio process such that, after consuming $(C_0, \ldots, C_{T-1})$, the investor has remaining wealth $W_T = C_T$. Thus, any consumption process satisfying (9.5) is feasible.

Because the feasible consumption processes are precisely those that satisfy (9.5), the investor's optimal consumption process must be the process that maximizes expected utility subject to (9.5). This optimization problem is called a static problem, because there is only a single budget constraint (as in a single-period model), the intertemporal budget constraints having been collapsed into (9.5).

As an example, consider CRRA utility with relative risk aversion ρ. The first-order condition implies

$$C_t^{-\rho} = C_0^{-\rho} \delta^{-t} M_t \quad \Leftrightarrow \quad C_t = C_0 \delta^{t/\rho} M_t^{-1/\rho}, \tag{9.6}$$

and substituting this into (9.5) yields

$$C_0 + C_0 \mathsf{E}\left[\sum_{t=1}^{T} \delta^{t/\rho} M_t^{1-1/\rho}\right] = W_0 + \sum_{t=1}^{T} \mathsf{E}_t[M_t Y_t],$$

which can be solved as

$$C_0 = \frac{W_0 + \sum_{t=1}^{\infty} \mathsf{E}_t[M_t Y_t]}{1 + \mathsf{E}\left[\sum_{t=1}^{T} \delta^{t/\rho} M_t^{1-1/\rho}\right]}.$$

Thus,

$$C_t = \frac{\left(W_0 + \sum_{t=1}^{T} \mathsf{E}_t[M_t Y_t]\right) \delta^{t/\rho} M_t^{-1/\rho}}{1 + \mathsf{E}\left[\sum_{t=1}^{T} \delta^{t/\rho} M_t^{1-1/\rho}\right]}. \tag{9.7}$$

The optimal wealth process is, for $t < T$,

$$W_t = C_t + \mathsf{E}_t\left[\sum_{s=t+1}^{T} \frac{M_s(C_s - Y_s)}{M_t}\right].$$

9.3 ORTHOGONAL PROJECTIONS FOR QUADRATIC UTILITY

Section 3.7 characterizes the optimal portfolio for an investor with quadratic utility in a single-period model. The investor shorts the portfolio that best approximates her labor income, shorts the portfolio with return $\tilde{R}_p$, and invests all of the proceeds from the short sales plus her initial wealth in the asset that best approximates a risk-free asset. In the presence of a risk-free asset, the optimum can be described thus: Short the best approximation to labor income and invest the proceeds of the short sale plus initial wealth in a portfolio on the mean-variance frontier. A similar characterization of the optimal portfolio can be made in a dynamic model. We will only consider the finite-horizon case. Denote the horizon by T.

Consider an investor with quadratic utility function $u(c) = -(c-\zeta)^2/2$ and no bequest motive. The Euler equation states that

$$\delta^t u'(C_t) = u'(C_0) M_t \quad \Leftrightarrow \quad C_t = \zeta - (\zeta - C_0)\delta^{-t} M_t \qquad (9.8)$$

for an SDF process M. Subtracting Y_t from both sides gives[1]

$$C_t - Y_t = \zeta - Y_t - (\zeta - C_0)\delta^{-t} M_t. \qquad (9.9)$$

At each date t, $C_t - Y_t$ is the amount the investor withdraws from her financial portfolio, so $C - Y$ is a marketed consumption process in the sense of Section 8.4. Set $X_t = C_t - Y_t$ and $\hat{M}_t = \delta^{-t} M_t$, so we can write (9.9) as

$$X = \zeta - Y - (\zeta - C_0)\hat{M}. \qquad (9.10)$$

Consider stochastic processes $A = (A_0, \ldots, A_T)$ with the property that $\mathsf{E}[A_t^2] < \infty$ for each t. Denote the set of such processes by $\mathcal{L}^2$. Use the investor's discount factor δ to define an inner product for $A, B \in \mathcal{L}^2$:

$$\langle A, B \rangle \stackrel{\text{def}}{=} \sum_{t=0}^{T} \delta^t \mathsf{E}[A_t B_t]. \qquad (9.11)$$

The set $\mathcal{L}^2$ equipped with this inner product is a Hilbert space (Section 3.8). Let $\mathcal{M}_0$ denote the subset of marketed consumption processes in $\mathcal{L}^2$. The set $\mathcal{M}_0$ is a linear subspace of $\mathcal{L}^2$. Let $\mathcal{M}$ denote its closure. For any $A \in \mathcal{L}^2$, the orthogonal projection of A onto $\mathcal{M}$ exists. It is $A_p \in \mathcal{M}$ such that $A - A_p$ is orthogonal to $\mathcal{M}$, in the sense that $\langle B, A - A_p \rangle = 0$ for every $B \in \mathcal{M}$. We will consistently use the subscript p to denote projections onto $\mathcal{M}$.

Assume the investor's labor income process Y and consumption process C belong to $\mathcal{L}^2$. Then, $X = C - Y \in \mathcal{L}^2$. As already remarked, X is marketed, so $X \in \mathcal{M}$. Therefore, it equals its own projection onto $\mathcal{M}$. Projections are linear, so the fact that $X = X_p$ combined with (9.10) gives

$$X = \zeta \, 1_p - Y_p - (\zeta - C_0)\hat{M}_p. \qquad (9.12)$$

---

1. Take $Y_0 = 0$ to be consistent with our usual convention that income at date 0 is already included in W_0.

Here 1_p denotes the projection onto $\mathcal{M}$ of the trivial stochastic process that equals 1 at every date and in every state of the world. Equation (9.12) shows that, as in a single-period model, the investor short sells the marketed consumption process that best approximates her labor income, short sells the consumption process that best approximates the SDF process, and invests her initial wealth plus the proceeds from the short sales into the consumption process that best approximates a constant.

The quantity C_0 in (9.12) is endogenous. We solve for it below using the budget constraint and show that

$$X = \zeta 1_p - Y_p - \left(\zeta \langle 1_p, \hat{M}_p \rangle - \langle Y_p, \hat{M}_p \rangle - W_0 \right) \frac{\hat{M}_p}{\langle \hat{M}_p, \hat{M}_p \rangle}. \qquad (9.13)$$

The consumption process $\hat{M}_p / \langle \hat{M}_p, \hat{M}_p \rangle$ is a unit-cost version of the projection $\hat{M}_p$, so it is analogous to $\tilde{R}_p$ in the single-period model. The formula (9.13) is a direct extension of the single-period formula (3.42).

Note that, for any $A \in \mathcal{M}$, we have

$$\langle A, \hat{M}_p \rangle = \langle A, \hat{M} \rangle = \sum_{t=0}^{T} \mathsf{E}[A_t M_t].$$

Thus, the inner product with $\hat{M}_p$ gives the fundamental value of a marketed consumption process. Using $C_T = W_T$ (no bequest motive), $Y_0 = 0$, and (8.17), we therefore have

$$\langle X, \hat{M}_p \rangle = W_0.$$

This is a reformulation of the investor's budget constraint. Substitute for X from (9.12) to obtain

$$\zeta \langle 1_p, \hat{M}_p \rangle - \langle Y_p, \hat{M}_p \rangle - (\zeta - C_0) \langle \hat{M}_p, \hat{M}_p \rangle = W_0.$$

This implies

$$\zeta - C_0 = \frac{\zeta \langle 1_p, \hat{M}_p \rangle - \langle Y_p, \hat{M}_p \rangle - W_0}{\langle \hat{M}_p, \hat{M}_p \rangle}.$$

Substitute this into (9.12) to obtain (9.13).

9.4 INTRODUCTION TO DYNAMIC PROGRAMMING

Many dynamic decision problems are easiest to solve when reduced to a series of single-period problems. This method is easiest to explain in a lattice (tree) model. The tree in Figure 9.1 represents a decision problem in which a person

Dynamic Portfolio Choice

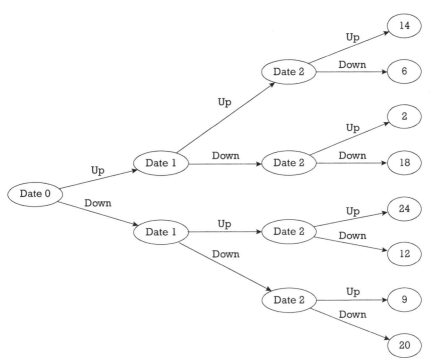

Figure 9.1 A decision tree. In this example, there are four dates 0, 1, 2, and 3, starting with date 0 on the left and ending with date 3 on the right. At dates 0, 1, and 2, the choice to be made is to go up or down. The rewards occur at date 3 and depend on the path taken. The rewards are shown in the date 3 nodes.

must decide at each of dates $t = 0, 1, 2$ whether to go up or down. The reward earned at date $t = 3$ depends on the sequence of decisions made and is shown at the right side of the graph. Clearly, 24 is the maximum possible reward, and the optimal sequence of decisions is Down-Up-Up.

To see how dynamic programming works in this simple problem, consider each of the four nodes at date $t = 2$. If we were to reach the top node, the optimal decision from that point is obviously Up, leading to a reward of 14. From the second-highest node at date $t = 2$, the optimal decision is Down, leading to a reward of 18. These calculations lead to the value function at date 2, which lists the maximum terminal reward that can be reached from each of the date–2 nodes.

Having computed the value function at date 2, we can compute the value function at date 1 by considering at each node whether Up or Down produces the highest date–2 value. For example, at the top node at date 1, we can choose between the values 14 and 18. Obviously, we would choose 18, meaning Down. We do not have to look forward to date 3, because the information we need to make an optimal decision at date 1 is already encoded in the date–2 values.

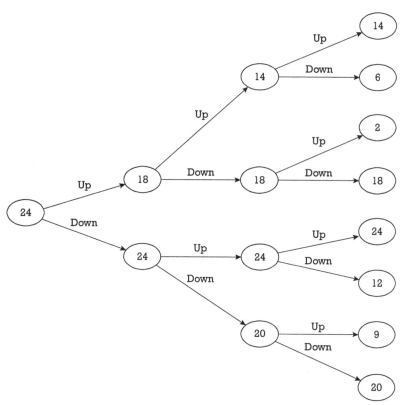

Figure 9.2 A value function. In this example, there are four dates 0, 1, 2, and 3, starting with date 0 on the left and ending with date 3 on the right. At dates 0, 1, and 2, the choice to be made is to go up or down. The rewards occur at date 3 and depend on the path taken. The rewards are shown in the date 3 nodes. The value shown at each node at dates 0, 1, and 2 is the maximum attainable reward starting at that node.

Likewise, we can compute the value at date 0 by considering whether Up or Down produces the highest date–1 value. This process of computing the value function is called backward induction or backward recursion. The complete set of values is shown in Figure 9.2.

State Transition Equation

To represent these calculations as a mathematical formula, note that there are 2^t nodes at each date t. Index the nodes, starting from the bottom, as $x = 1, \ldots, 2^t$. Let π represent the decision variable, with $\pi = 0$ meaning Up and $\pi = 1$ meaning Down. If x is the node at date t and the decision π is taken at date t, then $2x - \pi$ is the node at date $t + 1$, so we write

Dynamic Portfolio Choice

$$x_{t+1} = 2x_t - \pi_t. \tag{9.14}$$

An equation such as (9.14) is called a state transition equation, the state being here the node x.

Bellman Equation

Let $V_t(x)$ denote the maximum attainable value starting from node x at date t. The values at date 3 are the given rewards; for example, $V_3(2) = 9$. The values at dates $t = 0, 1, 2$ are the values we have computed by backward induction; for example, $V_2(1) = 20$. The backward induction process is expressed mathematically as

$$V_t(x) = \max_{\pi \in \{0,1\}} V_{t+1}(2x - \pi). \tag{9.15a}$$

This formula for V_t in terms of V_{t+1} is called the Bellman equation.

Intermediate Rewards

A variation of the decision problem is one in which there are rewards earned at each date and the objective is to maximize the sum of rewards over time. Consider a tree like that shown in Figure 9.1 but suppose there are rewards at each node. We can allow the reward earned at each node to depend on the decision (Up or Down) taken at that node. Denote the reward earned at date t at node x when decision π is taken by $u_t(x, \pi)$. There is obviously no decision to be made at the terminal date $t = 3$, so we can write $u_3(x)$ for the terminal reward at node x.

Denote the decision taken at date t at node x by $\pi_t(x)$. We want to choose the decisions $\pi_t(x)$ so as to maximize

$$\sum_{t=0}^{2} u_t(x_t, \pi_t(x_t)) + u_3(x_3),$$

where the path through the tree is determined by the decisions $\pi_t(x_t)$ and the state transition equation (9.14). Let $V_3(x) = u_3(x)$ and for $t < 3$, define

$$V_t(x) = \max_{\pi \in \{0,1\}} \left\{ u_t(x, \pi) + V_{t+1}(2x - \pi) \right\}. \tag{9.15b}$$

By this backward induction, we define the values at each node. Equation (9.15b) is the Bellman equation for this problem. To make the optimal decision at each date, it is again enough to look at the current reward and the values attainable at the next date, rather than looking forward to the end of the tree.

Dynamic Programming under Uncertainty

In dynamic programming under uncertainty, we use the maximum attainable *expected* utility as the value. This means that (9.15a) is replaced by

$$V_t(x) = \max_\pi E[V_{t+1}(X_{t+1}) \mid X_t = x], \qquad (9.16a)$$

where X_{t+1} denotes the random state (node) at date $t+1$, the distribution of which may depend on the decision π and the state x at date t. Also, (9.15b) is replaced by

$$V_t(x) = \max_\pi \left\{ u_t(x,\pi) + E[V_{t+1}(X_{t+1}) \mid X_t = x] \right\}. \qquad (9.16b)$$

In these equations, and in the statement of the Bellman equation throughout the book, the operator max means the least upper bound (supremum).

Obviously, in an infinite-horizon problem, we cannot calculate the value function by starting at T and working backward using the Bellman equation. Nevertheless, under certain conditions, we can find the value function by solving the Bellman equation, and the maximization in the Bellman equation produces the optimal decisions. This is discussed in the next section.

9.5 DYNAMIC PROGRAMMING FOR PORTFOLIO CHOICE

Dynamic programming is applicable only in a Markovian model, meaning that there are some variables the current values of which form a sufficient statistic for predicting future random variables (returns and labor income in our case). These variables play the role of the nodes in the decision tree shown earlier in the sense that they summarize the past in a manner that is sufficient for making optimal decisions. They are called state variables.

Assume the distribution of the vector of asset returns R_{t+1} depends on a vector of state variables X_t. Likewise, allow the distribution of the labor income Y_{t+1} to depend on the vector X_t. By "depend on," it is meant that the distribution of R_{t+1} and Y_{t+1} conditional on all information at date t is the same as the distribution conditional on X_t only. For this to be useful, the vector X must have the same property, namely, that the distribution of X_{t+1} conditional on all information at date t is the same as the distribution conditional on X_t only. This defines the sequence of random vectors X_t as a Markov process, and it implies that the sequence of random vectors (X_t, Y_t, R_t) is also a Markov process.[2]

It follows from these assumptions that the distributions of X_u, Y_u, and R_u for all $u > t$ conditional on information at date t are the same as the distributions conditional on X_t only. The properties assumed here are often expressed by saying that X_t is a sufficient statistic for predicting the future values of (X, Y, R), because all information at date t other than X_t can be ignored for making those predictions.

The decision variable π in the previous section is now the consumption-portfolio pair (c, π), and the state variable (node) x in the previous section is replaced by the state variables x and wealth w. Recall that we are including the return of the money market account, if it exists, as one of the returns in the vectors R_t. Therefore, a portfolio is a vector π such that $\iota'\pi = 1$. *Every maximization over π in the remainder of the chapter is to be understood as subject to the constraint $\iota'\pi = 1$.*

Bellman Equation for Portfolio Choice

If an investor has a finite horizon and seeks only to maximize the expected utility of terminal wealth $E[u(W_T)]$, then the Bellman equation is

$$V_t(x, w) = \max_{\pi} E\left[V_{t+1}\left(X_{t+1}, Y_{t+1} + w\pi'R_{t+1}\right) \mid X_t = x\right]. \tag{9.17}$$

Here, the conditional expectation is over the distributions of X_{t+1}, Y_{t+1}, and R_{t+1}, given $X_t = x$. The future value V_{t+1} depends on X_{t+1} and W_{t+1}. We have used the intertemporal budget equation to substitute for W_{t+1} so as to show how it depends on the decision π and wealth w at time t. The value function equals the utility function at the terminal date T, that is, $V_T(x, w) = u(w)$.

If there is consumption at each date as in (8.2a) and (8.2b), then the rewards u_t in the previous section are $\delta^t u(C_t)$. Therefore, the Bellman equation (9.16b) is

$$V_t(x, w) = \max_{c, \pi} \left\{ \delta^t u(c) + E\left[V_{t+1}\left(X_{t+1}, Y_{t+1} + (w-c)\pi'R_{t+1}\right) \mid X_t = x\right] \right\}. \tag{9.18}$$

When there is consumption at each date, it is convenient to define another value function $J_t(w, x) = \delta^{-t} V_t(w, x)$. The relation of J_t to V_t is that V_t is the maximum utility from date–t onward, discounted to date 0, whereas J_t is the same utility

---

2. A simple example of a Markov process is an AR(1) process:

$$X_{t+1} = a + AX_t + \varepsilon_{t+1},$$

for a vector a, a square matrix A, and a sequence of IID random vectors ε_t.

discounted only to date t. Substitute J_t for V_t in (9.18) and cancel the factor δ^t. This produces

$$J_t(x,w) = \max_{c,\pi} \left\{ u(c) + \delta \mathsf{E}\left[J_{t+1}\left(X_{t+1}, Y_{t+1} + (w-c)\pi' R_{t+1}\right) \mid X_t = x \right] \right\}. \tag{9.19}$$

We will use the value function J and the Bellman equation (9.19) whenever there is consumption at each date. If the horizon is finite and there is consumption at each date, then the value function J_t equals the utility function at the terminal date $t = T$, that is, $J_T(x,w) = u(w)$.

The Bellman equation (9.19) can be simplified in the infinite-horizon case (8.2a), due to stationarity. When there is an infinite horizon, the maximum value that can be achieved from date t onward discounted to date t and starting from $X_t = x$ and $W_t = w$ is the same for every t, given x and w. This is due to the fact there is always an infinite number of periods remaining at any date t and to the fact that X is Markov, so its distribution does not depend on the time index, given x. Thus, in the infinite-horizon case, J does not depend on t, and the Bellman equation is

$$J(x,w) = \max_{c,\pi} \left\{ u(c) + \delta \mathsf{E}\left[J\left(X_{t+1}, Y_{t+1} + (w-c)\pi' R_{t+1}\right) \mid X_t = x \right] \right\}. \tag{9.20}$$

Optimal Portfolio and Hedging Demands

In principle, given knowledge of the value function J, we could perform the maximization in the infinite-horizon Bellman equation (9.20) in two steps. First, we can compute the optimal consumption c^* and then find the optimal portfolio π^*.[3] In the second step, we are maximizing

$$\mathsf{E}\left[J\left(X_{t+1}, Y_{t+1} + (w-c^*)\pi' R_{t+1}\right) \mid X_t = x \right], \tag{9.21}$$

where c^* denotes the consumption attaining the maximum in (9.20). Maximizing (9.21) is similar to a single-period portfolio choice problem, which of course is the point of dynamic programming, but there is an important difference between (9.21) and the objective function in a single-period problem. The difference is that the function J in (9.21) depends on the state variables X_{t+1} in addition to date $t+1$ wealth.

---

3. In practice, we may want to solve for the consumption and portfolio simultaneously, or we may want to follow the opposite procedure, solving for the portfolio first. Examples are given in Section 9.6.

The same two-step procedure could be used for the Bellman equations (9.17) and (9.19), given knowledge of the value functions. So, in general, the optimal portfolio at any date t is the one that maximizes the expected value of the value function at $t+1$. In general, the value function at date $t+1$ is lower when the realized values of the state variables X_{t+1} imply a less favorable distribution for future returns $R_{t+2}, R_{t+3}, \ldots$, and/or a less favorable distribution for future income $Y_{t+2}, Y_{t+3}, \ldots$. Investors generally choose to hedge, to some extent, against such adverse changes in the state variables.[4] Investors typically face a trade-off between hedging and achieving high returns, in addition to the trade-off between risk and return that appears in a single-period model. In continuous-time models, a formula for the optimal portfolio (and for the hedging demands) can be given in terms of the partial derivatives of the value function, the covariance matrix of the returns, and the covariances between returns and state variables (Section 14.5).

Envelope Condition

The envelope theorem states that the partial derivative of a value function with respect to a parameter is the same whether a choice variable is varied when the parameter is varied or held fixed at the previous optimum. In the Bellman equation (9.20), we can regard the expression being maximized as a function of (x, w, π, c). The exogenous parameters are (x, w) and the choice variables are (π, c). The partial derivative of the maximum value with respect to w is the same whether c is varied with w or held fixed at the optimum, so we can suppose c is varied one-for-one with w, leaving $w - c$ fixed. The partial derivative of the maximum with respect to w—the left-hand side of (9.20)—is

$$\frac{\partial}{\partial w} J(x, w).$$

Taking $dc/dw = 1$ and holding $w - c$ fixed, the derivative with respect to w of the expression being maximized on the right-hand side of (9.20) is $u'(c)$. The envelope theorem implies the equality of these expressions:

$$\frac{\partial}{\partial w} J(x, w) = u'(c) \qquad (9.22)$$

when c is the optimal choice. Naturally, equation (9.22) is called the envelope condition. To repeat somewhat the discussion above, the interpretation of (9.22) is that, because the investor has optimized over consumption c and investment $w - c$, a small change in initial wealth can either be consumed or invested, with

---

4. An exception is when the investor has log utility (Exercise 9.2).

the value of either option being the same. Thus, the value of a small amount of additional wealth is the same as the value of a small amount of additional consumption. The envelope condition also holds for the finite-horizon Bellman equation (9.19).

CRRA Utility Implies a CRRA Value Function

Assume for the remainder of this section that there is no labor income. Then, the value function of an investor with CRRA utility has constant relative risk aversion for wealth with the same risk aversion that the utility function has for consumption.[5] For both log and power utility, the value function is separable in the state variables and in wealth. For log utility, the value function is additively separable. For power utility, the value function is multiplicatively separable.

If the horizon is finite, then the value function for log utility is $\gamma_t \log w + f_t(x)$ for some numbers γ_t and functions f_t. This is true when the objective is to maximize the expected utility of terminal wealth, and it is also true when there is consumption at each date. If the horizon is infinite, then $J(x,w) = \gamma \log w + f(x)$ for a constant γ and function f.

If the horizon is finite, then the value function for power utility is $f_t(x) w^{1-\rho}$ for some functions f_t, where ρ is the relative risk aversion of the utility function. This is true when the objective is to maximize the expected utility of terminal wealth, and it is also true when there is consumption at each date. If the horizon is infinite, then $J(x,w) = f(x) w^{1-\rho}$ for some function f.

The fact that the value function has constant relative risk aversion for wealth with risk aversion ρ is often expressed, when $\rho \neq 1$, by saying that the value function (like the utility function) is homogeneous of degree $1 - \rho$. The utility and value functions are also said to be homothetic, a homothetic function being a monotone transform of a homogeneous function.

When the horizon is finite, the above description of the value function can be verified by induction using the Bellman equation. Consider power utility with consumption at each date. The value function J_t is equal to the utility function at date $t = T$, so it is of the form $f_t(x) w^{1-\rho}$ with $f_t(x) = 1/(1-\rho)$ for all x. Suppose $J_s(x,w) = f_s(x) w^{1-\rho}$ for all $s = t+1, \ldots, T$ for some functions f_s. Then,

$$J_t(x,w) = \max_{c,\pi} \left\{ u(c) + \delta \mathsf{E}\left[J_{t+1}\left(X_{t+1}, (w-c)\pi' R_{t+1}\right) \mid X_t = x \right] \right\}$$

$$= \max_{c,\pi} \left\{ \frac{1}{1-\rho} c^{1-\rho} + \delta \mathsf{E}\left[f_{t+1}(X_{t+1}) \left((w-c)\pi' R_{t+1}\right)^{1-\rho} \mid X_t = x \right] \right\}.$$

---

5. Likewise, CARA utility implies a CARA value function (Exercise 9.3).

Dynamic Portfolio Choice

Let z denote the consumption rate c/w. We can maximize over z instead of c. This produces

$$J_t(x,w) = \max_{z,\pi} \left\{ \frac{1}{1-\rho} z^{1-\rho} w^{1-\rho} \right.$$
$$\left. + \delta \mathsf{E}\left[f_{t+1}(X_{t+1}) \left(w(1-z)\pi' R_{t+1} \right)^{1-\rho} \mid X_t = x \right] \right\}$$
$$= w^{1-\rho} \max_{z,\pi} \left\{ \frac{1}{1-\rho} z^{1-\rho} \right.$$
$$\left. + \delta \mathsf{E}\left[f_{t+1}(X_{t+1}) \left((1-z)\pi' R_{t+1} \right)^{1-\rho} \mid X_t = x \right] \right\}.$$

Setting

$$f_t(x) = \max_{z,\pi} \left\{ \frac{1}{1-\rho} z^{1-\rho} + \delta \mathsf{E}\left[f_{t+1}(X_{t+1}) \left((1-z)\pi' R_{t+1} \right)^{1-\rho} \mid X_t = x \right] \right\},$$

we see that $J_t(x,w) = f_t(x) w^{1-\rho}$ as claimed. By induction, the claim is true for all t. The argument is similar for log utility and for maximizing the expected utility of terminal wealth.

With an infinite horizon, a similar argument shows that there is a CRRA solution of the Bellman equation. However, this does not prove that the value function has constant relative risk aversion, because there can in general be multiple solutions of the Bellman equation in infinite-horizon dynamic programming problems, only one of which is the value function. A simple example is presented in Exercise 9.4. Therefore, we give a different proof below that the value function has constant relative risk aversion in the infinite-horizon model with CRRA utility and no labor income.

Adopt the constraint $C_t \leq W_t$ to rule out Ponzi schemes. This no-borrowing constraint is very sensible here, because we are assuming there is no future labor income Y to borrow against. Consider any consumption and portfolio processes. Set $Z_t = C_t/W_t$. The constraint $C_t \leq W_t$ is equivalent to $Z_t \leq 1$. Let R_{t+1} denote the portfolio return from t to $t+1$. The intertemporal budget constraint specifies that, for all dates s, $W_{s+1} = W_s(1-Z_s)R_{s+1}$. By induction, for any dates $t < u$,

$$W_u = W_t \prod_{s=t}^{u-1}(1-Z_s)R_{s+1}.$$

Therefore,

$$C_u = W_t Z_u \prod_{s=t}^{u-1}(1-Z_s)R_{s+1}. \tag{9.23}$$

We can use the same formula for $u = t$ by defining the product over an empty range to equal 1. It follows that, for any t,

$$\sum_{u=t}^{\infty} \delta^{u-t} \frac{1}{1-\rho} C_u^{1-\rho} = \frac{1}{1-\rho} W_t^{1-\rho} \sum_{u=t}^{\infty} \delta^{u-t} \left(Z_u \prod_{s=t}^{u-1} (1 - Z_s) R_{s+1} \right)^{1-\rho}.$$

Thus,

$$J(X_t, W_t) = \max_{C, \pi} \mathrm{E}_t \left[\sum_{u=t}^{\infty} \delta^{u-t} \frac{1}{1-\rho} C_u^{1-\rho} \right]$$

$$= \max_{Z, \pi} \mathrm{E}_t \left[\frac{1}{1-\rho} W_t^{1-\rho} \sum_{u=t}^{\infty} \delta^{u-t} \left(Z_u \prod_{s=t}^{u-1} (1-Z_s) R_{s+1} \right)^{1-\rho} \right]$$

$$= W_t^{1-\rho} \max_{Z, \pi} \mathrm{E}_t \left[\frac{1}{1-\rho} \sum_{u=t}^{\infty} \delta^{u-t} \left(Z_u \prod_{s=t}^{u-1} (1-Z_s) R_{s+1} \right)^{1-\rho} \right],$$

where the optimization is over $Z_u \le 1$ for all u and over the portfolio process that affects the returns R_{s+1}. The important fact is that

$$f(X_t) \stackrel{\mathrm{def}}{=} \max_{Z, \pi} \mathrm{E}_t \left[\frac{1}{1-\rho} \sum_{u=t}^{\infty} \delta^{u-t} \left(Z_u \prod_{s=t}^{u-1} (1-Z_s) R_{s+1} \right)^{1-\rho} \right]$$

does not depend on W_t, so the value function has the form claimed.

The Marginal Value of Wealth with CRRA Utility

Chapter 10 presents the Intertemporal Capital Asset Pricing Model (ICAPM), which is a factor model for asset risk premia that uses market wealth and the state variables X as the factors. The prices of risk for the state variables in the ICAPM are determined by how the state variables affect investors' marginal values of wealth. Chapter 10 derives the ICAPM as an approximate relation in discrete time, and Chapter 14 derives it as an exact relation in continuous time. Chapter 14 links the prices of risk of the state variables to the hedging demands. Briefly, if investors desire to hold assets because they help to hedge changes in state variables, then this increased demand will produce higher prices and lower expected returns, so hedging demands are in inverse relation to prices of risk.

Here, we explain how state variables affect the marginal value of wealth for an investor with CRRA utility and no labor income. This example is useful for interpreting the prices of risk in the ICAPM. The results are different for the three cases $\rho < 1$, $\rho = 1$, and $\rho > 1$. Specifically, changes in any state variable cause the investor's value and the investor's marginal value of wealth to change in the same direction if $\rho < 1$ and to change in opposite directions if $\rho > 1$. Changes in state variables have no effect on the marginal value of wealth if $\rho = 1$ (log utility). The case $\rho > 1$ seems the most intuitive, as is discussed below.

The proof is very simple. We present it for the value function in an infinite-horizon model to economize on notation, but, as can easily be seen, the calculation is exactly the same for a finite horizon, both with and without consumption prior to the terminal date. First, consider $\rho = 1$. Then,

$$J(x,w) = \gamma \log w + f(x) \quad \Rightarrow \quad \frac{\partial J(w,x)}{\partial w} = \frac{\gamma}{w}.$$

Thus, the marginal value of wealth is independent of all state variables. Now, consider $\rho \neq 1$. Then, $J(x,w) = f(x)w^{1-\rho}$, so for any $j = 1, \ldots, k$,

$$\frac{\partial J(x,w)}{\partial x_j} = w^{1-\rho}\frac{\partial f(x)}{\partial x_j}, \quad \frac{\partial^2 J(x,w)}{\partial w \partial x_j} = (1-\rho)w^{-\rho}\frac{\partial f(x)}{\partial x_j}$$

$$\Rightarrow \quad \frac{\partial^2 J(x,w)/\partial w \partial x_j}{\partial J(w,x)/\partial x_j} = \frac{1-\rho}{w}.$$

Thus, the signs of the first partial derivative and the cross partial derivative are the same if $\rho < 1$ and are opposite if $\rho > 1$.

To compare the different results for $\rho < 1$ and $\rho > 1$, consider a state variable X_j having the property that increases in X_j improve the investment opportunity set. Then, increases in X_j will be beneficial for the investor, meaning that $\partial J(w,x)/\partial x_j > 0$. The result we have just derived shows that if $\rho < 1$, then increases in X_j increase the marginal value of wealth, and if $\rho > 1$, then increases in X_j lower the marginal value of wealth. Thus, for aggressive investors ($\rho < 1$), having an additional unit of wealth is prized more highly when investment opportunities are better. It seems plausible that additional wealth would be more valuable when investment opportunities are poorer rather than when they are better, because when they are poorer, it is more difficult to increase wealth through investments. This is the case if $\rho > 1$.

9.6 CRRA UTILITY WITH IID RETURNS

The simplest dynamic portfolio choice problems are with IID returns and LRT utility. In this section, we analyze power utility. Log utility and CARA utility are considered in the exercises, and portfolio choice with shifted log and shifted power utility is discussed in the end-of-chapter notes.

Assume the investor has no labor income, and assume that the return vectors $R_1, R_2, \ldots$ are IID. Thus, in particular, the conditional distribution of R_{t+1} is independent of information at date t. The assumptions that returns are independent of conditioning information and there is no labor income imply that we do not need to keep track of conditioning information: The only state variable in the model is the investor's wealth. Because there is no dependence on state variables, the description of the value function in Section 9.5 simplifies. The functions $f_t(x)$ do not depend on x; instead, they depend only on time. Likewise, $f(x)$ is just a constant when the horizon is infinite.

The main lesson from this section is that there is not much new in a dynamic portfolio choice model with IID returns compared to a single-period model. The optimal portfolio for CRRA utility is the same in a dynamic model as in a single-period model—it is the portfolio that achieves the maximum in (9.24) when $\rho \neq 1$ and the portfolio that maximizes the expected log return when $\rho = 1$.[6] When there is consumption at each date, optimal consumption is proportional to wealth. The constant of proportionality is time dependent when the horizon is finite and independent of time (a single number) when the horizon is infinite.

Assume there exists a finite number $B > 0$ satisfying

$$\frac{1}{1-\rho} B^{1-\rho} = \max_{\pi} \mathsf{E}\left[\frac{1}{1-\rho} \left(\pi' R_t\right)^{1-\rho}\right]. \tag{9.24}$$

Thus, B is the certainty-equivalent end-of-period wealth for an investor with constant relative risk aversion ρ in a single-period problem with initial wealth $W_0 = 1$ and consumption only at the end of the period. The existence of B simply means that the maximum utility in this single-period problem is finite. In the infinite-horizon model, we will need the additional assumption

$$\delta B^{1-\rho} < 1, \tag{9.25}$$

where δ is the discount factor. This ensures that the maximum attainable lifetime utility is finite.

---

6. The optimal portfolio is constant over time, meaning that the fraction of invested wealth that is invested in each asset remains constant over time. However, the investor will usually trade all assets each period. She trades because her wealth changes—buying assets when wealth rises and selling them to finance consumption when wealth falls. Furthermore, she trades to rebalance her portfolio weights to the optimum. For example, if the price of an asset held by the investor rises relative to others, then the investor's portfolio weight on that asset will rise, and she will need to sell the asset to return the weight to the optimum.

Dynamic Portfolio Choice

Maximizing the Expected Utility of Terminal Wealth

Suppose that the investor does not consume at dates $t = 0, \ldots, T-1$ and seeks to maximize

$$\mathsf{E}\left[\frac{1}{1-\rho} W_T^{1-\rho}\right].$$

We will show that

$$V_t(w) = \frac{B^{(1-\rho)(T-t)}}{1-\rho} w^{1-\rho} \qquad (9.26)$$

for each t. The argument is by induction. Clearly, (9.26) is true for $t = T$. Suppose it is true for $t+1, \ldots, T$ for some t. Then, the Bellman equation gives us

$$V_t(w) = \max_\pi \mathsf{E}[V_{t+1}(w\pi' R_{t+1})]$$

$$= \max_\pi \mathsf{E}\left[\frac{B^{(1-\rho)(T-t-1)}}{1-\rho} (w\pi' R_{t+1})^{1-\rho}\right]$$

$$= B^{(1-\rho)(T-t-1)} w^{1-\rho} \max_\pi \mathsf{E}\left[\frac{1}{1-\rho} (\pi' R_{t+1})^{1-\rho}\right]$$

$$= B^{(1-\rho)(T-t-1)} w^{1-\rho} \cdot \frac{1}{1-\rho} B^{1-\rho}$$

$$= \frac{B^{(1-\rho)(T-t)}}{1-\rho} w^{1-\rho}.$$

Thus, (9.26) is true for all t. This proof also shows that the maximum value is attained at the portfolio π that maximizes the single-period objective function in (9.24).

Finite Horizon with Consumption at Each Date

Suppose now that the investor consumes at each date. Since we are assuming there is no bequest motive, the investor seeks to maximize

$$\frac{1}{1-\rho} \mathsf{E}\left[\sum_{t=0}^{T} \delta^t C_t^{1-\rho}\right].$$

Define A_t by

$$\frac{1}{A_t} = \sum_{s=0}^{T-t} (\delta B^{1-\rho})^{s/\rho}. \qquad (9.27)$$

Note that the A_t increase with t, culminating in $A_T = 1$. We will show that the optimal consumption is

$$C_t = A_t W_t, \qquad (9.28)$$

and the value function is

$$J_t(x, w) = \frac{A_t^{-\rho}}{1-\rho} w^{1-\rho}. \qquad (9.29)$$

The proof is again by induction and again shows that the optimal portfolio is the one that maximizes the single-period objective function in (9.24).

First, note that (9.27) implies $A_T = 1$ and, for $t < T$,

$$\frac{1}{A_t} = 1 + \sum_{s=1}^{T-t} (\delta B^{1-\rho})^{s/\rho}$$

$$= 1 + (\delta B^{1-\rho})^{1/\rho} \sum_{s=0}^{T-t-1} (\delta B^{1-\rho})^{s/\rho}$$

$$= 1 + \frac{(\delta B^{1-\rho})^{1/\rho}}{A_{t+1}}.$$

Thus,

$$A_t = \frac{A_{t+1}}{A_{t+1} + (\delta B^{1-\rho})^{1/\rho}}. \qquad (9.30)$$

Because the value function equals the utility at $t = T$ and $A_T = 1$, we have $J_t(x, w) = A_t w^{1-\rho}/(1-\rho)$ at $t = T$. Suppose $J_s(w) = A_s w^{1-\rho}/(1-\rho)$ for $s = t+1, \ldots, T$ for some t, where the A_s are defined in (9.27). Define $z = c/w$ and replace maximization over c in the Bellman equation with maximization over z. We have

$$J_t(w) = \max_{c,\pi} \left\{ \frac{1}{1-\rho} c^{1-\rho} + \delta E\left[J_{t+1}\left((w-c)\pi' R_{t+1}\right) \right] \right\}$$

$$= \max_{z,\pi} \left\{ \frac{1}{1-\rho} (zw)^{1-\rho} + \delta E\left[J_{t+1}\left(w(1-z)\pi' R_{t+1}\right) \right] \right\}$$

$$= \max_{z,\pi} \left\{ \frac{1}{1-\rho} (zw)^{1-\rho} + \delta E\left[A_{t+1}^{-\rho} \frac{1}{1-\rho} w^{1-\rho} (1-z)^{1-\rho} (\pi' R_{t+1})^{1-\rho} \right] \right\}$$

$$= w^{1-\rho} \max_{z,\pi} \left\{ \frac{1}{1-\rho} z^{1-\rho} + \delta E\left[A_{t+1}^{-\rho} (1-z)^{1-\rho} \frac{1}{1-\rho} (\pi' R_{t+1})^{1-\rho} \right] \right\}.$$

Because $A_{t+1}^{-\rho}(1-z)^{1-\rho}$ is a positive constant, the maximum in π is achieved at the portfolio that maximizes the single-period objective function in (9.24), and

Dynamic Portfolio Choice

the maximum value of that objective is $B^{1-\rho}/(1-\rho)$ by definition. Therefore,

$$J_t(w) = w^{1-\rho} \max_z \left\{ \frac{1}{1-\rho} z^{1-\rho} + \delta A_{t+1}^{-\rho}(1-z)^{1-\rho} \frac{1}{1-\rho} B^{1-\rho} \right\}. \quad (9.31)$$

The maximum here is achieved at

$$z = \frac{\gamma}{1+\gamma},$$

where

$$\gamma = A_{t+1} \left(\delta B^{1-\rho} \right)^{-1/\rho}.$$

Note that

$$\frac{\gamma}{1+\gamma} = \frac{A_{t+1} \left(\delta B^{1-\rho} \right)^{-1/\rho}}{1 + A_{t+1} \left(\delta B^{1-\rho} \right)^{-1/\rho}} = \frac{A_{t+1}}{(\delta B^{1-\rho})^{1/\rho} + A_{t+1}} = A_t,$$

using (9.30) for the last equality, so the optimal consumption is $A_t w$ as claimed in (9.28). Substitute the optimal z into (9.31) to obtain

$$J_t(w) = \frac{w^{1-\rho}}{1-\rho} \left[\left(\frac{\gamma}{1+\gamma} \right)^{1-\rho} + \gamma^{-\rho} \left(\frac{1}{1+\gamma} \right)^{1-\rho} \right] = \frac{w^{1-\rho}}{1-\rho} \left(\frac{1+\gamma}{\gamma} \right)^{\rho}.$$

Given that $\gamma/(1+\gamma) = A_t$, this verifies the formula (9.29) for the value function.

Infinite Horizon

Assume the horizon is infinite. Impose the no-borrowing constraint $C_t \leq W_t$ to preclude Ponzi schemes. As remarked previously, this is a very natural constraint, because we are assuming there is no labor income to borrow against. Assume (9.25) holds. We will show that the optimal consumption is

$$C_t = AW_t, \quad (9.32)$$

where

$$A = 1 - \left(\delta B^{1-\rho} \right)^{1/\rho}. \quad (9.33)$$

The value function is

$$J(w) = \frac{A^{-\rho}}{1-\rho} w^{1-\rho}. \quad (9.34)$$

Condition (9.25) is not very restrictive if $\rho > 1$. For example, it holds if $\rho > 1$ and there is a risk-free asset with return $R_f > 1$, because the maximum utility achievable in a single-period problem is at least as large as that achieved by investing everything in the risk-free asset. This implies $B \geq R_f$, so $B^{1-\rho} \leq R_f^{1-\rho} < 1$. Thus, (9.25) is an issue primarily when $\rho < 1$. In that case, if $\delta B^{1-\rho} > 1$, then the maximum expected lifetime utility is infinite. This can be seen from the fact that the value $J(w)$ is at least as large as the value of consuming nothing for n periods and then consuming all wealth, which is the value when maximizing the expected value of terminal wealth with n periods remaining and discounting by δ. Therefore, from (9.26),

$$J(w) \geq \frac{\delta^n B^{(1-\rho)n}}{1-\rho} w^{1-\rho} \to \infty$$

as $n \to \infty$ when $\delta B^{1-\rho} > 1$.

We write the Bellman equation here in terms of a generic function $\hat{J}$ (not necessarily equal to the true value function J):

$$\hat{J}(w) = \max_{c \leq w, \pi} \left\{ \frac{1}{1-\rho} c^{1-\rho} + \delta \mathsf{E}\left[\hat{J}\left((w-c)\pi' R_{t+1}\right)\right] \right\}. \qquad (9.35)$$

The value function J satisfies the Bellman equation (9.35), but there may in general be other functions $\hat{J}$ that also satisfy (9.35). In a finite-horizon model, the value of J_T is given, and the Bellman equation can be used to compute J_t by backward induction. However, in an infinite-horizon model, this procedure is not possible, so we must use some other method to ensure that a solution of the Bellman equation is actually the value function. See the end-of-chapter notes for further discussion and also Exercise 9.4.

With CRRA utility, the value function is a constant times $w^{1-\rho}$, as shown in Section 9.5. Because the value function has the same sign as the utility function, it is of the form (9.34) for some constant $A > 0$. We show below that (9.33) is the unique constant $A > 0$ such that (9.34) satisfies the Bellman equation. Therefore, (9.34) with A defined by (9.33) is the value function.

In a finite-horizon model, the choices (c, π) attaining the maximum in the Bellman equation, substituting the true value function $\hat{J} = J$ in the Bellman equation, are guaranteed to be optimal. This is also true in an infinite-horizon model if the transversality condition holds. The transversality condition is

$$\lim_{T \to \infty} \delta^T \mathsf{E}[J(W_T^*)] = 0, \qquad (9.36)$$

where W^* denotes the wealth process produced by the choices (c, π) attaining the maximum in the Bellman equation.[7] This condition automatically holds if the utility function is bounded above, as is the case when $\rho > 1$ (Exercise 9.5).

---

7. Note that this is not the same as the transversality condition discussed in Sections 8.5 and 9.1. The transversality condition discussed in Chapter 8 states that the contribution of consumption at dates $T, T+1, \ldots$ to the date–0 budget goes to zero as $T \to \infty$. Condition (9.36) states that the contribution of consumption at dates $T, T+1, \ldots$ to the date–0 expected lifetime utility goes to zero as $T \to \infty$.

Dynamic Portfolio Choice

It does not necessarily hold in general if the utility function is unbounded above (Exercise 9.4). We verify below that the transversality condition holds in this model and that the transversality condition and Bellman equation imply optimality.

As in the finite-horizon model, set $z = c/w$. We can maximize over $z \leq 1$ instead of over $c \leq w$. We want to find $A > 0$ satisfying

$$A^{-\rho}\left(\frac{1}{1-\rho}w^{1-\rho}\right)$$

$$= \max_{z,\pi}\left\{\frac{1}{1-\rho}(zw)^{1-\rho} + \delta A^{-\rho}\mathsf{E}\left[\frac{1}{1-\rho}\left(w(1-z)\pi'R_{t+1}\right)^{1-\rho}\right]\right\}$$

$$= w^{1-\rho}\max_{z,\pi}\left\{\frac{1}{1-\rho}z^{1-\rho} + \delta A^{-\rho}(1-z)^{1-\rho}\mathsf{E}\left[\frac{1}{1-\rho}\left(\pi'R_{t+1}\right)^{1-\rho}\right]\right\}. \tag{9.37}$$

Because $A^{-\rho}(1-z)^{1-\rho}$ is positive, the maximum in π is achieved at the portfolio that maximizes the single-period objective function in (9.24), and the maximum value of that objective is $B^{1-\rho}/(1-\rho)$ by definition. Making this substitution and canceling $w^{1-\rho}$, we see that (9.37) is equivalent to

$$\frac{1}{1-\rho}A^{-\rho} = \max_{z,\pi}\left\{\frac{1}{1-\rho}z^{1-\rho} + \frac{\delta A^{-\rho}B^{1-\rho}}{1-\rho}(1-z)^{1-\rho}\right\}.$$

The maximum here is achieved at

$$z = \frac{\gamma}{1+\gamma},$$

where

$$\gamma = A\left(\delta B^{1-\rho}\right)^{-1/\rho}.$$

Substituting the optimal z, we see that (9.37) is equivalent to

$$\frac{1}{1-\rho}A^{-\rho} = \frac{1}{1-\rho}\left[\left(\frac{\gamma}{1+\gamma}\right)^{1-\rho} + \gamma^{-\rho}\left(\frac{1}{1+\gamma}\right)^{1-\rho}\right] = \frac{1}{1-\rho}\left(\frac{1+\gamma}{\gamma}\right)^{\rho}.$$

Thus,

$$A = \frac{\gamma}{1+\gamma} = \frac{A\left(\delta B^{1-\rho}\right)^{-1/\rho}}{1+A\left(\delta B^{1-\rho}\right)^{-1/\rho}} = \frac{A}{(\delta B^{1-\rho})^{1/\rho}+A},$$

and the solution of this is (9.33). The condition $A = \gamma/(1+\gamma) = z$ also implies (9.32).

Let π^* denote the portfolio that maximizes the single-period objective function in (9.24), and let W^* denote the wealth process generated by this portfolio and consumption $C_t^* = AW_t^*$. We want to show that this portfolio and consumption are actually optimal. To this point, we have shown that they satisfy the Bellman equation, so

$$J(W_t^*) = u(C_t^*) + \delta E_t \left[J(W_{t+1}^*) \right]$$

for all t. Start at $t = 0$ and substitute this recursively to obtain

$$\begin{aligned} J(W_0) &= u(c_0^*) + \delta E \left[J(W_1^*) \right] \\ &= u(c_0^*) + \delta E \left[u(C_1^*) + \delta J(W_2^*) \right] \\ &\cdots \\ &= E \left[\sum_{t=0}^{T-1} \delta^t u(C_t^*) \right] + \delta^T E \left[J(W_T^*) \right]. \end{aligned}$$

From the monotone convergence theorem (Appendix A.5), we have

$$E \left[\sum_{t=0}^{T-1} \delta^t u(C_t^*) \right] \to E \left[\sum_{t=0}^{\infty} \delta^t u(C_t^*) \right]$$

as $T \to \infty$. Therefore,

$$J(W_0) = E \left[\sum_{t=0}^{\infty} \delta^t u(C_t^*) \right] + \lim_{T \to \infty} \delta^T E \left[J(W_T^*) \right]. \qquad (9.38)$$

Note that

$$W_T^* = W_0 (1-A)^T \prod_{t=0}^{T-1} (\pi' R_{t+1}).$$

Therefore,

$$\begin{aligned} \delta^T E \left[(W_T^*)^{1-\rho} \right] &= \delta^T W_0^{1-\rho} (1-A)^{(1-\rho)T} (1-\rho)^T \prod_{t=0}^{T-1} E \left[\frac{1}{1-\rho} (\pi' R_{t+1})^{1-\rho} \right] \\ &= W_0^{1-\rho} (1-A)^{(1-\rho)T} \left(\delta B^{1-\rho} \right)^T. \end{aligned}$$

Substitute

$$1 - A = (\delta B^{1-\rho})^{1/\rho}$$

from (9.33) to obtain

$$\delta^T E \left[(W_T^*)^{1-\rho} \right] = W_0^{1-\rho} \left(\delta B^{1-\rho} \right)^{T/\rho},$$

which converges to zero as $T \to \infty$ by virtue of (9.25). The formula (9.34) for the value function therefore implies the transversality condition (9.36). From the transversality condition and (9.38), we conclude that π^* and C^* are optimal.

9.7 NOTES AND REFERENCES

The portfolio choice method described for complete markets in Section 9.2 can also be applied, though less directly, when markets are incomplete or there are market frictions such as short sales constraints, margin requirements, or different borrowing and lending rates. Consider a finite-horizon finite-state model in which the market is incomplete. Add fictitious assets to complete the market. In the completed market, the investor can achieve expected utility at least as high as that achieved in the incomplete market, because not trading the new assets is always an option. If she can obtain higher expected utility in the completed market than in the incomplete market, then the consumption plan in the completed market must be infeasible in the incomplete market. The minimum expected utility attainable in any completed market is the expected utility attainable in the incomplete market, and the completed market solving the optimization problem "minimize the maximum attainable expected utility" is the market in which the new assets are priced by the SDF process $M_t = \delta^t u'(C_t)$, where C is the optimal consumption plan in the incomplete market. If we can solve the optimization problem to find what is called the least favorable fictitious completion, then we can find C and solve the incomplete market portfolio choice problem using the fictitious complete market as in Section 9.2. See He and Pearson (1991a).

The discussion of the optimal portfolio for a quadratic utility investor in Section 9.3 is based on Cochrane (2014). Mossin (1968) discusses maximizing the expected utility of terminal wealth with IID returns. He solves the problem with CRRA utility and shows that the optimal portfolio is the same in each period if and only if the investor has CRRA utility. Samuelson (1969) solves the finite-horizon problem with CRRA utility and consumption at each date. Hakansson (1970) solves the infinite-horizon problem with IID returns and CRRA utility (and also CARA utility).

Optimal portfolios for shifted log and shifted power utility can be deduced, in some circumstances, from the results for log and power utility. Consider shifted power utility with shift $\zeta > 0$ and with consumption only at the terminal date T. Assume there is a zero-coupon bond maturing at T. Recall that ζ can be interpreted as a subsistence level of consumption. An investor's total consumption equals the subsistence level plus a surplus: $c = \zeta + (c - \zeta)$. The investor can likewise separate her portfolio problem into two parts: She buys zero-coupon bonds maturing at T with a total face value of ζ,[8] and she invests

---

[8]. If there is consumption at each date, the investor buys zero-coupon bonds with face value of ζ for each maturity date $t = 1, \ldots, T$.

the remainder of her wealth in a portfolio to maximize power utility of the surplus consumption $c - \zeta$. Thus, the optimal portfolio for an investor with shifted power utility involves a position in a zero-coupon bond, and the remaining wealth is invested in the portfolio that is optimal for an investor with power utility. The same is true for a shift $\zeta < 0$, if the utility function is regarded as defined for all $c > \zeta$. In this case, the investor shorts zero-coupon bonds with a face value of ζ and invests the proceeds from the short sale plus her wealth in the portfolio that is optimal for an investor with power utility. However, it is not very sensible to allow negative consumption, and it is more natural to restrict the domain of the utility function to $c \geq 0$ when $\zeta < 0$.[9] When negative consumption is disallowed, the portfolio just described is infeasible. Back, Liu, and Teguia (2015) derive the solution to the problem when negative consumption is not allowed (in a continuous-time model).

The general properties of infinite-horizon stationary discounted dynamic programming problems are somewhat different for positive and negative utility functions, as mentioned earlier. If the utility function is bounded from below, we can add a constant and make it positive, or, if it is bounded from above, we can subtract a constant and make it negative, so the positive and negative cases include all utility functions that are bounded either from below or from above. Call these the positive (P) and negative (N) cases, respectively. If the utility function is bounded both from below and above, then it has the properties of both cases. Call this the bounded (B) case.

Here are important properties, with the cases in which they hold stated in parentheses. These properties are stated in terms of a portfolio choice problem, but they hold for general infinite-horizon stationary discounted dynamic programming problems (including those with state variables X_t). These results can be found in Hinderer (1970).

1. (P, N or B) The value function satisfies the Bellman equation.
2. (P, N, or B) Any optimal policy $(c(\cdot), \pi(\cdot))$ must attain the maximum in the Bellman equation (employing the true value function in the Bellman equation) with probability 1.
3. (P) The value function is the smallest positive solution of the Bellman equation.
4. (B) The value function is the unique bounded solution of the Bellman equation.

---

9. Recall that this is the LRT/DARA utility function that has increasing relative risk aversion. If $\rho > 1$, it is also bounded on the domain $[0, \infty)$ (Exercise 1.10).

5. (P or B) The value function can be computed by value iteration: Letting V_0^T denote the value function from a problem with intermediate consumption and horizon T, we have $V_0^T \to V_0$ as $T \to \infty$.
6. (N or B) If a policy $(c(\cdot), \pi(\cdot))$ attains the maximum in the Bellman equation with probability 1, using the true value function $\hat{J} = J$, then the policy is optimal.

Exercise 9.4 is an example of the positive case in which there are multiple solutions of the Bellman equation. It is also an example in which attaining the maximum in the Bellman equation, using the true value function in the Bellman equation, is not a sufficient condition for optimality, due to a failure of the transversality condition. Exercise 9.5 asks for a proof of no. 6 in cases N and B. Log utility is unbounded above and below, so it fits none of the cases listed above. Exercise 9.1 asks for a proof of the transversality condition with log utility and IID returns.

In general, regardless of the boundedness of the utility function, the following is true: If $\hat{J}$ is any solution of the Bellman equation and if

$$\lim_{T \to \infty} \mathsf{E}\left[\sum_{t=1}^{T} \delta^t u(C_t)\right] = \mathsf{E}\left[\sum_{t=1}^{\infty} \delta^t u(C_t)\right] \tag{9.39a}$$

and

$$\limsup_{T \to \infty} \delta^T \mathsf{E}\left[\hat{J}(W_T)\right] \geq 0 \tag{9.39b}$$

for every sequence of decisions (C_t, π_t), with W_t being the corresponding wealth process, and if

$$\lim_{T \to \infty} \delta^T \mathsf{E}\left[\hat{J}(W_T^*)\right] = 0, \tag{9.39c}$$

where W_t^* is the wealth process corresponding to the decisions that attain the maximum in the Bellman equation relative to $\hat{J}$, then (a) $\hat{J}$ is the value function, and (b) the decisions attaining the maximum in the Bellman equation are optimal. See Exercise 9.6. This is also true of general infinite-horizon stationary discounted dynamic programming problems. This fact immediately implies no. 4 above, because (9.39) holds whenever u and $\hat{J}$ are bounded. The lim sup equals the largest limit point of any subsequence, so (9.39b) holds if the limit is zero. Note that (9.39c) is the transversality condition (9.36), but relative to $\hat{J}$.

EXERCISES

9.1. Consider the infinite-horizon model with IID returns and no labor income. Assume
$$\max_\pi \mathsf{E}[\log \pi' R_{t+1}] < \infty.$$

(a) Calculate the unique constant γ such that
$$J(w) = \frac{\log w}{1-\delta} + \gamma$$
solves the Bellman equation.

(b) Show that the transversality condition
$$\lim_{T \to \infty} \delta^T \mathsf{E}[J(W_T^*)] = 0$$
holds.

(c) Show that the optimal portfolio is the one that maximizes $\mathsf{E}[\log \pi' R_{t+1}]$ and the optimal consumption is $C_t = (1-\delta)W_t$.

9.2. Consider the finite-horizon model with consumption at each date, state variables X_t, log utility, and no labor income. Assume $\max_\pi \mathsf{E}_t[\log(\pi' R_{t+1})]$ is finite for each t with probability 1. The value function at date T is
$$J_T(x,w) = \log w.$$

Define
$$\gamma_t = \frac{1 - \delta^{T+1-t}}{1-\delta}.$$

(a) Show that
$$J_t(x,w) = \gamma_t \log w + f_t(x)$$
for some functions f_t.

(b) Show that the optimal portfolio at each date t is the one that maximizes $\mathsf{E}_t[\log \pi' R_{t+1}]$ and that the optimal consumption at each date t is $C_t = W_t/\gamma_t$. Note: Similar results are true for power utility only when returns are IID. This exercise does not assume IID returns.

9.3. Consider the finite-horizon model with consumption at each date, IID returns, and no labor income. Suppose one of the assets is risk free with return R_f. Let **R** denote the vector of risky asset returns, let μ denote the expected value of **R**, and let Σ denote the covariance matrix of **R**. Assume Σ is nonsingular.

(a) For constants α, δ, κ_0, and γ_0, define

$$J_0(w) = -\kappa_0 e^{-\gamma_0 w}$$

and

$$J_1(w) = \max_{c,\phi} \left\{ -e^{-\alpha c} + \delta E[J_0((w-c)R_f + \phi'(R - R_f \iota))] \right\}.$$

Show that

$$J_1(w) = -\kappa_1 e^{-\gamma_1 w}$$

for constants κ_1 and γ_1.

(b) Using dynamic programming, deduce from the result of Part (a) that the value function J_t of an investor with CARA utility and horizon $T < \infty$ has constant absolute risk aversion. How does the risk aversion depend on the remaining time $T - t$ until the horizon?

(c) If the investor has an infinite horizon, then the value function is independent of t. A good guess would therefore be $J(w) = -\kappa_0 e^{-\gamma_0 w}$ where κ_0 and γ_0 are such that $\kappa_1 = \kappa_0$ and $\gamma_1 = \gamma_0$ in Part (a); that is, (κ_0, γ_0) is a fixed point of the map $(\kappa_0, \gamma_0) \mapsto (\kappa_1, \gamma_1)$ calculated in Part (a). Show that this implies

$$\gamma_0 = \left(\frac{R_f - 1}{R_f} \right) \alpha.$$

9.4. Suppose there is a single asset that is risk free with return $R_f > 1$. Consider an investor with an infinite horizon, utility function $u(c) = c$, and discount factor $\delta = 1/R_f$. Suppose she is constrained to consume $0 \leq C_t \leq W_t$.

(a) Show that the value function for this problem is $J(w) = w$.
(b) Show that the value function solves the Bellman equation.
(c) Show that $\hat{J}(w) = 2w$ also solves the Bellman equation.
(d) Show that, using the true value function $\hat{J}(w) = w$ in the Bellman equation, the suboptimal policy $C_t = 0$ for every t achieves the maximum for every value of w.

9.5. Consider the infinite-horizon model with IID returns and no labor income. Denote the investor's utility function by $u(c)$.

(a) Case B: Assume there is a constant K such that $-K \leq u(c) \leq K$ for each c. Show that the transversality condition (9.36) holds.
(b) Case N: Assume $u(c) \leq 0$ for each c and $J(w) > -\infty$ for each w. Show that the transversality condition (9.36) holds. Hint: Use (9.38) and the

definition of a value function to deduce that the limit in (9.36) is nonnegative.

9.6. Consider the infinite-horizon model with IID returns and no labor income. Denote the investor's utility function by $u(c)$. Let $\hat{J}$ be a function that solves the Bellman equation. Assume (9.39) holds. For arbitrary decisions (C_t, π_t), assume $\mathsf{E}[u(C_t)]$ and $\mathsf{E}[\hat{J}(W_t)]$ are finite for each t. Suppose (C_t^*, π_t^*) attain the maximum in the Bellman equation. Show that $\hat{J}$ is the value function and (C_t^*, π_t^*) are optimal.

10

Dynamic Asset Pricing

CONTENTS

10.1	CAPM, CCAPM, and ICAPM	234
10.2	Testing Conditional Models	246
10.3	Competitive Equilibria	247
10.4	Gordon Model and Representative Investors	249
10.5	Campbell-Shiller Linearization	251
10.6	Risk-Neutral Probabilities	254
10.7	Notes and References	256

The first two sections of this chapter describe dynamic factor models. As in a single-period model, dynamic factor pricing is equivalent to pricing via an SDF. However, there are at least four ways in which the study of dynamic factor pricing differs from that of static factor pricing:

- Consumption is different from wealth in a dynamic model, and the utility of consumption is different from the value of wealth. However, the marginal utility of consumption equals the marginal value of wealth (the envelope condition).
- An investor's value of wealth in a dynamic model typically depends on state variables that forecast future returns and/or labor income. Thus, the marginal value of wealth typically depends on state variables in addition to wealth.
- If the time periods in a dynamic model are short, then the change in marginal utility from one period to the next is approximately linear in consumption growth and the change in the marginal value of wealth is approximately linear in the changes in wealth and state variables. Thus, factor pricing with respect to marginal utility (= marginal value)

implies approximate factor pricing with respect to changes in consumption or changes in wealth and state variables. These relations can be aggregated over investors even if markets are incomplete and there is no representative investor, implying approximate factor pricing with respect to aggregate consumption or aggregate wealth and state variables. The approximate relations are exact in continuous time.
- In dynamic models, we can study the implications of exposures to risks and/or prices of risks changing over time. In particular, we can distinguish between conditional and unconditional models.

Sections 10.3 and 10.4 discuss representative investors in dynamic models. As in a single-period model, there is a representative investor in a dynamic model if the market is complete or if all investors have LRT utility functions with the same cautiousness parameter. Section 10.4 solves for the price of the market portfolio, the market return, and the risk-free return assuming a representative investor with CRRA utility and IID lognormal consumption growth. The results for the market return and risk-free return are the same as in a single-period model; in particular, we rederive the equity premium and risk-free rate puzzles. The new feature of the dynamic model is that the market dividend-price ratio (called the dividend yield) is constant over time. Thus, it cannot predict the market return. This is inconsistent with empirical evidence, which indicates that a high dividend yield predicts a low market return. Section 10.5 describes an approximate formula for the dividend-price ratio in terms of future returns and future dividend growth due to Campbell and Shiller (1988), which is useful for understanding return predictability.

10.1 CAPM, CCAPM, AND ICAPM

This section presents the Consumption-based Capital Asset Pricing Model (CCAPM) and the Intertemporal Capital Asset Pricing Model (ICAPM). The CCAPM states that risk premia depend on covariances with consumption growth. The ICAPM states that risk premia depend on covariances with market wealth and with state variables that determine investment opportunities and/or future labor income. The two models are consistent: Because optimal consumption depends on wealth and state variables, the covariances with market wealth and state variables in the ICAPM collapse to the covariance with consumption in the CCAPM. The CCAPM and ICAPM are based in discrete time on first-order Taylor series approximations—of the marginal utility of consumption and the marginal value of wealth, respectively. They are exact in continuous time (Chapter 14).

Aside from the CAPM, the CCAPM and ICAPM are the two most important factor models in asset pricing theory. The CCAPM fails to fit the data for reasons discussed in Section 7.3—the volatility of consumption growth is too small to fit the equity premium for reasonable values of risk aversion. Its failure is an important fact, because the CCAPM rests on minimal assumptions. Only time-additive utility is required; in particular, there is no need to assume a representative investor as in Section 7.3.

The ICAPM is an important organizing principle for interpreting the data. If it is found that an empirically motivated factor model seems to work, the immediate question is whether the factors are proxies for (projections of) some variables that determine investment opportunities; that is, whether the theory of the ICAPM can support the empirical results.

This section also discusses the conditional CAPM. In the conditional CAPM, risk premia are determined by betas with respect to the market return, as in the single-period CAPM, but betas and the market risk premium may depend on conditioning information. We distinguish it from the unconditional CAPM, which is the same formula but for unconditional expectations of excess returns and unconditional betas (unconditional covariances divided by the unconditional variance of the market return). Section 10.4 shows that the conditional CAPM holds when there is a representative investor and consumption growth is IID (it then follows from the CCAPM). The conditional CAPM also holds when there is no labor income and either returns are IID or investors have log utility (it then follows from the ICAPM). In general, the conditional CAPM does not imply the unconditional CAPM.

Conditional and Unconditional CAPM

Let $R_{m,t+1}$ denote the market return from t to $t+1$. The conditional CAPM is

$$(\forall i, t) \quad E_t[R_{i,t+1} - R_{z,t+1}] = \frac{\text{cov}_t(R_{i,t+1}, R_{m,t+1})}{\text{var}_t(R_{m,t+1})} E_t[R_{m,t+1} - R_{z,t+1}] \quad (10.1)$$

for some zero-beta return $R_{z,t+1}$ that is known at date t. The zero-beta return must equal the risk-free return $R_{f,t+1}$ if a money market account exists. Note that, because the zero-beta return is known at time t, we can replace the returns $R_{i,t+1}$ and $R_{m,t+1}$ in the covariance and variance in (10.1) by the excess returns $R_{i,t+1} - R_{z,t+1}$ and $R_{m,t+1} - R_{z,t+1}$ without affecting the covariance or variance. Thus, the conditional CAPM can be expressed entirely in terms of excess returns as

$$(\forall i,t) \quad E_t[R_{i,t+1} - R_{z,t+1}] = \frac{\text{cov}_t(R_{i,t+1} - R_{z,t+1}, R_{m,t+1} - R_{z,t+1})}{\text{var}_t(R_{m,t+1} - R_{z,t+1})}$$
$$\times E_t[R_{m,t+1} - R_{z,t+1}] \quad (10.2)$$

The unconditional CAPM is the formula (10.2) with substitution of unconditional expectations and the unconditional covariance and variance for the conditional versions. Specifically, the unconditional CAPM is

$$(\forall i,t) \quad E[R_{it} - R_{zt}] = \frac{\text{cov}(R_{it} - R_{zt}, R_{mt} - R_{zt})}{\text{var}(R_{mt} - R_{zt})} E[R_{mt} - R_{zt}]. \quad (10.3)$$

We subscript returns by time indices in (10.3), but the unconditional CAPM assumes that the risk premia, covariance, and variance are independent of time (the excess return processes are stationary). Note that we do *not* need to assume that the zero-beta return is the same for all t. If the excess returns $R_{it} - R_{zt}$ and $R_{mt} - R_{zt}$ are IID, then there is no difference between the unconditional and conditional versions of the CAPM.

The conditional CAPM does not imply the unconditional CAPM when excess returns are not IID. To see why, let β_{it} denote the conditional beta in (10.2). Take unconditional expectations of both sides of (10.2) to obtain

$$E[R_{i,t+1} - R_{z,t+1}] = E\left[\beta_{it} E_t[R_{m,t+1} - R_{z,t+1}]\right]$$
$$= E[\beta_{it}] E[R_{m,t+1} - R_{z,t+1}] + \text{cov}\left(\beta_{it}, E_t[R_{m,t+1} - R_{z,t+1}]\right). \quad (10.4)$$

The first term on the right-hand side of (10.4) is similar to the right-hand side of the unconditional CAPM (10.3), though

$$E[\beta_{it}] \neq \frac{\text{cov}(R_{it} - R_{zt}, R_{mt} - R_{zt})}{\text{var}(R_{mt} - R_{zt})}$$

in general. The primary difference between (10.3) and (10.4) is the covariance between the conditional beta and the conditional market risk premium that appears in (10.4).

As an example of the difference between the conditional and unconditional versions of the CAPM, suppose the market return $R_{m,t+1}$ is $\mu_t + \varepsilon_{t+1}$, where μ and ε are independent IID stochastic processes, with the mean of each ε_t being zero. Suppose $\mu_t = R_f - \Delta$ or $\mu_t = R_f + \Delta$ with probability $1/2$ each for some $\Delta > 0$, where R_f is the constant return of the money market account. Consider a trading strategy that invests 100% in the money market account when $\mu_t = R_f - \Delta$ and invests 100% in the market when $\mu_t = R_f + \Delta$. Let R_t denote the return of this strategy. The conditional CAPM clearly holds for this strategy. When

$\mu_t = R_f - \Delta$, the conditional beta of the strategy is 0, and its expected return is the risk-free return. When $\mu_t = R_f + \Delta$, the conditional beta is 1, and its expected return is the same as the expected market return. However, the unconditional CAPM does not hold. The unconditional expected market return is the risk-free return in this example, so the unconditional CAPM asserts that all unconditional expected returns should equal the risk-free return. However, the unconditional expected return of this strategy is $\mathsf{E}[R_t] = R_f + \Delta/2 > R_f$. The failure of the unconditional CAPM in this example can be traced to the covariance term in (10.4). The conditional beta of the strategy positively covaries with the conditional market risk premium. Consequently, the strategy has a higher expected return than predicted by the unconditional CAPM. This example is very special, but it illustrates a general principle. If the market risk premium varies over time (and there is evidence that it does), then a market-timing strategy that increases its exposure to the market when the risk premium is high and lowers its exposure when the risk premium is low will produce higher average returns than predicted by the unconditional CAPM, even if the conditional CAPM holds at each date.

We can write the covariance in (10.4) as

$$\frac{\mathrm{cov}(\beta_{it}, \mathsf{E}_t[R_{m,t+1} - R_{z,t+1}])}{\mathrm{var}(\mathsf{E}_t[R_{m,t+1} - R_{z,t+1}])} \times \mathrm{var}(\mathsf{E}_t[R_{m,t+1} - R_{z,t+1}]).$$

The first factor in this product is a beta of a beta: It is the unconditional beta of the conditional beta, with respect to the conditional market risk premium. The second factor is the variance of the conditional market risk premium. The second factor is independent of the asset i. When we make this substitution in (10.4), we see that the conditional CAPM implies that the unconditional risk premium is linear in the expected beta and in the beta of the conditional beta. Thus, the conditional CAPM implies an unconditional model that is similar to a two-factor model (it might be better to call it a two-characteristic model, the characteristics being the expected beta and the beta of the beta).

CCAPM

The CCAPM states that risk premia depend on covariances with aggregate consumption growth. It is an approximate relation in discrete time and an exact relation in continuous time. To derive it, we can start with the Euler equation (9.1), which states that an investor's MRS is a single-period SDF. Equation (8.8) shows that, as in a single-period model, risk premia are determined by covariances with any single-period SDF. Therefore, risk premia are determined by covariances with any investor's MRS. Specifically, for any investor with utility function u and at each date t,

$$(\forall i) \quad E_t[R_{i,t+1}] = \frac{u'(C_t)}{\delta E_t[u'(C_{t+1})]} - \frac{1}{E_t[u'(C_{t+1})]} \mathrm{cov}_t(R_{i,t+1}, u'(C_{t+1})). \tag{10.5}$$

This is a precise parallel of the single-period formula (7.13). In Chapter 7, we assume the investor is a representative investor, so that (7.13) expresses risk premia in terms of covariances with a function of aggregate consumption. Here, we take a different approach, which does not require the existence of a representative investor.

Assume the time periods are short, so the consumption changes $\Delta C_{t+1} = C_{t+1} - C_t$ are small. In this case, we can reasonably approximate the marginal utility $u'(C_{t+1})$ by a first-order Taylor series expansion around $u'(C_t)$; that is,

$$u'(C_{t+1}) \approx u'(C_t) + u''(C_t)\Delta C_{t+1}. \tag{10.6}$$

Substitute this into (10.5) to obtain

$$(\forall i) \quad E_t[R_{i,t+1}] \approx \frac{u'(C_t)}{\delta E_t[u'(C_{t+1})]} - \frac{u''(C_t)}{E_t[u'(C_{t+1})]} \mathrm{cov}_t(R_{i,t+1}, \Delta C_{t+1}). \tag{10.7}$$

This formula states that risk premia are approximately determined by covariances with any investor's consumption growth. There are two advantages to using covariances with consumption growth rather than covariances with marginal utility as in (10.5). First, consumption (at least, aggregate consumption) is observable, whereas the marginal utility of consumption is not. Second, consumption can be added across investors, whereas sums of marginal utilities are not especially meaningful. We discuss this further below.

In empirical implementations, it is important for variables to have stationary distributions. The change in aggregate consumption is not stationary, because $|\Delta C_t|$ tends to become larger over time, but the percent change probably is. So, we rewrite (10.7) as

$$(\forall i) \quad E_t[R_{i,t+1}] \approx \frac{u'(C_t)}{\delta E_t[u'(C_{t+1})]} - \frac{C_t u''(C_t)}{E_t[u'(C_{t+1})]} \mathrm{cov}_t\left(R_{i,t+1}, \frac{\Delta C_{t+1}}{C_t}\right). \tag{10.8}$$

If we want to assume a representative investor, then we can stop here. The formula (10.8) is one version of the CCAPM, in which C is aggregate consumption and u is the utility function of the representative investor. Note that if $E_t[u'(C_{t+1})] \approx u'(C_t)$, then the coefficient on the covariance is

$$-\frac{C_t u''(C_t)}{E_t[u'(C_{t+1})]} \approx -\frac{C_t u''(C_t)}{u'(C_t)},$$

which is the relative risk aversion of the representative investor. This is exactly true in continuous time. The term

$$\frac{u'(C_t)}{\delta E_t[u'(C_{t+1})]}$$

is the zero-beta return. Thus, this version of the CCAPM states that risk premia (using the zero-beta return as a proxy risk-free return) depend on covariances with the aggregate consumption growth rate, and the price of covariance risk is the representative investor's relative risk aversion.

The price of covariance risk in (10.8) is positive, so the CCAPM states that assets that have higher covariances with aggregate consumption have higher risk premia. An asset that has a high covariance with aggregate consumption is one that has its highest returns when consumption growth is high, which are "good times." Assets that are less positively correlated with consumption are more desirable for hedging purposes. According to the CCAPM, the more desirable assets with lower consumption covariances have higher prices and lower expected returns than do assets with higher consumption covariances.

As remarked before, we do not need to assume a representative investor to derive the CCAPM. We can simply add (10.7) across investors (after a slight rearrangement). Subscript each investor's utility and consumption by the index $h = 1 \ldots, H$. The formula (10.7) holds for each investor h. Divide both sides by the coefficient on the covariance, that is divide by

$$\frac{u''_h(C_{ht})}{E_t[u'_h(C_{h,t+1})]}.$$

Then, sum over h to obtain

$$\sum_h \frac{E_t[u'_h(C_{h,t+1})]}{u''_h(C_{ht})} E_t[R_{i,t+1}] \approx \sum_h \frac{u'_h(C_{ht})}{\delta u''_h(C_{ht})} - \text{cov}_t(R_{i,t+1}, \Delta C_{t+1}).$$

In this formula, C is aggregate consumption, as desired. Now, divide by the coefficient on $E_t[R_{i,t+1}]$ to obtain

$$E_t[R_{i,t+1}] \approx \sum_h \frac{u'_h(C_{ht})}{\delta u''_h(C_{ht})} \Big/ \sum_h \frac{E_t[u'_h(C_{h,t+1})]}{u''_h(C_{ht})}$$

$$- \left(1 \Big/ \sum_h \frac{E_t[u'_h(C_{h,t+1})]}{u''_h(C_{ht})} \right) \text{cov}_t(R_{i,t+1}, \Delta C_{t+1}).$$

The first term on the right-hand side is again the zero-beta return. To reduce the notation, denote the zero-beta return by $R_{z,t+1}$.[1] To obtain stationarity, multiply and divide the second term by C_t. This produces the general version of the CCAPM:

$$(\forall i) \quad E_t[R_{i,t+1}] \approx R_{z,t+1} - \left(C_t \bigg/ \sum_h \frac{E_t[u'_h(C_{h,t+1})]}{u''_h(C_{ht})} \right) \text{cov}_t\left(R_{i,t+1}, \frac{\Delta C_{t+1}}{C_t} \right). \quad (10.9)$$

It is possible to interpret the price of covariance risk in terms of relative risk aversion, even without a representative investor. Assume $E_t[u'_h(C_{h,t+1})] \approx u'_h(C_{ht})$ for each h. Making this substitution, the price of covariance risk in (10.9) is approximately aggregate consumption multiplied by aggregate absolute risk aversion (defined as always as the reciprocal of the sum of individual risk tolerances). Of course, consumption multiplied by absolute risk aversion is relative risk aversion, so it is sensible to say that the price of covariance risk is approximately aggregate relative risk aversion. This is exact in continuous time.

Letting ρ_t denote the price of covariance risk, the CCAPM (10.9) implies that conditional Sharpe ratios satisfy

$$\frac{E_t[R_{i,t+1}] - R_{z,t+1}}{\text{stdev}_t(R_{i,t+1})} \approx \rho_t \times \text{corr}_t\left(R_{i,t+1}, \frac{\Delta C_{t+1}}{C_t} \right) \times \text{stdev}_t\left(\frac{\Delta C_{t+1}}{C_t} \right).$$

Thus, they should approximately satisfy the following bound:

$$(\forall i, t) \quad \left| \frac{E_t[R_{i,t+1}] - R_{z,t+1}}{\text{stdev}_t(R_{i,t+1})} \right| \leq \rho_t \times \text{stdev}_t\left(\frac{\Delta C_{t+1}}{C_t} \right). \quad (10.10)$$

The same bound is derived in Exercise 7.2 by assuming lognormal consumption growth in a single-period model with a representative investor.

ICAPM

The ICAPM provides a theoretical foundation for macroeconomic variables to be pricing factors. If such variables are correlated with changes in investment opportunities (expected returns and risks) and/or changes in the distribution of labor income, then the ICAPM states that they are pricing factors. This foundation can be extended from macroeconomic variables to returns or excess returns, because we can replace any factors with their projections on returns or

---

1. The zero beta return $R_{z,t+1}$ is known at time t. The subscript $t+1$ indicates that it is a return (actually, an expected return) from time t to time $t+1$. This is the same convention we are using for the risk-free return $R_{f,t+1}$.

excess returns. Thus, if macroeconomic variables forecast (are correlated with the conditional distributions of) future returns and/or labor income, then the ICAPM implies that their projections on returns or excess returns are pricing factors. If we find empirically that certain returns or excess returns seem to work as pricing factors, then it is at least theoretically possible that they work because they are projections of such macroeconomic variables.

As a multifactor model, the ICAPM appears similar to the APT. However, the APT is silent on what the factors should be, whereas the ICAPM states that the factors should be market wealth and variables that predict future returns and labor income (or the projections of such variables). Moreover, the ICAPM provides some guidance as to what the factor risk premia should be, as we will see below.

The ICAPM can be derived in continuous time from the formula for optimal portfolios in terms of hedging demands, but here we will derive it as an approximate formula using the envelope condition (marginal value of wealth equals marginal utility of consumption), the Euler equation (the marginal utility of consumption is an SDF process), and the factor model (8.8) in terms of a single-period SDF. As with the CCAPM, the ICAPM is an approximate relation in discrete time and an exact relation in continuous time.

To reduce the notational burden, consider the infinite-horizon model.[2] Let $X = (X_1 \cdots X_k)'$ denote the vector of state variables. Consider an investor with utility function u and value function J. The Euler equation and SDF pricing model imply the pricing model (10.5) in which the factor is the marginal utility of consumption. Also, the envelope condition (9.22) states that $u'(C_t) = J_w(X_t, W_t)$ for all t, where we use subscripts on J to denote partial derivatives.

We use a first-order Taylor series approximation of the investor's marginal value of wealth at $t + 1$ around X_t and $W_t - C_t$. Recall that $W_t - C_t$ is the amount invested in a portfolio at date t. Include the risk-free return, if it exists, in the vector R_{t+1} of returns. Then, $W_{t+1} = (W_t - C_t)\pi_t' R_{t+1} + Y_{t+1}$. Use Δ again to denote changes ($\Delta X_{j,t+1} = X_{j,t+1} - X_{jt}$ and $\Delta W_{t+1} = W_{t+1} - W_t$). The approximation is

$$J_w(X_{t+1}, W_{t+1}) \approx J_w(X_t, W_t - C_t) + \sum_{j=1}^{k} J_{wx_j}(X_t, W_t - C_t)\Delta X_{j,t+1}$$

$$+ J_{ww}(X_t, W_t - C_t)(\Delta W_{t+1} + C_t).$$

---

2. This special case is considered only for simplicity: The ICAPM can also be derived in a finite-horizon model with or without intermediate consumption. We obviously do not have an envelope condition without intermediate consumption; nevertheless, the marginal value of wealth is proportional to an SDF process in that case also. This is demonstrated at the end of the section.

Substitute J_w for u' in (9.22) and use this approximation for $J_w(X_{t+1}, W_{t+1})$ to obtain

$$(\forall i) \quad \mathsf{E}_t[R_{i,t+1}] \approx \frac{J_w(X_t, W_t - C_t)}{\delta \mathsf{E}_t[J_w(X_{t+1}, W_{t+1})]}$$
$$- \sum_{j=1}^k \frac{J_{wx_j}(X_t, W_t - C_t)}{\mathsf{E}_t[J_w(X_{t+1}, W_{t+1})]} \mathrm{cov}_t(R_{i,t+1}, \Delta X_{j,t+1})$$
$$- \frac{J_{ww}(X_t, W_t - C_t)}{\mathsf{E}_t[J_w(X_{t+1}, W_{t+1})]} \mathrm{cov}_t(R_{i,t+1}, \Delta W_{t+1} + C_t). \quad (\mathbf{10.11})$$

To improve the chances for stationarity, convert to percent changes as

$$(\forall i) \quad \mathsf{E}_t[R_{i,t+1}] \approx \frac{J_w(X_t, W_t - C_t)}{\delta \mathsf{E}_t[J_w(X_{t+1}, W_{t+1})]}$$
$$- \sum_{j=1}^k \frac{X_{jt} J_{wx_j}(X_t, W_t - C_t)}{\mathsf{E}_t[J_w(X_{t+1}, W_{t+1})]} \mathrm{cov}_t\left(R_{i,t+1}, \frac{\Delta X_{j,t+1}}{X_{jt}}\right)$$
$$- \frac{(W_t - C_t) J_{ww}(X_t, W_t - C_t)}{\mathsf{E}_t[J_w(X_{t+1}, W_{t+1})]} \mathrm{cov}_t\left(R_{i,t+1}, \frac{\Delta W_{t+1} + C_t}{W_t - C_t}\right).$$
$$(\mathbf{10.12})$$

If there is a representative investor, then (10.12) holds for the representative investor. This is one version of the ICAPM.

When there is a representative investor, W is market wealth and C is aggregate consumption in (10.12). Also,

$$\frac{\Delta W_{t+1} + C_t}{W_t - C_t} = \frac{(W_t - C_t)(\pi'_{mt} R_{t+1} - 1) + Y_{t+1}}{W_t - C_t} = \pi'_{mt} R_{t+1} - 1 + \frac{Y_{t+1}}{W_t - C_t}.$$
$$(\mathbf{10.13})$$

This is the market return, including labor income as part of the payoff of the market portfolio. Thus, the ICAPM states that risk premia (using the zero-beta return as the risk-free return) depend on covariances with the market return and on covariances with changes in state variables.

To interpret the prices of risks (coefficients of the covariances) in (10.12), assume

$$\mathsf{E}_t[J_w(X_{t+1}, W_{t+1})] \approx J_w(X_t, W_t - C_t).$$

Then, the price of wealth risk in (10.12) is approximately the relative risk aversion of the representative investor's value function evaluated at $W_t - C_t$. Also, the price of risk of each state variable is approximately the elasticity of the marginal value of wealth with respect to the state variable. Specifically, the price of risk of state variable j is approximately

$$-\frac{X_{jt}J_{wx_j}(X_t, W_t - C_t)}{J_w(X_t, W_t - C_t)} = -\left.\frac{\partial \log J_w(x,w)}{\partial \log x_j}\right|_{x=X_t, w=W_t-C_t}. \qquad (10.14)$$

The price of risk is exactly the elasticity in continuous time.

The covariance with the market return (10.13) in the ICAPM (10.12) is the same as in the conditional CAPM. This is discussed further in the next subsection. Like the CAPM, the ICAPM states that assets that have higher covariances with the market return have higher expected returns. What is new in the ICAPM relative to the CAPM is of course the covariances with state variables. It is important to understand the prices of risk of the state variables. As a starting point, assume there is a representative investor with constant relative risk aversion $\rho > 1$ and no labor income. As shown in Section 9.5, the marginal value of wealth of such an investor moves in the opposite direction of the investor's value function in response to changes in state variables, meaning that the marginal value of wealth is low when investment opportunities are good. Consider a state variable for which an increase indicates "good times" in the sense of good investment opportunities. For such an investor and such a state variable, the elasticity (10.14) is positive, so the price of risk in (10.12) is positive. The ICAPM states that assets that covary highly with the state variable have low prices and high expected returns. They have low prices, because they have their highest payoffs when the state variable is high, which are good times; thus, they are not useful for hedging.

As shown in (10.14), the price of risk in the ICAPM of a state variable depends on how the state variable affects the marginal value of wealth, not how the state variable affects the value of wealth. For example, if there is a representative investor with log utility, then the prices of risk of state variables are zero, because, as shown in Section 9.5, the marginal value of wealth of a log utility investor does not depend on state variables. So, even if increases in a state variable indicate good times, assets that covary highly with the state variable do *not* have low prices and high risk premia if the representative investor has log utility. Log utility is the boundary between $\rho > 1$ and $\rho < 1$. It is not common to assume $\rho < 1$, because higher values of ρ seem more reasonable. But, another reason for not assuming $\rho < 1$ is that when there is a representative investor with constant relative risk aversion $\rho < 1$, the prices of risk of state variables seem to have the "wrong" signs. Consider a state variable for which increases indicate better investment opportunities (good times). If $\rho < 1$, then the ICAPM states that assets that covary highly with the state variable have high prices and low expected returns, because the investor has a higher marginal value of wealth when investment opportunities are better and hence is willing to pay more for assets that have their highest payoffs in those times.

As with the CCAPM, it is not necessary to assume a representative investor to derive the ICAPM. We can start with the formula (10.11) for each investor h.

In that formula, the covariance with wealth is the covariance with the investor's change in wealth. Divide both sides of (10.11) by

$$\frac{J_{hww}(X_t, W_{ht} - C_{ht})}{E_t[J_{hw}(X_{t+1}, W_{h,t+1})]}$$

for each investor h and then add over h, so that the covariance with wealth is the covariance with the change in aggregate wealth. Then, divide by the coefficient of the expected return on the left-hand side. These are analogous to the steps used to derive the CCAPM without a representative investor. Finally, convert to percent changes as in going from (10.11) to (10.12). The result is the ICAPM (10.12), but the price of wealth risk (the coefficient of the covariance with $(\Delta W_{t+1} + C_t)/(W_t - C_t)$) is aggregate relative risk aversion, that is, market wealth multiplied by the reciprocal of the aggregate risk tolerance for wealth. The prices of risk of the state variables (the coefficients of the covariances with $\Delta X_{j,t+1}/X_{jt}$) are approximately weighted averages of the investors' elasticities

$$-\frac{\partial \log J_{hw}(x, w)}{\partial \log x_j}\bigg|_{x=X_t, w=W_{ht}-C_{ht}}.$$

The weight on investor h's elasticity in this weighted average is the investor's risk tolerance for wealth divided by the aggregate risk tolerance for wealth.

We will demonstrate that the marginal value of wealth defines an SDF process even when an investor maximizes the expected utility of terminal wealth (in which case the envelope condition cannot be applied). In this case, the Bellman equation is

$$V_t(X_t, W_t) = \max_{\pi} E_t \left[V_{t+1}(X_{t+1}, Y_{t+1} + W_t\pi'R_{t+1})\right]. \tag{10.15}$$

Let V_{tw} denote the partial derivative of the value function V_t with respect to wealth. We will show that, at the optimum,

$$\frac{V_{tw}(X_t, W_t)}{V_{0w}(X_0, W_0)} \tag{10.16}$$

is an SDF process.

Assume there exists a solution π_t of the maximization problem in (10.15). The envelope theorem states that the partial derivative of (10.15) with respect to W_t is the same whether π is varied or held fixed at the optimum. Holding it fixed and differentiating both sides of (10.15) with respect to W_t (and making the mild assumption that differentiation and expectation can be interchanged) gives

$$V_{tw}(X_t, W_t) = E_t[V_{t+1,w}(X_{t+1}, W_{t+1})\pi'_t R_{t+1}]. \tag{10.17}$$

Moreover, the first-order condition for the maximization problem in (10.15), subject to the constraint $\iota'\pi = 1$, is

$$E_t[V_{t+1,w}(X_{t+1}, W_{t+1})W_t R_{t+1}] = \lambda \iota,$$

where λ is the Lagrange multiplier for the constraint. Equivalently,

$$E_t[V_{t+1,w}(X_{t+1}, W_{t+1})R_{t+1}] = \frac{\lambda}{W_t}\iota. \tag{10.18}$$

Hence,

$$\frac{\lambda}{W_t} = \pi'_t\left(\frac{\lambda}{W_t}\iota\right) = \pi'_t E_t[V_{t+1,w}(X_{t+1}, W_{t+1})R_{t+1}]$$
$$= V_{tw}(X_t, W_t),$$

using the constraint $\iota'\pi = 1$ for the first equality, (10.18) for the second, and (10.17) for the third. Combining this with (10.18) shows that

$$E_t\left[\frac{V_{t+1,w}(X_{t+1}, W_{t+1})}{V_{tw}(X_t, W_t)}R_{t+1}\right] = \iota,$$

which implies that (10.16) is an SDF process.

ICAPM and Conditional CAPM

The ICAPM implies the conditional CAPM in two circumstances. First, assume returns and labor income are IID. Then, there are no state variables. So, the only covariance in the ICAPM (10.12) is the covariance with the market return (10.13). Under this assumption, the ICAPM becomes

$$(\forall i) \quad E_t[R_{i,t+1}] \approx R_{z,t+1} - \rho_t \operatorname{cov}_t(R_{i,t+1}, R_{m,t+1}), \tag{10.19}$$

where we use $R_{z,t+1}$ to denote the zero-beta return and ρ_t to denote the (approximate) aggregate relative risk aversion for wealth.

The second circumstance in which the ICAPM implies the conditional CAPM is when investors have log utility and no labor income. In this environment, investors' marginal values of wealth do not depend on state variables (Section 9.5). Therefore, again, the only covariance in the ICAPM is the covariance with the market return. As in the previous paragraph, this implies an approximate version of the conditional CAPM. The approximations are exact in continuous time.

10.2 TESTING CONDITIONAL MODELS

In general, empirical tests of models are based on the assumption that the variables of interest have stationary and ergodic distributions, implying that sample averages converge to population means, that is, to unconditional expectations, as the number of observations increases. A model needs to be converted to some statement about unconditional expectations in order to be tested.

First, consider testing a hypothesis that a stochastic process Z is a sequence of single-period SDFs. Consider the case in which a money market account exists. Then, the hypothesis implies

$$(\forall i, t) \quad E_t[Z_{t+1}(R_{i,t+1} - R_{f,t+1})] = 0. \qquad (10.20)$$

By iterated expectations, we can replace the conditional expectation by the unconditional expectation:

$$(\forall i, t) \quad E[Z_{t+1}(R_{i,t+1} - R_{f,t+1})] = 0. \qquad (10.21)$$

Hence, we can test the hypothesis about Z by testing whether the sample mean of $Z_t(R_{it} - R_{ft})$ is zero for some set of returns R_i. If not, then the hypothesis is rejected. However, there are more powerful tests, because (10.21) does not use the full strength of the condition (10.20).

Condition (10.20) implies (is in fact equivalent to)

$$(\forall i, t) \quad E[Z_{t+1}(R_{i,t+1} - R_{f,t+1})\pi_{it}] = 0 \qquad (10.22)$$

for every random variable π_{it} that depends only on date–t information and has a finite mean.[3] Let R_{t+1} denote the vector of risky asset returns, and let π_t denote the vector with elements π_{it}. Adding (10.22) across assets implies

$$(\forall t) \quad E[Z_{t+1}\pi_t'(R_{t+1} - R_{f,t+1}\iota)] = 0. \qquad (10.23)$$

Equation (10.23) states that Z_{t+1} is orthogonal to the excess return of the portfolio π_t. The hypothesis that Z is a sequence of single-period SDFs is

---

3. This is a consequence of the law of iterated expectations. Calculate the expectation in (10.22) as

$$E[Z_{t+1}(R_{i,t+1} - R_{f,t+1})\pi_{it}] = E\left[E_t[Z_{t+1}(R_{i,t+1} - R_{f,t+1})\pi_{i,t}]\right]$$
$$= E\left[\pi_{it}E_t[Z_{t+1}(R_{i,t+1} - R_{f,t+1})]\right] = 0.$$

equivalent to (10.23) holding for each portfolio process π. The only restriction on portfolios is that each must be formed using information available at the beginning of the return period. This equivalence is called the managed portfolio theorem. It means that we can test the conditional hypothesis (10.20) for individual assets by performing unconditional tests on managed portfolios.

Now consider a conditional factor model

$$(\forall i, t) \quad \mathsf{E}_t[R_{i,t+1}] = R_{z,t+1} + \psi_t' \operatorname{Cov}_t(F_{t+1}, R_{i,t+1}). \tag{10.24}$$

Here, F is the vector of factors and $R_{z,t+1}$ is the zero-beta return (known at time t). Provided that the zero-beta return is nonzero (meaning that the zero-beta *rate* of return is not -100%), we can divide by it and define

$$Z_{t+1} = \frac{1}{R_{z,t+1}} - \frac{1}{R_{z,t+1}} \psi_t' \big(F_{t+1} - \mathsf{E}_t[F_{t+1}]\big). \tag{10.25}$$

The same algebra as in Section 6.2 shows that the hypothesis (10.24) with $R_{z,t+1} \neq 0$ is equivalent to Z_{t+1} defined in (10.25) being an SDF for the period from t to $t+1$. The managed portfolio theorem implies that testing the conditional factor model (10.24) is equivalent to testing whether Z_{t+1} is unconditionally orthogonal to all managed portfolio excess returns $\pi_t'(R_{t+1} - R_{z,t+1}\iota)$.

10.3 COMPETITIVE EQUILIBRIA

As in a single-period model, equilibrium is defined by markets clearing and investors optimizing, taking prices as given. However, in a dynamic model, instead of simply choosing a portfolio, each investor chooses a portfolio process. Moreover, the standard definition of competitive equilibrium (rational expectations equilibrium) is that, instead of taking only date–0 prices as given, each investor takes the price *stochastic process* for each asset as given, meaning that she anticipates perfectly how the price at each date depends on the state of the world (and believes that her actions have no effects on prices). Markets clear if the investors' plans (portfolio processes) are consistent in the sense that total demand for each asset equals total supply at each date and in each state of the world.

As explained in Section 4.3 for a single-period model, competitive equilibria in complete securities markets are equivalent to Arrow-Debreu equilibria. In a dynamic model, consumption in each date-state pair is considered a separate good in the Arrow-Debreu economy. The price of each such good is given by an

SDF process. Equilibrium in the Arrow-Debreu economy is defined by an SDF process such that the consumption good market clears, with optimal consumption being determined as described in Section 9.2. The SDF process determines equilibrium asset prices in the securities market. As in a single-period model, this perspective enables us to establish the Pareto optimality of competitive equilibria in complete securities markets.

As in a single-period model, there is a representative investor at a Pareto-optimal competitive equilibrium. The representative investor's utility function is defined at each date as

$$u_t(c) = \max \left\{ \sum_{h=1}^{H} \lambda_h \delta_h^t u_h(c_h) \,\Big|\, \sum_{h=1}^{H} c_h = c \right\}$$

for some positive weights λ_h. If the discount factor δ_h is the same for each investor h, say equal to δ, then

$$u_t(c) = \delta^t u(c),$$

where

$$u(c) = \max \left\{ \sum_{h=1}^{H} \lambda_h u_h(c_h) \,\Big|\, \sum_{h=1}^{H} c_h = c \right\}.$$

If the risk tolerance of each utility function u_h is $\tau_h(c) = A_h + Bc$, where the cautiousness parameter B is the same for each investor, then the representative investor's utility function has risk tolerance $\tau(c) = A + Bc$, where $A = \sum_{h=1}^{H} A_h$. This is shown in Section 7.2. For example, if all investors have CARA utility, then the representative investor has CARA utility, and if all investors have CRRA utility with the same coefficient of relative risk aversion, then the representative investor has CRRA utility with the same coefficient of relative risk aversion.

As in a single-period model, the envelope theorem implies that, at a Pareto-optimal competitive equilibrium, the representative investor's MRS

$$\frac{u_t'(C_t)}{u_0'(C_0)}$$

is an SDF process, where C denotes aggregate consumption. In particular, if all investors have the same discount factor δ, then

$$\frac{\delta^t u'(C_t)}{u'(C_0)}$$

is an SDF process.

10.4 GORDON MODEL AND REPRESENTATIVE INVESTORS

In this section, we analyze an economy with IID consumption growth and a representative investor having CRRA utility. We extend the results of Section 7.3 to this dynamic model. We will see that the market return is IID and of the same form as the market return in Section 7.3. The risk-free return is constant and equal to the risk-free return in Section 7.3. Thus, the equity premium puzzle and risk-free rate puzzles discussed in Section 7.3 arise in exactly the same way in this dynamic model. The new feature of the dynamic model is that we can show that the market price-dividend ratio is constant over time—this implies that the market return is IID.

Before beginning the analysis of the representative investor model, we consider a more general model, called the Gordon growth model. The assumptions of the Gordon growth model are that the vector stochastic process consisting of the single-period SDF and an asset's dividend growth rate is IID over time and that there is no bubble in the price of the asset. It yields a simple formula for the asset's price-dividend ratio in terms of the expected return and expected dividend growth rate.

Gordon Growth Model

We study a single asset in this subsection. It could be an individual asset, or it could be the market portfolio. Let P denote its price, D denote its dividend, and R denote its return. Let M be a strictly positive SDF process. Assume there is no bubble in the price of the asset, so the price at any date t is

$$P_t = \sum_{u=t+1}^{\infty} \mathsf{E}_t\left[\frac{M_u}{M_t}D_u\right] = D_t \sum_{u=t+1}^{\infty} \mathsf{E}_t\left[\left(\frac{M_u}{M_t}\right)\left(\frac{D_u}{D_t}\right)\right]$$

$$= D_t \sum_{u=t+1}^{\infty} \mathsf{E}_t\left[\prod_{i=t}^{u-1}\left(\frac{M_{i+1}}{M_i}\right)\left(\frac{D_{i+1}}{D_i}\right)\right]. \quad (10.26)$$

Assume the vector stochastic process consisting of the single-period SDF M_{t+1}/M_t and dividend growth D_{t+1}/D_t is IID over time. Set

$$\nu = \mathsf{E}\left[\left(\frac{M_{i+1}}{M_i}\right)\left(\frac{D_{i+1}}{D_i}\right)\right]. \quad (10.27)$$

Assume $v < 1$. By the IID hypothesis, the expected product in (10.26) is the product of the expectations, and each expectation equals v, so

$$\frac{P_t}{D_t} = \sum_{u=t+1}^{\infty} \prod_{i=t}^{u-1} \mathsf{E}\left[\left(\frac{M_{i+1}}{M_i}\right)\left(\frac{D_{i+1}}{D_i}\right)\right] = \sum_{u=t+1}^{\infty} v^{u-t} = \frac{v}{1-v}. \qquad (10.28)$$

The formula (10.28) for the price-dividend ratio implies that the asset return is

$$R_{t+1} = \frac{D_{t+1} + P_{t+1}}{P_t} = \frac{1 + P_{t+1}/D_{t+1}}{P_t/D_t}\left(\frac{D_{t+1}}{D_t}\right)$$

$$= \frac{1}{v}\left(\frac{D_{t+1}}{D_t}\right). \qquad (10.29)$$

Taking expectations on both sides and then rearranging yields

$$v = \frac{\mathsf{E}[D_{t+1}/D_t]}{\mathsf{E}[R_{t+1}]},$$

and substituting this into (10.28) produces

$$\frac{P_t}{D_t} = \frac{\mathsf{E}[D_{t+1}/D_t]}{\mathsf{E}[R_{+1}] - \mathsf{E}[D_{t+1}/D_t]}. \qquad (10.30)$$

Equation (10.30) is the Gordon formula. It states that the price-dividend ratio equals gross expected dividend growth divided by the difference between the expected rate of return and the expected dividend growth rate. Under our IID hypothesis, the expectations in (10.30) are the same for all t, and the price-dividend ratio is constant over time.

Representative Investor and IID Consumption Growth

We can apply the results of the previous subsection to calculate the market return when there is a representative investor and IID consumption growth. Define the market portfolio as the claim to aggregate consumption C. The dividend paid by the market portfolio is aggregate consumption. Assume consumption growth C_{t+1}/C_t is IID. Let u denote the utility function of the representative investor, and define

$$v = \delta \mathsf{E}\left[\frac{u'(C_{t+1})}{u'(C_t)} \frac{C_{t+1}}{C_t}\right].$$

There is an SDF process M with

$$\frac{M_{t+1}}{M_t} = \frac{\delta u'(C_{t+1})}{u'(C_t)},$$

Dynamic Asset Pricing

so (10.29) implies that the market return is

$$R_{m,t+1} = \frac{1}{\nu}\left(\frac{C_{t+1}}{C_t}\right). \tag{10.31}$$

Thus, the market return is proportional to consumption growth. Hence, the covariance with consumption growth in the CCAPM is proportional to the covariance with the market return. Therefore, the conditional CAPM holds when there is a representative investor and consumption growth is IID.

We can obtain more explicit formulas by assuming the representative investor has constant relative risk aversion ρ. In that case,

$$\frac{\delta u'(C_{t+1})}{u'(C_t)} = \delta\left(\frac{C_{t+1}}{C_t}\right)^{-\rho}.$$

Therefore,

$$\nu = \delta \mathsf{E}\left[\left(\frac{C_{t+1}}{C_t}\right)^{1-\rho}\right].$$

Equation (10.31) implies

$$R_{m,t+1} = \frac{1}{\delta} \cdot \frac{C_{t+1}/C_t}{\mathsf{E}[(C_{t+1}/C_t)^{1-\rho}]}. \tag{10.32}$$

This formula for the market return is a direct extension of (7.17). The usual formula for the risk-free return gives us

$$R_{f,t+1} = \frac{1}{\mathsf{E}[M_{t+1}/M_t]}$$
$$= \frac{1}{\delta} \cdot \frac{1}{\mathsf{E}[(C_{t+1}/C_t)^{-\rho}]}, \tag{10.33}$$

which is a direct extension of (7.22). If we make the additional assumption of lognormal consumption growth, then we obtain the equity premium and risk-free rate puzzles in exactly the same form as in Section 7.3.

10.5 CAMPBELL-SHILLER LINEARIZATION

The Gordon formula (10.30) establishes a link between the dividend-price ratio and the expected return and expected dividend growth of an asset. In log form, the relation is

$$\log\left(\frac{D_t}{P_t}\right) = \log\left(\mathsf{E}[R_{t+1}] - \mathsf{E}[D_{t+1}/D_t]\right) - \log\left(\mathsf{E}[D_{t+1}/D_t]\right).$$

This formula states that if the dividend-price ratio is high, then either the future return is expected to be high or dividend growth is expected to be low. This suggests that, when the dividend-price ratio increases, either future returns will be higher on average or future dividend growth will be lower on average. However, we cannot draw that conclusion from the Gordon formula, because in the Gordon model the dividend-price ratio is constant. In this section, we derive an approximate formula for the dividend-price ratio in a non-IID world due to Campbell and Shiller (1988) that shows that, as just conjectured, when the dividend-price ratio increases, then either future returns will be high or future dividend growth will be low. Thus, the dividend-price ratio must predict either future returns or future dividend growth (or both).

As in the discussion of the Gordon growth model, we are going to study a single asset. Let P denote its price and D denote its dividend. Let h denote the continuously compounded return, δ denote the log dividend-price ratio, and g denote the dividend growth rate; that is,

$$h_{t+1} = \log\left(\frac{P_{t+1} + D_{t+1}}{P_t}\right),$$

$$\delta_t = \log\left(\frac{D_t}{P_t}\right),$$

$$g_t = \log\left(\frac{D_t}{D_{t-1}}\right).$$

Assume δ is a stationary stochastic processes, and set $\delta^* = \mathsf{E}[\delta_t]$. The definitions of h, δ, and g imply

$$e^{h_{t+1} - g_{t+1}} = \frac{P_{t+1} + D_{t+1}}{P_t} \bigg/ \frac{D_{t+1}}{D_t} = \frac{D_t}{P_t}\left(1 + \frac{P_{t+1}}{D_{t+1}}\right)$$

and

$$e^{\delta_t - \delta_{t+1}} + e^{\delta_t} = \frac{D_t}{P_t} \bigg/ \frac{D_{t+1}}{P_{t+1}} + \frac{D_t}{P_t} = \frac{D_t}{P_t}\left(1 + \frac{P_{t+1}}{D_{t+1}}\right).$$

Therefore,

$$e^{h_{t+1} - g_{t+1}} = e^{\delta_t - \delta_{t+1}} + e^{\delta_t} \Leftrightarrow h_{t+1} = \log\left(e^{\delta_t - \delta_{t+1}} + e^{\delta_t}\right) + g_{t+1}.$$

Take a first-order Taylor series expansion of the function

$$(a, b) \mapsto \log\left(e^{a-b} + e^a\right)$$

around (δ^*, δ^*). Then, we obtain

$$h_{t+1} \approx \log(1+e^{\delta^*}) + (\delta_t - \delta^*) - \frac{1}{1+e^{\delta^*}}(\delta_{t+1} - \delta^*) + g_{t+1}$$
$$= \kappa_0 + \delta_t - \kappa_1 \delta_{t+1} + g_{t+1}, \qquad (10.34)$$

where we define

$$\kappa_0 = \log(1+e^{\delta^*}) - \frac{e^{\delta^*} \delta^*}{1+e^{\delta^*}},$$

$$\kappa_1 = \frac{1}{1+e^{\delta^*}}.$$

Formula (10.34) states that the continuously compounded return is approximately a constant minus the change in the log dividend-price ratio (with the end-of-period log dividend-price ratio δ_{t+1} discounted by κ_1) plus the growth rate of dividends.

Use κ_1 as a discount factor and add the discounted continuously compounded returns for any integer $n \geq 1$ as

$$\sum_{i=1}^{n} \kappa_1^{i-1} h_{t+i} \approx \kappa_0 \sum_{i=1}^{n} \kappa_1^{i-1} + \delta_t - \kappa_1^n \delta_{t+n} + \sum_{i=1}^{n} \kappa_1^{i-1} g_{t+i}.$$

We can rearrange this as

$$\delta_t \approx -\kappa_1^n \delta_{t+n} + \sum_{i=1}^{n} \kappa_1^{i-1} h_{t+i} - \sum_{i=1}^{n} \kappa_1^{i-1} g_{t+i} - \kappa_0 \sum_{i=1}^{n} \kappa_1^{i-1}.$$

If the infinite sums are convergent and δ_{t+n} does not grow too fast (so $\kappa_1^n \delta_{t+n} \to 0$), then we can take the limit as $n \to \infty$ and obtain

$$\delta_t \approx \sum_{i=1}^{\infty} \kappa_1^{i-1} h_{t+i} - \sum_{i=1}^{\infty} \kappa_1^{i-1} g_{t+i} - \frac{\kappa_0}{1-\kappa_1}. \qquad (10.35)$$

Thus, the log dividend-price ratio is approximately the discounted sum of future returns minus the discounted sum of future dividend growth rates minus a constant. When the dividend-price ratio is low, then either future returns will be low or future dividend growth will be high. In other words, the dividend-price ratio must (assuming the approximation is reasonable) predict either returns or dividend growth or both.

If the conditional expected return $E_t[h_{t+1}]$ is constant, then taking conditional expectations in (10.35) and using iterated expectations yields

$$\delta_t \approx -\sum_{i=1}^{\infty} \kappa_1^{i-1} E_t[g_{t+i}] + \frac{r - \kappa_0}{1-\kappa_1}, \qquad (10.36)$$

where $r = \mathsf{E}_t[h_{t+1}]$. In this case, the dividend-price ratio forecasts future dividend growth. Alternatively, suppose the risk-free rate $r_{ft} = \log(1+R_{ft})$ varies over time and the conditional expected return satisfies $\mathsf{E}_t[h_{t+1}] = r_{f,t+1} + c$ for a constant risk premium c. Then,

$$\delta_t \approx \sum_{i=1}^{\infty} \kappa_1^{i-1} \mathsf{E}_t[r_{f,t+i} - g_{t+i}] + \frac{c - \kappa_0}{1 - \gamma}. \tag{10.37}$$

In this case, the dividend-price ratio forecasts the difference between future risk-free rates and future dividend growth rates.

10.6 RISK-NEUTRAL PROBABILITIES

This section explains how risk-neutral probabilities—first introduced in Section 3.4—are defined in dynamic models. Suppose a money market account exists and denote the risk-free return from t to $t+1$ by $R_{f,t+1} = 1 + r_{f,t+1}$. Consider investing 1 in the money market account at date 0 and reinvesting all interest. Let V_t denote the amount you would have at time t. This is just the compounded money market return:

$$V_t \stackrel{\text{def}}{=} \prod_{s=1}^{t} R_{fs}.$$

Assume V is strictly positive (the rate of return of the money market account is never -100%). Denote the dividend-reinvested price of each risky asset i by S_i. The conditions in this section apply to each risky asset, and it is convenient to drop the i subscript.

Denote the probability under which we have been taking expectations by $\mathbb{P}$. As remarked in Section 3.4, this is called the physical probability (or the actual probability or something similar). A risk-neutral probability for the horizon $T < \infty$ is defined to be a probability $\mathbb{Q}$ on the events that can be distinguished by date T with the following properties:

(A) For any event A that can be distinguished by date T, $\mathbb{P}(A) = 0$ if and only if $\mathbb{Q}(A) = 0$.
(B) For each asset, the stochastic process S_t/V_t is a martingale on the time horizon $\{0, 1, \ldots, T\}$ relative to $\mathbb{Q}$.

Property (A) is expressed by saying that $\mathbb{Q}$ is equivalent to $\mathbb{P}$ (for the events distinguishable at date T). A risk-neutral probability is also called an equivalent martingale measure.

Property (B) means that

$$\frac{S_t}{V_t} = \mathsf{E}^*_t\left[\frac{S_{t+1}}{V_{t+1}}\right] \qquad (10.38)$$

for each $t < T$, where E^* denotes expectation with respect to $\mathbb{Q}$. Because V_{t+1} is known at t and $V_{t+1}/V_t = R_{f,t+1}$, this implies

$$S_t = \frac{\mathsf{E}^*_t[S_{t+1}]}{R_{f,t+1}} \qquad (10.39)$$

for each t. Thus, the dividend-reinvested price equals the expected future dividend-reinvested price discounted at the risk-free rate, when we take expectations under the risk-neutral probability. Multiplying both sides by $R_{f,t+1}/S_t$ and subtracting 1, we can express (10.39) as

$$\frac{\mathsf{E}^*_t[S_{t+1}] - S_t}{S_t} = r_{f,t+1}. \qquad (10.40)$$

This says that the expected rate of return of each asset equals the risk-free rate, when expectations are taken under $\mathbb{Q}$. As mentioned in Chapter 3, this would be the case in equilibrium if investors were risk neutral, whence the term "risk-neutral probability." The return S_{t+1}/S_t is equal to $(P_{t+1} + D_{t+1})/P_t$, so we can also write (10.39) as

$$P_t = \frac{\mathsf{E}^*_t[P_{t+1} + D_{t+1}]}{R_{f,t+1}}. \qquad (10.41)$$

This says that the price of any asset is the expectation of the future cum-dividend price discounted at the risk-free rate, when we take expectations under a risk-neutral probability.

Valuation via a risk-neutral probability is equivalent to valuing via an SDF process, and the martingale property (B) relative to a risk-neutral probability is equivalent to the martingale property discussed in Section 8.2. We can construct a risk-neutral probability from a strictly positive SDF according to the formula

$$\mathbb{Q}(A) = \mathsf{E}[V_T M_T 1_A] \qquad (10.42)$$

for each event A that is distinguishable at T (where, as usual, 1_A denotes the random variable that is equal to 1 on A and 0 on the complement of A).[4] We

---

4. A common way to write the formula (10.42) is

$$\frac{d\mathbb{Q}}{d\mathbb{P}} = M_T V_T,$$

and $M_T V_T$ is called the Radon-Nikodym derivative of $\mathbb{Q}$ with respect to $\mathbb{P}$ (both restricted to the events observable at T). This notation is useful when we write expectations as, for example, show below that the definition (10.42) produces a risk-neutral probability, that is, that $\mathbb{Q}$ has properties (A) and (B) stated above. In fact, $\mathbb{Q}$ defined in (10.42) is a risk-neutral probability if and only if M is a strictly positive SDF process.

Because $M_T V_T$ is strictly positive, the definition (10.42) implies that

$$\mathbb{Q}(A) = 0 \quad \Leftrightarrow \quad \mathsf{E}[1_A] = \mathbb{P}(A) = 0,$$

so property (A) holds. The importance of property (A) is that—according to (10.42)—$\mathbb{Q}(A)$ is the date–0 cost of receiving V_T at date T when A occurs. This cost should be nonzero if and only if the physical probability of A is nonzero, which is what (A) states.

In an infinite-horizon model, there is generally not a probability $\mathbb{Q}$ that has both properties (A) and (B). However, we can still use risk-neutral pricing as follows. Given a strictly positive SDF process M, we can define a probability $\mathbb{Q}$ that satisfies (10.42) for each T. This probability has the martingale property (10.38) for each t, and it is equivalent to $\mathbb{P}$ at all finite times, in the sense that if A is an event that can be distinguished in finite time, then $\mathbb{P}(A) = 0$ if and only if $\mathbb{Q}(A) = 0$. This suffices for valuation, because cash flows must occur in finite time.[5] See the end-of-chapter notes for further discussion.

We want to establish Property (B) for $\mathbb{Q}$ defined in (10.42). Set $\xi_t = \mathsf{E}_t[M_T V_T]$. A standard result, presented in Appendix A.12, is that a process X is a $\mathbb{P}$-martingale if and only if X/ξ is a $\mathbb{Q}$-martingale. Let S denote a dividend-reinvested asset price. Because M is an SDF process, MS is a $\mathbb{P}$-martingale. Therefore, MS/ξ is a $\mathbb{Q}$-martingale. Furthermore, because M is an SDF process, $\xi_t = M_t V_t$. Therefore,

$$\frac{M_t S_t}{\xi_t} = \frac{S_t}{V_t}.$$

Because this is a $\mathbb{Q}$-martingale, property (B) holds.

10.7 NOTES AND REFERENCES

The observation in Section 10.1 that the conditional CAPM implies unconditional risk premia are linear in expected betas and betas of betas is due to

$\mathsf{E}[1_A] = \int_A \mathrm{d}\mathbb{P}$. We can then write

$$\mathsf{E}^*[1_A] = \int_A \mathrm{d}\mathbb{Q} = \int_A \frac{\mathrm{d}\mathbb{Q}}{\mathrm{d}\mathbb{P}} \, \mathrm{d}\mathbb{P} = \int_A M_T V_T \, \mathrm{d}\mathbb{P} = \mathsf{E}[1_A M_T V_T].$$

---

5. An example of a random variable that cannot be observed in finite time is a bet on the limit as $t \to \infty$ of a stochastic process.

Jagannathan and Wang (1996). The ICAPM is due to Merton (1973a) and the CCAPM to Breeden (1979). The proofs in this chapter follow Grossman and Shiller (1982). Merton shows that the ICAPM implies the conditional CAPM if investors have log utility and the unconditional CAPM if returns are IID. Merton, Breeden, and Grossman and Shiller analyze continuous-time models (Chapter 14). The term "managed portfolio theorem" appears in Cochrane (2011). See also Cochrane (2001).

Valuation via a representative investor in an infinite-horizon pure exchange economy was pioneered by Rubinstein (1976) and Lucas (1978), and this type of economy is often called a Lucas economy. The assets are sometimes called trees, their dividends being fruit. This terminology is motivated by the fact that the production capacity of the economy is exogenously given in the model. Models of investments in production capacity are discussed in Chapter 20. Constantinides (1982) proves the existence of a representative investor in a complete dynamic market.

The Gordon model is so named in recognition of Gordon (1962). Of course, the derivation of the Gordon formula (10.30) in terms of an SDF process is more modern.

Mehra and Prescott (1985) derive the equity premium and risk-free rate puzzles in a dynamic model in which consumption growth is not IID. In their model, consumption growth follows a finite-state Markov chain. Other puzzling aspects of the data relative to a representative investor economy are that the market return appears excessively volatile (LeRoy and Porter, 1981; Shiller, 1981) and the dividend-price ratio predicts the market return (Fama and French, 1988a). The excess volatility puzzle is surveyed by LeRoy (1996), and all of these issues, including the equity premium and risk-free rate puzzles, are surveyed by Campbell (2003). The equity premium and risk-free rate puzzles are discussed further in Chapter 11.

The concept of competitive equilibrium that is described in Section 10.3 is called by Radner (1972) an equilibrium of plans, prices, and price expectations. Radner (1972) proves the existence of equilibrium assuming a finite horizon and a finite number of states of the world (allowing for incomplete markets and multiple consumption goods). The existence and Pareto optimality of equilibrium in a complete securities market with a finite horizon and finitely many states of the world follows from Arrow (1953).

Even though the equilibrium concept described in Section 10.3 is the standard concept in the literature, there is no apparent mechanism in real markets for equilibrating plans or price expectations. Grossman (1988) points out that date–0 markets for future (date, state)–contingent consumption, which obviate the need for dynamic trading plans, may perform better in practice than dynamic markets, because date–0 prices will equilibrate what may otherwise be inconsistent plans.

The concept of a risk-neutral probability is introduced by Cox and Ross (1976a,b) and developed further by Harrison and Kreps (1979). Dalang, Morton, and Willinger (1990) prove the existence of a strictly positive SDF process in an economy with no arbitrage opportunities. Specifically, they show that, in the absence of arbitrage opportunities, there is a single-period SDF Z_t for each t. This defines an SDF process M according to (8.7). See Delbaen and Schachermayer (2006) for a comprehensive survey of this topic.

Risk-neutral probabilities are defined as probabilities equivalent to $\mathbb{P}$ that make the ratios S_i/V martingales. This definition is convenient, but it is also frequently convenient to substitute some other asset for the money market account, so that S_i/S_j is a martingale for some j and each i. Such a probability can be constructed from an SDF process by substituting S_j for V in (10.42). This idea is developed by Geman, El Karoui, and Rochet (1995) and Schroder (1999). See Chapters 16 and 17 for some applications.

In an infinite-horizon model, we can construct a probability measure with the martingale property (10.38) as follows: Let $\mathcal{F}_t$ denote the σ-field of observable events at date t, and let $\mathcal{F}$ denote the smallest σ-field containing each of the $\mathcal{F}_t$. Note that $\cup_{t=1}^\infty \mathcal{F}_t$ is contained in $\mathcal{F}$ but is not equal to $\mathcal{F}$, because there are events in $\mathcal{F}$ that are not observable at any finite t (for example, events defined in terms of limits of average returns). For an event A that belongs to $\mathcal{F}_T$ for some T, define

$$\hat{\mathbb{Q}}(A) = \mathsf{E}[M_T V_T 1_A].$$

By the Carathéodory extension theorem (Shiryayev, 1984, p. 150) there exists a unique probability measure $\mathbb{Q}$ that extends $\hat{\mathbb{Q}}$ from $\cup_{t=1}^\infty \mathcal{F}_t$ to $\mathcal{F}$, that is, a unique $\mathbb{Q}$ defined on $\mathcal{F}$ such that

$$\mathbb{Q}(A) = \mathsf{E}[M_T V_T 1_A],$$

if A belongs to $\mathcal{F}_T$ for any T. The proof that the martingale property (10.38) holds is the same as the proof given in Section 10.6. This probability measure also permits risk-neutral valuation of nonnegative infinite-horizon consumption streams, because, using the monotone convergence theorem for the two outer equalities and the definition of $\mathbb{Q}$ for the middle equality,

$$\mathsf{E}^*\left[\sum_{t=1}^\infty \frac{C_t}{V_t}\right] = \lim_{T\to\infty} \mathsf{E}^*\left[\sum_{t=1}^T \frac{C_t}{V_t}\right] = \lim_{T\to\infty} \mathsf{E}\left[\sum_{t=1}^T M_t C_t\right] = \mathsf{E}\left[\sum_{t=1}^\infty M_t C_t\right],$$

where E^* denotes expectation with respect to $\mathbb{Q}$. However, $\mathbb{Q}$ typically does not have the same null sets as $\mathbb{P}$. See Exercise 10.3 for an example.

EXERCISES

10.1. Consider an investor with an infinite horizon in a market with a constant risk-free return and a single risky asset with returns

$$R_t = \frac{1}{\nu}e^{\mu+\sigma\varepsilon_t}$$

for a sequence of independent standard normals ε_t and a constant ν—as implied by (10.31) when the investor is a representative investor and consumption growth is IID lognormal. Assume it is optimal for the investor to choose $\pi = 1$—as is true when the investor is a representative investor. Show that the condition $\delta B^{1-\rho} < 1$ used in Section 9.6 to ensure the existence of an optimum for the investor implies $\nu < 1$, as assumed in the Gordon growth model.

10.2. In the model of Exercise 8.1, calculate the unique risk-neutral probability for any given horizon $T < \infty$, and show that the risk-neutral probability of any path depends on ν_t and the parameters R_f, k, λ_h, and λ_ℓ.

10.3. In the setting of Exercise 8.1, let $\mathbb{P}$ denote the physical probability and assume

$$\mathsf{E}\left[\frac{P_{t+1}+D_{t+1}}{P_t}\right] \neq R_f.$$

Suppose there is an infinite horizon. Show that there is no probability $\mathbb{Q}$ on the space of infinite paths that is (a) equivalent to $\mathbb{P}$, and (b) satisfies

$$\mathsf{E}^*_t\left[\frac{P_{t+1}+D_{t+1}}{P_t}\right] = R_f$$

for each t, where E^* denotes expectation with respect to $\mathbb{Q}$. Hint: Apply the strong law of large numbers to show that any $\mathbb{Q}$ satisfying (b) cannot be equivalent to $\mathbb{P}$.

11

Explaining Puzzles

CONTENTS

11.1 External Habits	260
11.2 Rare Disasters	266
11.3 Epstein-Zin-Weil Utility	268
11.4 Long-Run Risks	276
11.5 Uninsurable Labor Income Risk	279
11.6 Notes and References	283

The equity premium and risk-free rate puzzles discussed in Sections 7.3 and 10.4 have spawned large literatures, in which more general models have been analyzed to see if they are consistent with the data. In this chapter, we describe some of the more important models that have been developed to try to explain the puzzles. These models embody variations on the complete markets/CRRA utility/lognormal consumption model described in Chapters 7 and 10. The variations have other applications as well and may generally be helpful for understanding asset prices. Additional variations are discussed elsewhere in this book (see the end-of-chapter notes).

11.1 EXTERNAL HABITS

This section describes two representative investor models in which the representative investor's utility depends on consumption measured in relation to another variable. The other variable is called a habit, because of an analogy to models in which investors evaluate their consumption relative to their own past consumption. Utility is high and marginal utility is low when consumption is high relative to the habit. In the models of this section, all investors regard the habits as

Explaining Puzzles

exogenously determined. Thus, they are called external habits. The first model in this section is called a nonaddictive habit model, because consumption less than the habit is feasible. The second model is called an addictive habit model, because it is infeasible to consume less than the habit. Models in which the habit depends on the investor's own past consumption (internal habit models) are discussed in Chapter 14.

Abel's Model of Catching Up with the Joneses

In economics, the idea that people care about their consumption relative to that of others dates back at least to Veblen (1899), who describes "conspicuous consumption" as an effort to achieve social status. Concern for consumption relative to that of others is commonly described as a desire to "keep up with the Joneses." Let X_t denote aggregate consumption at date t. Assume there is a representative investor. This investor represents individual investors, all of whom regard aggregate consumption as being exogenous, since the consumption of each is a negligible part of the whole. Thus, the representative investor also regards aggregate consumption as being exogenous, even while she chooses to consume the same amount.

One way to model "keeping up with the Joneses" preferences is to take utility to be

$$\sum_{t=1}^{\infty} \delta^t u\left(\frac{C_t}{X_t}\right)$$

for some function u. Abel (1990) assumes utility is

$$\sum_{t=1}^{\infty} \delta^t u\left(\frac{C_t}{X_{t-1}}\right)$$

and calls this "catching up with the Joneses" preferences. Assume power utility. Then, the marginal utility is $X^{\rho-1}C^{-\rho}$, with X being lagged for catching up with the Joneses preferences. The representative investor consumes aggregate consumption, so $C = X$. Then, the date–t marginal utility is $C_t^{\rho-1}C_t^{-\rho} = C_t^{-1}$ in the keeping-up model and $C_{t-1}^{\rho-1}C_t^{-\rho}$ in the catching-up model. Marginal utility in the keeping-up model is the same as in a standard representative investor model with log utility. Therefore, asset pricing is as described in Section 10.4. The remainder of this section analyzes the catching-up model.

Assume consumption growth is IID. Define $\xi_t = C_t/C_{t-1}$. In the catching-up model, marginal utility at date t is $C_{t-1}^{\rho-1}C_t^{-\rho}$, so the Euler equation implies that there is an SDF process M with, for all t,

$$\frac{M_{t+1}}{M_t} = \frac{\delta C_t^{\rho-1} C_{t+1}^{-\rho}}{C_{t-1}^{\rho-1} C_t^{-\rho}} = \delta \xi_t^{\rho-1} \xi_{t+1}^{-\rho}. \qquad (11.1)$$

This formula shows that the conditional distribution of M_{t+1}/M_t at date t depends on the realization of ξ_t. Therefore, the single-period SDFs M_{t+1}/M_t are not IID, and we cannot directly apply the Gordon growth model from Section 10.4. Nevertheless, the calculations are straightforward.

The risk-free return is

$$R_{f,t+1} = \frac{1}{E_t[M_{t+1}/M_t]} = \frac{\xi_t^{1-\rho}}{\delta E[\xi_{t+1}^{-\rho}]}. \qquad (11.2)$$

The risk-free return changes from period to period, because it depends on the realization of ξ_t. However, it is stationary and has an unconditional mean equal to

$$E[R_{f,t+1}] = \frac{E[\xi_\tau^{1-\rho}]}{\delta E[\xi_\tau^{-\rho}]}. \qquad (11.3)$$

We replaced the subscripts t and $t+1$ on the right-hand side with τ to emphasize that the expectations are the same at all dates.

Define the market portfolio as the claim to aggregate consumption. Its dividend is aggregate consumption. Assume there is no bubble in the price of the market portfolio. Then, the market price-dividend ratio is

$$\frac{P_t}{C_t} = \frac{1}{C_t} E_t \left[\sum_{\tau=t+1}^{\infty} \frac{M_\tau}{M_t} C_\tau \right]. \qquad (11.4)$$

Using (11.1), we obtain

$$\frac{M_\tau}{M_t} \cdot \frac{C_\tau}{C_t} = \prod_{u=t}^{\tau-1} \frac{M_{u+1} C_{u+1}}{M_u C_u} = \prod_{u=t}^{\tau-1} \left(\delta \xi_u^{\rho-1} \xi_{u+1}^{1-\rho} \right) = \delta^{\tau-t} \xi_t^{\rho-1} \xi_\tau^{1-\rho}.$$

Hence, the price-dividend ratio (11.4) equals

$$\frac{P_t}{C_t} = \xi_t^{\rho-1} \sum_{\tau=t+1}^{\infty} \delta^{\tau-t} E[\xi_\tau^{1-\rho}] = \xi_t^{\rho-1} E[\xi_\tau^{1-\rho}] \frac{\delta}{1-\delta}. \qquad (11.5)$$

Explaining Puzzles

We used the fact that $E[\xi_\tau^{1-\rho}]$ is the same for every τ to obtain the last equality in (11.5). The return on the market portfolio is

$$R_{m,t+1} = \frac{P_{t+1}+C_{t+1}}{P_t} = \frac{C_{t+1}}{C_t}\left(\frac{1+P_{t+1}/C_{t+1}}{P_t/C_t}\right)$$

$$= \xi_{t+1}\left(\frac{1+\xi_{t+1}^{\rho-1}E[\xi_\tau^{1-\rho}]\left(\frac{\delta}{1-\delta}\right)}{\xi_t^{\rho-1}E[\xi_\tau^{1-\rho}]\left(\frac{\delta}{1-\delta}\right)}\right)$$

$$= \frac{1-\delta}{\delta}\cdot\frac{\xi_t^{1-\rho}\xi_{t+1}}{E[\xi_\tau^{1-\rho}]}+\xi_t^{1-\rho}\xi_{t+1}^{\rho}.$$

Because the ξ_t are IID, $E[\xi_t^{1-\rho}\xi_{t+1}] = E[\xi_\tau^{1-\rho}]E[\xi_t]$ and $E[\xi_t^{1-\rho}\xi_{t+1}^\rho] = E[\xi_\tau^{1-\rho}]E[\xi_t^\rho]$. Therefore, the unconditional mean of the market return is

$$E[R_{m,t+1}] = \frac{1-\delta}{\delta}E[\xi_\tau] + E[\xi_\tau^{1-\rho}]E[\xi_\tau^\rho].$$

This can easily be calculated if we assume that consumption growth is lognormal. Using the Mehra and Prescott (1985) consumption growth statistics, $\delta = 0.99$, and $\rho = 6$, we obtain an average risk-free rate of 2.0% and an average market risk premium of 4.6%. These do not quite match the historical numbers, but they are much closer than the results of the standard time-additive model with similar risk aversion and time preference. However, Abel (1990) observes that the volatility of the time series $(R_{f1}, R_{f2},\ldots)$ is much higher in this model than in the data.

Campbell-Cochrane Model

This subsection presents the model of Campbell and Cochrane (1999). Assume there is a representative investor who maximizes

$$E\left[\sum_{t=0}^\infty \delta^t u(C_t, X_t)\right], \qquad (11.6)$$

where

$$u(c,x) = \frac{1}{1-\rho}(c-x)^{1-\rho}. \qquad (11.7)$$

The stochastic process X is defined below. Note that the representative investor has CRRA utility for the excess of consumption over X_t. The desired interpretation of X is that it measures past and current consumption of other investors. As in the previous subsection, each investor (and hence the representative investor) is concerned with her consumption relative to that of others.

A key role is played by the percentage surplus consumption, which is

$$S_t \stackrel{\text{def}}{=} \frac{C_t - X_t}{C_t}. \tag{11.8}$$

For the representative investor to have finite utility, S must be a nonnegative process (C must be no smaller than X). We can interpret S as a business cycle indicator. In economic booms, consumption substantially exceeds the external habit (S is large), and in recessions consumption barely exceeds the external habit (S is small).

The marginal utility of the representative investor is

$$\delta^t (C_t - X_t)^{-\rho} = \delta^t S_t^{-\rho} C_t^{-\rho}. \tag{11.9}$$

Interpreting S as in the previous paragraph, marginal utility is low in booms and high in recessions. The representative investor's MRS

$$M_t = \delta^t \left(\frac{S_t}{S_0}\right)^{-\rho} \left(\frac{C_t}{C_0}\right)^{-\rho} \tag{11.10}$$

is an SDF process. Note that the MRS depends only on S and C and not on X. We can match any given SDF process M and consumption process C by using (11.10) to define S and then using (11.8) to define X as $X = (1 - S)C$. Thus, the model can "explain" any SDF process and hence any asset prices. To give the model empirical content, we make additional assumptions.

Assume IID lognormal consumption growth:

$$\log C_{t+1} - \log C_t = \mu + \sigma \varepsilon_{t+1}, \tag{11.11}$$

for a sequence $\varepsilon_1, \varepsilon_2, \ldots,$ of independent standard normals and constants μ and σ. Assume also that

$$\log S_{t+1} - \log S_t = (1 - \phi)(\zeta - \log S_t) + \lambda(S_t)\sigma \varepsilon_{t+1}, \tag{11.12}$$

where ϕ and ζ are constants, with $0 < \phi < 1$, and where $\lambda(\cdot)$ is a function to be specified. This makes S a mean-reverting process. If $\log S_t < \zeta$, then S increases on average ($E_t[\log S_{t+1}] > \log S_t$), and if $\log S_t > \zeta$, then S decreases on average ($E_t[\log S_{t+1}] < \log S_t$). The parameter ζ is called the steady-state value of $\log S$. Imposing (11.12) reduces the flexibility of the model. With this condition, it is possible in principle for the model to be rejected by the data. Additional constraints are presented below in (11.15).

The dynamics of the external habit X are determined by the dynamics of C and S and the fact that $X = (1 - S)C$. Note that the same random variable ε_{t+1} that determines aggregate consumption C_{t+1} in (11.11) also determines S_{t+1} in (11.12); consequently, it also determines X_{t+1}. This feature makes it possible to regard X as a measure of past and current consumption and hence as an external habit.

A single-period SDF is, from (11.10)–(11.12),

$$\frac{M_{t+1}}{M_t} = \delta \left(\frac{S_{t+1}}{S_t}\right)^{-\rho} \left(\frac{C_{t+1}}{C_t}\right)^{-\rho}$$

$$= \delta S_t^{\rho(1-\phi)} \exp\left(-\rho\mu - \rho\zeta(1-\phi) - \rho[1+\lambda(S_t)]\sigma\varepsilon_{t+1}\right). \quad (11.13)$$

The risk-free return is

$$R_{f,t+1} = \frac{1}{E_t[M_{t+1}/M_t]}$$

$$= \frac{1}{\delta} S_t^{-\rho(1-\phi)} \exp\left(\rho\mu + \rho\zeta(1-\phi) - \frac{1}{2}\rho^2\sigma^2[1+\lambda(S_t)]^2\right). \quad (11.14)$$

Conditional on date–t information, (11.13) is the exponential of a normally distributed random variable that has a variance of $\rho^2\sigma^2[1+\lambda(S_t)]^2$. Hence, as in Exercise 7.2, the Hansen-Jagannathan bound (3.35) implies that $\rho\sigma[1+\lambda(S_t)]$ is an approximate upper bound on the maximum Sharpe ratio. If we make λ a decreasing function of S, then this upper bound on Sharpe ratios will be countercyclical—higher in recessions than in booms. There is evidence for countercyclical Sharpe ratios in the data.

Campbell and Cochrane (1999) specify ζ by

$$e^\zeta = \sigma\sqrt{\frac{\rho}{1-\phi}}. \quad (11.15a)$$

They specify $\lambda(\cdot)$ by

$$\lambda(S) = \begin{cases} e^{-\zeta}\sqrt{1+2\zeta - 2\log S} - 1, & \text{if } \log S \leq \theta, \\ 0, & \text{otherwise}, \end{cases} \quad (11.15b)$$

where

$$\theta = \zeta + \frac{1-e^{2\zeta}}{2}. \quad (11.15c)$$

Note that the condition $\log S \leq \theta$ in (11.15b) is equivalent to the expression inside the square root sign being nonnegative. Condition (11.15b) defines λ as a decreasing function of S, decreasing to zero at $\log S = \theta$.

Assume

$$\rho\sigma^2 < 1 - \phi. \quad (11.16)$$

Under this condition, it follows from (11.15a) and (11.15c) that $\theta > \zeta$. Given the dynamics (11.12) of $\log S$, this implies that $\log S$ decreases toward ζ (with no uncertainty) whenever $\log S$ exceeds θ. Thus, $\log S$ will only rarely be above θ.

Assume $\log S_t \leq \theta$ and substitute (11.15) into (11.14) to obtain the continuously compounded risk-free rate:

$$\log R_{f,t+1} = -\log \delta + \rho\mu - \frac{1}{2}\rho(1-\phi). \tag{11.17}$$

Thus, the risk-free rate is constant when $\log S_t \leq \theta$. The constant (real) risk-free rate, the countercyclical maximum Sharpe ratio, and other considerations led Campbell and Cochrane to the specification (11.15).

In the representative investor model with standard CRRA utility, the continuously compounded risk-free rate has the same form as in (11.17) but with $\rho(1-\phi)$ replaced by $\rho^2\sigma^2$; see (7.25). The condition (11.16) allows for $\rho(1-\phi)$ to be larger than $\rho^2\sigma^2$ and therefore allows for a smaller risk-free return than in the standard model. Hence, it is better able to match the data than is the standard model. Campbell and Cochrane (1999) show that it is also better able to match various other features of the data.

11.2 RARE DISASTERS

A possible reason that the equity premium has been higher than predicted by theoretical models calibrated to U.S. consumption data is that the U.S. consumption sample does not match the distribution of consumption growth rates anticipated by investors. In particular, it is possible that the sample does not include some very negative outcomes that are anticipated to occur with low probability. The failure of a sample to include low probability (tail) events that have a substantial effect on market outcomes is called a peso problem. In this case, the possibility of very negative outcomes could have a substantial effect on the risk perceived by investors and consequently justify the high equity premium and low risk-free rate that are present in the data. These outcomes are called rare disasters. This explanation of the equity premium puzzle was first proposed by Rietz (1988). More recently, it has been studied by Barro (2006, 2009) and others. See the end-of-chapter notes for further references.

Assume a representative investor with CRRA utility and IID consumption growth. This is a special case of the model studied in Section 10.4. The market return is given in (10.31), and the risk-free return is given in (10.33). Assume

$$\Delta \log C_{t+1} = \mu + \sigma\xi_{t+1} + \chi_{t+1}\log(1-b_{t+1}), \tag{11.18}$$

where ξ, χ, and b are independent sequences of IID random variables, with each ξ_t having the standard normal distribution, each χ_t having a Bernoulli distribution

Explaining Puzzles

$$\chi_t = \begin{cases} 0 & \text{with probability } 1-p, \\ 1 & \text{with probability } p, \end{cases}$$

for some fixed $0 < p < 1$, and each b_t being distributed on the interval $(0,1)$. When $\chi_{t+1} = 1$, (11.18) implies that

$$\frac{C_{t+1}}{C_t} = e^{\mu + \sigma \xi_{t+1}}(1 - b_{t+1}).$$

Thus, b_{t+1} is the fraction by which consumption falls when $\chi_{t+1} = 1$. We assume p is small—Barro (2006) uses $p = 2\%$. Thus, usually consumption growth is lognormal, but occasionally there is a reduction by b_{t+1} (a disaster).

To calculate the expected market return in (10.31) and the risk-free return in (10.33), we have to evaluate $E[(C_{t+1}/C_t)^\gamma]$ for $\gamma = 1$, $\gamma = -\rho$, and $\gamma = 1 - \rho$. In this model, we have, for any γ,

$$\begin{aligned}
E\left[\left(\frac{C_{t+1}}{C_t}\right)^\gamma\right] &= E\left[e^{\gamma \Delta \log C_{t+1}}\right] \\
&= E\left[e^{\gamma \mu + \gamma \sigma \xi_{t+1}}\right] E\left[e^{\gamma \chi_{t+1} \log(1 - b_{t+1})}\right] \\
&= e^{\gamma \mu + \gamma^2 \sigma^2/2} \left\{(1-p)e^0 + p E\left[e^{\gamma \log(1-b_{t+1})}\right]\right\} \\
&= e^{\gamma \mu + \gamma^2 \sigma^2/2} \left\{1 - p + p E[(1 - b_{t+1})^\gamma]\right\}.
\end{aligned}$$

Substituting this into the formulas (10.31) and (10.33), we see that the expected market return and the risk-free return are the same in this model as for lognormal consumption growth, except for an additional factor

$$\frac{1 - p + p E[1 - b_{t+1}]}{1 - p + p E[(1 - b_{t+1})^{1-\rho}]} \tag{11.19}$$

multiplying the expected market return and an additional factor

$$\frac{1}{1 - p + p E[(1 - b_{t+1})^{-\rho}]} \tag{11.20}$$

multiplying the risk-free return.

The numerators of (11.19) and (11.20) are 1 or smaller. The denominators are larger than 1 when $\rho > 1$ (note that $1 - b_{t+1} \leq 1$ so a negative power of it is at least 1). Thus, this model shrinks both the expected market return and the risk-free return. Furthermore, it generally shrinks the risk-free return more, so the equity premium rises. Therefore, it fits both the equity premium and the risk-free rate better than does the same model without rare disasters. The lower risk-free rate and higher equity premium reflect the greater propensity of investors to save and the lower propensity of investors to hold risky assets when disasters are possible.

11.3 EPSTEIN-ZIN-WEIL UTILITY

The assumption of time-additive utility has no real theoretical foundation and has been adopted in economics and finance primarily for tractability. The assumption is restrictive in that it ties intertemporal substitution to risk aversion (Exercise 2.6). It implies that the utility and hence marginal utility of consumption at any date is independent of consumption at all other dates. A dependence between the marginal utility of consumption at one date and consumption at other dates is similar to an external habit in that it breaks the one-to-one link between marginal utility and contemporaneous consumption. This can be helpful, because it makes it possible to have a high standard deviation of marginal utility—as required by the Hansen-Jagannathan bound—without having a high standard deviation of consumption. In general, marginal utility could depend either on past consumption or on anticipated future consumption, or on both, in addition to depending on current consumption. Internal habit models (Section 15.5) allow marginal utility to depend on past consumption. In this section, we study recursive utility, which allows marginal utility to depend on anticipated future consumption. Recursive utility is defined by a certainty equivalent functional—which is based on preferences toward risks—and a time aggregator—which is based on intertemporal substitutability. We will focus on a CRRA certainty equivalent functional and constant elasticity of substitution (CES) time aggregator. This combination is called Epstein-Zin or Epstein-Zin-Weil (EZW) utility, in recognition of Epstein and Zin (1989) and Weil (1990).

Certainty Equivalents and Time Aggregators

It is useful to first discuss how standard time-additive utility can be expressed in terms of a certainty equivalent functional and time aggregator. Consider a consumption process $(C_0, C_1, \ldots)$. Given a utility function u, define the expected continuation utility at date t as

$$K_t = \mathsf{E}_t \left[\sum_{\tau=t}^{\infty} \delta^{\tau-t} u(C_\tau) \right]. \tag{11.21}$$

Let U_t measure the expected continuation utility in consumption good units, meaning $u(U_t) = K_t$ for each t. Note that maximizing K_t is equivalent to maximizing U_t, given monotonicity of u. Therefore, U_t serves as a measure of utility at date t for consumption processes $(C_t, C_{t+1}, \ldots)$. Iterated expectations implies that

$$K_t = u(C_t) + \delta \mathsf{E}_t[K_{t+1}].$$

Therefore,
$$u(U_t) = K_t = u(C_t) + \delta E_t[K_{t+1}] = u(C_t) + \delta E_t[u(U_{t+1})]. \qquad (11.22)$$

Let ξ_t denote the certainty equivalent at date t for future utility K_{t+1}, meaning
$$u(\xi_t) = E_t[K_{t+1}] = E_t[u(U_{t+1})]. \qquad (11.23)$$

Combine (11.22) and (11.23) to obtain
$$u(U_t) = u(C_t) + \delta u(\xi_t). \qquad (11.24)$$

Equivalently,
$$U_t = V(C_t, \xi_t), \qquad (11.25)$$

where we define
$$V(c,\xi) = u^{-1}(u(c) + \delta u(\xi)). \qquad (11.26)$$

The function V is called the time aggregator. It aggregates current consumption C_t and the certainty equivalent ξ_t of future utility to produce current utility U_t.

With time-additive CRRA utility, the certainty equivalent functional is linearly homogeneous, and the time aggregator has constant elasticity of substitution (CES). To see this, let $u(c) = c^{1-\rho}/(1-\rho)$. Then, (11.23) implies
$$\frac{1}{1-\rho}\xi_t^{1-\rho} = E_t\left[\frac{1}{1-\rho}U_{t+1}^{1-\rho}\right],$$

so we have
$$\xi_t = E_t\left[U_{t+1}^{1-\rho}\right]^{\frac{1}{1-\rho}}, \qquad (11.27)$$

which is linearly homogeneous as claimed. Moreover, (11.24) implies
$$\frac{1}{1-\rho}U_t^{1-\rho} = \frac{1}{1-\rho}C_t^{1-\rho} + \frac{\delta}{1-\rho}\xi_t^{1-\rho},$$

so $U_t = V(C_t, \xi_t)$ for the time aggregator
$$V(c,\xi) = \left(c^{1-\rho} + \delta \xi^{1-\rho}\right)^{\frac{1}{1-\rho}}, \qquad (11.28)$$

which is a CES utility function. Under certain conditions—see Epstein and Zin (1989)—there is a unique solution $U = (U_0, U_1, \ldots)$ of (11.25) and (11.27), given a consumption process C and the definition (11.28). Thus, rather than starting with the usual time-additive specification with CRRA utility, we can define U by the recursivity equation (11.25), the definition (11.27) of the certainty equivalent, and the definition (11.28) of the aggregator. Of course, this would be an unnecessarily complex way to define time-additive utility, but it provides a recipe for generalizing time-additive utility.

Definition of Epstein-Zin Weil Utility

EZW utility retains the linearly homogeneous certainty equivalent functional and the CES time aggregator but allows the parameters of the two to differ. We will retain (11.27) and take the time aggregator to be

$$V(c,\xi) = \left(c^{1-\alpha} + \delta \xi^{1-\alpha}\right)^{\frac{1}{1-\alpha}} \tag{11.29}$$

for $\alpha \neq \rho$. An interesting feature of EZW utility is that an investor with EZW utility has preferences for when uncertainty is resolved, even if the uncertainty is not decision relevant. The investor prefers early resolution of uncertainty if $\rho > \alpha$ and prefers late resolution if $\rho < \alpha$. If $\rho = \alpha$, then utility is time additive, and the investor is indifferent about the timing of uncertainty resolution whenever it is not decision relevant. See the end-of-chapter notes for further discussion.

We achieve some economy of notation in the following by setting $\theta = (1-\rho)/(1-\alpha)$. Note that $\theta = 1$ for time-additive CRRA utility.

Dynamic Programming for Epstein-Zin-Weil Utility

Consider an infinite-horizon Markovian portfolio choice problem with state variables X_t and with the certainty equivalent (11.27) and the time aggregator (11.29). Assume $\rho \neq 1$ and $\alpha \neq 1$ and assume there is no labor income Y_t. In the notation of Chapter 8, R_{t+1} is the vector of asset returns (including a risk-free return if one exists) between t and $t+1$, π_t is the portfolio held for the same period, and the intertemporal budget constraint is

$$W_{t+1} = (W_t - C_t)\pi_t' R_{t+1}.$$

A consumption process $C = (C_0, C_1, \ldots)$ and portfolio process $\pi = (\pi_0, \pi_1, \ldots)$ generates a utility process U satisfying the recursivity equation (11.25). The investor chooses C and π to maximize U_0.

The principle of optimality holds with recursive utility, meaning that optimal C and π are such that $(C_t, C_{t+1}, \ldots)$ and $(\pi_t, \pi_{t+1}, \ldots)$ maximize U_t with probability 1 for each t. The maximized value of U_t depends on W_t and X_t. Denote it by $J(W_t, X_t)$. As in the previous subsection, U_t is measured in units of the consumption good, so J is also in units of the consumption good. With V denoting the time aggregator, the Bellman equation holds in the form

$$J(w,x) = \max_{c,\pi} V(c,\xi) = \max_{c,\pi} \left\{c^{1-\alpha} + \delta \xi^{1-\alpha}\right\}^{1/(1-\alpha)}, \tag{11.30}$$

with the certainty equivalent ξ being defined by

$$\xi = \left\{\mathsf{E}\left[J((w-c)\pi'R_{t+1}, X_{t+1})^{1-\rho} \mid X_t = x\right]\right\}^{1/(1-\rho)}.$$

As with time-additive CRRA utility, the value function is homogeneous in w. Because we are measuring utility here in units of the consumption good, the value function is linearly homogeneous in w rather than being homogeneous of degree $1 - \rho$ as in Chapter 9.[1] Thus, $J(w,x) = wf(x)$ for some function f, and we can write the certainty equivalent as

$$\xi = (w - c)\left\{\mathsf{E}\left[f(X_{t+1})^{1-\rho}(\pi'R_{t+1})^{1-\rho} \mid X_t = x\right]\right\}^{1/(1-\rho)}.$$

The maximum in (11.30) occurs at the portfolio π that maximizes the certainty equivalent ξ. Thus, the optimal portfolio solves

$$\max_{\pi}\left\{\mathsf{E}\left[f(X_{t+1})^{1-\rho}(\pi'R_{t+1})^{1-\rho} \mid X_t = x\right]\right\}^{1/(1-\rho)}. \tag{11.31}$$

Let $\mu(x)$ denote the maximum in (11.31), so the maximum certainty equivalent is $\xi(x) = (w - c)\mu(x)$. Substitute this into the Bellman equation (11.30) to obtain

$$J(w,x) = \max_{c}\left\{c^{1-\alpha} + \delta(w-c)^{1-\alpha}\mu(x)^{1-\alpha}\right\}^{1/(1-\alpha)}.$$

Substitute $c = zw$ and maximize over the consumption-to-wealth ratio z instead of c to obtain

$$J(w,x) = w\max_{z}\left\{z^{1-\alpha} + \delta(1-z)^{1-\alpha}\mu(x)^{1-\alpha}\right\}^{1/(1-\alpha)}. \tag{11.32}$$

From (11.31) and (11.32), we see that the optimal portfolio is determined by risk aversion, and the optimal consumption is determined by intertemporal substitution (given the certainty equivalent of the optimal portfolio). Furthermore, (11.32) implies that $J(w,x) = wf(x)$, where

$$f(x) = \max_{z}\left\{z^{1-\alpha} + \delta(1-z)^{1-\alpha}\mu(x)^{1-\alpha}\right\}^{1/(1-\alpha)}. \tag{11.33}$$

Thus, our claim that $J(w,x) = wf(x)$ is consistent with the Bellman equation.

SDF for Epstein-Zin-Weil Utility

Let C, π, and W denote optimal consumption, portfolio, and wealth processes, and set $Z_\tau = C_\tau/W_\tau$ for $\tau = t, t+1$. By analyzing the first-order conditions for the optimization problems (11.31) and (11.32), we will derive an SDF process. The first-order condition for the portfolio optimization problem (11.31) is that

---

[1]. This is a consequence of the expression in braces on the right-hand side of the Bellman equation (11.30) being taken to the $1/(1-\alpha)$ power, which undoes what would otherwise turn out to be $w^{1-\alpha}$. We confirm below that the assumption $J(w,x) = wf(x)$ is consistent with the Bellman equation.

$$(\forall i) \qquad (1-\rho)E_t\left[f(X_{t+1})^{1-\rho}(\pi'R_{t+1})^{-\rho}R_{i,t+1}\right] = \lambda_t,$$

where λ_t is the Lagrange multiplier (depending on X_t) for the constraint $\sum \pi_i = 1$. This implies that

$$(\forall i,j) \qquad E_t\left[f(X_{t+1})^{1-\rho}(\pi'R_{t+1})^{-\rho}(R_{i,t+1} - R_{j,t+1})\right] = 0. \qquad (11.34)$$

To obtain a useful SDF from (11.34), we need a formula for $f(X_{t+1})$ in terms of observables. Use (11.33) at date $t+1$ to obtain

$$f(X_{t+1}) = \left\{Z_{t+1}^{1-\alpha} + \delta(1-Z_{t+1})^{1-\alpha}\mu(X_{t+1})^{1-\alpha}\right\}^{1/(1-\alpha)}. \qquad (11.35)$$

Now, we need a formula for $\mu(X_{t+1})$ in terms of observables. The first-order condition for the optimization problem (11.32) at date $t+1$ yields

$$Z_{t+1}^{-\alpha} = \delta(1-Z_{t+1})^{-\alpha}\mu(X_{t+1})^{1-\alpha}, \qquad (11.36)$$

which we can rearrange as

$$\mu(X_{t+1})^{1-\alpha} = \frac{1}{\delta}Z_{t+1}^{-\alpha}(1-Z_{t+1})^{\alpha}.$$

Substitute this into (11.35) to obtain

$$f(X_{t+1}) = \left\{Z_{t+1}^{1-\alpha} + \frac{Z_{t+1}^{-\alpha}}{(1-Z_{t+1})^{-\alpha}}(1-Z_{t+1})^{1-\alpha}\right\}^{1/(1-\alpha)} = Z_{t+1}^{-\alpha/(1-\alpha)}. \qquad (11.37)$$

Now, substitute this into the first-order condition (11.34) for the optimal portfolio to obtain

$$(\forall i,j) \qquad E_t\left[Z_{t+1}^{-\alpha\theta}(\pi'R_{t+1})^{-\rho}(R_{i,t+1} - R_{j,t+1})\right] = 0. \qquad (11.38)$$

Equation (11.38) is close to providing the SDF we want, but it is useful to replace wealth W_{t+1} in the denominator of the consumption-wealth ratio $Z_{t+1} = C_{t+1}/W_{t+1}$ using the budget equation $W_{t+1} = (W_t - C_t)\pi'_t R_{t+1}$. This yields

$$(\forall i,j) \qquad E_t\left[C_{t+1}^{-\alpha\theta}(W_t - C_t)^{\alpha\theta}(\pi'R_{t+1})^{\alpha\theta-\rho}(R_{i,t+1} - R_{j,t+1})\right] = 0.$$

Because $W_t - C_t$ is known at date t and $\alpha\theta - \rho = \theta - 1$, this implies

$$(\forall i,j) \qquad E_t\left[C_{t+1}^{-\alpha\theta}(\pi'R_{t+1})^{\theta-1}(R_{i,t+1} - R_{j,t+1})\right] = 0. \qquad (11.39)$$

Now, we are quite close to obtaining the SDF that we want. Equation (11.39) states that $C_{t+1}^{-\alpha\theta}(\pi'R_{t+1})^{\theta-1}$ is orthogonal to each excess return. We just need to scale it by the appropriate amount (depending on date-t information) so that it has a unit inner product with each return $R_{i,t+1}$. We show below that the appropriate scaling factor is $\delta^\theta C_t^{\alpha\theta}$. Thus,

$$\frac{M_{t+1}}{M_t} = \delta^\theta \left(\frac{C_{t+1}}{C_t}\right)^{-\alpha\theta}(\pi'_t R_{t+1})^{\theta-1} \qquad (11.40)$$

Explaining Puzzles

is a single-period SDF. If $\theta = 1$ (that is, if $\alpha = \rho$), then we have the usual SDF for time-additive CRRA utility, which depends only on consumption. When $\theta \neq 1$, then the portfolio return also enters into the SDF.

Multiply (11.39) by π_{tj} and add over assets j using $\sum_j \pi_{tj} R_{i,t+1} = (\sum_j \pi_{tj}) R_{i,t+1} = R_{i,t+1}$ and $\sum_j \pi_{tj} R_{j,t+1} = \pi'_t R_{t+1}$ to obtain

$$(\forall i) \qquad \mathsf{E}_t \left[C_{t+1}^{-\alpha\theta} (\pi' R_{t+1})^{\theta-1} (R_{i,t+1} - \pi' R_{t+1}) \right] = 0.$$

This implies

$$(\forall i) \qquad \mathsf{E}_t \left[C_{t+1}^{-\alpha\theta} (\pi' R_{t+1})^{\theta-1} R_{i,t+1} \right] = \mathsf{E}_t \left[C_{t+1}^{-\alpha\theta} (\pi' R_{t+1})^{\theta} \right]. \qquad (11.41)$$

We will show that the right-hand side of (11.41) is

$$\mathsf{E}_t \left[C_{t+1}^{-\alpha\theta} (\pi'_t R_{t+1})^{\theta} \right] = C_t^{-\alpha\theta} \delta^{-\theta}. \qquad (11.42)$$

Thus, dividing (11.41) by the right-hand side yields

$$(\forall i) \qquad \mathsf{E}_t \left[\delta^{\theta} \left(\frac{C_{t+1}}{C_t} \right)^{-\alpha\theta} (\pi' R_{t+1})^{\theta-1} R_{i,t+1} \right] = 1, \qquad (11.43)$$

confirming that (11.40) is a single-period SDF.

It remains to establish (11.42). The first-order condition for the optimization problem (11.32) at date t yields

$$Z_t^{-\alpha} = \delta(1 - Z_t)^{-\alpha} \mu(X_t)^{1-\alpha}. \qquad (11.44)$$

Also, $\mu(x)$ is defined as the maximum value of the portfolio choice problem (11.31), so

$$\mu(X_t) = \left\{ \mathsf{E}_t \left[f(X_{t+1})^{1-\rho} (\pi'_t R_{t+1})^{1-\rho} \right] \right\}^{1/(1-\rho)}.$$

Use the formula (11.37) for $f(X_{t+1})$ to obtain

$$\mu(X_t) = \left\{ \mathsf{E}_t \left[Z_{t+1}^{-\alpha\theta} (\pi'_t R_{t+1})^{1-\rho} \right] \right\}^{1/(1-\rho)}.$$

Substitute $Z_{t+1} = C_{t+1}/W_{t+1}$ and use the budget equation again to substitute for W_{t+1} to obtain

$$\mu(X_t) = [W_t(1 - Z_t)]^{\alpha/(1-\alpha)} \left\{ \mathsf{E}_t \left[C_{t+1}^{-\alpha\theta} (\pi'_t R_{t+1})^{\theta} \right] \right\}^{1/(1-\rho)}. \qquad (11.45)$$

Now substitute this into (11.44) to obtain

$$Z_t^{-\alpha} = \delta(1 - Z_t)^{-\alpha} [W_t(1 - Z_t)]^{\alpha} \left\{ \mathsf{E}_t \left[C_{t+1}^{-\alpha\theta} (\pi'_t R_{t+1})^{\theta} \right] \right\}^{1/\theta}$$

$$= \delta W_t^{\alpha} \left\{ \mathsf{E}_t \left[C_{t+1}^{-\alpha\theta} (\pi'_t R_{t+1})^{\theta} \right] \right\}^{1/\theta}.$$

Multiply by $W_t^{-\alpha}$ to obtain

$$C_t^{-\alpha} = \delta \left\{ \mathsf{E}_t \left[C_{t+1}^{-\alpha\theta} (\pi_t' R_{t+1})^\theta \right] \right\}^{1/\theta}.$$

This implies (11.42).

Representative Investor with Epstein-Zin-Weil Utility

If there is a representative investor with EZW utility, then the SDF (11.40) depends on aggregate consumption growth and on the market return. In continuous time, this produces a two-factor pricing model in which the factors are consumption growth and the market return (Duffie and Epstein, 1992a). Thus, it is a blend of the CCAPM and the CAPM. When consumption growth is IID, this two-factor model reduces to a single-factor model (with either consumption growth or the market return being the factor) because the market return is proportional to consumption growth when consumption growth is IID, as shown in (10.31).

EZW utility is a significant departure from time-additive CRRA utility in models in which consumption growth is not IID. Such a model is discussed in Section 11.4. To understand why it is more important in such models, consider a shock that affects expected consumption growth—so it affects future consumption—but does not affect current consumption. In a time-additive CRRA model, such a shock does not affect current marginal utility, because current utility depends only on current consumption, not on anticipated future consumption. Therefore, an asset's covariance with the shock does not affect its risk premium. This seems a bit unreasonable, because we would expect risk premia to depend on the ability of assets to hedge changes in the economic environment, including expected consumption growth. Indeed, this is precisely what the ICAPM states—risk premia depend on covariances with variables that determine investment opportunities and the distribution of labor income. However, all such risk premia in the ICAPM reduce to risk premia with consumption when utility is time additive, because of the response of consumption to the shocks in investment opportunities. Once consumption is included as a pricing factor, there is no role for other variables to play in determining risk premia in the time-additive model. With EZW utility, there is such a role, because the variables can affect the market return even if they do not affect consumption, and hence affect the SDF, because the SDF depends on the market return in addition to consumption.

Epstein-Zin-Weil Utility and IID Consumption Growth

The remainder of this section explains the effect of EZW utility with IID consumption growth. We will see that the effect is minimal. As just explained, the effect will be much larger in the model of Section 11.4, in which expected consumption growth is time varying.

Adopt the assumptions of Section 10.4 except for replacing CRRA utility with EZW utility. Because consumption growth is IID, the market return is also IID; thus, the single-period SDFs (11.40) are IID. Consequently, this is another example of the Gordon growth model. As in Section 10.4, the market price-dividend ratio is $v/(1-v)$, where

$$v = \mathsf{E}\left[\left(\frac{M_{t+1}}{M_t}\right)\left(\frac{C_{t+1}}{C_t}\right)\right]. \tag{11.46}$$

Here, the single-period SDF M_{t+1}/M_t is given by (11.40). Also, as in (10.29), the market return is

$$R_{m,t+1} = \frac{P_{t+1} + C_{t+1}}{P_t} = \frac{1}{v} \cdot \frac{C_{t+1}}{C_t}. \tag{11.47}$$

Substitute this into (11.40) to obtain

$$\frac{M_{t+1}}{M_t} = \delta^\theta \left(\frac{C_{t+1}}{C_t}\right)^{-\alpha\theta} v^{1-\theta} \left(\frac{C_{t+1}}{C_t}\right)^{\theta-1}$$

$$= \delta^\theta v^{1-\theta} \left(\frac{C_{t+1}}{C_t}\right)^{-\rho}. \tag{11.48}$$

Substitute (11.48) into (11.46) to obtain

$$v = \delta^\theta v^{1-\theta} \mathsf{E}_t\left[\left(\frac{C_{t+1}}{C_t}\right)^{1-\rho}\right].$$

Solving this for v gives

$$v = \delta \left\{\mathsf{E}_t\left[\left(\frac{C_{t+1}}{C_t}\right)^{1-\rho}\right]\right\}^{1/\theta}. \tag{11.49}$$

Substitute this into (11.47) to obtain

$$R_{m,t+1} = \frac{1}{\delta} \cdot \frac{C_{t+1}/C_t}{\mathsf{E}[(C_{t+1}/C_t)^{1-\rho}]^{1/\theta}}$$

$$= \frac{1}{\delta} \cdot \frac{C_{t+1}/C_t}{\mathsf{E}[(C_{t+1}/C_t)^{1-\rho}]} \cdot \mathsf{E}\left[\left(\frac{C_{t+1}}{C_t}\right)^{1-\rho}\right]^{\frac{\theta-1}{\theta}}. \tag{11.50}$$

To obtain the last line, we substituted $1/\theta = 1 - (\theta - 1)/\theta$ in the denominator of the previous line. Equation (11.50) shows that the market return is the same in the EZW model as in the CRRA model—compare (10.31)—except for the additional factor

$$E\left[\left(\frac{C_{t+1}}{C_t}\right)^{1-\rho}\right]^{\frac{\theta-1}{\theta}}. \tag{11.51}$$

The same additional factor appears in the risk-free return. To see this, substitute (11.49) into (11.48) to obtain

$$\frac{M_{t+1}}{M_t} = \delta E\left[\left(\frac{C_{t+1}}{C_t}\right)^{1-\rho}\right]^{\frac{1-\theta}{\theta}} \left(\frac{C_{t+1}}{C_t}\right)^{-\rho}.$$

Use the usual formula for the risk-free return and substitute to obtain

$$R_{f,t+1} = \frac{1}{E[M_{t+1}/M_t]} \tag{11.52}$$

$$= \frac{1}{\delta} \cdot \frac{1}{E[(C_{t+1}/C_t)^{-\rho}]} \cdot E\left[\left(\frac{C_{t+1}}{C_t}\right)^{1-\rho}\right]^{\frac{\theta-1}{\theta}}. \tag{11.53}$$

Comparing this to (10.33), we see that the risk-free return is the same in the EZW model as in the CRRA model except for the additional factor (11.51), as stated before. Because the additional factor is the same for the market return as for the risk-free return, the ratio of the expected market return to the risk-free return, or the difference in logs, is exactly the same in the EZW model as in the CRRA model. For example,

$$\log E[R_{m,t+1}] - \log R_{f,t+1} = \log E\left[\frac{C_{t+1}}{C_t}\right] + \log E\left[\left(\frac{C_{t+1}}{C_t}\right)^{-\rho}\right]$$
$$- \log E\left[\left(\frac{C_{t+1}}{C_t}\right)^{1-\rho}\right] \tag{11.54}$$

in both the CRRA model and the EZW model. Thus, EZW utility does not help to explain the equity premium puzzle when consumption growth is IID.

11.4 LONG-RUN RISKS

This section describes the model of Bansal and Yaron (2004) and Bansal, Kiku, and Yaron (2012), which incorporates several deviations from the CRRA lognormal consumption model of Section 10.4. The key elements of the model are

Explaining Puzzles

- The market portfolio is regarded as the claim to the aggregate dividend process, which is only part of consumption (consumption equals dividends plus labor income).
- Expected consumption growth and expected dividend growth each follow AR(1) processes.
- The volatilities of consumption and dividend growth are time varying.
- The representative investor has EZW utility.
- The EIS is larger than 1.

A shock to expected consumption growth might have no effect on contemporaneous consumption but could have a large effect on consumption over the long run, so this is called a model of long-run risks. With EZW utility, the marginal utility of consumption depends on the certainty equivalent of future utility in addition to depending on the level of consumption. Changes in expected consumption growth affect marginal utility through the certainty equivalent and therefore produce variability of marginal utility even when consumption does not vary. If the changes to expected consumption growth are sufficiently persistent, then we can obtain enough variability in marginal utility to match the equity premium, without requiring risk aversion to be too high. At the same time, the relatively high EIS—compared to that assumed by Weil (1989), for example—leads to a relatively low risk-free rate in equilibrium, avoiding the risk-free rate puzzle (see Exercise 2.7 for the relation between the EIS and the equilibrium risk-free rate in a simple model).

The model is

$$\Delta \log C_{t+1} = \mu_c + X_t + \sigma_t \eta_{t+1}, \tag{11.55a}$$

$$X_{t+1} = \lambda X_t + \varphi_e \sigma_t e_{t+1}, \tag{11.55b}$$

$$\sigma_{t+1}^2 = \overline{\sigma}^2 + \nu(\sigma_t^2 - \overline{\sigma}^2) + \sigma_w w_{t+1}, \tag{11.55c}$$

$$\Delta \log D_{t+1} = \mu_d + \phi X_t + \varphi \sigma_t u_{t+1} + \pi \sigma_t \eta_{t+1}, \tag{11.55d}$$

where η, e, w, and u are independent sequences of IID standard normal random variables and μ_c, λ, φ_e, $\overline{\sigma}$, ν, σ_w, μ_d, ϕ, φ, and π are constants.

We will use a subscript d to denote the claim to the dividend and a subscript c to denote the claim to consumption. Let P_i denote the price of the claim and R_i the return on the claim, for $i \in \{c, d\}$. Denote the price-consumption and price-dividend ratios by $Z_c = P_c/C$ and $Z_d = P_d/D$. We can derive an approximate solution to the model in which

$$\log Z_{it} = A_{i0} + A_{i1} X_t + A_{i2} \sigma_t^2 \tag{11.56}$$

for $i \in \{c,d\}$ and constants A_{ij} by using the Campbell-Shiller linearization. The Campbell-Shiller linearization (10.34) is

$$\log R_{c,t+1} = \kappa_{c0} - \log Z_{ct} + \kappa_{c1} \log Z_{c,t+1} + \Delta \log C_{t+1}, \quad (11.57a)$$

$$\log R_{d,t+1} = \kappa_{d0} - \log Z_{dt} + \kappa_{d1} \log Z_{d,t+1} + \Delta \log D_{t+1}. \quad (11.57b)$$

Set

$$\bar{z}_i = \mathrm{E}[\log Z_{it}] = A_{i0} + A_{i2}\bar{\sigma}^2. \quad (11.58)$$

The formulas for the κ's given in Section 10.5 can be expressed in terms of the $\bar{z}_i$ as

$$\kappa_{i0} = \log\left(1 + e^{-\bar{z}_i}\right) + \frac{e^{-\bar{z}_i}\bar{z}_i}{1 + e^{-\bar{z}_i}}, \quad (11.59a)$$

$$\kappa_{i1} = \frac{1}{1 + e^{-\bar{z}_i}}. \quad (11.59b)$$

Substituting (11.56) and (11.59) into (11.57) produces formulas for the returns in terms of the $\bar{z}_i$ and A_{ij} and the shocks. From (11.40), the SDF process M satisfies

$$\frac{M_{t+1}}{M_t} = \delta^\theta \left(\frac{C_{t+1}}{C_t}\right)^{-\alpha\theta} R_{c,t+1}^{\theta-1}, \quad (11.60)$$

where $\theta = (1-\rho)/(1-\alpha)$, ρ is relative risk aversion, and $1/\alpha$ is the EIS. Using the pricing equation

$$\mathrm{E}_t\left[\frac{M_{t+1}}{M_t} R_{i,t+1}\right] = 1 \quad (11.61)$$

and the formulas for the returns, we compute the A_{ij} in terms of the $\bar{z}_j$. Given the A_{ij}, we compute new values of the $\bar{z}_i$ from (11.58). We iterate until we obtain fixed points $\bar{z}_j$.

To explain how the A_{ij} are computed from the pricing equation (11.61), consider the claim to consumption. Substitute from (11.55a), (11.56), (11.57a), and (11.60) and note that

$$(-\alpha\theta + \theta - 1)\Delta \log C_{t+1} = -\rho(\mu_c + X_t + \sigma_t \eta_{t+1})$$

to obtain

$$\Delta \log M_{t+1} + \log R_{c,t+1} = \theta \log \delta - \rho(\mu_c + X_t + \sigma_t \eta_{t+1})$$
$$+ (\theta - 1)\{\kappa_{c0} - (A_{i0} + A_{i1}X_t + A_{i2}\sigma_t^2) + \kappa_{c1}(A_{i0} + A_{i1}X_{t+1} + A_{i2}\sigma_{t+1}^2)\}.$$

Write the pricing equation as

$$\log \mathrm{E}\left[e^{\Delta \log M_{t+1} + \log R_{c,t+1}}\right] = 0.$$

The expectation can be calculated explicitly, given the conditional normal distributions of η_{t+1}, X_{t+1}, and σ^2_{t+1}. The logarithm of the expectation is a linear function of the state variables X_t and σ_t^2. The logarithm must equal zero for all values of the state variables, so the coefficients of X_t and σ_t^2 must each be zero, and the other terms must sum to zero. This provides three equations that can be solved for A_{c0}, A_{c1}, and A_{c2}. The parameters A_{dj} can be computed likewise from the equation

$$\log \mathsf{E}\left[e^{\Delta \log M_{t+1} + \log R_{d,t+1}}\right] = 0.$$

Given the fixed points $\bar{z}_i$ (average log price-consumption and price-dividend ratios), the average equity premium and the average risk-free rate can be computed. These fit the data reasonably well in the calibration of Bansal and Yaron (2004). Furthermore, the model implies that the dividend yield predicts the market return and return volatility is time varying. See the end-of-chapter notes for further discussion.

11.5 UNINSURABLE LABOR INCOME RISK

One of the assumptions of the basic consumption-based asset pricing model in Sections 7.3 and 10.4 is that there is a representative investor. However, if markets are incomplete, then a representative investor may not exist. Therefore, the failure of the model to match the data (that is, the equity premium and risk-free rate puzzles) could be due to the nonexistence of a representative investor.

An important form of market incompleteness is labor income risk that cannot be hedged with assets. If shocks to labor income are short-lived (for example, if the level of labor income is IID), then it turns out that this incompleteness has small effects in theory, because investors can save enough to self-insure (Telmer, 1993; Heaton and Lucas, 1996). However, if shocks are persistent (for example, if labor income is a random walk), then this form of incompleteness matters more. This section describes a model of persistent uninsurable idiosyncratic income risk due to Constantinides and Duffie (1996). The model demonstrates the possibility of explaining asset returns via uninsurable labor income risk, but we will not make any attempt to calibrate the income risk to the data.

Let D denote aggregate dividends, let Y denote aggregate labor income, and let C denote aggregate consumption. Then, $C = D + Y$. Let ρ and δ be positive constants. Suppose asset prices are such that there is a strictly positive SDF process with the following properties:

$$(\forall t) \qquad \frac{M_{t+1}}{M_t} \geq \delta \left(\frac{C_{t+1}}{C_t}\right)^{-\rho}, \qquad (11.62a)$$

$$\sum_{t=1}^{\infty} \mathsf{E}[M_t Y_t] < \infty, \qquad (11.62b)$$

$$\sum_{t=1}^{\infty} \mathsf{E}[M_t D_t] < \infty, \qquad (11.62c)$$

$$(\forall i) \qquad \lim_{T \to \infty} \mathsf{E}[M_T P_{iT}] = 0. \qquad (11.62d)$$

Conditions (11.62b) and (11.62c) are minor regularity conditions. Condition (11.62d) is the transversality condition that rules out bubbles (Section 8.5). The significant assumption is (11.62a). If there were a representative investor with constant relative risk aversion ρ and discount factor δ, then (11.62a) would hold as an equality. Allowing the inequality in (11.62a) means that the model is more flexible and more likely to be able to fit the data. In particular, we can make the risk-free rate as low as desired (Exercise 11.3).

Define

$$Z_{t+1} = \sqrt{\frac{2}{\rho(1+\rho)} \left[\log\left(\frac{M_{t+1}}{M_t}\right) - \log \delta + \rho \log\left(\frac{C_{t+1}}{C_t}\right) \right]}. \qquad (11.63)$$

The importance of assumption (11.62a) is that it implies the expression inside braces in (11.63) is nonnegative; hence, it is possible to take the square root. The definition (11.63) implies

$$e^{\rho(\rho+1)Z_{t+1}^2/2} = \frac{1}{\delta} \cdot \frac{M_{t+1}}{M_t} \left(\frac{C_{t+1}}{C_t}\right)^{\rho}. \qquad (11.64)$$

To model individual income risk, it simplifies matters to have a large number of investors, so some aspects of individual income risk "diversify away" across individuals. It is conventional to model a "large number" of investors by assuming there is a continuum of investors and indexing the investors by $h \in [0,1]$. We sum over investors by integrating over $h \in [0,1]$. With a continuum of investors, individual income and consumption must be infinitesimal in order for aggregate income and consumption to be finite. We will denote the income of investor h by $Y_{ht} \, dh$ and the consumption of investor h by $C_{ht} \, dh$. Aggregate income and consumption are then

$$Y_t = \int_0^1 Y_{ht} \, dh \quad \text{and} \quad C_t = \int_0^1 C_{ht} \, dh, \qquad (11.65)$$

respectively.[2] Assume all investors have equal asset endowments at date 0. Hence, if there is no trade, each investor receives dividends equal to $D_t\,dh$ at each date t. Assume each investor's financial wealth is constrained to be nonnegative. This rules out Ponzi schemes.

We will show that the existence of an SDF process with the properties (11.62) implies that the asset prices are equilibrium prices in an economy in which all investors have CRRA utility with risk aversion ρ and discount factor δ. The proof involves defining labor income processes Y_h such that it is always optimal for each investor to hold her initial portfolio of assets. Thus, the equilibrium is autarkic, meaning that there is no trade. The equilibrium consumption of each investor h is the investor's labor income plus dividends $D_t\,dh$.

To define the labor income processes, let $\{\psi_{ht}\}$ be a family of IID standard normals that are independent of all other variables in the model. Define $\xi_{h0} = 1$ and, for $t \geq 0$, set

$$\xi_{h,t+1} = \xi_{ht}e^{\psi_{h,t+1}Z_{t+1} - Z_{t+1}^2/2},$$

where Z is defined in (11.63) and hence satisfies (11.64). The usual formula for the mean of an exponential of a normally distributed variable and (11.64) imply

$$\mathrm{E}\left[\left.\frac{\xi_{h,t+1}}{\xi_{ht}}\right|Z_{t+1}\right] = 1, \qquad (11.66)$$

$$\mathrm{E}\left[\left.\left(\frac{\xi_{h,t+1}}{\xi_{ht}}\right)^{-\rho}\right|Z_{t+1}\right] = e^{\rho(\rho+1)Z_{t+1}^2/2} = \frac{1}{\delta}\cdot\frac{M_{t+1}}{M_t}\left(\frac{C_{t+1}}{C_t}\right)^{\rho}. \qquad (11.67)$$

It is shown below that (11.66) and the strong law of large numbers imply[3]

$$\int_0^1 \xi_{ht}\,dh = 1 \qquad (11.68)$$

with probability 1 for each t. Therefore, we can define individual labor income as

$$Y_{ht} = \xi_{ht}(Y_t + D_t) - D_t.$$

This definition is consistent, because, by (11.68),

$$\int_0^1 Y_{ht}\,dh = (Y_t + D_t)\int_0^1 \xi_{ht}\,dh - D_t = Y_t.$$

---

2. We can alternatively view aggregate income and consumption as being infinite and express market clearing in terms of per capita variables. In this interpretation, Y_{ht} is the income and C_{ht} the consumption of investor h, and (11.65) defines per capita (i.e., average) rather than total income and consumption.

3. This requires some qualification. See the end-of-chapter notes.

In the absence of trade, $C_{ht} = Y_{ht} + D_t = \xi_{ht} C_t$. This implies

$$\frac{\delta u'_h(C_{h,t+1})}{u'_h(C_{ht})} = \delta \left(\frac{C_{t+1}}{C_t}\right)^{-\rho} \left(\frac{\xi_{h,t+1}}{\xi_{ht}}\right)^{-\rho}. \tag{11.69}$$

The key step in showing that autarky is an equilibrium is to show that this is a single-period SDF (that is, the Euler equation holds). This is a consequence of (11.67) and (11.69), as is shown below.

As remarked before, an important feature of the model is the persistence of the individual income shocks. We have

$$\log \xi_{h,t+1} = \log \xi_{ht} + \psi_{h,t+1} Z_{t+1} - \frac{1}{2} Z_{t+1}^2.$$

Thus, conditional on aggregate variables, including the Z process, $\log \xi_h$ is a random walk with independent normally distributed increments.

To establish (11.68), note that, because ξ_{ht} and $\xi_{h,t+1}/\xi_{ht}$ are independent conditional on the Z process, (11.66) implies

$$E[\xi_{h,t+1}|Z] = E\left[\xi_{ht} \frac{\xi_{h,t+1}}{\xi_{ht}} \bigg| Z\right]$$

$$= E[\xi_{ht}|Z] E\left[\frac{\xi_{h,t+1}}{\xi_{ht}} \bigg| Z\right]$$

$$= E[\xi_{ht}|Z]. \tag{11.70}$$

To deduce the last equality, use (11.66) and the fact that $\xi_{h,t+1}/\xi_{ht}$ depends on Z only via Z_{t+1}. It follows from (11.70), the fact that $\xi_{h0} = 1$, and induction that

$$E[\xi_{ht}|Z] = 1 \tag{11.71}$$

for all t. Note that, conditional on Z, the random variables ξ_{ht} for different h are IID. Hence, (11.71) and the strong law of large numbers imply (11.68).

Now, we show that it is optimal for each investor to hold her initial portfolio of assets. Section 9.1 shows that it is optimal if the Euler equation and the transversality condition hold. First, we establish the Euler equation. Conditional on Z_{t+1}, $\xi_{h,t+1}/\xi_{ht}$ depends only on $\psi_{h,t+1}$, so it is independent of $(C_{t+1}/C_t)^{-\rho} R_{i,t+1}$ for each asset i. Therefore, (11.69) implies

$$E_t\left[\frac{\delta u'_h(C_{h,t+1})}{u'_h(C_{ht})}R_{i,t+1}\bigg|Z_{t+1}\right]$$

$$= E_t\left[\delta\left(\frac{C_{t+1}}{C_t}\right)^{-\rho}R_{i,t+1}\bigg|Z_{t+1}\right]E\left[\left(\frac{\xi_{h,t+1}}{\xi_{ht}}\right)^{-\rho}\bigg|Z_{t+1}\right]$$

$$= E_t\left[\left(\frac{C_{t+1}}{C_t}\right)^{-\rho}R_{i,t+1}\bigg|Z_{t+1}\right]\frac{M_{t+1}}{M_t}\left(\frac{C_{t+1}}{C_t}\right)^{\rho}$$

$$= E_t\left[\frac{M_{t+1}}{M_t}R_{i,t+1}\bigg|Z_{t+1}\right].$$

The second equality is a consequence of (11.67). For the third equality, we used the fact that

$$\frac{M_{t+1}}{M_t}\left(\frac{C_{t+1}}{C_t}\right)^{\rho}$$

is a function of Z_{t+1} as shown in (11.67), so it can be moved through the conditional expectation operator. By iterated expectations,

$$E_t\left[\frac{\delta u'_h(C_{h,t+1})}{u'_h(C_{ht})}R_{i,t+1}\right] = E_t\left[\frac{M_{t+1}}{M_t}R_{i,t+1}\right] = 1.$$

Hence, the MRS of each investor h is an SDF (the Euler equation holds).

To complete the proof that it is optimal to consume $C_{ht} = Y_{ht} + D_t$, it remains to establish the transversality condition (9.4), where W in (9.4) is the wealth process corresponding to "no trade." This is equivalent to

$$\lim_{T\to\infty} E\left[\delta^T C_T^{-\rho}\xi_{hT}^{-\rho}P_{iT}\right] = 0$$

for each asset i. This follows from (11.62d), (11.64), and (11.67).

11.6 NOTES AND REFERENCES

The equity premium and risk-free rate puzzles were first described by Mehra and Prescott (1985). The early literature on the puzzles is surveyed by Kocherlakota (1996) and by Mehra and Prescott (2003). Some proposed resolutions of the puzzles discussed elsewhere in this book (perhaps briefly) are internal habits (Section 15.5), learning expected consumption growth (Section 23.4), prospect theory, generalized disappointment aversion, and ambiguity aversion. See Chapter 25 for the last three.

Chan and Kogan (2002) solve for equilibrium in a pure exchange economy with a continuum of investors having "catching up with the Joneses" preferences and different relative risk aversions. In their formulation, utility depends on a weighted geometric average of past per capita consumption (like the formulation of habit in the internal habit model) rather than just lagged per capita consumption. Variation over time in the distribution of wealth across agents produces variation in aggregate risk aversion, leading to a countercyclical maximum Sharpe ratio and other features that are consistent with the data.

DeMarzo, Kaniel, and Kremer (2004, 2008) endogenize "keeping up with the Joneses" preferences for wealth (not consumption) relative to a cohort group by introducing local goods markets in which prices depend on the wealth of other investors in the same cohort. In DeMarzo, Kaniel, and Kremer (2004), this results in investors in the same locality holding similar (perhaps undiversified) portfolios, consistent with the "home bias" puzzle. This is an example of "herding." In DeMarzo, Kaniel, and Kremer (2008), the cohorts are defined by age (the model is an overlapping generations model), the "goods" are the investment opportunities, and the herding produces asset price bubbles.

Santos and Veronesi (2010) show that external habit models of the Campbell-Cochrane form described in Section 11.1 do not fit the cross section of stock returns, because they imply a growth premium rather than the value premium found in the data. Bansal, Kiku, and Yaron (2012) show that the long-run risks model fits various features of the data better than does the external habits model. However, Beeler and Campbell (2012) argue that the long-run risks model is inconsistent with the data in several important ways. They make the following arguments:

- The serial correlation of consumption growth is negligible or perhaps even negative (depending on the time frame over which it is estimated), which is inconsistent with the persistent shocks to expected consumption growth that are at the heart of the model.
- In the model, changes to expected consumption growth produce changes in the price-dividend ratio, so changes in the price-dividend ratio forecast future dividend and consumption growth. However, the market price-dividend ratio is empirically a poor predictor of dividend and consumption growth.
- In the model, persistent changes in consumption volatility produce changes in the price-dividend ratio and produce changes in future stock price volatility, so the price-dividend ratio should forecast future stock price volatility. Again, there is little forecasting power in the data.
- In the model, long-term bonds provide good hedges to changes in expected consumption growth. For example, a decline in expected consumption growth leads to lower interest rates and therefore higher bond prices. This produces a downward-sloping term structure for real bond yields and very low real bond yields, which appear to be inconsistent with the data.

- The high assumed EIS implies that any variation in short-term real interest rates should be accompanied by variation in consumption growth. However, this is not present in the data.

Regarding the EIS, Weil (1989) uses 0.1 in his calibration, which was consistent with estimates at that time. More recently, Havránek (2015) examines more than 2,700 estimates reported in 169 published studies. He concludes that "the corrected mean of micro estimates of the EIS for asset holders is around 0.3–0.4. Calibrations greater than 0.8 are inconsistent with the bulk of the empirical evidence."

The original response to the rare disasters model of Rietz (1988) is that the required disasters are too large to reasonably assume; see Mehra and Prescott (1988). However, Barro (2006) catalogues disasters for various countries, calibrates the model using the empirical distribution of disasters, and finds that it fits the data reasonably well. In this work, it is important to use a distribution of disasters, because the left tail is important—the effects of disasters are not linear, so their effects are not well approximated by a disaster state that represents the average (Exercise 11.2). On the other hand, Backus, Chernov, and Martin (2011) calibrate the distribution of disasters implied by equity index options and conclude that the probabilities of disasters perceived by investors are smaller than in Barro's calibration. Furthermore, Julliard and Ghosh (2012) argue that the rare disasters model cannot fit the equity premium unless disasters occur much too frequently and also point out that the model worsens the ability of the CCAPM to fit the cross section of equity returns, because the presence of disasters in which all stocks fall together reduces the cross-sectional dispersion of consumption risk.

The model of a continuum of investors with IID income shocks presented in Section 11.5 is not quite correct. The difficulty is in deriving (11.68) from the strong law of large numbers. This is infeasible, because if the random variables ξ_{ht} are IID across h, then the realizations will not be measurable as a function of h, and hence the integral in (11.68) will not exist. Judd (1985) demonstrates this measurability issue, shows that it is surmountable, but shows that the strong law of large numbers may still not hold. Feldman and Gilles (1985) propose using a countable space of investors and a purely finitely additive measure on the space of investors. To do this, we simply replace the integral over [0, 1] in the definitions of aggregate income and consumption with an integral over the natural numbers with respect to the purely finitely additive measure. Under such a measure, any finite set of investors is a null set. Green (1994) and Sun (2006) propose other means of resurrecting the strong law of large numbers with an infinite number of investors. Most finance papers simply ignore this issue, with Constantinides and Duffie (1996) being one of the few exceptions.

EZW utility is due to Epstein and Zin (1989) and Weil (1989, 1990). Epstein and Zin give sufficient conditions for the existence of a solution U to the recursivity equation (11.25) when V is the CES aggregator (11.29), imposing only a continuity condition on the certainty equivalent function. They also allow for non-expected-utility certainty equivalents—betweenness preferences (Chapter 25). Weil (1989) analyzes a calibrated model as in Section 11.3 and also a calibrated model in which consumption growth follows a two-state Markov process, as in Mehra and Prescott (1985). He concludes that EZW utility does not by itself resolve the equity premium and risk-free rate puzzles.

EZW utility is a special case of recursive utility axiomatized by Kreps and Porteus (1978). Recursive utility is a dynamically consistent and history-independent generalization of time-additive utility. Recursive utility is developed in continuous time by Duffie and Epstein (1992b) and called stochastic differential utility. Duffie and Epstein (1992a) show that the CCAPM does not hold in general for stochastic differential utility but that there is a two-factor pricing model in which the factors are aggregate consumption and market wealth. Thus, the model combines the CCAPM with the CAPM. Schroder and Skiadas (1999) characterize optimal consumption and portfolios with stochastic differential utility.

The assumption of time-additive utility has no real theoretical foundation. It is unattractive for at least the following reasons:

(a) It ties intertemporal substitution to risk aversion.
(b) It does not allow consumption at nearby dates to be substitute goods. It seems reasonable (especially when time periods are short) that there could be some "carryover" effect in consumption, making consumption at nearby dates substitute goods. In other words, the marginal utility of consumption may be lower when past consumption has been higher.
(c) It does not allow consumption at different dates to be complementary goods. In reality, people become accustomed to a standard of living. Thus, consuming at a high rate for some period of time and at a lower rate later may be painful. In other words, the marginal utility of consumption may be higher when past consumption has been higher.
(d) It implies indifference about when information is revealed, unless the information is relevant for decision making.

The last three of these issues are illustrated by the following example presented by Duffie and Epstein (1992b). Let a and b be constants. Consider the following three consumption processes $(C_1, \ldots, C_T)$:

(i) At date 0, a single coin is tossed. If it comes up heads, then $C_t = a$ for all t; otherwise, $C_t = b$ for all t.
(ii) At date 0, T coins are tossed and the results observed. If the tth toss is a head, then $C_t = a$; otherwise, $C_t = b$.
(iii) At each date t, a coin is tossed. If it is a head, then $C_t = a$; otherwise, $C_t = b$.

The substitutability argument (b) suggests that (ii) and (iii) could be preferred to (i), due to the intertemporal diversification possessed by (ii) and (iii). On the other hand, the complementarity argument (c) suggests that (i) could be preferred to (ii) and (iii). Neither set of preferences could be termed irrational. Substitutability and complementarity are captured in internal habit models (Section 15.5). Finally, a person might or might not prefer to know in advance what consumption will be. However, time additivity implies indifference among (i)–(iii), because time additivity implies that the utility of each is

$$\sum_{t=1}^{T} \delta^t \left[\frac{1}{2} u(a) + \frac{1}{2} u(b) \right].$$

An investor with EZW utility prefers early resolution of uncertainty if $\rho > \alpha$ and prefers late resolution if $\rho < \alpha$ (Exercise 11.4). Thus, the three aspects of preferences—risk aversion, intertemporal substitution, and preference for the resolution of uncertainty—are still linked within EZW utility. If $\rho = \alpha$, then utility is time additive, so risk aversion equals the reciprocal of the EIS, and the investor is indifferent about the time at which uncertainty is resolved.

EXERCISES

11.1. Calculate the unconditional standard deviation of R_{ft} in the catching up with the Joneses model.

11.2. Calculate the expected market return and the risk-free return in the rare disasters model when

(a) b_{t+1} is uniformly distributed on $[0, b^*]$ for some constant $b^* < 1$.
(b) $b_{t+1} = b^*/2$ with probability 1 for some constant $b^* < 1$.

Explain why the ratio $E[R_{mt}]/R_{ft}$ is larger in the rare disasters model in case (a) than in case (b).

11.3. Let C denote aggregate consumption, and assume consumption growth C_{t+1}/C_t is IID. Assume

$$\frac{M_{t+1}}{M_t} \stackrel{\text{def}}{=} \delta \left(\frac{C_{t+1}}{C_t}\right)^{-\rho} + \alpha \left(\frac{C_{t+1}}{C_t}\right)^{-\gamma}$$

is an SDF process for some δ, ρ, α, and γ. For $\alpha > 0$, condition (11.62a) of the Constantinides-Duffie model is satisfied. Take $\delta = 0.99$ and $\rho = 10$ as in Mehra and Prescott (2003).

(a) Apply the Gordon growth model (Section 10.4) to derive formulas for the risk-free return and expected market return.
(b) Explain why it is possible to choose parameters α and γ such that the risk-free return is as low as desired.
(c) Are there limits as to how high you can make the equity premium? Explain.

11.4. Consider consumption processes (ii) and (iii) in Section 11.6. Take $T = 2$. Suppose consumption C_0 is known at date 0 (before any coins are tossed). Assume the power certainty equivalent and the CES aggregator.

(a) Assume two coins are tossed at date 0 determining C_1 and C_2. Calculate the utility U_0 of the person before the coins are tossed.
(b) Assume a coin is tossed at date 1 determining C_1, and a coin is tossed at date 2 determining C_2. Calculate the utility U_0.
(c) Show numerically that the utility is higher in part (a) than in part (b)—that is, early resolution of uncertainty is preferred—if $\rho > \alpha$. Show that late resolution is preferred if $\rho < \alpha$, and show that the person is indifferent about the timing of resolution of uncertainty if $\rho = \alpha$.

12

Brownian Motion and Stochastic Calculus

CONTENTS

12.1	Brownian Motion	290
12.2	Itô Integral and Itô Processes	292
12.3	Martingale Representation	298
12.4	Itô's Formula	299
12.5	Geometric Brownian Motion	303
12.6	Covariation of Itô Processes and General Itô's Formula	305
12.7	Conditional Variances and Covariances	308
12.8	Transformations of Models	309
12.9	Notes and References	311

This chapter explains how the calculus of Newton and Leibniz is extended to continuous-time models with uncertainty. We will see in subsequent chapters that many concepts regarding securities markets are essentially the same in continuous time as in discrete time. However, some results (for example, the ICAPM) that are true in continuous time hold only as approximations in discrete time. Furthermore, calculations are frequently simpler in continuous time. Many models that can be solved analytically in continuous time can only be solved numerically in discrete time. There are other cases in which numerical solutions are necessary, but it is easier to formulate a numerical solution for the continuous-time version of a model than for the discrete-time version.

A simple continuous-time model of a stock price S is that it evolves as

$$\frac{dS}{S} = \mu\, dt + \sigma\, dB, \qquad (12.1)$$

where B is a Brownian motion and μ and σ are constants. Here, μ is interpreted as the expected rate of increase of the stock price, and σ as the instantaneous standard deviation ("volatility") of the stock's rate of return. We start by explaining Brownian motion, and then we explain Equation (12.1) and related, more general models. For concreteness, take the unit in which time is measured to be years.

12.1 BROWNIAN MOTION

In continuous time, a stochastic process X is a collection of random variables X_t for $t \in [0, \infty)$ or for $t \in [0, T]$ for some $T < \infty$. The state of the world determines the value of X_t at each time t. Thus, it determines the path of X, which is the set of points (t, X_t) (that is, the graph of the function of time $t \mapsto X_t$) that shows how X evolves in the particular state of the world.

A Brownian motion is a continuous-time stochastic process B with the property that, for any dates $t < u$, and conditional on information at date t, the change $B_u - B_t$ is normally distributed with mean zero and variance $u - t$. Equivalently, B_u is conditionally normally distributed with mean B_t and variance $u - t$. In particular, the distribution of $B_u - B_t$ is the same for any conditioning information and hence is independent of conditioning information. This is expressed by saying that the Brownian motion has independent increments. We can regard $\Delta B = B_u - B_t$ as noise that is unpredictable by any date–t information. The starting value of a Brownian motion is typically not important, because only the increments ΔB are usually used to define the randomness in a model, so we can and will take $B_0 = 0$.

The definition of a Brownian motion involves both the random variables B_t and the conditioning information. Thus, a Brownian motion with respect to some information might not be a Brownian motion with respect to other information. For example, a stochastic process could be a Brownian motion for some investors but not for better-informed investors, who might be able to predict the increments to some degree. It is part of the definition of a Brownian motion that the past values B_s for $s \leq t$ are part of the information at each date t. Thus, in the preceding paragraph, conditioning on date–t information means conditioning on the history of B up to and including time t and on whatever other information there may be.

It can be shown that the paths of a Brownian motion must be continuous (with probability 1). However, the paths of a Brownian motion are almost everywhere nondifferentiable, again with probability 1.[1] The paths of a Brownian motion oscillate wildly, making many small up-and-down movements with extremely high frequency, so that the limits $\lim_{s \to t} (B_t - B_s)/(t - s)$ defining derivatives do

---

1. More precisely, with probability 1, the set of times t at which dB_t/dt exists has zero Lebesgue measure.

not exist. Thus, the seemingly innocuous assumption of independent zero-mean normal increments with variance equal to the length of the time interval has strong implications for the paths of the process. The name "Brownian motion" stems from the observations by the botanist Robert Brown of the erratic behavior of particles suspended in a fluid. Stock prices also fluctuate erratically, and Brownian motions, or related processes—particularly "geometric Brownian motions" as in (12.1)—are commonly used to model stock prices.

Quadratic and Total Variation

Let B be a Brownian motion. Consider a discrete partition

$$s = t_0 < t_1 < t_2 < \cdots < t_N = u$$

of a time interval $[s, u]$, and consider the sum of squared changes

$$\sum_{i=1}^{N}(B_{t_i} - B_{t_{i-1}})^2$$

in some state of the world. If we consider finer partitions (i.e., increase N) with the maximum length $t_i - t_{i-1}$ of the time intervals going to zero as $N \to \infty$, the limit of the sum is called the quadratic variation of B in that state of the world. For a Brownian motion, the quadratic variation over any interval $[s, u]$ is equal to $u - s$ with probability 1.

This is in sharp contrast to continuously differentiable functions of time. Consider, for example, a linear function: $f_t = at$ for some constant a. Taking $t_i - t_{i-1} = \Delta t = (u-s)/N$ for each i, the sum of squared changes over an interval $[s, u]$ is

$$\sum_{i=1}^{N}(f_{t_i} - f_{t_{i-1}})^2 = \sum_{i=1}^{N}(a\,\Delta t)^2 = Na^2\left(\frac{u-s}{N}\right)^2 = \frac{a^2(u-s)^2}{N} \to 0$$

as $N \to \infty$. A similar argument shows that the quadratic variation of any continuously differentiable function is zero, using the fact that such a function is approximately linear (can be approximated linearly by its derivative) at each time. This difference between Brownian motions and continuously differentiable functions results in a different calculus (the Itô calculus) for Brownian motions, which is described below and which is an essential tool for continuous-time finance.

A concept closely related to quadratic variation is total variation, which is defined in the same way as quadratic variation but with the squared changes replaced by the absolute values of the changes. A Brownian motion has infinite total variation (with probability 1).[2] This means that if we were to straighten out a path of a Brownian motion to measure it, its length would be infinite! This is true no matter how small the time period over which we measure the path.

Continuous Martingales and Levy's Theorem

A martingale is a stochastic process M with the property that $\mathsf{E}_t[M_u] = M_t$ for each $t < u$ (equivalently, $\mathsf{E}_t[M_u - M_t] = 0$). A continuous martingale is a martingale with continuous paths. Asset pricing theory inherently involves martingales. The most important example is the fact that MW is a martingale if M is an SDF process and W is a self-financing wealth process (see Chapters 8 and 13).[3] It simplifies calculations—and leads to stronger results—to assume that martingales are continuous.

A Brownian motion is an example of a continuous martingale. As we will see, Brownian motions can be used to construct other continuous martingales. The seemingly strange path properties of Brownian motions are unavoidable if we want to deal with continuous martingales. In fact, every continuous martingale that is not constant has infinite total variation, so Brownian motions are prototypical in that regard. Another important fact is Levy's theorem, which states that a continuous martingale is a Brownian motion if and only if its quadratic variation over each interval $[s, u]$ equals $u - s$.[4] Thus, if a stochastic process has (i) continuous paths, (ii) conditionally mean-zero increments, and (iii) quadratic variation over each interval equal to the length of the interval, then its increments must also be (iv) independent of conditioning information and (v) normally distributed. Having quadratic variation over any interval equal to the length of the interval is really just a normalization. Any stochastic process satisfying (i) and (ii)—that is, any continuous martingale—can be converted to a process also satisfying (iii), and hence having all of the properties (i)–(v), just by deforming the time scale (though that topic is not addressed in this book).

12.2 ITÔ INTEGRAL AND ITÔ PROCESSES

The basis of continuous-time finance is K. Itô's concept of integral. If θ is a stochastic process adapted to the information with respect to which B is a Brownian motion, is jointly measurable in (t, ω), and satisfies

---

2. In fact, this is true of any continuous function that has nonzero quadratic variation.

3. Actually, as is discussed in Section 13.3, in continuous time MW is in general only a "local martingale" and to ensure it is a martingale requires the imposition of a regularity condition.

4. In fact, Levy's theorem applies even to continuous local martingales.

$$\int_0^T \theta_t^2 \, dt < \infty \tag{12.2}$$

with probability 1, and if M_0 is a constant, then we can define the stochastic process

$$M_t = M_0 + \int_0^t \theta_s \, dB_s \tag{12.3}$$

for $t \in [0, T]$. The integral in (12.3) is called an Itô integral or stochastic integral. For each t, it can be approximated as (is a limit in probability of)

$$\sum_{i=1}^N \theta_{t_{i-1}} (B_{t_i} - B_{t_{i-1}})$$

given discrete partitions

$$0 = t_0 < t_1 < t_2 < \cdots < t_N = t$$

of the time interval $[0,t]$ with the maximum length $t_i - t_{i-1}$ of the time intervals going to zero as $N \to \infty$. Note that θ is evaluated in this sum at the beginning of each interval $[t_{i-1}, t_i]$ over which the change in B is computed.

The process M in (12.3) is a local martingale, local martingales being a class of stochastic processes that includes martingales as a proper subset. It may or may not be a martingale, as is discussed further below.

Given (12.3), we write

$$dM_t = \theta_t \, dB_t, \tag{12.3'}$$

or, more simply, $dM = \theta \, dB$. Heuristically, we can go from the differential form (12.3') and the initial condition M_0 to the integral form (12.3) by "summing" the changes dM as

$$M_t = M_0 + \int_0^t dM_s = M_0 + \int_0^t \theta_s \, dB_s.$$

Actually, the differential form (12.3') has no independent meaning—it is simply shorthand for the integral form (12.3). Nevertheless, we interpret M as changing by dM in each instant, with this change being equal to $\theta \, dB$. Because B has independent zero-mean increments, we interpret the instantaneous change dB as having a zero mean and hence also the instantaneous change dM as having a zero mean.

The stochastic process M defined by (12.3) has continuous paths and has quadratic variation equal to

$$\int_t^u \theta_s^2 \, ds \tag{12.4}$$

on each interval $[t, u]$. If the expected quadratic variation over $[0, T]$ is finite, that is, if

$$\mathsf{E}\left[\int_0^T \theta_t^2 \, dt\right] < \infty, \tag{12.5}$$

then M is a martingale—in fact, a continuous martingale with finite variance.[5] There are in general three possibilities: (i) condition (12.5) holds, in which case, as just said, M is a martingale with finite variance, (ii) condition (12.5) does not hold and M is a martingale with infinite variance, or (iii) condition (12.5) does not hold and M is not a martingale (but is still a local martingale). The relation between quadratic variation and variance is discussed further in Section 12.7.

Local Martingales and Doubling Strategies

The name "local martingale" reflects the fact that each instantaneous increment dM of the process M defined in (12.3) can be interpreted as having a zero mean; thus, M is at each instant (i.e., locally) like a martingale. However, because there are an infinite number of instants in any time interval, being locally a martingale at each instant does not necessarily imply that M is a martingale. This is a technical issue but one of some importance in finance.

The simplest example of a local martingale that is not a martingale is the wealth process from a doubling strategy in discrete time. Fix an infinite sequence of dates $0 = t_0 < t_1 < t_2 < \cdots < T$ within some interval $[0, T]$ for a finite T. For example, take $t_n = nT/(n+1)$. Suppose a casino allows you to gamble on the toss of a fair coin at each date t_n, with you winning if the coin comes up heads. Suppose the initial stake is \$1 and the stake doubles on each successive round. Suppose you can play the game until you win. Suppose the game ends when you win, meaning that your wealth is constant from then until T, so your wealth W_t is defined for $t \in [0, T]$. If you win on the second round, then you lost \$1 on the first and won \$2 on the second and hence won \$1 overall. With probability 1, you eventually win, and no matter the round on which you win, you win \$1 overall.[6] Each separate

---

5. We say that a martingale M on a time period $[0, T]$ has finite variance if M_T has a finite unconditional variance. This is equivalent to $\mathsf{E}[M_T^2] < \infty$. If M is a finite-variance martingale, then any increment $M_t - M_s$ for $0 \leq s < t \leq T$ has finite conditional and unconditional variance.

6. Note that if the game has not ended before the toss at t_n, then you have wealth (prior to the toss) of

$$W_{t_n} = W_0 - \sum_{i=0}^{n-1} 2^i = W_0 + 1 - 2^n,$$

and you have an equal chance of winning or losing 2^n at t_n.

coin toss is a fair game in the sense that your expected gain (expected change in wealth) is zero. Thus, your wealth process is a local martingale.[7] However, the overall game is not fair, because you win \$1 with probability 1, so your wealth process is not a martingale ($\mathsf{E}[W_T] = 1 + W_0 \neq W_0$). This distinction between the game being "locally fair" and fair overall arises only because the number of potential coin tosses is unbounded. In a continuous-time trading model, investors can always trade an unbounded number of times, so the distinction between local martingales and martingales is always an issue.

In finance applications, wealth processes typically are not martingales, because interest rates are positive and the stock market goes up on average. However, adjusted for the time value of money and risk, wealth processes in discrete time *are* martingales—that is, MW is a martingale if M is an SDF process and W is a self-financing wealth process. To ensure this is true in continuous time, we must prohibit strategies like the doubling strategy. In reality, such strategies are indeed impossible. To win for sure in the coin-tossing game above, you must start with infinite wealth or have a ready supply of credit if it takes too long for a winning toss to appear. No casino will supply the credit to play this game, and neither will financial markets.

As the previous paragraph suggests, one way to prohibit doubling strategies is to impose a lower bound on wealth. If W is a local martingale and bounded below, then it can be shown that $\mathsf{E}_t[W_u] \leq W_t$ for $u \geq t$ (see Appendix A.13), so a lower bound on wealth in the game above ensures that you do not win for sure (by forcing you to exit after a finite number of tosses). While the approach of imposing a lower bound on wealth is feasible, it is also feasible to require an analog of condition (12.5), which implies that the process M defined in (12.3) is actually a martingale. Condition (12.5) states that θ does not get too large over time and across states of the world. The interpretation of the analogous condition in the context of portfolio choice is that the amounts invested in the assets are not too large relative to the investor's wealth (as the bets are in the doubling game). This topic is discussed further in Chapter 13.

Itô Processes

The sum of an ordinary integral and a stochastic integral is called an Itô process. Such a process has the form

$$Y_t = Y_0 + \int_0^t \alpha_s \, ds + \int_0^t \theta_s \, dB_s, \qquad (12.6)$$

---

7. For the formal definition of a local martingale, see Appendix A.13. In this example, we can take the stopping times τ_n in Appendix A.13 to be $\tau_n = T$ if you have won by date t_n and $\tau_n = t_n$ otherwise.

which is also written as

$$dY_t = \alpha_t\, dt + \theta_t\, dB_t, \tag{12.6$'$}$$

or, more simply, as $dY = \alpha\, dt + \theta\, dB$. Heuristically, as before, we go from the differential form to the integral form (12.6) by "summing" the changes dY over time.

The first integral in (12.6) is random if α is random, but the integral is defined in each state of the world just as integrals are ordinarily defined. The process α is called the drift of Y, and in this book the term $\theta\, dB$ is called the stochastic part of dY. We can interpret $\alpha\, dt$ as the conditional mean of dY. As this interpretation suggests, it can be shown that Y is a local martingale (the conditional mean of dY is zero) if and only if the drift α is identically zero.[8] This simple fact, combined with Itô's formula (see below) to compute the drift of a function $f(t, Y_t)$, is the source of the partial differential equations (PDEs) that appear in finance—for example, the Hamilton-Jacobi-Bellman equation for optimal portfolio choice (Chapter 14) and the fundamental PDE for valuing derivative securities (Chapter 15 and Part III).

Asset and Portfolio Returns

Suppose that between dividend payments the price S of an asset satisfies

$$\frac{dS}{S} = \mu\, dt + \sigma\, dB \tag{12.7}$$

for a Brownian motion B and stochastic processes (or constants) μ and σ. We interpret dS/S as the instantaneous rate of return of the asset and $\mu\, dt$ as the expected rate of return. Equation (12.7) can be written equivalently as $dS = S\mu\, dt + S\sigma\, dB$, and the real meaning of it is that

$$S_u = S_0 + \int_0^u S_t \mu_t\, dt + \int_0^u S_t \sigma_t\, dB_t \tag{12.8}$$

for each u. Section 12.5 presents an explicit formula for the price S satisfying (12.7) when μ, and σ are constants.

Suppose there is also an asset that is locally risk free, meaning that its price R satisfies

$$\frac{dR}{R} = r\, dt \tag{12.9}$$

---

8. More precisely, Y is a local martingale if and only if, with probability 1, the set of times t at which $\alpha_t \neq 0$ has zero Lebesgue measure; equivalently, $\int_0^t \alpha_s^2\, ds = 0$ for all t with probability 1.

for some r (which can be a stochastic process). This equation can be solved explicitly as

$$R_u = R_0 \exp\left(\int_0^u r_t\, dt\right).$$

We interpret r_t as the interest rate at date t for an investment during the infinitesimal period $(t, t+dt)$. If the interest rate is constant, then

$$R_u = R_0 e^{ru},$$

meaning that interest is continuously compounded at the constant rate r. In general, we call r the instantaneous risk-free rate or the locally risk-free rate or the short rate ("short" referring to the infinitesimal maturity of the bond that pays it). Analogous to the discrete-time case, the asset with price R can be called a money market account (or the instantaneously risk-free asset or the locally risk-free asset).

A portfolio of these two assets is defined by the fraction π_t of wealth invested in the risky asset at each date t. If no funds are invested or withdrawn from the portfolio during a time period $[0, T]$ and the asset does not pay dividends during the period, then the wealth process W satisfies

$$\frac{dW}{W} = (1-\pi)r\, dt + \pi \frac{dS}{S}. \tag{12.10}$$

The meaning of (12.10) is that W_t equals W_0 plus an ordinary integral and a stochastic integral, exactly analogous to (12.8). An explicit formula for W when r, π, μ, and σ are constant is given in Section 12.5.

Equation (12.10) is called the intertemporal budget constraint or self-financing condition. It states that wealth grows only from interest earned and from the return on the risky asset. Obviously, it mirrors the discrete-time formula

$$(1-\pi_{it})R_{f,t+1} + \pi_{it}R_{i,t+1}$$

for the return on a portfolio of a risk-free asset and a single risky asset.

One possible concern about the continuous-time formula (12.10) is that we allow the portfolio π_t to depend on date–t information, whereas we also assume the price S_t is known at time t. Thus, it may appear that we allow the portfolio to be chosen after the return dS/S is known. However, this is not true. In the approximating sum

$$\sum_{i=1}^{N} \theta_{t_{i-1}}(B_{t_i} - B_{t_{i-1}})$$

for the Itô integral, the integrand θ is deliberately evaluated at t_{i-1} rather than at an arbitrary point in the interval $[t_{i-1}, t_i]$. Evaluating θ at different points in the interval leads to different stochastic integrals (different stochastic processes), though it does not matter for ordinary integrals. For example, choosing the midpoint $(t_{i-1}+t_i)/2$ produces what is called the Stratonovich integral. However, the Itô integral is the right integral for portfolio choice and asset pricing applications, because it means that, in the approximating sum and hence also in the limit defining the wealth process W in (12.10), portfolio returns are calculated based on the beginning-of-period portfolio $\pi_{t_{i-1}}$.

12.3 MARTINGALE REPRESENTATION

Consider a local martingale M that depends only on a Brownian motion B in the sense that, for each t, M_t depends only on the history of B through date t. This is expressed by saying that M is adapted to B. Any such local martingale has the property that

$$M_t = M_0 + \int_0^t \theta_s \, dB_s$$

for some θ. Thus, all local martingales (and hence all martingales) adapted to a Brownian motion are stochastic integrals with respect to the Brownian motion. This is a spanning property: A Brownian motion spans all of the local martingales M adapted to it in the sense that $dM = \theta \, dB$ for some θ. It is called the martingale representation theorem (or the predictable representation theorem). The theorem extends to vectors of Brownian motions: If M is a local martingale adapted to a vector $(B_1, \ldots, B_n)$ of Brownian motions, then there exist stochastic processes θ_i such that $dM = \sum_{i=1}^n \theta_i \, dB_i$.

The spanning property is very important. It underlies the Black-Scholes option pricing formula and much of the rest of continuous-time finance. It is analogous to the spanning property of a stock in a binomial model (see Exercise 3.1). That model is complete, because any derivative security can be replicated by a portfolio of the risk-free asset and the stock. Similar completeness properties follow from the martingale representation theorem in continuous time, as we will see.

As an example of this spanning, consider a constant σ and a Brownian motion B and define

$$M_t = \mathsf{E}_t\left[e^{\sigma B_1}\right]$$

for $t \leq 1$. The usual formula for expectations of exponentials of normally distributed variables yields

$$M_t = \exp\left(\sigma B_t + \frac{1}{2}\sigma^2(1-t)\right). \tag{12.11}$$

The martingale representation theorem tells us that $dM = \theta\, dB$ for some θ. Summing the changes dM and noting that $M_0 = e^{\sigma^2/2}$ yields

$$M_t = e^{\sigma^2/2} + \int_0^t \theta_s\, dB_s. \tag{12.12}$$

Thus, M_t, which is a nonlinear (exponential) function of B_t, is also an affine function of the changes dB_s, with coefficients θ_s. We will see in the next section how to calculate θ in this simple example.

12.4 ITÔ'S FORMULA

Before presenting Itô's formula, it is useful to review some basic facts of the ordinary calculus. Consider a deterministic (i.e., nonrandom) continuously differentiable function of time x_t and define $y_t = f(x_t)$ for some continuously differentiable function f. The chain rule of calculus expressed in differential form gives us[9]

$$dy = f'(x)\, dx. \tag{12.13}$$

The fundamental theorem of calculus states that we can "sum" the changes over an interval $[0, t]$ to obtain

$$y_t = y_0 + \int_0^t f'(x_s)\, dx_s. \tag{12.13'}$$

Of course, we can substitute $dx_s = x'_s\, ds$ in this integral.

Now consider a slightly more complicated example. Suppose $y_t = f(t, x_t)$ for some continuously differentiable function f of (t, x). Then, the chain rule of multivariate calculus states that

$$dy = \frac{\partial f}{\partial t}\, dt + \frac{\partial f}{\partial x}\, dx, \tag{12.14}$$

which implies

$$y_t = y_0 + \int_0^t \frac{\partial f(s, x_s)}{\partial s}\, ds + \int_0^t \frac{\partial f(s, x_s)}{\partial x}\, dx_s. \tag{12.14'}$$

Itô's formula (also called Itô's lemma) is the chain rule for stochastic calculus. Suppose a function $f(t, x)$ is continuously differentiable in t and twice continuously differentiable in x. Suppose $Y_t = f(t, B_t)$ for a Brownian motion B. Itô's formula states that

---

9. The ′ here denotes the derivative, not a transpose.

$$dY = \frac{\partial f}{\partial t} dt + \frac{\partial f}{\partial B} dB + \frac{1}{2} \frac{\partial^2 f}{\partial B^2} dt. \tag{12.15}$$

This shows that Y is an Itô process with

$$\frac{\partial f}{\partial t} + \frac{1}{2} \frac{\partial^2 f}{\partial B^2}$$

as its drift and $(\partial f / \partial B)\, dB$ as its stochastic part. The meaning of (12.15) is that, for each t,

$$Y_t = Y_0 + \int_0^t \left(\frac{\partial f(s, B_s)}{\partial s} + \frac{1}{2} \frac{\partial^2 f(s, B_s)}{\partial B^2} \right) ds + \int_0^t \frac{\partial f(s, B_s)}{\partial B} dB_s. \tag{12.15'}$$

Compared to the usual calculus, there is obviously an extra term in Itô's formula involving the second partial derivative $\partial^2 f / \partial B^2$. This is discussed further below.

Squared Differentials

A very convenient notation for applying Itô's formula, and for other purposes, is to write $(dB)^2 = dt$. This notation can be interpreted in terms of quadratic variation. Over any interval, the sum of squared changes $(\Delta B)^2 = (B_{t_i} - B_{t_{i-1}})^2$ converges to the length of the interval as the time periods within the interval get short and the number of time periods gets large. For an infinitesimal interval dt, we can think of this as $(\Delta B)^2 \to dt$ and hence as $(dB)^2 = dt$.[10] In terms of this notation, Itô's formula (12.15) can be written as

$$dY = \frac{\partial f}{\partial t} dt + \frac{\partial f}{\partial B} dB + \frac{1}{2} \frac{\partial^2 f}{\partial B^2} (dB)^2. \tag{12.15''}$$

---

10. Formally, the symbol $(dB)^2$ should be understood as the differential of the quadratic variation process. For a Brownian motion, the quadratic variation over any interval $[0, t]$ is t, so its differential is dt. Consistent with the notation $(dB)^2 = dt$, we say that dB is of order $\sqrt{dt}$. Of course, for small positive numbers x, $\sqrt{x} > x$ and in fact $\sqrt{x}/x \to \infty$ as $x \to 0$, so saying that dB is of order $\sqrt{dt}$ implies that dB is of larger order than dt. This is an informal but presumably intuitive way of expressing the fact that the paths of a Brownian motion have nonzero quadratic variation and infinite total variation.

Some Examples

As an example, let $Y_t = B_t^2$. Consider the increment $\Delta Y = Y_u - Y_s$ over an interval $[s, u]$. We have

$$\Delta Y = B_u^2 - B_s^2$$
$$= [B_s + \Delta B]^2 - B_s^2$$
$$= 2B_s \Delta B + (\Delta B)^2, \tag{12.16}$$

where $\Delta B = B_u - B_s$. Now apply Itô's formula to $Y_t = f(B_t)$ where f is the square function ($f(x) = x^2$). Using the notation $(dB)^2 = dt$, we have

$$dY = f'(B_t)\, dB + \frac{1}{2} f''(B_t)\, (dB)^2$$
$$= 2B_t\, dB + (dB)^2. \tag{12.17}$$

The similarity between (12.16) and (12.17) is evident.

For another example, consider the process M_t defined in (12.11). It can be written as

$$M_t = f(t, B_t)$$

by defining

$$f(t, x) = \exp\left(\sigma x + \frac{1}{2}\sigma^2(1 - t)\right). \tag{12.18}$$

The partial derivatives of f are

$$\frac{\partial f}{\partial t} = -\frac{1}{2}\sigma^2 f(t, x),$$
$$\frac{\partial f}{\partial x} = \sigma f(t, x),$$
$$\frac{\partial^2 f}{\partial x^2} = \sigma^2 f(t, x).$$

Notice that the $\partial f/\partial t$ term in (12.15) cancels in this instance with the term $(1/2)\partial^2 f/\partial B^2$. Itô's formula therefore implies

$$dM = \frac{\partial f}{\partial B}\, dB = \sigma M\, dB.$$

Hence, θ_s in (12.12) is σM_s.

Each of these examples illustrates why the ordinary calculus cannot be correct for a function of a Brownian motion. The ordinary calculus (12.13) applied to $Y_t = B_t^2$ would yield $dY = 2B\, dB$, implying that Y is a local martingale, but

$\mathsf{E}_s[B_u^2 - B_s^2] = u - s$, which is inconsistent with its being a local martingale.[11] On the other hand, M defined in (12.11) is a martingale, but the ordinary calculus (12.14) applied to the function f in (12.18) would yield $dM = -(1/2)\sigma^2 M\,dt + \sigma M\,dB$, implying that M has a nonzero drift, which is inconsistent with its being a martingale.

The only purpose of the following discussion is to provide some additional intuition for Itô's formula.

Consider the ordinary calculus again with $y_t = f(x_t)$. Given dates $t < u$, the derivative defines a linear approximation of the change in y over this time period; that is, setting $\Delta x = x_u - x_t$ and $\Delta y = y_u - y_t$, we have the approximation

$$\Delta y \approx f'(x_t)\,\Delta x.$$

A better approximation is given by the second-order Taylor series expansion

$$\Delta y \approx f'(x_t)\,\Delta x + \frac{1}{2}f''(x_t)\,(\Delta x)^2.$$

The fundamental theorem of calculus means that the linear approximation works perfectly for infinitesimal time periods dt, because we can compute the change in y in (12.13$'$) by "summing up" the infinitesimal changes $f'(x_t)\,dx_t$. In other words, the second-order term $\frac{1}{2}f''(x_t)\,(\Delta x)^2$ "vanishes" when we consider very small time periods.

Now consider $Y_t = f(B_t)$ for a twice continuously differentiable function f (choosing f to depend directly only on B rather than on t and B is for the sake of simplicity only). The second-order Taylor series expansion in the case of $Y = f(B)$ is

$$\Delta Y \approx f'(B_t)\,\Delta B + \frac{1}{2}f''(B_t)\,(\Delta B)^2.$$

For example, given a partition $s = t_0 < t_1 < \cdots < t_N = u$ of a time interval $[s, u]$, we have, with the same notation we have used earlier,

$$Y_u - Y_s = \sum_{i=1}^{N} \Delta Y_{t_i} \approx \sum_{i=1}^{N} f'(B_{t_{i-1}})\,\Delta B_{t_i} + \frac{1}{2}\sum_{i=1}^{N} f''(B_{t_{i-1}})\,(\Delta B_{t_i})^2. \quad (12.19)$$

If we make the time intervals $t_i - t_{i-1}$ shorter, letting $N \to \infty$, we cannot expect that the "extra" term here will disappear, leading to the result (12.13$'$) of the ordinary calculus, because we know that

---

11. As noted in Section 12.2, a local martingale Z that is bounded below (and B_t^2 is bounded below by zero) must satisfy $\mathsf{E}_s[Z_u - Z_s] \leq 0$. Such a process is an example of a "supermartingale." See Appendix A.13.

$$\lim_{N\to\infty} \sum_{i=1}^{N} (\Delta B_{t_i})^2 = u - s,$$

whereas for a continuously differentiable function x_t, the analogous limit is zero.

Taking the second-order Taylor series expansion and $(\Delta B)^2 \to dt$ yields Itô's formula: If we take the limit in (12.19), replacing the limit of $(\Delta B_{t_i})^2$ with $(dB)^2 = dt$, we obtain

$$Y_u = Y_s + \int_s^u f'(B_t)\, dB_t + \int_s^u \frac{1}{2} f''(B_t)\, dt,$$

or, in differential form,

$$dY = f'(B_t)\, dB + \frac{1}{2} f''(B_t)\, dt, \tag{12.20}$$

which is the same as (12.15) when f does not depend directly on t.

12.5 GEOMETRIC BROWNIAN MOTION

Now we are prepared to explain the formula (12.1), which is a relatively simple but fairly standard model of a stock price. Let

$$S_t = S_0 \exp\left(\mu t - \frac{1}{2}\sigma^2 t + \sigma B_t\right) \tag{12.21}$$

for constants S_0, μ, and σ, where B is a Brownian motion. Then, $S_t = f(t, B_t)$, where

$$f(t, x) = S_0 \exp\left(\mu t - \frac{1}{2}\sigma^2 t + \sigma x\right).$$

The partial derivatives of f are

$$\frac{\partial f}{\partial t} = \left(\mu - \frac{1}{2}\sigma^2\right) f(t, x),$$

$$\frac{\partial f}{\partial x} = \sigma f(t, x),$$

$$\frac{\partial^2 f}{\partial x^2} = \sigma^2 f(t, x).$$

Substituting S for $f(t, B)$ in these partial derivatives, Itô's formula (12.15″) implies

$$dS = \left(\mu - \frac{1}{2}\sigma^2\right) S\, dt + \sigma S\, dB + \frac{1}{2}\sigma^2 S\, (dB)^2$$

$$= \mu S\, dt + \sigma S\, dB.$$

This is equivalent to (12.1), that is,

$$\frac{dS}{S} = \mu \, dt + \sigma \, dB. \tag{12.21'}$$

Because (12.21) is an explicit formula for S, we say that (12.21) is the solution of (12.21').

The process S is called a geometric Brownian motion. Notice that the term μt in the exponent of (12.21) results in the term $\mu \, dt$ in (12.21'). As a result of this drift, S is not a martingale (or a local martingale), unless of course $\mu = 0$. As noted before, we interpret $\mu \, dt$ as the expected rate of change of S, and σ is called the volatility of S. The formula (12.21) and the usual rule for expectations of exponentials of normals shows that S grows at the average rate of μ, in the sense that

$$E_t[S_\tau] = e^{\mu(\tau-t)} S_t \tag{12.22}$$

for any dates $t < \tau$.

Taking the natural logarithm of (12.21) gives an equivalent formula for the solution:

$$\log S_t = \log S_0 + \left(\mu - \frac{1}{2}\sigma^2\right) t + \sigma B_t. \tag{12.23}$$

The differential of (12.23) is[12]

$$d\log S_t = \left(\mu - \frac{1}{2}\sigma^2\right) dt + \sigma \, dB_t. \tag{12.23'}$$

The geometric Brownian motion is an important process, and it should be committed to memory that (12.21') is equivalent to (12.23') and that the solution is the equivalent formulas (12.21) and (12.23). In fact, the differential versions (12.21') and (12.23') are equivalent even if μ and σ are stochastic processes; however, (12.21) and (12.23) are the solution and S is a geometric Brownian motion only if μ and σ are constants.

Because S is the exponential of its logarithm, S can never be negative. For this reason, a geometric Brownian motion is a better model for stock prices than is a Brownian motion. Taking S to be the price of a non-dividend-paying asset (or a dividend-reinvested price—see Section 13.1), so dS/S is the total return of the asset, the geometric Brownian motion model also has the convenient property of IID continuously compounded rates of return.

---

12. To derive (12.23') from (12.23), apply Itô's formula to $\log S_t = f(t, B_t)$, where

$$f(t,x) = \log S_0 + \left(\mu - \frac{1}{2}\sigma^2\right) t + \sigma x.$$

Because f is affine in (t,x), there are no second-order terms, and Itô's formula gives the result that we would get from the usual calculus.

Continuously Compounded Returns

To see the implications for continuously compounded rates of return, recall that the annualized continuously compounded rate of return of a non-dividend-paying asset over a time period $[t_1, t_2]$ of length $\Delta t = t_2 - t_1$ is r defined by

$$\frac{S_{t_2}}{S_{t_1}} = e^{r \Delta t}.$$

Setting $\Delta \log S = \log S_{t_2} - \log S_{t_1}$, this is equivalent to

$$r = \frac{\Delta \log S}{\Delta t}.$$

The geometric Brownian motion model—this is easiest to see from (12.23) or (12.23')—implies that

$$\Delta \log S = \left(\mu - \frac{1}{2}\sigma^2\right) \Delta t + \sigma \Delta B, \qquad (12.23'')$$

where $\Delta B = B_{t_2} - B_{t_1}$. Therefore, $\Delta \log S$ is conditionally normally distributed with mean $(\mu - \sigma^2/2) \Delta t$ and variance $\sigma^2 \Delta t$. Thus, the annualized continuously compounded rate of return r is conditionally normally distributed with mean $\mu - \sigma^2/2$ and standard deviation $\sigma \sqrt{\Delta t}/\Delta t = \sigma/\sqrt{\Delta t}$.

12.6 COVARIATION OF ITÔ PROCESSES AND GENERAL ITÔ'S FORMULA

To describe Itô's formula for functions of multiple Itô processes, we first need to explain the concept of covariation of Itô processes. Consider a discrete partition $s = t_0 < t_1 < t_2 < \cdots < t_N = u$ of a time interval $[s, u]$. For any two functions of time x and y, consider the sum of products of changes

$$\sum_{i=1}^{N} \Delta x_{t_i} \Delta y_{t_i},$$

where $\Delta x_{t_i} = x_{t_i} - x_{t_{i-1}}$ and $\Delta y_{t_i} = y_{t_i} - y_{t_{i-1}}$. The covariation (or joint variation) of x and y on the interval $[s, u]$ is defined as the limit of this sum as $N \to \infty$ and the lengths $t_i - t_{i-1}$ of the intervals go to zero. If $x = y$, then this is the same as the quadratic variation. As with quadratic variation, if x and y are continuously differentiable functions, then their covariation is zero.

The covariation of any two Brownian motions B_1 and B_2 is defined in each state of the world as the limit of the sum of products of increments $\Delta x = \Delta B_1$

and $\Delta y = \Delta B_2$ as above. It can be shown that the covariation over any interval $[s, u]$ equals

$$\int_s^u \rho_t \, dt$$

with probability 1, for some stochastic process (or constant) ρ with $-1 \leq \rho_t \leq 1$ for all t. Of course, if the two Brownian motions are the same ($B_1 = B_2$), then the covariation is the same as the quadratic variation, so $\rho = 1$. In general, ρ is called either the covariation process or, more commonly, correlation process of the two Brownian motions (see Section 12.7). Extending the notation $(dB)^2 = dt$, we write $(dB_1)(dB_2) = \rho \, dt$. Then, the covariation is

$$\int_s^u (dB_{1t})(dB_{2t}).$$

Now, consider two local martingales M_i with $dM_i = \theta_i \, dB_i$ for each i, where the B_i are Brownian motions. The covariation of M_1 and M_2 over any interval $[s, u]$ in each state of the world is defined in the same way as for two Brownian motions (and in the same way as for any functions x_t and y_t). It can be shown to equal

$$\int_s^u (dM_{1t})(dM_{2t})$$

with probability 1, where

$$(dM_1)(dM_2) = (\theta_1 \, dB_1)(\theta_2 \, dB_2) = \theta_1 \theta_2 \rho \, dt,$$

and where ρ is the correlation process of the Brownian motions B_1 and B_2.

Finally, consider two Itô processes $dX_i = \alpha_i \, dt + \theta_i \, dB_i$. It can be shown that the covariation of X_1 and X_2 over any interval $[s, u]$ is

$$\int_s^u \theta_{1t} \theta_{2t} \rho_t \, dt.$$

To remember this formula, it is convenient to introduce additional rules for multiplying differentials: Define $(dt)^2 = 0$ and $(dt)(dB) = 0$ for any Brownian motion B. Then, we can "compute" the covariation of X_1 and X_2 over an interval $[s, u]$ as

$$\int_s^u (dX_{1t})(dX_{2t}),$$

where

$$(dX_1)(dX_2) = (\alpha_1 \, dt + \theta_1 \, dB_1)(\alpha_2 \, dt + \theta_2 \, dB_2) = \theta_1 \theta_2 \rho \, dt.$$

We say that the two Itô processes are locally correlated if $(dX_1)(dX_2) \neq 0$, that is, if $\theta_1 \theta_2 \rho \neq 0$. These rules for multiplying differentials are also very convenient for stating and using the general version of Itô's formula.

General Itô's Formula

Recall that if $Y_t = f(t, B_t)$ where f is continuously differentiable in t and twice continuously differentiable in B, then Itô's formula is

$$dY = \frac{\partial f}{\partial t} dt + \frac{\partial f}{\partial B} dB + \frac{1}{2} \frac{\partial^2 f}{\partial B^2} (dB)^2.$$

More generally, suppose X is an Itô process: $dX = \alpha \, dt + \theta \, dB$ for stochastic processes α and θ. Let $Y = f(t, X_t)$. Itô's formula is

$$dY = \frac{\partial f}{\partial t} dt + \frac{\partial f}{\partial X} dX + \frac{1}{2} \frac{\partial^2 f}{\partial X^2} (dX)^2. \qquad (12.24)$$

By substituting $dX = \alpha \, dt + \theta \, dB$ and $(dX)^2 = \theta^2 \, dt$, we can write (12.24) more explicitly as

$$dY = \left(\frac{\partial f}{\partial t} + \frac{\partial f}{\partial X} \alpha + \frac{1}{2} \frac{\partial^2 f}{\partial X^2} \theta^2 \right) dt + \frac{\partial f}{\partial X} \theta \, dB. \qquad (12.24')$$

However, (12.24) should be easier to remember than (12.24').

A similar formula applies if Y depends on multiple Itô processes. Suppose $dX_i = \alpha_i \, dt + \theta_i \, dB_i$ for $i = 1, \ldots, n$. Let $Y_t = f(t, X_{1t}, \ldots, X_{nt})$, where f is continuously differentiable in t and twice continuously differentiable in the X_i. Itô's formula is:

$$dY = \frac{\partial f}{\partial t} dt + \sum_{i=1}^n \frac{\partial f}{\partial X_i} dX_i + \frac{1}{2} \sum_{i=1}^n \sum_{j=1}^n \frac{\partial^2 f}{\partial X_i \partial X_j} (dX_i)(dX_j). \qquad (12.25)$$

For example, if $n = 2$, then

$$dY = \frac{\partial f}{\partial t} dt + \sum_{i=1}^2 \frac{\partial f}{\partial X_i} dX_i + \frac{1}{2} \sum_{i=1}^2 \frac{\partial^2 f}{\partial X_i^2} (dX_i)^2 + \frac{\partial^2 f}{\partial X_1 \partial X_2} (dX_1)(dX_2).$$

This is exactly analogous to a second-order Taylor series expansion of f in the X_i (and a first-order expansion in t).

A special case of (12.25) that arises frequently is when $Y = X_1 X_2$. In this case, the second direct partial derivatives are 0 and the cross partial derivative is 1, so (12.25) implies

$$dY = X_1 \, dX_2 + X_2 \, dX_1 + (dX_1)(dX_2). \qquad (12.26)$$

This is called integration by parts.

12.7 CONDITIONAL VARIANCES AND COVARIANCES

There are close relations between quadratic variation and variance and between covariation and covariance. For any Brownian motion B and dates $s < u$, the variance of the increment $B_u - B_s$ is $u - s$, which is also the quadratic variation of the path of B between s and u (with probability 1). Thus,

$$\text{var}_s(B_u - B_s) = \mathsf{E}_s\left[\int_s^u (\mathrm{d}B_t)^2\right]. \tag{12.27}$$

Now consider a stochastic integral $\mathrm{d}M = \theta\,\mathrm{d}B$. Recall that the quadratic variation of M over any interval $[s, u]$ is

$$\int_s^u (\mathrm{d}M_t)^2 = \int_s^u \theta_t^2\,(\mathrm{d}B_t)^2 = \int_s^u \theta_t^2\,\mathrm{d}t.$$

Also, recall that M is a martingale on the time interval $[0, T]$ if its expected quadratic variation over the interval is finite—that is, if (12.5) holds. In this circumstance, the increment $M_u - M_s$ for $s \leq u \leq T$ has zero mean conditional on date–s information. Moreover, it can be shown that the conditional variance is equal to the expected quadratic variation over the interval:

$$\text{var}_s(M_u - M_s) = \mathsf{E}_s\left[\int_s^u (\mathrm{d}M_t)^2\right]. \tag{12.28}$$

These formulas motivate a refinement of the interpretations of $\mathrm{d}B$ and $\mathrm{d}M$ as follows: The instantaneous increment $\mathrm{d}B$ can be interpreted as having zero mean and variance equal to $\mathrm{d}t$ and the increment $\mathrm{d}M$ as having zero mean and conditional variance equal to $\theta^2\,\mathrm{d}t$.

Equation (12.28) is analogous to the properties of martingales in discrete time: If M is a martingale and $s = t_0 < t_1 < \cdots < t_N = u$ are any discrete dates, then the sequence of random variables

$$x_i = M_{t_i} - M_{t_{i-1}}$$

is called a martingale difference series. Because $M_u - M_s$ is the sum of the x_i, which have zero means and are uncorrelated,[13]

$$\text{var}_s(M_u - M_s) = \sum_{i=1}^N \text{var}_s(x_i) = \mathsf{E}_s\left[\sum_{i=1}^N x_i^2\right]. \tag{12.28'}$$

In continuous time, we simply write the sum of squared changes as $\int_s^u (\mathrm{d}M_t)^2$.

---

13. The conditional covariance of x_i and x_j for $j > i$ is

$$\text{cov}_s(x_i, x_j) = \mathsf{E}_s[x_i x_j] = \mathsf{E}_s[\mathsf{E}_{t_i}[x_i x_j]] = \mathsf{E}_s[x_i \mathsf{E}_{t_i}[x_j]] = 0.$$

Now consider two Brownian motions B_1 and B_2. It can be shown that, for any dates $s < u$,

$$\mathrm{cov}_s(B_{1u} - B_{1s}, B_{2u} - B_{2s}) = \mathsf{E}_s\left[\int_s^u (\mathrm{d}B_{1t})(\mathrm{d}B_{2t})\right]. \tag{12.29}$$

Let ρ be the correlation process of the two Brownian motions; that is, $(\mathrm{d}B_1)(\mathrm{d}B_2) = \rho\,\mathrm{d}t$. If ρ is constant, then (12.29) implies that the covariance of the increments is $(u-s)\rho$; hence, the correlation of the increments is ρ. This motivates the name "correlation process." Two normally distributed random variables are independent if and only if they are uncorrelated, and the same is true of Brownian motions—B_1 and B_2 are independent (knowing even the entire path of one provides no information about the distribution of the other) if and only if ρ is identically zero.

Now consider stochastic integrals $\mathrm{d}M_i = \theta_i\,\mathrm{d}B_i$, and suppose each θ_i satisfies (12.5), so M_i is a finite-variance martingale. Then, it can be shown that

$$\mathrm{cov}_s(M_{1u} - M_{1s}, M_{2u} - M_{2s}) = \mathsf{E}_s\left[\int_s^u (\mathrm{d}M_{1t})(\mathrm{d}M_{2t})\right]. \tag{12.30}$$

If θ_1, θ_2, and ρ are constants, then ρ is the correlation of the increments $M_{iu} - M_{is}$.[14] In general, the correlation process of two Brownian motions B_i is also said to be the correlation process of two local martingales $\mathrm{d}M_i = \theta_i\,\mathrm{d}B_i$. As with (12.28) and (12.28′), (12.30) is analogous to a property of discrete-time martingales (see Exercise 12.13). There are more general statements of (12.28) and (12.30) that also apply when the M_i are not finite-variance martingales (see Exercises 12.12 and 12.8).

12.8 TRANSFORMATIONS OF MODELS

We encounter models written in different ways, so it is important to understand when different forms are equivalent. Suppose we are given an Itô process

$$\mathrm{d}Y = \alpha\,\mathrm{d}t + \sum_{j=1}^{k} \sigma_j\,\mathrm{d}B_j$$

for independent Brownian motions B_j. We can always write the stochastic part as depending only on a single Brownian motion as follows: Define a stochastic

---

14. If θ_1, θ_2, and ρ are constants, then the covariance of the increments $\Delta M_i = M_{iu} - M_{is}$ is $(u-s)\theta_1\theta_2\rho$, and the standard deviation of ΔM_i is $\theta_i\sqrt{u-s}$.

process $\hat{B}$ by $\hat{B}_0 = 0$ and

$$d\hat{B} = \frac{1}{\sqrt{\sum_{j=1}^{k} \sigma_j^2}} \left(\sum_{j=1}^{k} \sigma_j \, dB_j \right). \qquad (12.31)$$

Then, $d\hat{B}$ has no drift, and

$$(d\hat{B})^2 = \frac{1}{\sum_{j=1}^{k} \sigma_j^2} \left(\sum_{j=1}^{k} \sigma_j^2 \, dt \right) = dt,$$

so Levy's theorem implies B is a Brownian motion. Obviously we have

$$dY = \alpha \, dt + \hat{\sigma} \, d\hat{B},$$

where

$$\hat{\sigma} = \sqrt{\sum_{j=1}^{k} \sigma_j^2}.$$

This does not mean that the other Brownian motions B_i are irrelevant. Unless α and $\hat{\sigma}$ are constants, they may not be adapted to $\hat{B}$; hence, the information in the other Brownian motions may be useful for forecasting the path of Y.

Given multiple Itô processes of this form, that is,

$$dY_i = \alpha_i \, dt + \sum_{j=1}^{k} \sigma_{ij} \, dB_j,$$

we can make this transformation to obtain

$$dY_i = \alpha_i \, dt + \hat{\sigma}_i \, d\hat{B}_i$$

for each i. The Brownian motions $\hat{B}_i$ will typically be correlated, because

$$\hat{\sigma}_i \hat{\sigma}_\ell (d\hat{B}_i)(d\hat{B}_\ell) = (dY_i)(dY_\ell) = \sum_{j=1}^{k} \sigma_{ij} \sigma_{\ell j} \, dt.$$

It is possible to reverse this process to transform a model written in terms of correlated Brownian motions into a model written in terms of independent Brownian motions. Moreover, it is possible to do so in such a way that the first process is locally correlated only with the first Brownian motion, the second process is locally correlated only with the first two Brownian motions, and so on. This is the Gram-Schmidt orthogonalization used for ordinary finite-variance random variables (Section 3.8). It is described below. The case $n = 2$ is treated in Exercises 12.10–12.11.

Suppose Y is an n-dimensional stochastic process with

$$dY_t = \alpha_t\, dt + A_t\, dB_t,$$

where α is an n-dimensional stochastic process, A is an $n \times n$ dimensional stochastic process, and B is an n-dimensional vector of possibly correlated Brownian motions. Let Σ denote the instantaneous covariance process of B, meaning $(dB_t)(dB_t)' = \Sigma_t\, dt$. The (i,j)th element of Σ is the correlation process of B_i and B_j, and the diagonal elements of Σ are 1's. The instantaneous covariance process of Y is $A\Sigma A'$, because

$$(dY_t)(dY_t)' = A_t(dB_t)(dB_t)'A_t' = A_t \Sigma_t A_t'\, dt.$$

Suppose that, with probability 1, $A_t \Sigma_t A_t'$ is nonsingular for all t. Let L_t be the Cholesky decomposition of $A_t \Sigma_t A_t'$, meaning the lower triangular matrix L_t with positive diagonal elements such that $L_t L_t' = A_t \Sigma_t A_t'$. Defining

$$dZ_t = L_t^{-1} A_t\, dB_t,$$

we have

$$(dZ_t)(dZ_t)' = L_t^{-1} A_t \Sigma_t A_t' (L_t')^{-1}\, dt = L_t^{-1} L_t L_t' (L_t')^{-1}\, dt = I\, dt,$$

so by Levy's theorem, Z is a vector of independent Brownian motions. Moreover,

$$dY_t = \alpha_t\, dt + L_t\, dZ_t.$$

Because L_t is lower triangular, Y_1 is locally correlated only with Z_1, Y_2 is locally correlated only with Z_1 and Z_2, and so on.

If the Brownian motions B_i are independent, then $\Sigma = I$, and L_t is the Cholesky decomposition of $A_t A_t'$. In this case, the matrix $L_t^{-1} A_t$ in the definition of dZ_t is idempotent, and Z is called a rotation of B (an idempotent matrix is a square matrix C such that $CC' = I$).

12.9 NOTES AND REFERENCES

Bachelier (1900) was the first to use continuous-time methods in finance, but Merton (1969) laid the foundation for most modern developments. Harrison and Kreps (1979) point out the necessity of prohibiting doubling strategies and also demonstrate the importance of the martingale representation theorem.

Note that the (very loose) definition of the Itô integral in Section 12.2 is similar to the definition of the usual (Riemann) integral. The important differences are (i) in the Itô integral, the integrand must be evaluated at the beginning of each interval rather than at an arbitrary point in the interval, and (ii) the limit does not exist state by state but rather only in probability (or in some related metric). As for ordinary integrals, the limit is independent of the approximating sequence of partitions: The Itô integral $\int_0^t \theta_s \, dB_s$ is a random variable X_t with the property that

$$\sum_{i=1}^{N} \theta_{t_{i-1}} (B_{t_i} - B_{t_{i-1}}) \to X_t$$

in probability, given any sequence (indexed by N) of partitions

$$0 = t_0 < t_1 < t_2 < \cdots < t_N = t$$

of the time interval $[0, t]$ having the property that the maximum length $t_i - t_{i-1}$ of any time interval converges to zero as $N \to \infty$. Obviously, the proof of existence of such a random variable is nontrivial, but it does exist provided θ is adapted and jointly measurable in (t, ω), and $\int_0^t \theta_s^2 \, ds < \infty$ with probability 1. Moreover, the stochastic process X_t can be taken to have continuous paths with probability 1. The continuous process X is unique up to indistinguishability, meaning that if Y is any other continuous process with the properties stated here, then, on a set of states ω having probability 1, $X_t(\omega) = Y_t(\omega)$ for all t. There are many texts that cover these issues and the Itô calculus, including Øksendal (2003) and Karatzas and Shreve (2004).

EXERCISES

12.1. Simulate the path of a Brownian motion over a year (using your favorite programming language or Excel) by simulating N standard normal random variables z_i and calculating $B_{t_i} = B_{t_{i-1}} + z_i \sqrt{\Delta t}$ for $i = 1, \ldots, N$, where $\Delta t = 1/N$ and $B_0 = 0$. (To simulate a standard normal random variable in a cell of an Excel worksheet, use the formula = NORMSINV(RAND()).)

(a) Plot the path—the set of points (t_i, B_{t_i}).
(b) Calculate the sum of the $(\Delta B_{t_i})^2$. Confirm that for large N the sum is approximately equal to 1.
(c) Calculate the sum of $|\Delta B_{t_i}|$. Confirm that this sum increases as N increases. Note: The sum converges to ∞ as $N \to \infty$ (because a Brownian motion has infinite total variation), but this may be difficult to see.
(d) Use the simulated Brownian motion to simulate a path of a geometric Brownian motion via the formula (12.23). Plot the path.

12.2. Assume X is an Itô process. Use Itô's formula to derive the following:

(a) Define $Y_t = e^{X_t}$. Show that
$$\frac{dY}{Y} = dX + \frac{1}{2}(dX)^2.$$

(b) Assume X is strictly positive. Define $Y_t = \log X_t$. Show that
$$dY = \frac{dX}{X} - \frac{1}{2}\left(\frac{dX}{X}\right)^2.$$

(c) Assume X is strictly positive. Define $Y_t = X_t^{-\lambda}$ for a constant λ. Show that
$$\frac{dY}{Y} = -\lambda \frac{dX}{X} + \frac{\lambda(1+\lambda)}{2}\left(\frac{dX}{X}\right)^2.$$

12.3. Assume X_1 and X_2 are strictly positive Itô processes. Use Itô's formula to derive the following:

(a) Define $Y_t = X_{1t}X_{2t}$. Show that
$$\frac{dY}{Y} = \frac{dX_1}{X_1} + \frac{dX_2}{X_2} + \left(\frac{dX_1}{X_1}\right)\left(\frac{dX_2}{X_2}\right).$$

(b) Define $Y_t = X_{1t}/X_{2t}$. Show that
$$\frac{dY}{Y} = \frac{dX_1}{X_1} - \frac{dX_2}{X_2} - \left(\frac{dX_1}{X_1}\right)\left(\frac{dX_2}{X_2}\right) + \left(\frac{dX_2}{X_2}\right)^2.$$

12.4. Assume S is a geometric Brownian motion:
$$\frac{dS}{S} = \mu\, dt + \sigma\, dB$$
for constants μ and σ and a Brownian motion B.

(a) Show that
$$\text{var}_t\left(\frac{S_{t+1}}{S_t}\right) = e^{2\mu}\left(e^{\sigma^2} - 1\right).$$
Hint: Compare Exercise 1.7.

(b) Use the result of the previous part, the formula (12.22), and the approximation $e^x \approx 1 + x$ to derive approximate formulas for $\text{var}_t(S_{t+1}/S_t)$ and $E_t[S_{t+1}/S_t]$.

12.5. Assume
$$X_t = \theta - e^{-\kappa t}(\theta - X_0) + \sigma \int_0^t e^{-\kappa(t-s)} dB_s$$
for a Brownian motion B and constants θ and κ. Show that
$$dX = \kappa(\theta - X) dt + \sigma dB.$$

Note: The process X is called an Ornstein-Uhlenbeck process. Assuming $\kappa > 0$, θ is called the long-run or unconditional mean, and κ is the rate of mean reversion. This is the interest rate process in the Vasicek model (Section 18.3).

12.6. Let X be an Ornstein-Uhlenbeck process with a long-run mean of zero; that is,
$$dX = -\kappa X dt + \sigma dB$$
for constants κ and σ. Set $Y = X^2$. Show that
$$dY = \hat{\kappa}(\hat{\theta} - Y) dt + \hat{\sigma} \sqrt{Y} dB$$
for constants $\hat{\kappa}$, $\hat{\theta}$ and $\hat{\sigma}$. Note: The squared Ornstein-Uhlenbeck process Y is a special case of the interest rate process in the Cox-Ingersoll-Ross model (Section 18.3) and a special case of the variance process in the Heston model (Section 17.4)—special because $\hat{\kappa}\hat{\theta} = \hat{\sigma}^2/4$.

12.7. Suppose $dS/S = \mu dt + \sigma dB$ for constants μ and σ and a Brownian motion B. Let r be a constant. Consider a wealth process W as defined in Section 12.2:
$$\frac{dW}{W} = (1 - \pi) r dt + \pi \frac{dS}{S},$$
where π is a constant.

(a) By observing that W is a geometric Brownian motion, derive an explicit formula for W_t.

(b) For a constant ρ and dates $s < t$, calculate $E_s[W_t^{1-\rho}]$. Hint: write $W_t^{1-\rho} = e^{(1-\rho)\log W_t}$.

(c) Consider an investor who chooses a portfolio process to maximize
$$E\left[\frac{1}{1-\rho} W_T^{1-\rho}\right].$$
Show that if a constant portfolio $\pi_t = \pi$ is optimal, then the optimal portfolio is
$$\pi = \frac{\mu - r}{\rho \sigma^2}.$$

12.8. Let B be a Brownian motion. Define $Y_t = B_t^2 - t$.

(a) Use the fact that a Brownian motion has independent zero-mean increments with variance equal to the length of the time interval to show that Y is a martingale.

(b) Apply Itô's formula to calculate dY and verify condition (12.5) to show that Y is a martingale. Hint: To verify (12.5) use the fact that

$$E\left[\int_0^T B_t^2\, dt\right] = \int_0^T E[B_t^2]\, dt.$$

(c) Let $dM = \theta\, dB$ for a Brownian motion B. Use Itô's formula to show that

$$M_t^2 - \int_0^t (dM_s)^2$$

is a local martingale.

(d) Let $dM_i = \theta_i\, dB_i$ for $i = 1, 2$, and Brownian motions B_1 and B_2. Use Itô's formula to show that

$$M_{1t}M_{2t} - \int_0^t (dM_{1s})(dM_{2s})$$

is a local martingale.

12.9. Let B_1 and B_2 be independent Brownian motions and let $\rho \in [-1, 1]$. Set $\hat{B}_1 = B_1$. Define $\hat{B}_2$ by $\hat{B}_{20} = 0$ and $d\hat{B}_2 = \rho\, dB_1 + \sqrt{1 - \rho^2}\, dB_2$.

(a) Use Levy's theorem to show that $\hat{B}_2$ is a Brownian motion.

(b) Show that ρ is the correlation process of the two Brownian motions $\hat{B}_1$ and $\hat{B}_2$.

12.10. Let $\rho \neq \pm 1$ be the correlation process of two Brownian motions B_1 and B_2. Set $\hat{B}_1 = B_1$. Define $\hat{B}_2$ by $\hat{B}_{20} = 0$ and

$$d\hat{B}_2 = \frac{1}{\sqrt{1-\rho^2}}(dB_2 - \rho\, dB_1).$$

Show that $\hat{B}_1$ and $\hat{B}_2$ are independent Brownian motions. Note: Obviously this reverses the process of the previous exercise. It gives us

$$dB_2 = \rho\, d\hat{B}_1 + \sqrt{1-\rho^2}\, d\hat{B}_2,$$

so $\rho\, dB_1$ can be viewed as the orthogonal projection of dB_2 on $dB_1 = d\hat{B}_1$.

12.11. Let B_1 and B_2 be independent Brownian motions and

$$dZ \stackrel{\text{def}}{=} \begin{pmatrix} dZ_1 \\ dZ_2 \end{pmatrix} = \begin{pmatrix} \sigma_{11} & \sigma_{12} \\ \sigma_{21} & \sigma_{22} \end{pmatrix} \begin{pmatrix} dB_1 \\ dB_2 \end{pmatrix} \stackrel{\text{def}}{=} A\, dB$$

for stochastic processes σ_{ij}, where A is the matrix of the σ_{ij}.

(a) Calculate a, b, and c with $a > 0$ and $c > 0$ such that $LL' = AA'$, where

$$L = \begin{pmatrix} a & 0 \\ b & c \end{pmatrix}.$$

(b) Define $\hat{B} = (\hat{B}_1 \ \hat{B}_2)'$ by $\hat{B}_{i0} = 0$ and $d\hat{B} = L^{-1} A\, dB$. Show that $\hat{B}_1$ and $\hat{B}_2$ are independent Brownian motions.

(c) Show that $dZ = L\, d\hat{B}$.

Note: This illustrates the implementation of Gram-Schmidt orthogonalization via the Cholesky decomposition discussed in Sections 3.9 and 12.8. The process can be applied for more than two Brownian motions.

12.12. Suppose $dM_i = \theta\, dB_i$ for $i = 1, 2$, where B_i is a Brownian motion and θ_i satisfies (12.5), so M_i is a finite-variance martingale.

(a) Show that the conditional variance formula (12.28) is equivalent to

$$M_{it}^2 - \int_0^t (dM_{is})^2 \qquad (12.32)$$

being a martingale.

(b) Show that the conditional covariance formula (12.30) is equivalent to

$$M_{1t} M_{2t} - \int_0^t (dM_{1s})(dM_{2s}) \qquad (12.33)$$

being a martingale.

Note: A more general fact, which does not require the finite-variance assumption, and which can be used as the definition of $(dM_i)^2$ and $(dM_i)(dM_j)$, is that $\int_0^t (dM_{is})^2$ is the finite-variation process such that (12.32) is a local martingale, and $\int_0^t (dM_{1s})(dM_{2s})$ is the finite-variation process such that (12.33) is a local martingale.

12.13. Let $dM_i = \theta_i\, dB_i$ for $i = 1, 2$ and Brownian motions B_1 and B_2. Suppose θ_1 and θ_2 satisfy condition (12.5), so M_1 and M_2 are finite-variance martingales.

Consider discrete dates $s = t_0 < t_1 < \cdots < t_N = u$ for some $s < u$. Show that

$$\operatorname{cov}_s(M_{1u} - M_{1s}, M_{2u} - M_{2s}) = \mathsf{E}_s\left[\sum_{j=1}^{N}(M_{1t_j} - M_{1t_{j-1}})(M_{2t_j} - M_{2t_{j-1}})\right].$$

Hint: This is true of discrete-time finite-variance martingales, and the assumption that the M_i are stochastic integrals is neither necessary nor helpful in this exercise. However, it is interesting to compare this to (12.30).

13

Continuous-Time Markets

CONTENTS

13.1 Asset Price Dynamics — 318
13.2 Intertemporal Budget Constraint — 322
13.3 Stochastic Discount Factor Processes — 323
13.4 Valuation via SDF Processes — 330
13.5 Complete Markets — 333
13.6 Markovian Model — 335
13.7 Real and Nominal SDFs and Interest Rates — 336
13.8 Notes and References — 337

We study a securities market that operates continuously over an infinite horizon. Assume there is an instantaneously risk-free asset (money market account) and n risky assets the prices of which are driven by Brownian motions. As in Section 12.2, denote the instantaneous risk-free rate at date t by r_t, and define

$$R_t = \exp\left(\int_0^t r_s \, ds\right). \qquad (13.1)$$

The price R is the price of the money market account.

13.1 ASSET PRICE DYNAMICS

We describe the model for the returns of the assets and then at the end of the section state technical conditions ensuring the existence of price processes with those returns. When working in continuous time, we use "return" as a synonym for "rate of return."

Dividend-Reinvested Asset Prices

As in discrete time, it is convenient to work with dividend-reinvested asset prices. The capital gain of a dividend-reinvested price is the total return of the asset. The simplest case is an asset that pays dividends continuously, at some rate D per unit of time. This means that the total of dividends paid by a share during a time interval $[s, u]$ is

$$\int_s^u D_t \, dt.$$

Let P denote the price of the asset and set

$$X_t = \exp\left(\int_0^t \frac{D_s}{P_s} \, ds\right). \qquad (13.2)$$

Then, $X_0 = 1$ and

$$dX_t = \frac{X_t D_t}{P_t} \, dt. \qquad (13.3)$$

Intuitively, in an instant dt, the asset pays dividends of $D \, dt$ per share. If we own X shares, then we receive $XD \, dt$ in dividends, which will purchase $(XD/P) \, dt$ new shares. Thus, the change in the number of shares from reinvesting dividends is given by (13.3), and X_t defined in (13.2) is the number of shares we would own at date t by starting with one share at date 0 and reinvesting dividends. The dividend-reinvested asset price is

$$S_t = P_t X_t. \qquad (13.4)$$

From Itô's formula (Exercise 12.3),

$$\frac{dS}{S} = \frac{dX}{X} + \frac{dP}{P} = \frac{D \, dt + dP}{P},$$

which is the total return of the asset, as desired.

If an asset pays a discrete (i.e., noninfinitesimal) dividend with the amount announced prior to payment—as, of course, real assets do—then, in the absence of tax issues, the price should drop by the amount of the dividend at the time of the dividend payment (more precisely, at the ex-dividend date). Therefore, the price process obtained by reinvesting dividends can be assumed to be continuous in that case also. So, we will assume in general that dividend-reinvested asset prices are Itô processes. This excludes price processes that jump due to discontinuous arrival of information. Jump processes are addressed briefly in Section 15.4.

Brownian Motion Model of Asset Prices

For $i = 1,\ldots,n$, let S_{it} denote the dividend-reinvested price of risky asset i. Let dS/S denote the n-dimensional column vector with dS_i/S_i as its ith component. Assume

$$\frac{dS_t}{S_t} = \mu_t\, dt + \sigma_t\, dB_t, \qquad (13.5)$$

where μ_t is the vector of expected returns, B is a k-vector of independent Brownian motions, and σ_t is an $n \times k$ matrix. Assume the Brownian motions $B_1,\ldots,B_k$ are the only sources of uncertainty. Every stochastic process that we consider is assumed to be adapted to the information provided by the Brownian motions. Because k can be large (larger than the number of assets), this is restrictive only in ruling out discontinuous information (and jumps in asset prices).

The correlations and covariances of the asset returns depend of course on the matrix σ. The important matrix is the instantaneous covariance matrix

$$(\sigma\, dB)(\sigma\, dB)' = \sigma(dB)(dB)'\sigma' = \sigma\sigma'\, dt,$$

the second equality here following from the rules for multiplying differentials, which imply that $(dB)(dB)'$ is the identity matrix times dt. Define the $n \times n$ covariance matrix

$$\Sigma = \sigma\sigma'.$$

For simplicity, assume there are no redundant assets (σ has full row rank), so Σ is nonsingular. This requires that we have at least as many Brownian motions as risky assets ($k \geq n$).

The square roots of the diagonal elements of Σ are called the volatilities of the assets. Specifically, the volatility of asset i is defined to be $\sqrt{e_i'\Sigma e_i}$, because

$$\left(\frac{dS_i}{S_i}\right)^2 = e_i'\Sigma e_i\, dt,$$

where e_i denotes the ith basis vector of $\mathbb{R}^n$ (having 1 in the ith place and 0 elsewhere).

Equivalent Models

We could write the asset returns in terms of correlated Brownian motions instead of independent Brownian motions, as discussed in Section 12.8. For $i = 1,\ldots,n$, define $Z_{i0} = 0$ and

$$dZ_i = \frac{1}{\sqrt{e_i'\Sigma e_i}} e_i'\sigma\, dB.$$

By Levy's theorem, Z_i is a Brownian motion. We have

$$\frac{dS_i}{S_i} = \mu_i\, dt + \sqrt{e_i'\Sigma e_i}\, dZ_i.$$

In this formulation, there is one Brownian motion for each risky asset. The Brownian motions are correlated:

$$(dZ_i)(dZ_j) = \frac{e_i'\Sigma e_j}{\sqrt{e_i'\Sigma e_i \times e_j'\Sigma e_j}}\, dt.$$

We call this correlation between the Brownian motions the correlation of the two asset returns also. An issue with this formulation is that the Brownian motions $Z_1, \ldots, Z_n$ may not carry the same information as $B_1, \ldots, B_k$. If $k > n$ (in which case the market is incomplete—see Section 13.5), we will need to specify that there are other Brownian motions besides $Z_1, \ldots, Z_n$ the histories of which may influence conditional expected returns, volatilities, and correlations. We can also reverse the above process: Given a model written in terms of correlated Brownian motions B_i, we can rewrite it in terms of independent Brownian motions Z_i (Section 12.9 and Exercise 12.11). Thus, the choice of using independent or correlated Brownian motions is just a question of convenience. We will usually work with the model (13.5) in which the Brownian motions are independent.

Mean-Variance Frontier

Given μ, r, and Σ, the mean-variance frontier of instantaneous returns in continuous time is defined as if the model were a single-period model. The tangency portfolio is defined as

$$\pi_{\text{tang}} = \frac{1}{\iota'\Sigma^{-1}(\mu - r\iota)}\Sigma^{-1}(\mu - r\iota), \qquad (13.6)$$

as in (5.17), provided $\iota'\Sigma^{-1}(\mu - r\iota) \neq 0$. The maximum Sharpe ratio is

$$\kappa = \sqrt{(\mu - r\iota)'\Sigma^{-1}(\mu - r\iota)}, \qquad (13.7)$$

as in (5.16). Of course, r, μ, and Σ—and hence the tangency portfolio π_{tang} and maximum Sharpe ratio κ—are in general stochastic processes (depend on t and the state of the world).

Some regularity conditions are needed to ensure the model is well defined. Assume that, with probability 1,

$$(\forall T < \infty) \qquad \int_0^T |r_t|\, dt < \infty, \qquad (13.8a)$$

$$(\forall T < \infty) \qquad \int_0^T |\mu_{it}|\, dt < \infty, \qquad (13.8b)$$

$$(\forall T < \infty) \qquad \int_0^T e_i' \Sigma_t e_i\, dt < \infty. \qquad (13.8c)$$

Then, the price R of the money market account is defined by (13.1). The risky asset prices are

$$S_{it} = S_{i0} \exp\left(\int_0^t \left(\mu_{is} - \frac{1}{2} e_i' \Sigma_s e_i \right) ds + \int_0^t \sqrt{e_i' \Sigma_s e_i}\, dZ_{is} \right) \qquad (13.9)$$

and are well defined when (13.8b) and (13.8c) hold.

13.2 INTERTEMPORAL BUDGET CONSTRAINT

Let π_{it} denote the fraction of wealth an investor holds in the ith risky security at date t and let π_t denote the n-dimensional column vector with ith component equal to π_{it}. Extending (12.10), in the absence of nonportfolio income (which we continue to call labor income) and in the absence of consumption, the intertemporal budget constraint is

$$\frac{dW}{W} = (1 - \iota'\pi) r\, dt + \pi' \left(\frac{dS}{S} \right)$$
$$= r\, dt + \pi'(\mu - r\iota)\, dt + \pi'\sigma\, dB. \qquad (13.10)$$

This means that the portfolio rate of return is the weighted average of the rates of return of the assets, the weight being π_{it} on risky asset i and therefore necessarily $1 - \iota'\pi_t$ on the risk-free asset. The constraint (13.10) is based on no cash being invested or withdrawn from the portfolio at date t. As in discrete time, we call wealth processes W satisfying (13.10) self-financing or non-dividend-paying. If there is a flow C of consumption and labor income Y, the intertemporal budget constraint is

$$\frac{dW}{W} = r\, dt + \pi'(\mu - r\iota)\, dt - \frac{C - Y}{W}\, dt + \pi'\sigma\, dB. \qquad (13.11)$$

We can also write the intertemporal budget constraint in terms of the amount of the consumption good invested in each asset, instead of the fraction of wealth invested. Let ϕ_i denote the amount of the consumption good invested in asset i and $\phi = (\phi_1 \cdots \phi_n)'$. A wealth process is self-financing if it satisfies

$$dW = Wr\,dt + \phi'(\mu - r\iota)\,dt + \phi'\sigma\,dB. \tag{13.12}$$

Actually, (13.12) is a more general statement of the self-financing condition than is (13.10). It is equivalent to (13.10) when $W > 0$, but, unlike (13.10), it is also meaningful when $W \leq 0$.

Some additional regularity conditions are needed to ensure that wealth processes are well defined. The self-financing wealth process W defined by (13.12) and a portfolio process ϕ is

$$W_T = e^{\int_0^T r_s\,ds}W_0 + \int_0^T e^{\int_t^T r_s\,ds}[(\phi'(\mu_t - r_t\iota)\,dt + \phi_t'\sigma_t\,dB_t]. \tag{13.13}$$

Given (13.8a), W is well defined on $[0, \infty)$ if the following hold:

$$(\forall T < \infty) \qquad \int_0^T e^{\int_t^T r_s\,ds} \left|\phi_t'(\mu_t - r_t\iota)\right| dt < \infty, \tag{13.14a}$$

$$(\forall T < \infty) \qquad \int_0^T e^{\int_t^T 2r_s\,ds} \phi_t'\Sigma_t\phi_t\,dt < \infty. \tag{13.14b}$$

If $W_0 > 0$, then W_T is also well defined by (13.10) and given by

$$W_T = W_0 \exp\left(\int_0^T \left(r_t + \pi_t'(\mu_t - r_t\iota) - \frac{1}{2}\pi_t'\Sigma_t\pi_t\right) dt + \int_0^T \pi_t'\sigma_t\,dB_t\right), \tag{13.15}$$

provided the portfolio process π satisfies the following conditions:

$$(\forall T < \infty) \qquad \int_0^T \left|\pi_t'(\mu_t - r_t\iota)\right| dt < \infty, \tag{13.16a}$$

$$(\forall T < \infty) \qquad \int_0^T \pi_t'\Sigma_t\pi_t\,dt < \infty. \tag{13.16b}$$

In this circumstance, $W_T > 0$ for all T with probability 1.

13.3 STOCHASTIC DISCOUNT FACTOR PROCESSES

It is convenient to employ a weaker definition of an SDF process M than the definition "MW is a martingale for each self-financing wealth process W" used in discrete time. Define M to be an SDF process if (i) $M_0 = 1$, (ii) $M_t > 0$ for all t with probability 1, and (iii) MR and MS_i are local martingales, for $i = 1,\ldots,n$, where the S_i are the dividend-reinvested asset prices. Section 13.4 examines the issue of when MS_i is actually a martingale and, more generally, when MW is a martingale for a self-financing wealth process W. The analysis of SDF processes in continuous time is very similar to that in a single-period model, as we see throughout this section.

Dynamics of SDF Processes

The local martingale properties imply specific dynamics for SDF processes, expressed in (13.17) and (13.20) below. We use the property relative to the money market account to derive the drift of an SDF process shown in (13.17). Then, we use the property relative to risky assets to derive the stochastic part of an SDF process, specifically, the property (13.20). These steps give us results that parallel properties of SDFs in single-period models.

Suppose M is an SDF process, and set $Y = MR$, where R is the price of the money market account. Itô's formula implies

$$\frac{dY}{Y} = \frac{dM}{M} + r\,dt.$$

Therefore, for Y to be a local martingale, the drift of dM/M must cancel the $r\,dt$ term; that is, the drift of dM/M must be $-r\,dt$. This is analogous to the fact that $E[\tilde{m}] = 1/R_f$ in a single-period model.

The martingale representation theorem tells us that the stochastic part of the SDF process is spanned by the Brownian motions. Therefore,

$$\frac{dM}{M} = -r\,dt - \lambda'\,dB \qquad (13.17)$$

for some stochastic process λ. The vector λ is called the vector of market prices of risk. As we will see, this is consistent with calling factor risk premia "prices of risk," where here the Brownian motions B_i are regarded as the factors. The minus sign we have placed in front of λ is arbitrary, but it implies that the elements of λ have convenient signs, as we will see in (13.31) below.

To characterize λ in (13.17), we use the local martingale property for the risky asset prices. Set $Y_i = MS_i$ for $i = 1,\ldots,n$. Apply Itô's formula and (13.17) to obtain

Continuous-Time Markets

$$\frac{dY_i}{Y_i} = \frac{dM}{M} + \frac{dS_i}{S_i} + \left(\frac{dM}{M}\right)\left(\frac{dS_i}{S_i}\right)$$

$$= -r\,dt - \lambda'\,dB + \frac{dS_i}{S_i} + \left(\frac{dM}{M}\right)\left(\frac{dS_i}{S_i}\right). \qquad (13.18)$$

By the definition of an SDF process, Y_i is a local martingale. This implies that the drift on the right-hand side of (13.18) is zero. We are writing the drift of dS_i/S_i as μ_i. Therefore,

$$(\mu_i - r)\,dt = -\left(\frac{dM}{M}\right)\left(\frac{dS_i}{S_i}\right). \qquad (13.19)$$

Equation (13.19) says that risk premia depend on covariances with dM/M. It is the continuous-time equivalent of the single-period formula $E[\tilde{R}] - R_f = -R_f\,\mathrm{cov}(\tilde{R}, \tilde{m})$.

Stack the equations (13.19) for $i = 1, \ldots, n$ to obtain

$$(\mu - r\iota)\,dt = -\left(\frac{dS}{S}\right)\left(\frac{dM}{M}\right)$$

$$= \sigma\,(dB)(dB')\lambda$$

$$= \sigma\lambda\,dt.$$

We used the formulas (13.5) and (13.17) to obtain the second equality and the fact that $(dB)(dB') = I\,dt$ to obtain the last equality. The conclusion is that the vector λ in the dynamics (13.17) of an SDF process must satisfy

$$\sigma\lambda = \mu - r\iota. \qquad (13.20)$$

In fact, M is an SDF process if and only if $M_0 = 1$ and (13.17) and (13.20) hold, because MR and the MS_i are local martingales if and only if (13.17) and (13.20) hold. Equation (13.20) is analyzed further below and is used extensively throughout this and following chapters.

Orthogonal Projections of SDF Processes

In a single-period model, the projection of any SDF $\tilde{m}$ onto the span of the assets equals the unique SDF $\tilde{m}_p$ that lies in the span of the assets. An analogous result is true in continuous time. A particular solution of (13.20) is

$$\lambda_p = \sigma'\Sigma^{-1}(\mu - r\iota). \qquad (13.21)$$

The general solution of (13.20) is

$$\lambda = \lambda_p + \zeta, \qquad \text{where} \qquad \sigma\zeta = 0. \qquad (13.22)$$

To see that λ_p is a solution of (13.20), premultiply by σ and recall that $\sigma\sigma' = \Sigma$. To see that (13.22) is the general solution, for any solution λ, define $\zeta = \lambda - \lambda_p$ and note that (13.20) implies $\sigma\zeta = 0$.

Define a stochastic process M_p by $M_{p0} = 1$ and[1]

$$\frac{\mathrm{d}M_p}{M_p} = -r\,\mathrm{d}t - \lambda_p'\,\mathrm{d}B. \tag{13.23}$$

The process M_p plays the role of the SDF $\tilde{m}_p$ that lies in the span of the assets in a single-period model. The process M_p is spanned by the assets in the sense of being perfectly correlated with a portfolio return, that is, $\lambda_p'\,\mathrm{d}B$ is the stochastic part of a portfolio return. Specifically,

$$\lambda_p'\,\mathrm{d}B = (\mu - rt)'\Sigma^{-1}\sigma\,\mathrm{d}B = \pi_p'\sigma\,\mathrm{d}B,$$

where

$$\pi_p = \Sigma^{-1}(\mu - rt). \tag{13.24}$$

Note that π_p is proportional to the tangency portfolio (13.6). In fact, it is the optimal portfolio for a log-utility investor and hence maximizes the expected continuously compounded rate of return (Exercises 14.3 and 14.4). For this reason, it is called the growth-optimal portfolio. Due to the minus sign in front of $\lambda_p'\,\mathrm{d}B$ in (13.23), the SDF process M_p is perfectly *negatively* correlated with the return of the growth-optimal portfolio. This is analogous to the fact that $\tilde{R}_p$ is on the inefficient part of the mean-variance frontier.

Now, let M be any SDF process. It has dynamics (13.17) where λ satisfies (13.20). Define $\zeta = \lambda - \lambda_p$. As stated in (13.22), $\sigma\zeta = 0$. Define a stochastic process ε by $\varepsilon_0 = 1$ and

$$\frac{\mathrm{d}\varepsilon}{\varepsilon} = -\zeta'\,\mathrm{d}B.$$

---

[1]. We have already assumed $\int_0^T |r_t|\,\mathrm{d}t < \infty$ for all T with probability 1. The stochastic process

$$M_{pt} = \exp\left(-\int_0^t r_s\,\mathrm{d}s - \frac{1}{2}\int_0^t \kappa_s^2\,\mathrm{d}t - \int_0^t \lambda_{ps}'\,\mathrm{d}B_s\right)$$

is well defined for all $t < \infty$ with probability 1 and is the unique solution of $M_{p0} = 1$ and (13.23) if $\int_0^T \lambda_{pt}^2\,\mathrm{d}t < \infty$ for all T with probability 1. Note that

$$\lambda_p^2 = (\mu - rt)'\Sigma^{-1}\sigma\sigma'\Sigma^{-1}(\mu - rt) = (\mu - rt)'\Sigma^{-1}(\mu - rt),$$

which is the square of the maximum Sharpe ratio. Thus, M_p is well defined if the integral of the squared maximum Sharpe ratio is finite for all T with probability 1. Adopt that assumption henceforth.

This enables us to write (13.17) as

$$\frac{dM}{M} = \frac{dM_p}{M_p} + \frac{d\varepsilon}{\varepsilon}. \tag{13.25}$$

The fact that $\sigma\zeta = 0$ implies $\lambda'_p \zeta = 0$. Hence,

$$\left(\frac{dM_p}{M_p}\right)\left(\frac{d\varepsilon}{\varepsilon}\right) = -\lambda'_p (dB)(dB)'\zeta = 0. \tag{13.26}$$

Furthermore, ε is orthogonal to every wealth process in the sense that, for any π,

$$(\pi'\sigma\, dB)\left(\frac{d\varepsilon}{\varepsilon}\right) = \pi'\sigma\zeta\, dt = 0. \tag{13.27}$$

Equation (13.25) is the orthogonal decomposition of SDF processes in continuous time. In fact, $M = M_p \varepsilon$, due to (13.25), (13.26), Itô's formula, and the initial conditions $M_0 = M_{p0} = \varepsilon_0 = 1$.

In a single-period complete market, $\tilde{m}_p$ is the unique SDF process. In our current model, if there are as many risky assets as Brownian motions ($k = n$), then M_p is the unique SDF process. This follows from our assumption that there are no redundant assets, which implies that σ is nonsingular when $k = n$. In this case, the only ζ satisfying $\sigma\zeta = 0$ is the zero vector. In fact, when σ is nonsingular, the unique solution to (13.20) is[2]

$$\lambda = \sigma^{-1}(\mu - r\iota). \tag{13.28}$$

For example, if there is a single risky asset and a single Brownian motion, then $\lambda = \lambda_p = (\mu - r)/\sigma$, which of course is the Sharpe ratio of the risky asset.

Absence of Arbitrage Opportunities

In a single-period model, the existence of a strictly positive SDF implies that there are no arbitrage opportunities. This is also true in continuous time. To see this, let M be an SDF process (which is strictly positive by definition) and let W be a self-financing wealth process. Then, MW is a local martingale. To see this, apply Itô's formula to $Y = MW$ and use (13.12) and (13.20) to obtain

$$dY = M(\phi'\sigma - W\lambda')\, dB.$$

---

2. This solution is λ_p defined in (13.21), because, when σ is nonsingular,

$$\lambda_p = \sigma'\Sigma^{-1}(\mu - r\iota) = \sigma'(\sigma\sigma')^{-1}(\mu - r\iota) = \sigma'(\sigma')^{-1}\sigma^{-1}(\mu - r\iota) = \sigma^{-1}(\mu - r\iota).$$

This verifies that Y has no drift and hence is a local martingale. Since $M > 0$, MW is actually a nonnegative local martingale when $W \geq 0$ and hence a supermartingale (Appendix A.13). This means that

$$M_t W_t \geq \mathsf{E}_t[M_T W_T] \qquad (13.29)$$

for each pair of dates $t < T$. We can rearrange (13.29) as

$$W_t \geq \mathsf{E}_t\left[\frac{M_T}{M_t} W_T\right]. \qquad (13.30)$$

Thus, the cost W_t of obtaining W_T is at least as great as the fundamental value of W_T, defined as the right-hand side of (13.30).[3] This precludes arbitrage opportunities—note that (13.30) implies for nonnegative W that $W_T = 0$ if $W_t = 0$; hence, it is impossible to make something from nothing.

Under the hypothesis of footnote 1 about the squared maximum Sharpe ratio, the stochastic process M_p is well defined and is an SDF process. Therefore, the hypothesis about the squared maximum Sharpe ratio implies that there are no arbitrage opportunities.

Hansen-Jagannathan Bound

In a single-period model with a risk-free asset, the SDF $\tilde{m}_p$ has the minimum variance, and its standard deviation equals the maximum Sharpe ratio divided by the risk-free return. This is the Hansen-Jagannathan bound. An analogous fact is true in continuous time. Conditions (13.25) and (13.26) give us

$$\left(\frac{\mathrm{d}M}{M}\right)^2 = \left(\frac{\mathrm{d}M_p}{M_p}\right)^2 + \left(\frac{\mathrm{d}\varepsilon}{\varepsilon}\right)^2,$$

which is interpreted as the conditional variance of $\mathrm{d}M/M$ being at least as large as the conditional variance of $\mathrm{d}M_p/M_p$. Moreover, the conditional variance of $\mathrm{d}M_p/M_p$ is

$$\left(\frac{\mathrm{d}M_p}{M_p}\right)^2 = \lambda_p' (\mathrm{d}B)(\mathrm{d}B)' \lambda_p = \lambda_p' \lambda_p \, \mathrm{d}t,$$

which is the square of the maximum Sharpe ratio (footnote 1). Thus, M_p has the smallest volatility of any SDF process, and its volatility is the maximum Sharpe ratio $\sqrt{\lambda_p' \lambda_p}$.

---

3. Strict inequality in (13.30) can be interpreted as a bubble. This is discussed further in the end-of-chapter notes.

Factor Pricing

As remarked before, the formula (13.19) for risk premia is an extension of the single-period formula $\mathsf{E}[\tilde{R}] - R_f = -R_f \operatorname{cov}(\tilde{R}, \tilde{m})$. The interpretation of (13.19) is that the risk premium equals minus the conditional covariance of the return with dM/M. It implies the same thing for portfolio returns:

$$\pi'(\mu - r\iota)\,dt = -\left(\pi'\frac{dS}{S}\right)\left(\frac{dM}{M}\right).$$

To see the interpretation of λ as the vector of market prices of risk, note that, given the form (13.17) for an SDF process, we can write (13.19) as

$$(\mu_i - r)\,dt = \lambda'\left(\frac{dS_i}{S_i}\right)(dB). \tag{13.31}$$

Thus, the risk premium is a linear combination of the conditional covariances of the rate of return with the changes dB_j of the Brownian motions, with λ_j being the risk premium for the factor dB_j.

In a single-period model, factor models are equivalent to SDFs in the sense that, disregarding trivial exceptions, a set of random variables serve as pricing factors if and only if an affine function of the variables is an SDF. An analogous result is true in continuous time. We say there is a factor model with Itô processes

$$dX_j = \alpha_j\,dt + \phi_j'\,dB, \tag{13.32}$$

for $j = 1, \ldots, \ell$, as the factors if, for some stochastic processes η_j,

$$(\mu_i - r)\,dt = \sum_{j=1}^{\ell} \eta_j \left(\frac{dS_i}{S_i}\right)(dX_j), \tag{13.33}$$

for each i. Of course, we interpret (13.33) as stating that risk premia are a linear combination of conditional covariances. Here, the η_j are the prices of risk.

The factor model (13.33) with X as the vector of factors holds if

$$\frac{dM}{M} = \text{something }dt - \sum_{j=1}^{\ell} \eta_j\,dX_j. \tag{13.34}$$

This follows from substituting into the SDF factor model (13.19) as in a single-period model. A very important fact is that (13.34) holds if M is any (sufficiently smooth) function of X: By Itô's formula, if $M_t = f(t, X_t)$, then (13.34) holds with $\eta_j = -\partial \log f / \partial x_j$.[4]

Conversely, (13.33) implies, subject to regularity conditions, that there is an SDF process M satisfying (13.34). Using (13.32), we can write (13.33) in vector form as

$$\mu - rt = \sigma \sum_{j=1}^{\ell} \eta_j \phi_j.$$

Setting

$$\zeta = \sum_{j=1}^{\ell} \eta_j \phi_j - \lambda_p,$$

we have $\sigma \zeta = 0$. Therefore, if $\lambda \stackrel{\text{def}}{=} \sum_{j=1}^{\ell} \eta_j \phi_j$ satisfies $\int_0^T \lambda_t^2 \, dt < \infty$ for all T with probability 1, then there is an SDF process M with

$$\frac{dM}{M} = -r \, dt - \left(\sum_{j=1}^{\ell} \eta_j \phi_j \right)' dB$$

$$= \left(-r + \sum_{j=1}^{\ell} \eta_j \alpha_j \right) dt - \sum_{j=1}^{\ell} \eta_j \, dX_j. \qquad (13.35)$$

13.4 VALUATION VIA SDF PROCESSES

The local martingale property is convenient for defining and characterizing SDF processes, but in order to perform valuation we really need local martingales to be martingales. If they are, then, as explained below, the value at date t of a payoff v at date $T > t$ is, for any SDF process M,

$$\mathrm{E}_t \left[\frac{M_T}{M_t} v \right]. \qquad (13.36)$$

Furthermore, the value at date t of a continuous stream of consumption C until $T > t$ is

$$\mathrm{E}_t \left[\int_t^T \frac{M_u}{M_t} C_u \, du \right]. \qquad (13.37)$$

---

4. Of course, (13.34) can also hold with M_t depending on the entire history of $X_1, \ldots, X_\ell$ through date t or on even more. The typical way to deduce (13.34) when there is such dependence is via the martingale representation theorem.

These formulas show that we can use M_u/M_t as an SDF at t for valuing payoffs at any future date u. Here, we are not claiming that v or C can be created by trading the securities (we are not claiming the market is complete). But, if they can be created, and the local martingales are martingales, then their costs are as shown in (13.36) and (13.36). Section 13.5 discusses market completeness.

Suppose first that v can be created by trading the securities, meaning that there is a self-financing wealth process W such that $W_T = v$. Then, MW is a local martingale. The martingale assumption we need is that MW is a martingale rather than just a local martingale. In that case,

$$M_t W_t = \mathsf{E}_t[M_T W_T] = \mathsf{E}_t[M_T v].$$

This implies that W_t equals (13.36). Thus, if MW is a martingale for *every* self-financing wealth process W satisfying $W_T = v$, then (13.36) is the unique cost at which v can be obtained.

Now assume that C can be created by trading the securities, meaning that there exist ϕ and W such that $W_T = 0$ and the intertemporal budget constraint

$$dW = rW\,dt + \phi'(\mu - r\iota)\,dt - C\,dt + \phi'\sigma\,dB \tag{13.38}$$

holds. Let M be an SDF process. Then, the stochastic process

$$\int_0^t M_s C_s\,ds + M_t W_t \tag{13.39}$$

is a local martingale (from Itô's formula, its differential is $M(\phi'\sigma - W\lambda')\,dB$). The martingale assumption we need is that (13.39) is a martingale. In that case, for any dates $t < T$, recalling that $W_T = 0$, we have

$$\int_0^t M_s C_s\,ds + M_t W_t = \mathsf{E}_t\left[\int_0^T M_s C_s\,ds\right]. \tag{13.40}$$

This implies that W_t equals (13.37). If (13.39) is a martingale for every wealth process W for which (i) $W_T = 0$ and (ii) the intertemporal budget constraint holds for some ϕ, then (13.37) is the unique cost at which C can be acquired. Often we want to take $T = \infty$ in (13.37). To do that, we need the local martingales to be martingales, and we also need to rule out bubbles as discussed in Section 8.5. See the end-of-chapter notes for further discussion.

Another useful consequence of MW being a martingale for a self-financing wealth process W is that we can calculate M_t in terms of M_T as

$$M_t = \mathsf{E}_t\left[\frac{W_T}{W_t}M_T\right].$$

For example,

$$M_t = \mathsf{E}_t\left[e^{\int_t^T r_u\,du}M_T\right].$$

Sufficient Conditions for Martingales

For valuation, we adopt additional assumptions of the form "assume MW is a martingale for all admissible portfolio processes." This is a restriction on the asset returns. Obviously, it also involves the definition of "admissible." Here, we discuss the types of assumptions that will work. Let M be an SDF process, and let W be a self-financing wealth process, so MW is a local martingale. A sufficient condition for MW to be a martingale on a time interval $[0, T]$ for any $T < \infty$ is Novikov's condition. In this setting (Exercise 13.4), Novikov's condition is

$$\mathsf{E}\left[\exp\left(\frac{1}{2}\int_0^T \lambda_p'\lambda_p + \zeta'\zeta + \pi'\Sigma\pi - 2\pi'(\mu - r\iota)\,\mathrm{d}t\right)\right] < \infty. \qquad (13.41)$$

This is a restriction on the asset returns, the orthogonal component ζ of the SDF process, and the portfolio process π that generates W. Another sufficient condition is that W be nonnegative and

$$W_0 = \mathsf{E}[M_T W_T]. \qquad (13.42)$$

Nonnegativity of W implies that the local martingale MW is a supermartingale, and (13.42) is a sufficient condition for the supermartingale MW to be a martingale (recalling that $M_0 = 1$). Yet another sufficient condition is presented in Section 15.3.

To see what conditions imply that (13.39) is a martingale, start by assuming W is strictly positive. If funds were retained in the portfolio instead of being consumed, then the wealth process would be the dividend-reinvested (or, rather, consumption-reinvested) price

$$W_t^\dagger = W_t \exp\left(\int_0^t \frac{C_s}{W_s}\,\mathrm{d}s\right). \qquad (13.43)$$

This wealth process is self-financing with the same portfolio process as W. In other words,

$$\frac{\mathrm{d}W^\dagger}{W^\dagger} = r\,\mathrm{d}t + \pi'(\mu - r\iota)\,\mathrm{d}t + \pi'\sigma\,\mathrm{d}B \qquad (13.44)$$

for the same π (this follows from applying Itô's formula to (13.43)). A useful fact is that (13.39) is a martingale if $MW^\dagger$ is a martingale (Exercise 13.5). Thus, restrictions on portfolio processes sufficient to ensure that $MW^\dagger$ is a martingale for self-financing wealth processes $W^\dagger$ are also sufficient to ensure that (13.39) is a martingale. A related result (based on investing dividends in the money market account) is presented in Exercise 13.6.

13.5 COMPLETE MARKETS

When investors are assumed to trade continuously, markets can be complete even with a finite number of assets and an infinite number of states of the world. This is an important distinction between discrete-time and continuous-time markets.

To obtain completeness, we must have as many risky assets as Brownian motions, so assume $k = n$. Because we are assuming there are no redundant assets, σ is invertible. Assume the squared maximum Sharpe ratio has a finite integral with probability 1, so, as discussed in footnote 1, the stochastic process M_p exists and is a strictly positive SDF process. Because σ is invertible, M_p is the unique SDF process. Recall that we are assuming that all of the uncertainty in the economy is generated by the vector Brownian motion B. Therefore, the martingale representation theorem (Section 12.3) implies that all local martingales are spanned by B.

In this circumstance, markets are complete in the following sense. Consider $T < \infty$, and let x be any random variable observable at time T. Let C be an adapted stochastic process. Assume

$$\mathsf{E}\left[\int_0^T |M_t C_t|\,dt + |M_T x|\right] < \infty. \tag{13.45a}$$

Then, there exists a portfolio process ϕ and wealth process W such that ϕ, W, and C satisfy the intertemporal budget constraint (13.38) and such that $W_T = x$. Likewise, in the infinite-horizon case, if

$$\mathsf{E}\left[\int_0^\infty |M_t C_t|\,dt\right] < \infty, \tag{13.45b}$$

then there exists a portfolio process ϕ and wealth process W such that π, W, and C satisfy the intertemporal budget constraint (13.38). Moreover, by construction,

$$\int_0^t M_s C_s\,ds + M_t W_t \tag{13.46}$$

is a martingale in both the finite- and infinite-horizon cases, and if C and x are nonnegative, then W is nonnegative. An alternate formulation of completeness, employing the risk-neutral probability, is presented in Exercise 15.7.

We provide the proof of completeness in two steps, each of which is sometimes useful in other circumstances. Consider the infinite-horizon case (the finite-horizon case is analogous). The first step is to define W and to show that (13.46) is a martingale. Given a consumption process C satisfying the regularity condition (13.45b), define

$$W_t = \frac{1}{M_t}\mathsf{E}_t\int_t^\infty M_u C_u\,du. \tag{13.47}$$

This definition is the infinite-horizon version of (13.37). Multiply by M_t and add $\int_0^t M_s C_s \, ds$ to both sides to obtain

$$\int_0^t M_s C_s \, ds + M_t W_t = \mathsf{E}_t \int_0^\infty M_u C_u \, du.$$

The right-hand side of this is the conditional expectation of a random variable that is independent of t; hence, the stochastic process is a martingale, so the left-hand side is a martingale. This completes the first step.

Now, consider any C and W such that (13.46) is a martingale. Actually, it is enough that (13.46) be a local martingale. Using only the local martingale property of (13.46), we will construct ϕ such that (C, W, ϕ) satisfies the intertemporal budget constraint. This is the second step and it finishes the proof of completeness.

The intertemporal budget constraint specifies that dW has a certain drift and a certain stochastic part. The stochastic part is $\phi' \sigma \, dB$. We will define ϕ so that the stochastic part of dW is $\phi' \sigma \, dB$. Having done so, W will always have the right drift, due to (13.46) being a local martingale. We verify that at the end of the proof.

To match the stochastic part of dW to $\phi' \sigma \, dB$, we first need to compute the stochastic part of dW. Let Z denote the local martingale (13.46), so

$$dZ = MC \, dt + M \, dW + W \, dM + (dW)(dM).$$

We can rearrange this as

$$dW = -C \, dt - W \frac{dM}{M} - (dW) \left(\frac{dM}{M} \right) + \frac{1}{M} dZ. \qquad (13.48)$$

For equation (13.48) to hold, the drifts on both sides must match, and the stochastic parts on both sides much match. Because M is an SDF process,

$$\frac{dM}{M} = -r \, dt - \lambda' \, dB, \qquad (13.49)$$

where $\sigma \lambda = \mu - r \iota$. Thus, the stochastic part of dW must be

$$W \lambda' \, dB + \frac{1}{M} dZ.$$

By the martingale representation theorem, there exists a stochastic process γ such that

$$\gamma' \, dB = W \lambda' \, dB + \frac{1}{M} dZ.$$

Set $\phi = (\sigma')^{-1} \gamma$. Then, $\gamma' = \phi' \sigma$, so the stochastic part of dW is $\phi' \sigma \, dB$ as desired.

It remains to show that the drift of W is $-C\,dt + Wr\,dt + \phi'(\mu - r\iota)\,dt$, which will confirm the intertemporal budget constraint. This follows from the fact that (13.46) is a local martingale—so (13.48) holds with Z being a local martingale—and from the fact that we have matched the stochastic parts. Equations (13.48) and (13.49) imply that the drift of W is

$$-C\,dt + Wr\,dt + (dW)(\lambda'dB).$$

Given that the stochastic part of dW is $\phi'\sigma\,dB$, we have

$$\begin{aligned}(dW)(\lambda'dB) &= (\phi'\sigma\,dB)(\lambda'dB)\\ &= (\phi'\sigma\,dB)(dB)'\lambda\\ &= \phi'\sigma\lambda\,dt\\ &= \phi'(\mu - r\iota)\,dt,\end{aligned}$$

using $(dB)(dB)' = I\,dt$ for the third equality and $\sigma\lambda = \mu - r\iota$ for the fourth. This completes the verification of the intertemporal budget constraint.

13.6 MARKOVIAN MODEL

It is often convenient to assume that any time variation in the investment opportunity set is due solely to variation in a finite set of state variables, which are themselves Markovian. This concept is introduced in discrete time in Section 9.5. This section explains the continuous-time version. We use this model in Chapter 14 to compute optimal consumption and portfolios by dynamic programming and to derive the ICAPM, we use it in Chapter 15 to calculate the values (13.36) and (13.37) as solutions of PDEs and to describe optimal portfolios in terms of the solution of a linear PDE, and we use it throughout Part III of the book.

Denote the state variables by $X_1, X_2, \ldots, X_\ell$. Let X denote the column vector $(X_1 \cdots X_\ell)'$. Assume the instantaneous risk-free rate r, the vector μ of expected returns and the matrix σ of volatilities depend on X in the sense that the instantaneous risk-free rate at date t is $r(X_t)$, the vector of expected returns is $\mu(X_t)$, and the matrix of volatilities is $\sigma(X_t)$ for some functions $r(\cdot)$, $\mu(\cdot)$, and $\sigma(\cdot)$. Let $\Sigma(X)$ denote the covariance matrix $\sigma(X)\sigma(X)'$. Under these assumptions, the dynamics of the SDF process M_p that is spanned by the assets also depend on X: The vector of prices of risk is

$$\lambda_p(X) = \sigma(X)'\Sigma(X)^{-1}[\mu(X) - r(X)\iota].$$

Assume there are functions $x \mapsto \phi(x) \in \mathbb{R}^\ell$ and $x \mapsto \nu(x) \in \mathbb{R}^\ell \times \mathbb{R}^k$ such that

$$dX_t = \phi(X_t)\,dt + \nu(X_t)\,dB_t, \qquad (13.50)$$

where B is the same k-vector of independent Brownian motions that determines the security returns.[5] Equation (13.50) implies that the vector X is a Markov process, meaning that, for any $u > t$, the distribution of X_u conditional on information at date t depends only on X_t. Any other information, in particular the history of X prior to t, is irrelevant, given knowledge of X_t. Because the investment opportunity set at date u is assumed to depend only on X_u, the vector X_t is also a sufficient statistic, relative to date–t information, for predicting the investment opportunity set at all dates $u > t$. Let $\nu_j(X)$ denote the jth row of $\nu(X)$ rearranged as a column vector, so we can write

$$dX_{jt} = \phi_j(X_t) + \nu_j(X_t)'\,dB_t \qquad (13.51)$$

for $j = 1, \ldots, \ell$.

13.7 REAL AND NOMINAL SDFS AND INTEREST RATES

Translating SDF processes between real and nominal prices and between two different currencies is the same in continuous time as in discrete time (Section 8.6). One new issue that arises in continuous-time models is how to model the price index. Let Z denote the price in dollars (or other currency) of the consumption good at date t. If M is an SDF process for assets denominated in real prices (units of the consumption good), then $M^* \stackrel{\text{def}}{=} Z_0 M/Z$ is an SDF process for assets denominated in nominal prices. The process Z is sometimes modeled as an absolutely continuous process and sometimes as an Itô process. In models of the former type, $dZ/Z = \pi\,dt$, where π is the inflation rate (which may be a stochastic process). In this case, we have

$$\frac{dM^*}{M^*} = -r\,dt - \lambda'\,dB - \pi\,dt,$$

showing that $r + \pi$ is the nominal interest rate. If Z is taken to be an Itô process, then there may be no instantaneously risk-free asset (money market account) in real terms, even if one exists in nominal terms. Note that taking Z to be an Itô process implies that deflation is roughly as likely as inflation over short time

---

5. It is without loss of generality to use the same vector of Brownian motions because we can take k as large as necessary and include as many zeros in σ and ν as necessary. In particular, we are not assuming the market to be complete here.

intervals, because the expected change of an Itô process is of order dt, whereas the standard deviation is of order $\sqrt{dt}$.

In the absence of a money market account, portfolios must satisfy $\iota'\pi = 1$, as in a single-period model without a risk-free asset, and self-financing wealth processes $W > 0$ satisfy

$$\frac{dW}{W} = \pi'\mu \, dt + \pi'\sigma \, dB. \qquad (13.52)$$

To define SDF processes without a money market account, we must obviously drop the requirement in Section 13.3 that MR be a local martingale. No other change is necessary. Thus, we can define an SDF process to be a strictly positive process M such that $M_0 = 1$ and MS_i is a local martingale for each risky asset i.

Given this definition, an SDF process must satisfy (13.17) for some drift, which we can call $-r$ as before. Thus, the characterization of SDF processes is the same as before, but r is now an arbitrary stochastic process. We call r a shadow risk-free rate. In the absence of a money market account, the market is incomplete, and, even if $n = k$ and σ is invertible, there is an infinite number of SDF processes, one for each imaginable r process (even $r < 0$). In this book, only Chapter 17 discusses continuous-time markets without a money market account.

13.8 NOTES AND REFERENCES

Use of the martingale representation theorem to deduce market completeness is due to Harrison and Pliska (1981). They also show that requiring $W \geq 0$ rules out arbitrage opportunities when there is an SDF process. On the latter point, see also Dybvig and Huang (1988). Harrison and Pliska call self-financing wealth processes W for which MW is a supermartingale but not a martingale "suicide strategies." The motivation for this name is that, if $M_t W_t > \mathsf{E}_t[M_T W_T]$, then, at least in a complete market, we can obtain W_T at cost $\mathsf{E}_t[M_T W_T/M_t] < W_t$. Thus, the strategy with wealth process W throws away money. Duffie and Huang (1985) prove the existence of a competitive equilibrium in a complete continuous-time securities market. For a comprehensive discussion of this topic, see Karatzas and Shreve (1998).

Novikov's condition can be found in many texts, including Øksendal (2003) and Karatzas and Shreve (2004). Finite-horizon bubbles, interpreted as MW being a local martingale but not a martingale, are studied by Lowenstein and Willard (2000), Cox and Hobson (2005), and Heston, Lowenstein, and Willard (2007). Heston et al. relate such bubbles to multiple solutions of the fundamental PDE for derivative security values (Chapter 16) and provide several examples, including multiple solutions of the fundamental PDE for option prices when the underlying asset has constant elasticity of variance or stochastic volatility.

To see the connection between local martingales and bubbles, suppose (13.39) is a local martingale but not a martingale. If W and C are nonnegative, then, because nonnegative local martingales are supermartingales,

$$W_t \geq E_t\left[\int_t^T \frac{M_s}{M_t} C_s\, ds\right] + E_t\left[\frac{M_T}{M_t} W_T\right]. \tag{13.53}$$

If there is strict inequality in (13.53), then we can say that there is a finite-horizon bubble, meaning that the price W_t of the finite-horizon consumption process $(C_s)_{t \leq s \leq T}$ and terminal value W_T is larger than its fundamental value, defined as the right-hand side of (13.53).

Naturally, the type of bubble discussed in Chapter 8 is also possible in this model. Specifically, note that if (13.39) is a martingale, $M \geq 0$, and $C \geq 0$, then we can take limits in (13.37) as $T \to \infty$. If the limits are finite,

$$W_t = E_t\left[\int_t^\infty \frac{M_s}{M_t} C_s\, ds\right] + \lim_{T \to \infty} E_t\left[\frac{M_T}{M_t} W_T\right]. \tag{13.54}$$

There is a bubble of the type discussed in Section 8.5 if (13.54) holds but

$$\lim_{T \to \infty} E_t\left[\frac{M_T}{M_t} W_T\right] > 0,$$

implying

$$W_t > E_t\left[\int_t^\infty \frac{M_s}{M_t} C_s\, ds\right].$$

The formula for the risk-free rate in Exercise 13.1 is due to Breeden (1986). Part (c) of Exercise 13.2 asks for a geometric Brownian motion aggregate consumption process to be calibrated to the historical moments of consumption growth reported by Mehra and Prescott (1985). Calibrating in the way requested is not quite the right thing to do. The data give aggregate consumption over a period of time, whereas C_t in the continuous-time model is the rate of consumption at time t. Thus, the data should be matched to the properties of the model time series $\int_0^1 C_t\, dt$, $\int_1^2 C_t\, dt$, ..., rather than to the properties of the model series $C_1, C_2, \ldots$. This is called the time aggregation issue.

EXERCISES

13.1. For constants $\delta > 0$ and $\rho > 0$, assume

$$M_t \stackrel{\text{def}}{=} e^{-\delta t}\left(\frac{C_t}{C_0}\right)^{-\rho} \tag{13.55}$$

is an SDF process, where C denotes aggregate consumption. Assume that

$$\frac{dC}{C} = \alpha \, dt + \theta' \, dB \tag{13.56}$$

for stochastic processes α and θ.

(a) Apply Itô's formula to calculate dM/M.
(b) Explain why the result of Part (a) implies that the instantaneous risk-free rate is

$$r = \delta + \rho\alpha - \frac{\rho(\rho+1)}{2}\theta'\theta \tag{13.57}$$

and the price of risk process is $\lambda = \rho\theta$.
(c) Explain why the risk premium of any asset with price S is

$$\rho\left(\frac{dS}{S}\right)\left(\frac{dC}{C}\right).$$

Note: This is a preview of the CCAPM (Section 14.6), which holds under more general assumptions.

13.2. Consider an asset paying dividends D over an infinite horizon. Assume D is a geometric Brownian motion:

$$\frac{dD}{D} = \mu \, dt + \sigma \, dB$$

for constants μ and σ and a Brownian motion B. Assume the instantaneous risk-free rate r is constant, and assume there is an SDF process M such that

$$\left(\frac{dD}{D}\right)\left(\frac{dM}{M}\right) = -\sigma\lambda \, dt \tag{13.58}$$

for a constant λ. Assume $\mu - \sigma\lambda < r$, and assume there are no bubbles in the price of the asset.

(a) Show that the asset price is

$$P_t = \frac{D_t}{r + \sigma\lambda - \mu}.$$

Show that the Sharpe ratio of the asset is λ. Note: This is a continuous-time version of the Gordon growth model (Section 10.4). This exercise is continued in Exercise 15.2.
(b) Assume (13.55) is an SDF for constants $\delta > 0$ and $\rho > 0$, where $C = D$. Show that (13.58) holds. What is λ? Referencing Exercise 12.4, calibrate to the following statistics reported by Mehra and Prescott (1985):

$r = \log 1.008$, $E_t[C_{t+1}/C_t] = 1.018$, $\text{stdev}_t(C_{t+1}/C_t) = 0.036$,
$E_t[(P_{t+1} + C_{t+1})/P_t] = 1.0698$, and $\text{stdev}_t(P_{t+1} + C_{t+1})/P_t) = 0.1654$.
Calculate ρ and δ.

13.3. Let r^d denote the instantaneous risk-free rate in the domestic currency, and let R^d denote the domestic currency price of the domestic money market account:

$$R_t^d = \exp\left(\int_0^t r_s^d \, ds\right).$$

As in Section 8.6, let X denote the price of a unit of a foreign currency in units of the domestic currency. Let r^f denote the instantaneous risk-free rate in the foreign currency, and let R^f denote the foreign currency price of the foreign money market account:

$$R_t^f = \exp\left(\int_0^t r_s^f \, ds\right).$$

Suppose M^d is an SDF process for the domestic currency, so $M^f \stackrel{\text{def}}{=} M^d X/X_0$ is an SDF process for the foreign currency. Assume

$$\frac{dX}{X} = \mu_x \, dt + \sigma_x \, dB$$

for a Brownian motion B.

(a) Show that $M^f R^f$ being a local martingale implies that

$$\frac{dM^f}{M^f} = -r^f \, dt + dZ$$

for some local martingale Z.

(b) Deduce from the previous result and Itô's formula that

$$\mu_x \, dt = (r^d - r^f) \, dt - \left(\frac{dX}{X}\right)\left(\frac{dM^d}{M^d}\right).$$

Note: This exercise is continued in Exercise 15.6.

13.4. For a local martingale Y satisfying $dY/Y = \theta' \, dB$ for some stochastic process θ, Novikov's condition is that

$$E\left[\exp\left(\frac{1}{2}\int_0^T \theta'\theta \, dt\right)\right] < \infty.$$

Under this condition, Y is a martingale on $[0, T]$. Consider $Y = MW$, where M is an SDF process and W is a self-financing wealth process.

(a) Show that $dY/Y = \theta' dB$, where $\theta = \sigma'\pi - \lambda_p - \zeta$ and $\sigma\zeta = 0$.
(b) Deduce that Novikov's condition is equivalent to (13.41).
(c) By specializing (13.41), state sufficient conditions for MS_i to be a martingale for $i = 1,\ldots,n$.

13.5. Suppose $W > 0$, C, and π satisfy the intertemporal budget constraint (13.38). Define the consumption-reinvested wealth process $W^\dagger$ by (13.43).

(a) Show that $W^\dagger$ satisfies the intertemporal budget constraint (13.44).
(b) Show that
$$W_t^\dagger - W_t = W_t^\dagger \int_0^t \frac{C_s}{W_s^\dagger} \, ds$$
for each t.

Hint: Define $Y = W/W^\dagger$ and use Itô's formula to show that
$$dY = -\frac{C}{W^\dagger} \, dt.$$

Conclude that
$$\frac{W_t}{W_t^\dagger} = 1 - \int_0^t \frac{C_s}{W_s^\dagger} \, ds$$
for all t.

(c) Let M be an SDF process and assume $MW^\dagger$ is a martingale. Use this assumption and iterated expectations to show that, for any $t < T$,
$$E_t\left[M_T W_T^\dagger \int_0^T \frac{C_s}{W_s^\dagger} \, ds\right] = M_t W_t^\dagger \int_0^t \frac{C_s}{W_s^\dagger} \, ds + E_t\left[\int_t^T M_s C_s \, ds\right].$$

(d) Let M be an SDF process and assume $MW^\dagger$ is a martingale. Use the results of the previous two parts to show that (13.39) is a martingale.

13.6. Suppose W, C, and π satisfy the intertemporal budget constraint (13.38). Define
$$W_t^\dagger = W_t + R_t \int_0^t \frac{C_s}{R_s} \, ds.$$
Note: This means consumption is reinvested in the money market account rather than in the portfolio generating the wealth process as in (13.43).

(a) Show that $W^\dagger$ satisfies the intertemporal budget constraint (13.12).
(b) Let M be an SDF process. Assume MR is a martingale and $MW^\dagger$ is a martingale. Deduce that (13.39) is a martingale.

14

Continuous-Time Portfolio Choice and Pricing

CONTENTS

14.1 Euler Equation   343
14.2 Representative Investor Pricing   343
14.3 Static Approach to Portfolio Choice   344
14.4 Introduction to Dynamic Programming   349
14.5 Markovian Portfolio Choice   352
14.6 CCAPM, ICAPM, and CAPM   357
14.7 Notes and References   360

This chapter studies investors who have time-additive utility. Such investors seek to maximize

$$\mathsf{E}\left[\int_0^T e^{-\delta t} u(C_t)\,\mathrm{d}t + U(W_T)\right] \qquad (14.1\text{a})$$

for some $T < \infty$, or

$$\mathsf{E}\left[\int_0^\infty e^{-\delta t} u(C_t)\,\mathrm{d}t\right]. \qquad (14.1\text{b})$$

In the finite-horizon case, U denotes the utility of a bequest. If $u = 0$ in the finite-horizon case, then the objective is to maximize the expected utility of terminal wealth. If $U = 0$, then there is no bequest motive. As before, the first-order condition for optimal consumption is called the Euler equation. It states that the investor's MRS is an SDF process. If there is a representative investor, then the MRS depends on aggregate consumption.

We describe two methods for optimizing (14.1). First, we describe the static approach, which works in some special models (it can be generalized to work in other models, but that is beyond the scope of this book; see the end-of-chapter notes for some references). Then, we explain dynamic programming in continuous time. Key results are the envelope condition and the decomposition of optimal portfolios into myopic and hedging demands. Section 14.6 derives implications of optimization for asset prices—in particular, the CCAPM and the ICAPM.

14.1 EULER EQUATION

The Euler equation (first-order condition for optimal consumption) is essentially the same in continuous time as in discrete time. The condition is that the MRS

$$\frac{e^{-\delta t} u'(C_t)}{u'(C_0)}$$

is an SDF process. This is a sufficient condition for optimality in a complete market (Section 14.3). We use it as a necessary condition for optimality to derive the CCAPM and the ICAPM in Section 14.6. Proving that it is a necessary condition for optimality is more delicate in continuous time than in discrete time, and we do not attempt it here (see the end-of-chapter notes).

14.2 REPRESENTATIVE INVESTOR PRICING

There is a representative investor if the market is complete or if all investors have LRT utility functions with the same cautiousness parameter. The proof of this fact is essentially the same as in discrete time or as in a single-period model. Pricing via the representative investor's MRS is also very much the same in continuous time as in discrete time, though calculations are sometimes easier in continuous time. For example, if there is a representative investor with constant relative risk aversion ρ and discount factor δ, then the Euler equation states that

$$M_t \stackrel{\text{def}}{=} e^{-\delta t} C_t^{-\rho} C_0^{\rho}$$

is an SDF process. This implies formulas for the risk-free rate and for the price of risk process (Exercise 13.1). If consumption growth is lognormal, then the market price-dividend ratio is constant, and the market Sharpe ratio is $\rho\sigma$ (Exercise 13.2). If instead,

$$\frac{dC}{C} = \alpha(X)\,dt + \theta(X)'\,dB,$$

where X is Markovian (Section 13.6), then the market price-dividend ratio is a function of X, say $f(X)$. In this case, the market risk premium depends on the partial derivatives $\partial \log f(x)/\partial x_i$ and the covariances between the elements of X and consumption growth (Exercise 15.1).

14.3 STATIC APPROACH TO PORTFOLIO CHOICE

If there is no labor income, and if consumption and wealth are nonnegative always, then an investor's consumption and wealth must satisfy the following static budget constraint for any SDF process M:

$$\mathsf{E}\left[\int_0^T M_t C_t \, dt + M_T W_T\right] \leq W_0 \qquad (14.2\text{a})$$

or

$$\mathsf{E}\left[\int_0^\infty M_t C_t \, dt\right] \leq W_0 \qquad (14.2\text{b})$$

if the horizon is infinite. The left-hand side of the constraint is of course the date–0 cost of consumption (and terminal wealth), using the SDF process for pricing, and the right-hand side is the given initial wealth.

We show below that, when the market is complete, we can find the optimal consumption process by maximizing expected utility subject to the static budget constraint. This is a relatively straightforward problem, because it does not involve the portfolio or the intertemporal budget constraint. The portfolio process is calculated from the optimal consumption process in a second step. We also show below that the same method works if the capital market line is constant over time. In particular, if returns are IID, then we can find the optimum by maximizing subject to a static budget constraint.

In other cases, we must deal either with a continuum of static budget constraints (corresponding to different SDF processes) or work directly with the portfolio process and intertemporal budget constraint. The latter approach is taken later in the chapter, using dynamic programming. First, we prove that the static budget constraint (14.2) must hold at any consumption process that, in conjunction with some portfolio process, satisfies the intertemporal budget constraint.

To derive (14.2), assume π, C, and W satisfy the intertemporal budget constraint (13.38). As observed in Section 13.3,

$$M_t W_t + \int_0^t M_s C_s \, ds \qquad (14.3)$$

is a local martingale. Under our current assumptions, it is a nonnegative local martingale and hence a supermartingale, implying

$$M_t W_t + \int_0^t M_s C_s \, ds \geq \mathsf{E}_t\left[M_\tau W_\tau + \int_0^\tau M_s C_s \, ds\right]$$

for any $t \leq \tau$. We can rearrange this as

$$M_t W_t \geq \mathsf{E}_t\left[M_\tau W_\tau + \int_t^\tau M_s C_s \, ds\right] \quad (14.4)$$

for each $t \leq T$.

The inequality (14.2a) is the special case of (14.4) obtained by taking $t = 0$ and $\tau = T$. In the infinite-horizon case, (14.2b) follows from (14.4) by taking $\tau \to \infty$ and using the monotone convergence theorem.

Complete Markets

Assume $n = k$ (as many risky assets as Brownian motions), so σ is invertible, and assume the squared maximum Sharpe ratio has a finite integral with probability 1. As shown in Section 13.3, there is a unique SDF process $M = M_p$. Assume investors have no labor income and are constrained to maintain nonnegative wealth.

The portfolio choice problem in this setting is equivalent to maximizing expected utility subject to the static budget constraint (14.2)—that is, the intertemporal budget constraint is equivalent to the static constraint. The previous section shows that the intertemporal budget constraint implies the static constraint (even in an incomplete market). In a complete market, the converse is true. In fact, Section 13.5 shows that if nonnegative C and W_T satisfy the static budget constraint, then there exists a portfolio process π in conjunction with which C and W_T satisfy the intertemporal budget constraint. Furthermore,

$$W_t = \mathsf{E}_t\left[\int_t^T \frac{M_u}{M_t} C_u \, du + \frac{M_T}{M_t} W_T\right]. \quad (14.5)$$

This implies that $W \geq 0$, as we are requiring here. Thus, the static and intertemporal budget constraints are equivalent in a complete market.

The static problem can in principle be solved by the same methods as a standard consumer choice problem. Any solution of the static budget constraint and first-order conditions

$$(\forall t \leq T) \quad e^{-\delta t} u'(C_t^*) = \gamma M_t \quad \text{and} \quad U'(W_T^*) = \gamma M_T \quad (14.6a)$$

or

$$(\forall t < \infty) \quad e^{-\delta t} u'(C_t^*) = \gamma M_t, \tag{14.6b}$$

in the infinite-horizon case, for some γ, is a solution of the static problem.[1] This follows from concavity, as usual. We invert the marginal utility functions and then attempt to solve the first-order conditions (14.6) for C (or C and W_T) in terms of γ.[2] We then find γ and hence C^* (or C^* and W_T^*) by imposing the static budget constraint (as an equality).

To illustrate this method for power utility, as in Section 9.2, consider maximizing

$$\mathsf{E}\left[\int_0^T \frac{e^{-\delta t}}{1-\rho} C_t^{1-\rho} \, dt + \frac{\beta}{1-\rho} W_T^{1-\rho}\right]$$

for some positive δ, β, and ρ, subject to the static constraint (14.2a). Let γ denote the Lagrange multiplier for the constraint. The first-order condition is

$$e^{-\delta t} C_t^{-\rho} = \gamma M_t \quad \text{and} \quad \beta W_T^{-\rho} = \gamma M_T.$$

Thus,

$$C_t = \left(e^{\delta t} \gamma M_t\right)^{-1/\rho} \quad \text{and} \quad W_T = \beta^{1/\rho} (\gamma M_T)^{-1/\rho}. \tag{14.7}$$

This implies that the left-hand side of the budget constraint (14.2a) is

$$\gamma^{-1/\rho} \mathsf{E}\left[\int_0^T e^{-(\delta/\rho)t} M_t^{1-1/\rho} \, dt + \beta^{1/\rho} M_T^{1-1/\rho}\right].$$

Equating this to W_0 defines γ. Then, the optimal consumption and terminal wealth are given by (14.7).

The solution of the static problem also yields the optimum in the presence of labor income, provided investors are allowed to borrow against the labor income, meaning that the nonnegative wealth constraint takes the form

$$W_t + \mathsf{E}_t\left[\int_t^T \frac{M_s}{M_t} Y_s \, ds\right] \geq 0 \tag{14.8}$$

in the finite-horizon case. See Exercise 14.7. However, there are typically limitations both on the extent to which an investor can borrow against labor income and on the terms at which she can borrow, due to moral hazard and adverse selection. These limitations may render the solution to the static problem infeasible.

---

[1]. Of course, the first half of (14.6a) can be ignored when $u = 0$ and the second half ignored when $U = 0$.

[2]. The first-order conditions have a solution C^* (or C^* and W_T^*) for any γ if $\lim_{c \to 0} u'(c) = \lim_{w \to 0} U'(w) = \infty$ and $\lim_{c \to \infty} u'(c) = \lim_{w \to \infty} U'(w) = 0$. These are called the Inada conditions.

Constant Capital Market Line

A second circumstance in which the static approach is feasible is when there is no labor income and the capital market line is constant over time. The capital market line is the line through the risk-free rate and the tangency portfolio in (standard deviation, mean) space, so for it to be constant means that the risk-free rate r and the maximum Sharpe ratio

$$\kappa = \sqrt{\lambda_p'\lambda_p} = \sqrt{(\mu - r\iota)'\Sigma^{-1}(\mu - r\iota)} \qquad (14.9)$$

are constants. This allows the investment opportunity set (defined by r, μ, and Σ) to be time varying and random to a limited extent. Also, the market can be incomplete. In this circumstance, each investor's optimal portfolio is proportional to the portfolio

$$\Sigma^{-1}(\mu - r\iota). \qquad (14.10)$$

Thus, two-fund separation holds. The portfolio (14.10) is the optimal portfolio for a log utility investor even when the capital market line is not constant (Exercises 14.3 and 14.4). Maximizing $E[\log W_T]$ means maximizing the expected continuously compounded return from 0 to T. For this reason, (14.10) is called the growth-optimal portfolio.

In a single-period model, two-fund separation follows from assumptions on preferences (quadratic utility) or assumptions on returns (separating distributions). Likewise, in continuous time, two-fund separation follows from assumptions on preferences (log utility) or assumptions on returns (constancy of the capital market line).

To be more precise about the assumptions producing two-fund separation, assume that investors have no labor income and are constrained to choose nonnegative consumption and maintain nonnegative wealth. Assume also that there is a solution to the static problem with $M = M_p$ that satisfies the first-order conditions (14.6). Then, as is shown below, the solution to the static problem is optimal, and the optimal portfolio is proportional to the tangency portfolio. This is true for all versions of the optimization problems (14.1) and (14.1b), including maximizing the expected utility of terminal wealth.

For concreteness, consider the finite-horizon case. The argument for the infinite-horizon case is the same. We are assuming there is a solution (C^*, W_T^*) to the static problem defined by M_p. Any feasible consumption process and terminal wealth must satisfy the constraints of the static problem and hence can provide no higher utility than (C^*, W_T^*). What remains is to show that (C^*, W_T^*) is feasible (satisfies the intertemporal budget constraint in conjunction with some portfolio process).

We will work with the SDF process M_p given by

$$\frac{dM_p}{M_p} = -r\,dt - \lambda'_p\,dB,$$

where $\lambda_p = \sigma'\Sigma^{-1}(\mu - r\iota)$ as defined in (13.21). To economize on notation, we write M for M_p. To construct the portfolio process, first define $X_0 = 0$ and

$$dX = \frac{1}{\kappa}\lambda'_p\,dB, \tag{14.11}$$

where κ is the maximum Sharpe ratio defined in (14.9). Then, $(dX)^2 = dt$, so by Levy's theorem, X is a Brownian motion. We have

$$\frac{dM}{M} = -r\,dt - \kappa\,dX, \tag{14.12}$$

and by assumption r and κ are constants. Therefore, M is a geometric Brownian motion with

$$M_t = e^{(r-\kappa^2/2)t + \kappa X_t} \tag{14.13}$$

for each t. The first-order condition (14.6a) for the static problem therefore defines C_t^* as a function of X_t and W_T^* as a function of X_T.

Define

$$W_t = \frac{1}{M_t}\mathsf{E}_t\left[\int_t^T M_u C_u^*\,du + M_T W_T^*\right]. \tag{14.14}$$

This implies $W_T = W_T^*$. Define

$$Z_t = \int_0^t M_u C_u^*\,du + M_t W_t = \mathsf{E}_t\left[\int_0^T M_u C_u^*\,du + M_T W_T^*\right].$$

This is a martingale adapted to X. We can follow the second step in Section 13.5, using the martingale representation theorem with B replaced by X, to deduce that the stochastic part of dW is $\gamma\,dX$ for some stochastic process γ. We have

$$\gamma\,dX = \frac{\gamma}{\kappa}\lambda'_p\,dB = \frac{\gamma}{\kappa}(\mu - r\iota)'\Sigma^{-1}\sigma\,dB = \phi'\sigma\,dB,$$

where we define

$$\phi = \frac{\gamma}{\kappa}\Sigma^{-1}(\mu - r\iota).$$

This portfolio is proportional to the growth-optimal portfolio. Furthermore, W satisfies (14.14) and the stochastic part of dW is $\phi'\sigma\,dB$, so the intertemporal budget constraint is satisfied by W, ϕ, and C^*, as explained in Section 13.5. Therefore, (C^*, W_T^*) is feasible.

14.4 INTRODUCTION TO DYNAMIC PROGRAMMING

In this section, we explain dynamic programming in continuous time in the relatively simple context of IID returns (that is, r, μ, and Σ are constants) and no labor income. Of course, the results of the preceding section apply to this case, but dynamic programming provides more detail about the solution than does the method of the previous section (which relies on the martingale representation theorem for the existence of the optimal portfolio). Section 14.5 discusses more general models.

Value Function

Consider the finite-horizon model. Because returns are IID and there is no labor income, the investor's value at any date t is only a function of time and wealth. Specifically,

$$V(t,w) = \max \mathsf{E}\left[\int_t^T e^{-\delta s} u(C_s)\,\mathrm{d}s + U(W_T)\,\bigg|\, W_t = w\right]. \qquad (14.15)$$

In particular, $V(T,\cdot) = U(\cdot)$. As in a discrete-time model, when there is consumption at each date, it is convenient to define $J(t,w) = e^{\delta t} V(t,w)$. Then, $J(T,\cdot) = e^{\delta T} U(\cdot)$.[3]

HJB Equation

Recall from Section 9.4 that the discrete-time Bellman equation (with state variable x, choice variable π, and utility function u_t) is

$$V_t(x) = \max_\pi \left\{ u_t(x,\pi) + \mathsf{E}_t[V_{t+1}(X_{t+1})] \right\}.$$

Subtract $V_t(x)$ from both sides to obtain

$$0 = \max_\pi \left\{ u_t(x,\pi) + \mathsf{E}_t[V_{t+1}(X_{t+1}) - V_t(x)] \right\}.$$

In continuous time, the reward at date t that substitutes for $u_t(x,\pi)$ is $e^{-\delta t} u(c)\,\mathrm{d}t$. The continuous-time Bellman equation is

$$0 = \max_{c,\pi} \left\{ e^{-\delta t} u(c)\,\mathrm{d}t + \mathsf{E}_t[\mathrm{d}V] \right\}, \qquad (14.16)$$

---

3. In (14.1), $U(w)$ denotes the utility of a bequest w discounted to date 0, so $e^{\delta T} U(w)$ is the utility at date T.

where $\mathsf{E}_t[dV]$ denotes the drift of the value function. This continuous-time version of the Bellman equation is usually called the Hamilton-Jacobi-Bellman equation, or the HJB equation. The intuition for (14.16), which applies also in discrete time, is that at the optimal action the current utility exactly offsets any expected decline in the value function, but at suboptimal actions the current utility is insufficient to offset the expected change in the value function.

When there is consumption at each date, we set $J(t,w) = e^{\delta t}V(t,w)$. This implies

$$dV = -\delta e^{-\delta t} J\, dt + e^{-\delta t}\, dJ.$$

Make this substitution in the HJB equation (14.16) and cancel the $e^{-\delta t}$ factor to obtain the equivalent form:

$$0 = \max_{c,\pi}\left\{u(c)\,dt - \delta J\,dt + \mathsf{E}_t[dJ]\right\}. \qquad (14.17)$$

Supermartingales and Martingales

To restate the explanation of (14.16) in a somewhat more rigorous way, note that the definition (14.15) of the value function implies, assuming there is an optimum,

$$V(t,W_t^*) = \mathsf{E}_t\left[\int_t^T e^{-\delta s} u(C_s^*)\,ds + U(W_T^*)\right],$$

where the asterisks denote the optimum. This implies that

$$\int_0^t e^{-\delta s} u(C_s^*)\,ds + V(t,W_t^*) = \mathsf{E}_t\left[\int_0^T e^{-\delta t} u(C_t^*)\,dt + U(W_T^*)\right],$$

which is a martingale. Thus, the accumulated utility plus remaining value

$$\int_0^t e^{-\delta s} u(C_s)\,ds + V(t,W_t) \qquad (14.18)$$

forms a martingale at the optimum. In particular, (14.18) is not expected to decrease over time, meaning that value is not expected to dissipate as time passes. However, if we consider a suboptimal policy, then $V(t,w)$ will exceed the conditional expectation in (14.15). This yields

$$\int_0^t e^{-\delta s} u(C_s)\,ds + V(t,W_t) > \mathsf{E}_t\left[\int_0^T e^{-\delta t} u(C_t)\,dt + U(W_T)\right],$$

implying that the left-hand side is a supermartingale, expected to decrease over time, reflecting a dissipation of value. The HJB equation simply restates this martingale/supermartingale distinction in terms of the drift of (14.18) being zero at the maximum and negative otherwise. See Section 15.6 for more details.

Calculating the Drift of the Value Function

We can compute the drift that is maximized in the HJB equation by assuming the value function is sufficiently smooth and applying Itô's formula. Consider the value function J and HJB equation (14.17). Using subscripts to denote partial derivatives, Itô's formula implies

$$dJ = J_t\, dt + J_w\, dW + \frac{1}{2}J_{ww}\,(dW)^2.$$

Substitute the intertemporal budget constraint to obtain

$$dJ = J_t\, dt + J_w[rW - C + W\pi'(\mu - r\iota)]\, dt + J_w W\pi'\sigma\, dB + \frac{1}{2}J_{ww}W^2\pi'\Sigma\pi\, dt.$$

Substitute the dt part of this into (14.16), writing c for an arbitrary value of C_t and w for an arbitrary value of W_t, and cancel the dt to obtain

$$0 = \max_{c,\pi}\left\{u(c) - \delta J + J_t + J_w[rw - c + w\pi'(\mu - r\iota)] + \frac{1}{2}J_{ww}w^2\pi'\Sigma\pi\right\}. \tag{14.19}$$

This is the HJB equation for portfolio choice with IID returns. It is to be solved for the maximizing c and π and then solved for J in conjunction with boundary conditions on J. These steps are explained below.

Optimal Consumption and Portfolio

The first-order condition for maximizing (14.19) in c is

$$u'(c) = J_w. \tag{14.20}$$

This is the envelope condition discussed in Chapter 9. The maximization over π in (14.19) is a quadratic optimization problem with solution

$$\pi = -\frac{J_w}{wJ_{ww}}\Sigma^{-1}(\mu - r\iota). \tag{14.21}$$

This is of course consistent with the result of the previous section that the optimal portfolio is proportional to the growth-optimal portfolio. The growth-optimal portfolio $\Sigma^{-1}(\mu - r\iota)$ is the optimal portfolio for an investor with log utility, even when the capital market line is not constant (see Exercise 14.3 for an example).

Thus, the optimal portfolio is the optimal portfolio for a log-utility investor scaled by the reciprocal of the relative risk aversion of the value function J.

The Nonlinear PDE

By inverting the marginal utility function, the optimal consumption can be computed from the envelope condition as

$$c = (u')^{-1}\left(J_w(t, w)\right).$$

Denote the function on the right-hand side of this as $f(t, w)$ so we can write $c = f(t, w)$. Of course, f depends on J. Substitute this and the optimal π (14.21) into the HJB equation (14.19) to obtain

$$0 = u(f(t, w)) - \delta J + J_t + wJ_w\left(r - \frac{\kappa^2}{2}\frac{J_w}{wJ_{ww}}\right). \tag{14.22}$$

This is a nonlinear PDE in the function J. It should be solved subject to the boundary condition $J(T, w) = e^{\delta T} U(w)$. In special cases (when u and U have the same linear risk tolerance), it can be solved explicitly. The case of CRRA utility is considered in the exercises. In other cases, there are standard methods that can be used to solve it numerically.

To derive (14.22), we assumed that r, μ, and Σ are constants. However, the only parameters of the market that appear in (14.22) are the risk-free rate r and the maximum Sharpe ratio κ. If they are constants—that is, if the capital market line is constant—then (14.22) is still meaningful as an equation in J. In fact, it is the correct equation to solve for J, and (14.20) and (14.21) give the optimal consumption and portfolio, whenever the capital market line is constant.

14.5 MARKOVIAN PORTFOLIO CHOICE

We now analyze portfolio choice when the investor's value function depends on a vector X of state variables. Assume the securities market is Markovian in the sense of Section 13.6. If the investor has labor income Y, assume $Y_t = \zeta(X_t)$ for some function ζ. Assume for now that the investor has a finite horizon (the infinite-horizon case is discussed at the end of the section). At any date t, the set of consumption processes $(C_s)_{t \leq s \leq T}$ and terminal wealths W_T attainable by an investor depends on the investor's wealth W_t and on the state vector X_t. Thus, the maximum attainable expected utility is a function of (t, X_t, W_t). Write the value function with discounting to date 0 as $V(t, x, w)$, and set $J = e^{\delta t} V$. Denote partial derivatives by subscripts again.

HJB Equation

The HJB equation is the same as before, except that now when computing $E_t[dJ]$ we have to include the following terms:

$$J_{x_j} E[dX_j] \quad \text{for } j = 1, \ldots, \ell, \tag{14.23a}$$

$$\frac{1}{2} J_{x_i x_j} (dX_i)(dX_j) \quad \text{for } i = 1, \ldots, \ell \text{ and } j = 1, \ldots, \ell, \tag{14.23b}$$

$$J_{wx_j} (dW)(dX_j) \quad \text{for } j = 1, \ldots, \ell. \tag{14.23c}$$

None of these additional terms involves the consumption rate c. Therefore, maximizing in c in the HJB equation produces the envelope condition as before.

The first two of the additional terms above do not involve the portfolio π either. Therefore, when maximizing with respect to π, the only change is that we must consider the terms in (14.23c). Assume dX is given by (13.50) and let $v_j(X)$ denote the jth row of $v(X)$ rearranged as a column vector, so (13.51) holds for each j. The stochastic part of dX_j is $v_j' \, dB$. Therefore,

$$(dW)(dX_j) = (\pi' \sigma \, dB)(dX_j) = \pi' \sigma v_j \, dt.$$

The terms in the HJB equation involving π, including the terms in (14.23c), are

$$J_w w \pi'(\mu - r\iota) + \frac{1}{2} J_{ww} w^2 \pi' \Sigma \pi + \sum_{j=1}^{\ell} J_{wx_j} w \pi' \sigma v_j.$$

Maximizing in π produces

$$\pi = -\frac{J_w}{w J_{ww}} \Sigma^{-1} (\mu - r\iota) - \sum_{j=1}^{\ell} \frac{J_{wx_j}}{w J_{ww}} \Sigma^{-1} \sigma v_j. \tag{14.24}$$

To compute the value function J, we need to write the HJB equation as a PDE by substituting the solutions for c and π (in terms of J). In this PDE, all of the terms in (14.23) appear. Numerical methods are usually required to solve the PDE.

Myopic and Hedging Demands

The first term in (14.24) is the optimal portfolio (14.21) when the capital market line is constant and is termed the myopic demand. The other terms in (14.24) are called the hedging demands. Consider the vector

$$\pi_j = \Sigma^{-1} \sigma v_j = (\sigma \sigma')^{-1} \sigma v_j.$$

The stochastic part of a portfolio return is $\pi'\sigma\,dB$. The stochastic part of the change in the jth state variable is $v_j'\,dB$. The portfolio having maximum instantaneous correlation with the change in X_j is the portfolio π for which the column vector $\sigma'\pi$ is closest to v_j. The vector π_j is this portfolio. To see this, note that

$$\sigma'\pi_j = \sigma'(\sigma\sigma')^{-1}\sigma v_j,$$

which is the orthogonal projection of v_j onto the linear span of the columns of σ'. Thus, the hedging demands consist of positions in the ℓ portfolios π_j that have maximum instantaneous correlation with the state variables. The formula (14.24) implies $(\ell + 2)$-fund separation: Investors combine the tangency portfolio and risk-free asset with the ℓ funds having maximum correlation with the state variables.

Because $J_{ww} < 0$ (due to concavity), the sign of the coefficient of $\Sigma^{-1}\sigma v_j$ in (14.24) is the same as the sign of J_{wx_j}. Thus, the hedging demand is positively correlated with the state variable if $J_{wx_j} > 0$, which occurs when an increase in the state variable increases the marginal value of wealth. The role of the hedging demands is to generate higher returns in states of the world in which wealth is more valuable at the margin.

It can be useful to rewrite (14.24) as

$$\pi = -\frac{J_w}{wJ_{ww}}\left[\Sigma^{-1}(\mu - r\iota) + \sum_{j=1}^{\ell}\frac{J_{wx_j}}{J_w}\Sigma^{-1}\sigma v_j\right]. \tag{14.24'}$$

The portfolio in braces is the sum of the growth-optimal portfolio $\Sigma^{-1}(\mu - r\iota)$ and scalar multiples of the ℓ portfolios having maximum instantaneous correlation with the state variables. The scalars equal

$$\frac{J_{wx_j}}{J_w} = \frac{\partial \log J_w}{\partial x_j},$$

which reflect the sensitivities of the marginal value of wealth to the respective state variables. Equation (14.24') demonstrates that the extent to which the optimal portfolio deviates from the growth-optimal portfolio depends on (i) the coefficient of relative risk aversion $-wJ_{ww}/J_w$, (ii) the portfolios $\pi_j = \Sigma^{-1}\sigma v_j$ having maximum correlation with the state variables, and (iii) the importance of the state variables, meaning how much and in what direction they affect the logarithm of the marginal value of wealth.

Infinite Horizon

The HJB equation simplifies in the infinite-horizon case, due to stationarity, as described in Section 9.5 in discrete time. The value function J is independent of time, so the term J_t disappears from the HJB equation. With IID returns (or a constant capital market line), the infinite-horizon HJB equation is

$$0 = \max_{c,\pi} \left\{ u(c) - \delta J + J_w[rw + w\pi'(\mu - r\iota) - c] + \frac{1}{2} J_{ww} w^2 \pi' \Sigma \pi \right\}. \quad (14.25)$$

Substituting the maximizing values of c and π converts (14.25) into an ordinary differential equation (ODE) in the function $J(w)$. In the more general Markovian case, the same reasoning leads to a PDE to be solved for the function $J(x, w)$.

CRRA Utility

In the absence of labor income and with CRRA utility, the value function inherits the risk aversion of the utility function, as discussed in discrete time in Chapter 9. With log utility and a finite horizon, the value function is

$$J(t, w, x) = \gamma(t) \log w + f(t, x),$$

for functions γ and f. Specifically, with $u(c) = a \log c$ and $U(w) = b \log w$, and with either a or b being possibly zero,

$$\gamma(t) = \frac{a}{\delta} \left(1 - e^{-\delta(T-t)}\right) + b e^{\delta t}.$$

When the horizon is infinite, the function γ is constant and equal to a/δ, and f is independent of t. With log utility, the marginal value of wealth does not depend on the state variables, so there are no hedging demands.

For power utility and a finite horizon,

$$J(t, w, x) = f(t, x) w^{1-\rho}$$

for some function f. If the horizon is infinite, then f is independent of t. In the infinite-horizon case with $\rho < 1$, the existence of a solution to the HJB equation and the existence of an optimum depend on the discount rate δ being sufficiently large. When the capital market line is constant over time, we need to assume

$$\delta > (1-\rho)\left(r + \frac{\kappa^2}{2\rho}\right), \qquad (14.26)$$

where κ is the maximum Sharpe ratio.[4] See Exercise 14.5.

The signs of hedging demands with CRRA utility differ depending on whether $\rho < 1$ or $\rho > 1$. This is exactly the same as discussed in Section 9.5 in discrete time. If $\rho < 1$, then J_{wx_j} has the same sign as J_{x_j}, meaning that the marginal value of wealth is higher when investment opportunities are better. Hence, when $\rho < 1$, the hedging demands produce higher returns in states of the world in which investment opportunities are better. In contrast, when $\rho > 1$, then J_{wx_j} and J_{x_j} have opposite signs, meaning that the marginal value for wealth is higher when investment opportunities are poorer. For such investors, the hedging demands produce higher returns in states of the world in which investment opportunities are worse.

The finite-horizon log-utility case is proven below. The other cases, including power utility, are demonstrated analogously.

Let Z denote the consumption wealth ratio C/W. Starting with wealth w at any date t and following a portfolio process π and consumption process Z produces wealth $W_s = w R_s^{\pi Z}$ at each date $s > t$, where

$$R_s^{\pi Z} = \exp\left(\int_t^s \left\{r + \pi'(\mu - r\iota) - \frac{1}{2}\pi'\Sigma\pi - Z\right\} da + \int_t^s \pi'\sigma\, dB\right).$$

This was seen earlier in (13.15) assuming $Z = 0$, in which case $R_s^{\pi Z}$ is the gross return on the portfolio between t and s. The expected utility produced by (π, Z) date t through date T is

$$\mathsf{E}_t\left[a \int_t^T e^{-\delta s} \log\left(Z_s w R_s^{\pi Z}\right) ds + b \log\left(w R_T^{\pi Z}\right)\right]$$

$$= \left(a \int_t^T e^{-\delta s} ds + b\right) \log w + \mathsf{E}_t\left[a \int_t^T e^{-\delta s} \log\left(Z_s R_s^{\pi Z}\right) ds + b \log\left(R_T^{\pi Z}\right)\right].$$

The value $V(t, w, x)$ is the maximum of this over (π, ξ). The first term is independent of (x, π, ξ) and the second term is independent of w. Thus, we have

---

4. To compare this to the restriction needed in discrete time, note that the discount factor denoted by δ in discrete time is now denoted by $e^{-\delta}$. In our current notation, condition (9.25) is $\delta > (1 - \rho) \log B$, where B is the certainty equivalent for a single-period problem defined in (9.24).

$$J(t,w,x) = e^{\delta t}V(t,w,x) = \left(a\int_t^T e^{-\delta(s-t)}\,ds + e^{\delta t}b\right)\log w$$

$$+ \max_{\pi,Z} \mathsf{E}_t\left[a\int_t^T e^{-\delta s}\log\left(Z_s R_s^{\pi Z}\right)ds + b\log\left(R_T^{\pi Z}\right)\right].$$

Letting $f(t,x)$ denote the last term establishes the claim.

14.6 CCAPM, ICAPM, AND CAPM

The CCAPM of Breeden (1979) states that risk premia depend on covariances with aggregate consumption growth. It is presented as an approximate relation in discrete time in Section 10.1. It is an exact relation in continuous time. As in Section 10.1, we derive it from the Euler equations. The Euler equation for investor h is that

$$\frac{e^{-\delta_h t}u'_h(C_{ht})}{u'_h(C_{h0})} = M_{ht} \tag{14.27}$$

for an SDF process M_h. We show below that the Euler equations imply

$$(\forall i) \quad (\mu_i - r)\,dt = \rho_t \left(\frac{dS_i}{S_i}\right)\left(\frac{dC}{C}\right), \tag{14.28}$$

where C is aggregate consumption and ρ_t is the (generally stochastic) coefficient of aggregate relative risk aversion (aggregate absolute risk aversion multiplied by aggregate consumption). The factor model (14.28) is the CCAPM.

Set

$$g_h(t,c) = \frac{e^{-\delta_h t}u'_h(c)}{u'_h(C_{h0})}, \tag{14.29a}$$

so we have

$$M_{ht} = g_h(t, C_{ht}). \tag{14.29b}$$

Assuming u_h is three times continuously differentiable, (14.29) implies—see the discussion following (13.34)—that, for each asset i,

$$(\mu_i - r)\,dt = \alpha_h\left(\frac{dS_i}{S_i}\right)(dC_h),$$

where

$$\alpha_{ht} = -\left.\frac{\partial \log g_h(t,c)}{\partial c}\right|_{c=C_{ht}} = -\frac{u''_h(C_t)}{u'_h(C_{ht})}.$$

Dividing by α_{ht}, summing over h, and then rearranging yields (14.28), where

$$\rho_t = \frac{C_t}{\sum_{h=1}^{H} 1/\alpha_{ht}}.$$

ICAPM

The ICAPM of Merton (1973a) states that risk premia depend on covariances with market wealth and state variables. It is presented as an approximate relation in discrete time in Section 10.1. It is an exact relation in continuous time. The risk premia of the state variables depend on how they affect the marginal value of wealth, as discussed in Section 10.1.

Assume that the market is Markovian as in Section 13.6. Assume there is consumption at each date. The envelope condition $J_{hw} = u'_h$ and the first-order condition (14.27) produce

$$\frac{e^{-\delta_h t} J_{hw}(t, X_t, W_{ht})}{J_{hw}(0, W_0, X_0)} = M_{ht} \qquad (14.30)$$

for an SDF process M_h and each investor h. It is shown below, following the same reasoning as for the CCAPM, that (14.30) holding for each h implies

$$(\mu_i - r)\,\mathrm{d}t = \rho_t \left(\frac{\mathrm{d}S_i}{S_i}\right)\left(\frac{\mathrm{d}W}{W}\right) + \sum_{j=1}^{\ell} \eta_{jt} \left(\frac{\mathrm{d}S_i}{S_i}\right)\left(\frac{\mathrm{d}X_j}{X_j}\right), \qquad (14.31\mathrm{a})$$

where $W = \sum_{h=1}^{H} W_h$ is market wealth, ρ is the aggregate relative risk aversion of investors' value functions, and η_j is a weighted average (weighted by risk tolerances) of the elasticities of the investors' marginal values of wealth with respect to the state variable X_j. Specifically, $\rho_t = \alpha_t W_t$, and

$$\alpha_t = 1 \bigg/ \sum_{h=1}^{H} \frac{1}{\alpha_{ht}}, \qquad \alpha_{ht} = -\frac{J_{hww}(t, X_t, W_{ht})}{J_{hw}(t, X_t, W_{ht})}, \qquad (14.31\mathrm{b})$$

$$\eta_{jt} = \alpha_t \sum_{h=1}^{H} \frac{\eta_{hjt}}{\alpha_{ht}}, \qquad \eta_{hjt} = \frac{-J_{hwx_j}(t, X_t, W_{ht}) X_{jt}}{J_{hw}(t, X_t, W_{ht})}. \qquad (14.31\mathrm{c})$$

The factor model (14.31) is the ICAPM.

An alternate proof of the ICAPM based on the portfolio formula (14.24) is considered in Exercise 14.2. One merit of the alternate proof is that it does not rely on the envelope condition and hence applies when investors maximize the expected utility of terminal wealth, without consumption at each date. We can also apply the proof given here when investors maximize the expected utility of terminal wealth by deducing (14.30) without using the envelope condition. See Section 10.1 for this proof in discrete time.

From (14.27), we have

$$M_{ht} = g_h(t, X_t, W_{ht}),$$

where

$$g_h(t, w, x) = \frac{e^{-\delta_h t} J_{hw}(t, w, x)}{J_{hw}(0, W_0, X_0)}.$$

As in the proof of the CCAPM, the SDF factor model (Section 13.3) implies

$$(\forall i) \quad (\mu_i - r)\,dt = \alpha_h \left(\frac{dS_i}{S_i}\right)(dW_h) + \sum_{j=1}^{\ell} \eta_{hj}\left(\frac{dS_i}{S_i}\right)\left(\frac{dX_j}{X_j}\right), \quad (14.32)$$

where

$$\alpha_{ht} = -\left.\frac{\partial \log g_h(t, x, w)}{\partial w}\right|_{x=X_t, w=W_{ht}} = -\frac{J_{hww}(t, X_t, W_{ht})}{J_{hw}(t, X_t, W_{ht})},$$

$$\eta_{hjt} = -\left.\frac{\partial \log g_h(t, x, w)}{\partial \log x_j}\right|_{x=X_t, w=W_{ht}} = -\frac{J_{hwx_j}(t, X_t, W_{ht}) X_{jt}}{J_{hw}(t, X_t, W_{ht})}.$$

Divide (14.32) by α_{ht}, sum over h, and rearrange to obtain (14.31a).

More on the Envelope Condition

The CCAPM and ICAPM use different factors to explain risk premia, so they may appear to be alternative models. However, they are tightly linked via the envelope condition. This has already been shown in the derivations of the models, but the relation of the covariances in the two models deserves further emphasis.

For simplicity, assume there is a representative investor and a single state variable. The argument can easily be generalized to heterogeneous investors and multiple state variables. Apply Itô's formula to both sides of the envelope condition $J_w(W, X) = u'(C)$ to obtain

$$J_{ww}(W, X)\,dW + J_{wx}(W, X)\,dX + \text{second-order terms}$$
$$= u''(C)\,dC + \text{second-order terms}.$$

Divide by $J_w = u'$ and convert to percent changes to obtain

$$\frac{J_{ww}(W,X)W}{J_w(W,X)}\left(\frac{\mathrm{d}W}{W}\right) - \frac{J_{wx}(W,X)X}{J_w(W,X)}\left(\frac{\mathrm{d}X}{X}\right)$$
$$+ \text{something } \mathrm{d}t = -\frac{u''(C)C}{u'(C)}\left(\frac{\mathrm{d}C}{C}\right) + \text{something } \mathrm{d}t.$$

It follows that the covariance of any stochastic process—for example, an asset return—with

$$-\frac{J_{ww}(W,X)W}{J_w(W,X)}\left(\frac{\mathrm{d}W}{W}\right) - \frac{J_{wx}(W,X)X}{J_w(W,X)}\left(\frac{\mathrm{d}X}{X}\right)$$

equals the covariance of that stochastic process with

$$-\frac{u''(C)C}{u'(C)}\left(\frac{\mathrm{d}C}{C}\right).$$

Thus, if one covariance explains asset risk premia, then the other must also. This is a consequence of the optimal response of consumption to changes in wealth and state variables in a model with time-additive utility. That response is captured in the envelope condition.

ICAPM and Conditional CAPM

The ICAPM simplifies to the conditional CAPM when $J_{hwx_j} = 0$ for each investor and each state variable j. This is true if there is no labor income and if either all investors have log utility or the capital market line is constant. If an investor has log utility, then, as discussed in Section 14.5, the investor's value function may depend on the state variables, but the marginal value of wealth does not; hence, $J_{hwx_j} = 0$ for each j. If the capital market line is constant, then the value functions do not depend on the state variables. The continuous-time conditional CAPM is discussed further in Exercise 14.1.

There is evidence that the maximum Sharpe ratio changes over time. Obviously, it is also doubtful that all investors have log utility. Therefore, the ICAPM seems a more plausible asset pricing relation than does the conditional CAPM.

14.7 NOTES AND REFERENCES

The static approach to portfolio choice described in Section 14.3 is due to Karatzas, Lehoczky, and Shreve (1987) and Cox and Huang (1989). A complete treatment is presented by Karatzas and Shreve (1998). Note that, at the solution to the static problem, the static budget constraint will hold as an equality, given strictly monotone utility. This implies that (14.3) is a martingale for the optimal consumption and wealth process (because a sufficient condition for a supermartingale X on a time horizon $[0, T]$ to be a martingale is that $\mathsf{E}[X_T] = X_0$). The fact that (14.3) is a supermartingale in general but a martingale at the optimum can be interpreted as follows: The nonnegative wealth constraint permits suicide strategies (Section 13.8), but the optimum can never involve a suicide strategy.

The static approach can also be used in incomplete markets or in markets with frictions, as discussed in Section 9.7, by finding the "least favorable fictitious completion." In the setting of continuous time, this method is due to He and Pearson (1991b), Karatzas, Lehoczky, Shreve, and Xu (1991), and Cvitanic and Karatzas (1992). Again, see Karatzas and Shreve (1998) for a full discussion.

The dynamic programming approach to portfolio choice in continuous time is due to Merton (1969). The result of Section 14.3 on portfolio choice with a constant capital market line appears in Nielsen and Vassalou (2006). It can also be derived from a more general result due to Chamberlain (1988). Chamberlain shows that if there is an SDF process M such that (i) all consumption processes adapted to M are marketed, and (ii) M is adapted to a vector of ℓ independent Brownian motions, then the optimal wealth process for each investor is spanned by the ℓ Brownian motions. The proof in Section 14.3 follows Chamberlain (1988) in showing that the optimal wealth process is spanned by the Brownian motion X because the SDF process M is adapted to X as a result of (14.12). Chamberlain's result implies $(\ell + 1)$–fund separation. Schachermayer, Sirbu, and Taflin (2009) prove separation theorems under conditions related to but distinct from those of Chamberlain (1988).

As discussed in Section 9.7, if zero-coupon bonds of all maturities are traded (in particular, if the market is complete), then shifted CRRA utility functions can be easily reduced to CRRA utility functions. Consider, for example, maximizing $\mathsf{E}[\log(W_T - \zeta)]$ for a constant $\zeta > 0$. The constant ζ can be interpreted as a subsistence level. To ensure subsistence, the investor should buy ζ units of the zero-coupon bond maturing at T. If the zero-coupon bond is not explicitly traded, the investor should purchase a self-financing portfolio having value ζ at date T. The cost at date t of ensuring subsistence at date T is $S_t = \zeta \mathsf{E}_t[M_T/M_t]$. Having allocated S_t to ensure subsistence, the investor should allocate the remaining wealth $W_t - S_t$ as a log-utility investor would allocate it. This means that the vector of investments (in consumption good units) in the risky assets should be $(W - S)\Sigma^{-1}(\mu - r\iota)$ plus the vector that replicates ζ units of the zero-coupon bond. If $\zeta < 0$ (and any $W_T \geq \zeta$ is allowed), then the investor should short ζ units of the zero-coupon bond, and the optimal vector of investments in the risky assets is $(W + S)\Sigma^{-1}(\mu - r\iota)$ minus the vector that replicates ζ units of the zero-coupon bond. It is unreasonable to allow negative terminal wealth, so allowing any $W_T \geq \zeta < 0$ is unreasonable. Back, Liu, and Teguia (2015) solve this problem with consumption at each date and a nonnegative consumption constraint.

Liu (2007) solves the HJB equation in quadratic models. In general, the HJB equation must be solved numerically. This is difficult with more than just a few state variables, because the time required increases exponentially with the number of state variables. This is called the curse of dimensionality. Brandt, Goyal, Santa-Clara, and Stroud (2005) analyze numerical methods for solving portfolio choice problems.

Conditions sufficient to imply that the Euler equation is a necessary condition for optimization can be found in Back (1991). That particular set of sufficient conditions is quite mild in complete markets but less so in incomplete markets. References for the ICAPM and the CCAPM are given in Section 10.7.

EXERCISES

14.1. Assume the continuous-time CAPM holds:
$$(\mu_i - r)\,dt = \rho \left(\frac{dS_i}{S_i}\right)\left(\frac{dW_m}{W_m}\right)$$
for each asset i, where W_m denotes the value of the market portfolio, $\rho = \alpha W_m$, and α denotes the aggregate absolute risk aversion. Define $\sigma_i = \sqrt{e_i' \Sigma e_i}$ to be the volatility of asset i, as described in Section 13.1, so we have
$$\frac{dS_i}{S_i} = \mu_i\,dt + \sigma_i\,dZ_i$$
for a Brownian motion Z_i. Likewise, the return on the market portfolio is
$$\frac{dW_m}{W_m} = \mu_m\,dt + \sigma_m\,dZ_m$$
for some μ_m, σ_m, and Brownian motion Z_m. Let ϕ_{im} denote the correlation process of the Brownian motions Z_i and Z_m.

(a) Using the fact that the market return must also satisfy the continuous-time CAPM, show that the continuous-time CAPM can be written as
$$\mu_i - r = \frac{\sigma_i \sigma_m \phi_{im}}{\sigma_m^2}(\mu_m - r).$$

(b) Suppose r, μ_i, μ_m, σ_i, σ_m, and ρ_i are constant over a time interval Δt, so both S_i and W_m are geometric Brownian motions over the time interval.

Define the annualized continuously compounded rates of return over the time interval:

$$r_i = \frac{\Delta \log S_i}{\Delta t} \quad \text{and} \quad r_m = \frac{\Delta \log W_m}{\Delta t}.$$

Let $\bar{r}_i$ and $\bar{r}_m$ denote the expected values of r_i and r_m. Show that the continuous-time CAPM implies

$$\bar{r}_i - r = \frac{\text{cov}(r_i, r_m)}{\text{var}(r_m)} (\bar{r}_m - r) + \frac{1}{2}[\text{cov}(r_i, r_m) - \text{var}(r_i)]\Delta t.$$

14.2. For each investor $h = 1, \ldots, H$, let π_h denote the optimal portfolio presented in (14.24). Using the notation of Section 14.6, set $\tau_h = 1/\alpha_h$ for each investor h. Then, (14.24) implies

$$W_h \pi_h = \tau_h \Sigma^{-1}(\mu - r\iota) - \sum_{j=1}^{\ell} \tau_h \eta_{hj} \Sigma^{-1} \sigma v_j.$$

(a) Deduce that

$$\mu - r\iota = \alpha W \Sigma \pi + \sum_{j=1}^{\ell} \eta_j \sigma v_j, \qquad (14.33)$$

where π denotes the market portfolio

$$\pi = \sum_{h=1}^{H} \frac{W_h}{W} \pi_h.$$

(b) Explain why (14.33) is the same as the ICAPM (14.31).

14.3. Consider an investor with initial wealth $W_0 > 0$ who seeks to maximize $E[\log W_T]$. Assume

$$E\left[\int_0^T |r_t|\, dt\right] < \infty \quad \text{and} \quad E\left[\int_0^T \kappa_t^2\, dt\right] < \infty,$$

where κ denotes the maximum Sharpe ratio. Assume portfolio processes are constrained to satisfy

$$E\left[\int_0^T \pi_t' \Sigma_t \pi_t\, dt\right] < \infty.$$

Recall that this constraint implies

$$E\left[\int_0^T \pi' \sigma\, dB\right] = 0.$$

(a) Using the formula (13.15) for W_t show that the optimal portfolio process is
$$\pi = \Sigma^{-1}(\mu - r\iota).$$
Hint: The objective function obtained by substituting the formula (13.15) for W_t can be maximized in π separately at each date and in each state of the world.

(b) Assume the market is Markovian. Show that the investor's value function is $V(t, w, x) = \log w + f(t, x)$, where
$$f(t,x) = \mathbb{E}\left[\int_t^T \left(r_s + \frac{1}{2}\kappa_s^2\right) ds \,\Big|\, X_t = x\right].$$

14.4. Consider an investor with log utility and an infinite horizon. Assume the capital market line is constant, so we can write $J(w)$ instead of $J(x, w)$ for the value function.

(a) Show that
$$J(w) = \frac{\log w}{\delta} + K$$
solves the HJB equation (14.25), where
$$K = \frac{\log \delta}{\delta} + \frac{r - \delta + \kappa^2/2}{\delta^2}.$$
Show that $c = \delta w$ and $\pi = \Sigma^{-1}(\mu - r\iota)$ achieve the maximum in the HJB equation.

(b) Show that the transversality condition
$$\lim_{T \to \infty} \mathbb{E}\left[e^{-\delta T} J(W_T^*)\right] = 0$$
holds, where W^* denotes the wealth process generated by the consumption and portfolio processes in Part (a).

14.5. Consider an investor with power utility and an infinite horizon. Assume the capital market line is constant, so we can write $J(w)$ instead of $J(x, w)$ for the value function.

(a) Define
$$\xi = \frac{\delta - (1-\rho)r}{\rho} - \frac{(1-\rho)\kappa^2}{2\rho^2}.$$
Assume (14.26) holds, so $\xi > 0$. Show that
$$J(w) = \xi^{-\rho}\left(\frac{1}{1-\rho} w^{1-\rho}\right)$$

solves the HJB equation (14.25). Show that $c = \xi w$ and $\pi = (1/\rho)\Sigma^{-1}(\mu - r\iota)$ achieve the maximum in the HJB equation.

(b) Show that, under the assumption $\xi > 0$, the transversality condition

$$\lim_{T \to \infty} \mathsf{E}\left[e^{-\delta t}J(W_T^*)\right] = 0$$

holds, where W^* denotes the wealth process generated by the consumption and portfolio processes in Part (a).

14.6. Consider an investor with power utility and a finite horizon. Assume the capital market line is constant and the investor is constrained to always have nonnegative wealth. Let $M = M_p$. Calculate the optimal portfolio as follows:

(a) Using (14.12), show that, for $s > t$,

$$\mathsf{E}_t\left[M_s^{1-1/\rho}\right] = M_t^{1-1/\rho} e^{\alpha(s-t)}$$

for a constant α.

(b) Define C_t and W_T from the first-order conditions (14.7) and set

$$W_t = \mathsf{E}_t\left[\int_t^T \frac{M_s}{M_t} C_s \, ds + \frac{M_T}{M_t} W_T\right].$$

Show that

$$W_t = g(t) M_t^{-1/\rho}$$

for some deterministic function g (which you could calculate).

(c) By applying Itô's formula to W in Part (b), show that the optimal portfolio is

$$\frac{1}{\rho}\Sigma^{-1}(\mu - r\iota).$$

14.7. This exercise demonstrates the equivalence between the intertemporal and static budget constraints in the presence of labor income when the investor can borrow against the income, as asserted in Section 14.3. Let M be an SDF process and Y a labor income process. Assume

$$\mathsf{E}\left[\int_0^T M_t |Y_t| \, dt\right] < \infty$$

for each finite T. The intertemporal budget constraint is

$$dW = rW \, dt + \phi'(\mu - r\iota) \, dt + Y \, dt - C \, dt + \phi'\sigma \, dB. \tag{14.34}$$

(a) Suppose that (C, W, ϕ) satisfies the intertemporal budget constraint (14.34), $C \geq 0$, and the nonnegativity constraint (14.8) holds.

(i) Suppose the horizon is finite. Show that (C, W) satisfies the static budget constraint

$$W_0 + \mathsf{E}\left[\int_0^T M_t Y_t \, dt\right] \geq \mathsf{E}\left[\int_0^T M_t C_t \, dt + M_T W_T\right] \quad (14.35)$$

by showing that

$$\int_0^t M_s(C_s - Y_s) \, ds + M_t W_t$$

is a supermartingale.

Hint: Show that it is a local martingale and at least as large as the martingale $-X_t$, where

$$X_t = \mathsf{E}_t\left[\int_0^T M_s Y_s \, ds\right].$$

This implies the supermartingale property (Appendix A.13.)

(ii) Suppose the horizon is infinite and $\lim_{T \to \infty} \mathsf{E}[M_T W_T] \geq 0$. Assume $Y \geq 0$. Show that the static budget constraint

$$W_0 + \mathsf{E}\left[\int_0^\infty M_t Y_t \, dt\right] \geq \mathsf{E}\left[\int_0^\infty M_t C_t \, dt\right]$$

holds.

(b) Suppose the horizon is finite, markets are complete, $C \geq 0$, and (C, W) satisfies the static budget constraint (14.35) as an equality. Show that there exists ϕ such that (C, W, ϕ) satisfies the intertemporal budget constraint (14.34).

15

Continuous-Time Topics

CONTENTS

15.1	Fundamental Partial Differential Equation	367
15.2	Fundamental PDE and Optimal Portfolio	369
15.3	Risk-Neutral Probabilities	370
15.4	Jump Risks	374
15.5	Internal Habits	380
15.6	Verification Theorem	387
15.7	Notes and References	390

This chapter addresses some miscellaneous topics in continuous-time models. The concepts discussed in Sections 15.1 and 15.3 are used extensively in the next part of the book. Section 15.4 introduces Poisson processes, which appear again in Chapters 23 and 24, and explains in general how jump risks are priced. Section 15.5 discusses internal habits, which are an important generalization of time-additive utility and provide a possible explanation of the equity premium and risk-free rate puzzles. Section 15.6 explains what assumptions are required to guarantee that the solution of an HJB equation is actually optimal in a continuous-time model.

15.1 FUNDAMENTAL PARTIAL DIFFERENTIAL EQUATION

This section describes a method for calculating the value (13.36) of a claim to a payoff at some future date T and the value (13.37) of a claim to a consumption or dividend stream. In some cases, the values can be calculated directly (for example, when they involve exponentials of normally distributed variables). In other cases, it is easier to express the value as the solution of a PDE and solve the PDE. The PDE is called the fundamental PDE. The fundamental PDE plays a very important role in Part III of this book (derivative securities).

Assume the securities market is Markovian with state process X as described in Section 13.6. Consider a payoff at some date T that depends on X_T. Denote the payoff by $g(X_T)$. Let M be an SDF process with price of risk process $\lambda(X_t)$. We want to compute the value (13.36) for $v = g(X_T)$. If $g(X_T)$ is a marketed payoff, then (13.36) is the unique price at which $g(X_T)$ can be obtained, provided MW is a martingale for all self-financing wealth processes W, as discussed in Section 13.4. The value (13.36) is some function of t and X_t. Denote it by $f(t, X_t)$. The formula (13.36) implies that $M_t f(t, X_t)$ is a martingale. Hence, it has no drift. We obtain the fundamental PDE by applying Itô's formula to Mf and equating the drift to zero.[1]

Assume the dynamics of X are given by (13.50), and let v_i denote the ith row of v arranged as a column vector. Itô's formula implies

$$d(Mf) = M\,df + f\,dM + (dM)(df).$$

Substituting from (13.50), the drift of this is

$$M\left[f_t + \sum_{i=1}^{\ell} f_{x_i}\phi_i + \frac{1}{2}\sum_{i=1}^{\ell}\sum_{j=1}^{\ell} f_{x_i x_j} v_i' v_j\right] - rMf\,dt - M\sum_{i=1}^{\ell} f_{x_i} v_i'\lambda.$$

Equate this to zero to obtain

$$f_t + \sum_{i=1}^{\ell} f_{x_i}(\phi_i - v_i'\lambda) + \frac{1}{2}\sum_{i=1}^{\ell}\sum_{j=1}^{\ell} f_{x_i x_j} v_i' v_j = rf. \tag{15.1}$$

This is the fundamental PDE. The PDE is the same for all terminal payoffs $g(X_T)$. The terminal payoff appears only in the boundary condition for the PDE, which is that $f(T, x) = g(x)$ for all values of x. Some examples of the fundamental PDE are given in the exercises.

Now consider a consumption stream C that depends on X, meaning $C_u = g(u, X_u)$ for some function g and all dates u. We want to compute the value (13.37). Denote the value at date t by $f(t, X_t)$. Then,

$$\int_0^t M_u g(u, X_u)\,du + M_t f(t, X_t)$$

is a martingale. Its drift is $Mg\,dt$ plus the drift of Mf. Use the calculation in the previous paragraph and equate the drift to 0 to obtain

$$g + f_t + \sum_{i=1}^{\ell} f_{x_i}(\phi_i - v_i'\lambda) + \frac{1}{2}\sum_{i=1}^{\ell}\sum_{j=1}^{\ell} f_{x_i x_j} v_i' v_j = rf. \tag{15.2}$$

---

[1]. The issue of local martingales versus martingale discussed in Section 13.4 is not important here. The drift would be zero even if Mf were only a local martingale. But, in fact, the definition of f as the value (13.36) implies directly that Mf is actually a martingale.

Continuous-Time Topics

This PDE should be solved subject to the boundary condition $f(T, x) = 0$ for all values of x. If $T = \infty$ and g does not depend on time, then the function f does not depend on time. It satisfies the PDE (15.2) with $f_t = 0$.

There are ways to derive the fundamental PDEs (15.1) and (15.2) that are equivalent to the calculation given here but which may be more intuitive. Consider (15.2), for example. It is equivalent to

$$\frac{g\,dt + \mathsf{E}[df]}{f} = r\,dt - \left(\frac{dM}{M}\right)\left(\frac{df}{f}\right).$$

The left-hand side is the expected rate of return. The right-hand side is the "required rate of return," meaning the risk-free rate plus the risk premium computed by the formula (13.19). Equating the expected rate of return to the required rate of return produces the fundamental PDE (Exercise 15.10). Yet another way to derive the fundamental PDEs (15.1) and (15.2) is to equate the expected rate of return under a risk-neutral probability to the risk-free rate (Section 15.3).

15.2 FUNDAMENTAL PDE AND OPTIMAL PORTFOLIO

We can extend the fundamental PDE to characterize the optimal portfolio of an investor in a complete market. This PDE is typically easier to solve than the PDE derived from the HJB equation, because of its linearity. Assume the market is Markovian with state process X as described in Section 13.6. Assume the market is complete. Then, there is a unique SDF process M and the price of risk process is $\lambda(X_t)$, where

$$\lambda(x) = \sigma(x)^{-1}[\mu(x) - r(x)\iota].$$

Assume the investor has a finite horizon (the infinite-horizon case is the same except for dealing with Ponzi schemes, etc.) The first-order conditions (14.6a) define optimal consumption and optimal terminal wealth as $C_t = g(t, M_t)$ and $W_T = h(M_T)$, where the functions g and h are the inverse marginal utilities. At the optimum, the investor's wealth process is given by (14.5), which we repeat here:

$$W_t = \mathsf{E}_t\left[\int_t^T \frac{M_u}{M_t} g(u, M_u)\,du + \frac{M_T}{M_t} h(M_T)\right].$$

The distribution of M_u for $u > t$, conditional on date–t information, depends on M_t and X_t. Therefore, $W_t = f(t, M_t, X_t)$ for some function f.

The previous section analyzes values that depend on X. Here, we have a value that depends on M and X. However, the augmented vector $(M\ X')'$ is also Markovian. So, f satisfies an extension of the PDE (15.1). Let m denote an arbitrary value of M_t. Following the same reasoning used to derive (15.1), we can see that the PDE satisfied by the function $f(t, m, x)$ is

$$g + f_t + \sum_{i=1}^{\ell} f_{x_i}(\phi_i - v_i'\lambda) - mf_m(r - \lambda'\lambda)$$

$$+ \frac{1}{2}\sum_{i=1}^{\ell}\sum_{j=1}^{\ell} f_{x_i x_j} v_i' v_j + \frac{1}{2} m^2 f_{mm} \lambda'\lambda - m\sum_{i=1}^{\ell} f_{mx_i} v_i'\lambda = rf. \quad (15.3)$$

A boundary condition that f must satisfy is $f(T, m, x) = h(m)$ for all x. For CRRA utility, the PDE (15.3) has a solution of the form $m^{-1/\rho}\hat{f}(t, x)$, where $\hat{f}$ satisfies a certain PDE (Exercise 15.3).

As explained in Section 14.3, the optimal portfolio is π such that $\pi'\sigma\,dB$ equals the stochastic part of df/f. The stochastic part of df is

$$\left(\sum_{i=1}^{\ell} f_{x_i} v_i - Mf_m \lambda\right)' dB.$$

Therefore, the optimal portfolio is $\pi(t, M_t, X_t)$, where

$$\pi(t, m, x)' = \left(\sum_{i=1}^{\ell} \frac{\partial \log f(t, m, x)}{\partial x_i} v_i(x) - \frac{\partial \log f(t, m, x)}{\partial m} m\lambda\right)' \sigma(x)^{-1}. \quad (15.4)$$

15.3 RISK-NEUTRAL PROBABILITIES

As in a discrete-time model (Section 10.6), a risk-neutral probability for the time horizon $[0, T]$ is a probability $\mathbb{Q}$ having the same null sets as the physical probability $\mathbb{P}$ and with the property that asset prices are expected discounted values, discounting at the instantaneous risk-free rate.

DEFINITION OF A RISK-NEUTRAL PROBABILITY

As before, let R denote the price of the money market account, so $R_0 = 1$ and $dR_t/R_t = r_t\,dt$. Let M be an SDF process, and assume MR is a martingale (a $\mathbb{P}$-martingale) instead of just a local martingale, so $\mathsf{E}[M_T R_T] = R_0 = 1$. Consider $T < \infty$. Define

$$\mathbb{Q}(A) = \mathsf{E}[M_T R_T 1_A] \quad (15.5)$$

for each event A that is distinguishable at date T, where 1_A denotes the random variable equal to 1 when the state of the world is in A and 0 otherwise. The construction (15.5) defines $M_T R_T$ as the Radon-Nikodym derivative $d\mathbb{Q}/d\mathbb{P}$ (see footnote 4 in Chapter 8). It follows from (15.5) that

$$E^*[X] = E[M_T R_T X] \tag{15.6}$$

for any random variable X depending on date–T information, where E^* denotes expectation with respect to the probability $\mathbb{Q}$. Different SDF processes M (for which MR is a martingale) define different risk-neutral probabilities.

Risk-Neutral Valuation

Let W be a self-financing wealth process such that MW is a $\mathbb{P}$-martingale. Because MR is a $\mathbb{P}$-martingale, $M_t R_t$ is the $\mathbb{P}$-conditional expectation at date t of the Radon-Nikodym derivative $M_T R_T$. Therefore, from a general result stated in Appendix A.12, MW is a $\mathbb{P}$-martingale if and only if $MW/(MR) = W/R$ is a $\mathbb{Q}$-martingale. Thus,

$$W_t = R_t E_t^* \left[\frac{W_T}{R_T} \right] = E_t^* \left[\exp\left(-\int_t^T r_u \, du \right) W_T \right]. \tag{15.7}$$

Therefore, as promised, asset values are expected discounted values, taking expectations with respect to a risk-neutral probability and discounting at the instantaneous risk-free rate. The formula (15.7) is an alternative to (13.36) for valuing a payoff $W_T = v$.

In an infinite-horizon model, for each SDF process M, there is a unique probability $\mathbb{Q}$ such that (15.5) holds for each event A that is observable at any finite date T. This probability has the risk-neutral pricing property (15.7) for each finite T. See the discussion in Section 8.7.

Girsanov's Theorem

An important result is Girsanov's theorem, which states that if

(a) B is a vector of independent $\mathbb{P}$-Brownian motions,
(b) ξ is a strictly positive $\mathbb{P}$-martingale,
(c) $\mathbb{Q}$ is a probability on the events observable at date T with $d\mathbb{Q}/d\mathbb{P} = \xi_T$, and
(d) $d\xi/\xi = -\lambda' dB$ for a vector stochastic process $\lambda = (\lambda_1 \cdots \lambda_n)'$,

then

$$B_t^* = B_t + \int_0^t \lambda_s \, ds \tag{15.8}$$

defines a vector of independent $\mathbb{Q}$–Brownian motions on $[0, T]$.[2] An exceedingly important consequence of Girsanov's theorem is that *changing probabilities changes drifts but does not affect volatilities or correlations*. To see this, let Z be an Itô process with

$$dZ = \mu \, dt + \sum_{i=1}^{n} \sigma_i \, dB_i.$$

Use (15.8) to substitute dB_i^* for dB_i. We obtain

$$dZ = \left(\mu - \sum_{i=1}^{n} \sigma_i \lambda_i \right) dt + \sum_{i=1}^{n} \sigma_i \, dB_i^*.$$

Thus, the drift of Z under $\mathbb{Q}$ differs from the drift of Z under $\mathbb{P}$ by the term $-\sum \sigma_i \lambda_i$. However, the coefficients σ_i are unchanged.

In the particular case of a risk-neutral probability $\mathbb{Q}$, the process ξ equals MR, so

$$\frac{d\xi}{\xi} = r \, dt + \frac{dM}{M} = -\lambda' \, dB,$$

where λ is the vector of prices of risk in the characterization (13.17) of an SDF process M. Thus, λ in the transformation (15.8) from Brownian motions under the physical probability to Brownian motions under a risk-neutral probability is the vector of prices of risk defining the SDF process.

Girsanov's theorem provides another condition sufficient to ensure that MW is a martingale for a self-financing wealth process W. Specifically, MW is a martingale on $[0, T]$ if

$$\mathsf{E} \int_0^T \frac{M}{R} \phi' \Sigma \phi \, dt < \infty, \tag{15.9}$$

where $\phi_i = W \pi_i$ is the amount of the consumption good invested in risky asset i (Exercise 15.8). We can also write (15.9) as

$$\mathsf{E}^* \int_0^T \eta' \Sigma \eta \, dt < \infty, \tag{15.9'}$$

where $\eta_i = \phi_i / R$ is the amount invested in asset i measured in units of the money market account.

---

2. Notice that $d\xi / \xi = -\lambda' dB$ implies $(dB_i)(d\xi / \xi) = -\lambda_i \, dt$ for each i. Thus, we can write the definition (15.8) of B^* as, for each i, $B_{i0}^* = 0$ and $dB_i^* = dB_i - (dB_i)(d\xi / \xi)$. This version of the formula extends as follows: if Z is any $\mathbb{P}$–Brownian motion, then $dZ^* = dZ - (dZ)(d\xi / \xi)$ is a $\mathbb{Q}$–Brownian motion.

Expected Returns under a Risk-Neutral Probability

Under a risk-neutral probability, the instantaneous expected rate of return on every portfolio is the risk-free rate. In other words, for every self-financing wealth process $W > 0$ and each risk-neutral probability,

$$\frac{dW}{W} = r\,dt + \pi'\sigma\,dB^*, \tag{15.10}$$

where B^* is a vector of independent Brownian motions under the risk-neutral probability. To deduce this, substitute (15.8) into the intertemporal budget constraint as

$$\frac{dW}{W} = r\,dt + \pi'(\mu - r\iota)\,dt + \pi'\sigma\,dB$$
$$= r\,dt + \pi'(\mu - r\iota)\,dt + \pi'\sigma\,(dB^* - \lambda\,dt)$$

and use the fact (13.20) that $\sigma\lambda = \mu - r\iota$.

Fundamental PDE Again

Consider the Markovian model of Section 13.6. Let M be an SDF process with price of risk process $\lambda(X_t)$. Assume MR is a martingale, so we can define a risk-neutral probability $\mathbb{Q}$ using M. Let B^* be the vector of $\mathbb{Q}$-Brownian motions (15.8). Then, the dynamics (13.50) of the Markov process X can be written as

$$dX_t = [\phi(X_t) - \nu(X_t)\lambda(X_t)]\,dt + \nu(X_t)\,dB^*_t. \tag{15.11}$$

Denote the $\mathbb{Q}$-drift of X by ϕ^*; that is, define $\phi^*(x) = \phi(x) - \nu(x)\lambda(x)$. Then, we can write the fundamental PDEs (15.1) and (15.3) as

$$f_t + \sum_{i=1}^{\ell} f_{x_i}\phi_i^* + \frac{1}{2}\sum_{i=1}^{\ell}\sum_{j=1}^{\ell} f_{x_ix_j}v_i'v_j = rf, \tag{15.12}$$

$$g + f_t + \sum_{i=1}^{\ell} f_{x_i}\phi_i^* + \frac{1}{2}\sum_{i=1}^{\ell}\sum_{j=1}^{\ell} f_{x_ix_j}v_i'v_j = rf. \tag{15.13}$$

These have simple interpretations. The left-hand side of (15.12) is the expected change in f under the risk-neutral probability. Equation (15.12) states that the expected rate of return of the asset with value f is the risk-free rate, when we take expectations under the risk-neutral probability. This is a statement that has to be true, given that all self-financing wealth processes have expected rates of return equal to the risk-free rate under any risk-neutral probability. Equation (15.13) also says that an expected rate of return is the risk-free rate. In this case, the return equals the dividend yield g/f plus the capital gain df/f.

15.4 JUMP RISKS

This section describes pricing via SDF processes when dividend-reinvested asset prices have jumps. The asset pricing results of this section are not used elsewhere in the book; however, Poisson processes do appear again in Chapters 23 and 24. Adopt the usual conditions regarding the information in the economy (Appendix A.14). We can take each local martingale X to have paths that are continuous from the right with limits from the left, meaning that, with probability 1, $X_t = \lim_{u \downarrow t} X_u$ for all t, and the limit

$$X_{t-} \stackrel{\text{def}}{=} \lim_{s \uparrow t} X_s$$

exists for all t. Adopt the notation $\Delta X_t = X_t - X_{t-}$. This is the jump of X at t. Every local martingale X can be written uniquely as $X = X^c + X^d$ where X^c is a continuous local martingale and X^d is a purely discontinuous local martingale (Dellacherie and Meyer, 1982, VIII.43). A purely discontinuous local martingale is also called a compensated sum of jumps. This terminology is explained below.[3]

ASSET PRICES AND SDF PROCESSES

Assume there is a money market account with price

$$\frac{dR_t}{R_t} = r_t \, dt$$

for a stochastic process r. Assume there is an asset with dividend-reinvested price S satisfying

$$\frac{dS_t}{S_{t-}} = \mu_t \, dt + dZ_t = \mu_t \, dt + dZ^c_t + dZ^d_t, \qquad (15.14)$$

for a stochastic process μ and a local martingale $Z = Z^c + Z^d$. There can be multiple risky assets as before. The formula (15.16) below applies to each of them.

As before, define an SDF process to be a stochastic process M such that $M_0 = 1$, $M_t > 0$ for all t with probability 1, assume MR is a local martingale,

---

3. The definition of a purely discontinuous local martingale is as follows: A local martingale X is purely discontinuous if its quadratic variation process is purely discontinuous, meaning that the quadratic variation through any date t is $\sum_{s \leq t} (\Delta X_s)^2$ (Dellacherie and Meyer, 1982, VIII.42, VI.52, VII.42).

Continuous-Time Topics

and assume MS is a local martingale (for each dividend-reinvested asset price S). By reasoning that is essentially the same as in the Brownian model, we can verify that the drift of an SDF process must be $-r$. This means that

$$\frac{dM_t}{M_{t-}} = -r_t\,dt + dY_t = -r_t\,dt + dY_t^c + dY_t^d \qquad (15.15)$$

for a local martingale $Y = Y^c + Y^d$.

Asset Risk Premia

The crucial formula (13.19) for risk premia in terms of covariances with an SDF process generalizes to models with jump risks as

$$(\mu - r)\,dt = -d\langle Y^c, Z^c \rangle - d\langle Y^d, Z^d \rangle. \qquad (15.16)$$

The sharp bracket $\langle Y^c, Z^c \rangle$ of the two continuous local martingales is the covariation of the local martingales (Section 12.6). In the Brownian model, all local martingales are continuous (the martingale representation theorem implies that they are stochastic integrals with respect to the Brownian motions), so in that model $Y^d = Z^d = 0$ and $d\langle Y^c, Z^c \rangle$ is what we have been writing as

$$\left(\frac{dM}{M}\right)\left(\frac{dS}{S}\right).$$

Thus, the result (15.16) generalizes (13.19). The definition of the sharp bracket process for discontinuous local martingales and some examples (Poisson and compound Poisson processes) are given below.

An important fact is that $\langle Y^d, Z^d \rangle = 0$ if Y and Z have no jump times in common. If there are no jumps that occur at the same time—that is, if there is an SDF process that never jumps when the asset price jumps—then the formula (15.16) implies that the risk premium for the jump risk is zero.

Poisson and Compensated Poisson Processes

A simple and common example of a jump process is a Poisson process. The paths of a Poisson process are step functions. Each path starts at 0, is constant between steps, and increases by 1 at each step. It is an example of a counting process. The value N_t of a Poisson process at date t is the number of steps (jumps) that have occurred by time t (including time t). We call each jump an arrival. So, the Poisson process counts arrivals. The feature that distinguishes a Poisson process N among other counting processes is that it has stationary and independent increments, meaning that each increment $N_u - N_t$ for $u > t$ is independent of date t–information, and two increments $N_u - N_t$ and $N_v - N_s$ have the same distribution if $u - t = v - s$. The feature of stationary and independent increments implies that there is some $\lambda > 0$ such that each increment $N_u - N_t$ has the Poisson distribution with parameter $(u - t)\lambda$. Thus, the mean and variance of each increment $N_u - N_t$ are each equal to $(u - t)\lambda$. Furthermore, the times between jumps (called interarrival times) are exponentially distributed with parameter λ. With probability 1, there are never two arrivals at the same time, so all jumps are of size 1. We say that the probability of a jump during an infinitesimal interval dt is $\lambda\,dt$.

A Poisson process is not a martingale, because it only goes up. However, if N is a Poisson process with parameter λ, then $N_t - \lambda t$ is a martingale. Intuitively, this is because $E[dN_t] = \lambda\,dt$. The process $N_t - \lambda t$ is called a compensated Poisson process. It is the simplest example of a purely discontinuous (local) martingale. As remarked before, purely discontinuous local martingales are also called compensated sums of jumps.

Compound Poisson Processes

Let N be a Poisson process with parameter λ, and let $y_1, y_2, \ldots$ be a sequence of IID finite-variance random variables that are independent of N. Define

$$Y_t = \sum_{i=1}^{N_t} y_i.$$

The paths of Y are constant between arrivals of N. At the ith arrival of N, Y jumps by y_i (jumps up if $y_i > 0$ and down if $y_i < 0$). The process Y is called a compound Poisson process. The process

$$Y_t - E[y_i]\lambda t$$

is a purely discontinuous local martingale (compensated sum of jumps).

Let $z_1, z_2, \ldots$ be another sequence of IID finite-variance random variables that are independent of N. In fact, assume $(y_1, z_1), (y_2, z_2), \ldots$ is an IID sequence of random vectors. Define

$$Z_t = \sum_{i=1}^{N_t} z_i.$$

Then, $\langle Y, Z \rangle_t = E[y_i z_i]\lambda t$ and

$$d\langle Y, Z \rangle = E[y_i z_i]\lambda\,dt.$$

This is a simple example of the sharp bracket process of purely discontinuous local martingales Y and Z. If instead

$$Z_t = \sum_{i=1}^{\hat{N}_t} z_i,$$

where $\hat{N}$ is a Poisson process that is independent of N, then $\langle Y, Z \rangle = 0$. This is because N and $\hat{N}$ have no jump times in common (with probability 1). These two examples can be viewed as examples of the jump risk premium $\mathrm{d}\langle Y^d, Z^d \rangle$ in (15.16). They illustrate the general fact that jump risks are priced based on contemporaneous jumps of the asset price and SDF process.

Square Brackets and Sharp Brackets

The conventional notation for quadratic variation and covariation is a square bracket. We denote the quadratic variation of a local martingale X over $[0, t]$ by $[X, X]_t$ and the covariation of two local martingales X and Y over $[0, t]$ by $[X, Y]_t$. The square bracket process is what shows up in Itô's formula as the second-order term. For example, if Y and Z are local martingales, then (Dellacherie and Meyer, 1982, VIII.18)

$$\mathrm{d}(YZ)_t = Y_{t-}\,\mathrm{d}Z_t + Z_{t-}\,\mathrm{d}Y_t + \mathrm{d}[Y, Z]_t. \qquad (15.17)$$

Thus, $\mathrm{d}[Y, Z]$ is what we denoted as $(\mathrm{d}Y)(\mathrm{d}Z)$ previously.

The sharp bracket process that appears in (15.16) is sometimes called the conditional quadratic variation or the conditional covariation. For continuous local martingales, the sharp and square brackets are the same. In general, the sharp bracket process of a local martingale X is defined to be the predictable finite-variation process[4] $\langle X, X \rangle$ such that $X^2 - \langle X, X \rangle$ is a local martingale, if such a process exists.[5] Likewise, the sharp bracket process of two local martingales X and Y is the predictable finite-variation process $\langle X, Y \rangle$ such that $XY - \langle X, Y \rangle$ is a local martingale, if such a process exists.

The Poisson process provides a good example of the difference between the sharp and square bracket processes for discontinuous local martingales. The

---

4. Finite variation means that the paths have finite variation on each bounded interval, with probability 1. Stochastic processes with left-continuous paths (in particular, continuous processes) are predictable. More generally, a stochastic process is said to be predictable if it is measurable with respect to the σ-field on $\Omega \times [0, \infty)$ generated by the left-continuous adapted processes.

5. The sharp bracket process exists under mild regularity conditions. For example, it exists if the jumps of X are bounded. More generally, it exists if the process $\sup_{s \leq t}(\Delta X_s)^2$ is "locally integrable" (Dellacherie and Meyer, 1982, VII.25).

quadratic variation of a Poisson process N over $[0,t]$ is the sum of squared jumps during $[0,t]$. Because each jump size is 1, the squared jumps equal the jumps, so the quadratic variation is N_t. Thus, $[N,N]_t = N_t$. On the other hand, the sharp bracket process is the compensator $\langle N,N \rangle_t = \lambda t$. Recall that $\lambda t = \mathsf{E}[N_t]$. In general, we can interpret $d\langle X,Y\rangle_t$ as the conditional expectation of $d[X,Y]_t$ given information prior to t.[6]

There is a characterization of the square bracket process that parallels the definition of the sharp bracket process. For two local martingales X and Y, the square bracket $[X,Y]$ is the unique finite-variation process such that $XY - [X,Y]$ is a local martingale and $\Delta[X,Y] = \Delta X \Delta Y$ (Dellacherie and Meyer, 1982, VII.42). This characterization makes it easy to see why the square bracket equals the sharp bracket for continuous local martingales. For continuous local martingales, $\Delta[X,Y] = \Delta X \Delta Y$ implies that $[X,Y]$ is continuous, hence predictable. Thus, for continuous local martingales, $[X,Y]$ is a predictable finite-variation process such that $XY - [X,Y]$ is a local martingale. This by definition is the sharp bracket process.

We want to verify (15.15) and (15.16). For (15.15), write

$$M = (MR)\frac{1}{R}$$

---

6. More formally, let X be a square-integrable martingale, fix t, and consider a sequence of partitions

$$0 < \frac{t}{2^n} < \frac{2t}{2^n} < \frac{3t}{2^n} < \frac{(2^n-1)t}{2^n} < t.$$

The square bracket is the quadratic variation, so

$$[X,X]_t = \lim_{n\to\infty} \sum_{i=0}^{2^n-1} [X_{2^{-n}(i+1)t} - X_{2^{-n}it}]^2.$$

On the other hand,

$$\langle X,X \rangle_t = \lim_{n\to\infty} \sum_{i=0}^{2^n-1} \mathsf{E}\left[[X_{2^{-n}(i+1)t} - X_{2^{-n}it}]^2 \mid \mathcal{F}_{2^{-n}it-}\right],$$

where $\mathcal{F}_{s-}$ denotes the smallest σ-field that contains all of the information known prior to s (that is, that contains all of the σ-fields $\mathcal{F}_u$ for $u < s$). See Dellacherie and Meyer (1982, VII.43).

Continuous-Time Topics

and integrate by parts to obtain[7]

$$dM_t = (M_{t-}-R_t)\, d\frac{1}{R_t} + \frac{1}{R_t}\, d(MR)_t$$

$$= -M_{t-}r_t\, dt + \frac{M_{t-}}{M_{t-}R_{t-}}\, d(MR)_t. \quad (15.18)$$

By assumption, MR is a local martingale, so

$$Y_t \stackrel{\text{def}}{=} \int_0^t \frac{1}{M_{s-}R_{s-}}\, d(MR)_s$$

is a local martingale. Thus, (15.18) implies (15.15).

To derive (15.16), use (15.14) and (15.15) and integrate by parts to obtain

$$M_t S_t = S_0 + \int_0^t M_{u-}S_{u-}\left\{(\mu_u - r_u)\, du + dY_u + dZ_u + d[Y,Z]_u\right\}.$$

Thus,

$$M_t S_t - S_0 - \int_0^t M_{u-}S_{u-}\left\{dY_u + dZ_u\right\} = \int_0^t M_{u-}S_{u-}\left\{(\mu_u - r_u)\, du + d[Y,Z]_u\right\}.$$

The left-hand side is a local martingale. Hence, the right-hand side is also. This implies that

$$Z_t \stackrel{\text{def}}{=} \int_0^t (\mu_u - r_u)\, du + [Y,Z]_t$$

is a local martingale. Use this definition of Z to write

$$Y_t Z_t - [Y,Z]_t = Y_t Z_t - Z_t + \int_0^t (\mu_u - r_u)\, du.$$

By the characterization stated above of the square bracket process, the left-hand side is a local martingale. Hence, the right-hand side is also. The difference of two local martingales is a local martingale, so

$$Y_t Z_t + \int_0^t (\mu_u - r_u)\, du$$

is a local martingale. The second term is a predictable finite-variation process, so by the definition of the sharp bracket process,

$$\langle Y, Z \rangle_t = -\int_0^t (\mu_u - r_u)\, du.$$

---

7. Integration by parts is the formula (15.17). It is Itô's formula applied to the function $f(x,y) = xy$. In this case, there is no second-order term because R has paths that are continuous and have finite variation on each bounded interval.

The sharp bracket is bilinear and $Y = Y^c + Y^d$ and $Z = Z^c + Z^d$, so

$$\langle Y^c, Z^c \rangle_t + \langle Y^c, Z^d \rangle_t + \langle Y^d, Z^c \rangle_t + \langle Y^d, Z^d \rangle_t = -\int_0^t (\mu_u - r_u)\,du. \quad (15.19)$$

Finally, the sharp bracket of any continuous local martingale with any purely discontinuous local martingale is zero (Dellacherie and Meyer, 1982, VIII.43). Hence, (15.19) implies (15.16).

15.5 INTERNAL HABITS

As remarked in Chapter 11, there is no real theoretical foundation for the assumption that utility is time additive. A very reasonable variation of time-additive utility is to allow utility at each date to depend on past consumption. One interpretation is that an investor becomes habituated to a certain standard of living and evaluates her present consumption in light of that standard of living. This is called an internal habit model. Like external habits (Section 11.1), internal habits break the link between marginal utility and contemporaneous consumption. They can produce high variability in marginal utility without requiring high variability in consumption. Hence, like external habits, they provide a possible explanation of the equity premium puzzle. We discuss this in a continuous-time model because calculations are simpler in continuous time.

The first part of this section analyzes portfolio choice with an internal habit via the static approach. We assume the habit is an exponentially weighted average of past consumption, utility depends on the difference between current consumption and the habit, and the market is complete. The last part of the section calculates the equity premium and risk-free rate when there is a representative investor with an internal habit.

EXPONENTIALLY DECAYING HABITS IN COMPLETE MARKETS

Denote the investor's habit by X. For simplicity, consider a finite-horizon model without a bequest motive. Other models can be analyzed similarly. Assume the investor seeks to maximize

$$\mathsf{E}\int_0^T e^{-\delta t} u(C_t - X_t)\,dt \quad (15.20)$$

for some utility function u. Assume the habit evolves as

$$dX_t = -aX_t\,dt + bC_t\,dt \quad (15.21)$$

for constants $a > 0$ and b. This equation has the explicit solution[8]

$$X_t = e^{-at}X_0 + b\int_0^t e^{-a(t-s)}C_s\,ds. \qquad (15.22)$$

If $b > 0$, then consumption at any two different dates are complementary goods: Increasing C_s increases the habit X_t for $t > s$ and increases the marginal utility of consumption at $t > s$. This is consistent with the "standard of living" interpretation given above. However, we can also allow $b < 0$. In that case, consumption at any two different dates are substitute goods: Consumption at any date is less important (marginal utility is lower) if consumption was high previously. The parameter a determines the persistence of the effect of consumption on future habits; that is, it determines the rate at which the habit decays.

By adding and subtracting $bX_t\,dt$ to the right-hand side of (15.21), we can write it as

$$dX_t = -(a-b)X_t\,dt + b(C_t - X_t)\,dt. \qquad (15.23)$$

This has the solution (footnote 8):

$$X_t = e^{-(a-b)t}X_0 + b\int_0^t e^{-(a-b)(t-s)}(C_s - X_s)\,ds. \qquad (15.24)$$

Assume the market is complete, and let $M = M_p$ denote the unique SDF process. If the utility function is only defined for $C_t \geq X_t$, then we need to assume

$$W_0 \geq X_0 \,\mathrm{E}\int_0^T M_t e^{-(a-b)t}\,dt. \qquad (15.25)$$

This means that the investor can afford to pay for her initial habit; that is, she can afford to consume $C_t = X_t$ for all t, in which case the habit decays at rate $a - b$ as shown in (15.24) and the cost of C is the right-hand side of (15.25).[9]

Because the market is complete, the intertemporal budget constraint is equivalent to the static budget constraint, as discussed in Section 14.3. We will define

---

8. To see that (15.22) satisfies (15.21), write (15.22) as

$$X_t = e^{-at}\left(X_0 + b\int_0^t e^{as}C_s\,ds\right)$$

and calculate

$$\frac{dX}{dt} = -ae^{-at}\left(X_0 + b\int_0^t e^{as}C_s\,ds\right) + e^{-at}be^{at}C_t = -aX_t + bC_t.$$

9. The most natural assumption is $a > b$, in which case the habit does decay when consumption is the minimum possible ($C_t = X_t$). However, the results stated in this section for the finite-horizon model are also true when $b \geq a$.

a stochastic process $\hat{M}$ and a constant $\hat{W}_0$ such that

$$\mathsf{E}\int_0^T M_t C_t \, dt = W_0 \quad \Leftrightarrow \quad \mathsf{E}\int_0^T \hat{M}_t (C_t - X_t) \, dt = \hat{W}_0.$$

Setting $\hat{C}_t = C_t - X_t$, this equivalence enables us to write the optimization problem as

$$\max \quad \mathsf{E}\int_0^T e^{-\delta t} u(\hat{C}_t) \, dt \quad \text{subject to} \quad \mathsf{E}\int_0^T \hat{M}_t \hat{C}_t \, dt = \hat{W}_0. \quad (15.26)$$

Thus, the habit model is equivalent to a standard time-additive model in which the choice variable is $\hat{C}$, the SDF process is $\hat{M}$, and initial wealth is $\hat{W}_0$.

The optimization problem (15.26) can be solved for $\hat{C}$ as in Section 14.3. Given the optimal $\hat{C}$, the optimal C is $C = \hat{C} + X$. Using (15.24), we can write the optimal consumption process C explicitly in terms of $\hat{C}$ as

$$C_t = \hat{C}_t + e^{-(a-b)t} X_0 + b \int_0^t e^{-(a-b)(t-s)} \hat{C}_s \, ds. \quad (15.27)$$

As in a time-additive model, the optimal wealth process is

$$W_t = \mathsf{E}_t \int_t^T \frac{M_u}{M_t} C_u \, du,$$

and the optimal portfolio process is found by equating $\phi' \sigma \, dB$ to the stochastic part of dW (Section 13.5).

The stochastic process $\hat{M}$ is defined in (15.30) below. It has a simple interpretation. For each t, $\hat{M}_t$ is the date–0 price (state price divided by probability) of the following amount of consumption at date t:

$$1 + b \int_t^T e^{-(a-b)(u-t)} P_t(u) \, du,$$

where $P_t(u)$ is the price at t of a zero-coupon bond maturing at u. Whereas M_t is the price of a single unit of consumption, $\hat{M}_t$ is the price of one unit plus an amount of consumption that will pay for bonds to provide $be^{-(a-b)(u-t)}$ units of consumption at each date $u > t$. This amount of consumption at $u > t$ is the increment to the habit at u created by a single unit of consumption at t. Therefore, $\hat{M}_t$ is an "all in" price of consumption—it covers consumption plus the future increment to habit created by the consumption.

Substitute from (15.24) to obtain

$$\mathsf{E}\int_0^T M_t C_t \, dt = \mathsf{E}\int_0^T M_t (C_t - X_t) \, dt + \mathsf{E}\int_0^T M_t X_t \, dt$$

$$= \mathsf{E}\int_0^T M_t (C_t - X_t) \, dt + X_0 \mathsf{E}\int_0^T M_t e^{-(a-b)t} \, dt$$

$$+ b \mathsf{E}\int_0^T M_t \int_0^t e^{-(a-b)(t-s)} (C_s - X_s) \, ds \, dt.$$

Define

$$\hat{W}_0 = W_0 - X_0 \mathsf{E}\int_0^T M_t e^{-(a-b)t} \, dt. \qquad (15.28)$$

Then, $\mathsf{E}\int_0^T M_t C_t \, dt = W_0$ if and only if

$$\mathsf{E}\int_0^T M_t (C_t - X_t) \, dt + b \mathsf{E}\int_0^T M_t \int_0^t e^{-(a-b)(t-s)} (C_s - X_s) \, ds \, dt = \hat{W}_0. \qquad (15.29)$$

What we need to do now is to define $\hat{M}$ so that the left-hand side of this equals $\mathsf{E}\int_0^T \hat{M}_t (C_t - X_t) \, dt$. To do this, consider the second term on the left-hand side. Change variables (from t to u and from s to t) and interchange integration and expectation (which we can do by Fubini's theorem) to write it as

$$b \int_0^T \int_0^u e^{-(a-b)(u-t)} \mathsf{E}[M_u (C_t - X_t)] \, dt \, du.$$

By iterated expectations,

$$\mathsf{E}[M_u (C_t - X_t)] = \mathsf{E}[(C_t - X_t) \mathsf{E}_t [M_u]].$$

Moreover, $\mathsf{E}_t[M_u] = M_t P_t(u)$, where $P_t(u)$ denotes the price at t of a zero-coupon bond maturing at u. Therefore, the second term on the left-hand side of (15.29) equals

$$b \mathsf{E}\int_0^T \int_0^u e^{-(a-b)(u-t)} M_t P_t(u) (C_t - X_t) \, dt \, du.$$

Now, change the order of integration to write it as

$$b \mathsf{E}\int_0^T M_t (C_t - X_t) \int_t^T e^{-(a-b)(u-t)} P_t(u) \, du \, dt.$$

We are finished by defining

$$\hat{M}_t = M_t + b M_t \int_t^T e^{-(a-b)(u-t)} P_t(u) \, du. \qquad (15.30)$$

A Representative Investor with an Internal Habit in a Production Economy

Assume there is a risk-free production technology with constant rate of return r and a single risky production technology. Assume that a unit of consumption invested into the risky technology at any time t grows (or declines) to S_u/S_t units of consumption at any date $u > t$, where S is a geometric Brownian motion:

$$\frac{dS}{S} = \mu\, dt + \sigma\, dB.$$

Assume both technologies are in infinitely elastic supply, meaning that consumption can be frictionlessly invested or withdrawn from either technology at any time, and the amounts invested do not affect the returns. Assume B is the only source of uncertainty in the economy, so the market is complete, and the unique SDF process M satisfies

$$\frac{dM}{M} = -r\, dt - \frac{\mu - r}{\sigma}\, dB. \tag{15.31}$$

Assume there is a representative investor with an infinite horizon, habit (15.21), and power utility for consumption in excess of the habit:

$$u(c - x) = \frac{1}{1 - \rho}(c - x)^{1-\rho}.$$

Take $b > 0$, so this is a "standard of living" model as described at the beginning of the section.

With a finite horizon, the optimal consumption and portfolio for this model are given in Exercise 15.9. Under parameter restrictions described below, these solutions extend to an infinite horizon. The infinite-horizon solution is

$$C_t = X_t + K e^{-\delta t/\rho}[(1 + \gamma)M_t]^{-1/\rho}, \tag{15.32a}$$

$$\pi_t = \left(1 - \frac{\gamma X_t}{W_t}\right)\frac{\mu - r}{\rho \sigma^2}, \tag{15.32b}$$

for a constant K, where

$$\gamma = \frac{1}{r + a - b}. \tag{15.32c}$$

In the absence of a habit and with a constant investment opportunity set and no labor income, the optimal portfolio for a CRRA investor is

$$\frac{\mu - r}{\rho \sigma^2}.$$

See Exercise 14.6. The formula (15.32b) shows that an investor with an internal habit invests less in the risky asset than would an investor without a habit. This is intuitive. The investor with a habit must put some funds in the risk-free asset to ensure she can afford the habit she has already acquired. As the formula (15.32b) shows, the allocation to the risky asset declines when the habit increases in relation to wealth.

The dependence of the optimal portfolio on the habit-to-wealth ratio (in particular, the dependence on wealth W) causes the optimal portfolio to be an Itô process with a nonzero stochastic part. This is the reason for assuming the assets are in infinitely elastic supply, instead of analyzing a pure exchange economy as in previous parts of the book. In a pure exchange economy with a risk-free asset in zero net supply, a representative investor would hold the portfolio $\pi_t = 1$ for all t. However, it is impossible for π defined in (15.32b) to satisfy $\pi_t = 1$ for all t, due to the stochastic part.

We want to relate the risk-free rate r and the equity premium $\mu - r$ to consumption growth, as we have done in other discussions of the equity premium and risk-free rate puzzles. In the other discussions, we analyzed pure exchange economies, so consumption growth was taken as exogenous and r and $\mu - r$ were computed. In this model, consumption growth is endogenous. We will solve for its dynamics in terms of r and μ. As in the previous subsection, set $\hat{C} = C - X$. Define

$$\alpha = \frac{r - \delta}{\rho} + \frac{(1+\rho)(\mu-r)^2}{2\rho^2\sigma^2}. \tag{15.33}$$

From (15.32a), we obtain

$$\frac{d\hat{C}}{\hat{C}} = -\frac{\delta}{\rho} dt + \frac{dM^{-1/\rho}}{M^{-1/\rho}}$$

$$= \alpha \, dt + \frac{\mu - r}{\rho\sigma} dB. \tag{15.34}$$

Define $Z_t = X_t/C_t$, so $\hat{C}/C = 1 - Z$. Use (15.34) and the definition (15.21) of dX to obtain

$$\frac{dC}{C} = \frac{\hat{C}}{C}\left(\frac{d\hat{C}}{\hat{C}}\right) + \frac{dX}{C}$$

$$= [\alpha(1-Z) - aZ + b] dt + (1-Z)\frac{\mu-r}{\rho\sigma} dB. \tag{15.35}$$

Note that (15.35) is also true in the absence of an internal habit, in which case $a = b = Z = 0$.

This production model cannot directly provide an explanation of the equity premium, because we are taking the market return as exogenous. However, it can explain the low volatility of consumption, given the market return. Recall that in a pure exchange economy with IID lognormal consumption growth and CRRA utility, the low volatility of consumption implies both a lower equity premium and a lower market volatility than is observed in the data, given reasonable risk aversion (Section 10.4). In other words, in that model, a high equity premium and high market volatility can only be explained by high risk aversion, given the low volatility of consumption. On the other hand, in the internal habit model, a low volatility of consumption is consistent with a high equity premium and high market volatility and reasonable risk aversion.

Constantinides (1990) gives sufficient conditions for Z to have a stationary distribution and computes the mean drift and volatility of dC/C in (15.35) for various parameter values. A rough summary of his results is as follows: Suppose the mean value of Z is 0.8 (this is approximately true for various parameter values). Then, the formula for the volatility in (15.35) shows that the risk aversion ρ required to match any historical consumption volatility is only 20% of what it is in the same model without habit. Using the return and consumption growth statistics of Mehra and Prescott (1985)—see Section 7.3—we obtain $\rho = 2.2$, which is within the range generally regarded as reasonable.[10] In the same model without habit, the drift of dC/C is α and matching α to the historical statistics ($\alpha = 0.0178$) produces a negative rate of time preference (Exercise 13.2). Here, we have $0.2\alpha - 0.8a + b = 0.0178$. Taking δ to be such that $e^{-\delta} = 0.99$ and taking $\rho = 2.2$, we obtain $-0.8a + b = 0.0081$. For example, taking $a = 0.5$ and $b = .4081$ matches the historical statistics and satisfies all the constraints necessary for there to be a unique optimum for the investor and a stationary distribution for Z.

One interpretation of the ability of a habit model to resolve the equity premium and risk-free rate puzzles is that it breaks the link between risk aversion and intertemporal substitution, like EZW utility but in a different way. Constantinides (1990) defines risk aversion as the risk aversion of the value function for wealth, given the habit level, and defines the EIS as the partial derivative of the drift of dC/C with respect to r, holding the risk premium $\mu - r$ constant. With CRRA utility, a constant investment opportunity set, and no habit, the risk aversion of the value function is ρ, the drift of dC/C is α, and $\partial \alpha / \partial r = 1/\rho$. This is the familiar equality between risk aversion and the reciprocal of the EIS in a time-additive model with CRRA utility. This equality does not hold in the habit model. Constantinides (1990) computes the mean values of risk aversion and the EIS in the habit model and shows

---

10. However, to ensure $\pi_t \leq 1$ for all t, which means that the representative investor does not short the risk-free technology, we require $\rho \geq (\mu - r)/\sigma^2$ (for example, $\rho = 2.53$).

that they seem reasonable for parameter values that are consistent with the historical return and consumption growth statistics. For various parameter values, relative risk aversion is on average only about 25% of the reciprocal of the EIS.

The parameter restrictions needed in the infinite-horizon model are

$$r + a > b, \tag{15.36a}$$

$$W_0 \geq \frac{X_0}{r+a-b}, \tag{15.36b}$$

$$\delta > (1-\rho)\left(r + \frac{(\mu-r)^2}{2\rho\sigma^2}\right). \tag{15.36c}$$

The condition $r + a > b$ implies that γ_t in Exercise 15.9 converges—to γ defined in (15.32c)—for each fixed t as the horizon T goes to infinity. Condition (15.36b) implies that the problem is feasible (the investor can afford her initial habit). Condition (15.36c) is condition (14.26) that is needed in the infinite-horizon model without a habit.

15.6 VERIFICATION THEOREM

In a portfolio choice application, we want to find (either analytically or numerically) the function V (or, equivalently, J) satisfying the HJB equation. We also need to verify that the solution of the HJB equation is the true value function and that the consumption C and portfolio π attaining the maximum in the HJB equation are optimal. Doing this is called "proving the verification theorem." The theorem is based on the supermartingale/martingale interpretation of the HJB equation discussed in Section 14.5. The reason we need a verification theorem is that the HJB equation states that a drift is zero at the optimum, which is not really a martingale condition, but instead is only a local martingale condition. The solution of the HJB equation can fail to be optimal if the local martingale is not a martingale. A sketch of the proof of the verification theorem when r, μ, and Σ are constant is as follows: The proof follows the same general lines in any continuous-time optimization problem.

Consider the finite-horizon case. Suppose that V satisfies the boundary condition $V(T, w) = U(w)$ and the HJB equation for each (t, w). Assume further that V is sufficiently smooth to apply Itô's formula. Consider an arbitrary consumption process C and portfolio process π. In each state of the world,

$$U(W_T) = V(T, W_T) = V(0, W_0) + \int_0^T dV(t, W_t).$$

Apply Itô's formula to calculate dV and use the intertemporal budget constraint to conclude that

$$U(W_T) = V(0, W_0) + \int_0^T \{V_t + V_w[rW + W\pi'(\mu - r\iota) - C]$$
$$+ \frac{1}{2}V_{ww}W^2\pi'\Sigma\pi\} \, dt + \int_0^T V_w W\pi'\sigma \, dB. \tag{15.37}$$

Suppose that consumption and portfolio processes are constrained to satisfy

$$\mathsf{E}\int_0^T V_w^2 W^2 \pi'\Sigma\pi \, dt < \infty. \tag{15.38}$$

It is shown below how this condition can be relaxed. The condition implies that the stochastic integral

$$\int_0^t V_w W\pi'\sigma \, dB \tag{15.39}$$

is a martingale. Therefore, taking expectations in (15.37) and adding the integral of $e^{-\delta t}u(C)$ to both sides yields

$$\mathsf{E}\left[\int_0^T e^{-\delta t}u(C) \, dt + U(W_T)\right]$$
$$= V(0, W_0) + \mathsf{E}\left[\int_0^T \left\{e^{-\delta t}u(C) + V_t\right.\right.$$
$$\left.\left.+ V_w[rW + W\pi'(\mu - r\iota) - C] + \frac{1}{2}V_{ww}W^2\pi'\Sigma\pi\right\} dt\right]. \tag{15.40}$$

Note that the integrand on the right-hand side (in curly braces) is what, according to the HJB equation (14.19), must have a maximum value of zero. For the arbitrary consumption process C and portfolio process π, the expression in curly braces in (15.40) must therefore be negative or zero everywhere, implying

$$\mathsf{E}\left[\int_0^T e^{-\delta t}u(C) \, dt + U(W_T)\right] \leq V(0, W_0). \tag{15.41a}$$

However, for the consumption C_t^* and portfolio π_t^* attaining the maximum, the HJB equation (14.19) establishes that the expression in curly braces is everywhere zero. Thus, for this consumption process and portfolio process and associated terminal wealth W_T^*, we have

$$\mathsf{E}\left[\int_0^T e^{-\delta t}u(C^*) \, dt + U(W_T^*)\right] = V(0, W_0). \tag{15.41b}$$

It follows from (15.41) that C^* and π^* are optimal and that $V(0, W_0)$ is the maximum attainable expected utility. Therefore, solving the HJB equation produces the optimal C and π in a finite-horizon model when C and π are constrained to satisfy (15.38).

Suppose now that the horizon is infinite. Assume (15.38) holds for each finite T. Then, the same reasoning leading to (15.41) produces

$$\mathsf{E}\left[\int_0^T e^{-\delta t} u(C_t)\,\mathrm{d}s + V(T, W_T)\right] \leq V(0, W_0) \qquad (15.42)$$

for each finite T, with equality at the consumption and portfolio processes C^* and π^* that attain the maximum in the HJB equation. Now, we need to take limits as $T \to \infty$.

The simplest case is when the utility function is nonnegative. Then, we should seek a nonnegative solution V of the HJB equation. Given that $V \geq 0$, (15.42) implies

$$\mathsf{E}\left[\int_0^T e^{-\delta t} u(C_t)\,\mathrm{d}s\right] \leq V(0, W_0),$$

and the monotone convergence theorem (Appendix A.5) then implies

$$\mathsf{E}\left[\int_0^\infty e^{-\delta t} u(C_t)\,\mathrm{d}s\right] \leq V(0, W_0). \qquad (15.43)$$

There is equality in (15.42) at the consumption and portfolio processes C^* and π^* that attain the maximum in the HJB equation. Letting W^* denote the associated wealth process, suppose the transversality condition

$$\lim_{T \to \infty} \mathsf{E}[V(T, W_T^*)] = 0 \qquad (15.44)$$

holds. Equality in (15.42), the monotone convergence theorem, and the transversality condition imply

$$\mathsf{E}\left[\int_0^\infty e^{-\delta t} u(C_t^*)\,\mathrm{d}s\right] = V(0, W_0). \qquad (15.45)$$

It follows from (15.43) and (15.45) that C^* and π^* are optimal. To summarize, when the utility function is nonnegative, the HJB equation produces the optimal policy if consumption and portfolio processes are constrained to satisfy (15.38) and if the transversality condition (15.44) holds.

For utility functions that can be negative, the discussion in Section 9.7 about positive, negative, and bounded dynamic programming applies. In particular, if the utility function is always negative, then we do not need to verify the transversality condition, but we must use some other method to guarantee that the solution of the HJB equation is the true value function. For example, with CRRA utility, the true value function is known to be homogeneous, so the unique homogeneous solution of the HJB equation is the true value function; furthermore, the policies that achieve the maximum in the HJB equation are optimal, if consumption and portfolio processes are constrained to satisfy (15.38).

The objective here is to explain how to relax the condition (15.38). Consider the finite-horizon case. Suppose we adopt the weaker constraint that C and π be such that

$$\int_0^T V_w^2 W^2 \pi' \Sigma \pi \, dt < \infty$$

with probability 1. Then, the stochastic integral (15.39) exists and is a local martingale. By the definition of a local martingale, there is a sequence of stopping times $\tau_k \uparrow T$ with probability 1 such that the stopped process

$$\int_0^{\min(t,\tau_k)} V_w W \pi' \sigma \, dB$$

is a martingale for each k. Taking expectations and reasoning as above yields (15.41) with T replaced by τ_k. Now, we need to take limits as $k \uparrow \infty$, and the issue is the same as in the infinite-horizon model. Given monotonicity or uniform integrability (Appendix A.5), we obtain (15.41) in the limit.

15.7 NOTES AND REFERENCES

The characterization of an investor's optimal wealth in a complete market as the solution of the linear PDE (15.3) is due to Cox and Huang (1989). An advantage of the approach is that a linear PDE is easier to solve than is the nonlinear PDE derived from the HJB equation (that equation is nonlinear because of the nonlinearity of marginal utility). An example of this approach appears in Wachter (2002), who solves for the optimal portfolio in a model with a single risky asset, assuming the Sharpe ratio of the asset follows an Ornstein-Uhlenbeck process. Exercise 15.3 shows that the characterization of the optimal portfolio simplifies when the investor has CRRA utility. In the usual terminology of differential equations, what Exercise 15.3 shows is that the PDE (15.3) in the CRRA case can be solved by separation of variables as $f(t, m, x) = a(m)b(t, x)$ for functions a and b.

Harrison and Kreps (1979) were the first to give a sufficient condition ("viability") for the existence of a risk-neutral probability and were also the first to point out the role of Girsanov's theorem for risk-neutral valuation. Girsanov's theorem can be found in many texts, including Øksendal (2003) and Karatzas and Shreve (2004). Girsanov's theorem for an infinite-horizon model as stated in Section 15.3 can be found in Revuz and Yor (1991, Theorem VIII.1.4).

Suppose the squared maximum Sharpe ratio has a finite integral over $[0, T]$ with probability 1, so M_p is a strictly positive SDF process over $[0, T]$. Define an arbitrage opportunity to be a self-financing wealth process W such that (i) $W_0 = 0$, (ii) W/R is bounded below, and (iii) for some $t \leq T$, $W_t \geq 0$ with probability 1 and $W_t > 0$ with positive probability. The purpose of condition (ii) is to rule out doubling strategies. Assume the market is complete. Then, Levental and Skorohod (1995, Corollary 2) show that there are no arbitrage opportunities if and only if $M_p R$ is a martingale. Thus, there are no arbitrage opportunities if and only if there is a risk-neutral probability.

Now drop the assumption on the squared maximum Sharpe ratio and drop the market completeness assumption. Define a free lunch with vanishing risk to consist of (a) a nonzero nonnegative bounded random variable x that is observable at date T, (b) a sequence of bounded random variables x_i that are observable at date T and that converge to x in the L^∞ norm,[11] and (c) a sequence of self-financing wealth processes W_i, each of which satisfies conditions (i) and (ii) above and which satisfy $W_{iT}/R_T \geq x_i$ with probability 1 for each i. If some x_i in the sequence were a nonzero nonnegative random variable, then the corresponding wealth process W_i would be an arbitrage opportunity. In general, each x_i may be negative with positive probability, so a free lunch with vanishing risk does not make something from nothing. There is some risk of loss. The term "vanishing risk" refers to the convergence of the x_i to x in the L^∞ norm, meaning that the convergence is uniform across states, so the value at risk diminishes to zero as i increases. Delbaen and Schachermayer (1994) show that there is a risk-neutral probability if and only if there is no free lunch with vanishing risk. This applies even in non-Brownian markets. See Delbaen and Schachermayer (2006) for a comprehensive treatment of this subject.

The square-root interest-rate model of Cox, Ingersoll, and Ross (1985) is an example of a market for which, for some parameter values, there is a strictly positive SDF process M but MR is not a martingale. Thus, again, there is no risk-neutral probability. This is discussed further in Chapter 18. See also Back (2010) for a more extensive survey of these issues.

The formula (15.16) for risk premia of jump processes is derived by Madan (1988) and Back (1991). They study somewhat more general models in that they allow the drifts of the money market account price and the risky asset prices to be predictable finite-variation processes instead of assuming them to be Lebesgue integrals. Theorem 1 of Back (1991) assumes that, in the notation of Section 15.4, MY is a special semimartingale, but, as the proof in Section 15.4 shows, that assumption is unnecessary.

---

11. Convergence in the L^∞ norm means that there exist numbers $K_i \downarrow 0$ such that $|x_i - x| \leq K_i$ with probability 1 for each i.

The approach to linear habits in complete markets described in Section 15.5 is due to Detemple and Zapatero (1991) and Schroder and Skiadas (2002). Detemple and Zapatero (1991) provide characterizations of the optimal portfolio in a very general complete markets setting. They prove the existence of equilibrium in a pure exchange economy with a representative investor having an internal habit. They also derive a general representation of the SDF process and derive general formulas for risk premia and the risk-free rate. Schroder and Skiadas (2002) extend their work and show how to transform the optimization problem with habits into an equivalent problem without habits. They also show that the approach can also be applied to recursive utility (Section 11.6) with habits. Hindy and Huang (1993) model complementarity and substitutability of consumption at different dates by taking the period utility function to depend only on the habit X_t. Schroder and Skiadas (2002) show that the Hindy-Huang model can be solved by the approach described in Section 15.5 via the solution of a model in which the decision maker is intolerant of any decline in consumption, due to Dybvig (1995).

Exercise 15.9 asks for the finite-horizon version of the Constantinides model in Section 15.5 to be solved by the method described in Section 15.5. Constantinides (1990) solves the HJB equation instead. The optimal portfolio process has the property that the allocation to the risky asset is lower when the ratio of habit to wealth is higher. This is intuitive, because a high habit necessitates high consumption at future dates; thus, it limits a person's ability and inclination to take on risk. This can potentially explain the standard advice of financial planners that older investors (who typically have higher habits) should hold safer portfolios (in contrast to a model with time-additive CRRA utility and IID returns, which implies a constant optimal portfolio). Research on internal habits and life-cycle investing includes Gomes and Michaelides (2003), Bodie, Detemple, Otruba, and Walter (2004), Polkovnichenko (2007), and Munk (2008). Gomes and Michaelides (2003) and Polkovnichenko (2007) study habits with uninsurable labor income, borrowing constraints, and short sales constraints. Bodie, Detemple, Otruba, and Walter (2004) include a labor/leisure choice within the optimization problem. Munk (2008) analyzes optimal portfolios for an investor with habits in a complete market for some specific types of stochastic investment opportunities.

The coefficient b in the linear habit model (15.21) can be either positive (complements) or negative (substitutes) as discussed in Section 15.5. With $b > 0$, it is possible for consumption C_t to be so high that the utility created at date t by the marginal unit of consumption is less than the subsequent disutility it creates via its contribution to X_s for $s > t$. In other words, these preferences are not strictly monotone. In fact, a very high C_t may be infeasible, in the sense that the person cannot maintain $C_s \geq X_s$ for all $s > t$. This is not an issue for portfolio/consumption choice when asset prices are taken as given, because such high levels of consumption will simply be avoided. However, it is an issue for general equilibrium. In particular, in a pure exchange economy, equilibrium may not exist (due to infeasibility), or it may involve negative state prices (due to negative marginal utilities). Chapman (1998) shows that the ability of a habit model to resolve the equity premium and risk-free rate puzzles in a pure exchange economy is significantly constrained if we impose the natural condition that state prices be positive.

The calibration of the representative investor model with an internal habit in Section 15.5 suffers from the time aggregation issue discussed in Section 13.8. Chapman (2002) corrects for time aggregation and concludes that it has an economically insignificant effect on the results.

Exercise 15.5 is based on the "two trees" model of Cochrane, Longstaff, and Santa-Clara (2008). An implication of their model is that there can be momentum in returns. An asset that has a positive dividend shock in the two trees model will see its price rise, but, because it becomes a larger part of the market portfolio, its absolute covariation with the SDF process also increases. Hence, the risk premium increases, forecasting a higher future return. This should be an empirically important channel for momentum only for aggregates, like stock indices, rather than for individual stocks. Martin (2013b) extends the model to more than two trees (a "Lucas orchard"). A good reference for the verification theorem (and much more about dynamic programming) is Øksendal and Sulem (2007).

EXERCISES

15.1. Assume there is a representative investor with constant relative risk aversion ρ. Assume aggregate consumption C satisfies

$$\frac{dC}{C} = \alpha(X)\,dt + \theta(X)'\,dB$$

for functions α and θ, where X is the Markov process (13.50).

(a) Explain why the market price-dividend ratio is a function of X_t.
(b) Denote the market price-dividend ratio by $f(X_t)$. Explain why the market risk premium is

$$\rho\theta'\theta + \rho\theta' \left(\sum_{j=1}^{\ell} \left. \frac{\partial \log f(x)}{\partial x_j} \right|_{x=X_t} \nu_j \right).$$

How does this compare to the geometric Brownian motion model of consumption in Exercise 13.2?

15.2. Adopt the assumptions of Part (a) of Exercise 13.2. Assume $e^{rt}M_t$ is a martingale.

(a) Using Girsanov's theorem, show that
$$\frac{dD}{D} = (\mu - \sigma\lambda)\,dt + \sigma\,dB^*,$$
where B^* is a Brownian motion under the risk-neutral probability associated with M.

(b) Calculating under the risk-neutral probability, show that the asset price is
$$P_t \stackrel{\text{def}}{=} \frac{D_t}{r + \sigma\lambda - \mu}.$$
Verify that the expected rate of return of the asset under the risk-neutral probability is the risk-free rate.

(c) Define $\delta = r + \sigma\lambda - \mu$, so we have $D/P = \delta$ (in other words, δ is the dividend yield). Verify that
$$\frac{dP}{P} = (r - \delta)\,dt + \sigma\,dB^*.$$

(d) Write down the fundamental PDE—which is here actually an ODE—for the asset value $P_t = f(D_t)$. Verify that the ODE is satisfied by $f(D) = D/(r + \sigma\lambda - \mu)$.

15.3. Adopt the assumptions of Section 15.2. Assume the investor has constant relative risk aversion ρ. Define optimal consumption C and terminal wealth W_T from the first-order conditions (14.7), and define W_t from (14.5).

(a) Show that $W_t = M_t^{-1/\rho} f(t, X_t)$ for some function f.
(b) Derive a PDE for f.
(c) Explain why the optimal portfolio is $\pi(t, X_t)$, where π satisfies
$$\pi(t,x)' = \left(\frac{1}{\rho}\lambda(x)' + \sum_{i=1}^{\ell} \frac{\partial \log f(t,x)}{\partial x_i} v_i(x)'\right)\sigma(x)^{-1}.$$
How does this compare to the optimal portfolio derived in Exercise 14.6 for a constant investment opportunity set? How does it compare to the optimal portfolio (14.24) derived from dynamic programming?

15.4. Assume aggregate consumption C and its expected growth rate μ satisfy
$$\frac{dC}{C} = \mu\,dt + \sigma\,dB_1$$
$$d\mu = \kappa(\theta - \mu)\,dt + \gamma\left[\rho\,dB_1 + \sqrt{1-\rho^2}\right]dB_2$$

for constants σ, κ, θ, ρ, and γ and independent Brownian motions B_1 and B_2. Then, the vector process (C, μ) is Markovian. Assume

$$M_t \stackrel{\text{def}}{=} e^{-\delta t} \left(\frac{C_t}{C_0} \right)^{-\rho}$$

is an SDF process for constants δ and ρ.

(a) Show that the price of risk vector λ for the SDF process M is constant.
(b) Explain why the market price-dividend ratio is a function of μ.
(c) Let $f(\mu)$ denote the market price-dividend ratio, that is,

$$f(\mu_t) = \mathsf{E}_t \int_t^\infty \frac{M_u}{M_t} \cdot \frac{C_u}{C_t} \, du.$$

Write down the fundamental ODE for f.

15.5. Assume two dividend processes D_i are independent geometric Brownian motions:

$$\frac{dD_i}{D_i} = \mu_i \, dt + \sigma_i \, dB_i$$

for constants μ_i and σ_i and independent Brownian motions B_i. Define $C_t = D_{1t} + D_{2t}$. Assume

$$M_t \stackrel{\text{def}}{=} e^{-\delta t} \frac{C_0}{C_t}$$

is an SDF process. (This is the MRS for a log-utility investor.) Define $X_t = D_{1t}/C_t$.

(a) Show that X is a Markov process.
(b) Show that the value

$$\mathsf{E}_t \int_t^\infty \frac{M_u}{M_t} D_{1u} \, du$$

equals $C_t f(X_t)$ for some function f.
(c) Write down the ODE that f must satisfy.

15.6. Adopt the notation of Exercise 13.3. Suppose $M^d R^d$ is a martingale and define the risk-neutral probability corresponding to M^d. Assume $M^d X R^f$ is also a martingale. Show that

$$\frac{dX}{X} = (r^d - r^f) \, dt + \sigma_x \, dB^*,$$

where B^* is a Brownian motion under the risk-neutral probability. Note: This is called uncovered interest parity under the risk-neutral probability. Suppose, for example, that $r^f < r^d$. Then, it may appear profitable to borrow in the foreign currency and invest in the domestic currency money market. The result states that, under the risk-neutral probability, the cost of the foreign currency is expected to increase so as to exactly offset the interest rate differential.

15.7. Assume the market is complete, and let M denote the unique SDF process. Assume MR is a martingale. Consider $T < \infty$, and define the probability $\mathbb{Q}$ in terms of $\xi_T = M_T R_T$ by (15.5). Define B^* by (15.8). Let x be a random variable that depends only on the path of the vector process B^* up to time T and let C be a process adapted to B^*. Assume

$$\mathsf{E}^*\left[\int_0^T \frac{C_s}{R_s}\,ds + \frac{x}{R_T}\right] < \infty.$$

For $t \leq T$, define

$$W_t^* = \mathsf{E}_t^*\left[\int_t^T \frac{C_s}{R_s}\,ds + \frac{x}{R_T}\right].$$

Set $W_t = R_t W_t^*$ (so, in particular, $W_T = x$). Use martingale representation under $\mathbb{Q}$ and the fact that

$$\int_0^t \frac{C_s}{R_s}\,ds + W_t^* = \mathsf{E}_t^*\left[\int_0^T \frac{C_s}{R_s}\,ds + \frac{x}{R_T}\right]$$

is a $\mathbb{Q}$-martingale to prove that there is a portfolio process ϕ such that W, C, and ϕ satisfy the intertemporal budget constraint (13.38).

15.8. This exercise verifies that, as asserted in Section 15.3, condition (15.9) is sufficient for MW to be a martingale. Let M be an SDF process such that MR is a martingale. Define B^* by (15.8). Let W be a positive self-financing wealth process. Define $W^* = W/R$.

(a) Use Itô's formula, (13.10), (13.20), and (15.8) to show that

$$dW^* = \frac{1}{R}\phi'\sigma\,dB^*.$$

(b) Explain why the condition

$$\mathsf{E}^*\left[\int_0^T \frac{1}{R^2}\phi'\Sigma\phi\,dt\right] < \infty$$

implies that W^* is a martingale on $[0, T]$ under the risk-neutral probability defined from M, where E^* denotes expectation with respect to the risk-neutral probability.

(c) Deduce from the previous part that (15.9) implies MW is a martingale on $[0, T]$ under the physical probability.

15.9. Consider the continuous-time portfolio choice problem with exponentially decaying habit described in Section 15.5. Assume (15.25) holds with strict inequality. Repeating the argument at the end of Section 15.5 shows that, for any date t,

$$E_t \int_t^T M_u C_u \, du = E_t \int_t^T \hat{M}_u \hat{C}_u \, du + X_t E_t \int_t^T M_u e^{-(a-b)(u-t)} \, du.$$

(Section 15.5 considers $t = 0$.) Assume power utility: $u(c - x) = \frac{1}{1-\rho}(c - x)^{1-\rho}$. Assume the information in the economy is generated by a single Brownian motion B, there is a constant risk-free rate r, and there is a single risky asset with constant expected rate of return μ and constant volatility σ.

(a) Show that the optimal $\hat{C}$ is

$$\hat{C}_t = K e^{-(\delta/\rho)t} \hat{M}_t^{-\frac{1}{\rho}}$$

for a constant K.

(b) Define

$$\gamma(t) = \frac{1}{r + a - b} \left[1 - e^{-(r+a-b)(T-t)} \right].$$

Show that

$$E_t \int_t^T \frac{M_u}{M_t} e^{-(a-b)(u-t)} \, du = \gamma(t),$$

$$\hat{M}_t = [1 + b\gamma(t)] M_t,$$

$$\frac{1}{M_t} E_t \int_t^T \hat{M}_u \hat{C}_u \, du = \beta(t) M_t^{-\frac{1}{\rho}},$$

for a nonrandom function β.

(c) Define

$$W_t = E_t \int_t^T \frac{M_u}{M_t} C_u \, du.$$

Show that

$$dW_t = \beta(t) M_t^{-\frac{1}{\rho}} \left(\frac{\mu - r}{\rho \sigma} \right) dB_t + \text{something } dt$$

$$= [W_t - \gamma(t) X_t] \left(\frac{\mu - r}{\rho \sigma} \right) dB_t + \text{something } dt.$$

(d) Show that the optimal portfolio is

$$\pi_t = \left(1 - \frac{\gamma(t) X_t}{W_t} \right) \frac{\mu - r}{\rho \sigma^2}.$$

15.10. Derive the fundamental PDEs in Section 15.1 from the fact that the expected rate of return of an asset must equal its required rate of return, as discussed at the end of Section 15.1. Specifically,

(a) Derive (15.1) from the fact that
$$\frac{\mathsf{E}[df]}{f} = r\,dt - \left(\frac{dM}{M}\right)\left(\frac{df}{f}\right).$$

(b) Derive (15.2) from the fact that
$$\frac{g\,dt + \mathsf{E}[df]}{f} = r\,dt - \left(\frac{dM}{M}\right)\left(\frac{df}{f}\right).$$

15.11. In the Markov model of Section 15.1, consider valuing an asset that pays an infinite stream of dividends D, where
$$\frac{dD_t}{D_t} = \gamma(X_t)\,dt + \theta(X_t)'\,dB_t$$
for functions γ and θ. Assume there is no bubble. Then, the price of the asset is
$$P_t = \mathsf{E}_t \int_t^\infty \frac{M_u}{M_t} D_u\, du = D_t \mathsf{E}_t \int_t^\infty \frac{M_u D_u}{M_t D_t}\, du = D_t f(X_t)$$
for some function f.

(a) Deduce that the rate of return is
$$\frac{D\,dt + dP}{P} = \frac{1}{f}\,dt + \frac{dD}{D} + \frac{df}{f} + \left(\frac{dD}{D}\right)\left(\frac{df}{f}\right).$$

(b) Use the result of Part (a) to calculate the expected rate of return.
(c) Use the result of Part (b) and the fact that the expected rate of return must equal
$$r\,dt - \left(\frac{dM}{M}\right)\left(\frac{dP}{P}\right)$$
to derive a PDE for f.

15.12. Derive the PDE in Exercise 15.11 by working under the risk-neutral probability corresponding to M. Use Girsanov's theorem and the fact that the expected rate of return of the asset under the risk-neutral probability must be the risk-free rate.

PART THREE

Derivative Securities

16

Option Pricing

CONTENTS

16.1 Uses of Options and Put-Call Parity 403
16.2 "No Arbitrage" Assumptions 406
16.3 Changing Probabilities 407
16.4 Black-Scholes Formula 409
16.5 Fundamental PDE 413
16.6 Delta Hedging and Greeks 415
16.7 American Options and Smooth Pasting 419
16.8 Dividends 423
16.9 Notes and References 424

A derivative security is a security the value of which depends on another ("underlying") security. For the sake of brevity, the word "underlying" is commonly used as a noun, and we refer to the underlying security simply as the underlying. The basic derivative securities are calls, puts, forwards, and futures. This chapter addresses the valuation and hedging of derivative securities, focusing primarily on calls and puts. Forwards and futures are defined in Chapter 17. Throughout this part of the book, we work with a continuous-time model, assuming dividend-reinvested asset prices are Itô processes.

It is convenient to depart somewhat from the notation of previous chapters. As noted in Section 13.1, it is possible to write each asset price as driven by its own Brownian motion, with the Brownian motions corresponding to different assets typically being correlated. Let S denote the price of the underlying asset. The price process S is written in this chapter as

$$\frac{dS}{S} = \mu\,dt + \sigma\,dB \qquad (16.1)$$

for real-valued stochastic processes μ and σ and a single Brownian motion B. We call σ the volatility of the asset. We are *not* assuming that there is only a single risky asset. There may be other risky assets and other Brownian motions in the background.

Investors' preferences are irrelevant in this part of the book—pricing is by arbitrage. Therefore, there is no reason to assume that prices and cash flows are denominated in units of a consumption good. For concreteness, prices will be taken to be denominated in dollars, though any other currency could be substituted instead.

Assume the existence of a money market account. As before, denote its price at date t by R_t, taking $R_0 = 1$, so we have $dR/R = r\,dt$ and

$$R_t = \exp\left(\int_0^t r_u\,du\right),$$

where r_u is the instantaneous risk-free rate. We assume, except in Section 16.8, that the underlying asset does not pay dividends prior to the date T at which an option matures. We say that such an asset is non-dividend-paying, though of course it must pay dividends eventually or else it is worthless. Section 16.8 extends the results of previous sections using the dividend-reinvested asset price. Beginning in Section 16.4, we assume that the volatility σ and the instantaneous risk-free rate r are constants. Chapter 17 presents extensions to stochastic interest rates and to stochastic volatilities.

We sometimes assume that a zero-coupon bond is traded or can be created by trading. A zero-coupon bond is a security that pays one dollar at some fixed date and has no other cash flows. The date at which the payoff occurs is the bond's maturity date. Such a bond is also called a pure discount bond, or, more briefly, a discount bond. At date $t \leq T$, a discount bond maturing at T has time $T - t$ remaining before maturity, and its price is denoted by $P_t(T)$, or by P_t when the maturity date T is understood. The yield of a discount bond is defined to be the annualized interest rate such that continuous discounting at that rate produces the bond price. This means that the yield at date 0 of the discount bond maturing at T is y defined by

$$e^{-yT} = P_0(T) \quad \Leftrightarrow \quad y = -\frac{\log P_0(T)}{T}. \tag{16.2}$$

If the instantaneous risk-free rate r is nonrandom, then investing

$$\exp\left(-\int_t^T r_u\,du\right) \tag{16.3}$$

in the money market account at date t (and reinvesting interest) produces \$1 at date $T > t$. Therefore, the price $P_t(T)$ of a discount bond must equal (16.3) when r is nonrandom.

16.1 USES OF OPTIONS AND PUT-CALL PARITY

A call option is the right to buy ("call") an asset at a fixed price, called the exercise price or the strike price or simply the strike. A put option is the right to sell ("put") an asset at a fixed price. The price paid to acquire an option is called its premium. An option is called European if it can only be exercised at its maturity date and American if it can be exercised at any date prior to and including its maturity.

Letting T denote the maturity of the options and K the strike price, a call has value $\max(0, S_T - K)$ at maturity and a put has value $\max(0, K - S_T)$ at maturity. The "max" operator reflects the fact that the owner of the option has discretion and will choose to exercise the option (buy the asset in the case of a call and sell it in the case of a put) only if doing so has a positive value. In that case, the option is said to be "in the money." If exercising would have a negative value (the option is "out of the money"), the owner will allow it to expire unexercised and obtain a value of zero. The values of long and short calls and puts are illustrated in Figure 16.1.

The following are some basic uses of options. Other standard option portfolios are presented in Exercises 16.1 and 16.2.

Protective Puts An owner of an asset may buy a put option on it. The payoff of the asset combined with a put is

$$S_T + \max(0, K - S_T) = \max(S_T, K).$$

This reflects the fact that the owner can obtain at least K by exercising the put; thus, the put is insurance for the underlying asset. Similarly, someone who is short the underlying asset can buy a call option to insure the short position.

Covered Calls An owner of an asset may sell a call option on it. The payoff of the asset combined with a short call is

$$S_T - \max(0, S_T - K) = S_T + \min(0, K - S_T) = \min(S_T, K).$$

This means that the upside is capped at K. In exchange for accepting this cap, the seller of the call receives income (the option premium). The call is said to be covered because the seller owns the underlying and hence will not have to buy it in the market to deliver if the call is exercised. If the seller does

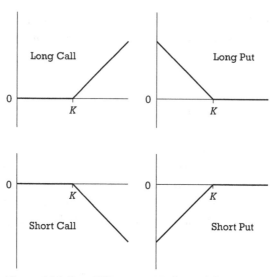

Figure 16.1 Payoff diagrams. Each panel depicts the value of an option position at the option maturity. The underlying asset price is plotted on the horizontal axis, and the strike price K is indicated. The value of a long call is $\max(0, S - K)$, where K denotes the underlying asset price. The value of a long put is $\max(0, K - S)$. The value of a short call is minus the value of a long call, and the value of a short put is minus the value of a long put.

not own the underlying, the call is said to be naked. Similarly, someone who is short the underlying asset might sell a put for income.

Leverage Buying a call is similar to buying the underlying on margin. This is shown explicitly in Section 16.6. Very high leverage (and very high risk) can be created by buying options. Unlike someone who buys on margin, the buyer of a call has no default risk. In practice, this often makes it possible to obtain higher leverage by buying calls than by buying the underlying asset on margin. Similarly, buying a put is an alternative to short selling, and higher leverage can be obtained by buying puts than by short selling the underlying asset.

Trading Volatility A straddle is a put and a call with the same strike. It will profit if the underlying moves in either direction from the strike. Someone who buys straddles is buying volatility, and someone who sells straddles is selling volatility.

As the protective put example illustrates, options play the economic role of insurance contracts. The language of option markets reflects this fact: The price of an option is called its premium, and someone who sells an option is said to write it. To appreciate the benefits and costs of options, it may help to keep in mind that options are bought for insurance or to speculate, and they are sold for income.

The returns of portfolios containing options may have extreme skewness and kurtosis. For example, selling deep out of the money options (puts with strikes considerably below the current value of the asset or calls with strikes considerably above the current value of the asset) is likely to generate income with no loss, because the options are likely to finish out of the money. However, such a portfolio involves a risk of catastrophic loss. The loss can even exceed the value of the portfolio. A time series of returns from such a strategy is likely to show little volatility until a significant loss occurs, so the returns have negative skewness and high kurtosis. Such strategies are commonly called "picking up nickels before a steamroller." As is shown in Section 16.4, under certain assumptions, there are dynamic trading strategies in the underlying asset and money market account that have the same payoffs as options. Thus, dynamic trading strategies can also produce returns of the "picking up nickels" variety. This is an important practical issue in performance evaluation.

Put-Call Parity and Option Bounds

As discussed in the protective put example, the payoff of the underlying asset plus a European put option is $\max(K, S_T)$. The same payoff is obtained with K units of a discount bond maturing at T and a European call option, namely,

$$K + \max(0, S_T - K) = \max(K, S_T).$$

Thus, the value at T of the following portfolios are the same: (i) K units of a discount bond maturing at T plus a European call option and (ii) one unit of the underlying asset plus a European put option.[1] In the absence of arbitrage opportunities, the values of the portfolios at each date $t < T$ must be also the same; therefore,

$$\text{European Call Price} + KP_t(T) = \text{European Put Price} + S_t. \qquad (16.4)$$

This formula is called put-call parity. It is implied by the absence of arbitrage opportunities and does not depend on any particular model of the underlying asset price (it is model independent).

---

1. If the options are American, these portfolios need not be equivalent, and put-call parity (16.4) need not hold (Section 16.7).

Because the put price must be nonnegative, put-call parity implies the following lower bound on a European call option price:

$$\text{Call Price} \geq S_t - KP_t(T). \tag{16.5a}$$

Likewise, put-call parity and the nonnegativity of call prices implies

$$\text{Put Price} \geq KP_t(T) - S_t. \tag{16.5b}$$

These bounds are also model independent.

American options are worth at least as much as their European counterparts, so the bounds (16.5) must also hold for American options. Furthermore, the value of an American option at any date must be at least the value of exercising it at that date (the "intrinsic value"). For American puts, this implies that the intrinsic value $K - S_t$ is a lower bound, which improves on (16.5b), assuming a positive time value of money ($P_t(T) \leq 1$). On the other hand, the bound (16.5a) on a call price is higher than the intrinsic value. This implies that an American call on a non-dividend-paying asset should not be exercised early and hence has the same value as its European counterpart. This is commonly expressed as "calls are better alive than dead."

16.2 "NO ARBITRAGE" ASSUMPTIONS

Each SDF process M satisfies

$$\frac{dM}{M} = -r\,dt - \frac{\mu - r}{\sigma}\,dB + \frac{d\varepsilon}{\varepsilon}, \tag{16.6}$$

where ε is a local martingale uncorrelated with B (i.e., it is spanned by independent Brownian motions). This is effectively established in Section 13.3 (in the "Orthogonal Projections" subsection). Actually, in Section 13.3, the stochastic part of an SDF process is shown to consist of a local martingale spanned by the assets and an orthogonal local martingale. However, repeating that argument focusing only on the asset with price S gives (16.6). For a more formal verification, see Exercise 16.8.

Assume there is an SDF process M such that MR and MS are martingales. From Itô's formula and (16.6),

$$\frac{d(MR)}{MR} = -\frac{\mu - r}{\sigma}\,dB + \frac{d\varepsilon}{\varepsilon}, \tag{16.7a}$$

$$\frac{d(MS)}{MS} = -\frac{\mu - r - \sigma^2}{\sigma}\,dB + \frac{d\varepsilon}{\varepsilon}. \tag{16.7b}$$

Sufficient conditions for MR and MS to be martingales on a time interval $[0, T]$ are (i) μ, r, and σ are adapted to B, (ii) ε is a martingale that is independent of B, and (iii) Novikov's condition:

$$\mathsf{E}\left[\exp\left(\frac{1}{2}\int_0^T (\mu_t - r_t)^2/\sigma_t^2 \, dt\right)\right] < \infty, \qquad (16.8a)$$

$$\mathsf{E}\left[\exp\left(\frac{1}{2}\int_0^T (\mu_t - r_t - \sigma_t^2)^2/\sigma_t^2 \, dt\right)\right] < \infty. \qquad (16.8b)$$

In the remainder of this chapter, we simply assume that MR and MS are martingales for some SDF process M. This rules out some models, but it includes all models satisfying the sufficient conditions (i)–(iii). It also includes some other models, because the sufficient conditions are not necessary conditions.

We also assume through the remainder of the chapter that the class of allowed portfolio processes is constrained in such a way that MW is a martingale for each self-financing wealth process. Thus, for any self-financing wealth process W, there is a unique cost at date t of achieving W_T at date $T > t$, and the cost is $W_t = \mathsf{E}[M_T W_T / M_t]$. All of the assumptions made here are usually expressed rather imprecisely but conveniently as "assume there are no arbitrage opportunities." The link between "no arbitrage" assumptions and the existence of an SDF process M such that MR is a martingale is discussed in Section 15.7.

16.3 CHANGING PROBABILITIES

Let M denote any SDF process such that MR is a martingale on $[0, T]$. There is a risk-neutral probability associated to M for the events observable at T. In Chapter 15, the risk-neutral probability is denoted by $\mathbb{Q}$. In this chapter, the notation prob^R is used, the R denoting the money market account. We use E^R to denote expectation with respect to the risk-neutral probability. There can be multiple SDF processes in the economy, but they differ only by the component $d\varepsilon/\varepsilon$ in (16.6), which is orthogonal to S. Hence, they all give the same valuation for derivative securities written on S. In particular, the different risk-neutral probabilities they define all give the same valuation for derivative securities written on S.

Recall that S/R is a martingale on the interval $[0, T]$ with respect to the risk-neutral probability. Thus,

$$S_t = \mathsf{E}_t^R\left[\frac{R_t}{R_T} S_T\right] = \mathsf{E}_t^R\left[\exp\left(-\int_t^T r_u \, du\right) S_T\right]$$

for each $t < T$. Making calculations of this sort is said to be changing measures or changing probabilities or changing numeraires. The last phrase is motivated by the fact that S/R is a value in units of the money market account.

We can just as easily use any other non-dividend-paying (or dividend-reinvested) asset as numeraire. For example, we can use the asset with price S as the numeraire. To do so, assume MS is a martingale and use S in place of R in the construction of Section 15.3. Specifically, set $\xi_T = M_T S_T/S_0$ and define

$$\text{prob}^S(A) = \mathsf{E}[\xi_T 1_A] \tag{16.9}$$

for each event A that is distinguishable at date T, where 1_A denotes the random variable equal to 1 when the state of the world is in A, and 0 otherwise. This means that $M_T S_T/S_0$ is the Radon-Nikodym derivative $d\,\text{prob}^S/d\mathbb{P}$. As in Section 15.3, it follows that Y/S is a prob^S-martingale for any stochastic process Y such that MY is a $\mathbb{P}$-martingale. Denote expectation with respect to prob^S by E^S.

Some important facts are

(A) The expected rate of return of S under prob^R is r.
(B) The expected rate of return of S under prob^S is $r + \sigma^2$.

Property (A) is established in (15.10). The proof of (B) is very similar and is given below. Note that facts (A) and (B) are true even when r, μ, and σ are stochastic processes.

Set $\xi_t = M_t S_t/S_0$. Because MS is a $\mathbb{P}$-martingale, $\xi_t = \mathsf{E}_t[\xi_T]$. Therefore, Girsanov's theorem implies that B^* is a prob^S-Brownian motion, where $B_0^* = 0$ and

$$dB^* = dB - \left(\frac{d\xi}{\xi}\right)(dB).$$

The formula (16.1) for dS/S and the formula (16.6) for dM/M imply

$$\frac{d\xi}{\xi} = \frac{dM}{M} + \frac{dS}{S} + \left(\frac{dM}{M}\right)\left(\frac{dS}{S}\right)$$

$$= \frac{\sigma^2 + r - \mu}{\sigma}\,dB.$$

Therefore,

$$dB^* = dB + \frac{\mu - r - \sigma^2}{\sigma}\,dt.$$

Use this to substitute for dB in (16.1). This produces

$$\frac{dS}{S} = \mu\,dt + \sigma\left(dB^* - \frac{\mu - r - \sigma^2}{\sigma}\,dt\right)$$

$$= (r + \sigma^2)\,dt + \sigma\,dB^*. \tag{16.10}$$

16.4 BLACK-SCHOLES FORMULA

This section derives the Black-Scholes formulas for the values of European calls and puts. In this section and the remainder of the chapter, we assume that the risk-free rate r and volatility σ are constants. These assumptions do not imply that the market is complete, but (in conjunction with the assumption that MR is a martingale) they do imply that the market is complete for payoffs that depend only on the history of S, in the following sense.

Let x be a random variable that depends only on the history of S through date T and satisfies $\mathsf{E}[M_T |x|] < \infty$ for some SDF process M. Then, there is a self-financing wealth process W such that $W_T = x$. Furthermore, the portfolio process involves positions only in the money market account and in the asset with price S. Also,

$$(\forall t \leq T) \quad W_t = \mathsf{E}_t\left[\frac{M_T}{M_t}x\right] = e^{-r(T-t)}\mathsf{E}_t^R[x]. \quad (16.11)$$

These claims are verified at the end of the section.

First, we derive the value of a European call option. Under our maintained assumptions (Section 16.2) and given the market completeness result, there is a unique date–0 cost of replicating the call, and that cost is $\mathsf{E}[M_T \max(0, S_T - K)]$.[2] Let A denote the set of states of the world such that $S_T \geq K$, and let 1_A denote the random variable that equals 1 when $S_T \geq K$ and 0 otherwise. Then,

$$\max(0, S_T - K) = S_T 1_A - K 1_A. \quad (16.12)$$

The random variable $K 1_A$ is the payoff of what is called a digital or binary option, because it is either "on" or "off," being on (paying K) when $S_T \geq K$ and off (paying zero) when $S_T < K$. The random variable $S_T 1_A$ is the payoff of what can be called a share digital. It is on, and pays one share worth S_T, when $S_T \geq K$, and off, paying zero, when $S_T < K$. A European call option is equivalent to a long position in the share digital and a short position in the digital—this is the meaning of (16.12).

From (16.12), we have

$$\text{Call Price} = \mathsf{E}[M_T S_T 1_A] - K\mathsf{E}[M_T 1_A]. \quad (16.13)$$

This states that the value of a European call is the difference between the values of the share digital and the digital.

---

2. It is without loss of generality to conduct the valuation only at date 0. We can translate the time axis to consider any other date as "date 0." The option pricing formulas remain valid if the remaining time to maturity is used in place of T and if the price of the underlying asset at the date of valuation is input instead of the date–0 price.

To value the share digital and digital, some simplification is obtained by changing numeraires. To value the share digital, it is convenient to use the underlying asset as numeraire. Recall that $M_T S_T / S_0$ is the Radon-Nikodym derivative in the change of probability to prob^S. Therefore,

$$\mathsf{E}[M_T S_T 1_A] = S_0 \mathsf{E}\left[\frac{M_T S_T}{S_0} 1_A\right] = S_0 \mathsf{E}^S[1_A] = S_0 \text{prob}^S(A).$$

The last equality is due to the fact that the expectation of a variable that is 1 on some event and 0 elsewhere is just the probability of the event. To value the digital, it is convenient to use the risk-free asset as numeraire. $M_T R_T = M_T e^{rT}$ is the Radon-Nikodym derivative for the risk-neutral probability, so

$$\mathsf{E}[M_T 1_A] = e^{-rT} \mathsf{E}[M_T e^{rT} 1_A] = e^{-rT} \mathsf{E}^R[1_A] = e^{-rT} \text{prob}^R(A).$$

Combining these facts with (16.13) yields

$$\text{Call Price} = S_0 \text{prob}^S(A) - e^{-rT} K \text{prob}^R(A). \qquad (16.14)$$

It remains to compute the probabilities $\text{prob}^S(A)$ and $\text{prob}^R(A)$, which are the probabilities of the option finishing in the money ($S_T \geq K$) under the two changes of numeraire. To do this, we can use facts (A) and (B) of Section 16.3, namely, S has expected rate of return r under prob^R and expected rate of return $r + \sigma^2$ under prob^S. As is shown below, these facts imply

(A') $\text{prob}^R(A) = N(d_2)$,
(B') $\text{prob}^S(A) = N(d_1)$,

where N denotes the cumulative distribution function of a standard normal random variable and d_1 and d_2 are defined as

$$d_1 = \frac{\log(S_0/K) + \left(r + \tfrac{1}{2}\sigma^2\right) T}{\sigma \sqrt{T}}, \qquad (16.15a)$$

$$d_2 = \frac{\log(S_0/K) + \left(r - \tfrac{1}{2}\sigma^2\right) T}{\sigma \sqrt{T}}$$

$$= d_1 - \sigma \sqrt{T}. \qquad (16.15b)$$

Substituting (A') and (B') in (16.14) yields

$$\text{Call Price} = S_0 N(d_1) - e^{-rT} K N(d_2). \qquad (16.15c)$$

This is the Black-Scholes formula for the value of a European call on a non-dividend-paying asset that has time T to maturity.

To value a European put, we can use put-call parity, or we can reason as follows: The payoff of a European put is the difference in the values of a digital and a share digital but with the digital and share digital paying on the complement of A (i.e., when $S_T < K$). Therefore, the value of a European put is

$$\text{Put Price} = e^{-rT}K[1 - \text{prob}^R(A)] - S_0[1 - \text{prob}^S(A)].$$

Using facts (A′) and (B′) above and the facts that $1 - N(d_2) = N(-d_2)$ and $1 - N(d_1) = N(-d_1)$ (due to the symmetry of the standard normal distribution function), this implies the Black-Scholes formula:

$$\text{Put Price} = e^{-rT}KN(-d_2) - S_0 N(-d_1). \qquad (16.15d)$$

In reality, the short-term interest rate is not constant as assumed in the derivation above. Chapter 17 shows that the Black-Scholes formula is nevertheless valid, provided we make two substitutions. First, we should use the yield of a zero-coupon bond maturing at the date the option matures as the risk-free rate r. Second, we should substitute the volatility of the forward price of the asset for the volatility of the asset price itself. Fortunately, the volatility of the forward price is close to the volatility of the underlying asset price if the time to maturity is short. Also, the volatility of the forward price is exactly equal to the volatility of the underlying asset if the risk-free rate is constant or if it varies nonrandomly. This is discussed further in Section 17.3.

An important fact about the Black-Scholes formula is that the expected return μ of the asset does not appear in the formula. This is surprising, because we might expect calls to be worth more and puts to be worth less if the price is expected to grow more (that is, if μ is higher). However, we cannot change μ without also changing some other feature of the model. For example, suppose we want to hold all of the inputs (S_0, K, r, σ, and T) to the Black-Scholes formula fixed and change μ. Then, the SDF process must change, because the formula $S_0 = E[M_T S_T]$ can continue to hold when there is a shift in the distribution of S_T (due to the change in μ) only if there is also a shift in the distribution of M_T. The precise change in the SDF process is given by (16.6). Intuitively, states in which S_T is large must become cheaper if it becomes more likely that S_T is large (due to an increase in μ) but S_0 does not change. Thus, an increase in μ shifts the distributions of terminal option values but, due to changes in state prices, does not change date–0 values of options. However, this does not mean that the expected return μ is entirely irrelevant. There are some stochastic processes μ for which MR and MS are not martingales. The "no arbitrage" assumptions of Section 16.2 rule those processes out.[3]

---

3. A sufficient condition for MR and MS to be martingales is (16.8), substituting constant r and σ.

First, we establish the market completeness result stated at the beginning of the section. Then, we prove (A′) and (B′). Recall from Section 15.3 (Girsanov's theorem) that B^* defined by $B_0^* = 0$ and

$$dB^* = dB + \frac{\mu - r}{\sigma} dt \qquad (16.16)$$

is a Brownian motion under the risk-neutral probability. Define

$$W_t = e^{-r(T-t)} E_t^R [x], \qquad (16.17)$$

so the second equality in (16.11) holds. It follows from (16.17) that $e^{-rt} W_t$ is a prob^R-martingale, and this implies that MW is a $\mathbb{P}$-martingale (Appendix A.12). Thus, the first equality in (16.11) also holds.

Under our assumptions, S is a geometric Brownian motion under the risk-neutral probability, driven by B^*. In particular, S is adapted to B^*. Because x depends only on the path of S, it must depend only on the path of B^*. Therefore, the prob^R-martingale $e^{-rt} W_t$ is adapted to the prob^R-Brownian motion B^*. By the martingale representation theorem, there exists a stochastic process γ such that

$$d(e^{-rt} W_t) = \gamma \, dB^*.$$

This implies

$$\frac{dW}{W} = r \, dt + \frac{\gamma e^{rt}}{W} dB^*.$$

Define

$$\pi = \frac{\gamma e^{rt}}{\sigma W}.$$

This yields

$$\frac{dW}{W} = r \, dt + \pi \sigma \, dB^*.$$

Now, substitute from (16.16) to obtain

$$\frac{dW}{W} = r \, dt + \pi (\mu - r) \, dt + \pi \sigma \, dB.$$

This shows that W is a self-financing wealth process.

(A′) Condition (A) in Section 16.3 states that

$$\frac{dS}{S} = r \, dt + \sigma \, dB^*$$

Option Pricing

for a Brownian motion B^* under prob^R. Therefore, we have the geometric Brownian motion formula

$$\log S_T = \log S_0 + \left(r - \frac{1}{2}\sigma^2\right)T + \sigma B_T^*.$$

The condition $S_T \geq K$ is equivalent to $\log S_T \geq \log K$. Thus, $\text{prob}^R(A)$ is the risk-neutral probability of the event

$$\log S_0 + \left(r - \frac{1}{2}\sigma^2\right)T + \sigma B_T^* \geq \log K.$$

This can be rearranged as

$$-\frac{B_T^*}{\sqrt{T}} \leq d_2.$$

The random variable on the left-hand side is a standard normal under the risk-neutral probability. Therefore, $\text{prob}^R(A) = N(d_2)$.

(B') Condition (B) in Section 16.3 states that

$$\frac{dS}{S} = (r + \sigma^2)\,dt + \sigma\,dB^*$$

for a Brownian motion B^* under prob^S. Therefore, we have the geometric Brownian motion formula

$$\log S_T = \log S_0 + \left(r + \frac{1}{2}\sigma^2\right)T + \sigma B_T^*.$$

Thus, $\text{prob}^S(A)$ is the probability under prob^S of the event

$$\log S_0 + \left(r + \frac{1}{2}\sigma^2\right)T + \sigma B_T^* \geq \log K.$$

This can be rearranged as

$$-\frac{B_T^*}{\sqrt{T}} \leq d_1.$$

The random variable on the left-hand side is a standard normal under prob^S. Therefore, $\text{prob}^S(A) = N(d_1)$.

16.5 FUNDAMENTAL PARTIAL DIFFERENTIAL EQUATION

For the value V of a path-independent derivative security (which is defined below), the fact that MV is a local martingale is equivalent to a PDE for V. The PDE is the fundamental PDE discussed in Section 15.1. Here, the underlying asset price S plays the role of the Markov process X in Section 15.1.[4] The PDE can be transformed into the heat equation, which is a standard PDE in physics. This is the route by which Black and Scholes originally derived their option pricing formulas. See the end-of-chapter notes for further discussion.

All path-independent derivative securities written on S satisfy the same PDE. Different derivative securities have different values because of different boundary conditions. A path-independent European derivative security is a random variable x that depends only on the value of S_T—it *does not* depend on the path of S prior to T. The derivative security pays x at date T. European calls and puts are examples, as are digital options and digital share options. There are many more possibilities. As in (16.11), the value of the derivative security at each date $t \leq T$ is

$$\mathsf{E}_t\left[\frac{M_T}{M_t}x\right] = e^{-r(T-t)}\mathsf{E}_t^R[x]. \qquad (16.18)$$

As noted in the proof of (A′) in the previous section—and as is stated again in (16.19) below—S is a geometric Brownian motion under the risk-neutral probability, so $\mathsf{E}_t^R[x]$ depends only on t and S_t. Hence, the value of the derivative security is $V(t, S_t)$ for some function V. This means in particular that $V(T, S_T) = x$. Assume V is sufficiently smooth to apply Itô's formula.

We can take either of two routes to derive the PDE. We can use the fact that

$$V(t, S_t) = \mathsf{E}_t\left[\frac{M_T}{M_t}x\right] \quad \Leftrightarrow \quad M_t V(t, S_t) = \mathbb{P}\text{-martingale}$$

or the fact that

$$V(t, S_t) = e^{-r(T-t)}\mathsf{E}_t^R[x] \quad \Leftrightarrow \quad e^{-rt}V(t, S_t) = \text{prob}^R\text{-martingale}.$$

We take the second route here (see Section 15.1 for the first). Recall that every self-financing wealth process has expected return equal to r under a risk-neutral probability, and changing probabilities does not change volatilities (Section 15.3), so

$$\frac{dS}{S} = r\,dt + \sigma\,dB^* \qquad (16.19)$$

---

4. Actually, because we have allowed the expected return process to be a general stochastic process, S is not necessarily Markovian under the physical probability. However, when we equate the drift of MV to zero, the drift of S under the physical probability drops out. Hence, we still end up with a PDE for V. Furthermore, S is Markovian under the risk-neutral probability, and we can derive the PDE by working exclusively under the risk-neutral probability.

for a Brownian motion B^* under prob^R. Apply Itô's formula to $e^{-rt}V(t, S_t)$ and use subscripts to denote partial derivatives. The differential is

$$-re^{-rt}V\,dt + e^{-rt}\left\{V_t\,dt + V_s\left[rS\,dt + \sigma S\,dB^*\right] + \frac{1}{2}V_{ss}\sigma^2 S^2\,dt\right\}.$$

This cannot have a drift. Equate the drift to 0 and cancel the e^{-rt} factor to obtain

$$rV = V_t + rSV_s + \frac{1}{2}\sigma^2 S^2 V_{ss}.$$

Here, we have dropped time subscripts, and S is shorthand for S_t. Because S_t can take any positive value, the equation must hold for $S_t = s$ and any positive s. So, more explicitly,

$$rV(t,s) = V_t(t,s) + rsV_s(t,s) + \frac{1}{2}\sigma^2 s^2 V_{ss}(s,t). \qquad (16.20)$$

This is the fundamental PDE. It must hold for all $t < T$ and $s > 0$. The fundamental PDE is equivalent to the statement that $e^{-rt}V$ is a local martingale under the risk-neutral probability (that is, $e^{-rt}V$ has no drift under the risk-neutral probability) and also equivalent to the statement that MV is a local martingale under the physical probability. The formula (16.18) for $V(t, S_t)$ is a solution of the fundamental PDE and the boundary condition $V(T, S_T) = x$.

Section 13.5 shows (writing now V instead of W and taking $C = 0$) that if MV is a local martingale and the stochastic part of dV equals $\phi'\sigma\,dB$ for some stochastic process ϕ, then V is a self-financing wealth process with portfolio process ϕ. Thus, if we can match the stochastic parts, then the local martingale property implies that the drifts match in the intertemporal budget constraint. The fundamental PDE plays the same role as the local martingale property—not surprisingly, since it is equivalent to the local martingale property. If we match the stochastic parts and the fundamental PDE holds, then the intertemporal budget constraint holds. This plays an important role in the topic of the next section.

16.6 DELTA HEDGING AND GREEKS

We continue in this section to study a European path-independent derivative security with value x at date T. We specialize to calls and puts later in the section. The delta is the (calculus!) derivative of the derivative security's value with respect to the underlying asset price. It is V_s in the notation of the previous section. The other option "greeks" are the theta $\Theta = V_t$, the gamma $\Gamma = V_{ss}$, the vega $\mathcal{V} = \partial V/\partial \sigma$, and the rho $\rho = \partial V/\partial r$. For the last two greeks, we view V as a function of the parameters of the model as well as its arguments t and S_t. We discuss the last two greeks further at the end of the section.

The delta of a derivative security is the number of shares of the underlying asset that we should hold in order to replicate the derivative. In other words, it is the number of shares we should hold at each date to obtain a portfolio worth x at date T. To hedge a short (written) derivative, we replicate the long derivative. Thus, the delta is the number of shares we should hold to hedge a written derivative. This is called delta hedging.

To verify that holding delta shares replicates the derivative, consider a portfolio with value V that includes V_s shares of the underlying asset and a position in the money market account. To have value V, the investment in the money market account must be $V - SV_s$. The change in the portfolio value in an instant dt is

$$V_s\,dS + (V - SV_s)r\,dt. \tag{16.21}$$

Note that the stochastic part comes from the term $V_s\,dS$. In order to replicate the derivative, we need the change (16.21) to equal dV. By Itô's formula,

$$dV = V_t\,dt + V_s\,dS + \frac{1}{2}\sigma^2 S^2 V_{ss}\,dt. \tag{16.22}$$

The stochastic part of this also comes from the term $V_s\,dS$. Thus, delta hedging matches the stochastic part of the change in the portfolio value (16.21) to the stochastic part of the change in the derivative security value (16.22). Because the $V_s\,dS$ terms match, the changes (16.21) and (16.22) match exactly if and only if

$$(V - SV_s)r = V_t + \frac{1}{2}\sigma^2 S^2 V_{ss}.$$

This is the fundamental PDE. It is satisfied by the derivative security value V, so holding delta shares of the underlying asset replicates the derivative security.

The delta of a European call option is $N(d_1)$, where N is the standard normal distribution function and d_1 is defined in (16.15a). The delta of a European put option is $-N(-d_1)$. These facts are verified at the end of the section. The variable d_1 changes when the time to maturity changes and when the underlying asset price changes, so a delta hedge must be adjusted continuously. From the Black-Scholes formula, it is evident that $SN(d_1) > V$ for a European call option, so the replicating portfolio involves a short position in the money market account. In other words, to replicate a European call, we need to buy the underlying asset on margin. The delta of a European put is negative. Therefore, to replicate a European put, we should short the underlying asset and invest the proceeds plus some additional funds in the money market account.

From the calculations above, the change in value of a portfolio consisting of a short option and a long replicating portfolio is

$$-dV + V_s\,dS + (V - SV_s)r\,dt = -V_t - \frac{1}{2}\sigma^2 S^2 V_{ss} - (SV_s - V)r\,dt.$$

The first term on the right-hand side is minus the theta of the option, the second term involves the gamma of the option, and the final term is interest expense. The theta of a call option is negative, meaning that the value decreases over time if the underlying asset price does not change. This is called time decay. The gamma is positive, because the value V is a strictly convex function of the underlying asset price. The delta-hedged short call portfolio is said to be short gamma or to have negative convexity. Thus, the delta-hedged short call portfolio gains from time decay, loses from gamma, and loses due to interest expense. The fundamental PDE states that the gain from time decay offsets the losses due to gamma and due to interest expense.

Figure 16.2 depicts how a delta hedge of a European call option would perform if the underlying asset price changed without any adjustment in the hedge. Interpret the change in the asset price as being an instantaneous discrete change, because interest on the borrowed funds in the delta hedge is ignored. As the figure illustrates, the delta hedge would underperform, regardless of the direction in which the underlying asset price changes. This is a result of the fact that the delta-hedged short call portfolio is short gamma. Even if the hedge is adjusted continuously (as it must be in order to maintain a perfect hedge), gamma matters because of the nonzero quadratic variation of an Itô process—roughly speaking, S changes by a relatively large amount even in an infinitesimal time interval; that is, dS is "of order $\sqrt{dt}$." Though Figure 16.2 indicates that the delta-hedged short call would lose in response to a discrete change in the underlying asset price, it does not lose when the asset price evolves continuously, because of the gain from time decay.

In reality, there are transactions costs (commissions, bid-ask spreads, and market impact costs) that make it impossible to continuously trade. Therefore, it is impossible in practice to perfectly delta hedge. Note that, to replicate an option with a positive gamma, like a call option, we must buy shares when the underlying asset price rises and sell shares when it falls, because a positive gamma means that the delta increases when the underlying asset price increases. This is a "liquidity demanding" strategy that may have relatively high market impact costs if carried out in scale.

The greeks vega and rho do not appear in a delta hedge. This is because we are assuming σ and r are constants. In reality, they are random. As mentioned in Section 16.4 and as is discussed further in Section 17.3, a random r can easily be accommodated by substituting the yield of a discount bond. The discount bond yield does not have to be constant in order for the Black-Scholes formula to be valid. Part of the change in the derivative security value is due to the changing discount bond yield, and the rho determines how much the derivative security value changes in response. A delta hedge in this case consists of positions in the

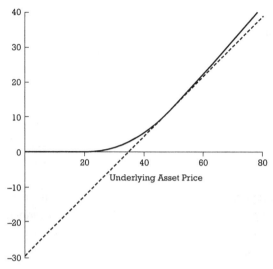

Figure 16.2 Delta hedge of a call option. The solid line is the graph of a European call option value as a function of the underlying asset price when $r = 0.05$, $\sigma = 0.3$, $K = 40$, and the time to maturity is $T = 1$. When $S_0 = 50$, the call option value is $V(0, 50) = 13.61$, and the delta at $S_0 = 50$ is $N(d_1) = 0.856$. The dashed line graphs the value, as a function of the underlying asset price, of a delta hedge consisting of 0.856 shares purchased with $0.856 \times 50 - 13.61 = 29.16$ in borrowed funds.

underlying asset and the discount bond. Enough units of the discount bond must be held to hedge the rho part of dV.

The Black-Scholes formula is not valid if σ is random. This is discussed in Section 17.4. Because it is not valid when σ changes, how it changes when sigma changes (the vega) may seem irrelevant. However, practitioners may use the Black-Scholes vega to approximate the sensitivity of a call or put value to changing volatility. They worry a lot about changing volatility, so vega (of the Black-Scholes formula or some other formula) matters a lot.

Let n denote the standard normal density function:

$$n(x) = \frac{1}{\sqrt{2\pi}} e^{-x^2/2}.$$

It is easy to verify directly that

$$S_0 \, n(d_1) = e^{-rT} K \, n(d_2) \quad \text{and} \quad \frac{\partial d_1}{\partial S_0} = \frac{\partial d_2}{\partial S_0}.$$

The delta of a call option can be computed using the chain rule and these facts as

$$N(d_1) + S_0 n(d_1)\frac{\partial d_1}{\partial S_0} - e^{-rT}K n(d_2)\frac{\partial d_2}{\partial S_0} = N(d_1).$$

We can perform similar calculations to compute the delta of a put, or we can use put-call parity, which states that

$$\text{European Put Price} = \text{European Call Price} + e^{-rT}K - S_0.$$

Differentiating both sides of this with respect to S_0 shows that the delta of a put equals the delta of a call minus 1. Hence, the delta of a put is

$$N(d_1) - 1 = -N(-d_1).$$

16.7 AMERICAN OPTIONS AND SMOOTH PASTING

As discussed in Section 16.1, calls are better alive than dead when the underlying asset does not pay dividends, so American calls have the same values as European calls when the underlying asset does not pay dividends. This section discusses the valuation of American puts. Some of what is discussed applies also to American calls on dividend-paying assets.

It is optimal to exercise an American put early when it is sufficiently far in the money, meaning that the underlying asset price S is sufficiently low. In general, American options must be considered in conjunction with exercise policies. For an American put, with constant r and σ, it is sufficient to consider exercise boundaries, meaning continuous functions $f : [0, T] \to [0, K]$. The interpretation is that the option is exercised at the first time that S falls to the boundary, that is, at the first time t such that $S_t \leq f(t)$. Of course, we would never exercise an out-of-the money option, so we can assume $f(t) \leq K$ for all t.

Continue to assume there are no arbitrage opportunities (Section 16.2) and that r and σ are constant. Given an exercise boundary, let A denote the set of states of the world such that $S_t \leq f(t)$ for some $t \leq T$, and on the event A define $\tau = \min\{t \mid S_t \leq f(t)\}$. The American put with this exercise policy pays $K - S_\tau$ at the random date τ in the event A and pays 0 on the complement of A. There is a self-financing trading strategy in the underlying and risk-free assets with the same payoff. The cost at date 0 of the replicating portfolio is[5]

$$E^R[e^{-r\tau}(K - S_\tau)1_A]. \qquad (16.23)$$

---

5. Exercising the put at date τ and investing the gain in the money market account until date T produces the payoff $e^{r(T-\tau)}(K - S_\tau)$ at date T. The market completeness result of Section 16.4 shows that there is a trading strategy in the underlying asset and money market account with this payoff at date T. The same strategy has value $K - S_\tau$ at date τ (otherwise, there would be an

Optimal Exercise Boundary and Arbitrage-Free Price

There is an exercise boundary f^* that maximizes (16.23). Denote the maximum value by V_0. This is the unique arbitrage-free value of the American put at date 0. To see this, suppose first that the put were sold at a price $p < V_0$. In that case, we should buy the put at price p, short the replicating strategy corresponding to the exercise boundary f^*, producing proceeds V_0, and then exercise the put at the boundary f^*. The exercise of the put covers the short replicating strategy, so we obtain $V_0 - p$ at date 0 without any further expense. This is an arbitrage opportunity. Therefore, an arbitrage-free price must be V_0 or greater.

Now, suppose the put were sold at a price $p > V_0$. To arbitrage, we want to short the put at price p and buy ("go long") a replicating strategy. The appropriate replicating strategy seems to depend on the exercise policy followed by the buyer of the put, which the seller may not know. However, this turns out not to be a problem, because it can be shown that there is a trading strategy with initial cost V_0 for which the portfolio value is at least $K - S_t$ at each $t \leq T$.[6] Thus, regardless of the buyer's exercise policy, selling the put at p and employing this strategy produces income $p - V_0$ at date 0 without any further expense (and with possibly some further income). This is again an arbitrage opportunity. Hence, the maximum value of (16.23) is the unique arbitrage-free price for the put.

We would like to calculate the maximum value V_0 and the optimal exercise boundary. However, there is no simple formula for V_0 like the Black-Scholes formula for European options (though it is fairly simple to compute V_0 numerically). The optimal exercise boundary is likewise unknown, though it is known to be an increasing continuously differentiable function f^* with the property that $f^*(T) = K$. A typical case is depicted in Figure 16.3.

Smooth Pasting

Consider an exercise boundary f, with exercise event A and exercise time τ as before. For any t and $s \geq f(t)$, define

$$V(t,s;f) = \mathsf{E}^R_t \left[e^{-r(\tau-t)}(K - S_\tau) 1_A \mid S_t = s, S_u > f(u) \text{ for all } u < t \right]. \quad (16.24)$$

---

arbitrage opportunity using the strategy and the money market account from τ to T), so it replicates the put payoff with exercise at τ. The cost of the strategy at date 0 is $e^{-rT} \mathsf{E}^R[e^{r(T-\tau)}(K - S_\tau)]$, which is the same as (16.23).

6. Because the portfolio's value is at least $K - S_t$ at each $t \leq T$ (and possibly more), this is called a super-replicating strategy.

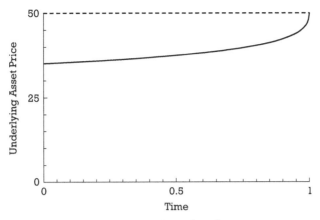

Figure 16.3 Optimal exercise boundary for an American put. The dashed line is the strike price, and the solid line is the optimal exercise boundary for an American put with maturity date $T = 1$ and strike $K = 50$ when the risk-free rate is 5% and the underlying asset is non-dividend-paying and has a volatility of 30%. It is optimal to exercise at the first time the underlying asset price falls to the boundary. The boundary lies below the strike and increases to the strike as the option approaches maturity.

This is the value at t of an American put exercised at the boundary f, conditional on it not having been exercised by t. Like other derivative security values, $e^{-rt}V(t, S_t; f)$ is a local martingale under the risk-neutral probability; hence, the fundamental PDE holds prior to exercise, which means that it holds on the region $\{(t,s) \mid s > f(t)\}$. This is called the continuation or inaction region (associated to f). The put value also satisfies $V(t,s;f) = K - s$ when $s = f(t)$, which is called the value matching condition.

The optimal exercise boundary f^* and associated value $V(t,s;f^*)$ satisfy (i) the fundamental PDE in the continuation region, (ii) the value matching condition, (iii) $V(t,s;f^*) \geq K - s$ in the continuation region, and (iv) smooth pasting:

$$V_s(t,s;f^*)\big|_{s=f^*(t)} = -1.$$

Condition (iii) states that the put value is at least its intrinsic value, which is clearly true. The partial derivative in the smooth pasting condition (iv) is defined as the right-hand derivative, that is,

$$\lim_{\varepsilon \downarrow 0} \frac{V(t,s+\varepsilon;f^*) - V(t,s;f^*)}{\varepsilon}.$$

The name "smooth pasting" refers to the fact that the value $V(t,s;f^*)$ of the put above the exercise boundary f^* and the value $K - s$ of immediate exercise below the exercise boundary "paste" together in a smooth (i.e., continuously differentiable) way at the boundary. For a call option on a dividend-paying asset, the smooth pasting condition is $V_s = 1$ at the boundary.

The necessity of the smooth-pasting condition can be seen as follows: For $s > f^*(t)$ and any ε such that $s \geq f^*(t) + \varepsilon$, set $J(t,s,\varepsilon) = V(t,s;f^* + \varepsilon)$. The optimality of f^* implies J is maximized in ε at $\varepsilon = 0$, so, assuming J is differentiable in ε, we have

$$J_\varepsilon(t,s,0) = 0. \tag{16.25}$$

By the value matching condition,

$$J(t,f^*(t)+\varepsilon,\varepsilon) = K - f^*(t) - \varepsilon$$

for all ε, so totally differentiating with respect to ε gives

$$J_s(t,f^*(t)+\varepsilon,\varepsilon) + J_\varepsilon(t,f^*(t)+\varepsilon,\varepsilon) = -1$$

for all ε. Substitute (16.25) for $\varepsilon = 0$ to obtain

$$J_s(t,f^*(t),0) = -1.$$

This is the smooth pasting condition. A sketch of the proof that (i)–(iv) are sufficient conditions for f^* to be the optimal exercise boundary follows.

Suppose V and f^* satisfy (i)–(iv) above and V is continuously differentiable in t and twice continuously differentiable in s on the inaction region. For convenience, omit the f^* argument from $V(\cdot)$. Define $V(t,s) = K - s$ for $s < f^*(t)$. By the smooth pasting condition, V is continuously differentiable in s across the boundary and hence continuously differentiable in s for all $s > 0$. Assume V is twice continuously differentiable in s for $s > 0$, so we can apply Itô's formula (this can be justified by an approximation argument). From the fundamental PDE and the definition of V on the region $s < f^*(t)$, we have

$$-rV + V_t + rsV_s + \frac{1}{2}\sigma^2 s^2 V_{ss} = \begin{cases} 0 & \text{if } s > f^*(t), \\ -rK & \text{if } s < f^*(t). \end{cases} \tag{16.26}$$

Consider any exercise time $\tau \leq T$. Apply Itô's formula to $e^{-rt}V(t,S_t)$, using the dynamics (16.19) of S under the risk-neutral probability, and substitute the extension (16.26) of the fundamental PDE to obtain

$$e^{-r\tau}V(\tau,S_\tau) = V(0,S_0) + \int_0^\tau d\left(e^{-rt}V(t,S_t)\right)$$

$$= V(0,S_0) + \int_0^\tau e^{-rt}SV_s\sigma\, dB^* - rK\int_0^\tau 1_{\{S_t \leq f^*(t)\}}e^{-rt}\, dt$$

$$\leq V(0,S_0) + \int_0^\tau e^{-rt}SV_s\sigma\, dB^*. \qquad (16.27)$$

Note that there is equality in the last line if $S_t > f^*(t)$ for all $t < \tau$, which is true if τ is the exercise time associated with f^*. If the stochastic integral in (16.27) is a martingale under the risk-neutral probability, then taking expectations yields

$$\mathsf{E}^R\left[e^{-r\tau}V(\tau,S_\tau)\right] \leq V(0,S_0)$$

with equality if $S_t > f^*(t)$ for all $t < \tau$. If the stochastic integral is merely a local martingale, then using a localizing sequence and taking limits, using the dominated convergence theorem (Appendix A.5) for the left-hand side, yields the same fact. Furthermore,

$$K - S_\tau \leq V(\tau,S_\tau)$$

with equality if $S_\tau \leq f^*(\tau)$. Therefore,

$$\mathsf{E}^R\left[e^{-r\tau}(K - S_\tau)\right] \leq \mathsf{E}^R\left[e^{-r\tau}V(\tau,S_\tau)\right] \leq V(0,S_0)$$

with equality when τ is the exercise time associated with f^*. This shows that τ is the optimal exercise time and V is the maximum value.

16.8 DIVIDENDS

It is easy to extend the Black-Scholes formula to dividend-paying assets when the dividend yield is constant. We break here with the convention of previous chapters and use S to denote the actual price of the asset, not the dividend-reinvested price. A constant dividend yield means that the dividend rate (dividend per unit time) at date t is $D_t = qS_t$ for a constant q. In this circumstance, the dividend-reinvested price defined in Section 13.1 is $e^{qt}S_t$. This is the value of a portfolio that starts with one share of the asset and reinvests dividends. If we start instead with e^{-qT} shares and reinvest dividends, then we will have a non-dividend-paying portfolio with value $Z_t = e^{qt}e^{-qT}S_t$ at each date t. In particular, $Z_T = S_T$, so a European option on the asset with price S is equivalent to a European option on the non-dividend-paying portfolio with price Z. Therefore, we can apply the Black-Scholes formula to value calls and puts on the asset with price S by using $Z_0 = e^{-qT} S_0$ as the initial asset price. This produces

$$\text{Call Price} = e^{-qT} S_0 N(d_1) - e^{-rT} K N(d_2), \qquad (16.28a)$$

$$\text{Put Price} = e^{-rT} K N(-d_2) - e^{-qT} S_0 N(-d_1), \qquad (16.28b)$$

where

$$d_1 = \frac{\log(S_0/K) + \left(r - q + \frac{1}{2}\sigma^2\right) T}{\sigma \sqrt{T}}, \qquad (16.28c)$$

$$d_2 = d_1 - \sigma \sqrt{T}. \qquad (16.28d)$$

In this circumstance, the delta of a call option is $e^{-qT} N(d_1)$, and the delta of a put is $e^{-qT} N(-d_1)$. The drift of dZ/Z under the risk-neutral probability is r, so the drift of dS/S under the risk-neutral probability is $r - q$. The fundamental PDE for the value $V(t, S_t)$ of an option is

$$V_t + (r - q) s V_s + \frac{1}{2} \sigma^2 s^2 V_{ss} = r V. \qquad (16.29)$$

The fundamental PDE is again equivalent to the statement that $e^{-rt} V(t, S_t)$ is a local martingale under the risk-neutral probability.

The assumption of continuous dividend payments and a constant dividend yield is frequently made when valuing options on stock indices, because the dividend return for a stock index consists of frequent small dividends. This is a consequence of different companies paying dividends on different dates. However, it is not reasonable to assume a constant dividend yield for options on individual stocks, because owners of individual stocks usually receive dividends only four times per year. Section 17.3 discusses the valuation of options on assets that pay discrete dividends.

16.9 NOTES AND REFERENCES

The bounds on option prices and the fact that "calls are better alive than dead" are due to Merton (1973b). The Black-Scholes formula is frequently called the Black-Scholes-Merton formula in recognition of Merton's contributions, especially the generalization discussed in Section 17.3. Use of the martingale representation theorem to deduce market completeness, as in Section 16.4 and elsewhere in this book, is due to Harrison and Pliska (1981). Valuation with risk-neutral probabilities is due to Cox and Ross (1976a,b) and Harrison and Kreps (1979). The use of other numeraires/probabilities, such as in Sections 16.3 and 16.4, is developed systematically by Geman, El Karoui, and Rochet (1995) and Schroder (1999).

Option Pricing

The value $E[M_T \max(0, S_T - K)]$ of a European call can be calculated by other means than that used in Section 16.4. For example, we can work entirely under the risk-neutral probability and calculate it as

$$e^{-rT} E^R[S_T 1_A] - e^{-rT} K E^R[1_A],$$

where A is the event $S_T > K$. The second expectation is shown in Section 16.4 to equal $e^{-rT} K N(d_2)$. The first expectation can be computed by noting that, for a non-dividend-paying asset,

$$\log S_T = \log S_0 + \left(r - \frac{1}{2}\sigma^2\right) T + \sigma \sqrt{T} \tilde{x},$$

where $\tilde{x} \stackrel{\text{def}}{=} B_T^* / \sqrt{T}$ is a standard normal under the risk-neutral probability. The event A is the event $\tilde{x} > -d_2$, so

$$E^R[S_T 1_A] = \int_{-d_2}^{\infty} \frac{1}{\sqrt{2\pi}} S_0 e^{(r-\sigma^2/2)T + \sigma\sqrt{T}x} e^{-x^2/2} \, dx$$

$$= e^{rT} S_0 \int_{-d_2}^{\infty} \frac{1}{\sqrt{2\pi}} e^{-(x-\sigma\sqrt{T})^2/2} \, dx$$

$$= e^{rT} S_0 \int_{-d_1}^{\infty} \frac{1}{\sqrt{2\pi}} e^{-y^2/2} \, dy$$

$$= e^{rT} S_0 [1 - N(-d_1)]$$

$$= e^{rT} S_0 N(d_1).$$

We used the change of variables $y = x - \sigma\sqrt{T}$ to obtain the third equality. Also, we can compute integrals to directly calculate $E[M_T S_T 1_A] - K E[M_T 1_A]$ without any changes of probabilities.

With constant r and σ, the expected returns of calls and puts under the physical probability are given in Exercise 16.6. Because calls are levered long positions in the underlying and puts are short positions, the risk premia of calls exceed the risk premium on the underlying, and the risk premia of puts are negative, assuming the underlying has a positive risk premium. The negative risk premium of a put can be understood as an insurance premium, because puts are insurance contracts for the underlying. The risk premia of calls and puts can be quite large in absolute value, as Exercise 16.6 illustrates. This point is emphasized by Broadie, Chernov, and Johannes (2009).

The fundamental PDE (16.20) can be put in a more standard form by defining

$$J(t, x) = V\left(t, S_0 e^{\sigma x + (r - \sigma^2/2)t}\right). \tag{16.30}$$

The fundamental PDE is equivalent to

$$J_t + \frac{1}{2} J_{xx} = rJ,$$

and the terminal condition $V(T,S) = \max(0, S - K)$ for a call is equivalent to

$$J(T,x) = \max\left(0, S_0 e^{\sigma x + (r - \sigma^2/2)T} - K\right).$$

The risk-neutral martingale $e^{-r(T-t)} E_t^R[V(T, S_T)]$ is the unique solution of the fundamental PDE (16.20) satisfying the terminal condition at T and for which J defined in (16.30) satisfies the condition

$$(\forall x) \quad \max_{0 \leq t \leq T} J(T,x) \leq A e^{ax^2} \qquad (16.31)$$

for any constants A and $0 < a < T/2$. This is a special case of the Feynman-Kac theorem—see Karatzas and Shreve (2004). Condition (16.31) is called a growth condition. Another standard formulation is obtained by defining

$$L(t,x) = e^{rt} J(T - t, x) = e^{rt} V\left(T - t, S_0 e^{\sigma x + (r - \sigma^2/2)(T-t)}\right).$$

The fundamental PDE is equivalent to

$$L_t = \frac{1}{2} L_{xx}.$$

This is the heat equation. The terminal condition $V(T,S) = \max(0, S - K)$ for a call becomes an initial condition for L, namely,

$$L(0,x) = \max\left(0, S_0 e^{\sigma x + (r - \sigma^2/2)T} - K\right).$$

As opposed to the HJB equation studied in Chapter 14, the fundamental PDE is a linear equation in the function V, meaning that a solution is a root of a linear operator on functions

$$V \mapsto V_t + rSV_S + \frac{1}{2}\sigma^2 S^2 V_{SS} - rV.$$

This reflects the linearity of valuation (the value of the sum of two payoffs is the sum of the values of the respective payoffs, and the value of a scalar multiple of a payoff is the scalar multiple of the value of the payoff) as opposed to the nonlinearity of utility functions.

Without assuming that portfolio processes and associated self-financing wealth processes are constrained to be such that MW is a martingale, we can say the following: Among all nonnegative self-financing wealth processes satisfying $W_T = \max(0, S_T - K)$, the Black-Scholes formula gives the smallest W_0. This is because MW is a supermartingale for any nonnegative self-financing wealth process; hence, $W_0 \geq \mathsf{E}[M_T W_T]$, and the right-hand side of this inequality is the Black-Scholes formula.

Important contributions to the theory of American options include McKean (1965), van Moerbeke (1976), Bensoussan (1984), and Karatzas (1988). For more thorough surveys and additional references, see Myneni (1992), Karatzas and Shreve (1998), Duffie (2001), or Shreve (2004).

A warrant is a call option issued by the firm on whose stock it is written. If exercised, the firm receives the exercise price and issues a new share. To value warrants, some assumption must be made about what the firm does with the proceeds from exercise, because that affects the value of the stock received by the exerciser. The values of warrants and the optimal exercise strategy also depend on whether all of the warrants are held by a single investor, by a small number of strategic investors, or by a large number of investors who act competitively. These considerations apply to the valuation of convertible bonds also. See Emanuel (1983), Constantinides (1984), and Spatt and Sterbenz (1988).

EXERCISES

16.1. A bull spread consists of a long call with strike K and a short call with strike $K' > K$. A collar consists of a long put with strike K and a short call with strike $K' > K$. Draw the payoff diagram (meaning the value at maturity as a function of the underlying asset price, as in Figure 16.1) of a bull spread, and draw the payoff diagram of a long position in the underlying asset combined with a collar. Use put-call parity to explain why the two payoff diagrams have the same shape.

16.2. A butterfly spread consists of a long call with strike $K - \Delta$, two short calls with strikes K, and a long call with strike $K + \Delta$, for $\Delta > 0$.

(a) Draw the payoff diagram of a butterfly spread. What is it a bet on?
(b) As explained in Section 16.1, selling straddles is a way to sell volatility, but it is clearly a high-risk strategy. To obtain insurance against the tail risks, we can add a long put and a long call. Draw the payoff diagram of a short straddle with strike K combined with a long put with strike $K - \Delta$ and a long call with strike $K + \Delta$.
(c) What principle underlies the relation between the payoff diagrams in Parts (a) and (b)? Explain.

16.3. Calculate the theta and gamma for the Black-Scholes call option formula and verify that it satisfies the fundamental PDE.

16.4. Use put-call parity to show that a European put and a European call with the same strike and time to maturity have the same gamma.

16.5. Assume constant r and σ and the "no arbitrage" assumptions of Section 16.2. Calculate the value and delta of a European derivative security that pays S_T^2 at date T.

16.6. Consider an asset with a constant dividend yield q. Assume the price S of the asset satisfies
$$\frac{dS}{S} = (\mu - q)\,dt + \sigma\,dB,$$
where B is a Brownian motion under the physical probability, and μ and σ are constants. Consider a European call and a European put with strike K on the asset. Assume the risk-free rate is constant, and adopt the assumptions of Section 16.2.

(a) Let A denote the event $S_T > K$. Show that $E[S_T 1_A] = e^{(\mu-q)T} S_0 N(d_1^*)$, where
$$d_1^* = \frac{\log(S_0/K) + \left(\mu - q + \frac{1}{2}\sigma^2\right)T}{\sigma\sqrt{T}}.$$
Hint: This can be computed directly under the physical probability or by changing probabilities using $e^{(q-\mu)T} S_T/S_0$ as the Radon-Nikodym derivative.

(b) Show that $E[K 1_A] = K N(d_2^*)$, where $d_2^* = d_1^* - \sigma\sqrt{T}$.

(c) It follows from the previous parts that the expected return of the European call under the physical probability, if held to maturity, is
$$\frac{e^{(\mu-q)T} S_0 N(d_1^*) - K N(d_2^*)}{e^{-qT} S_0 N(d_1) - e^{-rT} K N(d_2)},$$
where d_1 and d_2 are defined in (16.28). Assuming $T = 1$, $\mu = 0.12$, $r = 0.04$, $q = 0.02$, and $\sigma = 0.20$, show that the expected *rate* of return of a European call that is 20% out of the money $(S_0/K = 0.8)$ exceeds 100%.

(d) Show that the expected return of the European put under the physical probability, if held to maturity, is
$$\frac{K N(-d_2^*) - e^{(\mu-q)T} S_0 N(-d_1^*)}{e^{-rT} K N(-d_2) - e^{-qT} S_0 N(-d_1)}.$$
Assuming $T = 1$, $\mu = 0.12$, $r = 0.04$, $q = 0.02$, and $\sigma = 0.20$, show that the expected *rate* of return of a European put that is 20% out of the money $(S_0/K = 1.2)$ is less than -50%.

16.7. Consider a second risky non-dividend-paying asset with price Z. Assume

$$\frac{dZ}{Z} = \mu_z \, dt + \sigma_z \, dB_z,$$

where B_z is a Brownian motion under the physical probability. Let ρ denote the correlation process of B_z and B.

(a) Define $\xi_t = M_t S_t = E_t[M_T S_T]$. Show that

$$\frac{d\xi}{\xi} = r \, dt + \sigma \, dB + \frac{dM}{M}.$$

(b) Prove that

$$(dB_z)\left(\frac{d\xi}{\xi}\right) = \left(\rho\sigma - \frac{\mu_z - r}{\sigma_z}\right) dt.$$

(c) Use Girsanov's theorem to show that

$$\frac{dZ}{Z} = (r + \rho\sigma\sigma_z) \, dt + \sigma_z \, dB_z^*,$$

where B_z^* is a Brownian motion under prob^S.

16.8. This exercise verifies the assertion (16.6) regarding the dynamics of an SDF process. Suppose the information in the economy is given by independent Brownian motions $B_1, \ldots, B_k$. Consider a non-dividend-paying asset with price S satisfying

$$\frac{dS}{S} = \mu \, dt + \sum_{i=1}^{k} \sigma_i \, dB_i$$

for stochastic processes μ and σ_i.

(a) Define a stochastic process σ and Brownian motion B^s such that

$$\frac{dS}{S} = \mu \, dt + \sigma \, dB^s.$$

(b) Consider an SDF process M. We have

$$\frac{dM}{M} = -r \, dt - \sum_{i=1}^{k} \lambda_i \, dB_i$$

for some stochastic processes λ_i. Define a stochastic process λ and Brownian motion B^m such that

$$\frac{dM}{M} = -r \, dt - \lambda \, dB^m.$$

(c) Let ρ denote the correlation process of B^s and B^m. Show that
$$\lambda \rho = (\mu - r)/\sigma.$$
(d) Show that there is a local martingale Z such that
$$\frac{dM}{M} = -r\,dt - \frac{\mu - r}{\sigma}\,dB^s + dZ$$
and $(dB^s)(dZ) = 0$.

16.9. A compound option is an option on an option. Suppose there is a constant risk-free rate, and the underlying asset price has a constant volatility. Consider a European call option with strike K' maturing at T'. Consider now a European option maturing at $T < T'$ to purchase the first call option at price K. The purpose of this exercise is to value the "call on a call." The value at T of the underlying call is given by the Black-Scholes formula with $T' - T$ being the time to maturity. Denote this value by $V(T, S_T)$. The value at T of the call on a call is
$$\max(0, V(T, S_T) - K).$$
Let S^* denote the value of the underlying asset price such that $V(T, S^*) = K$. The call on a call is in the money at its maturity T if $S_T > S^*$, and it is out of the money otherwise. Let A denote the set of states of the world such that $S_T > S^*$, and let 1_A denote the random variable that equals 1 when $S_T > S^*$ and 0 otherwise. The value at its maturity T of the call on a call is
$$V(T, S_T)1_A - K1_A.$$

(a) What is the value at date 0 of receiving the payoff $K1_A$ at date T?
(b) To value receiving $V(T, S_T)1_A$ at date T, let C denote the set of states of the world such that $S_{T'} > K'$, and let 1_C denote the random variable that equals 1 when $S_{T'} > K'$ and 0 otherwise. Recall that $V(T, S_T)$ is the value at T of receiving $S_{T'}1_C - K'1_C$ at date T'. Hence, the value at date 0 of receiving $V(T, S_T)1_A$ at date T must be the value at date 0 of receiving $(S_{T'}1_C - K'1_C)1_A$ at date T'.

(i) Show that the value at date 0 of receiving $V(T, S_T)1_A$ at date T is
$$S_0 \operatorname{prob}^S(D) - K' e^{-rT'} \operatorname{prob}^R(D),$$
where $D = A \cap C$.
(ii) Show that $\operatorname{prob}^S(D)$ is the probability that
$$-\frac{B_T^*}{\sqrt{T}} < d_1 \quad \text{and} \quad -\frac{B_{T'}^*}{\sqrt{T'}} < d_1', \qquad (16.32)$$

where
$$d_1 = \frac{\log(S_0/S^*) + \left(r + \tfrac{1}{2}\sigma^2\right)T}{\sigma\sqrt{T}},$$
$$d_1' = \frac{\log(S_0/K') + \left(r + \tfrac{1}{2}\sigma^2\right)T'}{\sigma\sqrt{T'}},$$

and where B^* denotes a Brownian motion under prob^S. Note that the random variables in (16.32) are standard normals under prob^S with a correlation equal to $\sqrt{T/T'}$. Therefore, $\text{prob}^S(D)$ can be computed from the bivariate normal cumulative distribution function.

(iii) Show that $\text{prob}^R(D)$ is the probability that

$$-\frac{B_T^*}{\sqrt{T}} < d_2 \quad \text{and} \quad -\frac{B_{T'}^*}{\sqrt{T'}} < d_2', \qquad (16.33)$$

where
$$d_2 = d_1 - \sigma\sqrt{T},$$
$$d_2' = d_1' - \sigma\sqrt{T'},$$

and where B^* now denotes a Brownian motion under prob^R.

16.10. Calculate the value of a call on a put assuming a constant risk-free rate and a constant volatility for the underlying asset price.

17

Forwards, Futures, and More Option Pricing

CONTENTS

17.1 Forward Measures 432
17.2 Forwards and Futures 433
17.3 Margrabe, Black, and Merton Formulas 437
17.4 Implied and Local Volatilities 443
17.5 Stochastic Volatility 445
17.6 Notes and References 449

This chapter explains forward and futures contracts and presents the spot-forward parity formula and expectations hypothesis. It presents the formula of Margrabe (1978) for exchange options, the formula of Black (1976) for options on forwards and futures, and the formula of Merton (1973b) for options when there is no money market account, or when the risk-free rate varies randomly. The last part of the chapter deals with volatilities: estimating volatilities from option prices and valuing options when the volatility of the underlying asset changes randomly over time. As in Section 16.8, we break with the convention of previous chapters and use S to denote the actual price of the underlying asset, not the dividend-reinvested price. If there is a constant dividend yield q, then $e^{qt}S_t$ or $e^{-q(T-t)}S_t$ is the dividend-reinvested price. We start by defining forward measures, which are used in the chapter and are of independent interest.

17.1 FORWARD MEASURES

A probability created by using a discount bond as the numeraire is called a forward probability or, more commonly, a forward measure. We create the forward measure using the same steps used to create the risk-neutral probability and the probability prob^S in Chapter 16. Let P denote the price of a discount bond maturing at T, so $P_T = 1$. Let M be an SDF process. Assume MP is a martingale. The Radon-Nikodym derivative for the forward measure is $M_T P_T / P_0 = M_T / P_0$. In other words, we define

$$\text{prob}^P(A) = \frac{1}{P_0} \mathsf{E}[M_T 1_A] \tag{17.1}$$

for each event A observable at date T. Denote expectation with respect to prob^P by E^P. By a standard result cited several times before (Appendix A.12), if Y is a stochastic process such that MY is a martingale under the physical probability, then $MY/(MP) = Y/P$ is a martingale under the forward measure prob^P.

The forward measure equals the risk-neutral probability when the short rate is nonrandom. They are equal because $1/P_0$ in (17.1) equals $1/R_T$ in the definition (15.5) of the risk-neutral probability when r is nonrandom. Thus, the two probabilities are the same. When interest rates are random, the forward measure is sometimes more useful than the risk-neutral probability.

17.2 FORWARDS AND FUTURES

A forward contract is a contract to make an exchange at a future date. One party (the "long") agrees at date t to purchase an asset at date $u > t$ from another party (the "short") at a price F_t. The price F_t is called the forward price, and it is fixed at date t as part of the contract. When the additional specificity seems convenient, we write $F_t(u)$ for the forward price at t for a contract maturing at u. We assume neither party to a forward contract makes a payment to the other at date t or at any date prior to u (though in reality, some collateral may be required). Let S denote the price of the underlying asset. Then, the value to the long party at the maturity u of a forward contract is $S_u - F_t(u)$, and the value to the short party at maturity is $F_t(u) - S_u$.

It is useful to compute how the value of a forward contract evolves over time. Suppose we enter into a forward as the long at date s at some forward price K (for example, $K = F_s$). The value at $t > s$ can be seen as follows. We could unwind the long position at t by selling a forward at the market forward price $F_t(u)$. The delivery/receipt obligations for the underlying asset of the short and long forwards would cancel. On the long forward, we are obligated to pay K at u. On the short forward, we receive $F_t(u)$ at u. Thus, we would receive a net cash flow of $F_t(u) - K$ at u. A cash flow of $F_t(u) - K$ at date u is worth $[F_t(u) - K]P_t(u)$ at date t, where, as before, $P_t(u)$ denotes the price at t of a discount bond maturing at u. Therefore, the value of the long forward at t must be $[F_t(u) - K]P_t(u)$.

Another useful observation is that $F_t(u)P_t(u)$ is the value of a non-dividend-paying portfolio for $t \leq u$. Specifically, we can go long the forward contract at date 0 at price $F_0(u)$ and simultaneously buy $F_0(u)$ units of the discount bond that matures at u. The previous paragraph shows that the value of the forward at each date $t \in [0, u]$ is $[F_t(u) - K]P_t(u)$ with $K = F_0(u)$. The value of the bonds at date t is $KP_t(u)$. So, the value of the portfolio consisting of the long forward and the bonds is $F_t(u)P_t(u)$. This portfolio is a way to own the underlying asset at date u by making a payment at date 0 and having no further cash flows between date 0 and date u. The cost of buying the asset without interim cash flows is the portfolio value at date 0, namely, $F_0(u)P_0(u)$. Date 0 is arbitrary in this discussion. The forward and discount bonds can be used in this way to buy the underlying asset without interim cash flows at any date $t < u$ at cost $F_t(u)P_t(u)$.

Because FP is the value of a non-dividend-paying portfolio, MFP is a local martingale for any SDF process M. Assume MFP is actually a martingale. Then, as discussed in Section 17.1, $MFP/(MP) = F$ is a martingale under the forward measure. To summarize, given regularity conditions, the forward price $F_t(u)$ is a martingale under the forward measure defined by the discount bond with the same maturity u.

Spot-Forward Parity and Convergence

It is easy to replicate a forward contract if the underlying asset has a constant dividend yield. Replicating a forward is called creating a synthetic forward. To do so, we simply buy the asset with borrowed funds. Dividends paid by the asset before the forward matures are reinvested in the asset. This produces a portfolio that has cash flows only at the forward maturity. Let q denote the dividend yield, and let u denote the date at which the forward matures. At any date $t < u$, we can buy $e^{-q(u-t)}$ units of the underlying asset and finance the purchase by shorting $e^{-q(u-t)}S_t/P_t(u)$ units of the discount bond maturing at u. Through reinvestment of dividends, we will have a full share of the underlying asset at date u. We will have to pay $e^{-q(u-t)}S_t/P_t(u)$ at u to cover the short of the discount bond. Note that this is a value known at date t. Because this strategy replicates the forward, the market forward price at t must be

$$F_t(u) = \frac{e^{-q(u-t)}S_t}{P_t(u)}. \qquad (17.2)$$

Equation (17.2) is called spot-forward parity (the price S_t is the "spot price" of the asset at date t). It is not strictly necessary to assume that the asset pays continuous dividends. Exercise 17.1 presents the spot-forward parity formula for an underlying asset that pays discrete known dividends at a finite number of known dates.

A special case of spot-forward parity is that $F_u(u) = S_u$. This is called spot-forward convergence, because it reflects the fact that spot and forward prices must converge as the forward approaches maturity. Of course, at maturity, the forward contract is actually a spot contract, whence the convergence. Spot-forward convergence must hold even for assets that do not have constant dividend yields and even for assets for which dividends are not known at the time the forward is initiated. Hence, spot-forward convergence is more general than spot-forward parity.

Futures Contracts

A futures contract is an exchange-traded forward. The exchange requires collateral (margin) and marking of the contract to market. Marking to market means that when the futures price increases, the long party receives the change in the futures price (as a deposit to her margin account) and the short party pays the change in the futures price (as a debit to her margin account). The delivery price on the contract is simultaneously adjusted to the market futures price. The reverse occurs when the futures price falls. Marking to market causes a futures contract to always have a zero value. It could be canceled with no further cash flows by making an offsetting trade (selling if long and buying if short). The marking-to-market cash flows occur at the end of each trading day on an actual exchange, but it is simpler to model them as occurring continuously.

Let S denote the price of the underlying asset, and let $\hat{F}_t(u)$ denote the futures price at t for a contract maturing at u. We sometimes drop the u argument when the maturity date is understood. We must have $\hat{F}_u(u) = S_u$. In other words, there must be spot-futures convergence, just as there is spot-forward convergence. Assume $\hat{F}$ is an Itô process. In this model, the Itô differential $d\hat{F}$ is the cash flow that is received by the long party and paid by the short party. Assume the cash flows are paid to or debited from the money market account; that is, they earn interest at the instantaneous risk-free rate. Thus, if we purchase at date s a futures contract with maturity u and hold it until date $t \in [s, u]$, we will have

$$V_t \stackrel{\text{def}}{=} R_t \int_s^t \frac{1}{R_\tau} \, d\hat{F}_\tau \tag{17.3}$$

at date t. The process V is a self-financing wealth process.

Expectations Hypotheses

The expectations hypothesis for forward prices states that the forward price is the best predictor of the future price of the underlying asset in the sense that $F_t(u) = \mathsf{E}_t[S_u]$. There is also an expectations hypothesis for futures prices which states that $\hat{F}_t(u) = \mathsf{E}_t[S_u]$. To see the motivation for the hypothesis, consider the forward price and suppose for simplicity that there is a constant risk-free rate r and the underlying asset has a constant dividend yield q. As is shown below, forward prices equal futures prices when there is a constant risk-free rate. Under these hypotheses, the spot-forward parity formula (17.2) is

$$\hat{F}_t(u) = F_t(u) = \mathrm{e}^{(r-q)(u-t)} S_t. \tag{17.4}$$

Given spot-forward parity, the expectations hypothesis is equivalent to

$$S_t = \mathrm{e}^{-r(u-t)} \mathsf{E}_t \left[\mathrm{e}^{q(u-t)} S_u \right].$$

The right-hand side is how the underlying asset would be valued at t if investors were risk neutral. Thus, the expectations hypothesis is a hypothesis about risk-neutral valuation.

The expectations hypothesis has been shown to be invalid empirically. This is not surprising, because risk neutrality is an unreasonable assumption.[1] With risk-averse investors, we can derive the expectations hypothesis only by changing probabilities. The expectations hypothesis for the forward price should be true under the forward measure. The expectations hypothesis for the futures price should be true under the risk-neutral probability.

The expectations hypothesis for the forward price should be true under the forward measure, because, as explained above, the forward price is a martingale under the forward measure, given regularity conditions. When we combine the martingale property with spot-forward convergence, we obtain

$$F_t(u) = \mathsf{E}_t^P[S_u]. \tag{17.5}$$

The expectations hypothesis for the futures price should be true under the risk-neutral probability, because V defined in (17.3) is a self-financing wealth process, so MV is a local martingale. Under regularity conditions, it is a martingale. In this case,

$$\frac{V}{R} = \int_s^t \frac{1}{R_\tau} \, \mathrm{d}\hat{F}_\tau$$

is a martingale under the risk-neutral probability. For this to be true, it must be that $\hat{F}$ is a martingale or at least a local martingale under the risk-neutral

---

1. There are related empirical results that are surprising. This is discussed in the end-of-chapter notes.

probability. Assume it is a martingale and apply spot-futures convergence to obtain

$$\hat{F}_t = \mathsf{E}^R_t[S_u]. \qquad (17.6)$$

Equality of Forwards and Futures Prices

As noted in Section 17.1, if the short rate is nonrandom, then the risk-neutral probability and the forward measure are the same. Thus, the two expectations hypotheses (17.5) and (17.6) imply that forward prices equal futures prices when the short rate is nonrandom.

17.3 MARGRABE, BLACK, AND MERTON FORMULAS

This section derives three extensions of the Black-Scholes formula. Each extension builds on the previous one, and each builds on and generalizes the Black-Scholes formula. We price exchange options (Margrabe, 1978), options on forwards and futures (Black, 1976), and options on assets that can pay either continuous or discrete dividends, allowing for random interest rates (Merton, 1973b).

Exchange Options

An exchange option is an option to exchange two assets. The party who is long an exchange option has the choice whether to exercise it and will do so when the value of the asset she receives is greater than the value of the asset she delivers. Call the asset she receives asset 1 and the asset she delivers asset 2. Let S_1 and S_2 denote the prices of the assets. The value of the option at the date T that it matures is

$$\max(0, S_{1T} - S_{2T}).$$

The value to the seller of the option is the negative of this (as with puts and calls, the values for the short and long sum to zero). Calls and puts are special cases of exchange options. A call option is an exchange option in which cash is exchanged for the underlying asset. A put option is an exchange option in which the underlying asset is exchanged for cash. Here, cash means a position in the money market account or in the discount bond that matures at the option maturity date.

Assume the two assets have constant dividend yields q_1 and q_2, constant volatilities σ_1 and σ_2, and a constant correlation ρ. Then, the volatility of S_1/S_2

and also the volatility of S_2/S_1 is equal to σ, defined as $\sqrt{\sigma_1^2 - 2\rho\sigma_1\sigma_2 + \sigma_2^2}$.[2] Assume there is an SDF process M such that MS_1 and MS_2 are martingales (sufficient conditions, including Novikov's condition, can be adapted from Section 16.2). We do *not* need to assume that there is a money market account, nor do we need to assume that there are traded discount bonds. The existence of those assets is neither necessary nor useful for valuing or hedging an exchange option.

We want to show that there is a self-financing wealth process W for which $W_T = \max(0, S_{1T} - S_{2T})$. We then want to calculate $E[M_T \max(0, S_{1T} - S_{2T})]$. We can do these two things directly (Exercises 17.3–17.4). However, we can also derive them from the Black-Scholes formula. The argument is based on a change of numeraire. Assume for simplicity that the assets do not pay dividends ($q_1 = q_2 = 0$). The more general case is obtained from this special case by using the fact that $e^{-q_i(T-t)}S_{it}$ is the price of a non-dividend-paying asset, as in Section 16.8. Use asset 2 as the numeraire. The price of asset 2 in this numeraire is $S_2/S_2 = 1$. Thus, asset 2 is risk free in this numeraire with rate of return equal to 0. Define $R^* = 1$. The price of asset 1 in this numeraire is $S^* = S_1/S_2$. For valuing assets in the new numeraire, $M^* = MS_2/S_{20}$ is an SDF process. This is analogous to the relation between nominal and real SDF processes discussed in Section 13.7. To see that M^* is an SDF process in the new numeraire, it suffices to observe that (i) $M_0^* = M_0 = 1$ and (ii) for any non-dividend-paying asset with price S in the original numeraire and price S/S_2 in the new numeraire, we have $M^*S/S_2 = MS/S_{20}$, which is a local martingale. We are assuming MS_1 and MS_2 are martingales, so $M^*S^* = MS_1/S_{20}$ is a martingale, and $M^*R^* = MS_2/S_{20}$ is a martingale. Hence, with an appropriate constraint on portfolio processes, the assumptions of Section 16.2 are satisfied for the underlying asset price S^*, money market price R^*, and SDF process M^*.

The payoff of the exchange option is

$$S_{2T} \max\left(0, \frac{S_{1T}}{S_{2T}} - 1\right) = S_{2T} \max(0, S_T^* - 1). \tag{17.7}$$

2. By Itô's formula,

$$\frac{d(S_1/S_2)}{S_1/S_2} = \frac{dS_1}{S_1} - \frac{dS_2}{S_2} - \left(\frac{dS_1}{S_1}\right)\left(\frac{dS_2}{S_2}\right) + \left(\frac{dS_2}{S_2}\right)^2.$$

Therefore,

$$\left(\frac{d(S_1/S_2)}{S_1/S_2}\right)^2 = \left(\frac{dS_1}{S_1}\right)^2 - 2\left(\frac{dS_1}{S_1}\right)\left(\frac{dS_2}{S_2}\right) + \left(\frac{dS_2}{S_2}\right)^2 = (\sigma_1^2 - 2\rho\sigma_1\sigma_2 + \sigma_2^2)\,dt = \sigma^2\,dt.$$

Forwards, Futures, and More Option Pricing

In the new numeraire, the payoff of the exchange option is $\max(0, S^*_T - 1)$. The market completeness result of Section 16.4 shows that there is a trading strategy in the underlying asset with price S^* and money market account with price R^* that has payoff $\max(0, S^*_T - 1)$. When we convert to the original numeraire, the payoff is the payoff of the exchange option. Thus, the payoff of the exchange option can be replicated using the assets with prices S_1 and S_2 (Exercise 17.2). Furthermore,

$$E[M_T \max(0, S_{1T} - S_{2T})] = S_{20}E[M^*_T \max(0, S^*_T - 1)],$$

and the expectation on the right-hand side is given by the Black-Scholes call option formula. Multiplying the Black-Scholes call formula (for the underlying with price S^* and strike equal to 1) by S_{20} shows that the value of the exchange option is the Black-Scholes call option formula (16.28), where we input

- underlying asset price $S_0 = S_{10}$,
- dividend yield $q = q_1$,
- interest rate $r = q_2$,
- strike price $K = S_{20}$,
- volatility $\sigma = \sqrt{\sigma_1^2 - 2\rho\sigma_1\sigma_2 + \sigma_2^2}$.

This is called Margrabe's formula.

Options on Forwards

Options on forwards work slightly differently from standard options. When an option on a forward is exercised, a forward contract is created with forward price equal to the option strike. The strike is paid and the underlying asset is delivered when the forward matures. Exercise of a call on a forward creates a long forward for the party that exercises the option. Exercise of a put on a forward creates a short forward for the party that exercises the option. The option counterparty becomes the counterparty to the forward contract.

Consider European options on a forward contract. Let T denote the date at which the options mature and $u \geq T$ the date at which the forward matures. Let K denote the strike. Assume there is a traded discount bond maturing at u with price $P_t(u)$. Let $F_t(u)$ denote the forward price. Section 17.2 shows that FP is a self-financing wealth process (the portfolio consists of a long forward and discount bonds maturing at the forward maturity date). Hence, MP and MFP are local martingales for any SDF process M. Assume there is an SDF process M such that MP and MFP are martingales, and assume the volatility σ of the forward price is constant.

If an option is exercised at T, then the value of the long forward that is created is shown in Section 17.2 to equal $F_T(u)P_T(u) - KP_T(u)$. Therefore, the value of the call option at its maturity is

$$\max(0, F_T(u)P_T(u) - KP_T(u)). \tag{17.8a}$$

Likewise, the value of a European put at its maturity is

$$\max(0, KP_T(u) - F_T(u)P_T(u)). \tag{17.8b}$$

Calls and puts on forwards are equivalent to exchange options. A call on a forward is equivalent to an exchange option in which the asset with price KP (which is K units of the discount bond maturing at u) is exchanged for the asset with price FP. A put on a forward is equivalent to the reverse exchange option. Thus, the values of calls and puts on forwards are given by Margrabe's formula. Specifically, they are given by the Black-Scholes formula (16.28), where we input

- underlying asset price $S_0 = F_0(u)P_0(u)$,
- dividend yield $q = 0$,
- interest rate $r =$ yield y of the discount bond maturing when the forward matures,
- strike price K,
- volatility σ of the forward price.

By the definition of the discount bond yield, the underlying asset price can be written as $S_0 = e^{-yu}F_0(u)$.

Options on Futures

Futures options are similar to options on forwards except for marking to market. When a futures option is exercised, a futures contract is created with futures price equal to the option strike. The futures contract is immediately marked to market, so the futures price is reset to the market futures price and a cash flow equal to the difference between the option strike and the market futures price is transferred. The remainder of this subsection assumes that the short rate is nonrandom (we must acknowledge that this is an unreasonable assumption for valuing options on bond futures). Then, the futures price equals the forward price. Hence, the values of futures options differ from the values of forward options only because of the marking to market of the futures contract created by exercise of a futures option.

Consider European options on a futures contract. Let T denote the date at which the options mature, $u \geq T$ the date at which the futures matures, $\hat{F}_t(u)$ the futures price, and K the option strike. The value of a call on a futures at the call's maturity date T is $\max(0, \hat{F}_T(u) - K)$, and the value of a put on a futures is $\max(0, K - \hat{F}_T(u))$. These equal the values of forward options given in (17.8) divided by $P_T(u)$. Therefore, the values at date 0 of futures options equal the values at date 0 of forward options divided by $P_T(u)$. Consequently, with a nonrandom short rate, the values of calls and puts on futures are given by the Black-Scholes formula (16.28), where we input

- underlying asset price $S_0 = \hat{F}_0(u)P_0(T)$,
- dividend yield $q = 0$,
- interest rate $r =$ yield y of the discount bond maturing when the option (not the futures) matures,
- strike price K,
- volatility σ of the futures price.

By the definition of the discount bond yield, the underlying asset price can be written as $S_0 = e^{-yT}\hat{F}_0(u)$.

Arbitrary Dividends and Random Interest Rates

Now, consider European options on an underlying asset that may pay either continuous or discrete dividends. We do not assume that the interest rate is constant. We do not even need to assume the existence of a money market account. However, we do assume that there is a traded discount bond that matures when the options mature. Also, we assume there is a traded forward contract that matures when the options mature.[3] Assume the forward price has a constant volatility, and assume there is an SDF process M such that MP and MFP are martingales.

We use the discount bond and the forward to create the portfolio with price FP discussed in Section 17.2. As explained there, the portfolio is a way to own the asset at the date the bond and forward mature without receiving dividends in the interim. The process FP is a self-financing wealth process with value $F_T(T)P_T(T) = F_T(T) = S_T$ at date T, using successively the facts that the discount bond pays 1 at its maturity and spot-forward convergence. Thus, an option on the portfolio with price FP that matures at T is equivalent to an option on the underlying asset. To avoid assuming a constant risk-free rate, consider the option as an exchange option (as in the discussion of forward options) where K units of the discount bond maturing at T are exchanged for the portfolio with price FP. Then, option values are given by the formula for options on forwards,

---

3. We do not need to assume the discount bond is traded directly if there is a money market with a nonrandom interest rate, because then the discount bond can be created from the money market account. Also, we do not need to assume that the forward is traded directly if it can be created synthetically using the underlying asset as discussed in connection with the spot-forward parity formula.

substituting $u = T$ for the maturity of the forward and discount bond. Specifically, option values are given by the Black-Scholes formula (16.28), where we input

- underlying asset price $S_0 = F_0(T)P_0(T)$,
- dividend yield $q = 0$,
- interest rate $r =$ yield y of the discount bond maturing when the option matures,
- strike price K,
- volatility σ of the forward price.

By the definition of the discount bond yield, the underlying asset price can be written as $S_0 = e^{-yT} F_0(T)$.

An obvious difficulty with applying this formula is that there may not be a traded forward contract. This is not a problem if the underlying asset has a constant dividend yield or pays known dividends at known dates, because in those cases the forward can be created synthetically (see Exercise 17.1 for an example with known discrete dividends paid at known dates). Suppose for the remainder of this section that the underlying asset has a constant dividend yield q. Then, the spot-forward parity formula (17.2) implies that the underlying asset price above is $S_0 = F_0(T)P_0(T) = e^{-qT} S_0$. Inspection of the Black-Scholes formula shows that inputting this initial asset price and a dividend yield of 0 is equivalent to inputting an initial asset price of S_0 and a dividend yield of q (this is how the formula (16.28) is derived from (16.15)). Thus, with a constant dividend yield but a possibly random short rate (or a nonexistent money market account), the values of European puts and calls are given by the Black-Scholes formula (16.28) with two substitutions: We should use the yield of the discount bond maturing when the option matures as the interest rate r, and we should use the volatility of the forward price as the volatility σ. When the short rate is nonrandom, the spot-forward parity formula implies that the volatility of the forward price equals the volatility of the underlying asset price. In general, the two volatilities differ, with the difference depending on the volatility of the discount bond price. The difference is small for options with short maturities, because discount bonds have small volatilities when they are near maturity. Exercise 17.5 presents an example.

The market completeness result for exchange options (Exercises 17.2 and 17.3) shows that the value V of an exchange option satisfies

$$\frac{dV}{V} = \pi \frac{dS_1}{S_1} + (1-\pi) \frac{dS_2}{S_2}$$

for some π. In our current context, $S_1 = FP$ and $S_2 = KP$. Because K is a constant, $dS_2/S_2 = dP/P$. With a constant dividend yield, spot-forward parity implies that

$dS_1/S_1 = dS/S$, where S is the underlying asset price as before. Thus,

$$\frac{dV}{V} = \pi \frac{dS}{S} + (1-\pi)\frac{dP}{P}.$$

This shows that, with random interest rates, the value of a European option on an asset with a constant dividend yield can be replicated using the underlying asset and the discount bond that matures when the option matures.

17.4 IMPLIED AND LOCAL VOLATILITIES

Volatilities must be estimated before option pricing formulas can be applied. A common way to estimate a volatility for use in an option pricing formula (and for other purposes) is to imply it from the price of a liquid (that is, frequently and easily traded) option. The implied volatility is the volatility that, when input into the option pricing formula, produces the market price. It is the volatility that matches the model to the market. In practice, implied volatilities will be different for options with different strikes and for options with different maturities. There is an easy extension to the Black-Scholes formula that allows us to interpret different implied volatilities for different maturities as the result of the actual volatility changing nonrandomly over time. There are further extensions that accommodate different implied volatilities at both different strikes and different maturities. We first consider matching implied volatilities at different maturities.

Term Structure of Volatilities

Let T denote the maturity date of an option. Suppose

$$\frac{dS_t}{S_t} = \mu_t \, dt + \sigma(t) \, dB_t,$$

where $\sigma(\cdot)$ is a nonrandom function of time. The Black-Scholes formula is valid in this circumstance if we replace the volatility σ in the formula by

$$\sigma_{\text{avg}}(T) \overset{\text{def}}{=} \sqrt{\frac{1}{T}\int_0^T \sigma(t)^2 \, dt}. \qquad (17.9)$$

The proof of the formula with this extension is virtually unchanged. For example, we obtain

$$\log S_T = \log S_t + \left(r - \frac{1}{2}\sigma_{\text{avg}}(T)^2\right)T + \int_0^T \sigma(t) \, dB_T^*,$$

where B^* is a Brownian motion under the risk-neutral probability. The random variable

$$\int_0^T \sigma(t)\,dB_T^*$$

is normally distributed with mean zero and variance $T\sigma_{\text{avg}}(T)^2$ under the risk-neutral probability. Therefore, the derivations of (A′) and (B′) in Section 16.4 go through unchanged. Using this extension, we can usually fit different implied volatilities at different maturities (for example, for the at-the-money option at each maturity) by constructing a nonrandom function $\sigma(t)$ such that, for each T, $\sigma_{\text{avg}}(T)$ equals the implied volatility for the option with time T to maturity. The mapping $t \mapsto \sigma(t)$ is called the term structure of implied volatilities.

Smiles and Smirks

When we fix the maturity and consider options with different strikes, we usually find that implied volatilities are higher for low strikes and for high strikes than they are for strikes near the current value of the underlying asset. In other words, implied volatilities are typically higher for deep out-of-the-money and for deep in-the-money options than they are for options that are near the money. When we graph implied volatilities against strikes, plotting strikes on the horizontal axis and implied volatilities on the vertical axis, the plot looks like a smile, and the pattern is commonly called an option smile. Implied volatilities for equity index options are commonly higher for low strikes than for high strikes, so the smile appears twisted. It is called a smirk. The smirk means that there are especially high prices for deep out-of-the money puts compared to the Black-Scholes formula. This may reflect high demand for insurance against crashes, which would correspond to a risk-neutral distribution for $\log S_T$ that is negatively skewed instead of normal.

Local Volatility

A local volatility model is a model in which volatility changes both with time and with the underlying asset price. In other words, volatility is a function $\sigma(t,s)$ of time t and the underlying asset price s. A function of this form is called a volatility surface. Subject to MR and MS being martingales, the market completeness result of Section 16.4 is still valid in this context, and the pricing formula

$$e^{-qT} S_0 \operatorname{prob}^S(A) - e^{-yT} K \operatorname{prob}^R(A)$$

for a call and corresponding formula for a put are still valid (A is the event $S_T > K$ for a call, as before). There are different proposals in the literature for functional forms for σ to match the model to market prices. In all such models, the probabilities $\text{prob}^S(A)$ and $\text{prob}^R(A)$ are computed numerically.

Practitioners typically construct different volatility surfaces each day (or more frequently) to match market option prices. There is an inconsistency in doing this, because if the function $\sigma(t,s)$ changes each day, then it must depend on more than time and the underlying asset price. There must be an omitted factor. Stochastic volatility models described in the next section do not suffer from this inconsistency and are at least somewhat useful for producing smiles and smirks of the sort seen in market data, though they generally have too few parameters to match a given smile or smirk exactly.

17.5 STOCHASTIC VOLATILITY

If the volatility σ is a stochastic process not locally perfectly correlated with the asset price S, then options cannot be replicated using the underlying asset and the risk-free asset. In this circumstance, we cannot price an option unless we make some assumption about the market price of risk for the volatility process. This is called equilibrium asset pricing—as opposed to arbitrage pricing—because it involves an assumption about which SDF process is appropriate for pricing the option. The assumption might be justified, for example, by assuming there is a representative investor with a specific utility function and specific aggregate consumption process.

The model of Heston (1993) is quite tractable (because it is an affine model—see Chapter 18). Set $V_t = \sigma_t^2$. Heston assumes

$$dV_t = \kappa(\theta - V_t)\,dt + \gamma\sqrt{V_t}\,dB_{1t}, \tag{17.10a}$$

$$\frac{dS_t}{S_t} = \mu\,dt + \sqrt{V_t}\left[\rho\,dB_{1t} + \sqrt{1-\rho^2}\,dB_{2t}\right], \tag{17.10b}$$

where μ, $\kappa > 0$, $\theta > 0$, γ and ρ are constants, and B_1 and B_2 are independent Brownian motions under the physical probability. This implies that ρ is the instantaneous correlation of dS/S and dV. It is known that the solution of (17.10a) starting from $V_0 > 0$ has the property that, with probability 1, $V_t \geq 0$ for all t, as a variance must be. Intuitively, the reason for the nonnegativity of V is that the instantaneous standard deviation $\gamma\sqrt{V_t}$ in (17.10a) vanishes as $V \downarrow 0$; hence, the drift dominates as $V \downarrow 0$, pulling V up toward θ.

The market is not complete, because there are two Brownian motions in (17.10) and only a single underlying asset. There are many SDF processes. The condition $\sigma\lambda = \mu - r$ means that the two-dimensional vector λ of prices of risk satisfies a single equation.[4] Thus, there is one degree of indeterminacy. Assume the risk-free rate is constant, and assume the asset does not pay dividends. Any SDF process M satisfies

$$\frac{dM_t}{M_t} = -r\,dt - \lambda_{1t}\,dB_{1t} - \lambda_{2t}\,dB_{2t} + \frac{d\varepsilon_t}{\varepsilon_t}, \qquad (17.11)$$

where ε is a local martingale uncorrelated with B_1 and B_2 (it is spanned by Brownian motions independent of B_1 and B_2). From

$$\sigma\lambda = \mu - r \quad \Leftrightarrow \quad -\left(\frac{dM}{M}\right)\left(\frac{dS}{S}\right) = (\mu - r)\,dt,$$

we obtain the restriction[5]

$$\left(\rho\lambda_{1t} + \sqrt{1-\rho^2}\,\lambda_{2t}\right)\sqrt{V_t} = \mu - r. \qquad (17.12)$$

This is a single equation in the two prices of risk λ_1 and λ_2, so, as remarked before, there is one degree of indeterminacy.

In the remainder of this section, we choose to use the prices of risk that satisfy

$$\lambda_{1t} = \phi\sqrt{V_t} \qquad (17.13a)$$

for some constant ϕ. This is the "equilibrium assumption" mentioned above. We now regard ϕ as another parameter of the model. Conditions (17.12) and (17.13a) imply

$$\lambda_{2t} = \frac{\mu - r - \rho\phi V_t}{\sqrt{(1-\rho^2)V_t}}. \qquad (17.13b)$$

Thus, the equilibrium assumption identifies a unique SDF process M (ignoring the orthogonal component $d\varepsilon/\varepsilon$), given a value for the parameter ϕ. This SDF process has the property that MR and MS are martingales (see the end-of-chapter notes).

The value of a European call option is

$$\mathsf{E}\left[M_T \max(0, S_T - K)\right].$$

---

[4]. Here, we use the notation of Chapter 13, and σ is the row vector $(\sqrt{V}\rho, \sqrt{V(1-\rho^2)})$ that premultiplies the column vector of Brownian motion differentials in (17.10b).

[5]. If $\mu \neq r$, then V must be always strictly positive in order for (17.12) to hold—that is, the asset must always be risky ($V_t > 0$) in order to earn a risk premium. A necessary and sufficient condition to have, with probability 1, $V_t > 0$ for all t, is that $\kappa\theta \geq \gamma^2/2$. For the remainder of this section, consider only parameters satisfying this restriction.

The derivation of (16.14) is still valid in this context, so we have

$$\mathsf{E}\left[M_T \max(0, S_T - K)\right] = S_0 \operatorname{prob}^S(A) - e^{-rT} K \operatorname{prob}^R(A), \qquad (17.14)$$

where A is the event $S_T > K$, and prob^S and prob^R are the probabilities defined in terms of M. To calculate these probabilities of the event A, we need to know the dynamics of S and V under the two probabilities. Changing probabilities does not change volatilities, so what we need to know are the drifts. As in Section 16.3, the expected rate of return of S is r under prob^R and r plus the squared volatility under prob^S. Here, we are denoting the squared volatility by V. Hence,

$$\frac{dS}{S} = (r + \delta^i V)\, dt + \sqrt{V}\left[\rho\, dB_1^i + \sqrt{1 - \rho^2}\, dB_2^i\right] \qquad (17.15a)$$

for $i \in \{R, S\}$, where B_1^i and B_2^i are independent Brownian motions under prob^i and where $\delta^R = 0$ and $\delta^S = 1$. We verify this below and also show that

$$dV = \kappa^i(\theta^i - V)\, dt + \gamma \sqrt{V}\, dB_1^i, \qquad (17.15b)$$

where

$$\kappa^R = \kappa + \gamma\phi,$$
$$\kappa^S = \kappa + \gamma(\phi - \rho),$$
$$\theta^R = \frac{\kappa\theta}{\kappa^R},$$
$$\theta^S = \frac{\kappa\theta}{\kappa^S}.$$

The result that $\kappa^R \theta^R = \kappa^S \theta^S = \kappa\theta$ is important. Under the assumption that $\kappa\theta \geq \gamma^2/2$ (footnote 5), we have $V_t > 0$ for all t with probability 1 under the physical probability, under prob^R, and under prob^S. This is discussed further in the end-of-chapter notes.

To compute the probabilities in the option pricing formula (17.14), it may be helpful to represent them as solutions of PDEs. Denote the conditional probabilities given date–t information by $\operatorname{prob}_t^R(A) = Q^R(t, S_t, V_t)$ and $\operatorname{prob}_t^S(A) = Q^S(t, S_t, V_t)$. These conditional probabilities are martingales under the respective probabilities (because they are conditional expectations of 1_A), so their drifts are zero. Calculate the drifts from Itô's formula using (17.15) and equate the drifts to zero to obtain

$$Q_t^i + (r + \delta^i V)S Q_s^i + \kappa^i(\theta^i - V)Q_v^i + \frac{1}{2}VS^2 Q_{ss}^i + \frac{1}{2}\gamma^2 V Q_{vv}^i + \gamma\rho V S Q_{sv}^i = 0$$
$$(17.16)$$

for $i \in \{R, S\}$. These equations should be solved subject to the condition that the solutions lie between 0 and 1 and subject to the terminal condition

$$Q^R(T, a, b) = Q^S(T, a, b) = \begin{cases} 1 & \text{if } a > K, \\ 0 & \text{otherwise}. \end{cases}$$

Heston (1993) presents the solutions of the PDEs as integrals. This is a closed-form solution for the option price in the same sense that the Black-Scholes formula is a closed-form solution (the cumulative normal distribution function in the Black-Scholes formula is computed by integrating the normal density function).

The Radon-Nikodym derivative of the risk-neutral probability relative to the physical probability is $\xi_T = M_T R_T / R_0$. Set $\xi_t = E_t[\xi_T]$. Because MR is a martingale, we have $\xi_t = M_t R_t / R_0$. Therefore,

$$\frac{d\xi}{\xi} = -\lambda_1 \, dB_1 - \lambda_2 \, dB_2 + \frac{d\varepsilon}{\varepsilon}.$$

Girsanov's theorem implies that

$$dB_j^R = dB_j - \left(\frac{d\xi}{\xi}\right)(dB_j) = dB_j + \lambda_j \, dt$$

is the differential of a Brownian motion under prob^R for $j = 1, 2$. Substitute the formulas for the λ_i in (17.13) and substitute the dB_j^R for the dB_j in (17.10) to obtain (17.15) for $i = R$.

The argument for $i = S$ is very similar. The Radon-Nikodym derivative of prob^S relative to the physical probability is $\xi_T = M_T S_T / S_0$. Set $\xi_t = E_t[\xi_T]$. Because MS is a martingale, we have $\xi_t = M_t S_t / S_0$. Use Itô's formula and the fact that $(dS/S)(dM/M) = (r - \mu) \, dt$ to obtain

$$\frac{d\xi}{\xi} = \left(\rho\sqrt{V} - \lambda_1\right) dB_1 + \left(\sqrt{(1-\rho^2)V} - \lambda_2\right) dB_2 + \frac{d\varepsilon}{\varepsilon}.$$

Girsanov's theorem implies that

$$dB_1^S = dB_1 - \left(\frac{d\xi}{\xi}\right)(dB_1) = dB_1 + \left(\lambda_1 - \rho\sqrt{V}\right) dt$$

$$dB_2^S = dB_2 - \left(\frac{d\xi}{\xi}\right)(dB_2) = dB_2 + \left(\lambda_2 - \sqrt{(1-\rho^2)V}\right) dt$$

are differentials of independent Brownian motions under prob^S. Substitute the formulas for the λ_i in (17.13) and substitute the dB_j^S for the dB_j in (17.10) to obtain (17.15) for $i = S$.

17.6 NOTES AND REFERENCES

The forward measure concept first appears in Jamshidian (1989). The derivation of the exchange option formula from the Black-Scholes formula in Section 17.3 is attributed by Margrabe (1978) to Steve Ross. Cox and Ross (1976b) discuss local volatility models in which the volatility is S^γ for a constant γ. These are called constant elasticity of variance (CEV) models. Seminal local volatility models designed to match market option prices are the binomial tree of Rubinstein (1994) and the trinomial tree of Derman and Kani (1998). Jackwerth and Rubinstein (1996) use nonparametric methods to construct a risk-neutral distribution for the underlying asset that produces option prices matching market prices. Hobson and Rogers (1998) propose a class of models in which the volatility is adapted to S but not a function of the contemporaneous value of S. These models are related to discrete-time GARCH option pricing models (Duan, 1995; Heston and Nandi, 2000).

If the market is incomplete, as in the case of stochastic volatility, then the introduction of a zero-net-supply nonspanned asset will generally (unless, for example, all investors have LRT utility functions with the same cautiousness parameter) change equilibrium prices, so attempting to compute what the price of a nonspanned option would be if it were traded is somewhat problematic. Detemple and Selden (1991) provide a general equilibrium analysis of the effect of introducing a nontraded option on the price of the underlying asset.

The process (17.10a) for the variance in the Heston model is called a square-root process. In the special case $\kappa\theta = \gamma^2/4$, the solution of (17.10a) is the square of an Ornstein-Uhlenbeck process (Exercise 12.6). In general, the solution of (17.10a) can be represented as a time-changed squared Bessel process. If $\kappa\theta \geq \gamma^2/2$, then V_t is not just nonnegative but in fact strictly positive for all t (with probability 1). This condition is equivalent to the "dimension" of the Bessel process being at least 2. See, for example, Back (2005, Appendix B3). If $\kappa\theta \geq \gamma^2/2$, we say that the boundary ($V = 0$) is inaccessible, whereas the boundary is accessible if the inequality does not hold.

The volatility process in the Heston model can be normalized by defining $Y = V/\gamma^2$. The dynamics (17.10a) for V imply

$$dY = \left(\frac{\kappa\theta}{\gamma^2} - \kappa Y\right) dt + \sqrt{Y}\, dB_1.$$

The condition for the boundary ($V = 0 \Leftrightarrow Y = 0$) being inaccessible is that the constant $\kappa\theta/\gamma^2$ in the drift of Y be at least $1/2$. This constant is the same under prob^R and prob^S as under the physical probability in the Heston model. Thus, the boundary being inaccessible under the physical probability implies that it is also inaccessible under prob^R and prob^S. This equivalence under the various probabilities of the boundary being accessible is a necessary condition for them to be equivalent as probability measures.

The assumption in Section 17.4 that $\kappa\theta \geq \gamma^2/2$ implies that the Heston model is a member of the extended affine class of models defined by Cheridito, Filipović, and Kimmel (2007). This ensures that MR and MS are strictly positive martingales if we assume B_1 and B_2 are the only sources of uncertainty (or more generally that ε is a martingale independent of B_1 and B_2). The univariate process V with the price of risk specification $\lambda_1 = \phi\sqrt{V}$ is a member of the completely affine class, but the form (17.13b) for λ_2 implies that the joint process $(V, \log S)$ is only extended affine. Exercise 17.9 illustrates that the price of risk assumption (17.13) can be generalized while retaining the extended affine property. See Section 18.7 for further discussion. Exercise 17.7 is adapted from Hull and White (1987).

EXERCISES

17.1. Consider a forward contract on an asset that pays a single known discrete dividend x at a known date $T < u$, where u is the date the forward matures. Suppose there are traded discount bonds maturing at T and u. Let S denote the price of the asset. Prove the following spot-forward parity formula for $t < T$:

$$F_t(u) = \frac{S_t - P_t(T)x}{P_t(u)}.$$

17.2. Suppose V^* is a self-financing wealth process with risky asset price $S^* = S_1/S_2$ and money market price $R^* = 1$, meaning

$$\frac{dV^*}{V^*} = \pi^* \frac{dS^*}{S^*} + (1-\pi^*)\frac{dR^*}{R^*}$$

for some π^*. Define $V = S_2 V^*$. Show that V is a self-financing wealth process with risky asset prices S_1 and S_2 and no money market account.

17.3. Suppose the prices of two non-dividend-paying assets are given by

$$\frac{dS_i}{S_i} = \mu_i\,dt + \sigma_i\,dB_i,$$

where the B_i are Brownian motions with correlation ρ. The μ_i, σ_i, and ρ can be stochastic processes. However, assume the volatility of S_1/S_2 is a constant σ. Assume there is an SDF process such that MS_1 and MS_2 are martingales. Let $X_T = S_{2T}f(S_{1T}/S_{2T})$ for some nonnegative function f, for example, $f(a) = \max(0, a-1)$ as in (17.7). Assume $\mathsf{E}[M_T X_T] < \infty$. Define $Z = S_1/S_2$.

(a) Show that
$$\frac{dZ}{Z} = \sigma\, dB^*,$$
where B^* is a Brownian motion under the probability prob^{S_2} and satisfies
$$\sigma\, dB^* = (\mu_1 - \mu_2 - \rho\sigma_2\sigma_2 + \sigma_2^2)\, dt + \sigma_1\, dB_1 - \sigma_2\, dB_2.$$

(b) Show that
$$\mathsf{E}_t^{S_2}[f(Z_T)] = \mathsf{E}^{S_2}[f(Z_T)] + \int_0^t \psi_s\, dB_s^*$$
for some stochastic process ψ and all $0 \le t \le T$.

(c) Define $W_t = S_{2t}\mathsf{E}_t^{S_2}[f(Z_T)]$. Note that $W_T = X_T$. Show that W is a self-financing wealth process generated by the portfolio process in which the fraction
$$\pi = \frac{S_2 \psi}{\sigma W}$$
of wealth is invested in asset 1 and $1 - \pi$ is invested in asset 2.

17.4. Adopt the assumptions of Exercise 17.3. Let A denote the event $S_{1T} > S_{2T}$.

(a) Show that, for $i = 1, 2$,
$$\mathsf{E}[M_T S_{iT} 1_A] = S_{i0} \text{prob}^{S_i}(A).$$
Conclude that the value at date 0 of an option to exchange asset 2 for asset 1 at date T is
$$S_{10} \text{prob}^{S_1}(A) - S_{20} \text{prob}^{S_2}(A).$$

(b) Define $Y = S_2/S_1$. Show that
$$d\log Y = -\frac{1}{2}\sigma^2\, dt + \sigma\, dB^*,$$
where B^* is a Brownian motion under the probability measure prob^{S_1}. Use this fact and the fact that A is the event $\log Y_T < 0$ to show that $\text{prob}^{S_1}(A) = N(d_1)$, where
$$d_1 = \frac{\log(S_{10}/S_{20}) + \tfrac{1}{2}\sigma^2 T}{\sigma\sqrt{T}}.$$

(c) Define $Z = S_1/S_2$. Show that
$$d\log Z = -\frac{1}{2}\sigma^2\, dt + \sigma\, dB^*,$$

where B^* is a Brownian motion under the probability measure prob^{S_2}. Use this fact and the fact that A is the event $\log Z_T > 0$ to show that $\text{prob}^{S_2}(A) = N(d_2)$, where

$$d_2 = d_1 - \sigma\sqrt{T}.$$

17.5. Suppose the price S of a non-dividend-paying asset has a constant volatility σ. Assume the volatility at date t of a discount bond maturing at $T > t$ is

$$\frac{\phi}{\kappa}\left(1 - e^{-\kappa(T-t)}\right) \qquad (17.17)$$

for constants $\kappa > 0$ and $\phi > 0$. Assume the discount bond and stock have a constant correlation ρ. Let P denote the price of the discount bond.

(a) Calculate the volatility of S/P as a function $\hat{\sigma}(t)$.
(b) Define

$$\sigma_{\text{avg}} = \sqrt{\frac{1}{T}\int_0^T \hat{\sigma}(t)^2\,dt}.$$

Show that

$$\sigma_{\text{avg}}^2 = \sigma^2 + \frac{1}{\kappa^2}\left[\phi^2 - 2\kappa\rho\sigma\phi - (2\phi^2 - 2\kappa\rho\sigma\phi)\left(\frac{1 - e^{-\kappa T}}{\kappa T}\right)\right.$$
$$\left. + \phi^2\left(\frac{1 - e^{-2\kappa T}}{2\kappa T}\right)\right].$$

(c) Apply results from Section 17.3 to derive a formula for the value of a European call option on the asset with price S that matures at T.
(d) Use l'Hôpital's rule to show that $\sigma_{\text{avg}} \approx \sigma$ for small T.
(e) Show that $\sigma_{\text{avg}} > \sigma$ for large T if ρ is sufficiently small.

Note: The bond volatility (17.17) arises in the Vasicek term structure model, with κ being the rate of mean reversion of the short rate process and ϕ being the (absolute) volatility of the short rate process (Section 18.3).

17.6. Consider a European call option on an asset that pays a single known discrete dividend x at a known date $T < u$, where u is the date the option expires. Assume the asset price S drops by x when it goes ex-dividend at date T (i.e., $S_T = \lim_{t\uparrow T} S_t - x$) and otherwise is an Itô process. Suppose there are traded discount bonds maturing at T and u. Assume the volatility of the process

$$Z_t = \begin{cases} [S_t - xP_t(T)]/P_t(u) & \text{if } t < T \\ S_t/P_t(u) & \text{if } T \leq t \leq u \end{cases}$$

is a constant σ during $[0, u]$.

(a) Show that the value at date 0 of the call option is

$$[S_0 - xP_0(T)]N(d_1) - e^{-yu}KN(d_2),$$

where y is the yield at date 0 of the discount bond maturing at u and

$$d_1 = \frac{\log[(S_0 - xP_0(T))/K] + \left(y + \frac{1}{2}\sigma^2\right)u}{\sigma\sqrt{u}},$$

$$d_2 = d_1 - \sigma\sqrt{u}.$$

(b) Interpret the process Z.

17.7. Let S denote the price of a non-dividend-paying asset. Assume

$$\frac{dS_t}{S_t} = \mu_t\, dt + \sigma_t\, dB_{1t}$$

$$d\sigma_t = \phi(\sigma_t)\, dt + \gamma(\sigma_t)\, dB_{2t}$$

for some functions $\phi(\cdot)$ and $\gamma(\cdot)$, where B_1 and B_2 are independent Brownian motions and μ may be a stochastic process. Assume the price of risk for B_2 equals $\lambda(\sigma_t)$ for some function $\lambda(\cdot)$. Assume there is a constant risk-free rate r.

(a) Show that

$$\frac{dS_t}{S_t} = r\, dt + \sigma_t\, dB_{1t}^*,$$

$$d\sigma_t = \phi^*(\sigma_t)\, dt + \gamma(\sigma_t)\, dB_{2t}^*,$$

for some function $\phi^*(\cdot)$, where B_1^* and B_2^* are independent Brownian motions under the risk-neutral probability.

(b) Use iterated expectations to show that the date–0 value of a call option equals

$$E^R\left[S_0 N(d_1) - e^{-rT}KN(d_2)\right], \qquad (17.18)$$

where

$$d_1 = \frac{\log(S_0/K) + \left(r + \frac{1}{2}\sigma_{\text{avg}}^2\right)T}{\sigma_{\text{avg}}\sqrt{T}},$$

$$d_2 = d_1 - \sigma_{\text{avg}}\sqrt{T},$$

and the risk-neutral expectation in (17.18) is taken over the random "average" volatility

$$\sigma_{avg} = \sqrt{\frac{1}{T}\int_0^T \sigma_t^2\, dt}$$

on which d_1 and d_2 depend.

(c) Implement the Black-Scholes formula numerically. Plot the value of an at-the-money ($S_0 = K$) call option as a function of the volatility σ. Observe that the option value is approximately an affine (linear plus constant) function of σ.

(d) Explain why the value of an at-the-money call option on an asset with random volatility is approximately given by the Black-Scholes formula with

$$E^R\left[\sqrt{\frac{1}{T}\int_0^T \sigma_t^2\, dt}\right]$$

input as the volatility.

(e) Part (c) should indicate that the Black-Scholes value is not exactly linear in the volatility. Neither is it uniformly concave nor uniformly convex; instead, it has different shapes in different regions. Explain why if it were concave (convex) over the relevant region, then the Black-Scholes formula with

$$E^R\left[\sqrt{\frac{1}{T}\int_0^T \sigma_t^2\, dt}\right]$$

input as the volatility would overstate (understate) the value of the option.

17.8. Set $V_t = \log \sigma_t$, where σ_t is the volatility of a non-dividend-paying asset with price S. Assume

$$\frac{dS_t}{S_t} = \mu_t\, dt + \sigma_t\, dB_{1t},$$

$$dV_t = \kappa(\theta - V_t)\, dt + \gamma\left[\rho\, dB_{1t} + \sqrt{1-\rho^2}\, dB_{2t}\right],$$

where μ, κ, θ, γ, and ρ are constants and B_1 and B_2 are independent Brownian motions under the physical probability measure. Assume there is a constant risk-free rate.

(a) Show that any SDF process must satisfy

$$\frac{dM_t}{M_t} = -r\,dt - \frac{\mu_t - r}{\sigma_t}\,dB_{1t} - \lambda_t\,dB_{2t} + \frac{d\varepsilon_t}{\varepsilon_t} \qquad (17.19)$$

for some stochastic process λ, where ε is a local martingale uncorrelated with B_1 and B_2.

(b) Assume that λ in the previous part is a constant. Show that

$$\frac{dS_t}{S_t} = r\,dt + \sigma_t\,dB_{1t}^*$$

$$dV_t = \kappa^*(\theta^* - V_t)\,dt - \frac{\gamma\rho(\mu_t - r)}{\sigma_t}\,dt + \gamma\left[\rho\,dB_{1t}^* + \sqrt{1-\rho^2}\,dB_{2t}^*\right]$$

for some constants κ^* and θ^*, where B_1^* and B_2^* are independent Brownian motions under the risk-neutral probability corresponding to M.

(c) Let $W(t, S_t, V_t)$ denote the conditional probability $\operatorname{prob}_t^R(S_T > K)$ for a constant K. Show that W must satisfy the PDE

$$W_t + rSW_S + \left[\kappa^*\theta^* - \kappa^*V - \gamma\rho(\mu - r)e^{-V}\right]W_V + \frac{1}{2}e^{2V}S^2W_{SS}$$

$$+ \frac{1}{2}\gamma^2 W_{VV} + \gamma\rho e^V SW_{SV} = 0.$$

17.9. In the Heston model (17.10), define $Y_1 = V/\gamma^2$ and $Y_2 = \log S - \rho V/\gamma$.

(a) Derive the constants a_i, b_{ij}, and β such that

$$\begin{pmatrix} dY_1 \\ dY_2 \end{pmatrix} = \begin{pmatrix} a_1 \\ a_2 \end{pmatrix} dt + \begin{pmatrix} b_{11} & b_{12} \\ b_{21} & b_{22} \end{pmatrix}\begin{pmatrix} Y_1 \\ Y_2 \end{pmatrix} dt + \begin{pmatrix} \sqrt{Y_1} & 0 \\ 0 & \sqrt{\beta Y_1} \end{pmatrix}\begin{pmatrix} dB_1 \\ dB_2 \end{pmatrix}.$$

(b) Consider a price of risk specification

$$\lambda_{1t} = \frac{c_{10} + c_{11}Y_{1t}}{\sqrt{Y_{1t}}},$$

$$\lambda_{2t} = \frac{c_{20} + c_{21}Y_{1t}}{\sqrt{\beta Y_{1t}}},$$

for constants c_{ij}. Derive c_{20} and c_{21} as functions of c_{10} and c_{11} from the fact that (17.12) must hold for all V. Note: The specification in Section 17.4 is the special case $c_{10} = 0$.

(c) Assume that M defined in terms of c_{10} and c_{11} is such that MR is a martingale. Derive constants a_i^* and b_{ij}^* in terms of c_{10} and c_{11} such that

$$\begin{pmatrix} dY_1 \\ dY_2 \end{pmatrix} = \begin{pmatrix} a_1^* \\ a_2^* \end{pmatrix} dt + \begin{pmatrix} b_{11}^* & b_{12}^* \\ b_{21}^* & b_{22}^* \end{pmatrix} \begin{pmatrix} Y_1 \\ Y_2 \end{pmatrix} dt + \begin{pmatrix} \sqrt{Y_1} & 0 \\ 0 & \sqrt{\beta Y_1} \end{pmatrix} \begin{pmatrix} dB_1^* \\ dB_2^* \end{pmatrix},$$

where the B_i^* are independent Brownian motions under the risk-neutral probability.

(d) Assume $\kappa \theta / \gamma^2 \geq 1/2$. Under what condition on c_{10} is $a_1^* \geq 1/2$?

17.10. Consider an *American* call option with strike K on an asset that pays a single known discrete dividend x at a known date $T < u$, where u is the date the option expires. Assume the asset price S drops by x when it goes ex-dividend at date T (i.e., $S_T = \lim_{t \uparrow T} S_t - x$) and otherwise is an Itô process. Assume there is a constant risk-free rate r.

(a) Show that if $x < \left(1 - e^{-r(u-T)}\right) K$, then the call should not be exercised early.

For the remainder of the exercise, assume $x > \left(1 - e^{-r(u-T)}\right) K$. Assume the volatility of the process

$$Z_t = \begin{cases} S_t - e^{-r(T-t)} x & \text{if } t < T \\ S_t & \text{if } T \leq t \leq u \end{cases}$$

is constant over $[0, u]$. Let $V(t, S_t)$ denote the value of a European call on the asset with strike K maturing at u. Let S^* denote the value of the stock price just before T such that the holder of the American option would be indifferent about exercising just before the stock goes ex-dividend. This value is given by $S^* - K = V(T, S^* - x)$. Exercise is optimal just before T if $\lim_{t \uparrow T} S_t > S^*$, and equivalently, if $S_T > S^* - x$. Let A denote the event $S_T > S^* - x$ and let C denote the set of states of the world such that $S_T \leq S^* - x$ and $S_u > K$. The cash flows to a holder of the option who exercises optimally are $(S_T + x - K) 1_A$ at (or, rather, "just before") date T and $(S_u - K) 1_C$ at date u.

(b) Show that the value at date 0 of receiving $(S_T + x - K) 1_A$ at date T is

$$\left(S_0 - e^{-rT} x\right) N(d_1) - e^{-rT} (K - x) N(d_2),$$

where

$$d_1 = \frac{\log(S_0 - e^{-rT} x) - \log(S^* - x) + \left(r + \tfrac{1}{2} \sigma^2\right) T}{\sigma \sqrt{T}},$$

$$d_2 = d_1 - \sigma \sqrt{T}.$$

(c) Show that the value at date 0 of receiving $(S_u - K)1_C$ at date u is

$$\left(S_0 - e^{-rT}x\right) M(-d_1, d'_1, -\sqrt{T/u}) - e^{-ru}KM(-d_2, d'_2, -\sqrt{T/u}),$$

where $M(a, b, \rho)$ denotes the probability that $\xi_1 < a$ and $\xi_2 < b$ when ξ_1 and ξ_2 are standard normal random variables with correlation ρ, and where

$$d'_1 = \frac{\log(S_0 - e^{-rT}x) - \log K + \left(r + \frac{1}{2}\sigma^2\right)u}{\sigma\sqrt{u}},$$

$$d'_2 = d'_1 - \sigma\sqrt{u}.$$

18

Term Structure Models

CONTENTS

18.1	Forward Rates	459
18.2	Factor Models and the Fundamental PDE	460
18.3	Affine Models	461
18.4	Quadratic Models	469
18.5	Expectations Hypotheses	469
18.6	Fitting the Yield Curve and HJM Models	474
18.7	Notes and References	477

This chapter addresses the modeling of default-free discount bonds. There are generally two objects of interest: bond prices and expected returns. For bond prices, it suffices to work entirely under a risk-neutral probability. For expected returns, we need the physical probability. Of course, the two probabilities are related via the SDF process. We only consider two probabilities in this chapter: the physical probability and the risk-neutral probability. So, *we use the standard notation E^* for expectation under the risk-neutral probability instead of the notation E^R used in the previous two chapters.*

As before, we write $P_t(T)$ for the price at $t \leq T$ of a discount (zero-coupon) bond paying \$1 at date T. Recall from (16.2) that

$$\frac{-\log P_t(T)}{T-t}$$

is called the yield at t of a discount bond maturing at T. Denote this yield by $Y_t(T)$. The yield curve at date t is the function $\tau \mapsto Y_t(t+\tau)$ specifying the yield of each discount bond as a function of its time to maturity τ. This is also called the term structure of interest rates.

Term Structure Models

Assume there is a money market account and denote its price as before by

$$R_t = \exp\left(\int_0^t r_s \, ds\right),$$

where r is the instantaneous risk-free rate, also called the short rate. Assume there is an SDF process M such that MR is a martingale and such that $MP(T)$ is a martingale on $[0, T]$ for each T. Define a risk-neutral probability on the horizon $[0, T]$ in terms of M as in Section 15.3 to obtain

$$P_t(T) = \mathsf{E}_t\left[\frac{M_T}{M_t}\right] = \mathsf{E}_t^*\left[\exp\left(-\int_t^T r_u \, du\right)\right]. \tag{18.1}$$

This shows that bond prices can be computed either by modeling an SDF process under the physical probability or by modeling the short rate under a risk-neutral probability. The latter approach is generally taken in this chapter, though we do model the SDF process when we want to compute bond risk premia. Coupon bonds are portfolios of discount bonds (one for each coupon payment and one for the face value), so the pricing formulas for default-free discount bonds also give prices of default-free coupon bonds.

18.1 FORWARD RATES

The forward rate curve at date t is the function $\tau \mapsto F_t(t+\tau)$ for $\tau \geq 0$ such that

$$\exp\left(-\int_t^T F_t(u) \, du\right) = P_t(T) \tag{18.2}$$

for every $T \geq t$. Taking logs and then differentiating with respect to u shows that this is equivalent to

$$F_t(T) = -\frac{d \log P_t(T)}{dT}. \tag{18.3}$$

From (18.2), the relation between yields and forward rates is

$$Y_t(T) = \frac{1}{T-t} \int_t^T F_t(u) \, du. \tag{18.4}$$

Thus, the yield at t of the bond maturing at T is the average of the forward rates between t and T.

To understand the term "forward rate," consider how to lock in at date t the return on an investment of $\$1$ from u to T, for $t < u < T$. To do this, at date t, short sell one unit of the discount bond maturing at u and invest the proceeds in the bond maturing at T, buying $P_t(u)/P_t(T)$ units of that bond. The $\$1$ investment that you plan to make at date u will cover the short sale, and the return on the investment is the $P_t(u)/P_t(T)$ dollars received at the maturity T of the bonds that you buy. The annualized continuously compounded rate of return on this investment is the rate z such that

$$e^{z(T-u)} = \frac{P_t(u)}{P_t(T)}.$$

Solving for z gives

$$z = -\frac{\log P_t(T) - \log P_t(u)}{T-u}.$$

The limit of this difference quotient as $u \uparrow T$ is the definition of the derivative on the right-hand side of (18.3). Thus, $F_t(T)$ is interpreted as the rate of return that we can lock in at t on an investment of infinitesimal duration to be made at the subsequent date T.

18.2 FACTOR MODELS AND THE FUNDAMENTAL PDE

A factor model of the term structure is a model in which the short rate depends on a vector of state variables X that is Markovian under the risk-neutral probability. Consider the Markovian model of Section 13.6 but operate under the risk-neutral probability. In other words, assume there are functions r, ϕ, and ν such that $r_t = r(X_t)$ and

$$dX_t = \phi(X_t)\,dt + \nu(X_t)\,dB_t^*, \qquad (18.5)$$

where B^* is a vector of independent Brownian motions under the risk-neutral probability. As before, let ν_i denote the ith row of ν arranged as a column vector. Let n denote the dimension of X and B^*, so $\nu(x)$ is an $n \times n$ matrix for each value x of X_t.[1] Equation (18.1) shows that a discount bond price depends on the distribution of the future path of the short rate under a risk-neutral probability, so a discount bond price in a factor model is a function of time and the state vector X_t.

Fix a maturity date T and let $f(t, X_t)$ denote the price at $t \leq T$ of the discount bond maturing at T. The stochastic process $f(t, X_t)/R_t$ is a martingale—hence its drift is zero—under the risk-neutral probability. Use Itô's formula to compute the drift of $R_t f(t, X_t)$ and equate it to zero to obtain

$$f_t + \sum_{i=1}^{n} \phi_i f_{x_i} + \frac{1}{2}\sum_{i=1}^{n}\sum_{j=1}^{n} f_{x_i x_j} \nu_i' \nu_j = rf. \qquad (18.6)$$

---

[1]. The distribution of X is determined by the drift and the covariance matrix $\nu\nu'$. There is no loss of generality in assuming that the number of Brownian motions equals the dimension of X, as explained in Section 12.8.

This is the fundamental PDE (15.12). It should be solved subject to the terminal condition $f(T,x) = 1$ for all x.

To model bond risk premia, we need to work under the physical probability. Assume the price of risk process is a function $\lambda(X_t)$. By Girsanov's theorem, the vector of risk-neutral Brownian motions B^* in (18.5) is related to a vector B of Brownian motions under the physical probability as $dB_t^* = dB_t + \lambda(X_t)\,dt$.[2] Therefore,

$$dX = [\phi(X) + \nu(X)\lambda(X)]\,dt + \nu(X)\,dB. \tag{18.7}$$

The relation between (18.5) and (18.7) is the same as the relation between (15.11) and (13.50). The only difference is that here we started with the risk-neutral probability (and we use ϕ to denote the drift of X under the risk-neutral probability rather than under the physical probability).

18.3 AFFINE MODELS

A factor model is affine if

- $r(x)$ is an affine function of x,
- $\phi(x)$ is an affine function of x,
- each element of the covariance matrix $\nu(x)\nu(x)'$ is an affine function of x.

In any single-factor affine model, the single factor can be taken to be the short rate (Exercise 18.2). Instantaneous variances must be nonnegative, so the diagonal elements of $\nu(X_t)\nu(X_t)'$ must be nonnegative in any factor model. When the variances are nonconstant affine functions of X, this implies that the solution X of (18.5) must lie in a restricted domain. A model is said to be admissible if a unique solution to (18.5) exists.

In an affine model, discount bond yields are affine functions of the factors. This means that

$$Y_t(t+\tau) = a(\tau) + \sum_{i=1}^{n} b_i(\tau) X_{it} \tag{18.8}$$

---

2. The SDF process from which the risk-neutral probability is defined satisfies

$$\frac{dM_t}{M_t} = -r(X_t)\,dt - \lambda(X_t)'\,dB_t + \frac{d\varepsilon_t}{\varepsilon_t},$$

where ε is a local martingale uncorrelated with the vector B.

for some real-valued functions $a(\cdot)$ and $b_i(\cdot)$, for all $\tau > 0$. This is established at the end of the section by analyzing the fundamental PDE. In an affine model,

$$P_t(t+\tau) = \exp\left(-\tau a(\tau) - \sum_{i=1}^{n} \tau b_i(\tau) X_{it}\right). \tag{18.9}$$

We say that bond prices are exponential-affine. Equation (18.9) implies that forward rates are

$$F_t(t+\tau) = \alpha'(\tau) + \sum_{i=1}^{n} \beta_i'(\tau) X_{it}, \tag{18.10}$$

where $\alpha(\tau) = \tau a(\tau)$, $\beta_i(\tau) = \tau b_i(\tau)$, and the primes denote derivatives. Thus, forward rates are also affine in the factors.

A useful feature of affine models is that we can usually take a vector of yields to be the factors. Fix n positive numbers τ_i. According to (18.8), the yield at t of a discount bond maturing at $t + \tau_i$ is $a(\tau_i) + b(\tau_i)' X_t$. Stack the yields to form a column vector Y_t, the constants $a(\tau_i)$ to form a column vector $\mathcal{A}$, and the row vectors $b(\tau_i)'$ to form a matrix $\mathcal{B}$. Then, we have

$$Y_t = \mathcal{A} + \mathcal{B} X_t. \tag{18.11a}$$

Provided $\mathcal{B}$ is invertible, we have

$$X_t = \mathcal{B}^{-1}(Y_t - \mathcal{A}). \tag{18.11b}$$

Substituting (18.11b) and $dX = \mathcal{B}^{-1} dY$ in (18.5) produces an affine model in which the vector Y of yields (at fixed times-to-maturity τ_i) is the vector of factors. We can also use only $n - 1$ yields and use the short rate as the other factor, replacing the equation $Y_{nt} = a(\tau_n) + b(\tau_n)' X_t$ in the above construction with $r_t = r(X_t)$.

We can transform from the vector X to a vector $Y = \mathcal{A} + \mathcal{B} X$ for any $\mathcal{A}$ and nonsingular $\mathcal{B}$. It is not at all necessary to take Y to be yields. Examples are in Exercises 18.3–18.4. The model based on X and the model based on Y have identical implications for the short rate and bond prices. Frequently, factors are regarded as unobservable ("latent"), in which case the factor values and their dynamics can only be inferred from bond yields (including a proxy for the short rate). Nonsingular affine transformations of latent factor models produce observationally equivalent models. Also, we can rotate the Brownian motions as discussed in Section 12.8 without altering the implications for bond prices. Because these transformations produce observationally equivalent models, the issue of which parameter values or combinations of parameter values can be identified empirically is complex. See the end-of-chapter notes for some discussion.

GAUSSIAN AFFINE MODELS

If $v(x)$ in an affine model is a constant matrix—that is, the same for all x—then the model is said to be Gaussian, because the distribution of (r_u, X_u) conditional on X_t is multivariate normal for all $u > t$. The volatility of the short rate is constant when $v(x)$ is a constant matrix. In any two-factor Gaussian term structure model, the two factors can be taken to be the short rate and its drift (Exercise 18.3). We observe below that if a Gaussian model is completely affine, then bond risk premia depend only on times to maturity and do not depend on the factors.

Completely Affine Models

An affine factor model is completely affine if

$$v(x) = \sigma S(x), \qquad (18.12a)$$

$$\lambda(x) = S(x)\zeta, \qquad (18.12b)$$

for a constant vector ζ, a constant $n \times n$ matrix σ, and a function S having the property that $S(x)$ is an $n \times n$ diagonal matrix with each squared diagonal element being an affine function of x. A completely affine model is also affine under the physical probability. To see this, substitute (18.12) into (18.7) to obtain

$$dX_t = [\phi(X_t) + \sigma S(X_t) S(X_t) \zeta] \, dt + \sigma S(X_t) \, dB_t. \qquad (18.13)$$

The matrix $S(x)S(x)$ is a diagonal matrix with affine functions of x on the diagonal. Hence, each element of the vector $\phi(x) + \sigma S(x) S(x) \zeta$ is an affine function of x.

Bond risk premia are affine in x in a completely affine model. To derive the risk premia, observe that (18.9) implies the stochastic part of the return at t of the bond maturing at $t + \tau$ is

$$-\tau b(\tau)' \sigma S(X_t) \, dB_t^*.$$

Therefore, the price of risk specification (18.12b) implies that the risk premium is

$$-\left(\frac{dP}{P}\right)\left(\frac{dM}{M}\right) = -\tau b(\tau)' \sigma S(X_t) S(X_t) \zeta, \qquad (18.14)$$

which is affine in X_t. In a Gaussian completely affine model, $S(x)$ is a constant matrix, so the risk premium of a bond depends only on its time to maturity and does not depend on the factors.

The specification (18.12b) of the price of risk process can be generalized while still obtaining affine dynamics under the physical probability and affine risk premia (Exercises 18.7 and 18.8). In making such a generalization, the key issue is whether MR is a martingale for the specified price of risk process (that is, whether the risk-neutral probability exists). See the end-of-chapter notes for further discussion.

Vasicek Model

The single-factor Gaussian term structure model is called the Vasicek model in recognition of Vasicek (1977). As remarked before, the single factor can be taken to be the short rate. In the Vasicek model, the short rate r satisfies

$$dr = \kappa(\theta - r)\,dt + \sigma\,dB^*, \tag{18.15}$$

where B^* is a Brownian motion under a risk-neutral probability and θ, $\kappa > 0$, and $\sigma > 0$ are constants. The stochastic process (18.15) is called an Ornstein-Uhlenbeck process. .

The solution of (18.15)—see Exercise 12.5—is, for any $u \geq t \geq 0$,

$$r_u = \theta - e^{-\kappa(u-t)}(\theta - r_t) + \sigma \int_t^u e^{-\kappa(u-s)}\,dB_s^*. \tag{18.16}$$

Thus, conditional on date–t information (that is, conditional on r_t), the short rate at date $u > t$ is normally distributed with mean

$$\theta - e^{-\kappa(u-t)}(\theta - r_t) \tag{18.17a}$$

and variance

$$\sigma^2 \int_t^u e^{-2\kappa(u-s)}\,ds = \frac{\sigma^2}{2\kappa}\left(1 - e^{-2\kappa(u-t)}\right). \tag{18.17b}$$

As $u \to \infty$, the conditional mean of r_u converges to θ, which is the unconditional mean of r. The parameter κ is called the rate of mean reversion. We can see from (18.15) that r always drifts toward θ, because the drift is positive when $r < \theta$ and negative when $r > \theta$. As $u \to \infty$, the conditional variance (18.17b) increases to $\sigma^2/2\kappa$, which is the unconditional variance of the short rate process.

As stated in (18.8), discount bond yields are affine in the factor r. In the Vasicek model, the functions a and b are

$$a(\tau) = \theta - \frac{\sigma^2}{2\kappa^2} + \left(\frac{\sigma^2 - \theta\kappa^2}{\kappa^3}\right)\left(\frac{1-e^{-\kappa\tau}}{\tau}\right) - \frac{\sigma^2}{4\kappa^3}\left(\frac{1-e^{-2\kappa\tau}}{\tau}\right), \tag{18.18a}$$

$$b(\tau) = \frac{1}{\kappa}\left(\frac{1-e^{-\kappa\tau}}{\tau}\right). \tag{18.18b}$$

This can be shown by solving the fundamental PDE or by a direct calculation using the fact that $\int_t^T r_u \, du$ is normally distributed given information at date t (which implies the risk-neutral expectation (18.1) is the expectation of an exponential of a normally distributed variable). The direct calculation is provided at the end of the section. Exercise 18.5 asks for the solution of the completely affine version of the Vasicek model.

Conditional on information at any date t, bond yields at any date $u > t$ in the Vasicek model are normally distributed, due to the normality of r_u. Thus, negative yields occur with positive probability. This is unattractive if we are modeling nominal interest rates (nominal interest rates should be nonnegative if it is costless to store cash).

Cox-Ingersoll-Ross Model

The single-factor affine model in which the variance of the factor is proportional to the factor is called the Cox-Ingersoll-Ross (CIR) model in recognition of Cox, Ingersoll, and Ross (1985). In the CIR model,

$$dr = \kappa(\theta - r)\, dt + \sigma \sqrt{r}\, dB^*, \qquad (18.19)$$

where B^* is a Brownian motion under a risk-neutral probability and κ, θ, and σ are positive constants.[3] The fundamental PDE for the CIR model is

$$f_t + \kappa(\theta - r)f_r + \frac{1}{2}\sigma^2 r f_{rr} = rf. \qquad (18.20)$$

We show at the end of the section that there is an exponential-affine solution of (18.20) subject to the terminal condition $f(T, r) = 1$ for which yields satisfy (18.8) with

$$a(\tau) = -\frac{2\kappa\theta}{\sigma^2}\left[\frac{\kappa + \gamma}{2} + \frac{1}{\tau}\log\frac{2\gamma}{c(\tau)}\right], \qquad (18.21\text{a})$$

$$b(\tau) = \frac{2(e^{\gamma\tau} - 1)}{\tau c(\tau)}, \qquad (18.21\text{b})$$

$$c(\tau) = 2\gamma + (\kappa + \gamma)(e^{\gamma\tau} - 1), \qquad (18.21\text{c})$$

$$\gamma = \sqrt{\kappa^2 + 2\sigma^2}. \qquad (18.21\text{d})$$

---

3. The variance process in the model of Heston (1993) discussed in Section 17.5 is the same as the CIR short rate process (18.19). As observed in Sections 17.5 and 17.6, the solution r of (18.19) is nonnegative for all t—hence, it is possible to take the square root in (18.19)—and, if $\kappa\theta \geq \sigma^2/2$, then r is strictly positive for all t, with probability 1.

Multifactor CIR Models

A simple multifactor model is obtained by assuming the short rate is the sum of independent square-root processes, that is, processes of the form (18.19). Such a model is called a multifactor CIR model, because the procedure is suggested by Cox, Ingersoll, and Ross (1985). For example, adding two independent square-root processes produces the following affine model:

$$r_t = X_{1t} + X_{2t}, \qquad (18.22a)$$

where

$$dX_i = \kappa_i(\theta_i - X_i)\,dt + \sigma_i\sqrt{X_i}\,dB_i^* \qquad (18.22b)$$

for positive constants κ_i, θ_i, and σ_i, with the B_i^* being independent Brownian motions under a risk-neutral probability. This is an easy model to solve, given the solution of the single-factor CIR model. Independence implies that the expectation of a product is the product of the expectations, so

$$\mathsf{E}_t^*\!\left[\exp\!\left(-\int_t^{t+\tau} r_u\,du\right)\right]$$

$$= \mathsf{E}_t^*\!\left[\exp\!\left(-\int_t^{t+\tau} X_{1u}\,du\right)\right] \mathsf{E}_t^*\!\left[\exp\!\left(-\int_t^{t+\tau} X_{2u}\,du\right)\right]$$

$$= \exp\!\left(-\tau a(\tau) - \tau b_1(\tau) X_{1t} - \tau b_2(\tau) X_{2t}\right), \qquad (18.23)$$

where $a(\tau) = a_1(\tau) + a_2(\tau)$, and the functions a_i and b_i depend on the parameters κ_i, θ_i, and σ_i as in (18.21), for $i = 1, 2$. Exercise 18.4 shows that the factors in a two-factor CIR model can be taken to be the short rate and its instantaneous variance. Exercise 18.6 addresses the completely affine version of a multifactor CIR model.

We can define the short rate as the sum of other independent processes. If a bond pricing formula exists for each process, then multiplying as above produces a bond pricing formula for the sum. A simple example is taking one of the processes to be nonrandom (Section 18.6).

We first analyze the fundamental PDE for the general affine model. We show that an exponential-affine solution $f(t,x)$ of the fundamental PDE is equivalent to a system of ODEs for the functions a and b in (18.8). We then solve the ODEs for the CIR model. Finally, we use the direct method to solve the Vasicek model.

Instead of working with the functions a and b in (18.8), it is slightly more convenient to work with α and β defined as $\alpha(\tau) = \tau a(\tau)$ and $\beta(\tau) = \tau b(\tau)$.

Term Structure Models

Guess a solution of the fundamental PDE of the form

$$f(t,x) = \exp\left(-\alpha(T-t) - \sum_{i=1}^{n}\beta_i(T-t)x_i\right).$$

The boundary condition $f(T,x) = 1$ implies $\alpha(0)$ and $\beta(0) = 0$. We have

$$f_t(t,x) = \left[\alpha'(T-t) + \sum_{i=1}^{n}\beta'_i(T-t)x_i\right]f(t,x),$$

$$f_{x_i}(t,x) = -\beta_i(T-t)f(t,x),$$

$$f_{x_ix_j}(t,x) = \beta_i(T-t)\beta_j(T-t)f(t,x).$$

Hence, the fundamental PDE (18.6) is equivalent to

$$\alpha'(T-t) + \sum_{i=1}^{n}\beta'_i(T-t)x_i - \sum_{i=1}^{n}\beta_i(T-t)\phi_i(x)$$

$$+ \frac{1}{2}\sum_{i=1}^{n}\sum_{j=1}^{n}\beta_i(T-t)\beta_j(T-t)v_i(x)'v_j(x) = r(x). \quad (18.24)$$

Both sides of this equation are affine in x in an affine model. For the equation to hold for all values of x, the coefficients of each x_i on the two sides must match, and the constant terms must match. Equating the coefficients of the x_i produces a system of n ODEs for the β_i functions. They are called Riccati equations because they are affine in the β'_i and quadratic in the β_i. Matching the constant terms, given the β_i, determines α', which can be integrated starting from $\alpha(0) = 0$ to compute α. Thus, a solution of the Riccati equations produces an exponential-affine solution of the fundamental PDE.

For the CIR model, the PDE (18.24) is

$$\alpha'(T-t) + \beta'(T-t)r - \kappa(\theta - r)\beta(T-t) + \frac{1}{2}\sigma^2\beta(T-t)^2r = r. \quad (18.25)$$

This holds for all r if and only if the following pair of ODEs is satisfied:

$$\beta'(T-t) + \kappa\beta(T-t) + \frac{1}{2}\sigma^2\beta(T-t)^2 = 1, \quad (18.26)$$

$$\alpha'(T-t) - \kappa\theta\beta(T-t) = 0. \quad (18.27)$$

The unique solution of (18.26) subject to the boundary condition $\beta(0) = 0$ is $\beta(\tau) = \tau b(\tau)$, where b is given in (18.21). We can see that this function b is a solution of the ODE simply by differentiating the function. Uniqueness follows from standard results on uniqueness of solutions to ODEs. Given β, we can integrate (18.27) from 0 to τ, starting from $\alpha(0) = 0$, to obtain the formula for $\alpha(\tau) = \tau a(\tau)$ in (18.21).

Now, we solve the Vasicek model by directly calculating the expected value in (18.1). We have

$$\int_t^T r_u \, du = (T-t)\theta - (\theta - r_t) \int_t^T e^{-\kappa(u-t)} \, du + \sigma \int_t^T \left\{ \int_t^u e^{-\kappa(u-s)} \, dB_s^* \right\} du$$

$$= (T-t)\theta - \frac{1}{\kappa}\left(1 - e^{-\kappa(T-t)}\right)(\theta - r_t) + \sigma \int_t^T \left\{ \int_t^u e^{-\kappa(u-s)} \, dB_s^* \right\} du.$$

We can change the order of integration in the remaining integral to obtain

$$\sigma \int_t^T \left\{ \int_t^u e^{-\kappa(u-s)} \, dB_s^* \right\} du = \sigma \int_t^T \left\{ \int_s^T e^{-\kappa(u-s)} \, du \right\} dB_s^*$$

$$= \frac{\sigma}{\kappa} \int_t^T \left(1 - e^{-\kappa(T-s)}\right) dB_s^*. \quad (18.28)$$

Because the integrand in (18.28) is nonrandom, the stochastic integral (18.28) is normally distributed with mean zero and variance

$$\frac{\sigma^2}{\kappa^2} \int_t^T \left(1 - e^{-\kappa(T-s)}\right)^2 ds$$

$$= (T-t)\frac{\sigma^2}{\kappa^2} - \frac{2\sigma^2}{\kappa^3}\left(1 - e^{-\kappa(T-t)}\right) + \frac{\sigma^2}{2\kappa^3}\left(1 - e^{-2\kappa(T-t)}\right).$$

Thus, $-\int_t^T r_u \, du$ is normally distributed with mean

$$-(T-t)\theta + \frac{1}{\kappa}\left(1 - e^{-\kappa(T-t)}\right)(\theta - r_t)$$

and variance

$$(T-t)\frac{\sigma^2}{\kappa^2} - \frac{2\sigma^2}{\kappa^3}\left(1 - e^{-\kappa(T-t)}\right) + \frac{\sigma^2}{2\kappa^3}\left(1 - e^{-2\kappa(T-t)}\right).$$

Using the usual rule for expectations of exponentials of normals, it follows that

$$\log \mathrm{E}_t^* \left[\exp\left(-\int_t^T r_u \, du\right) \right] = -(T-t)\theta + \frac{1}{\kappa}\left(1 - e^{-\kappa(T-t)}\right)(\theta - r_t)$$

$$+ \frac{1}{2}\left[(T-t)\frac{\sigma^2}{\kappa^2} - \frac{2\sigma^2}{\kappa^3}\left(1 - e^{-\kappa(T-t)}\right) + \frac{\sigma^2}{2\kappa^3}\left(1 - e^{-2\kappa(T-t)}\right) \right].$$

This implies (18.18).

18.4 QUADRATIC MODELS

A quadratic (or quadratic-Gaussian) model is a factor model in which $r(x)$ is quadratic in x, $\phi(x)$ is affine in x, and $v(x)$ is a constant matrix. For r to be quadratic means that

$$r(x) = \delta_0 + \delta'x + x'\Delta x, \qquad (18.29)$$

for a constant δ_0, a constant vector δ, and a constant matrix Δ. If Δ is nonsingular, then

$$r(x) = \hat{\delta}_0 + (X - \xi)'\Delta(X - \xi), \qquad (18.29')$$

where $\xi' = -\delta'\Delta^{-1}/2$ and $\hat{\delta}_0 = \delta_0 - \delta'\Delta^{-1}\delta/4$. Consequently, the short rate is nonnegative if Δ is positive definite and $\hat{\delta}_0 \geq 0$, that is, if $4\delta_0 \geq \delta'\Delta^{-1}\delta$.

In a quadratic model, discount bond prices are exponential-quadratic, meaning that yields are

$$Y_t(t+\tau) = a(\tau) + b(\tau)'X_t + X_t'c(\tau)X_t,$$

for some real-valued function a, vector-valued function b, and matrix-valued function c. By guessing an exponential-quadratic solution of the fundamental PDE, ODEs can be derived for a, b, and c, similar to the analysis of affine models.

To derive bond risk premia, it is convenient to assume that the price-of-risk vector is affine in the state variables—this is part of the definition of a quadratic model in Ahn, Dittmar, and Gallant (2002). In that case, risk premia are quadratic functions of the factors with coefficients depending on the time to maturity.

18.5 EXPECTATIONS HYPOTHESES

Expectations hypotheses have a long tradition in finance and are based on the idea that investors should respond to differences in expected returns by buying assets with high expected returns and selling assets with low expected returns until prices moved to equalize expected returns. This would be true if investors were risk neutral. This reasoning motivates the expectations hypotheses for forwards and futures discussed in Section 17.2. However, risk neutrality is an uninteresting assumption when analyzing the term structure of interest rates, because it implies (when there is an interior optimum) that the short rate is constant and all bond yields equal the short rate. To see this, consider an investor who maximizes

$$\mathrm{E}\int_0^\infty e^{-\delta t} C_t \, dt$$

subject to the constraint $C_t \geq 0$ and assume that the investor has an optimum with $C_t > 0$ for all t with probability 1 (or, more simply, suppose that there is no nonnegativity constraint). Then, $M_t = e^{-\delta t}$ is an SDF process, so bond prices are $P_t(T) = e^{-\delta(T-t)}$. Thus, the term structure of interest rates is constant and flat at $r = \delta$. Consequently, it is unreasonable to appeal to risk neutrality to motivate expectations hypotheses of the term structure. Nevertheless, the various expectations hypotheses of the term structure are useful as simple conjectures about how the current term structure forecasts future term structures.

We first describe what are called pure versions of the expectations hypotheses and explain the "impure" versions later. There are three different pure versions: for all $t < T$,

(A) The return from t to T of the discount bond maturing at T equals the expected return on the money market account from t to T, that is,

$$\frac{1}{P_t(T)} = E_t\left[e^{\int_t^T r_u\, du}\right]. \qquad (18.30)$$

(B) The instantaneous expected return at t of the discount bond maturing at T equals the short rate at t, that is,

$$E_t\left[\frac{dP_t(T)}{P_t(T)}\right] = r_t\, dt. \qquad (18.31)$$

(C) The yield at t of the discount bond maturing at T equals the expected average short rate between t and T, that is,

$$Y_t(T) = \frac{1}{T-t} E_t \int_t^T r_u\, du. \qquad (18.32)$$

Simple arguments show that there are essentially equivalent versions of (B) and (C):

(B′) Discount bond prices can be computed using $e^{-\int_0^t r_s\, ds}$ as an SDF process:

$$P_t(T) = E_t\left[e^{-\int_t^T r_u\, du}\right]. \qquad (18.33)$$

(C′) Forward rates equal expected spot rates:

$$F_t(T) = E_t[r_T]. \qquad (18.34)$$

To see the relation between (B) and (B′), note that (B) is equivalent to $e^{-\int_0^t r_s\, ds} P_t(T)$ being a local martingale and (B′) is equivalent to it being a martingale. To see that (C) and (C′) are equivalent, use the fact (18.4) that yields are averages of forward rates to write (18.32) as

$$\int_t^T F_t(u)\,du = \mathsf{E}_t \int_t^T r_u\,du$$

and then differentiate with respect to T to obtain (C′).

We can write all three hypotheses compactly in terms of the random variable $\xi = \mathrm{e}^{-\int_t^T r_u\,du}$:

(A) $1/P_t(T) = \mathsf{E}_t[1/\xi]$.
(B) $P_t(T) = \mathsf{E}_t[\xi]$.
(C) $\log P_t(T) = \mathsf{E}_t[\log \xi]$.

Clearly, the three hypotheses cannot be simultaneously true when the short rate is random, because of Jensen's inequality. We show below that (A) implies

$$\mathsf{E}_t\left[\frac{\mathrm{d}P}{P}\right] = r\,\mathrm{d}t + \left(\frac{\mathrm{d}P}{P}\right)^2, \qquad (18.35)$$

and (C) implies

$$\mathsf{E}_t\left[\frac{\mathrm{d}P}{P}\right] = r\,\mathrm{d}t + \frac{1}{2}\left(\frac{\mathrm{d}P}{P}\right)^2, \qquad (18.36)$$

where we have suppressed the arguments t and T of the discount bond price P. Thus, (A)–(C) imply three different formulas for bond risk premia. We can calculate bond risk premia based on covariances with any SDF process. Thus, (A)–(C) imply three different formulas for covariances between bond returns and SDF processes. If (A) is true, then the covariance of every bond return with the SDF process must equal the negative of the bond return's variance. If (C) is true, then the covariance of every bond return with the SDF process must equal minus one-half of the bond return's variance. We explain at the end of the section why neither (A) nor (C) can be true in a single-factor model.

The impure versions of the expectations hypotheses generalize the pure versions by allowing the left and right-hand sides of the hypotheses to differ by some nonrandom amount that depends on the time to maturity. In the following, we denote these differences by $\gamma_i(\tau)$ with $\tau = T - t$:

(A*) $1/P_t(T) = \mathsf{E}_t\left[\mathrm{e}^{\int_t^T r_u\,du}\right] + \gamma_a(\tau)$.
(B*) $\mathsf{E}_t[\mathrm{d}P_t(T)/P_t(T)] = [r_t + \gamma_b(\tau)]\,\mathrm{d}t$.
(C*) $F_t(T) = \mathsf{E}_t[r_T] + \gamma_c(\tau)$.

As observed earlier, condition (B*) is true in any Gaussian completely affine model.

Hypothesis (C*) is what is usually tested in empirical work on the expectations hypothesis. Integrating (C*) and using the fact that yields are averages of forward rates shows that (C*) is equivalent to

$$Y_t(T) = \frac{1}{\tau}\mathsf{E}_t\int_t^T r_u\,du + \gamma(\tau), \qquad (18.37)$$

where we define

$$\gamma(\tau) = \frac{1}{\tau}\int_0^\tau \gamma_c(x)\,dx.$$

Consider dates $s < t < u$. Using (18.37) and iterated expectations, it is straightforward to show that

$$\mathsf{E}_s[Y_t(u) - Y_s(u)] = \frac{t-s}{u-t}[Y_s(u) - Y_s(t)] + \kappa(s,t,u), \qquad (18.38)$$

where

$$\kappa(s,t,u) \stackrel{\text{def}}{=} \frac{(t-s)\gamma(t-s) + (u-t)\gamma(u-t) - (u-s)\gamma(u-s)}{u-t}.$$

Call the bond that matures at t the short bond and the bond that matures at u the long bond (here, long and short refer to maturities, not to trades). The left-hand side of (18.38) is the expected change between s and t of the yield of the long bond. The first term on the right-hand side of (18.38) is a multiple of the difference at s between the yields of the long and short bonds. The equation states that the slope of the term structure (the difference in yields at long and short maturities) predicts changes in yields. It suggests that if we regress changes in yields on differences in yields, then we should find coefficients equal to $(t-s)/(u-t)$. The term $\kappa(s,t,u)$ should be the intercept in the regression. We do not have a theory about what value $\kappa(s,t,u)$ should take. The empirical question is whether the coefficient of the difference in yields equals $(t-s)/(u-t)$.

To explain the intuition for why differences in yields should forecast changes in yields, we first need to discuss how bond returns are related to changes in yields. By the definition of yields, the return of the long bond between s and t is

$$\frac{P_t(u)}{P_s(u)} = \exp\Big((u-s)Y_s(u) - (u-t)Y_t(u)\Big).$$

Set $y = Y_s(u)$. If the yield of the bond remains constant between s and t, then $Y_t(u) = y$ and the return of the bond is $e^{(t-s)y}$. Hence, the continuously compounded rate of return is the yield if the yield remains constant. If the yield rises (falls) between s and t, then the rate of return is less (more) than y. Now, suppose the term structure is upward sloping, meaning that the long bond has a higher yield than the short bond. Then, the longer bond might appear to be a better investment for the holding period from s to t. As just explained, if the long bond yield remains constant, it will indeed turn out to be a better investment, because it will have earned its yield between s and t. Also, the return on the long bond will be even higher if its yield falls between s and t. If we conjecture that all bonds are equally good investments in equilibrium, then we might conjecture that the yield on the long bond will rise by t, reducing its return between s and t. Thus, by this reasoning, the upward sloping term structure should forecast that the long bond yield will rise. This is the expectations hypothesis. Of course, this reasoning is based on risk neutrality and is unreliable, as discussed at the beginning of the section.

First we show show that (A) implies (18.35). The argument that (C) implies (18.36) is similar and is omitted. Fix T and write P for $P_t(T)$. Itô's formula implies that

$$\frac{d(1/P)}{1/P} = -\frac{dP}{P} + \left(\frac{dP}{P}\right)^2.$$

Therefore,

$$\frac{dP}{P} = -\frac{d(1/P)}{1/P} + \left(\frac{dP}{P}\right)^2.$$

To prove (18.35), it suffices to show that the drift of the first term on the right-hand side is the short rate. Condition (A) implies that

$$e^{\int_0^t r_u \, du} \frac{1}{P_t} = E_t \left[e^{\int_0^T r_u \, du} \right],$$

which is a martingale and hence has no drift. Applying Itô's formula to compute the drift, we see that

$$E_t \left[\frac{d(1/P)}{1/P} \right] = -r \, dt.$$

This completes the proof of (18.35).

Now, suppose there is a single-factor model and hence a single Brownian motion. Let λ denote the price of risk, and let σ denote the standard deviation of a bond return. Let μ denote the expected bond return. As usual, we have $\lambda = (\mu - r)/\sigma$. Suppose Condition (A) holds. The formula (18.35) means that $\mu - r = \sigma^2$. Therefore, $\lambda = \sigma$. This is true for bonds of all maturities, so all bonds have the same risk σ. Consequently, all bonds have the same risk premium $\mu - r$. This implies that all bond returns are identical. Likewise, if Condition (C) holds, then we can deduce from (18.36) that all bond returns must be identical. However, all bond returns cannot be identical, unless they are risk free, because each bond price must converge to 1 at its maturity.

18.6 FITTING THE YIELD CURVE AND HJM MODELS

A term structure model will not exactly fit market bond prices if there are more prices to be fit than there are parameters in the model. When estimating models, academics take the view that some or all of the market prices are measured with error, so the model can be true even if it does not fit perfectly. Practitioners, when valuing derivatives based on the term structure, such as caps, floors, or swaptions, want the model to exactly fit market bond prices, or, equivalently, to exactly fit market yields. To do that, we essentially need to add extra parameters. This section briefly describes some ways that this is done.

The simplest approach is to add a deterministic function of time to the short rate process. For convenience, call the date at which the model is being fit date 0. Let $Q_0(u)$ denote the market price at date 0 of a discount bond maturing at u. Suppose we have a model implying that discount bond prices are $P_0(u) = f(u, X_0)$ for some function f and factors X. Denote the short rate process in the model as r. Let $g(\cdot)$ be a deterministic function of time with $g(0) = 0$ and define a new model with short rate process $\hat{r}_t = r_t + g(t)$. Discount bond prices in the new model are

$$\hat{P}_0(u) = \mathsf{E}^* \left[\exp\left(-\int_0^u \hat{r}_t \, dt \right) \right]$$

$$= \exp\left(-\int_0^u g(t) \, dt \right) \mathsf{E}^* \left[\exp\left(-\int_0^u r_t \, dt \right) \right]$$

$$= \exp\left(-\int_0^u g(t) \, dt \right) P_0(u). \qquad (18.39)$$

We want the new model to fit market prices, which can be accomplished by setting

$$\exp\left(-\int_0^u g(t) \, dt \right) P_0(u) = Q_0(u) \quad \Leftrightarrow \quad \int_0^u g(t) \, dt = \log P_0(u) - \log Q_0(u).$$

This should hold for every u. Differentiating in u yields

$$g(u) = \frac{d \log P_0(u)}{du} - \frac{d \log Q_0(u)}{du} \qquad (18.40)$$

for every u. From (18.39), forward rates in the new model are

$$-\frac{d \log \hat{P}_0(u)}{du} = g(u) - \frac{d \log P_0(u)}{du},$$

so (18.40) is equivalent to matching forward rates in the new model to market forward rates:

$$-\frac{d \log \hat{P}_0(u)}{du} = -\frac{d \log Q_0(u)}{du}.$$

Another approach is to let some of the parameters in the original model vary with time. For example, in the Vasicek model or single-factor CIR model, we can let θ, κ, or σ vary with time. Allowing the long-run mean θ in the Vasicek model to vary with time is equivalent to adding a deterministic function g to the short rate as just described (Exercise 18.10). By allowing more than one of the parameters to vary with time, we can fit market bond prices and other prices as well, for example, market cap prices.

Models that fit the yield curve can be described as models of forward rates in which market forward rates are taken as initial conditions. Models of this type are called Heath-Jarrow-Morton (HJM) models in recognition of Heath, Jarrow, and Morton (1992). The basic form of an HJM model is as follows. Let $B_1^*, \ldots, B_n^*$ be independent Brownian motions under a risk-neutral probability. For each fixed u, assume the forward rate evolves for $s < u$ as

$$\mathrm{d}f_s(u) = \alpha_s(u)\,\mathrm{d}s + \sum_{i=1}^{n} \sigma_{is}(u)\,\mathrm{d}B_{is}^*, \qquad (18.41)$$

where $s \mapsto \alpha_s(u)$ and $s \mapsto \sigma_{is}(u)$ are adapted stochastic processes on $[0, u]$. Take the short rate at date s to be $r_s = f_s(s)$. Heath, Jarrow, and Morton (1992) show that

$$\alpha_s(u) = \sum_{i=1}^{n} \sigma_{is}(u) \int_s^u \sigma_{is}(t)\,\mathrm{d}t. \qquad (18.42)$$

This equation is derived below. It implies that the drifts of forward rates under the risk-neutral probability are determined by the forward rate volatilities, so a model is completely determined by specifying the forward rate volatilities.

Define

$$\Sigma_{is}(u) = \int_s^u \sigma_{is}(t)\,\mathrm{d}t.$$

Then,

$$\frac{\mathrm{d}}{\mathrm{d}u}\Sigma_{is}(u) = \sigma_{is}(u),$$

so the HJM equations (18.41)–(18.42) can be written as

$$\mathrm{d}f_s(u) = \sum_{i=1}^{n} \Sigma_{is}(u)\frac{\mathrm{d}\Sigma_{is}(u)}{\mathrm{d}u}\,\mathrm{d}s + \sum_{i=1}^{n} \frac{\mathrm{d}\Sigma_{is}(u)}{\mathrm{d}u}\,\mathrm{d}B_{is}^*. \qquad (18.43)$$

Any model that is fit to the current yield curve can be written in the HJM form, that is, as a model of forward rates with the initial forward rate curve as an input. See Exercise 18.11 for the Vasicek model. Note that in a general HJM model, forward rates, yields, and bond prices at any date t can depend on the entire histories of the Brownian motions B_i^* prior to t. Factor models such as are discussed elsewhere in this chapter are generally more tractable in that forward rates, yields, and bond prices depend only on the contemporaneous values of the factors.

Fix any u and consider $s < u$. Define $Y_s = (u-s)y_s(u)$, so $P_s(u) = e^{-Y_s}$, and $Y_s = \int_s^u f_s(t)\, dt$. From Itô's formula,

$$\frac{dP(u)}{P(u)} = -dY + \frac{1}{2}(dY)^2. \tag{18.44}$$

Also,

$$dY_s = d\int_s^u f_s(t)\, dt$$

$$= -f_s(s)\, ds + \int_s^u df_s(t)\, dt$$

$$= -r_s\, ds + \left(\int_s^u \alpha_s(t)\, dt\right) ds + \sum_{i=1}^n \left(\int_s^u \sigma_{is}(t)\, dt\right) dB_{is}^*. \tag{18.45}$$

The interchange of differentials in the last line is justified by the Fubini theorem for stochastic integrals—for example, Protter (1990).

Equation (18.45) implies

$$(dY)^2 = \sum_{i=1}^n \left(\int_s^u \sigma_{is}(t)\, dt\right)^2 ds.$$

Combine this with (18.44) and (18.45) to obtain

$$\frac{dP(u)}{P(u)} = r_s\, ds - \left(\int_s^u \alpha_s(t)\, dt\right) ds + \frac{1}{2}\sum_{i=1}^n \left(\int_s^u \sigma_{is}(t)\, dt\right)^2 ds$$

$$- \sum_{i=1}^n \left(\int_s^u \sigma_{is}(t)\, dt\right) dB_{is}^*.$$

Because the expected rate of return under a risk-neutral probability must be the short rate, this implies

$$-\left(\int_s^u \alpha_s(t)\, dt\right) + \frac{1}{2}\sum_{i=1}^n \left(\int_s^u \sigma_{is}(t)\, dt\right)^2 = 0.$$

This must hold for each u. Differentiating in u gives

$$-\alpha_s(u) + \sum_{i=1}^N \left(\int_s^u \sigma_{is}(t)\, dt\right) \sigma_{is}(u) = 0.$$

18.7 NOTES AND REFERENCES

There are generally three important technical issues regarding affine models (or factor models in general):

(i) Is there a state process X satisfying the stochastic differential equation that defines the model and is it unique?
(ii) Is there a solution of the fundamental PDE and is it unique (subject to an appropriate growth condition)?
(iii) If starting from the physical measure and a specification of the prices of risk, is MR a martingale (i.e., is there a risk-neutral probability as specified)?

Duffie and Kan (1996) give sufficient conditions for (i). Levendorskiĭ (2004) gives sufficient conditions for (ii). Cheridito, Filipović, and Kimmel (2007) answer (iii) for the extended affine class (which includes the completely affine class and the essentially affine class defined by Duffee (2002)). For additional analysis of affine models, including affine jump-diffusions, see Duffie, Pan, and Singleton (2000) and Duffie, Filipović, and Schachermayer (2003).

Though we can usually take yields to be the factors in an affine factor model, this is not always possible, and the exceptions may be interesting. Collin-Dufresne and Goldstein (2002) show that in a non-Gaussian affine model with three or more factors, there are parameter values such that the rank of the matrix $\mathcal{B}$ in (18.11a) mapping factors to yields is less than the number of factors, regardless of the maturities τ_i; hence, $\mathcal{B}$ is not invertible. In this circumstance, there are factors that affect the volatilities of bond returns that cannot be hedged with bond portfolios ("unspanned stochastic volatility").

Consider a completely affine model. Stacking the squared diagonal elements of $S(X)$, we have

$$\begin{pmatrix} S_{11}(X)^2 \\ \vdots \\ S_{nn}(X)^2 \end{pmatrix} = C + DX$$

for some constant vector C and constant $n \times n$ matrix D. Let m denote the rank of D. In the notation of Dai and Singleton (2000), the model belongs to the class $A_m(n)$. Suppose the factors are latent. As remarked in Section 18.3, affine transformations of the factors and rotations of the Brownian motions produce observationally equivalent models. Rescalings of the elements of σ, $S(x)$, and the vector λ in the price of risk specification that leave $\sigma S(x)$ and $S(x)\lambda$ unchanged also produce observationally equivalent models. In addition, we can obviously permute (change the order of) the factors without changing the implications for bond prices and expected returns. Dai and Singleton (2000) state that by such transformations we can transform any model of class $A_m(n)$ into what they call a canonical form for that class. The canonical form has the maximum number of identifiable parameters. Any particular model of class $A_m(n)$ may impose additional restrictions on the parameters. Because such restrictions are generally not motivated by theory, they can be regarded as unnecessary constraints on the goodness of fit of the model. Collin-Dufresne, Goldstein, and Jones (2008) take a different approach to identification, recommending the use of observable factors.

The term "completely affine" is due to Duffee (2002). Duffee introduces a generalization of the price of risk specification that also has the property that affine dynamics under the physical probability imply affine dynamics under the risk-neutral probability and vice versa. Models in this larger class are called "essentially affine." The essentially affine price of risk specification is the same as in (18.49) in Exercise 18.7, except that each element of the matrix $S(X)^{-1}$ is replaced by a zero if it is unbounded (equivalently, if the corresponding element of $S(X)$ is not bounded away from zero). This enables use of Novikov's condition to prove that MR is a martingale. See Exercise 18.8 for an example from Duffee (2002). Cheridito, Filipović, and Kimmel (2007) show that this replacement by zero is not always necessary in order to ensure that MR is a martingale. The key requirement that they identify is that the drifts of square-root processes under both the physical and potential risk-neutral probability should be such that the processes stay strictly positive with probability 1 (for a square root process $dX = \kappa(\theta - X)\,dt + \sigma\sqrt{X}\,dB$, this condition is that $\kappa\theta \geq \sigma^2/2$). This produces a larger class of models called the "extended affine" class.

Multifactor Gaussian term structure models are studied by Langetieg (1980). The Vasicek model with time-dependent parameters mentioned in Section 18.6 and addressed in Exercise 18.10 is frequently called the Hull-White model, because it is studied by Hull and White (1990). It is possible to analyze the Vasicek model without mean reversion, that is, with $\kappa = 0$. We do that by writing it in the form $dr = \phi\,dt + \sigma\,dB^*$. This model with a nonrandom function of time added to the short rate process is called the continuous-time Ho-Lee model, because it is the continuous-time limit of the binomial model of Ho and Lee (1986).

Exercise 18.4 asks for proof that the factors in a two-factor CIR model can be taken to be the short rate and its instantaneous variance. This idea is due to Longstaff and Schwartz (1992).

Ahn, Dittmar, and Gallant (2002) and Leippold and Wu (2002) study quadratic models. Ahn, Dittmar, and Gallant (2002) observe that the quadratic model nests the SAINTS model of Constantinides (1992); see Exercise 18.9. Constantinides (1992) formulates the SAINTS model in terms of an SDF process, and Ahn, Dittmar, and Gallant (2002) formulate quadratic models in terms of an SDF process, computing discount bond prices as $P_t(T) = \mathsf{E}_t[M_T/M_t]$. Rogers (1997) discusses other models of this type.

The discussion of the three types of pure expectations hypotheses in Section 18.5 is based on Cox, Ingersoll, and Ross (1981). Section 18.5 points out that the hypotheses (A) and (C) cannot hold in a single-factor model. Cox, Ingersoll, and Ross show more generally that they cannot hold in any factor model. Campbell (1986) argues that they are nevertheless worthy of study as approximate relations. Campbell also points out that the impure hypothesis (B*) is true in the Vasicek model. There is a large empirical literature regarding (C*). Campbell and Shiller (1991) test (C*) and conclude that a high yield spread between long and short bonds predicts a falling yield for the long bond, rather than the rising yield implied by the hypothesis. Bekaert, Hodrick, and Marshall (1997) show that the bias due to autocorrelation in the regression tests is large. Bekaert and Hodrick (2001) analyze alternative tests. See Piazzesi (2006) and Singleton (2006) for extensive surveys of term structure empirics.

With the slight exception of Exercise 18.12, this book does not cover the pricing of term structure derivatives, such as caps, floors, swaptions, and bond options. There are many books on that subject. An introduction is given in Back (2005).

EXERCISES

18.1. In the Vasicek model, set $f(t,r) = \exp(-\alpha(T-t) - \beta(T-t)r)$ for fixed T.

(a) Show that f satisfies the fundamental PDE and the boundary condition $f(T,r) = 1$ if and only if

$$\beta' + \kappa\beta = 1,$$

$$\alpha' - \kappa\theta\beta + \frac{1}{2}\sigma^2\beta^2 = 0,$$

and $\alpha(0) = \beta(0) = 0$.

(b) Verify that the formula (18.18) for yields with $\beta(\tau) = \tau b(\tau)$ and $\alpha(\tau) = \tau a(\tau)$ satisfies these conditions.

18.2. Consider a single-factor affine model.

(a) Use the fact that the short rate r is an affine function of the state variable to show that

$$dr = \phi\,dt - \kappa r\,dt + \sqrt{\alpha + \beta r}\,dB^* \qquad (18.46)$$

for constants ϕ, κ, α, and β.

(b) Assume $\beta > 0$ in (18.46). Show that r is a translation of a square-root process—that is, there exist η and Z such that

$$r_t = \eta + Z_t, \tag{18.47a}$$

$$dZ = \hat{\phi}\, dt - \hat{\kappa} Z\, dt + \hat{\sigma}\sqrt{Z}\, dB^* \tag{18.47b}$$

for constants $\hat{\kappa}, \hat{\theta}$, and $\hat{\sigma}$.

(c) The condition $\hat{\phi} > 0$ is necessary and sufficient for $Z_t \geq 0$ for all t in (18.47b) and hence for the square root to exist. Assuming $\beta > 0$, what are the corresponding conditions on the coefficients in (18.46) that guarantee $\alpha + \beta r_t \geq 0$ for all t?

18.3. Consider a two-factor Gaussian affine model. Show that the two factors can be taken to be the short rate and its drift in the sense that

$$dr_t = Z_t\, dt + \sigma\, d\hat{B}^*_{1t}, \tag{18.48a}$$

$$dZ_t = (a + br_t + cZ_t)\, dt + \zeta_1\, d\hat{B}^*_{1t} + \zeta_2\, d\hat{B}^*_{2t} \tag{18.48b}$$

for constants σ, a, b, c, ζ_1, and ζ_2 and independent Brownian motions $\hat{B}^*_i$ under the risk-neutral probability.

18.4. Suppose r is the sum of two independent square-root processes X_1 and X_2. Define $Z_t = \sigma_1^2 X_{1t} + \sigma_2^2 X_{2t}$. Note that the instantaneous variance of $r = X_1 + X_2$ is

$$\left(\sigma_1\sqrt{X_{1t}}\, dB^*_{1t} + \sigma_2\sqrt{X_{2t}}\, dB^*_{2t}\right)^2 = Z_t\, dt\,.$$

Assume $\sigma_1 \neq \sigma_2$. Show that

$$\begin{pmatrix} dr \\ dZ \end{pmatrix} = \Phi\, dt + K \begin{pmatrix} r \\ Z \end{pmatrix} dt + \Sigma S(r, Y)\, dB^*$$

for a constant vector Φ and constant matrices K and Σ, where $S(r, Z)$ is a diagonal matrix the squared elements of which are affine functions of (r, Z).

18.5. Assume the short rate is an Ornstein-Uhlenbeck process under the physical probability, that is,

$$dr = \hat{\kappa}(\hat{\theta} - r)\, dt + \sigma\, dB,$$

for constants $\hat{\kappa}, \hat{\theta}$, and σ, where B is a Brownian motion under the physical probability. Assume there is an SDF process M with

$$\frac{dM}{M} = -r\, dt - \lambda\, dB + \frac{d\varepsilon}{\varepsilon},$$

where λ is a constant and ε is a local martingale uncorrelated with B.

(a) Show that the short rate is an Ornstein-Uhlenbeck process under the risk-neutral probability corresponding to M (that is, the Vasicek model holds).

(b) Calculate the risk premium of a discount bond.

18.6. Assume $r = X_1 + X_2$, where the X_i are independent square-root processes under the physical probability; that is,

$$dX_i = \hat{\kappa}_i(\hat{\theta}_i - X_i)\,dt + \sigma_i\sqrt{X_i}\,dB_i$$

for constants $\hat{\kappa}_i$, $\hat{\theta}_i$, and σ_i, where the B_i are independent Brownian motions under the physical probability. Assume there is an SDF process M with

$$\frac{dM}{M} = -r\,dt - \lambda_1\sqrt{X_1}\,dB_1 - \lambda_2\sqrt{X_2}\,dB_2 + \frac{d\varepsilon}{\varepsilon},$$

where λ_1 and λ_2 are constants and ε is a local martingale uncorrelated with B.

(a) Show that the X_i are independent square-root processes under the risk-neutral probability corresponding to M.

(b) Calculate the risk premium of a discount bond.

18.7. Assume there is an SDF process M with

$$\frac{dM}{M} = -r\,dt - \left[S(X_t)\lambda + S(X_t)^{-1}\Lambda X_t\right]'dB + \frac{d\varepsilon}{\varepsilon}, \qquad (18.49)$$

where B is a vector of independent Brownian motions under the physical probability, ε is a local martingale uncorrelated with B, $S(X)$ is a diagonal matrix the squared elements of which are affine functions of X, $S(X)^{-1}$ denotes the inverse of $S(X)$, λ is a constant vector, and Λ is a constant matrix. Assume MR is a martingale, so there is a risk-neutral probability corresponding to M. (*Warning: This assumption is not valid in general. See the end-of-chapter notes.*)

(a) Assume $r = \delta_0 + \delta'X$ and $dX = (\phi + KX)\,dt + \sigma S(X)\,dB^*$, where B^* is a vector of independent Brownian motions under the risk-neutral probability, δ_0 is a constant, δ and ϕ are constant vectors, and K and σ are constant matrices. Show that

$$dX = (\hat{\phi} + \hat{K}X)\,dt + \sigma S(X)\,dB$$

for a constant vector $\hat{\phi}$ and constant matrix $\hat{K}$.

(b) Using the fact that bond prices are exponential-affine, calculate

$$-\left(\frac{dP}{P}\right)\left(\frac{dM}{M}\right)$$

to show that the risk premium of a discount bond is affine in X.
(c) Consider the Vasicek model with the price of risk specification (18.49). Show that, in contrast to the completely affine model considered in Exercise 18.5, the risk premium of a discount bond can depend on the short rate.

18.8. Assume

$$\begin{pmatrix} dr \\ dY \end{pmatrix} = \begin{pmatrix} \kappa(\theta - r) \\ \gamma(\phi - Y) \end{pmatrix} dt + \begin{pmatrix} \sigma & 0 \\ 0 & \eta \end{pmatrix} \begin{pmatrix} 1 & 0 \\ 0 & \sqrt{Y} \end{pmatrix} \begin{pmatrix} dB_1^* \\ dB_2^* \end{pmatrix} \qquad (18.50)$$

for constants $\kappa, \theta, \sigma, \gamma, \phi,$ and η, where the B_i^* are independent Brownian motions under a risk-neutral probability.

(a) Given the completely affine price of risk specification (18.12b), where $X = (r \ Y)'$ and

$$S(X) = \begin{pmatrix} 1 & 0 \\ 0 & \sqrt{Y_t} \end{pmatrix},$$

show that

$$dr = \kappa(\hat{\theta} - r) \, dt + \sigma \, dB_1$$

for some constant $\hat{\theta}$, where B_1 is a Brownian motion under the physical probability. Show that the risk premium of a discount bond depends only on its time to maturity and does not depend on r or Y.

(b) Consider the price of risk specification (18.49), replacing $S(X)^{-1}$ by

$$\begin{pmatrix} 1 & 0 \\ 0 & 0 \end{pmatrix}.$$

Show that the risk premium of a discount bond can depend on r and Y. Note: This is an essentially affine model. It is an example given by Duffee (2002).

18.9. Assume $M_t = \exp(X_{1t} + (X_{2t} - a)^2)$ is an SDF process, where

$$dX_1 = \mu \, dt + \sigma \, dB_{1t},$$

$$dX_2 = -\kappa X_{2t} \, dt + \phi \, dB_{2t},$$

with $\mu, \sigma, \kappa,$ and ϕ being constants and with B_1 and B_2 being independent Brownian motions under the physical probability.

(a) Derive dM/M, and deduce that r is a quadratic function of X_1 and X_2.
(b) Given the prices of risk calculated in the previous part, find Brownian motions B_1^* and B_2^* under the risk-neutral probability and show that the

dX satisfy (18.5) for an affine μ and a constant σ (thus, the model is quadratic).

18.10. Consider the Vasicek model with time-dependent parameters:
$$dr_t = \kappa(t)\big(\theta(t) - r_t\big)\,dt + \sigma(t)\,dB_t^*, \qquad (18.51)$$
where B^* is a Brownian motion under a risk-neutral probability. Define
$$\hat{r}_t = \exp\left(-\int_0^t \kappa(s)\,ds\right) r_0 + \int_0^t \exp\left(-\int_u^t \kappa(s)\,ds\right) \sigma(u)\,dB_u^* \qquad (18.52a)$$
$$g(t) = \int_0^t \exp\left(-\int_u^t \kappa(s)\,ds\right) \kappa(u)\theta(u)\,du. \qquad (18.52b)$$

(a) Show that $\hat{r}$ defined in (18.52a) satisfies
$$d\hat{r}_t = -\kappa(t)\hat{r}_t\,dt + \sigma(t)\,dB_t^*.$$
(b) Define $r_t = \hat{r}_t + g(t)$. Show that r satisfies (18.51).
(c) Given any functions $\kappa(\cdot)$ and $\sigma(\cdot)$, explain how to choose $\theta(\cdot)$ to fit the current yield curve.

18.11. Assume the short rate is $r_t = \hat{r}_t + g(t)$, where
$$d\hat{r} = -\kappa \hat{r}\,dt + \sigma\,dB^*$$
for constants κ and σ and $g(\cdot)$ is chosen to fit the current yield curve.

(a) Calculate the forward rates $f_s(u)$ using the Vasicek bond pricing formula.
(b) Calculate $\alpha_s(u)$ and $\sigma_s(u)$ such that, as s changes,
$$df_s(u) = \alpha_s(u)\,ds + \sigma_s(u)\,dB_s^*.$$
(c) Prove that
$$\alpha_s(u) = \sigma_s(u) \int_s^u \sigma_s(t)\,dt.$$

18.12. Assume the short rate is $r_t = \hat{r}_t + g(t)$, where
$$d\hat{r} = -\kappa \hat{r}\,dt + \sigma\,dB^*$$
and $g(\cdot)$ is chosen to fit the current yield curve.

(a) Consider a forward contract maturing at T on a discount bond maturing at $u > T$. Let F_t denote the forward price for $t \leq T$. What is the volatility of dF_t/F_t?

(b) What is the average volatility between 0 and T of dF_t/F_t in the sense of (17.9)?

(c) Consider a call option maturing at T on a discount bond that matures at $u > T$. Derive a formula for the value of the call option at date 0.

19

Perpetual Options and the Leland Model

CONTENTS

19.1	Perpetual Options	486
19.2	More Time-Independent Derivatives	492
19.3	Perpetual Debt with Endogenous Default	494
19.4	Optimal Static Capital Structure	498
19.5	Optimal Dynamic Capital Structure	500
19.6	Finite Maturity Debt	505
19.7	Notes and References	509

Option pricing theory is applicable to corporate debt and equity. Black and Scholes (1973) point out that shareholders can be viewed as owning a call option on the assets of the firm. Consider the simple case of a company with a single maturity of zero-coupon bonds outstanding. When the bonds mature, the shareholders can pay off the debt (exercising the call) and become the sole claimants to the firm. On the other hand, if the firm is worth less than the outstanding debt, then the call is out of the money and shareholders should choose not to exercise it. By put-call parity, shareholders can also be viewed as owning a put rather than a call. Shareholders own the firm (the underlying asset), are short cash (the face value of the debt), and have an option to put the firm to the bondholders for the face value of the debt by declaring bankruptcy, which will cover their short cash position. Bondholders own cash (the face value of the debt) and are short the put. If the value of the firm's assets has a lognormal distribution, then the Black-Scholes formulas for the values of put and call options can be directly applied to value debt and equity. This model of corporate debt is usually called the Merton model in recognition of Merton (1974).

This chapter analyzes coupon-paying debt and American options, in contrast to the zero-coupon debt and European options in the Merton model. We study the option of shareholders to default and also the option of shareholders to refinance outstanding debt. There are other options in corporate debt securities not discussed here—for example, convertible debt and callable convertible debt. The model studied in this chapter is due to Hayne Leland and co-authors. The Leland model is useful for analyzing credit risk, which is a topic in asset pricing, and it is also a workhorse model of the trade-off theory of optimal capital structure, which is an important topic in corporate finance. The trade-off theory assumes that firms choose their capital structure by trading off the tax advantages of debt against the deadweight costs of bankruptcy.

The Leland model builds on the theory of perpetual options, which is useful in its own right. We use the theory of perpetual options in Chapter 20 to analyze corporate investment. As in the previous chapter, denote expectation with respect to a risk-neutral probability by E^*.

19.1 PERPETUAL OPTIONS

This section derives the values of perpetual calls and puts. Assume the underlying asset pays a constant dividend yield $\delta > 0$ and has a constant volatility σ, and assume there is a constant risk-free rate r. Then, the underlying asset value S is a geometric Brownian motion under a risk-neutral probability:

$$\frac{\mathrm{d}S}{S} = (r - \delta)\,\mathrm{d}t + \sigma\,\mathrm{d}B^*, \qquad (19.1)$$

where B^* is a Brownian motion under the risk-neutral probability.[1] Because the claims we study are perpetual, the value of each is independent of time. The value depends only on the underlying asset value S. Thus, the fundamental PDE for the value as a function of (t, S) is actually an ODE for the value as a function of S. The ODE has a simple analytic solution. Moreover, there are simple formulas for optimal exercise times.

The Fundamental Ordinary Differential Equation

Consider any security for which the value at each date t is a function of S_t alone. Denote the value by $f(S_t)$. The expected rate of return under the risk-neutral

---

[1]. The dividend-reinvested price $e^{\delta t} S_t$ has expected rate of return r under a risk-neutral probability, so the expected growth rate of S under a risk-neutral probability is $r - \delta$. As always, the volatility σ does not change when we change probabilities.

probability must be the risk-free rate. We can compute the expected return using Itô's formula and the dynamics (19.1) of S. Equating the expected rate of return to the risk-free rate shows that f must satisfy the ODE

$$(r-\delta)sf'(s)+\frac{1}{2}\sigma^2 s^2 f''(s)=rf(s). \tag{19.2}$$

Trying a power solution $f(s)=s^\beta$ for the ODE (19.2) for a constant β, we see that the s^β factor cancels throughout, and we have a solution if and only if

$$(r-\delta)\beta+\frac{1}{2}\sigma^2\beta(\beta-1)=r,$$

or, equivalently,[2]

$$\frac{1}{2}\sigma^2\beta^2+\left(r-\delta-\frac{1}{2}\sigma^2\right)\beta-r=0. \tag{19.3}$$

In fact, the same reasoning shows that the ODE is satisfied by

$$f(s)=as^\gamma+bs^\beta \tag{19.4}$$

for any constants a, b, γ, and β if and only if γ and β are the two roots of the quadratic equation (19.3). The function (19.4) is the general solution of the ODE (19.2). From the quadratic formula, one root of (19.3) is negative and one is positive. Furthermore, the positive root is equal to 1 if $\delta=0$ and is greater than 1 if $\delta>0$.

Henceforth, we write the general solution (19.4) of the ODE (19.2) as

$$f(s)=as^{-\gamma}+bs^\beta, \tag{19.5}$$

where γ and β are both positive and $\beta>1$ if $\delta>0$. The parameter γ is the absolute value of the negative root of the quadratic equation. The constants a and b are different for different securities and are determined by boundary conditions specific to the security being valued.

Valuing a Cash Flow at a Hitting Time

To derive the Black-Scholes formula, we first valued digital options and share digitals. We follow a similar route to value perpetual calls and puts. Instead of claims that are either on or off depending on whether we are above or below the strike price at maturity, we consider claims that turn on when we first reach some critical value of the underlying asset price. For any number $s^*>0$, consider a

---

2. The ODE (19.2) is called a Cauchy-Euler differential equation, and the quadratic equation (19.3) is called its characteristic equation.

claim that pays $1 at the first time S_t hits s^*. There are two different possibilities, depending on whether S starts above or below s^*. In both cases, the value prior to the hitting time is $f(S_t)$, where f satisfies (19.5). A boundary condition is $f(s^*) = 1$.

Suppose the security pays when S falls to s^* (that is, at the first time τ such that $S_\tau \leq s^*$). In this case, an additional boundary condition is obtained by noting that, if the value of S at any time is taken to be $s \to \infty$, then the time at which s^* is hit will become indefinitely far away, so the value of the security converges to zero. Hence, f should satisfy the boundary condition $\lim_{s \to \infty} f(s) = 0$. This implies that the coefficient b on the positive power of s in (19.5) must be zero, so $f(s) = as^{-\gamma}$, where γ is the absolute value of the negative root of (19.3). Furthermore, the condition $f(s^*) = 1$ implies $a = (s^*)^\gamma$. Thus, the value of the security is

$$\left(\frac{S_t}{s^*}\right)^{-\gamma} \qquad (19.6)$$

for all $S_t \geq s^*$.

Now, suppose the security pays when S rises to s^*. In this case, the value goes to zero as S approaches zero, because the hitting time then becomes indefinitely far away. Thus, $\lim_{s \to 0} f(s) = 0$. This implies that the coefficient a on the negative power of s in (19.5) must be zero, so $f(s) = bs^\beta$, where β is the positive root of (19.3). As in the previous case, the boundary condition $f(s^*) = 1$ gives us $f(s) = (s/s^*)^\beta$. Thus, the value of the security is

$$\left(\frac{S_t}{s^*}\right)^\beta \qquad (19.7)$$

for all $S_t \leq s^*$.

Perpetual Calls

A perpetual call is a call that never expires. Let K denote the strike price. If the price of the underlying asset rises high enough, then it is worthwhile to exercise the call in order to capture future dividends. Let s^* denote a potential exercise price. If the call is exercised at s^*, then the investor receives value $s^* - K$ at the time the underlying rises to s^*. Thus, the value of the call prior to the hitting time is $s^* - K$ times the price of receiving $1 at the hitting time. From (19.7), the value of the call prior to the hitting time, that is, when $S_t < s^*$, is $(s^* - K)(S_t/s^*)^\beta$. The optimal exercise policy is the one that maximizes this value. The value is concave in s^*, and solving the first-order condition gives

$$s^* = \frac{\beta K}{\beta - 1}. \qquad (19.8)$$

This is the optimal exercise boundary. If the price of the underlying asset is initially larger than (19.8), then it is optimal to exercise immediately. Assuming

S_t is less than (19.8), substituting (19.8) into the call value $(s^* - K)(S_t/s^*)^\beta$ produces a call value at date t of

$$\left(\frac{\beta-1}{K}\right)^{\beta-1}\left(\frac{S_t}{\beta}\right)^\beta. \tag{19.9}$$

An example of the call value (19.9) is shown in Figure 19.1.

Perpetual Puts

Let K denote the strike price of a perpetual put. If the price of the underlying asset falls far enough, then it is optimal to exercise the put to capture the strike. Let s^* denote a potential exercise price. If the put is exercised at s^*, then the investor receives value $K - s^*$ at the time the underlying falls to s^*. Applying the formula (19.6) for the value of \$1 at the time S falls to s^*, the value of the put with this exercise policy is $(K - s^*)(S_t/s^*)^{-\gamma}$. The optimal exercise policy is the one that maximizes this value. Again, the value is concave in s^*, and solving the first-order condition gives

$$s^* = \frac{\gamma K}{1+\gamma}. \tag{19.10}$$

This is the optimal exercise boundary. If the price of the underlying asset is initially less than (19.10), then it is optimal to exercise immediately. Assuming S_t is greater than (19.10), substituting (19.10) into the put value $(K - s^*)(s^*/S_t)^{-\gamma}$ produces a put value of

$$\left(\frac{K}{1+\gamma}\right)^{1+\gamma}\left(\frac{S_t}{\gamma}\right)^{-\gamma}. \tag{19.11}$$

Smooth Pasting

An alternative way to derive the optimal boundary for a perpetual call or put is to use the smooth pasting condition. Consider a call, and let

$$V(s,s^*) = (s^* - K)\left(\frac{s}{s^*}\right)^\beta \tag{19.12}$$

for all $s < s^*$. This is the call value at any date t when $S_t = s$ and the exercise boundary is s^*. We derived the optimal s^* by maximizing V in s^*. That is, we solved

$$\frac{\partial V(s,s^*)}{\partial s^*} = 0. \tag{19.13}$$

As we discuss further below, the solution s^* to this equation is invariant with respect to s, for $s \leq s^*$.

The smooth pasting condition for a call is

$$\left.\frac{\partial V(s,s^*)}{\partial s}\right|_{s=s^*} = 1. \qquad (19.14)$$

This means that the derivative with respect to s is continuous at the boundary s^* (the value above the boundary is the intrinsic value $s - K$, so the derivative is 1 above the boundary). From (19.12), the partial derivative in (19.14) equals

$$\beta\left(\frac{s^*-K}{s}\right)\left(\frac{s}{s^*}\right)^\beta.$$

Setting $s = s^*$ and equating this partial derivative to 1 as in (19.14) gives the boundary (19.8) again.

It is not surprising that smooth pasting and direct optimization over the boundary give the same result, because the general justification for the smooth pasting condition is precisely that it gives the same result as the optimization. See Section 16.7.[3] The virtue of smooth pasting is that it can be used when we do not have a tractable formula for the value as a function of the boundary. When such a formula is available, as it is here, direct optimization over the boundary is often simpler. Figure 19.1 illustrates that the call value pastes smoothly with the intrinsic value when the optimal boundary is used, but it does not paste smoothly when suboptimal boundaries are used.

Ex Ante and Ex Post Optima

Note that, for both the call and the put, the optimal boundary s^* is independent of the starting value S_t. This is a general property of optimal stopping problems and follows from dynamic programming. Consider a call and consider $s_1 < s_2$. If we start at s_1, then we can choose a boundary $s^* \leq s_2$ or a boundary $s^* > s_2$. If it is optimal to choose $s^* > s_2$, then we will reach s_2 before reaching s^*, so the

---

3. The equivalence of the two methods can be seen in the current time–independent model as follows: For any s^*, we have $V(s^*,s^*) = s^* - K$. This simply states that if we have reached the boundary, then we will exercise, so the value is the intrinsic value (this is value matching). Take the total derivative of both sides of $V(s,s^*) = s^* - K$ to obtain

$$\left.\frac{\partial V(s,s^*)}{\partial s}\right|_{s=s^*} + \left.\frac{\partial V(s,s^*)}{\partial s^*}\right|_{s=s^*} = 1.$$

Direct optimization sets the partial with respect to s^* equal to 0. Because of this equation, setting the partial with respect to s^* equal to 0 is equivalent to setting the partial with respect to s equal to 1, which is smooth pasting.

Perpetual Options and the Leland Model

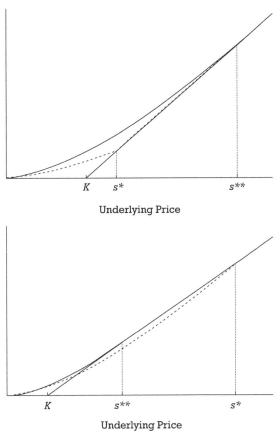

Figure 19.1 Smooth pasting for a perpetual call. The solid curved line is the optimal value (19.9) of a perpetual call. The optimal exercise boundary is s^{**}. In the upper panel, the dashed line is the value $V(s, s^*) = (s^* - K)(s/s^*)^\beta$ plotted as a function of s for an exercise boundary s^* smaller than the optimal boundary. In the bottom panel, the dashed line is the value $V(s, s^*)$ plotted as a function of s for an exercise boundary s^* larger than the optimal boundary. The optimal value pastes smoothly with the intrinsic value of the option, but the suboptimal values do not.

optimal boundary $s^* > s_2$ must be the boundary that is optimal when starting at s_2. Furthermore, if we conclude that is optimal to choose $s^* \leq s_2$, then we are saying that we should exercise immediately at all $s \geq s^*$; that is, the boundary s^* is optimal when starting at s_2 also. This is sometimes described as the ex ante (starting at s_1) optimum being the same as the ex post (after reaching some other value s_2) optimum.

19.2 MORE TIME-INDEPENDENT DERIVATIVES

The previous section values the receipt of a cash flow at a hitting time and values perpetual calls and puts. Here, we value some other time-independent claims. First, we consider knock-out boundaries for cash flows at hitting times. Then, we consider receiving a cash flow until a hitting time—either a constant cash flow or the dividends of the asset. We retain the assumptions of the previous section of a constant dividend yield, constant volatility, and constant risk-free rate. As a result of these assumptions, all of the perpetual derivative values have the general form (19.5).

Knock-Out Boundaries

Consider two numbers s_L and s_H with $s_L < S_0 < s_H$. Consider a security that pays \$1 when S reaches s_L if S reaches s_L before reaching s_H. Suppose the security expires with zero value if s_H is reached first. Thus, s_H is a "knock-out" boundary. The value $f(S_t)$ of such a security is of the form (19.5) before either boundary is reached. Here, the boundary conditions are that $f(s_L) = 1$ and $f(s_H) = 0$. The coefficients a and b in (19.5) are therefore the solutions of the linear equations

$$as_L^{-\gamma} + bs_L^{\beta} = 1, \tag{19.15a}$$

$$as_H^{-\gamma} + bs_H^{\beta} = 0. \tag{19.15b}$$

Symmetrically, we can also value a security that pays \$1 when S reaches s_H if S reaches s_H before reaching s_L and is knocked out if s_L is reached first. The value of that security is of the form (19.5) where a and b are the solutions of the linear equations

$$as_L^{-\gamma} + bs_L^{\beta} = 0, \tag{19.16a}$$

$$as_H^{-\gamma} + bs_H^{\beta} = 1. \tag{19.16b}$$

Valuing Constant Cash Flows Prior to a Hitting Time

Consider a security that pays \$1 per unit of time until the first time S hits some constant s^*. The value of this security is the value of receiving \$1 per unit of time forever (which is $1/r$) minus the value of receiving \$1 per unit of time forever after the hitting time. In other words, letting τ denote the random hitting time and assuming $t < \tau$, the value of receiving \$1 during (t, τ) is the value of receiving \$1 during (t, ∞) minus the value of receiving \$1 during (τ, ∞). The latter value is the value of receiving the single cash flow $1/r$ at time τ. We already know how to value receiving a cash flow at the hitting time τ. Applying (19.6), we see that the value of receiving \$1 per unit of time until the first time S falls to $s^* < S_t$ is

$$\frac{1}{r}\left[1 - \left(\frac{S_t}{s^*}\right)^{-\gamma}\right]. \qquad (19.17)$$

Applying (19.7), we see that the value of receiving \$1 per unit of time until the first time S rises to $s^* > S_t$ is

$$\frac{1}{r}\left[1 - \left(\frac{S_t}{s^*}\right)^{\beta}\right]. \qquad (19.18)$$

Valuing Dividends Prior to a Hitting Time

Consider owning the underlying asset and receiving the dividends δS until the first time S hits some constant s^*. After the hitting time, no further dividends are received. We can value receiving the stream of dividends δS until the hitting time by reasoning similar to that in the previous subsection. Let τ denote the hitting time and consider $t < \tau$. Note that the value at date t of receiving the dividend stream forever after time t is simply S_t. Also, the value of receiving δS during (t, τ) is the value of receiving δS during (t, ∞) minus the value of receiving δS during (τ, ∞). The latter value is the value of receiving S_τ at τ. Because τ is a hitting time, we know the value S_τ in advance (it is the boundary s^*). Thus, again we can use the formulas for valuing a known cash flow at a hitting time. Applying (19.6), we see that the value of receiving δS until the first time S falls to $s^* < S_t$ is

$$S_t - s^*\left(\frac{S_t}{s^*}\right)^{-\gamma}. \qquad (19.19)$$

Applying (19.7), we see that the value of receiving δS until the first time S rises to $s^* > S_t$ is

$$S_t - s^*\left(\frac{S_t}{s^*}\right)^{\beta}. \qquad (19.20)$$

19.3 PERPETUAL DEBT WITH ENDOGENOUS DEFAULT

In this and the following two sections, we study a model of corporate debt in which all of the debt is perpetual with a fixed interest rate (called consols). The firm's earnings before interest and taxes (EBIT) is a geometric Brownian motion under a risk-neutral probability. The firm chooses to default when EBIT falls to a boundary chosen by shareholders. In the first part of this section, we consider an arbitrary fixed boundary. At the end of the section, we compute the boundary that is optimal for shareholders. The optimal boundary trades off the option value to shareholders of keeping the firm alive versus the costs to shareholders of contributing additional capital to pay coupons to bondholders when the firm's EBIT is insufficient to pay them. In subsequent sections, we analyze how the amount of debt is chosen.

Assume for the remainder of the chapter that markets are complete, so there is a unique SDF process. This implies shareholder unanimity—all shareholders agree that the firm should maximize the market value of equity computed according to the unique SDF process. Absent completeness, different shareholders could have different opinions due to their different marginal rates of substitution.

The formulas in this section are all straightforward applications of the formulas (19.6), (19.17), and (19.19) for the value of receiving \$1 at the first time a geometric Brownian motion falls to a given level, the value of receiving \$1 per unit of time until the geometric Brownian motion falls to the given level, and the value of receiving a constant fraction (constant dividend yield) of the geometric Brownian motion until it falls to a given level.

Taxes

Let c denote the total coupons per unit time. The firm pays coupons, taxes, and net dividends from EBIT, which is an exogenous stochastic process denoted by Y. Assume that corporate income taxes are a fixed proportion τ_c of earnings before taxes (EBT = EBIT minus interest). This means that taxes paid by the firm in an instant dt are $\tau_c(Y_t - c)\,dt$. The residual

$$(1 - \tau_c)(Y_t - c)\,dt$$

is paid to shareholders. Note that we are allowing $Y_t < c$. Our assumption that taxes are a fixed proportion of EBT therefore implies that the firm receives immediate tax credits when it incurs losses. This is a simplification that is inconsistent with actual tax codes, which imply convexity in tax schedules due to progressivity and due to limitations on tax credits from losses.

When $Y_t < c$, shareholders receive negative dividends. The interpretation is that additional equity must be issued in the amount of $(1 - \tau_c)(c - Y_t)$ to supplement the tax credit in order to pay the coupons to bondholders. Current shareholders either contribute the capital themselves by buying the new shares, or they experience dilution. They can avoid this by choosing to default; however, some negative dividends will be acceptable due to the option value to shareholders of keeping the firm alive.

Assume there are personal taxes on coupon income at rate τ_i and personal taxes on dividends at rate τ_d. As with corporate taxes, it is unreasonable to assume personal taxes are levied at the same rate on negative income as on positive income (negative dividends represent capital contributions by shareholders that should be nontaxable) but we do so for the sake of simplicity. Thus, bondholders receive $(1 - \tau_i)c$ in coupon income net of taxes, and shareholders receive $(1 - \tau_d)(1 - \tau_c)(Y_t - c)$ in dividends net of taxes. Define the effective dividend tax rate τ_{eff} by $1 - \tau_{\text{eff}} = (1 - \tau_d)(1 - \tau_c)$; so we can write after-tax dividend income as $(1 - \tau_{\text{eff}})(Y_t - c)$.

Gordon Growth Model

We use a version of the Gordon growth model as our model of EBIT Y and its value. Specifically, assume each of the following:

- There is a constant risk-free rate r.
- $dY/Y = \mu\, dt + \sigma\, dB$ for constants μ and σ and a Brownian motion B under the physical probability.
- The unique SDF process M is such that $e^{rt}M_t$ is a martingale.
- $(dY/Y)(dM/M) = -\sigma \lambda\, dt$ for a constant λ.
- $\mu < r + \sigma \lambda$.
- There is no bubble in the value of the claim to EBIT.

Set $\delta = r + \sigma\lambda - \mu$. The asset value of the firm—meaning the sum of the claims of shareholders, bondholders, and the government (via taxes) and the value of possible bankruptcy costs—is the value of future EBIT. The value of future EBIT Y_u at any date t is X_t defined as Y_t/δ (Exercises 13.2 and 15.2). Furthermore, the asset value X satisfies

$$\frac{dX}{X} = (r - \delta)\, dt + \sigma\, dB^*, \tag{19.21}$$

where B^* is a Brownian motion under the risk-neutral probability associated with the SDF process M.

Default and Bankruptcy Costs

Assume the firm defaults when its EBIT first falls to some level y_D; equivalently, when the asset value X_t first falls to x_D defined as y_D/δ. Assume $x_D < X_0$, so the firm does not default at time 0. To simplify, assume that after default the firm is run as an all-equity firm forever. The market value at the time of default of the all-equity firm is $(1 - \tau_{\text{eff}})x_D$. Assume further that some fraction α of this value is lost to bankruptcy costs at the time of default, so bondholders receive $(1 - \alpha)(1 - \tau_{\text{eff}})x_D$ at the time of default.

Values of Corporate Claims

There are four claims to the company's EBIT. The cash flows associated with those claims are as follows. We use the fact that EBIT is $Y_t = \delta X_t$.

- Debt:
    - $(1 - \tau_i)c$ until default
    - $(1 - \alpha)(1 - \tau_{\text{eff}})x_D$ at default
- Equity: $(1 - \tau_{\text{eff}})(\delta X_t - c)$ until default
- Bankrucpty costs: $\alpha(1 - \tau_{\text{eff}})x_D$ at default
- Taxes:
    - $\tau_{\text{eff}}(\delta X_t - c)$ until default
    - $\tau_i c$ until default
    - $\tau_{\text{eff}} \delta X_t$ after default

Each of these can be valued using (19.6), (19.17), and (19.19). The calculations are left to the exercises. Let γ denote the absolute value of the negative root of the quadratic equation (19.3). Let D_t denote the debt value, E_t the equity value, G_t the value of bankruptcy costs, and H_t the value of tax payments. The results are as follows. These sum to the value X_t of the firm's assets.

$$D_t = \frac{(1-\tau_i)c}{r}\left[1 - \left(\frac{x_D}{X_t}\right)^\gamma\right] + (1-\alpha)(1-\tau_{\text{eff}})x_D\left(\frac{x_D}{X_t}\right)^\gamma, \quad (19.22)$$

$$E_t = (1-\tau_{\text{eff}})\left[X_t - x_D\left(\frac{x_D}{X_t}\right)^\gamma\right] - \frac{(1-\tau_{\text{eff}})c}{r}\left[1 - \left(\frac{x_D}{X_t}\right)^\gamma\right]. \quad (19.23)$$

$$G_t = \alpha(1 - \tau_{\text{eff}})x_D \left(\frac{x_D}{X_t}\right)^\gamma, \qquad (19.24)$$

$$H_t = \tau_{\text{eff}}X_t + \frac{(\tau_i - \tau_{\text{eff}})c}{r}\left[1 - \left(\frac{x_D}{X_t}\right)^\gamma\right]. \qquad (19.25)$$

Note that

$$(1 - \tau_{\text{eff}})\left(X_t - \frac{c}{r}\right)$$

would be the value of equity if the firm had no option to declare bankruptcy—it is the value of receiving $(1 - \tau_{\text{eff}})(Y_t - c)$ forever. Therefore, the remaining piece of E_t, namely,

$$(1 - \tau_{\text{eff}})\left(\frac{c}{r} - x_D\right)\left(\frac{x_D}{X_t}\right)^\gamma,$$

is the value to shareholders of the option to declare bankruptcy; that is, it is the value of the put option to sell the firm to the bondholders in exchange for extinguishing the debt.

The total market value of the firm is the sum of the debt and equity values. It can be arranged as

$$D_t + E_t = (1 - \tau_{\text{eff}})X_t + \frac{(\tau_{\text{eff}} - \tau_i)c}{r}\left[1 - \left(\frac{x_D}{X_t}\right)^\gamma\right] - (1 - \tau_{\text{eff}})\alpha x_D\left(\frac{x_D}{X_t}\right)^\gamma. \qquad (19.26)$$

The first term is the all-equity firm value. The second term is the value of net interest tax shields. The last term is the market value of bankruptcy costs. Thus, the market value of the firm is the all-equity value plus the value of tax shields minus the cost of bankruptcy.

Optimal Default Boundary

If the firm acts in the interests of shareholders, then it chooses the default boundary x_D to maximize the value of equity. Maximizing E_t over x_D gives

$$x_D = \frac{\gamma}{1+\gamma} \cdot \frac{c}{r}. \qquad (19.27)$$

Thus, the optimal default boundary is a fraction of c/r. The firm defaults earlier (that is, x_D is larger) when the debt burden c/r is higher. If the firm starts with $X_0 < x_D$, then it should default immediately. As discussed in Section 19.1, the fact that the optimal default boundary is independent of the current value of X_t is a general property. Section 19.1 also explains why the same default boundary is obtained if we compute it using the smooth pasting condition rather than directly optimizing over the boundary.

19.4 OPTIMAL STATIC CAPITAL STRUCTURE

This section derives the firm's optimal capital structure assuming it has only one opportunity to issue debt. This is a static model in the sense that the amount of debt is unchanging (prior to default). Assume that it is costly to issue debt. The firm should choose the coupon to maximize the sum of the debt and equity values, less the cost of issuance.[4] Assume issuance costs are a proportion q of the issue price (the post-issue debt value). Then, the firm chooses c at date 0 to maximize

$$(1-q)D_0 + E_0 = (1-\tau_{\text{eff}})X_0 + \frac{[(1-q)(1-\tau_i) - (1-\tau_{\text{eff}})]c}{r}\left[1 - \left(\frac{x_D}{X_0}\right)^\gamma\right]$$

$$- (1-\tau_{\text{eff}})[1 - (1-q)(1-\alpha)]x_D\left(\frac{x_D}{X_0}\right)^\gamma. \quad (19.28)$$

Agency Problem

In (19.28), x_D is the optimal default boundary, conditional on the coupon; that is, x_D is given by (19.27). Note that x_D is *not* chosen to maximize (19.28). Instead, x_D is chosen solely to maximize the equity value as described in the previous section. The difference between the optimization problems for c and x_D is that the debt is priced at issue after c is fixed but before the firm makes any commitment regarding x_D. Thus, the price at which debt is issued depends directly on c, but it does not depend directly on x_D—instead, it depends on the anticipated value of x_D. This assumes, as is reasonable, that the firm cannot credibly commit to a default boundary prior to the debt issue. In the absence of commitment, bondholders rationally anticipate that shareholders will choose x_D to maximize equity value as described previously. This agency issue between shareholders and bondholders—that is, the fact that shareholders are self-interested rather than maximizing debt plus equity value—ultimately acts to the detriment of shareholders, because it would be better for shareholders if they could commit

---

4. Shareholders care about the debt value because it is the price at which they can issue debt. We have assumed that the firm's asset value is exogenously given, independent of the leverage decision. For this to be the case, it must be that the proceeds of the debt issue are used to repurchase shares. In this case, the value to shareholders is the net cash from the debt issue plus the post-issue equity value.

to x_D and maximize (19.28), thereby increasing the issue price of debt. In fact, provided only that it is optimal to have debt—see the next paragraph—the first-best policy for shareholders prior to debt issuance is to commit to never default. That would avoid the deadweight costs of bankruptcy. Clearly, such a commitment is not feasible.

Optimal Coupon

Substitute the optimal default boundary (19.27) into the objective function (19.28), with (19.27) written as $x_D = \theta c/r$, where

$$\theta \stackrel{\text{def}}{=} \frac{\gamma}{1+\gamma}.$$

Also, to simplify, define

$$A_0 = (1-q)(1-\tau_i) - (1-\tau_{\text{eff}})$$

and

$$A_1 = (1-\tau_{\text{eff}})[1-(1-q)(1-\alpha)].$$

With these substitutions, and dropping the constant term $(1-\tau_{\text{eff}})X_0$, the objective function (19.28) is

$$\frac{A_0 c}{r}\left[1-\left(\frac{\theta}{X_0}\right)^\gamma \left(\frac{c}{r}\right)^\gamma\right] - A_1 \theta \left(\frac{\theta}{X_0}\right)^\gamma \left(\frac{c}{r}\right)^{1+\gamma}. \tag{19.29}$$

If $A_0 \leq 0$, then (19.29) is maximized at $c=0$. In this case, the net tax savings $\tau_{\text{eff}} - \tau_i$ is too small compared to the issuance cost to make debt worthwhile. In the more realistic case $A_0 > 0$, the optimal coupon is

$$c = \frac{rX_0}{\theta}\left(\frac{A_0}{(1+\gamma)(A_0 + \theta A_1)}\right)^{1/\gamma}. \tag{19.30}$$

Scaling

Note that the optimal coupon (19.30) is proportional to the asset value X_0. Furthermore, from (19.27), the optimal default boundary is proportional to the coupon ($x_D = \theta c/r$) and hence proportional to the asset value. If, for example, the firm asset value were twice as large at the time of debt issuance, then both the amount of debt and the default boundary would be doubled. Furthermore, with both c and x_D being proportional to X_0, the date–0 debt and equity values D_0 and E_0 are both proportional to X_0. This scaling property simplifies the analysis in the next section significantly.

19.5 OPTIMAL DYNAMIC CAPITAL STRUCTURE

The assumption that the firm has only one opportunity to issue debt is obviously unrealistic. Furthermore, it has a significant effect on the results. With only one opportunity to issue debt, the firm will issue more debt than is observed empirically; that is, the leverage ratio $D/(D+E)$ implied by the optimal coupon (19.30) is larger than typical firm leverage ratios, given realistic parameter values in (19.30). Much more realistic numbers are obtained when we give the firm the opportunity to issue debt whenever it wishes.

Denote the asset value at which a second issue of debt is made as x_R. Thus, there are now two boundaries: the default boundary x_D and the refinancing boundary x_R. Of course, we will have $x_D < x_R$. Set $\phi_D = x_D/X_0$, $\phi_R = x_R/X_0$, and $\phi_C = c/X_0$. Thus, ϕ_D is the fraction of initial firm value at which the firm defaults, ϕ_R is the multiple of initial firm value at which the firm refinances, and the coupon is $\phi_C X_0$.

The scaling property discussed in the previous section is very convenient. We want to preserve and extend it. Suppose X_t reaches $\phi_R X_0$ before reaching $\phi_D X_0$, so the firm refinances. It chooses a new coupon and new default and refinancing boundaries. We want these to be the same proportions of the new asset value $\phi_R X_0$ as the initial coupon and boundaries were of the original asset value X_0. In other words, we want the new coupon to be $\phi_C \phi_R X_0$, the new default boundary to be $\phi_D \phi_R X_0$, and the new refinancing boundary to be $\phi_R^2 X_0$. If such choices are optimal, then the dynamic problem can be solved simply by finding the optimal three numbers ϕ_D, ϕ_R, and ϕ_C.

The scaling described in the previous paragraph will be optimal if everything in the firm's problem is linearly homogeneous with respect to the asset value. Since the firm starts by assumption with zero debt prior to the first issue, a requirement for the scaling is that the firm must have zero debt prior to subsequent refinancings. This will be true if the firm is forced to retire all existing debt each time it refinances. While this is not generally the case, there are sometimes covenants that prohibit issuance of any additional debt of the same or greater seniority. Such covenants could require the original debt to be retired before any additional debt is issued. To simplify the analysis, assume such covenants exist and assume that the debt is callable at par (at D_0), so the firm can issue additional debt by first retiring the existing debt and then reissuing that debt plus the incremental debt desired.

The firm pays the issuance cost on all of the debt issued. This creates "lumpy" costs, which lead to infrequent debt issues—it is not optimal to increase debt frequently by small amounts, since issuance costs must be paid on the entirety of the issue rather than only on the amount of incremental debt.

Denote x_D by $\phi_D X_0$ and x_R by $\phi_R X_0$ in the remainder of the section. In fact, because all values scale by X_0, there is no loss in taking $X_0 = 1$ in the remainder of the section. Thus, the lower boundary is ϕ_D and the upper boundary is ϕ_R.

Valuing Cash Flows at and Prior to Hitting Times

If neither default nor refinancing has occurred prior to some date t, then let P_t^D denote the value at t of receiving \$1 at the time of default if default occurs before any refinancing. In this case, refinancing is the knock-out boundary described in Section 19.2, and, as shown there,

$$P_t^D = a_D X_t^{-\gamma} + b_D X_t^\beta, \tag{19.31}$$

where γ is the absolute value of the negative root and β is the positive root of the quadratic equation (19.3), and where a_D and b_D are constants that solve the linear equations

$$a_D \phi_D^{-\gamma} + b_D \phi_D^\beta = 1, \tag{19.32a}$$

$$a_D \phi_R^{-\gamma} + b_D \phi_R^\beta = 0. \tag{19.32b}$$

Likewise, let P_t^R denote the value at t of receiving \$1 at the time of the first refinancing if refinancing occurs before default, assuming neither refinancing nor default has occurred by t. As shown in Section 19.2,

$$P_t^R = a_R X_t^{-\gamma} + b_R X_t^\beta, \tag{19.33}$$

where a_R and b_R are constants that solve the linear equations

$$a_R \phi_D^{-\gamma} + b_R \phi_D^\beta = 0, \tag{19.34a}$$

$$a_R \phi_R^{-\gamma} + b_R \phi_R^\beta = 1. \tag{19.34b}$$

Consider a security that pays \$1 per unit of time prior to any default or refinancing. Reasoning as in Section 19.2, the value of receiving \$1 during (t, ξ), where ξ is the first time that either ϕ_D or ϕ_R is hit, is the value of receiving \$1 during (t, ∞) minus the value of receiving \$1 during (ξ, ∞), and the latter is the value of receiving $1/r$ at ξ. Receiving \$1 at ξ means receiving \$1 at the time of default if default occurs before refinancing and receiving \$1 at the time of refinancing if refinancing occurs before default. Thus, it has value $P_t^D + P_t^R$. Collecting these facts, the value of receiving \$1 per unit of time prior to any default or refinancing is

$$\frac{1}{r}\left(1 - P_t^D - P_t^R\right). \tag{19.35}$$

Finally, consider a security that pays $Y = \delta X$ prior to any default or refinancing. The value of receiving Y forever is X, and the value of receiving Y during (t,ξ) is the value of receiving Y during (t,∞) minus the value of receiving Y during (ξ,∞). The latter is the value of receiving X_ξ at time ξ, which means receiving ϕ_D at ξ if ξ is the default time and receiving ϕ_R at ξ if ξ is the refinancing time (ξ is either the default time or the refinancing time, depending on which occurs first). Thus, the value of receiving Y prior to any default or refinancing is

$$X_t - \phi_D P_t^D - \phi_R P_t^R.$$

Debt Value

The holders of the bonds issued at date 0 receive three types of cash flows:

- $(1-\tau_i)c$ until either default or refinancing occurs,
- $(1-\alpha)(1-\tau_{\text{eff}})\phi_D$ at the time of default if default occurs prior to refinancing,
- D_0 at the time of refinancing if refinancing occurs prior to default.

Using (19.35) to value the first, we see that the value of debt prior to default or refinancing is

$$D_t = \frac{(1-\tau_i)c}{r}\left(1 - P_t^D - P_t^R\right) + (1-\alpha)(1-\tau_{\text{eff}})\phi_D P_t^D + D_0 P_t^R. \quad (19.36)$$

This must be true at all dates prior to default or refinancing, including date 0. Equating this at time $t = 0$ to D_0 and rearranging gives

$$D_0 = \frac{(1-\tau_i)(c/r)\left(1 - P_0^D - P_0^R\right) + (1-\alpha)(1-\tau_{\text{eff}})\phi_D P_0^D}{1 - P_0^R}. \quad (19.37)$$

Computing D_0 as a fixed point of (19.36) resolves the following circularity: the issue price of debt depends on the call price paid if the firm refinances, and the call price is assumed to equal the issue price.

Deadweight Costs of Bankruptcy

The bankruptcy cost is $\alpha(1-\tau_{\text{eff}})\phi_D$ if default occurs before any refinancing. If refinancing occurs, then the market value at the time of refinancing of possible future bankruptcy costs must be the multiple ϕ_R of the date–0 market value of possible future bankruptcy costs. So, the market value of bankruptcy costs prior to default or refinancing is G_t defined by

$$G_t = \alpha(1-\tau_{\text{eff}})\phi_D P_t^D + \phi_R G_0 P_t^R. \tag{19.38}$$

Use this equation at $t=0$ and solve for G_0 to obtain

$$G_0 = \frac{\alpha(1-\tau_{\text{eff}})\phi_D P_0^D}{1-\phi_R P_0^R}. \tag{19.39}$$

We will encounter this division by $1-\phi_R P_0^R$ again in the values of tax payments, issuance costs, and equity. It is similar to (19.37), but the foundation for (19.39) is somewhat different from the foundation for (19.37). To understand the formula (19.39), notice that there is a potentially infinite sequence of refinancings. At date 0, the value of possible bankruptcy costs incurred if the firm defaults before the first refinancing is the first term in (19.38), evaluated at $t=0$. That is, it is

$$\alpha(1-\tau_{\text{eff}})\phi_D P_0^D. \tag{19.40}$$

If there is a refinancing, then the value of possible bankruptcy costs incurred if the firm defaults before a second refinancing, and with the value calculated as of the time of the first refinancing, is (19.40) scaled by ϕ_R, because the firm grows by ϕ_R and bankruptcy costs are assumed to be proportional to firm value. If there is a second refinancing, then the value of possible bankruptcy costs incurred if the firm defaults before a third refinancing, and with the value calculated as of the time of the second refinancing, is (19.40) multiplied by ϕ_R^2, and so on. To put all of these values in date–0 terms, we need to multiply each by the value at date 0 of receiving \$1 at the refinancing date, which is P_0^R for the first, $(P_0^R)^2$ for the second, and so forth. Thus, the value at date 0 of possible bankruptcy costs is

$$\alpha(1-\tau_{\text{eff}})\phi_D P_0^D \left[1 + \phi_R P_0^R + (\phi_R P_0^R)^2 + (\phi_R P_0^R)^3 + \cdots\right].$$

This infinite sum equals (19.39).[5]

Values of Tax Payments and Issuance Costs

Using the reasoning in the previous subsection about scaling at refinancing, we can show that the value at date 0 of tax payments is

$$H_0 \stackrel{\text{def}}{=} \frac{(\tau_i - \tau_{\text{eff}})(c/r)(1-P_0^D-P_0^R) + \tau_{\text{eff}} - \tau_{\text{eff}}\phi_R P_0^R}{1-\phi_R P_0^R}. \tag{19.41}$$

---

5. The same infinite sum logic applies to the value (19.37) of debt when we consider the value of all debt that might eventually be issued rather than only the debt issued at date 0. In that calculation, we consider bondholders to receive D_0 at refinancing and to pay $\phi_R D_0$ at the same time for newly issued debt, thereby continuing to be claimants of the firm. Continuing to be claimants does not change the value, because bondholders pay $\phi_R D_0$ at refinancing for bonds worth $\phi_R D_0$.

Likewise, the value at date 0 of future issuance costs (excluding date-0 issuance costs) is

$$I_0 \stackrel{\text{def}}{=} \frac{q\phi_R D_0 P_0^R}{1 - \phi_R P_0^R}. \qquad (19.42)$$

The details are left to the exercises.

Equity Value

We could compute the equity value E_t as the residual $E_t = X_t - D_t - G_t - H_t - I_t$. However, it may be instructive to consider the separate components of the equity cash flows. Shareholders receive $(1 - \tau_{\text{eff}})(Y_t - c)$ until default or refinancing. If the firm refinances, shareholders pay D_0 to call the existing debt and shareholders receive $(1 - q)\phi_R D_0$ in net proceeds from the new debt issue. As with other claims, the equity value must scale by ϕ_R at refinancing. Thus, E_t is the sum of the following values:

- value of $(1 - \tau_{\text{eff}})Y_t$ until default or financing:
  $(1 - \tau_{\text{eff}})(X_t - \phi_D P_t^D - \phi_R P_t^R)$,
- value of $-(1 - \tau_{\text{eff}})c$ until default or refinancing:
  $-\frac{(1-\tau_{\text{eff}})c}{r}(1 - P_t^D - P_t^R)$,
- value of $-D_0$ at refinancing if it occurs first: $-D_0 P_t^R$,
- value of $(1 - q)\phi_R D_0$ at refinancing if it occurs first: $(1 - q)\phi_R D_0 P_t^R$,
- value of $\phi_R E_0$ at refinancing if it occurs first: $\phi_R E_0 P_t^R$.

Using this at $t = 0$ and solving for E_0 shows that

$$E_0 = \frac{(1 - \tau_{\text{eff}})\left(1 - \phi_D P_0^D - \phi_R P_0^R\right)}{1 - \phi_R P_0^R} - \frac{(1 - \tau_{\text{eff}})(c/r)\left(1 - P_0^D - P_0^R\right)}{1 - \phi_R P_0^R}$$
$$+ \frac{-D_0 P_0^R + (1 - q)\phi_R D_0 P_0^R}{1 - \phi_R P_0^R}. \qquad (19.43)$$

Commitment, Optimization, and Equilibrium

We discussed before why commitment to a default boundary is not feasible (if shareholders could commit before issuing debt, then they would commit to never default). Commitment to the refinancing boundary might be possible, because refinancing requires that existing debt be called, and the debt covenants could specify that the debt can be called only when $X_t = \phi_R$. Nevertheless, a more realistic assumption is that the firm can commit to neither the default nor the refinancing boundary.

Without commitment, the firm chooses the boundaries ϕ_D and ϕ_R to maximize E_0, given c. In the formula (19.43) for E_0, P_0^D and P_0^R should be regarded as functions of ϕ_D and ϕ_R—namely, the functions (19.31) and (19.33)—because they are values to shareholders of the decision made by shareholders. However, D_0 in (19.43) should be regarded as fixed rather than as a function of ϕ_D and ϕ_R. The reason is that, if the firm refinances, then the proceeds from the debt issue depend on what bondholders anticipate future boundaries will be, rather than on the actual boundaries chosen. Thus, in the formula (19.43) for D_0, the variables ϕ_D and ϕ_R and the corresponding P_0^D and P_0^R should be regarded as fixed, equal to the anticipated values, when the optimization is conducted. This causes D_0 to be fixed. However, there should be rational expectations in equilibrium. Therefore, the optimal boundaries ϕ_D and ϕ_R (which will depend on D_0 and hence will depend on the anticipated boundaries) should equal the anticipated boundaries in equilibrium. This fixed point condition determines the equilibrium boundaries.

When optimizing and determining the equilibrium ϕ_D and ϕ_R, we take the coupon c as given. The equilibrium ϕ_D and ϕ_R (and therefore also P_0^D and P_0^R) depends on the coupon. The coupon c is chosen to maximize $(1 - q)D_0 + E_0$. This expression depends directly on c and also indirectly on c through the dependence of ϕ_D, ϕ_R, P_0^D, and P_0^R on c. The total derivative should be set to zero when maximizing; that is, all of the dependence should be considered.

Example

As an example, take $\delta = 0.03$, $r = 0.05$, $\sigma = 0.25$, $\alpha = 0.05$, $q = 0.05$, $\tau_c = 0.35$, $\tau_d = 0.10$, and $\tau_i = 0.35$. Then, in the static model, the optimal coupon is 26.9% of X_0, the default boundary is 28.1% of X_0, and the leverage ratio is 49.5%. The dynamic model must be solved numerically. For the same parameter values in the dynamic model, when the firm can commit to neither the default nor the refinancing boundary, the optimal coupon is 15.8% of X_0, the default boundary is 16.4% of X_0, the refinancing boundary is 292.8% of X_0, and the leverage ratio is 30.8%. This illustrates the lower leverage implied by the dynamic model compared to the static model.

19.6 FINITE MATURITY DEBT

The model with perpetual debt is relatively simple, because the security values depend only on the firm's asset value and do not depend on time. This section briefly describes the finite maturity debt model of Leland and Toft (1996). The assumptions of the model imply that, even though the price of each bond is time dependent, the total value of the outstanding debt of the firm is independent of time.

Assume the firm continuously issues debt of a constant maturity T to replace maturing debt, which also had initial maturity T. Assume all of the debt pays the same coupon, has the same face value, and has the same priority in bankruptcy. Assume that the debt is issued at a constant rate per unit of time. Thus, the firm has a constant term structure of debt. Let C denote the aggregate coupon. The aggregate coupon is the integral over maturities of the coupon paid by each maturity, so, since all maturities pay the same coupon, the coupon paid by each issue is $c = C/T$. Let p denote the principal (face) value of each issue. Thus, the cash flow paid to bondholders is $C + p$ per unit of time.

Ordinarily, firms issue debt at par, meaning they would set the coupon so that the issue price is p. However, the coupon is fixed here, so the issue price fluctuates over time as the firm's asset value and hence its creditworthiness fluctuates. Let $\pi(X_t)$ denote the issue price at time t.[6] If the proceeds from issue exceed the principal of maturing debt—that is, if $\pi(X_t) > p$—then the excess is paid to shareholders. On the other hand, if the proceeds from issue fall short of the maturing principal, then the deficit is paid via a reduction in shareholder cash flows. Ignore taxes on imputed interest that arise from issuing bonds below par. Then, the income to shareholders in an instant dt is $\tau_{\text{eff}}(Y_t - C)\,dt + [\pi(X_t) - p]\,dt$. The component $[\pi(X_t) - p]\,dt$ of shareholder cash flows is called the rollover gain or loss.

Adopt the assumptions of Section 19.3 regarding the dynamics of the EBIT process and market pricing, and assume again that the firm is operated as an all-equity firm after bankruptcy and that bankruptcy costs are a proportion α of the all-equity asset value $(1 - \tau_{\text{eff}})x_D$. As in Section 19.3, there are four claims to the firm's cash flows: debt, equity, taxes, and bankruptcy costs. The values of tax payments and bankruptcy costs are the same as in Section 19.3. We describe the valuation of debt and then calculate the equity value as the residual.

Debt Value

Consider a bond issued at date 0 and maturing at T. We want to value it at date $t < T$, assuming the firm has not yet defaulted at t. The value depends on the remaining time to maturity $T - t$ and the firm's asset value at t. The same formula

---

6. The issue price is given in formula (19.46), taking the time to maturity $T - t$ to be T.

will apply to a bond issued at any date s that matures at $s+T$ and that is being valued at $s+t$.

Let $\xi = \min\{u \mid X_u \leq x_D\}$ denote the hitting time of the default boundary. If $\xi > T$, then the holders of bonds issued at date 0 receive the coupon c during (t, T) and receive the face value p at T. If $\xi \leq T$, then the bondholders receive the coupon c during (t, ξ) and receive $(1-\alpha)(1-\tau_{\text{eff}})x_D/T$—their share of the value of the firm after bankruptcy costs, which is paid equally to all outstanding maturities—at the default date ξ.[7] Denote the minimum of ξ and T by $\xi \wedge T$. To summarize, the holders of bonds issued at date 0 receive the coupon c during $(t, \xi \wedge T)$, receive p at T if $T > \xi$, and receive $(1-\alpha)(1-\tau_{\text{eff}})x_D/T$ at ξ if $\xi \leq T$.

In the perpetuity model with no refinancing option, the key is to value receiving \$1 at the default time. In this model, there are two values that must be calculated: the value of receiving \$1 at T if $\xi > T$, and the value of receiving \$1 at ξ if $\xi \leq T$. The former value at date $t < T$ is

$$G_1(T-t, X_t) \stackrel{\text{def}}{=} \mathsf{E}^*\left[e^{-r(T-t)}1_{\{\xi > T\}} \mid \xi > t, X_t\right] = e^{-r(T-t)}\mathbb{Q}(\xi > T \mid \xi > t, X_t), \quad (19.44)$$

and the latter value is

$$G_2(T-t, X_t) \stackrel{\text{def}}{=} \mathsf{E}^*\left[e^{-r(\xi-t)}1_{\{\xi < T\}} \mid \xi > t, X_t\right], \quad (19.45)$$

where $\mathbb{Q}$ denotes the risk-neutral probability. The value of receiving the coupon c during $(t, \xi \wedge T)$ is the value of receiving the coupon during (t, ∞) minus the value of receiving the coupon during $(\xi \wedge T, \infty)$. The latter value is $(G_1 + G_2)c/r$. Thus, the value of the coupon to the bondholder is $(1 - G_1 - G_2)c/r$. Add in the values of the face and the proceeds from bankruptcy if the firm defaults before T to obtain the total bond value

$$\frac{c}{r}(1 - G_1 - G_2) + pG_1 + \frac{(1-\alpha)(1-\tau_{\text{eff}})x_D}{T}G_2. \quad (19.46)$$

Let u denote the time to maturity $T-t$. At any time prior to default, the firm has bonds outstanding with time to maturity u for all $0 < u < T$. The aggregate value of debt is the sum of the values of all of the outstanding maturities; that is,

---

7. Assume the firm at date 0 is already in the steady state in the sense that it already has bonds outstanding of all maturities $t \leq T$.

it is the integral of (19.46) over maturities $0 < u < T$. This equals

$$D(X_t) \stackrel{\text{def}}{=} \frac{C}{r} + \left(p - \frac{c}{r}\right) \int_0^T G_1(u, X_t)\, du + \left(\frac{(1-\alpha)(1-\tau_{\text{eff}})x_D}{T} - \frac{c}{r}\right)$$

$$\int_0^T G_2(u, X_t)\, du. \tag{19.47}$$

Note that G_1 and G_2 are values of non-dividend-paying assets and therefore must satisfy the fundamental PDE (for $x > x_D$):

$$\frac{\partial}{\partial t} G_i(T-t, x) + (r - \delta)x \frac{\partial}{\partial x} G_i(T-t), x)$$

$$+ \frac{1}{2}\sigma^2 x^2 \frac{\partial^2}{\partial x^2} G_i(T-t), x) = rG_i(T-t, x).$$

The boundary conditions are (for $x > x_D$, and $u > 0$): $G_1(0, x) = 1$, $G_1(u, x_D) = 0$, $G_2(0, x) = 0$, $G_2(u, x_D) = 1$, $\lim_{x \to \infty} G_1(u, x) = e^{-ru}$, and $\lim_{x \to \infty} G_1(u, x) = 0$. Leland and Toft (1996) give explicit formulas for G_1 and G_2 and for the integrals of G_1 and G_2 appearing in (19.47).[8]

Equity Value

The equity value is the asset value X minus the values of bankruptcy costs, taxes, and debt. Using the formulas in Section 19.3 for taxes and bankruptcy costs, the equity value is, when $X_t = x > x_D$,[9]

$$E(x) = x - \alpha(1-\tau_{\text{eff}})x_D \left(\frac{x_D}{x}\right)^\gamma - \tau_{\text{eff}}x - \frac{(\tau_i - \tau_{\text{eff}})C}{r}\left[1 - \left(\frac{x_D}{x}\right)^\gamma\right] - D(x). \tag{19.48}$$

The optimal default boundary can be found by maximizing this in x_D or by using smooth pasting ($E'(x) = 0$ when evaluated at $x = x_D$). Leland and Toft (1996) give an explicit formula for the optimal boundary.

---

8. Our notation is somewhat different from that of Leland and Toft (1996). Using our notation on the left-hand side and that of Leland and Toft (1996) on the right-hand side, the correspondence is $G_1(u, x) = e^{-ru}[1 - F(u)]$ and $G_2(u, x) = G(u)$.

9. Leland and Toft use the all-equity firm value as the fundamental process. The all-equity value is $v = (1 - \tau_{\text{eff}})x$. Set $v_D = (1 - \tau_{\text{eff}})x_D$ and denote the net tax advantage of debt by $\tau = \tau_{\text{eff}} - \tau_i$. Then, the equity value (19.48) can be written, as in equation (9) of Leland and Toft (1996), as

$$E(v) = v - \alpha v_D \left(\frac{v_D}{v}\right)^\gamma + \frac{\tau C}{r}\left[1 - \left(\frac{v_D}{v}\right)^\gamma\right] - D^*(v),$$

where we set $D^*(v) = D(v/(1 - \tau_{\text{eff}}))$.

19.7 NOTES AND REFERENCES

The Leland model first appears in Leland (1994). The optimal dynamic capital structure model is due to Goldstein, Ju, and Leland (2001). The model of finite maturity debt is due to Leland and Toft (1996).

The Merton and Leland models are called structural models of capital structure or structural models of credit risk, to contrast them with reduced-form models. Reduced-form models of capital structure are statistical models, usually positing linear relationships between variables of interest that are only loosely grounded in theory. Reduced-form models of credit risk treat default as an exogenous process and are largely concerned with statistically modeling correlations of defaults across companies and the linkages between default probabilities and credit ratings. For a survey of structural models of capital structure, see Strebulaev and Whited (2011). For a textbook treatment of the Leland model compared to other models of credit spreads and default risk, see Lando (2004).

Leland (1998) analyzes a model of finite maturity debt that is simpler in some respects than the Leland-Toft model. In Leland's model a fraction m of the outstanding debt is called per unit of time and the same amount of debt reissued. Each outstanding bond has the same probability of being called at each instant, regardless of when it was issued. The bonds in this model do not have any fixed maturity, but the average (expected) maturity of bonds is $1/m$. A faster call rate (higher m) implies a shorter average maturity. In the Leland model and in the Leland-Toft model, firms commit to bond maturity, meaning that they do not adjust maturities as circumstances change, that is, as X changes. He and Milbradt (2015) analyze a version of the Leland (1998) model of finite maturity debt in which the firm chooses between two maturities at each instant.

An important analysis of the Leland model is made by Strebulaev (2007). He extends the model by allowing for asset sales to reduce leverage prior to default, and he simulates the model and compares the simulation to reported empirical results. He shows, for example, that the negative relation between profitability and leverage that is found empirically and that is inconsistent with static models of the trade-off theory is consistent with the Leland model, because profitable firms that have not yet reached the refinancing boundary will have low leverage (which is the value of debt divided by the sum of debt and equity values) due to having high market equity. The debates about structural versus reduced-form models of capital structure and about the trade-off theory versus the pecking-order theory constitute large literatures within corporate finance. Those literatures are much too large to survey here, but the articles by Welch (2013) and Strebulaev and Whited (2013) are worth noting.

Credit spreads are differences between risky debt yields (of a particular credit rating and maturity) and Treasury yields (of the same maturity). Structural models of credit risk, including the Merton and Leland models, under-estimate credit spreads for highly rated bonds and especially underestimate credit spreads of highly rated bonds of short maturities (Huang and Huang, 2012). This is called the credit spread puzzle. Leland (2004) shows that structural models perform much better for estimating default probabilities—except again at short maturities, for which actual default probabilities exceed the predictions of structural models. The basic issue is that if the asset value is above the default boundary and a bond has only a short time to maturity, then it is highly unlikely that the asset value will fall all the way to the default boundary, so the default probability is very low, lower than appears in the data. An obvious remedy is to include jump risk in the asset value (Chen and Kou, 2009). An alternative explanation for high credit spreads at short maturities is that the market cannot observe the asset value continuously, so it cannot be sure of the distance from the default boundary. This explanation is developed by Duffie and Lando (2001).

Extensions of the Leland model not discussed in this chapter include

- asset substitution: Leland (1998),
- manager-shareholder conflicts: Morellec, Nikolov, and Schürhoff (2012),
- renegotiation of debt prior to default: François and Morellec (2004), Hackbarth, Hennesy, and Leland (2007), Broadie, Chernov, and Sundaresan (2007),
- illiquidity in the debt market: He and Xiong (2012) and He and Milbradt (2014).

EXERCISES

19.1. Calculate the value of a perpetual call that is knocked out when the underlying asset price falls to a boundary s_L. What is the optimal exercise boundary for the call? How does the optimal boundary compare to the optimal boundary when there is no knock-out feature and why? Answer the same questions for a perpetual put that is knocked out when the underlying asset price rises to a boundary s_H.

19.2. Explain how (19.6), (19.17), and (19.19) are used to compute the values (19.22)–(19.25) of corporate claims in the Leland model.

19.3. Derive the value of tax payments (19.41) and the value of issuance costs (19.42) in the model of dynamic capital structure.

19.4. Consider the value $f(S_t)$ of receiving a cash flow c per unit of time until the asset price S hits a boundary s^*. Assume the dividend yield, volatility, and risk-free rate are constant. The value is derived in Section 19.2 by reducing the valuation problem to valuing the receipt of $1 at the hitting time. The valuation can be done more directly as follows:

(a) Write down the ODE that the function f satisfies, taking into account that part of the return from the asset is the cash flow $c\,dt$.
(b) Show that c/r satisfies the ODE.
(c) Show that
$$\frac{c}{r} + as^{-\gamma} + bs^{\beta}$$
satisfies the ODE for any constants a and b, where $-\gamma$ and β are the roots of the quadratic equation (19.3). (The ODE is called nonhomogeneous, because it includes the constant term c—a term that is not linear in f. The general method for solving a nonhomogeneous version of (19.2) is to value the nonhomogeneous part as a cash flow to be received forever and then to add that value to the general solution (19.4) of the homogeneous part.)
(d) Use boundary conditions to calculate a and b in these two cases: (i) the cash flow terminates when S_t first falls to s^*, and (ii) the cash flow terminates when S_t first rises to s^*.

19.5. Repeat the previous exercise for the value of receiving κS_t until S hits a boundary, for a constant $\kappa > 0$. In lieu of Part (b) of the previous exercise, show that $f(s) = \kappa s/\delta$ satisfies the ODE.

19.6. In the Leland (1998) model, the firm issues at date 0 debt with principal P that pays aggregate coupons C. Prior to bankruptcy, a fraction $m\,dt$ of the outstanding debt is called each instant and redeemed at par. Therefore, $e^{-mt}P$ of principal remains outstanding at each $t > 0$ from the debt issued at date 0, if the firm has not declared bankruptcy prior to t. The expected cash flow received by this outstanding principal in an instant dt is $e^{-mt}(C + mP)\,dt$. This is the sum of the coupon $e^{-mt}C\,dt$ and the principal payments $e^{-mt}mP\,dt$ on the debt called during dt. Assume the firm declares bankruptcy when X first falls to some boundary x_D, where X is the geometric Brownian motion (19.21). Let $f(t, X_t)$ denote the value of the debt that is issued at date 0 and remains outstanding at date t if bankruptcy has not been declared by t.

(a) Explain why f should satisfy the PDE
$$f_t + (r - \delta)xf_x + \frac{1}{2}\sigma^2 x^2 f_{xx} + e^{-mt}(C + mP) = rf.$$
(b) Define $g = e^{mt}f$. Derive an ODE satisfied by g.

(c) Using the results of Section 19.2 or Exercise 19.4, solve the ODE for g. (In the Leland model, the firm continuously issues new debt to replace the debt being redeemed. Furthermore, the value of any outstanding bond is independent of the time at which it was issued, because all bonds have the same terms and the same probability of being called. Therefore, $g(X_t)$ is the value of all outstanding debt at any time prior to bankruptcy.)

20

Real Options and q Theory

CONTENTS

20.1 An Indivisible Investment Project	515
20.2 q Theory	518
20.3 Irreversible Investment as a Series of Real Options	524
20.4 Dynamic Programming for Irreversible Investment	530
20.5 Irreversible Investment and Perfect Competition	535
20.6 Berk-Green-Naik Model	541
20.7 Notes and References	546

The purpose of this chapter is to show how to apply methods developed earlier—dynamic programming and option pricing—to study optimal corporate investment. Some corporate investment problems are most easily solved by dynamic programming and some are most easily solved by option pricing. Sections 20.3 and 20.4 solve a single problem by the two different methods to show that they give the same result when both are applicable and to illustrate the relationship between them.

Corporate investment is a central topic in corporate finance and has also been of increasing interest in asset pricing research in recent years, as researchers attempt to endogenize firm characteristics, firm risks, and expected stock returns so as to study the cross-sectional relations between characteristics and returns in theoretical models with optimizing and rational firms and investors. The interest in such models stems largely from empirical findings of cross-sectional anomalies such as size, value, and momentum (Section 6.6). One possible explanation of the empirical results is that firm characteristics such as market capitalization and book-to-market ratio are related to priced risk factors that have been omitted from the asset pricing models being tested. Another possible explanation is that errors in estimating betas with respect to risk factors leave explanatory power for firm characteristics. Market value is always related to risk; hence, as Berk (1995) observes, variables based on market value—such as size or book-to-market—can be expected to have explanatory power for returns whenever risk factors are omitted or betas misestimated. To determine whether these explanations for anomalies are plausible, we need a model in which firm characteristics and risks are jointly and endogenously determined. Another motivation for such models is the desire to understand average returns preceding and following corporate financing decisions, for example equity issues or dividend changes. Such decisions are naturally related to investment decisions. For example, the empirical observation that returns are lower on average following seasoned equity issues may be due to the fact that investment converts a growth option into assets in place, lowering risk (Carlson, Fisher, and Giammarino, 2006).

The basic prescription for optimal investment is to invest up to the point that the marginal cost of investment equals the marginal value of capital. The marginal value of capital divided by the price of capital is called marginal q. The marginal cost of investment includes both the price of capital and any adjustment costs. Throughout the chapter, we take capital as the numeraire, so its price is always 1. If there are no adjustment costs, then the optimal investment rule is to invest up to the point that marginal q equals 1. If adjustment costs are quadratic, then the optimal investment rate is an affine function of marginal q. These and related results are presented in Section 20.2, using the method of dynamic programming.

The simplest application of option pricing theory to corporate investment is to determine the optimal time for a company to make an investment, when the investment is indivisible and irreversible. This is the topic of Section 20.1. If there is no fixed horizon for making the investment, then the investment option is a perpetual call option, and, under certain assumptions, the optimal exercise time is given by the results of Section 19.1. The essential lesson from this model is that if a firm can delay investment—that is, if it has an investment option that is not yet at its maturity—then the simple net present value (NPV) rule of investing if the present value of project cash flows exceeds the project cost is not the correct rule. Instead, the firm should wait until the option is sufficiently far in the money, and how far it should be in the money is given by the formula of Section 19.1. Projects may include abandonment options, which can be analyzed as perpetual puts, and expansion options, which are additional calls. The method described in Section 20.1 can easily accommodate these variations.

Continuous (perfectly divisible) corporate investment can sometimes be analyzed by considering each marginal unit of capital as a separate investment option and applying option pricing theory. This is illustrated in Section 20.3, which uses option pricing theory to solve an optimal investment problem when there is no depreciation of capital and investment is irreversible. The same problem is solved by dynamic programming in Section 20.4. Irreversible investment is a particular type of adjustment cost. When investment is irreversible, then the optimal investment rule is to invest only when marginal q equals 1 and to do nothing when marginal q is less than 1.

Section 20.5 analyzes a perfectly competitive industry with irreversible investment and free entry. Because of free entry, growth options are exercised as soon as they reach the money. Therefore, growth options have no value, and the value of each firm is the value of its assets in place. The equilibrium industry output price is a reflected geometric Brownian motion. The equilibrium industry investment process is the singular process that causes the price to be reflected. Section 20.4 explains singular processes, and Section 20.5 explains reflected geometric Brownian motion. An interesting feature of the model is that the risk of a company's stock is small when the output price is high, because, at the reflection point, increases in demand are absorbed by increases in supply rather than by price increases. Below the reflection point, demand shocks produce price shocks and corresponding shocks to stock returns. In fact, risk is a decreasing function of the output price and consequently an increasing function of the book-to-market ratio: The model predicts that value (high book-to-market) industries have higher risks and higher expected returns than do growth (low book-to-market) industries.

Section 20.6 presents the model of Berk, Green, and Naik (1999), which is a seminal contribution to the literature that endogenizes firm characteristics and risks to determine whether empirical relationships between characteristics and returns are consistent with optimal investment and rational asset pricing. The model is a discrete-time model in which a new investment opportunity appears to each firm each period. The different opportunities have different risks, and the risk of a firm at a point in time depends on the risks of the projects it has taken. It also depends on the values of the projects taken (which determines the value of assets in place) versus the value of growth options (stemming from future investment opportunities). As a consequence, in the cross section of firms, risks and hence expected returns are related to market capitalizations and to book-to-market ratios.

Assume there is no debt and no taxes, so the value of the firm is the value of equity. Assume that markets are complete, so there is a unique SDF process. As discussed in Section 19.3, this implies that all shareholders agree that the firm should maximize the market value of equity computed according to the unique SDF process. For simplicity, assume also that there is a constant risk-free rate.

20.1 AN INDIVISIBLE INVESTMENT PROJECT

In this section, we consider a single option to make a discrete investment in a project. Let K denote the investment cost, and let Y_t denote the cash flow that the project would generate at date t if the project were adopted prior to t. As in Section 19.3, use a version of the Gordon growth model for Y and its value. Assume the following:

- There is a constant risk-free rate r.
- $dY/Y = \mu\,dt + \sigma\,dB$ for constants μ and σ and a Brownian motion B under the physical probability.
- The unique SDF process M is such that $e^{rt}M_t$ is a martingale.
- $(dY/Y)(dM/M) = -\sigma\lambda\,dt$ for a constant λ.
- $\mu < r + \sigma\lambda$.
- There are no bubbles.

Set $\delta = r + \sigma\lambda - \mu$. The value of the project's future cash flows at any date t (if the investment has been made by t) is X_t defined as Y_t/δ (Exercise 13.2). Furthermore, the value X satisfies

$$\frac{dX}{X} = (r-\delta)\,dt + \sigma\,dB^*, \tag{20.1}$$

where B^* is a Brownian motion under the risk-neutral probability associated with the SDF process M (Exercise 15.2).

Optimal Exercise and Value

The investment option is a perpetual call option on an asset with price X that has a constant dividend yield δ (the cash flow stream of the asset is $Y = \delta X$). Therefore, from (19.8), the optimal exercise boundary is

$$x^* = \frac{\beta K}{\beta - 1}, \tag{20.2}$$

where β is the positive root of the quadratic equation (19.3). From (19.9), the value of the option, given optimal exercise, is

$$\left(\frac{\beta-1}{K}\right)^{\beta-1}\left(\frac{X_t}{\beta}\right)^{\beta}. \tag{20.3}$$

The Net Present Value Rule

Consider an alternative scenario in which the option can only be exercised at date 0. In this case, the option should be exercised if it is in the money, meaning $X_0 > K$. The NPV of the project in this scenario is $X_0 - K$, and the NPV rule is to invest if the NPV is positive. Clearly it is inappropriate (and costly) to follow this rule if the firm has the flexibility to delay investment. If the firm has the flexibility to delay investment, then it should not invest until

$$X_t \geq \frac{\beta K}{\beta - 1} = K + \frac{K}{\beta - 1},$$

as shown in (20.2). Firms usually have the flexibility to delay investment, so the NPV rule is usually inappropriate.

The NPV rule can be modified so that it applies correctly to perpetual options. The key is to recognize that adopting the project imposes two separate costs. One is the exercise price K. The other is an opportunity cost: taking the project extinguishes the investment option. The correct investment rule can be described as follows: invest when the adjusted NPV is greater than or equal to zero, where the adjusted NPV is defined as $X_t - K - V_t$, with V_t denoting the value of the option. The adjusted NPV is never strictly positive. It is usually negative, because the value of the option is usually greater than the option's intrinsic value (that is, $V_t > X_t - K$). At the optimal exercise time, we have value matching ($V_t = X_t - K$), which means that the adjusted NPV is zero. Thus, investing when the adjusted NPV is greater than or equal to zero produces the correct decision. This perspective about the opportunity cost of investing is useful for understanding marginal q in the irreversible investment model discussed in Sections 20.3–20.5.

An Abandonment Option

Now suppose the firm has an option to abandon the project after it is taken. Suppose that abandoning the project produces a positive cash flow L (if the cash flow L were negative, then the project would never be abandoned, given our assumption that the project cash flow Y is always positive). The abandonment option is a put option, giving the firm the right to sell the asset with value X in exchange for receiving L. From (19.11), the value of the put option, when optimally exercised, is

$$\left(\frac{L}{1+\gamma}\right)^{1+\gamma} \left(\frac{X_t}{\gamma}\right)^{-\gamma},$$

where γ is the absolute value of the negative root of the quadratic equation (19.3). Let $f(X_t)$ denote the project value including the abandonment option; that is,

$$f(X_t) = X_t + \left(\frac{L}{1+\gamma}\right)^{1+\gamma} \left(\frac{X_t}{\gamma}\right)^{-\gamma}.$$

To determine when the project with the abandonment option should be taken, consider exercising the call option on the project when X first rises to a boundary

x^*. The value at exercise is the intrinsic value $f(x^*) - K$, and the value of receiving $1 at the exercise time is $(X_t/x^*)^\beta$, so the value of the call option for $X_t < x^*$ with this exercise policy is

$$(f(x^*) - K)\left(\frac{X_t}{x^*}\right)^\beta = \left[x^* - K + \left(\frac{L}{1+\gamma}\right)^{1+\gamma}\left(\frac{x^*}{\gamma}\right)^{-\gamma}\right]\left(\frac{X_t}{x^*}\right)^\beta,$$

where β is the positive root of the quadratic equation (19.3). Maximizing this in x^* produces the optimal exercise boundary for the investment option when the project includes the abandonment option. Other project features, such as expansion options, can be addressed similarly by defining f to equal the value of the project including the additional features.

20.2 q THEORY

This section derives the optimal investment rule for a firm that can make continuous investments. The rule is expressed in terms of marginal q. The market value of the firm depends on its capital stock K and an exogenous stochastic process X. Let $J(K_t, X_t)$ denote the market value of the firm at date t. Use subscripts to denote partial derivatives. Marginal q is defined to be $J_k(K_t, X_t)$, which is the marginal value of capital. Average q is the average value of capital, which is $J(K_t, X_t)/K_t$.[1] Average q is often proxied empirically by the sum of the market values of debt and equity divided by the sum of their book values. The investment rule is in terms of marginal q, but, under circumstances described below, marginal q equals average q.

The basic model is as follows. A firm has a given initial capital stock K_0. It chooses an investment rate I_t at each date t, and its capital stock K_t evolves as

$$\dot{K}_t = I_t - \rho K_t, \tag{20.4}$$

where the dot notation denotes the derivative with respect to time. The constant ρ is the depreciation rate. The firm's operating cash flow depends on an exogenous stochastic process X and is given by a function $\pi(K_t, X_t)$. Assume X is a Markov process relative to the risk-neutral probability with dynamics

$$dX = \mu(X)\,dt + \sigma(X)\,dB^* \tag{20.5}$$

---

[1] More precisely, marginal q is the marginal value of capital divided by the price of capital, but we take capital to be the numeraire, so its price is 1. Likewise, average q is the value of the firm $J(K_t, X_t)$ divided by the replacement cost of capital. The replacement cost of capital equals K_t, again because we take capital as the numeraire.

for functions μ and σ, where B^* is a Brownian motion relative to the risk-neutral probability. For example, we could have $\mu(x) = \mu x$ and $\sigma(x) = \sigma x$ for constants μ and σ, in which case X is a geometric Brownian motion relative to the risk-neutral probability. Let E^* denote expectation with respect to the risk-neutral probability.

The firm's net cash flow is its operating cash flow minus the cost of investment. The cost of investment is given by a function $\theta(K_t, I_t)$. The cost of investment includes the price of capital plus any adjustment costs. Examples are given below.[2] The market value of the firm is the maximum over investment policies of the risk-neutral expectation of discounted cash flows:

$$J(K_0, X_0) \stackrel{\text{def}}{=} \max_{I} \mathsf{E}^* \int_0^\infty e^{-rt} [\pi(K_t, X_t) - \theta(K_t, I_t)] \, dt. \tag{20.6}$$

HJB Equation and the First-Order Condition

The HJB equation is

$$0 = \max_i \left[\pi(k, x) - \theta(k, i) - rJ(k, x) + J_x(k, x)\mu(x) + J_k(k, x)(i - \rho k) \right. $$
$$\left. + \frac{1}{2} J_{xx}(k, x) \sigma(x)^2 \right].$$

This is based on the principles explained in Section 14.5, using the dynamics (20.4) of K, the dynamics (20.5) of X, and the fact that there is discounting in (20.6) at rate r. Note that there are no second-order terms in the HJB equation with respect to the k argument, because dK is of order dt. Assume θ is differentiable in the investment rate. Then, the first-order condition for the maximum in the HJB equation is that $\theta_i = J_k$. Thus, the optimal investment rate is the rate that equates the marginal cost of investment θ_i to the marginal value of capital J_k (which is marginal q).

Examples of Investment Costs

Most of the examples we will discuss have the property that θ is linearly homogeneous in (k, i). In that case, it is convenient to define a function ϕ by

---

2. An equivalent model is sometimes studied. In that model, investment is measured by its cost rather than by its contribution to capital. So, using a $\hat{}$ to distinguish the models, the cost of investment is called $\hat{I}_t$ rather than $\theta(K_t, I_t)$, and the capital evolution equation is $\dot{K}_t = f(K_t, \hat{I}_t) - \rho K_t$ for some function f that should be assumed to be strictly increasing in $\hat{I}$. Given such a model, we can define $I_t = f(K_t, \hat{I}_t)$ and $\theta(K_t, I_t) = \hat{I}_t = f(K_t, \cdot)^{-1}(I_t)$ to write the model in the form studied here.

$\phi(y) = \theta(1, y)$. We then have $\theta(k, i) = k\theta(1, i/k) = k\phi(i/k)$. This implies that $\theta_i(k, i) = \phi'(i/k)$. Therefore, the first-order condition for the maximum in the HJB equation is $\phi'(i/k) = J_k(k, x)$.

An example of a linearly homogeneous θ is $\theta(k, i) = ai + bi^2/k$ for $a, b > 0$. This quadratic case is quite natural. The linear term can be interpreted as the price of capital and the quadratic term as an adjustment cost. As remarked before, we take capital to be the numeraire. This means that we take $a = 1$ and consider quadratic cost functions $\theta(k, i) = i + bi^2/k$ for some b. Note that if $i < 0$ then the "cost" i is actually a cash inflow, representing the receipts from selling capital. The quadratic term implies that it is more costly to install capital quickly than it is to install capital slowly. For example, installing at rate $i = 1$ for a unit period of time costs $1 + b/k$, but installing at rate $i = 2$ for half of a period costs $1 + 2b/k$. The extra cost could represent higher prices paid to suppliers for quicker delivery, overtime costs for employees, and so on. Notice that if $i < 0$ then the quadratic term implies that receipts fall below $|i|$. This could represent higher labor costs for dismantling and moving capital or depressed prices for capital as a result of a "fire sale."

Any smooth strictly convex function θ can be interpreted similarly to the quadratic function. We can represent such a θ by a Taylor series expansion, and the first-order term $\theta_i(0, k)$ can be interpreted as the price of capital. Strict convexity implies that the second-order term $\theta_{ii}(0, k)$ is positive, like the coefficient $b > 0$ in the quadratic case.

A linear θ represents costless adjustment. In this case, capital can be bought and sold at a fixed price. Costless adjustment is unrealistic but nevertheless interesting as a benchmark. With $\theta(k, i) = i$, the first-order condition in the HJB equation states that $J_k = 1$. This is an equation that can be solved for k as a function of x. Thus, with costless adjustment, capital will be adjusted instantaneously in response to changes in X in such a way as to maintain marginal q equal to 1. In contrast, with quadratic investment costs $\theta(k, i) = i + bi^2/k$, the first-order condition implies that marginal q satisfies $J_k = 1 + bi/k > 1$ when the firm is investing ($i > 0$) and satisfies $J_k = 1 + bi/k < 1$ when the firm is disinvesting ($i < 0$). Capital is not adjusted instantly to make marginal q equal to 1, because of the adjustment costs.

Another interesting special case is $\theta(k, i) = i^+ - ci^-$, where $c < 1$ and where $i^+ = \max(i, 0)$ and $i^- = \max(-i, 0)$. In this case, c is the resale price of capital when the firm disinvests. This function is linearly homogeneous. It is also convex, but it is not strictly convex, and it is not differentiable at $i = 0$. In this case, the first-order condition for the maximization in the HJB equation is that $J_k = 1$ when the firm is investing, $J_k = c$ when the firm disinvests, and, when $c < J_k < 1$, the firm neither invests nor disinvests.

A special case of the preceding is $\theta(k,i) = i^+$. In this case, there is no market for the firm's used capital. If the capital produces nonnegative operating cash flows, then the firm will never dispose of capital at a zero price, so, for practical purposes, investment is irreversible.

Another possibility not yet considered is that there could be fixed costs of investment. Fixed costs are costs that are independent of the scale of investment. For example, we could have $\theta(i,k) = i^+ + F1_{\{i>0\}}$ for a constant F. In this case, the resale price of capital is zero, and there is a fixed cost F to be paid whenever investment is positive. This cost function is not linearly homogeneous and is not studied further in this chapter.

Main Results of q Theory

Assume the investment cost function $\theta(k,i)$ is linearly homogeneous, and define $\phi(y) = \theta(1,y)$. Recall that marginal q is defined as J_k.

(i) If ϕ is differentiable and strictly convex, then the optimal investment-to-capital ratio is a strictly increasing function of marginal q. Specifically, $i/k = (\phi')^{-1}(J_k)$.
(ii) In the quadratic case—meaning $\theta(k,i) = i + bi^2/k$ for a constant $b > 0$—the optimal investment-to-capital ratio is an affine function of marginal q. Specifically, $i/k = (J_k - 1)/(2b)$.
(iii) If $\pi(k,x)$ is linear in k, then marginal q equals average q; that is, $J_k(k,x) = J(k,x)/k$.

As is discussed below, operating cash flows are linear in capital when there is perfect competition and constant returns to scale in production. Together, (ii) and (iii) imply that with linear operating cash flows and quadratic investment costs, the optimal investment-to-capital ratio is an affine function of average q. That relationship has been tested extensively.

The assumption that ϕ is differentiable and strictly convex excludes the cost function $\theta(i,k) = i^+ - ci^-$. We discuss that case later in this section and again with $c = 0$ in Sections 20.3–20.5.

To prove (i) and (ii), we use the first-order condition $J_k(k,x) = \phi'(i/k)$ from the HJB equation. Strict convexity of ϕ implies that ϕ' is strictly increasing, hence invertible. Therefore, $i/k = (\phi')^{-1}(J_k)$. When $\theta(k,i) = i + bi^2/k$, we have $\phi(y) = y + by^2$, and $\phi'(y) = 1 + 2by$. Hence, $(\phi')^{-1}(z) = (z-1)/(2b)$, and the first-order condition implies $i/k = (J_k - 1)/(2b)$.

To establish (iii), we show that the market value of the firm is linear in k; that is, $J(k,x) = kq(x)$ for some function q. Our assumptions are that $\pi(k,x)$ is linear in k, meaning that $\pi(k,x) = kf(x)$ for some function f, and that θ is linearly homogeneous. Given those assumptions, the date–0 market value of net cash flows for any investment process I is

$$\mathsf{E}^* \int_0^\infty e^{-rt} K_t [f(X_t) - \phi(I_t/K_t)]\, dt = K_0 \cdot \mathsf{E}^* \int_0^\infty e^{-rt} \frac{K_t}{K_0} [f(X_t) - \phi(I_t/K_t)]\, dt.$$

Make the change of variables $Y_t = K_t/K_0$ and $Z_t = I_t/K_0$. Then, the date–0 market value of the firm is

$$J(K_0, X_0) = K_0 \cdot \max_Z \mathsf{E}^* \int_0^\infty e^{-rt} Y_t [f(X_t) - \phi(Z_t/Y_t)]\, dt. \qquad (20.7)$$

Here, the maximization takes as given that $Y_0 = 1$ and that the dynamics of Y are $\dot{Y}_t = Z_t - \rho Y_t$. The value of the maximization problem in (20.7) is independent of K_0. Denote it as $q(X_0)$. Then, (20.7) states that $J(K_0, X_0) = K_0 q(X_0)$. Because K_0 and X_0 are arbitrary, we have $J(k,x) = kq(x)$ for all k and x. Thus, $J_k(k,x) = q(x) = J(k,x)/k$.

Resale Price of Capital, Marginal q, and Inactivity

It is very reasonable to assume that the firm can sell capital only at a discount to the price it pays for capital. This motivates the investment cost function $\theta(k,i) = i^+ - ci^-$ with $c < 1$ that is discussed above. It can be generalized to include adjustment costs, as, for example, $\theta(k,i) = i^+ - ci^- + bi^2/k$. This cost function is strictly convex but not differentiable at zero, due to the kink at the origin caused by the lower resale price of capital. Consequently, the optimal investment rate is not a strictly increasing function of marginal q. With the cost function $\theta(k,i) = i^+ - ci^- + bi^2/k$, the optimal investment rate is

$$i/k = \begin{cases} (J_k - 1)/(2b) & \text{if } J_k \geq 1, \\ 0 & \text{if } c < J_k < 1, \\ (J_k - c)/(2b) & \text{if } J_k \leq c. \end{cases}$$

Thus, the investment rate is insensitive to marginal q when marginal q is greater than the resale price of capital but below the price of capital. This region of inactivity weakens the correlation between marginal q and the investment rate. This issue is discussed further in Section 20.3.

Examples of Operating Cash Flows

A common model is to assume that the firm has a Cobb-Douglas production function in capital and labor: $y = Ak^\alpha \ell^\psi$ with $A, \alpha, \psi > 0$ and $\alpha + \psi \le 1$. Labor ℓ is hired in a perfectly competitive market. Suppose first that the firm is a price taker in the output market. Then, given the output price p, wage rate w, and capital stock k, the firm chooses labor ℓ to maximize its operating cash flow $Apk^\alpha \ell^\psi - w\ell$. Solving for the optimal ℓ and substituting, the firm's operating cash flow is $xk^{\alpha/(1-\psi)}$, where x is a function of the output price, the wage rate, and the productivity factor A. Setting $\eta = \alpha/(1-\psi)$, the operating cash flow is $\pi(k, x) = xk^\eta$. If there are decreasing returns to scale, meaning $\alpha + \psi < 1$, then $\eta < 1$ and π is strictly concave in k. If there are constant returns to scale, meaning $\alpha + \psi = 1$, then $\eta = 1$ and π is linear in k. In this model, the exogenous stochastic process X is the price or the wage rate or the productivity factor A, or a combination of the three.

Alternatively, assume the firm is a monopolist in the output market, and assume the industry demand curve satisfies $p = by^{-1/\varepsilon}$ for a constant $\varepsilon > 1$ (ε is the elasticity of demand). Maximizing over the labor input as before, we can compute the operating cash flow as xk^η, where x is a function of the demand parameter b, the wage rate w, and the productivity factor A, and where $\eta = \hat{\alpha}/(1 - \hat{\psi})$ with $\hat{\alpha} = \alpha(\varepsilon - 1)/\varepsilon$ and $\hat{\psi} = \psi(\varepsilon - 1)/\varepsilon$. In this case, $\eta < 1$ and the operating cash flow is strictly concave in k even if there are constant returns to scale in production.

In the Cobb-Douglas model, marginal q is less than average q if there are decreasing returns to scale or if the firm is a monopolist. Marginal q equals average q in the Cobb-Douglas model if and only if there are constant returns to scale and perfect competition.

It is also possible for marginal q to be greater than average q. An example is $\pi(k, x) = xk - c$ for a constant c. In this example, c represents fixed costs that are independent of the capital stock. Such costs are called operating leverage and could be due to overhead expenses in the form of executive salaries, for example.

The Investment Return in Discrete Time

Assuming θ is differentiable, the ratio J_k/θ_i is the marginal return on investment spending. To see this, note that the marginal unit of investment generates a cost equal to θ_i. So, spending θ_i creates a unit of investment, and spending 1 creates $1/\theta_i$ units of investment. Each unit of investment creates a unit of capital, which contributes J_k to the value of the firm. Therefore, spending 1 contributes J_k/θ_i to the value of the firm. The ratio J_k/θ_i is therefore the marginal return on investment spending, or, more simply, the investment return.

The first-order condition in the continuous-time model is that the investment return is equal to 1. This means that exactly one dollar of value is created by the marginal dollar spent. In discrete time models, it is natural to assume there is a delay between the time of investment spending and the time that the spending affects the stock of capital: investment at t affects capital at $t+1$. In this case, the value of the marginal dollar spent at t is not known until $t+1$, because the value of capital at $t+1$ depends on X_{t+1}. Consequently, the first-order condition for optimal investment is a bit different.

Assume capital evolves as $K_{t+1} = (1-\rho)K_t + I_t$. At each date t, the firm chooses investment I_t to maximize

$$E_t\left[\frac{M_{t+1}}{M_t}J\big((1-\rho)K_t + I_t, X_{t+1}\big)\right] - \theta(K_t, I_t).$$

The first term is the date–t market value of the cum-dividend value of the firm at $t+1$; that is, it is the date–t market value of owning the firm at $t+1$ prior to distributions of cash flows at $t+1$. Assuming we can interchange differentiation and expectation, the first-order condition for the optimization can be arranged as

$$E_t\left[\frac{M_{t+1}}{M_t}\cdot\frac{J_k(K_{t+1}, X_{t+1})}{\theta_i(K_t, I_t)}\right] = 1. \qquad (20.8)$$

Thus, in a discrete-time model, the first-order condition is that the investment return be priced like an asset return. Consequently, investment returns can be used instead of, or in conjunction with, asset returns to identify the SDF process.

20.3 IRREVERSIBLE INVESTMENT AS A SERIES OF REAL OPTIONS

This section studies the q theory model of Section 20.2 with the investment cost function $\theta(k,i) = i^+$. This means that the resale price of capital is zero, and there are no adjustment costs, except for the fact that downward adjustments occur at the zero price. In the model of this section, the marginal value of capital is always positive, so the firm will never sell capital at a zero price. Thus, investment is effectively irreversible. In this section, we analyze irreversible investment using the real options approach developed in Section 20.1. We analyze the same model using dynamic programming in Section 20.4.

The investment cost function $\theta(k,i) = i^+$ is linearly homogeneous, but it is not differentiable at $i = 0$, and it is not strictly convex. Thus, the q theory described in Section 20.2 does not directly apply. However, the model with $\theta(k,i) = i^+$

is quite tractable, and we can even solve it in closed form. We will solve it in this section by considering each marginal unit of capital as a separate investment option. We will see that the firm's marginal q equals 1 at the optimal exercise times for the investment options and is less than 1 at all other times. Thus, exercising the investment options optimally can be described thus: Invest when marginal q equals 1.

Assume there is no depreciation of capital ($\rho = 0$). Depreciation and other extensions of the model can be analyzed by dynamic programming as discussed in Section 20.4.

Operating Cash Flows

Assume the operating cash flow at each date t is $Y_t K_t^\eta$, where $\eta < 1$, and Y satisfies the Gordon growth model assumptions in Section 20.1. Then, for fixed capital k, the value at any date t of all future operating cash flows $Y_u k^\eta$ is $X_t k^\eta$, where $X = Y/\delta$ is a geometric Brownian motion under the risk-neutral probability with dynamics (20.1). Write the operating cash flow function as $\pi(k, x) = \delta x k^\eta$. As discussed in Section 20.2, the condition $\eta < 1$ follows from Cobb-Douglas production and monopoly power in the product market, with either constant or decreasing returns to scale. Perfect competition is discussed in Section 20.5.

Capital Stock Process and the Firm's Objective

Denote the initial capital stock of the firm by k_0. The investment cost function $\theta(k, i) = i^+$ means that there are no adjustment costs for positive investment (beyond the fixed price of capital). Therefore, it could be optimal to make a discrete investment in an instant of time. Such discrete investments represent jumps in the capital stock process. Let K_t denote the firm's capital stock at each date t, including any possible discrete investment at time t. If there is a discrete investment at date 0, then $K_0 > k_0$. There will be a discrete investment at date 0 if the given initial capital stock k_0 is too small. However, it is not optimal to make discrete investments at any other times.

In order to accommodate jumps in the capital stock process, we do not assume that the investment rate is well defined, and we drop the evolution equation $\dot{K}_t = I_t - \rho K_t$. The issue with the investment rate being undefined actually goes beyond jumps, as will be explained. Under our assumptions, the paths of the capital stock process are nondecreasing (investment is irreversible and there is no depreciation) and the firm chooses the capital stock process to maximize

$$\mathsf{E}^* \int_0^\infty e^{-rt}[\pi(K_t, X_t)\,dt - dK_t]. \tag{20.9}$$

Here, the integral $\int_0^\infty e^{-rt}\, dK_t$ adds up each increment to capital discounted back to date 0 at rate r. These are the discounted investment costs. Formally, the integral is a Lebesgue-Stieltjes integral.

Optimal Exercise and Value of Investment Options

The firm has a continuum of investment options, one for each infinitesimal unit of capital. At a capital stock level of k, the cash flow produced by the marginal unit of investment is the marginal operating cash flow $\pi_k(k,x) = \eta \delta x/k^{1-\eta}$. The market value of receiving δX is X, so the market value of the marginal cash flow stream at date t is $\eta X_t/k^{1-\eta}$.

Consider a perpetual call option with a strike equal to 1 on an asset with market value $\eta X_t/k^{1-\eta}$. The market value inherits the dynamics of X; thus, it is a geometric Brownian motion with drift $r - \delta$ under the risk-neutral probability. Because the strike is equal to 1, it follows from (19.8) that the call should be exercised when the underlying value $\eta X_t/k^{1-\eta}$ reaches $\beta/(\beta - 1)$, or equivalently, when X_t reaches

$$\frac{\beta k^{1-\eta}}{(\beta-1)\eta}, \qquad (20.10)$$

where β is the positive root of the quadratic equation (19.3). From (19.9), the value of the option prior to exercise is

$$(\beta - 1)^{\beta-1} \left(\frac{\eta X_t}{\beta k^{1-\eta}} \right)^\beta. \qquad (20.11)$$

Assets in Place and Growth Options

The value of the firm at any date t is the value of the cash flow stream produced by the firm's existing capital K_t plus the sum of the values of the infinitesimal investment options. The value of the cash flow stream produced by the firm's existing capital K_t is called the value of assets in place. The cash flow at $u > t$ from capital K_t is $\delta X_u K_t^\eta$, so the value of assets in place at date t is $X_t K_t^\eta$. The sum of the values of the infinitesimal investment options is called the value of growth options. It equals the integral of (20.11) over all capital stock levels greater than K_t. This is

$$\int_{K_t}^\infty (\beta-1)^{\beta-1} \left(\frac{\eta X_t}{\beta k^{1-\eta}}\right)^\beta dk = (\beta-1)^{\beta-1} \left(\frac{\eta X_t}{\beta}\right)^\beta$$

$$\times \left(\frac{1}{\beta(1-\eta)-1}\right) K_t^{1-\beta(1-\eta)}. \qquad (20.12)$$

Actually, (20.12) is correct only when $\beta(1-\eta) > 1$. Otherwise, the left-hand side of (20.12)—the value of growth options—is infinite. Given that $\eta < 1$, the condition $\beta(1-\eta) > 1$ holds if and only if $\beta > 1/(1-\eta)$. Using the quadratic formula to calculate β as the positive root of the quadratic equation (19.3) and employing some algebra, it can be seen that the condition $\beta > 1/(1-\eta)$ is equivalent to

$$\frac{\eta\sigma^2}{2(1-\eta)^2} + \frac{r-\delta}{1-\eta} < r. \qquad (20.13)$$

Assume this condition holds for the remainder of this section.[3]

Optimal Capital Stock Process

From (20.10), the optimal investment policy is to invest whenever X_t reaches

$$\frac{\beta K_t^{1-\eta}}{(\beta-1)\eta}. \qquad (20.14)$$

This means that the capital stock process K should be the smallest nondecreasing process such that, for all t,

$$X_t \leq \frac{\beta K_t^{1-\eta}}{(\beta-1)\eta}.$$

This is equivalent to

$$K_t \geq \left(\frac{(\beta-1)\eta X_t}{\beta}\right)^{\frac{1}{1-\eta}}. \qquad (20.15)$$

Of course, we also must have $K_t \geq k_0$ for all t, where k_0 is the given capital stock at date 0. The smallest nondecreasing process satisfying these conditions for all t is given by

$$K_t = \max\left(k_0, \max_{s \leq t}\left(\frac{\eta(\beta-1)X_s}{\beta}\right)^{\frac{1}{1-\eta}}\right). \qquad (20.16)$$

---

3. Condition (20.13) is violated when the standard deviation σ of X is large or the risk-neutral expected growth rate $r - \delta$ of X is large or the interest rate r is small or when η is large (close to 1). It is quite intuitive that high risk or high expected growth of X or a low interest rate produces a high value for growth options. To see how η is related to the value of growth options, recall that, in the monopoly model with Cobb-Douglas production, the parameter η depends on returns to scale and on the elasticity of demand. A high value for η means that returns to scale diminish slowly and/or demand is close to being perfectly elastic. Either condition implies that marginal operating cash flows decrease slowly as capital increases.

This is the optimal capital stock process. According to this formula,

$$k_0 < \left(\frac{(\beta-1)\eta X_0}{\beta}\right)^{\frac{1}{1-\eta}} \quad \Rightarrow \quad K_0 = \left(\frac{(\beta-1)\eta X_0}{\beta}\right)^{\frac{1}{1-\eta}} > k_0.$$

Thus, there is a discrete investment at date 0 when k_0 is smaller than is optimal at date 0.

The paths of K are continuous (except for the possible jump from k_0 to K_0 at date 0) and nondecreasing, but their derivatives with respect to time are zero at almost all times t (that is, except for times in a set having zero Lebesgue measure) with probability 1. This is a consequence of the properties of Brownian motion paths. In particular, the paths of K are not equal to the integrals of their derivatives! We could regard the investment rate as being zero almost all of the time but infinite on the zero Lebesgue measure set of times at which investment occurs. However, it is safer to simply regard the investment rate as being undefined. The paths of K are called singular with respect to the Lebesgue measure, and the process K is called a singular process. An example of a path of K is shown in Figure 20.1.

Marginal q

The value of the firm is the value $X_t K_t^\eta$ of assets in place plus the value (20.12) of growth options. Differentiate the sum with respect to capital k to obtain

$$J_k(k,x) = \frac{\eta x}{k^{1-\eta}} - (\beta-1)^{\beta-1}\left(\frac{\eta x}{\beta k^{1-\eta}}\right)^\beta. \qquad (20.17)$$

This is marginal q (because capital is the numeraire). The investment boundary defined by (20.10), namely,

$$\left\{(k,x) \;\middle|\; x = \frac{\beta k^{1-\eta}}{(\beta-1)\eta}\right\}, \qquad (20.18)$$

is the level curve $\{(k,x) \mid J_k(k,x) = 1\}$. Furthermore, $J_k < 1$ for x less than the boundary (20.18).[4] Therefore, investing at the boundary is equivalent to

---

4. To see this, set $s = \eta x/k^{1-\eta}$ and $s^* = \beta/(\beta-1)$. Then,

$$J_k(k,x) = s - (s/s^*)^\beta/(\beta-1).$$

At the boundary, we have $s = s^*$ and $J_k(k,x) = s^* - 1/(\beta-1) = 1$. Below the boundary, $s < s^*$, and

$$\frac{\partial}{\partial s}\{s - (s/s^*)^\beta/(\beta-1)\} = 1 - (s/s^*)^\beta > 0.$$

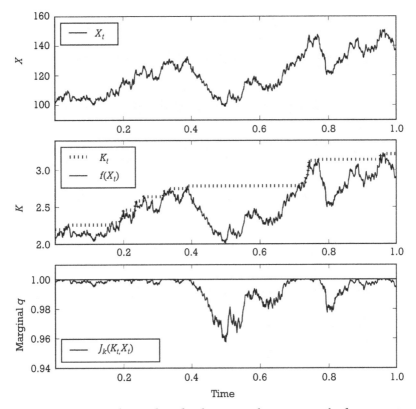

Figure 20.1 Optimal capital stock. The top panel presents a path of a geometric Brownian motion X. The middle panel plots $f(X_t) = [(\beta - 1)\eta X_t/\beta]^{1/(1-\eta)}$. The dotted line in the middle panel is $K_t = \max(k_0, \max_{s \leq t} f(X_s))$, which is the optimal capital process (20.16). The bottom panel plots marginal q. In this example, $r = 0.05, \delta = 0.02, \sigma = 0.20, \eta = 0.10, X_0 = 100$, and $k_0 = 2.2$.

investing when marginal q is equal to 1. The boundary is plotted in Figure 20.2 in Section 20.4.

To understand the formula (20.17) for marginal q, note that the first term is the value of the future cash flows contributed by the marginal unit of capital. To understand the second term, it is useful to recall the previous discussion of adjusted NPV. There are two consequences of investment: the first is that investment generates additional future cash flows; the second is that investment extinguishes the option to invest. The second term in (20.17) reflects the value of the option that is extinguished by investing the marginal unit of capital, which

Therefore, $J_k(k,x)$ is increasing in x below the boundary, increasing up to its value of 1 at the boundary.

is the option value (20.11). Thus, marginal q is the difference between the value of the marginal cash flows and the value of the option. According to the adjusted NPV rule, the firm should invest when $X - V \geq K$, where $X - V$ is the value of future cash flows minus the value of the option. In the current setting, the value of future cash flows minus the value of the option is marginal q (J_k), and the strike is the price of capital (1). Therefore, the adjusted NPV rule is to invest when $J_k \geq 1$. Adjusted NPV is never strictly positive, and it is never the case that $J_k > 1$. We always have $J_k \leq 1$ and investment is optimal only when $J_k = 1$. This is illustrated in Figure 20.1.

Hysteresis and the Correlation between q and Investment

Hysteresis means path dependence. The optimal capital stock process in this model is dependent on the history of the exogenous process X, not just on the current value of X. Consider Figure 20.1 and time 0.5. By time 0.5, the process X has dipped from its previous high. Consequently, the firm has too much capital at time 0.5, compared to the optimum at that time. If the firm could choose its capital stock freely—understanding that it would be irreversible going forward—then it would choose the amount denoted as $f(X_t)$ in the figure, which is the right-hand side of (20.15). At that level, marginal q would equal 1, whereas it is actually much lower than 1 at time 0.5. Hysteresis implies that we cannot predict the firm's capital stock at any point in time based on the firm's circumstances at that time. We have to know how the firm arrived at where it is. This occurs in any model with costly adjustment of capital.

Hysteresis is due to costly adjustment. In this model, in which the investment cost function is not differentiable at zero, another phenomenon also occurs. As Figure 20.1 illustrates, there are many fluctuations in marginal q that do not produce changes in investment, because q remains below 1. For example, the uptick in marginal q between times 0.5 and 0.6 does not lead to any increase in investment. This implies that empirical regressions of investment on proxies for marginal q or for changes in marginal q may produce weaker results than we might otherwise expect.

20.4 DYNAMIC PROGRAMMING FOR IRREVERSIBLE INVESTMENT

Dynamic programming gives the same result as the option pricing method applied in the previous section. The option pricing method is perhaps simpler, but dynamic programming is more general. To highlight the equivalence of the two methods, we use dynamic programming to solve the model of the previous section.

HJB Equation and First-Order Condition

The method of dynamic programming for this model is basically the same as for portfolio choice. The only new element is that the control process K is singular. Itô's formula still allows us to calculate $dJ(K_t, X_t)$ even when K is singular. Because K is continuous and has finite total variation, it has zero quadratic variation. Consequently, Itô's formula and the rules for multiplying differentials apply just as if dK were "something dt." Specifically,

$$dJ = J_k\, dK + J_x\, dX + \frac{1}{2} J_{xx} (dX)^2.$$

Therefore, the HJB equation is

$$\max_{dK \geq 0}\left\{\pi\, dt - dK - rJ\, dt + J_k\, dK + J_x(r-\delta)x\, dt + \frac{1}{2}J_{xx}\sigma^2 x^2\, dt\right\} = 0. \tag{20.19}$$

The maximization in the HJB equation is a constrained maximization, because, heuristically, $dK \geq 0$ as a result of K being nondecreasing (irreversible investment). The first-order condition is that $J_k \leq 1$ and that $(J_k - 1)\, dK = 0$. The latter condition (called the complementary slackness condition in the Kuhn-Tucker theory of constrained optimization) states that the capital stock increases only when marginal q is equal to 1.[5]

Fundamental ODE

The first-order condition implies that the dK terms cancel in the objective function in the HJB equation (20.19). Therefore, the HJB equation implies

$$\pi - rJ + J_x(r-\delta)x + \frac{1}{2}J_{xx}\sigma^2 x^2 = 0. \tag{20.20}$$

Even though J is a function of k and x, the derivatives in this equation are only with respect to x. It is an ODE for a function $x \mapsto J(k, x)$ that should hold for

---

5. The real meaning of the condition $(J_k - 1)\, dK = 0$ is that, with probability 1,

$$(\forall t) \qquad \int_0^t [J_k(K_s, X_s) - 1]\, dK_s = 0.$$

each fixed k (for a range of x values depending on k that is described as the inaction region below). As such, it is a nonhomogeneous version of the fundamental ODE (19.2). It states that, for a fixed capital stock k, the expected return of the asset paying dividends π and with value $J(k,x)$ is equal to the risk-free rate under the risk-neutral probability.

Depreciation

For this paragraph, let I_t denote cumulative investment through date t (instead of the rate of investment at date t as in Section 20.2). Without depreciation, $K_t = k_0 + I_t$. We could include depreciation by specifying instead that $dK = -\rho K\,dt + dI$ for a constant ρ. In this case, the HJB equation would be

$$\max_{dI \geq 0} \left\{ \pi\,dt - dI - rJ\,dt + J_k(-\rho K\,dt + dI) + J_x(r-\delta)x\,dt + \frac{1}{2}J_{xx}\sigma^2 x^2\,dt \right\} = 0.$$

The first-order condition is still that investment occur only when $J_k = 1$. The fundamental ODE would be a PDE:

$$\pi - rJ - \rho k J_k + (r-\delta)x J_x + \frac{1}{2}\sigma^2 x^2 J_{xx} = 0.$$

In the remainder of this section, we continue to study the model without depreciation.

Action and Inaction Regions

Figure 20.2 plots the optimal investment boundary (20.18), which we know from the analysis in Section 20.3, though we will calculate it independently in this section. The ODE (20.20) must hold in what is called the inaction region. This is the region in which X is below the investment boundary. The region above the investment boundary is called the action region. It is never reached, except possibly at date 0. Because the action region is not reached after date 0, the ODE (20.20) does not need to hold in the action region.

It is simple to calculate the value function in the action region based on the value function in the inaction region. If the firm is in the action region at date 0, then the optimal action is to add capital so as to move immediately to the boundary of the region. Figure 20.2 illustrates an increase from $k = k_0$ to K_0 based on the value of x_0. The value function at the point (k, x_0) in the action region is the value at (K_0, x_0) minus the cost of moving from k to K_0, which is $K_0 - k$. Note that investing to increase capital beyond K_0 would be suboptimal, because $J_k < 1$ in

Real Options and q Theory

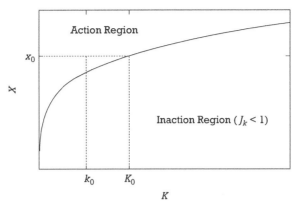

Figure 20.2 Action region, inaction region, and investment boundary. The curve is the investment boundary (20.18). The area below the curve is the inaction region, and the area above is the action region. If the initial condition (k_0, x_0) is in the action region, then the firm should invest a discrete amount (shown as $K_0 - k_0$ in the plot) to move immediately to the investment boundary.

the inaction region. Thus, we can describe the value function in the action region as

$$J(k, x_0) = J(K_0, x_0) - (K_0 - k),$$

where K_0 is such that (K_0, x_0) is on the investment boundary. Given x_0, this K_0 is the same for all $k < K_0$. Thus, J is linear in k in the action region with coefficient equal to 1. We therefore obtain the following derivatives in the action region: $J_k(k, x) = 1$ and $J_{kk}(k, x) = 0$.

Smooth Pasting and Super Contact

To solve the problem using only dynamic programming methods, we need to find the investment boundary and the function J below the boundary (in the inaction region). We do that by solving the ODE (20.20) with the requirement that the solution paste smoothly at the boundary with the value in the action region described in the previous paragraph. That is, we require $J_k(k, x) = 1$ and $J_{kk}(k, x) = 0$ at the boundary. The first of these conditions is commonly called smooth pasting, and the second is called super contact. However, the first condition is value matching and the second is smooth pasting in the option framework. To see that, we can use the formula (20.17) for J_k. Set $s = \eta x / k^{1-\eta}$. The condition $J_k(k, x) = 1$ is equivalent to

$$(\beta-1)^{\beta-1}\left(\frac{s}{\beta}\right)^{\beta} = s - 1. \qquad (20.21)$$

The left-hand side of (20.21) is the value (20.11) of the infinitesimal investment option. The right-hand side is the value of the underlying asset for the investment option minus its exercise price. Thus, the condition $J_k(k,x) = 1$ is equivalent to value matching: the value of the option equals its intrinsic value. Using the same formula for J_k, we can calculate J_{kk} by using the chain rule as

$$J_{kk}(k,x) = \frac{\partial}{\partial s}\left\{(\beta-1)^{\beta-1}\left(\frac{s}{\beta}\right)^{\beta}\right\} \cdot \frac{\partial s}{\partial k} - \frac{\partial s}{\partial k}.$$

Therefore, the condition $J_{kk}(k,x) = 0$ is equivalent to

$$\frac{\partial}{\partial s}\left\{(\beta-1)^{\beta-1}\left(\frac{s}{\beta}\right)^{\beta}\right\} = 1,$$

which is smooth pasting for the option value.

Solving the ODE with the Boundary Conditions

The ODE (20.20) is a nonhomogeneous version of the ODE (19.2). As discussed in Exercise 19.4, it can be solved by valuing the nonhomogeneous part as a cash flow to be received forever and adding that to the general solution (19.4) of the homogeneous part. Here, the nonhomogeneous part is $\pi(k,x) = \delta xk^\eta$. This is the cash flow from assets in place. Its value is the value xk^η of assets in place. Thus, the general solution of the ODE (20.20) is $xk^\eta + ax^{-\gamma} + bx^\beta$, where γ is the absolute value of the negative root of the quadratic equation (19.3) and β is the positive root. The coefficients a and b can depend on k. Because the value converges to zero as x converges to zero, the coefficient of $x^{-\gamma}$ must be zero. Hence, it must be that

$$J(k,x) = xk^\eta + b(k)x^\beta$$

for a function b. The term $b(k)x^\beta$ is the value of growth options. The function $b(k)$ and the investment boundary can be derived from the smooth pasting and super contact conditions $J_k = 1$ and $J_{kk} = 0$. We leave the details as an exercise. The resulting formula $b(k)x^\beta$ for the value of growth options is the same as the formula (20.12) obtained in Section 20.3.

20.5 IRREVERSIBLE INVESTMENT AND PERFECT COMPETITION

This section describes industry equilibrium assuming perfect competition with free entry and irreversible investment. The definition of industry equilibrium is that firms invest optimally, and growth options never get strictly in the money. When they reach the money—that is, when the intrinsic value reaches zero—entry of new firms or expansion of existing firms exhausts the options that are at the money. Therefore, growth options have no value, and the value of each firm is the value of its assets in place. As in the previous section, optimality implies that investment occurs when and only when marginal q is equal to 1. At the end of the section, we explain how fluctuations in q result in risky returns for the firm's shareholders.

Depreciation and Investment

For the remainder of the chapter, let I_t denote cumulative investment through date t (instead of the rate of investment at date t as in Section 20.2). We allow for depreciation. Assume $dK = -\rho K\, dt + dI$ for a constant ρ. Require I to be a nondecreasing process (investment is irreversible). If there is a discrete investment at date 0, then I_0 denotes the amount of the investment and $K_0 = k_0 + I_0$; otherwise, $I_0 = 0$.

Output and Operating Cash Flow

Assume Cobb-Douglas production with constant returns to scale[6] as discussed in Section 20.2; that is, $y = A k^\alpha \ell^{1-\alpha}$. Take A to be constant and the wage rate w to be constant. From each firm's point of view, the output price P is exogenous. We leave it as an exercise to show that, at the optimal labor input, output is

$$c_q P_t^{(1-\alpha)/\alpha} K_t \qquad (20.22)$$

and operating cash flow is $c_\pi P_t^{1/\alpha} K_t$ for constants c_q and c_π.

---

6. If there were decreasing returns to scale, then free entry would imply that there should be infinitely many firms of infinitely small scale, which is certainly unrealistic. It may be more realistic to assume increasing returns to scale up to some point and then decreasing returns thereafter, yielding a U-shaped average cost curve, but constant returns to scale is more tractable.

Industry Output Price

The output price P depends on supply and demand. Supply is determined by the industry capital stock. Denote the industry capital stock by K^{ind}. By summing (20.22) over firms, we see that industry output is

$$Q_t = c_q P_t^{(1-\alpha)/\alpha} K_t^{\text{ind}}. \tag{20.23}$$

Assume the industry demand curve is $P_t = Z_t Q_t^{-1/\varepsilon}$, where Z is a geometric Brownian motion under the risk-neutral probability. Demand is higher when Z is higher. The parameter ε is the elasticity of demand.

Let P_t denote the price at which supply equals demand, and define $Y_t = c_\pi P_t^{1/\alpha}$. Then, operating cash flow is

$$c_\pi P_t^{1/\alpha} K_t = Y_t K_t. \tag{20.24}$$

We leave it as an exercise to show that

$$\frac{\mathrm{d}Y}{Y} = (r - \delta)\,\mathrm{d}t + \sigma\,\mathrm{d}B^* - \kappa \left(\frac{\mathrm{d}I^{\text{ind}}}{K^{\text{ind}}} \right) \tag{20.25}$$

for constants δ, σ, and κ, where B^* is a Brownian motion under the risk-neutral probability and where I^{ind} denotes cumulative industry investment.

Reflected Geometric Brownian Motion

We solve for equilibrium in two steps. First, we define I^{ind} in such a way that Y defined by (20.25) is a geometric Brownian motion reflected from above at some constant y^*. This construction has the property that firms in the industry invest (I^{ind} increases) if and only if Y is at y^*. Then, we calculate y^* by requiring marginal q to equal 1 if and only if Y is at y^*. This produces the equilibrium price process $P = (Y/c_\pi)^\alpha$ and the equilibrium industry capital stock process.

Before beginning the construction of I^{ind}, it is useful to explain reflected Brownian motions and reflected geometric Brownian motions. Figure 20.3 illustrates how the subtraction of a singular process can cause a Brownian motion to be reflected. Figure 20.3 shows a path of a Brownian motion that is reflected from above at 0. The reflected Brownian motion is the process N defined by $N_t = B_t - \max\{B_s \mid s \leq t\}$. The process $\max\{B_s \mid s \leq t\}$ is called the running maximum of the Brownian motion B. Denote it by U, so we have $N = B - U$. Like the optimal capital stock process in Sections 20.3 and 20.4, the process U is a singular process—it has continuous and nondecreasing paths, but the derivative of each path with respect to time is zero at almost all times (with probability 1). The reflected Brownian motion evolves as $\mathrm{d}N = \mathrm{d}B - \mathrm{d}U$. At almost all times, $\mathrm{d}U = 0$,

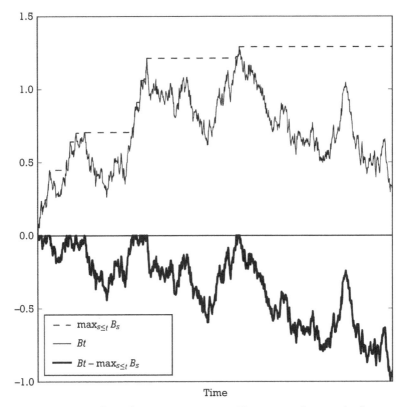

Figure 20.3 A reflected Brownian motion. The upper path is a path of a Brownian motion B. The dashed line is the running maximum of B. Subtracting the running maximum from the Brownian motion produces a Brownian motion reflected from above at 0.

and the dynamics of N are the same as the dynamics of B. However, subtraction of the singular process U from the Brownian motion B leads to reflection at 0.

The same construction allows us to reflect a (μ, σ)–Brownian motion from above at 0. If Z is such a Brownian motion (so $dZ = \mu\, dt + \sigma\, dB$ for a standard Brownian motion B), then N defined as $N_t = Z_t - \max\{Z_s \mid s \leq t\}$ is a (μ, σ)–Brownian motion reflected from above at 0. Furthermore, we can easily modify the construction to reflect Z from above at any number a. We simply take $N_t = Z_t - \max\{(Z_s - a)^+ \mid s \leq t\}$. This ensures that

$$N_t \leq Z_t - (Z_t - a)^+ = a + (Z_t - a) - (Z_t - a)^+ \leq a.$$

Finally, we can construct a reflected geometric Brownian motion as e^N.

Industry Investment and Price Processes

Now we define an industry investment process I^{ind} that causes Y to be a geometric Brownian motion reflected from above at y^*. Here, we consider any $y^* \geq Y_0$. Later, we compute the equilibrium y^*. Define

$$Z_t = \log Y_0 + \left(r - \delta - \frac{1}{2}\sigma^2\right)t + \sigma B_t^* \qquad (20.26\text{a})$$

and

$$U_t = \max_{s \leq t}(Z_s - \log y^*)^+, \qquad (20.26\text{b})$$

and set $N = Z - U$. Then, N is an $(r - \delta - \sigma^2/2, \sigma)$–Brownian motion under the risk-neutral probability reflected from above at $\log y^*$. Define I^{ind} by

$$dI^{\text{ind}} = \frac{K^{\text{ind}}}{K} dU. \qquad (20.26\text{c})$$

Then, Y defined as

$$Y_t = e^{N_t} \qquad (20.26\text{d})$$

satisfies (20.25) and is a geometric Brownian motion under the risk-neutral probability reflected from above at y^*. Furthermore, U and I^{ind} increase only when Y is at y^*.

A reflected geometric Brownian motion is a Markov process. We are taking the risk-free rate to be constant. Hence, the value at date t of a claim to a reflected geometric Brownian motion cash flow process Y_u for $u \geq t$ depends on Y_t. We use X_t to denote this value below. The value X_t is a function of the contemporaneous cash flow Y_t.

The industry output price is $P = (Y/c_\pi)^\alpha$. Consequently, it is also a geometric Brownian motion reflected from above at p^* defined as $(y^*/c_\pi)^\alpha$. The industry investment process (20.26c) is the smallest nondecreasing process that maintains the industry output price P below p^*. We will calculate the equilibrium reflection point y^* and hence the equilibrium p^*. In equilibrium, firms invest only when the industry output price reaches p^*, and they invest just enough to keep the price from ever rising above p^*.

Marginal and Average q

The value of a firm is the value of assets in place plus the value of growth options. Part of the definition of industry equilibrium is that growth options have no value, because investment in the industry occurs whenever growth options reach the money, due to free entry. Thus, the value of each firm is the value of its assets in place. The operating cash flows are $Y_t K_t$, so the value of a firm's assets in place at date t is $X_t K_t$, where we define

$$X_t = \mathsf{E}_t^* \int_t^\infty e^{-r(u-t)} Y_u \, du. \tag{20.27}$$

Because the value of a firm is XK, both marginal and average q are equal to X. We will calculate X assuming Y is a geometric Brownian motion reflected at $y^* \geq Y_0$. Then, we will determine the equilibrium y^* from the requirement that $X_t = 1$ when $Y_t = y^*$.

As remarked before, X_t depends on Y_t. For $y \leq y^*$, set

$$f(y) = \mathsf{E}^* \left[\int_t^\infty e^{-r(u-t)} Y_u \, du \,\bigg|\, Y_t = y \right],$$

assuming Y is a geometric Brownian motion reflected from above at y^*. Then, $f(Y_t)$ is the value at date t of an asset that pays Y as a dividend. Calculating f is similar to but not quite the same as calculating the value of an asset that pays a cash flow until a hitting time as in Section 19.1. If cash flows end at the hitting time, then $f(y^*) = 0$. However, in the problem here, the cash flows continue, but risk vanishes at the hitting time, because of the reflection. Thus, $f'(y^*) = 0$. We derive this fact in the following paragraph.

Assume $\delta > 0$. Equate the expected rate of return under the risk-neutral probability of the asset that pays Y to the risk-free rate. This yields

$$Y \, dt + \mathsf{E}^* \left[f'(Y) \, dY + \frac{1}{2} f''(Y) \, (dY)^2 \right] = rf(Y) \, dt. \tag{20.28}$$

There are dt terms and dI^{ind} terms in dY, and they must separately match on the two sides of (20.28). Equating the dt terms, using the dynamics (20.25) of Y, gives

$$y + (r - \delta) y f'(y) + \frac{1}{2} \sigma^2 y^2 f''(y) = rf(y). \tag{20.29}$$

This is a nonhomogeneous version of the fundamental ODE (19.2). The only remaining term on either side of (20.28) is the singular part of dY in $f'(Y) \, dY$. It is equal to zero if and only if $f'(y^*) = 0$.

Previously, we solved nonhomogeneous versions of the fundamental ODE by valuing a claim to the nonhomogeneous part as a cash flow to be received forever. We can do the same here, ignoring reflection in the valuation. The value of receiving Y forever if it is a geometric Brownian motion under the risk-neutral probability with drift $r - \delta$ is Y/δ. Clearly, $f(y) = y/\delta$ satisfies the ODE (20.29). The general solution of the ODE is $y/\delta + ay^{-\gamma} + by^\beta$, where γ is the absolute value of the negative root and β is the positive root of the quadratic equation (19.3).

The boundary condition $\lim_{y\to 0} f(y) = 0$ implies $a = 0$, so $f(y) = y/\delta + by^\beta$. The boundary condition $f'(y^*) = 0$ implies

$$b = -\frac{1}{\delta\beta(y^*)^{\beta-1}}.$$

Therefore, when there is reflection at y^*, marginal and average q are

$$X_t = \frac{Y_t}{\delta} + bY_t^\beta = \frac{Y_t}{\delta}\left[1 - \frac{1}{\beta}\left(\frac{Y_t}{y^*}\right)^{\beta-1}\right]. \qquad (20.30)$$

When Y_t reaches y^*, marginal and average q (20.30) are equal to

$$\frac{(\beta-1)y^*}{\delta\beta}. \qquad (20.31)$$

Equilibrium Reflection Point

In the construction (20.26), there is investment in the industry when and only when $Y_t = y^*$. We know from general theory that investment is optimal when and only when marginal q equals 1. Therefore, the construction (20.26) is consistent with industry equilibrium if and only if marginal q equals 1 when $Y_t = y^*$. Equating (20.31) to 1, we see that the equilibrium reflection point is

$$y^* = \frac{\delta\beta}{\beta-1}. \qquad (20.32)$$

The equilibrium output price is $P = (Y/c_\pi)^\alpha$, so P is a geometric Brownian motion reflected at

$$p^* = \left(\frac{\delta\beta}{(\beta-1)c_\pi}\right)^\alpha. \qquad (20.33)$$

Irreversible Investment, q, and Risk

The value of a firm is $J(K_t, X_t) = X_t K_t$. From this fact and the fact that the capital process K has zero quadratic variation, we obtain

$$\frac{dJ(K_t, X_t)}{J(K_t, X_t)} = \frac{dX_t}{X_t} + \frac{dK_t}{K_t} = \frac{dX_t}{X_t} - \rho\, dt + \frac{dI_t}{K_t}.$$

The rate of return earned by the firm's shareholders is

$$\frac{Y_t K_t\, dt - dI_t + dJ(K_t, X_t)}{J(K_t, X_t)} = \frac{Y_t K_t\, dt}{J(K_t, X_t)} - \frac{dI_t}{J(K_t, X_t)} + \frac{dX_t}{X_t} - \rho\, dt + \frac{dI_t}{K_t}.$$

Investment occurs only when $X_t = 1$, and this implies $J(K_t, X_t) = K_t$. Therefore, the dI_t terms in the shareholders' return cancel, and the return is

$$\frac{Y_t K_t \, dt}{J(K_t, X_t)} + \frac{dX_t}{X_t} - \rho \, dt. \tag{20.34}$$

If investment were perfectly reversible ($\theta(k, i) = i$), then we would always have marginal q equal to 1, and the return would be the risk-free return in equilibrium. However, with irreversible investment, fluctuations in marginal q add the risk dX/X for the firm's shareholders.

From the formula (20.30) for X and the dynamics (20.25) of Y, we can calculate that the volatility of X (and hence the risk of the stock return) is

$$\frac{Y_t}{\delta}\left[1 - \left(\frac{Y_t}{y^*}\right)^{\beta-1}\right]\sigma. \tag{20.35}$$

See Exercise 20.3. The volatility is decreasing in Y_t, decreasing to zero at $Y_t = y^*$. Equivalently, the volatility is decreasing in the industry output price P_t, decreasing to zero at $P_t = p^*$. Equivalently, the volatility is decreasing in X_t, decreasing to zero at $X_t = 1$. The book-to-market ratio of a firm in this model is $K_t/J(K_t, X_t) = 1/X_t$. All firms in the industry have the same book-to-market ratio. If we consider different industries that are ex ante identical, then firms in industries with higher book-to-market ratios (lower values of X) will have higher risks and higher expected returns. The reason for the correlation between output price (or book-to-market ratio) and risk is that increases in demand are absorbed by increases in supply when the price is at the reflection point. However, below the reflection point, demand shocks produce price shocks, which produce shocks to stock prices.

20.6 BERK-GREEN-NAIK MODEL

This section presents a model due to Berk, Green, and Naik (1999) of a firm that confronts a sequence of indivisible investment options. The options are for projects with varying risk; consequently, the risk of the firm changes over time. The model is a tractable model in which we can derive relationships between firm risks, expected returns, book-to-market ratios, and size (market capitalization). It is an example of an equilibrium model with rational agents in which book-to-market and size predict returns in the cross section of stocks: As in the data, high book-to-market (value) stocks and small stocks have higher average returns than do low book-to-market (growth) stocks and large stocks.

Projects and Capital

Unlike the other models discussed in this chapter, there are no timing options in this model. A project is available only at the date it appears, so the NPV rule determines optimal investment. A single new project arrives at each date, investment in a project can only be made at the date it arrives, and investment in each project is irreversible. The first project arrives at date 0. If the firm invests in the project, its operating cash flows begin at date 1. Capital in all projects depreciates at a common rate δ. There is a maximum feasible investment I in any project. There are constant returns to scale in operation in each project up to the feasible scale. The capital at date u in a project that arrived at date $t < u$ is $\delta^{u-t-1} I_t$, where I_t is the investment made at date t (which will be either 0 or I).

The Net Present Value Rule

The operating cash flow generated at date u by a project that arrived at $t < u$, per unit investment at t, is $\delta^{u-t-1} C_{tu}$, where C_{tu} is an exogenous random variable observable at date u, and where δ is the depreciation rate mentioned in the previous paragraph. The NPV of a unit investment in the project that arrives at t is

$$\mathrm{E}_t \left[\sum_{u=t+1}^{\infty} \frac{M_u}{M_t} \delta^{u-t-1} C_{tu} \right] - 1, \tag{20.36}$$

where M is the SDF process. Define $\chi_t = 1$ if the NPV is nonnegative and $\chi_t = 0$ if the NPV is negative. An optimal investment process is $I_t = \chi_t I$. Assuming optimal investment, the capital stock of the firm at date t is

$$K_t = I \sum_{s=0}^{t-1} \delta^{t-s-1} \chi_s.$$

SDF Process

Assume

$$\log M_{t+1} = \log M_t - r_t - \frac{1}{2}\lambda^2 - \lambda \varepsilon_{t+1}, \tag{20.37a}$$

for a constant λ (the market price of risk) and a stochastic process r, where ε is a sequence of independent standard normal random variables. This implies

$$\mathrm{E}_t \left[\frac{M_{t+1}}{M_t} \right] = e^{-r_t},$$

Real Options and q Theory

so r_t is the continuously compounded risk-free rate from date t to date $t+1$. To obtain precise formulas, some assumption about the distribution of the risk-free rate process must be made. However, for our purposes, the risk-free rate process can be quite general.

Project Cash Flows

Assume for each s that $(C_{s,s+1}, C_{s,s+2}, \ldots)$ is an IID lognormal sequence. Specifically, assume

$$\log C_{st} = \log \hat{C} - \frac{1}{2}\phi_s^2 + \phi_s \xi_{st}, \qquad (20.37b)$$

for $t > s$, where $\hat{C}$ is a constant, ϕ_s is observable at time s, and $(\xi_{s,s+1}, \xi_{s,s+2}, \ldots)$ is a sequence of independent standard normal random variables. From the usual rule for means of exponentials of normals, we have, for $t \geq s$,

$$E_t \left[\frac{M_{t+1}}{M_t} C_{s,t+1} \right] = e^{-r_t - \beta_s} \hat{C}, \qquad (20.38)$$

where we set $\beta_s = \lambda \phi_s \, \mathrm{corr}_t(\varepsilon_{t+1}, \xi_{s,t+1})$ (assume the correlation depends only on information available at s). Thus, the value at t of the cash flow $C_{s,t+1}$ is its expected value discounted continuously at the risk-adjusted rate $r_t + \beta_s$. Assume $(\beta_1, \beta_2, \ldots)$ is an IID sequence that is independent of the ε's and ξ's. This completes the assumptions of the model.

Value of Assets in Place

By iterated expectations, (20.38) generalizes as follows:[7] for $s \leq t < u$,

$$E_t \left[\frac{M_u}{M_t} C_{su} \right] = e^{-\beta_s} \hat{C} P_t(u), \qquad (20.39)$$

---

7. By (20.38) and iterated expectations,

$$E_t \left[\frac{M_u}{M_t} C_{su} \right] = E_t \left[\frac{M_{u-1}}{M_t} E_{u-1} \left[\frac{M_u}{M_{u-1}} C_{su} \right] \right] = e^{-\beta_s} \hat{C} E_t \left[\frac{M_{u-1}}{M_t} e^{-r_{u-1}} \right]$$

$$= e^{-\beta_s} \hat{C} E_t \left[\frac{M_{u-1}}{M_t} E_{u-1} \left[\frac{M_u}{M_{u-1}} \right] \right]$$

$$= e^{-\beta_s} \hat{C} E_t \left[\frac{M_u}{M_t} \right].$$

where $P_t(u)$ denotes the price at t of a discount bond maturing at u. Define

$$D_t = \sum_{u=t+1}^{\infty} \delta^{u-t-1} P_t(u).$$

Note that D_t is the value of a perpetual bond with coupons declining at rate $1 - \delta$. It follows from (20.39) that the value at t of the cash flows produced at $t + 1$, $t + 2$, ..., by a project that arrived at $s \leq t$ is

$$E_t \left[\sum_{u=t+1}^{\infty} \frac{M_u}{M_t} \delta^{u-s-1} \chi_s IC_{su} \right] = \delta^{t-s} \chi_s I e^{-\beta_s} \hat{C} D_t. \tag{20.40}$$

Set

$$\gamma_t = -\log \left(\frac{\sum_{s=0}^{t} \delta^{t-s} \chi_s I e^{-\beta_s}}{K_{t+1}} \right). \tag{20.41}$$

This means that $e^{-\gamma_t}$ is a weighted average of the risks of projects in which the firm has invested by date t, with the weight being the fraction $\delta^{t-s} \chi_s I / K_{t+1}$ of capital invested in the project when operations begin at date $t + 1$ (recall that K_{t+1} is known at date t, because it depends on investment decisions made at date t and before).

Let A_t denote the value at t of the cash flows produced at $t + 1, t + 2, ...,$ by all projects that arrived at t or before. This is the value of assets in place at date t. From (20.40) and (20.41), we have

$$A_t = \hat{C} D_t \sum_{s=0}^{t} \delta^{t-s} \chi_s I e^{-\beta_s} = e^{-\gamma_t} K_{t+1} \hat{C} D_t. \tag{20.42}$$

Note that $K_{t+1} \hat{C}$ is the expected cash flow at date $t + 1$, conditional on date–t information. Thus, the value of assets in place at date t is the expected cash flow at $t + 1$ multiplied by a risk-adjusted value of the perpetual bond.

Value of Growth Options

Setting $s = t$ in (20.40) gives the value of (20.36). Thus, the firm invests at t ($\chi_t = 1$) if and only if

$$e^{-\beta_t} \hat{C} D_t > 1.$$

This is true when interest rates are low, implying that the value D_t of the perpetual bond is high, or when project risk β_t is low. The value at t of the option to invest at t is

$$I \left(e^{-\beta_t} \hat{C} D_t - 1 \right)^+,$$

and the value at t of the option to invest at $u > t$ is

$$\mathrm{I}\mathrm{E}_t\left[\frac{M_u}{M_t}\left(e^{-\beta_u}\hat{C}D_u - 1\right)^+\right].$$

The total value of growth options at t is G_t defined by

$$G_t = I \sum_{u=t+1}^{\infty} \mathrm{E}_t\left[\frac{M_u}{M_t}\left(e^{-\beta_u}\hat{C}D_u - 1\right)^+\right]. \tag{20.43}$$

Expected Return

The total value of the firm after the distribution of cash flows at date t is $A_t + G_t$. Denote this by S_t. So, we have

$$S_t = e^{-\gamma t} K_{t+1} \hat{C} D_t + G_t. \tag{20.44}$$

The main purpose of this exercise is to compute the expected return and determine how it relates to characteristics of the firm. The conditional expectation at date t of the operating cash flow at date $t+1$ is $K_{t+1}\hat{C} = e^{\gamma t} A_t/D_t$, and the conditional expectation of investment cash flow is $-\mathrm{IE}_t[\chi_{t+1}]$. The expected return from t to $t+1$ for an owner of the firm is therefore

$$\frac{e^{\gamma t}}{D_t} \cdot \frac{A_t}{S_t} + \frac{\mathrm{E}_t[A_{t+1}] - \mathrm{IE}_t[\chi_{t+1}] + \mathrm{E}_t[G_{t+1}]}{S_t}. \tag{20.45}$$

The value at $t+1$ of assets in place equals the value at $t+1$ of the assets in place at t plus the value at $t+1$ of assets added at $t+1$. This decomposition is

$$A_{t+1} = \delta A_t \frac{D_{t+1}}{D_t} + \chi_{t+1} I e^{-\beta_{t+1}} \hat{C} D_{t+1}.$$

It follows that

$$\mathrm{E}_t[A_{t+1}] - \mathrm{IE}_t[\chi_{t+1}] = \delta A_t \mathrm{E}_t\left[\frac{D_{t+1}}{D_t}\right] + \mathrm{E}_t\left[\left(Ie^{-\beta_{t+1}}\hat{C}D_{t+1} - I\right)\chi_{t+1}\right]$$

$$= \delta A_t \mathrm{E}_t\left[\frac{D_{t+1}}{D_t}\right] + \mathrm{E}_t\left[I\left(e^{-\beta_{t+1}}\hat{C}D_{t+1} - 1\right)^+\right].$$

Substituting this into (20.45) shows that the expected return equals

$$\frac{e^{\gamma t} + \delta \mathrm{E}_t[D_{t+1}]}{D_t} \cdot \frac{A_t}{S_t} + \frac{\mathrm{E}_t\left[G_{t+1} + I\left(e^{-\beta_{t+1}}\hat{C}D_{t+1} - 1\right)^+\right]}{G_t} \cdot \frac{G_t}{S_t}. \tag{20.46}$$

The factor multiplying A_t/S_t is the expected return on assets in place. Note that $(1 + \delta \mathrm{E}_t[D_{t+1}])/D_t$ is the expected return on the perpetual bond. The factor multiplying G_t/S_t is the expected return on growth options, including the growth option that matures at $t+1$. The sequence of project risk realizations $\beta_0,\ldots,\beta_t$ and the interest rates faced by the firm through date t determine the firm risk γ_t and the relative importance of assets in place versus growth options. In conjunction with the interest rate environment at date t, these firm characteristics determine the expected return.

Expected Returns, Book-to-Market, and Size

Given a sample of ex-ante identical firms in this model, the firms will be distinguished at date t by their risks γ_t and capital stocks K_{t+1} (because these depend on the project risk realizations $\beta_0,\ldots,\beta_t$). The firm risk γ_t is not directly observable. We can rewrite the expected return by substituting $K_{t+1}\hat{C}$ in place of $e^{\gamma_t}A_t/D_t$ as the expected operating cash flow and substituting $A_t = S_t - G_t$. Make these substitutions in (20.46) to obtain

$$\frac{K_{t+1}\hat{C}}{S_t}+\delta\mathsf{E}_t\left[\frac{D_{t+1}}{D_t}\right]+\frac{\mathsf{E}_t\left[G_{t+1}+I\left(e^{-\beta_{t+1}}\hat{C}D_{t+1}-1\right)^+\right]-\delta G_t\mathsf{E}_t[D_{t+1}/D_t]}{S_t}.$$

(20.47)

The interesting feature of this formula is that the first term is proportional to the book-to-market ratio K_{t+1}/S_t, the second term depends only on the distribution of interest rates, and the numerator of the third term depends only on the distribution of interest rates, so the third term is inversely proportional to market value S_t. Thus, given a sample of ex-ante identical firms, expected returns will vary across the sample depending on book-to-market and size.

20.7 NOTES AND REFERENCES

Tobin (1969) uses the symbol q to denote the market value of a firm divided by the replacement cost of its capital. This is now called average q. The q theory in Section 20.2 is due to Hayashi (1982) and Abel (1985). Formula (20.8) for the investment return in discrete time is due to Cochrane (1991).

Important early work on real options includes Brennan and Schwartz (1985), McDonald and Siegel (1986), and Dixit and Pindyck (1994). Many of the issues discussed in this chapter are covered by Dixit and Pindyck. The indivisible project model of Section 20.1 has been widely applied and is covered in textbooks such as Trigeorgis (1996).

The effect of asymmetric information on the optimal exercise time in the indivisible project model of Section 20.1 is studied by Morellec and Schürhoff

(2011), Grenadier and Malenko (2011), and Bustamante (2012). Morellec and Schürhoff and Bustamante assume the project is financed by raising equity, and firms wish to be viewed as having high project quality because that increases the price at which shares can be issued, minimizing the dilution of existing shareholders. They show that firms with good projects will generally exercise investment options earlier in order to signal project quality, though Bustamante shows that all managers will exercise early (there is a pooling equilibrium) in "hot markets." Grenadier and Malenko (2011) assume that managers wish to maximize their own utility rather than shareholder value and with a specific type of utility function show that managers with good projects will sometimes delay investment.

The monopoly model presented in Sections 20.3 and 20.4 is a special case of the model studied by Abel and Eberly (1996). Cooper (2006) solves a version of that model in which there is also a fixed cost to invest and relates the risks and hence expected returns of firms to their book-to-market ratios. The model of perfect competition with irreversible investment studied in Section 20.5 is due to Leahy (1993). The discussion of irreversibility, marginal q, and risk at the end of Section 20.5 is based on Kogan (2004). Kogan analyzes a model in which investment is a singular process, as in Section 20.5, and he also analyzes a model in which there is a maximum feasible investment rate. In that model, marginal q can exceed 1—when it is above 1, firms invest at the maximum rate. In that model, risk increases as q decreases below 1, as in Section 20.5, and it also increases as q rises above 1.

For oligopoly versions of the irreversible investment model, see Baldursson (1998), Grenadier (2002), Aguerrevere (2009), Back and Paulsen (2009), and Steg (2012). In the oligopoly model, firms play a continuous-time stochastic game with singular strategies. The theory of such games has so far been developed only for open-loop strategies, meaning that the investment of each firm depends on the anticipated investment of other firms (via the Nash equilibrium condition) but does not depend on the realized investment of other firms. Open-loop equilibria are not subgame perfect. The difference between open-loop and closed-loop equilibria in deterministic investment games is explained in the textbook by Fudenberg and Tirole (1992).

The singular control problem of monopoly investment studied in Sections 20.3 and 20.4—in which there is no depreciation—is an example of what is called a monotone follower problem in the optimal control literature. The relation between the dynamic programming method and optimal stopping (which is optimal exercise in the option context) for monotone follower problems is developed by Karatzas and Shreve (1984). A more practitioner-oriented exposition of the equivalence between option pricing and dynamic programming (or decision tree analysis) is given by Smith and Nau (1995).

To apply the condition "invest when marginal q equals 1" in the irreversible investment model, we first have to calculate marginal q, which is the partial derivative J_k of the value function J. An equivalent condition that does not require prior calculation of the value function is given by Bank (2005) for the case in which there is no depreciation. Bank shows that, given a capital stock process K, we can define a stochastic process D with the property that

$$D_\tau = \mathsf{E}^*_\tau \int_\tau^\infty e^{-rt} \pi_k(K_t, X_t)\, dt - e^{-r\tau}$$

for all stopping times τ. Given some technical conditions, necessary and sufficient conditions for K to be optimal are that (i) $D \leq 0$ and (ii) $\int_0^\infty D_t\, dK_t = 0$. These are versions of the statements that (i) marginal q never exceeds 1, and (ii) investment occurs only when marginal q equals 1. Steg (2012) uses Bank's result to analyze the oligopoly model with irreversible investment.

The theory of singular control is relevant for issues in finance other than corporate investment. For example, it applies to portfolio choice with proportional transactions costs (Davis and Norman, 1990) and has been applied to international finance (Dumas, 1992). Textbook treatments of singular control include Harrison (1985, 2013), Øksendal and Sulem (2007), and Stokey (2009).

Gomes, Kogan, and Zhang (2003) solve a model similar to that of Berk, Green, and Naik (1999) while endogenizing the SDF process by assuming a representative investor with CRRA preferences. In their model, the conditional CAPM holds, but empirical tests of the CAPM on simulated samples show that size and book-to-market have additional explanatory power for average returns, due to misestimation of time-varying betas. In the Gomes-Kogan-Zhang model, projects differ by productivity (rather than by covariance with an exogenously specified SDF process as in the Berk-Green-Naik model). However, the Gomes-Kogan-Zhang model shares the feature of the Berk-Green-Naik model that all firms have equal growth options. Thus, in the Gomes-Kogan-Zhang model, a firm with more productive projects is what is generally defined to be a growth firm because it has a lower book-to-market ratio, yet growth options constitute a lower fraction of its value. The model generates a value premium, because growth options are a greater fraction of the value of a high book-to-market firm and growth options are riskier than assets in place. However, this implies that value firms have higher cash flow durations than growth firms, which is inconsistent with the data (see Section C.4 of Zhang, 2005).

Zhang (2005) analyzes industry equilibrium in a perfectly competitive model with an exogenously specified SDF process, assuming asymmetric quadratic adjustment costs (with a higher cost for disinvesting than for investing) and fixed operating costs. The SDF process has a countercyclical market price of risk. The adjustment costs produce higher risk for firms with excess capital, which generally are firms with higher book-to-market ratios. In combination with the countercyclical market price of risk, this produces a value premium. Li, Livdan, and Zhang (2009) and Livdan, Sapriza, and Zhang (2009) extend the Zhang model by incorporating equity issues, dividend changes, and capital structure choice.

Carlson, Fisher, and Giammarino (2004) solve a monopoly model with irreversible investment in which there is only a discrete set $(K_0 < K_1 < K_2)$ of feasible capital stocks. They incorporate quasi-fixed operating costs (operating costs that depend on the capital stock). This "operating leverage" produces additional risk for assets in place beyond the additional risk already due to irreversibility and thereby generates a value premium. Carlson, Fisher, and Giammarino (2006) extend the model to analyze risk and expected returns preceding and following equity issuance. Kogan and Papanikolaou (2012) survey investment models and their implications for firm characteristics and expected returns.

EXERCISES

20.1. Consider the optimization problem

$$\max_{\ell} \ pAk^\alpha \ell^{1-\alpha} - w\ell$$

that arises in the perfect competition model of Section 20.5. Compute constants c_q and c_π such that, at the optimal ℓ,

$$Ak^\alpha \ell^{1-\alpha} = c_q p^{(1-\alpha)/\alpha} k,$$
$$pAk^\alpha \ell^{1-\alpha} - w\ell = c_\pi p^{1/\alpha} k.$$

20.2. As in Section 20.5, assume

$$\frac{dZ}{Z} = \mu_z \, dt + \sigma_z \, dB^*$$

for constants μ_z and σ_z and a Brownian motion B^* under the risk-neutral probability. Match industry supply (20.23) to industry demand $Q_t = (Z_t/P_t)^\varepsilon$ to compute the equilibrium output price P_t. Define $Y_t = c_\pi P_t^{1/\alpha}$. Compute constants δ, σ, and κ such that Y satisfies (20.25). Using the formula for the constant c_π derived in the previous exercise, specify a condition on the parameters A, α, w, μ_z, and σ_z that is equivalent to $\delta > 0$.

20.3. Use the dynamics (20.25) of Y and the definition (20.30) of X to verify the formula (20.35) for the risk of a stock return in the model of perfect competition in Section 20.5.

Beliefs, Information, and Preferences

21

Heterogeneous Beliefs

CONTENTS

21.1	State-Dependent Utility Formulation	554
21.2	Aggregation in Single-Period Markets	555
21.3	Aggregation in Dynamic Markets	558
21.4	Short Sales Constraints and Overpricing	562
21.5	Speculative Trade and Bubbles	564
21.6	Notes and References	565

This part of the book provides brief introductions to several important extensions of the theory. There is a large literature on each of the topics covered. Rather than trying to survey the literatures, we try to provide careful treatments of some foundational issues and models within each literature. The goal is to provide enough background to make the literatures accessible.

This chapter relaxes the assumption made in most previous chapters that investors have identical beliefs. We assume in this chapter that investors hold their beliefs dogmatically, in the sense that they are not persuaded to revise their beliefs when they realize that others hold different beliefs. It is very natural to assume different individuals have different beliefs and hold them dogmatically if we take the subjectivist view of probability (Ramsey, 1931; Savage, 1954). Chapter 22 examines differences in beliefs arising from differences in information. In that case, each investor learns something about other investors' information from the terms of trade they are willing to accept, and they regard this information as useful, revising their own beliefs in consequence. Differences in beliefs due to differences in information are differences in *posterior* beliefs—that is, *after* observing information. In this chapter, we study differences in *prior* beliefs.

The two main topics discussed in this chapter are aggregation (existence of a representative investor) and equilibria in markets with short sales constraints.

If all investors have log utility, then there is a representative investor in either a single-period model or a dynamic model. If all investors have CARA utility, then there is a representative investor in a single-period model.

If investors agree on which events have zero probability—in the sense that if an investor h assesses the probability of any event to be zero, then so do all other investors—then their beliefs are said to be mutually absolutely continuous. Mutual absolute continuity is necessary for the existence of a Pareto optimum, because if investor h assesses the probability of an event A to be zero and investor j does not, then adding 1_A to the consumption of investor j and subtracting 1_A from the consumption of investor h leads to a Pareto improvement. Furthermore, if investors' beliefs are not mutually absolutely continuous and there are no constraints on portfolios, then a competitive equilibrium typically does not exist. If some investor regards an event as having zero probability and another investor regards it as having positive probability and an Arrow security for the event exists, then the investor who views it as having zero probability will want to sell an infinite amount of the Arrow security to the other investor. Short sales constraints make equilibria possible in this circumstance.

21.1 STATE-DEPENDENT UTILITY FORMULATION

When beliefs are heterogeneous, it is frequently useful to transform the model to one with homogeneous beliefs and state-dependent utility. Let $\mathbb{P}_h$ denote the beliefs of investor h, for $h = 1, \ldots, H$, and let E_h denote the corresponding expectation operator. Define $\mathbb{P}$ to be the average beliefs; that is, for each event A, set

$$\mathbb{P}(A) = \frac{1}{H} \sum_{h=1}^{H} \mathbb{P}_h(A).$$

Let E denote expectation with respect to $\mathbb{P}$. For each h, $\mathbb{P}_h$ is absolutely continuous with respect to $\mathbb{P}$ (meaning that if $\mathbb{P}(A) = 0$, then $\mathbb{P}_h(A) = 0$, for each event A). Hence, there exists a nonnegative random variable $\tilde{z}_h$ (the Radon-Nikodym derivative of $\mathbb{P}_h$ with respect to $\mathbb{P}$—see Appendix A.10) such that, for each random variable $\tilde{x}$,

$$\mathsf{E}_h[\tilde{x}] = \mathsf{E}[\tilde{z}_h \tilde{x}].$$

In particular, for a random wealth $\tilde{w}$ and utility function u_h,

$$\mathsf{E}_h[u_h(\tilde{w})] = \mathsf{E}[\tilde{z}_h u_h(\tilde{w})].$$

Therefore, we can interpret all investors as having the same beliefs (the average beliefs) and $\tilde{z}_h u_h(w)$ as being the state-dependent utility of wealth of investor h.

If investors' beliefs are mutually absolutely continuous, then $\mathbb{P}$ is absolutely continuous with respect to each $\mathbb{P}_h$, and each $\tilde{z}_h$ is strictly positive. An example of $\tilde{z}_h$ in the case of normal distributions is given in Exercise 21.1.

If there are only finitely many states of the world, then the definitions of the previous paragraph are very simple: For each state ω,

$$\mathbb{P}(\{\omega\}) = \frac{1}{H} \sum_{h=1}^{H} \mathbb{P}_h(\{\omega\}),$$

and, for each h,

$$\tilde{z}_h(\omega) = \frac{\mathbb{P}_h(\{\omega\})}{\mathbb{P}(\{\omega\})}.$$

Note that the set of possible states of the world should be defined as the union of the sets regarded as possible by the various investors.

The first-order condition for an investor in a single-period model is, as always, that marginal utility evaluated at the optimal wealth is proportional to an SDF: $u_h'(\tilde{w}_h) = \gamma_h \tilde{m}_h$, where u_h is the utility function of investor h, $\tilde{w}_h$ is her optimal wealth, γ_h is a constant, and $\tilde{m}_h$ is an SDF. The meaning of an SDF is the same as before—equation (3.1)—but now specifying that the expectation operator is E_h; that is, for all assets i, $\mathsf{E}_h[\tilde{m}_h \tilde{x}_i] = p_i$. The expectation operator is E_h, because it is with respect to E_h that investor h maximizes her expected utility. Transforming to the average beliefs, we have

$$\frac{1}{\gamma_h} \mathsf{E}_h[u_h'(\tilde{w}_h)\tilde{x}_i] = p_i \quad \Rightarrow \quad \frac{1}{\gamma_h} \mathsf{E}[\tilde{z}_h u_h'(\tilde{w}_h)\tilde{x}_i] = p_i.$$

Thus, at the optimal wealth $\tilde{w}_h$ of investor h, $\tilde{z}_h u_h'(\tilde{w}_h)$ is proportional to an SDF relative to the average beliefs.

21.2 AGGREGATION IN SINGLE-PERIOD MARKETS

This section establishes that there is a representative investor in a complete single-period market if all investors have log utility or if all investors have CARA utility. Investors can differ in beliefs and in endowments. Assume the beliefs of investors are mutually absolutely continuous. Except for the heterogeneity of beliefs, we model the securities market as in Chapter 4. In particular, we assume short sales are unconstrained. We assume the market is at a competitive equilibrium, and we exploit the Pareto optimality of competitive equilibria in complete markets, that is, the First Welfare Theorem. The First Welfare Theorem is true under very general assumptions on preferences—it is not necessary that investors' preferences be represented by expected utilities nor, if they are, that the beliefs underlying the expectations be the same.

With heterogeneous beliefs in a single-period model, the social planner's objective function (4.2) is replaced by

$$\sum_{h=1}^{H} \lambda_h \mathsf{E}_h[u_h(\tilde{w}_h)] = \sum_{h=1}^{H} \lambda_h \mathsf{E}[\tilde{z}_h u_h(\tilde{w}_h)]. \tag{21.1}$$

We can interpret the social planner as a representative investor having the average beliefs and a state-dependent utility function, but the state dependence renders the concept much less useful, compared to the model with homogeneous beliefs. With either log utility or CARA utility, there is a representative investor with a state-*independent* utility function.

Log Utility

Suppose each investor h has logarithmic utility. A competitive equilibrium in a complete market maximizes the social planner's utility function for some positive weights λ_h. Take the weights $\lambda_1, \ldots, \lambda_H$ to sum to 1 (which we can always do by dividing each weight by the sum). The solution of the social planning problem

$$\max \sum_{h=1}^{H} \lambda_h \mathsf{E}[\tilde{z}_h \log \tilde{w}_h] \quad \text{subject to} \quad \sum_{h=1}^{H} \tilde{w}_h = \tilde{w}_m \tag{21.2}$$

is

$$\tilde{w}_h = \tilde{w}_m \left(\frac{\lambda_h \tilde{z}_h}{\sum_{h=1}^{H} \lambda_h \tilde{z}_h} \right). \tag{21.3}$$

Substitute this into the objective function in (21.2) to see that the social planner's utility is

$$\sum_{h=1}^{H} \lambda_h \mathsf{E}[\tilde{z}_h \log \tilde{w}_h] = \mathsf{E}\left[\left(\sum_{h=1}^{H} \lambda_h \tilde{z}_h \right) \log \tilde{w}_m \right]$$

$$+ \mathsf{E}\left[\sum_{h=1}^{H} \lambda_h \tilde{z}_h \log \left(\frac{\lambda_h \tilde{z}_h}{\sum_{h=1}^{H} \lambda_h \tilde{z}_h} \right) \right]. \tag{21.4}$$

Define the weighted-average beliefs

$$\mathbb{P}_m(A) = \sum_{h=1}^{H} \lambda_h \mathbb{P}_h(A).$$

The expectation operator E_m corresponding to $\mathbb{P}_m$ satisfies

$$\mathsf{E}_m[\tilde{x}] = \sum_{h=1}^{h} \lambda_h \mathsf{E}_h[\tilde{x}]$$

$$= \mathsf{E}\left[\left(\sum_{h=1}^{H} \lambda_h \tilde{z}_h\right)\tilde{x}\right].$$

Therefore, the social planner's utility (21.4) can be written as

$$\mathsf{E}_m[\log \tilde{w}_m] + \mathsf{E}\left[\sum_{h=1}^{H} \lambda_h \tilde{z}_h \log\left(\frac{\lambda_h \tilde{z}_h}{\sum_{h=1}^{H} \lambda_h \tilde{z}_h}\right)\right]. \tag{21.4'}$$

Because the second term in (21.4′) is a constant that does not depend on the $\tilde{w}_h$, (21.4′) implies that there is a representative investor with log utility and beliefs $\mathbb{P}_m$ (Section 7.1).

Even with log utility, there is an important distinction between homogeneous beliefs and heterogeneous beliefs. The sharing rules (21.3) typically cannot be implemented unless markets are complete, because of the state dependence introduced through the $\tilde{z}_h$. This is in contrast to log utility with homogeneous beliefs, in which case the sharing rules are affine, and, in the absence of labor income, an equilibrium allocation in incomplete markets is Pareto optimal.

Constant Absolute Risk Aversion

Assume now that each investor h has utility $u_h(w) = -e^{-\alpha_h w}$. Set $\tau_h = 1/\alpha_h$, $\tau = \sum_{h=1}^{H} \tau_h$, and $\alpha = 1/\tau$, so α is the aggregate absolute risk aversion. The solution of the social planning problem

$$\max \; -\sum_{h=1}^{H} \lambda_h \mathsf{E}[\tilde{z}_h e^{-\alpha_h \tilde{w}_h}] \quad \text{subject to} \quad \sum_{h=1}^{H} \tilde{w}_h = \tilde{w}_m \tag{21.5}$$

is

$$\tilde{w}_h = \frac{\tau_h}{\tau}\tilde{w}_m + \tau_h\left[\log(\lambda_h \alpha_h \tilde{z}_h) - \sum_{j=1}^{H} \frac{\tau_j}{\tau}\log(\lambda_j \alpha_j \tilde{z}_j)\right]. \tag{21.6}$$

Substitute this into the objective function in (21.5) to see that the social planner's utility is

$$-\tau\left(\prod_{j=1}^{H}(\lambda_j \alpha_j)^{\tau_j/\tau}\right)\mathsf{E}\left[e^{-\alpha \tilde{w}_m}\prod_{j=1}^{H}\tilde{z}_j^{\tau_j/\tau}\right]. \tag{21.7}$$

Define

$$\tilde{z} = \frac{\prod_{j=1}^{H} \tilde{z}_j^{\tau_j/\tau}}{\mathsf{E}\left[\prod_{j=1}^{H} \tilde{z}_j^{\tau_j/\tau}\right]}.$$

Note that the numerator in the definition of $\tilde{z}$ is a weighted geometric average of the $\tilde{z}_j$. Each $\tilde{z}_j$ has mean equal to 1 under the average beliefs, but the geometric average does not. This motivates the division by the mean in the definition of $\tilde{z}$. The random variable $\tilde{z}$ is nonnegative and has an expected value equal to 1, so it can serve as a Radon-Nikodym derivative. For each event A, define

$$\mathbb{P}_m(A) = \mathsf{E}[\tilde{z} 1_A],$$

where 1_A is the random variable equal to 1 when the state of the world is in A and equal to 0 otherwise. Let E_m denote expectation with respect to $\mathbb{P}_m$. The social planner's utility (21.7) is proportional to

$$-\mathsf{E}\left[-e^{-\alpha \tilde{w}_m} \tilde{z}\right] = \mathsf{E}_m\left[-e^{-\alpha \tilde{w}_m}\right].$$

Therefore, there is a representative investor with constant absolute risk aversion α and beliefs $\mathbb{P}_m$.

As in the case of log utility, markets typically must be complete in order for the sharing rules (21.6) to be implementable in the securities market and for the existence of a representative investor. However, in one special case, the sharing rule (21.6) is affine and hence can be implemented (if there is a risk-free asset and no labor income). This special case is the case of investors who agree that aggregate wealth $\tilde{w}_m$ is normally distributed and agree on its variance. If investors disagree about the variance, then it is sufficient to have an asset the payoff of which is a quadratic function of $\tilde{w}_m$ (Exercise 21.1).

21.3 AGGREGATION IN DYNAMIC MARKETS

There is a representative investor with log utility in a dynamic complete market if all investors have log utility and discount utility at the same rate. Investors can differ in beliefs and endowments. However, aggregation in dynamic markets—with state-independent utility—is not possible for CARA utility or for other LRT utility functions.

Consider a discrete-time model with horizon T. Suppose the market is complete, all investors have the same discount factor δ, and investors' beliefs are mutually absolutely continuous. Let C denote aggregate consumption. Scale the weights λ_h in the social planner's objective function so that they sum to 1.

Log Utility

The social planning problem is

$$\max \sum_{h=1}^{H}\sum_{t=0}^{T} \lambda_h \delta^t \mathsf{E}[\tilde{z}_h \log C_{ht}] \quad \text{subject to} \quad (\forall t) \sum_{h=1}^{H} C_{ht} = C_t.$$

We cannot solve this pointwise (in each date and state) because doing so would produce C_{ht} that depend on $\tilde{z}_h$ and hence are not measurable with respect to information at date t. However, it is easy to modify the problem so it can be solved pointwise. For any C_{ht} that is measurable with respect to date–t information, iterated expectations implies $\mathsf{E}[\tilde{z}_h \log C_{ht}] = \mathsf{E}[Z_{ht} \log C_{ht}]$, where $Z_{ht} = \mathsf{E}_t[\tilde{z}_h]$. The social planning problem can be restated as

$$\max \sum_{h=1}^{H}\sum_{t=0}^{T} \lambda_h \delta^t \mathsf{E}[Z_{ht} \log C_{ht}] \quad \text{subject to} \quad (\forall t) \sum_{h=1}^{H} C_{ht} = C_t. \quad (21.8)$$

The solution is

$$C_{ht} = \left(\frac{\lambda_h Z_{ht}}{Z_t}\right) C_t, \quad (21.9)$$

where

$$Z_t = \sum_{h=1}^{H} \lambda_h Z_{ht}. \quad (21.10)$$

Substitute this into the objective function in (21.8) to see that the social planner's utility is

$$\sum_{t=0}^{T} \delta^t \sum_{h=1}^{H} \lambda_h \mathsf{E}[Z_{ht} \log C_{ht}] = \sum_{t=0}^{T} \delta^t \mathsf{E}[Z_t \log C_t]$$

$$+ \sum_{t=0}^{T} \delta^t \mathsf{E}\left[\sum_{h=1}^{H} \lambda_h Z_{ht} \log\left(\frac{\lambda_h Z_{ht}}{Z_t}\right)\right]. \quad (21.11)$$

As in Section 21.2, define the weighted-average beliefs

$$\mathbb{P}_m(A) = \sum_{h=1}^{H} \lambda_h \mathbb{P}_h(A),$$

and let E_m denote the expectation operator. Use iterated expectations and the fact that the Radon-Nikodym derivative of $\mathbb{P}_m$ with respect to $\mathbb{P}$ is $\sum_{h=1}^{H} \lambda_h \tilde{z}_h$ to obtain

$$E[Z_t \log C_t] = E\left[\left(\sum_{h=1}^{H} \lambda_h \tilde{z}_h\right) \log C_t\right]$$
$$= E_m[\log C_t].$$

Therefore, the social planner's utility (21.11) is a constant plus

$$E_m\left[\sum_{t=0}^{T} \delta^t \log C_t\right].$$

Thus, the social planner is a representative investor with log utility and beliefs $\mathbb{P}_m$.

Linear Risk Tolerance

Assume first that all investors have CARA utility. As for log utility, substitute $Z_{ht} = E_t[\tilde{z}_h]$ for $\tilde{z}_h$ in the social planning problem and solve the optimization problem pointwise to calculate the social planner's utility. The result is that the social planner's utility is

$$-\tau \left(\prod_{h=1}^{H}(\lambda_h \alpha_h)^{\tau_h/\tau}\right) \sum_{t=0}^{T} \delta^t E\left[Z_t e^{-\alpha C_t}\right], \qquad (21.12\text{a})$$

where

$$Z_t = \prod_{h=1}^{H} Z_{ht}^{\tau_h/\tau}. \qquad (21.12\text{b})$$

This is a straightforward extension of (21.7). This would imply the existence of a representative investor with state-independent CARA utility and beliefs $\mathbb{P}_m$ if it were true that

$$E\left[Z_t e^{-\alpha C_t}\right] = \gamma E_m\left[e^{-\alpha C_t}\right] \qquad (21.13)$$

for some probability $\mathbb{P}_m$ and a positive constant γ (which we could drop). By iterated expectations,

$$\gamma E_m\left[e^{-\alpha C_t}\right] = \gamma E\left[\frac{d\mathbb{P}_m}{d\mathbb{P}} e^{-\alpha C_t}\right] = \gamma E\left[e^{-\alpha C_t} E_t\left[\frac{d\mathbb{P}_m}{d\mathbb{P}}\right]\right].$$

Therefore, (21.13) holds for all consumption processes C if and only if

$$Z_t = \gamma E_t\left[\frac{d\mathbb{P}_m}{d\mathbb{P}}\right] \qquad (21.14)$$

Heterogeneous Beliefs

for each t. Taking $t = T$ here yields $d\mathbb{P}_m/d\mathbb{P} = Z_T/\gamma$, so (21.14) implies $Z_t = \mathsf{E}_t[Z_T]$. Thus, a necessary and sufficient condition for (21.13) is that Z be a martingale relative to $\mathbb{P}$.

In the log case, Z defined in (21.10) is a martingale, so we obtain a representative investor with state-independent utility. However, Z defined in (21.12b) is a supermartingale but not a martingale (see below). Thus, the most we can say in the CARA case is that there is a representative investor with state-dependent utility

$$\mathsf{E}\left[-\sum_{t=0}^{T}\delta^t Z_t e^{-\alpha C_t}\right]. \tag{21.15}$$

An analogous result is true for general LRT utility functions in complete dynamic markets. For general LRT utility, whether Z is a supermartingale or a submartingale depends on the cautiousness parameter (Exercise 21.2).

The supermartingale/submartingale property of Z has economic implications. We can interpret Z as a random discounting factor. A supermartingale decreases on average, so when Z is a supermartingale, the future is discounted more on average, producing a higher risk-free rate in equilibrium.

The fact that Z is a supermartingale in the case of CARA utility follows from the fact that a geometric average is smaller than an arithmetic average. To see the relation between geometric and arithmetic averages, apply Jensen's inequality to the logarithm function to obtain

$$\prod_{h=1}^{H}\left(\frac{Z_{h,t+1}}{Z_{ht}}\right)^{\tau_h/\tau} = \exp\left(\sum_{h=1}^{H}\frac{\tau_h}{\tau}\log\left(\frac{Z_{h,t+1}}{Z_{ht}}\right)\right)$$

$$\leq \exp\left(\log\left(\sum_{h=1}^{H}\frac{\tau_h}{\tau}\left(\frac{Z_{h,t+1}}{Z_{ht}}\right)\right)\right)$$

$$= \sum_{h=1}^{H}\frac{\tau_h}{\tau}\left(\frac{Z_{h,t+1}}{Z_{ht}}\right).$$

This implies

$$\mathsf{E}_t\left[\frac{Z_{t+1}}{Z_t}\right] = \mathsf{E}_t\left[\prod_{h=1}^{H}\left(\frac{Z_{h,t+1}}{Z_{ht}}\right)^{\tau_h/\tau}\right] \leq \mathsf{E}_t\left[\sum_{h=1}^{H}\frac{\tau_h}{\tau}\left(\frac{Z_{h,t+1}}{Z_{ht}}\right)\right] = 1.$$

The supermartingale property also follows directly from Hölder's inequality:

$$\mathsf{E}_t\left[\prod_{h=1}^{H}\left(\frac{Z_{h,t+1}}{Z_{ht}}\right)^{\tau_h/\tau}\right] \leq \prod_{h=1}^{H}\mathsf{E}_t\left[\frac{Z_{h,t+1}}{Z_{ht}}\right]^{\tau_h/\tau} = 1.$$

21.4 SHORT SALES CONSTRAINTS AND OVERPRICING

When investors have heterogeneous beliefs, investors who are optimistic about an asset should be long the asset in equilibrium, and investors who are pessimistic should be short. There are many investors who cannot short sell and many others who find it costly to short (because they do not obtain use of the proceeds and must post additional margin on which they earn no interest). In the presence of short sales constraints, optimistic investors will hold the asset in equilibrium, and pessimistic investors may be on the sideline. Short selling increases the supply of an asset available to those who want to buy it. Curtailing short selling limits the available supply, and, of course, any limitation of supply good should increase the price. Thus, in the presence of short sales constraints, prices may be too high (relative to average beliefs).

To illustrate this, suppose there is a single risky asset (the market portfolio) and all investors have CARA utility. The payoff of the risky asset is market wealth $\tilde{w}_m$. Suppose the risky asset cannot be sold short. Normalize the shares of the risky asset so that the total supply is 1 share, and assume each investor is endowed with $1/H$ shares. Assume there is a risk-free asset in zero net supply. Assume investors have CARA utility with the same absolute risk aversion α, agree that $\tilde{w}_m$ is normally distributed, and agree on the variance σ^2 of $\tilde{w}_m$. Let μ_h denote the mean of $\tilde{w}_m$ perceived by investor h.

Given a price P for the risky asset, it follows from (2.19) that the optimal number of shares of the risky asset for investor h to hold, if she faced no short sales constraints, is

$$\frac{\mu_h - PR_f}{\alpha \sigma^2}.$$

When short sales are not allowed, investor h's optimal demand is

$$\theta_h = \begin{cases} \frac{\mu_h - PR_f}{\alpha \sigma^2} & \text{if } \mu_h \geq PR_f, \\ 0 & \text{otherwise}. \end{cases}$$

For a given price P, aggregate demand is

$$\theta_m \stackrel{\text{def}}{=} \sum_{\{h \mid \mu_h \geq PR_f\}} \frac{\mu_h - PR_f}{\alpha \sigma^2}. \tag{21.16}$$

Market clearing requires $\theta_m = 1$.

As usual, since we have not introduced a date–0 consumption good and normalized prices by taking the price of date–0 consumption to be 1, there is one degree of indeterminacy in equilibrium prices. It is convenient to normalize prices by taking $R_f = 1$ (that is, by taking a quantity of the risk-free asset that pays 1 unit of the consumption good at date 1 to be the numeraire at date 0). The equation $\theta_m = 1$ can then be solved for the equilibrium price P of the risky asset.

To simplify the solution of the equation $\theta_m = 1$, it is convenient to modify the model by assuming there is a continuum of investors, of total mass equal to 1, and that μ_h is uniformly distributed across investors on some interval $(\mu^* - \Delta, \mu^* + \Delta)$. The parameter μ^* is the average belief. In the following formulas, we should interpret α as aggregate absolute risk aversion: It equals individual risk aversion because the mass of investors is normalized to equal 1.

In the modified model, in the absence of short sales constraints, aggregate demand would be

$$\theta_m = \frac{1}{2\Delta} \int_{\mu^* - \Delta}^{\mu^* + \Delta} \frac{\mu - P}{\alpha \sigma^2} \, d\mu = \frac{\mu^* - P}{\alpha \sigma^2},$$

just as if all investors agreed that μ^* is the mean of $\tilde{w}_m$. The market clearing condition $\theta_m = 1$ would imply $P = \mu^* - \alpha \sigma^2$. This is the average expectation minus a discount for risk.

If the market clearing price in the unconstrained case is below the expectation of the most pessimistic investor—that is, if $\alpha \sigma^2 \geq \Delta$—then all investors are long the asset in the unconstrained case. Hence, the imposition of a short sale constraint has no effect. However, if $\Delta > \alpha \sigma^2$, then some investors short sell the asset in the unconstrained case, and constraining short sales affects their demands and hence affects the equilibrium price.

Assume $\Delta > \alpha \sigma^2$, so the short sales constraint is binding on some investors. Consider any price $P \geq \mu^* - \Delta$. Generalizing (21.16) and using the condition $R_f = 1$, aggregate demand is

$$\theta_m = \frac{1}{2\Delta} \int_{P}^{\mu^* + \Delta} \frac{\mu - P}{\alpha \sigma^2} \, d\mu = \frac{(\mu^* + \Delta - P)^2}{4\Delta \alpha \sigma^2}.$$

Hence, the market clearing condition $\theta_m = 1$ implies

$$P = \mu^* + \Delta - 2\sqrt{\Delta \alpha \sigma^2}.$$

The difference between the price with and without the short sales constraint is

$$\Delta - 2\sqrt{\Delta \alpha \sigma^2} + \alpha \sigma^2 = (\sqrt{\Delta} - \sqrt{\alpha \sigma^2})^2 > 0.$$

This confirms that constraining short selling increases the asset price when $\Delta > \alpha \sigma^2$. Furthermore, the price is increasing in Δ:

$$\frac{\partial P}{\partial \Delta} = 1 - \sqrt{\frac{\alpha \sigma^2}{\Delta}} > 0$$

when $\Delta > \alpha \sigma^2$. Therefore, greater dispersion of beliefs (greater Δ) leads to higher prices.

21.5 SPECULATIVE TRADE AND BUBBLES

The previous section presents an example in which, in the presence of short sales constraints, only relatively optimistic investors hold the risky asset, and the asset price is higher than it would be if all investors possessed the average beliefs. Even more interesting phenomena arise in dynamic models. For example, it need not be that optimistic investors always hold the asset. Instead, at any given point in time, pessimistic investors may value the asset more, because of the right to resell it later to the optimistic investors. Buying an asset when we regard its fundamental value as low in order to resell it later to others with higher valuations is speculative trading. Due to speculative trading, asset prices can be even higher than they would be if all investors possessed optimistic beliefs. This is a type of bubble in the asset price.

To illustrate this, consider the following discrete-time example. Suppose there are two investors (or two classes of investors) $h = 1, 2$ who are risk neutral and have the same discount factor δ. Suppose the horizon T is finite. Assume there is a risk-free asset in each period. Consider a risky asset (not necessarily the market portfolio) that pays a dividend D_t in period t. Assume there are no margin requirements for purchasing the risky asset, but short sales of the risky asset are prohibited.

Let E_{ht} denote conditional expectation at date t, given the beliefs of investor h. The equilibrium price in the penultimate period must be

$$P_{T-1} = \delta \max_h \mathsf{E}_{h,T-1}[D_T],$$

and in other periods it must satisfy

$$P_t = \delta \max_h \mathsf{E}_{ht}[D_{t+1} + P_{t+1}]. \tag{21.17}$$

If P_t (or P_{T-1}) were more than this, then neither investor would be willing to hold the asset, preferring to consume more in period t (or $T-1$). If it were less, then one of the investors would want to buy an infinite amount on margin.

The fundamental value of the asset at date t for investor h is

$$V_{ht} \stackrel{\text{def}}{=} \mathsf{E}_{ht}\left[\sum_{s=t+1}^{T} \delta^{s-t} D_t\right].$$

We must have $P_t \geq V_{ht}$, because otherwise investor h would want to buy an infinite amount of the asset, planning to buy and hold. Set $V_t = \max\{V_{1t}, V_{2t}\}$. We might expect, based on the single-period model, that the price at t is set by the most optimistic investor, meaning $P_t = V_t$. However, as remarked before, the equilibrium price can exceed the fundamental value of even the optimistic investors in a dynamic model, due to the value inherent in the opportunity to resell the asset. To see this, suppose that, in some state of the world, investor 2 is the most optimistic about the fundamental value at date t, that is, $V_t = V_{2t} > V_{1t}$, but investor 1 is the most optimistic about investor 2's future valuation in the sense that

$$\mathsf{E}_{1t}[D_{t+1} + V_{2,t+1}] > \mathsf{E}_{2t}[D_{t+1} + V_{2,t+1}].$$

Because $P_{t+1} \geq V_{2,t+1}$, this implies

$$P_t \geq \delta \mathsf{E}_{1t}[D_{t+1} + P_{t+1}] \geq \delta \mathsf{E}_{1t}[D_{t+1} + V_{2,t+1}]$$
$$> \delta \mathsf{E}_{2t}[D_{t+1} + V_{2,t+1}] = V_{2t} = V_t.$$

A specific (infinite-horizon) numerical example from Harrison and Kreps (1978) is presented in Exercise 21.3.

21.6 NOTES AND REFERENCES

Rubinstein (1974) proves the existence of a representative investor in a single-period model with log or CARA utility. The discussion in Section 21.3 of representative investors in dynamic models with heterogeneous beliefs is based on Jouini and Napp (2006). The existence of a representative investor depends on the market being frictionless. Detemple and Murthy (1997) and Basak and Croitoru (2000) study dynamic models with log utility and heterogeneous beliefs in the presence of margin requirements and other portfolio constraints. Cao and Ou-Yang (2009) is the source for Part (c) of Exercise 21.1, which establishes that the existence of an asset with a payoff that is quadratic in market wealth is sufficient to implement Pareto optima when investors have CARA utility and agree that market wealth is normally distributed.

The idea that short sales constraints increase the prices of assets when investors have heterogeneous beliefs is due to Lintner (1969) and Miller (1977). It is commonly called Miller's model. The model in Section 21.4 is due to Chen, Hong, and Stein (2002), who emphasize that overpricing is increasing in the dispersion of beliefs. Gallmeyer and Hollifield (2008) study the effect of a market-wide short sales constraint and show that it may either raise or lower asset prices, depending on investors' elasticities of intertemporal substitution. They also show that the imposition of a short sales constraint increases the equilibrium interest rate. Hong and Stein (2003) develop a theory of market crashes based on heterogeneous beliefs and short-sale constraints.

The idea that speculative trade can cause prices to be above the fundamental values of even optimistic investors is due to Harrison and Kreps (1978), who analyze an infinite-horizon version of the model presented in Section 21.5. Exercise 21.3 presents a numerical example given by Harrison and Kreps (1978).

The role of the short sales constraint in the Harrison-Kreps model is to ensure the existence of equilibrium: In its absence, risk-neutral investors with heterogeneous beliefs would want to go infinitely short and long. Cao and Ou-Yang (2005) show that the price can be above the fundamental value of optimistic investors and can in other times be below the fundamental value of pessimistic investors when risk-averse investors have heterogeneous beliefs.

Scheinkman and Xiong (2003) analyze a continuous-time version of the Harrison-Kreps model. In the Scheinkman-Xiong model, investors observe processes that forecast future dividends. They disagree on the precisions with which the various signal processes forecast dividends, which Scheinkman and Xiong interpret as reflecting overconfidence of investors. There are many other papers that model heterogeneous beliefs with learning about fundamentals. Those papers are surveyed in Section 23.6.

It is generally regarded as a puzzle that the volume of trading in financial markets is as high as it is. Speculative trading is one possible explanation for the magnitude of observed volume. Harris and Raviv (1993), Kandel and Pearson (1995), and Cao and Ou-Yang (2009) present models of volume with heterogeneous beliefs.

Anderson, Ghysels, and Juergens (2005) and David (2008a) ask whether heterogeneity in beliefs can explain the equity premium puzzle. With different models, they reach different conclusions. Anderson, Ghysels, and Juergens (2005) also test whether heterogeneity in beliefs is a priced risk factor. Banerjee, Kaniel, and Kremer (2009) show that stock returns can exhibit momentum when investors have "higher order" differences in beliefs. For a survey of the implications of heterogeneous beliefs for asset prices and trading volume, see Hong and Stein (2007).

EXERCISES

21.1. Suppose each investor h has CARA utility with absolute risk aversion α_h. Assume the information in the economy is generated by $\tilde{w}_m$. Assume investor h believes $\tilde{w}_m$ is normally distributed with mean μ_h and variance σ^2, where σ is the same for all investors.

(a) Show that the Radon-Nikodym derivative of investor h's probability $\mathbb{P}_h$ with respect to the average probability $\mathbb{P}$ is

$$\tilde{z}_h = \frac{\exp\left(-\frac{(\tilde{w}_m - \mu_h)^2}{2\sigma^2}\right)}{\frac{1}{H}\sum_{h=1}^{H} \exp\left(-\frac{(\tilde{w}_m - \mu_h)^2}{2\sigma^2}\right)}.$$

(b) Show that the sharing rule (21.6) is equivalent to

$$\tilde{w}_h = \tau_h \sum_{h=1}^{H} \frac{\tau_h}{\tau} \left[\log\left(\frac{\lambda_h \alpha_h}{\lambda_h \alpha_h}\right) + \frac{\mu_h^2 - \mu_h^2}{2\sigma^2} \right]$$

$$+ \frac{\tau_h}{\tau} \tilde{w}_m + \tau_h \left(\sum_{h=1}^{H} \frac{\tau_h(\mu_h - \mu_h)}{\tau \sigma^2} \right) \tilde{w}_m.$$

(c) Show that if investors also disagree about the variance of $\tilde{w}_m$, then the sharing rule (21.6) is quadratic in $\tilde{w}_m$.

21.2. Assume all investors have constant relative risk aversion ρ and the same discount factor δ. Solve the social planning problem in a finite-horizon discrete-time model to show that the social planner's utility is

$$\mathsf{E}\left[\sum_{t=0}^{T} \delta^t Z_t \frac{C_t^{1-\rho}}{1-\rho} \right]$$

for some stochastic process Z. Show that Z is a supermartingale relative to the average beliefs if $\rho > 1$. Hint: For the last statement, use a conditional version of the Minkowski inequality. The Minkowski inequality states that for random variables $\tilde{x}_h$ and any $\rho > 1$,

$$\mathsf{E}\left[\left(\sum_{h=1}^{H} \tilde{x}_h \right)^\rho \right]^{1/\rho} \le \sum_{h=1}^{H} \mathsf{E}[\tilde{x}_h^\rho]^{1/\rho}.$$

21.3. Consider an infinite-horizon version of the model in Section 21.5 in which both investors agree the dividend process is a two-state Markov chain, with states $D = 0$ and $D = 1$. Suppose the investors' beliefs $\mathbb{P}_h$ satisfy, for all $t \ge 0$,

$$\mathbb{P}_1(D_{t+1} = 0 | D_t = 0) = 1/2, \quad \mathbb{P}_1(D_{t+1} = 1 | D_t = 0) = 1/2,$$
$$\mathbb{P}_1(D_{t+1} = 0 | D_t = 1) = 2/3, \quad \mathbb{P}_1(D_{t+1} = 1 | D_t = 1) = 1/3,$$

$$\mathbb{P}_2(D_{t+1} = 0 | D_t = 0) = 2/3, \quad \mathbb{P}_2(D_{t+1} = 1 | D_t = 0) = 1/3,$$
$$\mathbb{P}_2(D_{t+1} = 0 | D_t = 1) = 1/4, \quad \mathbb{P}_2(D_{t+1} = 1 | D_t = 1) = 3/4.$$

Assume the discount factor of each investor is $\delta = 3/4$. For $s = 0$ and $s = 1$, set

$$V_h(s) = \mathsf{E}_h\left[\sum_{t=1}^{\infty} \delta^t D_t \,|\, D_0 = s \right].$$

For each h, use the pair of equations

$$\frac{V_h(s)}{\delta} = \mathbb{P}_h(D_{t+1} = 0 | D_t = s) V_h(0) + \mathbb{P}_h(D_{t+1} = 1 | D_t = s)[1 + V_h(1)]$$

to calculate $V_h(0)$ and $V_h(1)$. Show that investor 2 has the highest fundamental value in both states $[V_2(0) > V_1(0)$ and $V_2(1) > V_1(1)]$ but investor 1 is the most optimistic in state $D = 0$ about investor 2's future valuation, in the sense that

$$\mathbb{P}_1(D_{t+1} = 0 \mid D_t = 0)V_2(0) + \mathbb{P}_1(D_{t+1} = 1 \mid D_t = 0)[1 + V_2(1)]$$
$$> \mathbb{P}_2(D_{t+1} = 0 \mid D_t = 0)V_2(0) + \mathbb{P}_2(D_{t+1} = 1 \mid D_t = 0)[1 + V_2(1)].$$

Rational Expectations Equilibria

CONTENTS

22.1 No-Trade Theorem	570
22.2 Normal-Normal Updating	573
22.3 Fully Revealing Equilibria	577
22.4 Grossman-Stiglitz Model	578
22.5 Hellwig Model	583
22.6 Notes and References	586

This chapter discusses securities markets when investors have common (homogeneous) priors but heterogeneous information. Conditioning on different information causes investors to have different beliefs, but investors do not hold those beliefs dogmatically. Instead, they learn from security prices about the information of other investors, and they revise their beliefs in response. The concept that investors understand how prices depend on information, and hence can make correct inferences from prices, is called rational expectations.

The extent to which prices reveal information is a fundamental issue in finance and in economics in general. If prices are fully revealing, then markets are said to be strong-form efficient. Strong-form efficiency is paradoxical in that, if prices are fully revealing, then private information is of no benefit in equilibrium. Yet, it is presumably costly to acquire private information. Of course, if no one acquires information—because it is costly to do so and of no benefit in equilibrium—then there is no information for prices to reveal. This is called the Grossman-Stiglitz paradox. To avoid this paradox, prices must be less than fully revealing.

One reason for partial revelation is that demands for assets depend both on information that is relevant to asset payoffs and on other information that is irrelevant but which nonetheless causes investors to trade. Such motives for trading could include a need for cash or a surplus of cash due to some personal liquidity shock. Trading based on such motives is commonly called liquidity trading. In Section 22.1, we see that some non-information-based motive for trade—such as risk sharing—must be present for asymmetrically informed investors to trade.

22.1 NO-TRADE THEOREM

Strictly risk-averse investors with common priors do not bet on events that are independent of asset payoffs and endowments (labor income), even if they have different information about the likelihoods of such events occurring. For example, they do not bet on sports events. The key to this conclusion is that the terms of the bet are observable to both parties. If one party is agreeable to a bet, then the other party should learn enough from this fact that she will not agree to it. This applies to security trades as well. If there is no risk sharing benefit to a trade, then risk-averse investors with common priors will not make the trade, even if the trade appears profitable to both given their different private information.

Example

For a simple example, consider a binary random variable $\tilde{x} \in \{0, 1\}$ on which two individuals might bet. The prior probabilities of the outcomes 0 and 1 are 1/2 each. Suppose each individual observes a signal $\tilde{s}_h$ prior to the bet. Conditional on the outcome of the event $\tilde{x}$, the signals are independently distributed with 3/4 probability that $\tilde{s}_h = \tilde{x}$ and 1/4 probability that $\tilde{s}_h = 1 - \tilde{x}$. The posterior probabilities of the events conditional on a signal are calculated as, for example,

$$\text{prob}(\tilde{x}=0\,|\,\tilde{s}_h=0) = \frac{\text{prob}(\tilde{x}=0, \tilde{s}_h=0)}{\text{prob}(\tilde{s}_h=0)}$$

$$= \frac{\text{prob}(\tilde{s}_h=0\,|\,\tilde{x}=0)\,\text{prob}(\tilde{x}=0)}{\text{prob}(\tilde{s}_h=0)} = \frac{3}{4}.$$

Suppose the realizations of the signals observed by the two individuals are $\tilde{s}_1 = 0$ and $\tilde{s}_2 = 1$. So, the first individual believes there is a 3/4 probability that $\tilde{x} = 0$, and the second individual believes there is a 3/4 probability that $\tilde{x} = 1$. If this difference in beliefs were due to differences in priors, then the two individuals would bet. However, in the present example, the difference is due to differences in information, and the two individuals will not bet. Consider, for example, a bet in which the first individual receives \$1 if $\tilde{x} = 0$ and pays \$2 if $\tilde{x} = 1$. This bet is acceptable to both given their private information, provided they are not too risk averse. For the first individual, the expected gain from the bet is

$$\frac{3}{4} \times \$1 - \frac{1}{4} \times \$2 = \$0.25.$$

For the second individual, the expected gain is

$$-\frac{1}{4} \times \$1 + \frac{3}{4} \times \$2 = \$1.25.$$

However, for this bet to take place, the first individual must realize that it is acceptable to the second individual, which is possible only if the second individual has observed $\tilde{s}_2 = 1$. Given this information, the first individual realizes that the events $\tilde{x} = 0$ and $\tilde{x} = 1$ are equally likely, so the expected gain from the bet to her is actually −\$0.50. She will therefore refuse the bet. Groucho Marx famously said that he would refuse to join any club that would admit him. There is a parallel here: An investor should refuse to accept any bet on $\tilde{x}$ that the other investor is willing to make.

Information about Exogenous Risks

For a more general treatment of the above example, suppose two investors have signals $\tilde{s}_h$ about the value of some random variable $\tilde{x}$ that is independent of endowments and asset payoffs, both conditionally on the $\tilde{s}_h$ and unconditionally. Suppose for the sake of argument that $\tilde{x}$ is traded at price $p(\tilde{s}_1, \tilde{s}_2)$ from investor 2 to investor 1, meaning that investor 1 adds $\tilde{x} - p(\tilde{s}_1, \tilde{s}_2)$ to her terminal wealth, and investor 2 subtracts the same amount. The independence assumption implies that neither investor receives any hedging benefits from this trade. Given strict risk aversion, a necessary condition for the trade to be acceptable to both investors is that each investor view her gain from the trade as having a positive conditional expectation (Section 1.5). Each investor h conditions on the signal $\tilde{s}_h$ that she observes directly and on the price $p(\tilde{s}_1, \tilde{s}_2)$ of the trade. Conditioning on the price means that the investor knows how the price depends on signals—that is, she knows the function p. As remarked before, this is called rational expectations. Because of rational expectations, observing a realization of the price provides some information about the realized signal of the other investor. The positive conditional expectations are expressed as

$$\mathsf{E}[\tilde{x} - p(\tilde{s}_1, \tilde{s}_2) \mid \tilde{s}_1, p(\tilde{s}_1, \tilde{s}_2)] > 0,$$
$$\mathsf{E}[p(\tilde{s}_1, \tilde{s}_2) - \tilde{x} \mid \tilde{s}_2, p(\tilde{s}_1, \tilde{s}_2)] > 0.$$

They imply

$$\mathsf{E}[\tilde{x} \mid \tilde{s}_1, p(\tilde{s}_1, \tilde{s}_2)] > p(\tilde{s}_1, \tilde{s}_2) > \mathsf{E}[\tilde{x} \mid \tilde{s}_2, p(\tilde{s}_1, \tilde{s}_2)].$$

Take the expectation conditional on $p(\tilde{s}_1,\tilde{s}_2)$ throughout to obtain

$$E[\tilde{x}|p(\tilde{s}_1,\tilde{s}_2)] > p(\tilde{s}_1,\tilde{s}_2) > E[\tilde{x}|p(\tilde{s}_1,\tilde{s}_2)],$$

which is a contradiction. Thus, given rational expectations, there is no price function p at which strictly risk-averse investors will trade a payoff-irrelevant random variable.

If investors are risk neutral, then they *may* make bets on events unrelated to endowments and payoffs of positive net supply assets, but they do not expect to gain from such bets. For risk-neutral investors, the strict inequalities in the previous paragraph are replaced by weak inequalities, leading to the conclusion[1]

$$E[\tilde{x}|\tilde{s}_1,p(\tilde{s}_1,\tilde{s}_2)] = p(\tilde{s}_1,\tilde{s}_2) = E[\tilde{x}|\tilde{s}_2,p(\tilde{s}_1,\tilde{s}_2)]. \tag{22.1}$$

Pareto Optima and Information about Payoff Relevant Events

The no-trade theorem also applies to trading on events that are relevant to endowments and asset payoffs, if we start from a Pareto optimum. Suppose the economy has reached a Pareto-optimal allocation of assets, perhaps through a round of trade in complete markets. Suppose investors then receive new private information about asset payoffs. Will they retrade the assets based on the new information? The answer is "no." Even though, based solely on each investor's private information, there may appear to be trades that would benefit each investor, the perceived benefits to some investors must disappear when the investors condition on the willingness of all parties to make the trades.

---

1. To derive (22.1), start from

$$E[\tilde{x}|\tilde{s}_1,p(\tilde{s}_1,\tilde{s}_2)] \geq p(\tilde{s}_1,\tilde{s}_2) \geq E[\tilde{x}|\tilde{s}_2,p(\tilde{s}_1,\tilde{s}_2)]$$

to obtain

$$E[\tilde{x}|p(\tilde{s}_1,\tilde{s}_2)] \geq p(\tilde{s}_1,\tilde{s}_2) \geq E[\tilde{x}|p(\tilde{s}_1,\tilde{s}_2)],$$

which is possible only if

$$E[\tilde{x}|p(\tilde{s}_1,\tilde{s}_2)] = p(\tilde{s}_1,\tilde{s}_2).$$

If two random variables $\tilde{y}$ and $\tilde{z}$ satisfy $\tilde{y} \geq \tilde{z}$ and $E[\tilde{y}] = E[\tilde{z}]$, then it must be that $\tilde{y} = \tilde{z}$ with probability 1. Apply this fact to $\tilde{y} = E[\tilde{x}|\tilde{s}_1,p(\tilde{s}_1,\tilde{s}_2)]$ and $\tilde{z} = p(\tilde{s}_1,\tilde{s}_2)$ and to expectation conditional on $p(\tilde{s}_1,\tilde{s}_2)$ to obtain the first equality in (22.1). The second equality is derived by the same reasoning.

22.2 NORMAL-NORMAL UPDATING

Most of the models in the remainder of the chapter are based on what is called normal-normal updating. This is a special case of Bayes' rule. We assume that a variable $\tilde{x}$ to be estimated and a signal $\tilde{s}$ are joint normally distributed.

Conditional Mean

Because of normality, the expectation of $\tilde{x}$ conditional on $\tilde{s}$ is the orthogonal projection of $\tilde{x}$ on the space spanned by $\tilde{s}$ and a constant.[2] This projection is

$$\mathsf{E}[\tilde{x}|\tilde{s}] = \mathsf{E}[\tilde{x}] + \beta(\tilde{s} - \mathsf{E}[\tilde{s}]), \tag{22.2}$$

where

$$\beta = \frac{\mathrm{cov}(\tilde{x},\tilde{s})}{\mathrm{var}(\tilde{s})}. \tag{22.3}$$

Conditional on $\tilde{s}$, $\tilde{x}$ is normally distributed with (22.2) as its mean.

Conditional Variance

Given $\tilde{s}$, the unknown part of $\tilde{x}$ is the residual from the projection (22.2). The variance of $\tilde{x}$ conditional on $\tilde{s}$ is the variance of the residual. The residual is $\tilde{u}$ defined as $\tilde{u} = \tilde{x} - \mathsf{E}[\tilde{x}|\tilde{s}]$. Equivalently,

$$\tilde{x} = \mathsf{E}[\tilde{x}] + \beta(\tilde{s} - \mathsf{E}[\tilde{s}]) + \tilde{u}. \tag{22.4}$$

Because $\tilde{u}$ and $\tilde{s}$ are uncorrelated,

$$\mathrm{var}(\tilde{x}) = \beta^2 \mathrm{var}(\tilde{s}) + \mathrm{var}(\tilde{u})$$
$$= \frac{\mathrm{cov}(\tilde{x},\tilde{s})^2}{\mathrm{var}(\tilde{s})} + \mathrm{var}(\tilde{u}),$$

which implies

$$\mathrm{var}(\tilde{u}) = \left[1 - \frac{\mathrm{cov}(\tilde{x},\tilde{s})^2}{\mathrm{var}(\tilde{s})\mathrm{var}(\tilde{x})}\right]\mathrm{var}(\tilde{x})$$
$$= [1 - \mathrm{corr}(\tilde{x},\tilde{s})^2]\mathrm{var}(\tilde{x}). \tag{22.5}$$

2. Normality implies that the residual $\tilde{\varepsilon}$ in the projection (which is always uncorrelated with $\tilde{s}$) is independent of $\tilde{s}$, hence mean independent of $\tilde{s}$. The fact that $\mathsf{E}[\tilde{\varepsilon} \mid \tilde{s}] = 0$ implies that $\mathsf{E}[\tilde{x} \mid \tilde{s}]$ equals the projection.

The squared correlation in this formula is the fraction of the variance of $\tilde{x}$ that is attributable to its correlation with $\tilde{s}$. It is called the R^2 of the projection of $\tilde{x}$ on $\tilde{s}$.

Truth-Plus-Noise Signals

There is no loss of generality in assuming that the signal is "truth plus noise," meaning that $\tilde{s} = \tilde{x} + \tilde{\varepsilon}$, where $\tilde{\varepsilon}$ is normally distributed and independent of $\tilde{x}$. To see that there is no generality lost in this assumption, consider an arbitrary $\tilde{s}$ that is joint normally distributed with $\tilde{x}$. Project $\tilde{s}$ on $\tilde{x}$ as

$$\tilde{s} = \alpha + \frac{\text{cov}(\tilde{x},\tilde{s})}{\text{var}(\tilde{x})}\tilde{x} + \tilde{\xi}.$$

In this projection, $\tilde{\xi}$ is uncorrelated with—and therefore, because of normality, independent of—$\tilde{x}$. The same information is carried in the following affine transform of $\tilde{s}$:

$$\frac{\text{var}(\tilde{x})}{\text{cov}(\tilde{x},\tilde{s})}(\tilde{s} - \alpha). \tag{22.6}$$

The affine transform equals $\tilde{x} + \tilde{\varepsilon}$, where

$$\tilde{\varepsilon} \stackrel{\text{def}}{=} \frac{\text{var}(\tilde{x})}{\text{cov}(\tilde{x},\tilde{s})}\tilde{\xi}.$$

Thus, by working with the affine transform, we obtain a truth-plus-noise signal. Note that $E[\tilde{s}] = E[\tilde{x}]$ when the signal is truth plus noise, so the formula (22.2) for the conditional mean can be written as

$$E[\tilde{x}|\tilde{s}] = (1-\beta)E[\tilde{x}] + \beta\tilde{s}. \tag{22.7}$$

R^2 and Beta for Truth-Plus-Noise Signals

When the signal is truth plus noise, then $\text{cov}(\tilde{x},\tilde{s}) = \text{var}(\tilde{x})$, so $\beta = \text{var}(\tilde{x})/\text{var}(\tilde{s})$. This implies further that

$$\text{corr}(\tilde{x},\tilde{s})^2 = \frac{\text{var}(\tilde{x})^2}{\text{var}(\tilde{x})\text{var}(\tilde{s})} = \beta.$$

In conjunction with (22.5), this produces the following formula for the conditional variance:

$$\text{var}(\tilde{x}|\tilde{s}) = (1-\beta)\text{var}(\tilde{x}), \tag{22.8}$$

where, because $\tilde{s} = \tilde{x} + \tilde{\varepsilon}$,

$$\beta = \frac{\text{cov}(\tilde{x}, \tilde{s})}{\text{var}(\tilde{s})} = \frac{\text{var}(\tilde{x})}{\text{var}(\tilde{x}) + \text{var}(\tilde{\varepsilon})}. \tag{22.9}$$

Increase in Precision

The reciprocal of a variance is called a precision. For a truth-plus-noise signal, there is an alternative to formula (22.8) for the conditional variance that is often useful. The alternative formula is expressed in terms of precisions. First, note that (22.9) implies

$$1 - \beta = \frac{\text{var}(\tilde{\varepsilon})}{\text{var}(\tilde{x}) + \text{var}(\tilde{\varepsilon})}.$$

Substitute this and take reciprocals of both sides of (22.8) to obtain

$$\frac{1}{\text{var}(\tilde{x}|\tilde{s})} = \left(1 + \frac{\text{var}(\tilde{x})}{\text{var}(\tilde{\varepsilon})}\right) \frac{1}{\text{var}(\tilde{x})}$$

$$= \frac{1}{\text{var}(\tilde{x})} + \frac{1}{\text{var}(\tilde{\varepsilon})}. \tag{22.10}$$

Thus, the precision of the estimate of $\tilde{x}$ increases when the signal is observed, and the increase in the precision equals the precision of the signal's noise term.

Multivariate Signals

If either $\tilde{x}$ or $\tilde{s}$ is a vector with $(\tilde{x}, \tilde{s})$ joint normal, then, using the notation of Section 3.5, the conditional expectation of $\tilde{x}$ given $\tilde{s}$ is given by the multivariate projection formula (3.32):

$$\mathsf{E}[\tilde{x}|\tilde{s}] = \mathsf{E}[\tilde{x}] + \text{Cov}(\tilde{s}, \tilde{x})' \, \text{Cov}(\tilde{s})^{-1} (\tilde{s} - \mathsf{E}[\tilde{s}]). \tag{22.11}$$

Following the same reasoning leading to (22.5) yields the following formula for the conditional covariance matrix of $\tilde{x}$:

$$\text{Cov}(\tilde{x}|\tilde{s}) = \text{Cov}(\tilde{x}) - \text{Cov}(\tilde{s}, \tilde{x})' \, \text{Cov}(\tilde{s})^{-1} \text{Cov}(\tilde{s}, \tilde{x}). \tag{22.12}$$

Sequential Projections

Suppose $\tilde{x}$ is a scalar, and there are signals $\tilde{s}_1, \ldots, \tilde{s}_n$. An alternative to using the multivariate formulas (22.11) and (22.12) is to use sequential conditional projections. The projection of $\tilde{x}$ on a random variable $\tilde{s}_m$ conditional on other random variables $\tilde{s}_1, \ldots, \tilde{s}_{m-1}$ is defined as in (22.2) except that means, variances, and covariances are replaced by means, variances, and covariances conditional on $\tilde{s}_1, \ldots, \tilde{s}_{m-1}$.

Suppose the signals are truth-plus-noise $\tilde{s}_i = \tilde{x} + \tilde{\varepsilon}_i$ with independent noise terms $\tilde{\varepsilon}_i$. The formula (22.7) for the conditional mean generalizes as

$$E[\tilde{x}|\tilde{s}_1,\ldots,\tilde{s}_m] = (1-\beta_m)E[\tilde{x}|\tilde{s}_1,\ldots,\tilde{s}_{m-1}] + \beta_m \tilde{s}_m \qquad (22.13)$$

for $m \leq n$, where

$$\beta_m = \frac{\mathrm{cov}(\tilde{x},\tilde{s}_m|\tilde{s}_1,\ldots,\tilde{s}_{m-1})}{\mathrm{var}(\tilde{s}_m|\tilde{s}_1,\ldots,\tilde{s}_{m-1})} = \frac{\mathrm{var}(\tilde{x}|\tilde{s}_1,\ldots,\tilde{s}_{m-1})}{\mathrm{var}(\tilde{x}|\tilde{s}_1,\ldots,\tilde{s}_{m-1}) + \mathrm{var}(\tilde{\varepsilon}_m)}. \qquad (22.14)$$

Formula (22.13) states that the change in the conditional expectation when the signal $\tilde{s}_m$ is observed is proportional (with proportionality constant β_m) to the "innovation" in the signal, meaning the difference between $\tilde{s}_m$ and its conditional mean $E[\tilde{s}_m|\tilde{s}_1,\ldots,\tilde{s}_{m-1}] = E[\tilde{x}|\tilde{s}_1,\ldots,\tilde{s}_{m-1}]$.

The formula (22.10) for the precision generalizes as

$$\frac{1}{\mathrm{var}(\tilde{x}|\tilde{s}_1,\ldots,\tilde{s}_m)} = \frac{1}{\mathrm{var}(\tilde{x}|\tilde{s}_1,\ldots\tilde{s}_{m-1})} + \frac{1}{\mathrm{var}(\tilde{\varepsilon}_m)}$$

for $m \leq n$. Successively substituting this formula yields

$$\frac{1}{\mathrm{var}(\tilde{x}|\tilde{s}_1,\ldots,\tilde{s}_m)} = \frac{1}{\mathrm{var}(\tilde{x})} + \sum_{i=1}^{m} \frac{1}{\mathrm{var}(\tilde{\varepsilon}_i)}. \qquad (22.15)$$

Thus, the precision of the estimate of $\tilde{x}$ increases by the precision of the signal noise term each time we observe an additional signal (under our assumption that the noise terms are independent).

We can use (22.14) and (22.15) to obtain a formula for β_m directly in terms of the variances of $\tilde{x}$ and the $\tilde{\varepsilon}_i$. Namely,

$$\frac{1}{\beta_m} = 1 + \frac{\mathrm{var}(\tilde{\varepsilon}_m)}{\mathrm{var}(\tilde{x}|\tilde{s}_1,\ldots,\tilde{s}_{m-1})}$$

$$= 1 + \frac{\mathrm{var}(\tilde{\varepsilon}_m)}{\mathrm{var}(\tilde{x})} + \sum_{i=1}^{m-1} \frac{\mathrm{var}(\tilde{\varepsilon}_m)}{\mathrm{var}(\tilde{\varepsilon}_i)}. \qquad (22.14')$$

Finally, divide both sides of (22.14') by $\mathrm{var}(\tilde{\varepsilon}_m)$, substitute (22.15), and take reciprocals of both sides to obtain the following alternative to (22.15):

$$\beta_m \mathrm{var}(\tilde{\varepsilon}_m) = \mathrm{var}(\tilde{x}|\tilde{s}_1,\ldots,\tilde{s}_m). \qquad (22.15')$$

22.3 FULLY REVEALING EQUILIBRIA

This section presents an example of a fully revealing equilibrium. Consider a single-period market with a risk-free asset in zero net supply. Assume all investors have CARA utility. Let $\tilde{x}$ denote the vector of risky asset payoffs, and let $\tilde{s} = (\tilde{s}_1, \ldots, \tilde{s}_H)$ denote the vector of signals observed by investors before trade. Assume $(\tilde{x}, \tilde{s})$ has a joint normal distribution. The standard device for constructing a fully revealing equilibrium is to consider an artificial economy in which each investor observes the entire vector $\tilde{s}$. We will compute the equilibrium of this artificial economy and then show that equilibrium prices reveal all that investors need to know about $\tilde{s}$. Hence, these prices in the actual economy with rational expectations produce the same demands as in the artificial economy and are therefore equilibrium prices of the actual economy.

Because of joint normality, the distribution of $\tilde{x}$ conditional on $\tilde{s}$ is normal, and the covariance matrix of $\tilde{x}$ conditional on $\tilde{s}$ is constant. Let $\mu(\tilde{s})$ denote $E[\tilde{x}|\tilde{s}]$, and let Σ denote the covariance matrix of $\tilde{x}$ conditional on $\tilde{s}$.[3] The random vector $\mu(\tilde{s})$ is a sufficient statistic for $\tilde{s}$ in terms of predicting $\tilde{x}$: The distribution of $\tilde{x}$ conditional on $\tilde{s}$ depends on $\tilde{s}$ only via $\mu(\tilde{s})$.

Equilibrium of the Artificial Economy

In the artificial economy, equilibrium prices can be computed for each realization of $\tilde{s}$ by using $\mu(\tilde{s})$ as the vector of expected payoffs and Σ as the covariance matrix of the payoffs in the model of Part I of this book. Thus, from Exercise 4.1, the risk-free return $R_f(\tilde{s})$ and equilibrium price vector $p(\tilde{s})$ are

$$R_f(\tilde{s}) = \frac{1}{\delta} \exp\left(\alpha \left[\bar{\theta}' \mu(\tilde{s}) - \bar{c}_0\right] - \frac{1}{2}\alpha^2 \bar{\theta}' \Sigma \bar{\theta}\right), \qquad (22.16a)$$

$$p(\tilde{s}) = \frac{1}{R_f(\tilde{s})}[\mu(\tilde{s}) - \alpha \Sigma \bar{\theta}], \qquad (22.16b)$$

where δ is a weighted geometric average of the investors' discount factors, α is the aggregate absolute risk aversion, $\bar{\theta}$ is the vector of supplies of the risky assets, and $\bar{c}_0$ is aggregate date–0 consumption.

---

3. If the matrix $\mathrm{Cov}(\tilde{s})$ is nonsingular, then the covariance matrix of $\tilde{x}$ conditional on $\tilde{s}$ is shown in (22.12) to be $\Sigma = \mathrm{Cov}(\tilde{x}) - \mathrm{Cov}(\tilde{s},\tilde{x})' \mathrm{Cov}(\tilde{s})^{-1} \mathrm{Cov}(\tilde{s},\tilde{x})$.

Revelation

From the equilibrium prices (22.16), investors can compute $\mu(\tilde{s})$ as

$$\mu(\tilde{s}) = \alpha \Sigma \bar{\theta} + R_f(\tilde{s}) p(\tilde{s}). \tag{22.17}$$

Thus, equilibrium prices are fully revealing in the sense of revealing a sufficient statistic for predicting $\tilde{x}$. In particular, each investor can compute the portfolio of risky assets that would be optimal if $\tilde{s}$ were known (Exercise 2.2) simply by observing equilibrium prices; that is, each investor h can compute

$$\frac{1}{\alpha_h} \Sigma^{-1} [\mu(\tilde{s}) - R_f(\tilde{s}) p(\tilde{s})]. \tag{22.18}$$

Thus, the equilibrium in the artificial economy is a fully revealing equilibrium in the actual economy.

Grossman-Stiglitz Paradox

Fully revealing equilibria suffer from the Grossman-Stiglitz paradox as remarked before. Notice that no investor h needs to use her private signal $\tilde{s}_h$ to compute $\mu(\tilde{s})$, because $\mu(\tilde{s})$ is fully revealed by the equilibrium prices. Thus, no investor benefits from her private information in equilibrium.

Diamond-Verrecchia Paradox

A related paradox is that the equilibrium demand (22.18), when $\mu(\tilde{s})$ is inferred from prices as in (22.17), is

$$\frac{\alpha}{\alpha_h} \bar{\theta}.$$

Thus, each investor's demand is just a fraction of the aggregate supply and independent of all signals. Even if investors acquire information before trade, it is not clear how their information could get into prices when all investors express constant demands to the market.

22.4 GROSSMAN-STIGLITZ MODEL

One circumstance in which equilibrium prices are only partially revealing is when the date–0 supply of the risky assets is random. This would occur if there were traders other than the H investors being modeled who trade for exogenous reasons, perhaps due to liquidity shocks. Such traders are called noise traders or liquidity traders. We can also regard the noisy supply as being due to random endowments of the H investors being modeled. With CARA utility, the endowments do not affect demands (no wealth effects), so equilibrium prices are the same with random endowments as with liquidity trades.

Model

Assume CARA utility and normal distributions as in the preceding section. Suppose $\tilde{s}$ is a scalar instead of a vector, and adopt the following simplifying assumptions:

- There is a single risky asset.
- The supply of the risky asset is a normally distributed random variable $\tilde{z}$ that is independent of $\tilde{x}$ and $\tilde{s}$.
- The equilibrium risk-free return R_f is exogenously given (see the end-of-chapter notes for discussion).
- $H_I < H$ investors observe $\tilde{s}$, and $H_U = H - H_I$ investors have no information other than the equilibrium price.

Linear Equilibrium and Revelation

We will look for an equilibrium in which $p(\tilde{s},\tilde{z}) = a_0 + a_1\tilde{s} + a_2\tilde{z}$ with $a_1 \neq 0$. By observing the price, uninformed investors can calculate

$$\frac{p(\tilde{s},\tilde{z}) - a_0}{a_1} = \tilde{s} + b\tilde{z},$$

where $b = a_2/a_1$. The solution for a_0, a_1, and a_2 is presented below.

Let α_I denote the aggregate absolute risk aversion of the informed investors. As usual, this means the reciprocal of the aggregate risk tolerance of the informed investors. If the informed investors all have the same absolute risk aversion α, then $\alpha_I = \alpha/H_I$. A useful observation about this model is that in equilibrium observing $p(\tilde{s},\tilde{z})$ is equivalent to observing

$$\mu(\tilde{s}) - \alpha_I \sigma^2 \tilde{z},$$

where $\mu(\tilde{s})$ denotes the mean of $\tilde{x}$ conditional on $\tilde{s}$ as before, and σ^2 denotes the variance of $\tilde{x}$ conditional on $\tilde{s}$. In the fully revealing equilibrium of the previous section, all investors infer $\mu(\tilde{s})$ from equilibrium prices. Here, investors who are ex ante uninformed only observe $\mu(\tilde{s})$ perturbed by noise. Thus, there is an advantage to being an informed investor in this model. This advantage can be large enough to justify the acquisition of information, so the model does not suffer from the Grossman-Stiglitz paradox.

Note that the noise $-\alpha_I \sigma^2 \tilde{z}$ disappears in the limit as $\alpha_I \to 0$ or $\sigma^2 \to 0$. If the aggregate risk aversion of informed traders is small ($\alpha_I \to 0$), informed traders push the price close to the expected discounted payoff, conveying this information to uninformed traders. The same is true if their information is nearly perfect ($\sigma^2 \to 0$), because they bear very little risk in that circumstance.

Risk Premium

The equilibrium price in this model is a weighted average of the informed and uninformed investors' conditional expectations of $\tilde{x}/R_f$ minus a risk premium term (see (22.20) below). The discount of the price for risk shown in (22.20) is

$$\frac{\tilde{z}}{(\tau_I \phi_I + \tau_U \phi_U) R_f},$$

where τ_i is the aggregate risk tolerance of investor class i, and ϕ_i is the precision of the information of investor class i (the reciprocal of the conditional variance of $\tilde{x}$ given the information obtained in equilibrium). The unconditional expectation of the price is

$$\frac{E[\tilde{x}]}{R_f} - \frac{E[\tilde{z}]}{(\tau_I \phi_I + \tau_U \phi_U) R_f}.$$

Assuming the expected supply $E[\tilde{z}]$ is positive, we have

$$\frac{E[\tilde{z}]}{(\tau_I \phi_I + \tau_U \phi_U) R_f} > \frac{E[\tilde{z}]}{(\tau_I + \tau_U) \phi_I R_f}. \tag{22.19}$$

The right-hand side of (22.19) would be the expected discount for risk if all investors observed $\tilde{s}$. Thus, on average, the price is lower and the expected return higher due to the presence of uninformed investors.

Risk-Neutral Investors

If the informed investors were risk neutral, then the only possible equilibrium would be fully revealing. This is suggested by the result for $\alpha_I \to 0$, but a more direct argument is based on the fact that such investors do not have optima unless

$$p(\tilde{s}, \tilde{z}) = \frac{\mu(\tilde{s})}{R_f}.$$

Thus, the price reveals $\mu(\tilde{s})$. Note that the random supply of the asset does not affect the price in this setting, because the risk-neutral investors are content to absorb whatever supply is offered when the asset is priced at its expected discounted value. Because the equilibrium is fully revealing, the Grossman-Stiglitz paradox *does* apply to the model with risk-neutral investors.

We will solve for an equilibrium price of the form $p(\tilde{s}, \tilde{z}) = a_0 + a_1 \tilde{s} + a_2 \tilde{z}$ for constants a_0, $a_1 \neq 0$, and a_2. In such an equilibrium, each investor can calculate $\tilde{s} + b\tilde{z}$, where $b = a_2/a_1$.

Let σ_I^2 denote the variance of $\tilde{x}$ conditional on $\tilde{s}$. Denote the variance of $\tilde{x}$ conditional on $\tilde{s} + b\tilde{z}$ by σ_U^2. The number of shares demanded by an informed investor h is

$$\frac{\mathsf{E}[\tilde{x}|\tilde{s}] - R_f p(\tilde{s}, \tilde{z})}{\alpha_h \sigma_I^2},$$

so the aggregate demand of informed investors is

$$\frac{\mathsf{E}[\tilde{x}|\tilde{s}] - R_f p(\tilde{s}, \tilde{z})}{\sigma_I^2} \left(\sum_{\text{informed } h} \frac{1}{\alpha_h} \right) = \frac{\mathsf{E}[\tilde{x}|\tilde{s}] - R_f p(\tilde{s}, \tilde{z})}{\alpha_I \sigma_I^2}.$$

The number of shares demanded by an uninformed investor h is

$$\frac{\mathsf{E}[\tilde{x}|\tilde{s} + b\tilde{z}] - R_f p(\tilde{s}, \tilde{z})}{\alpha_h \sigma_U^2},$$

so the aggregate demand of uninformed investors is

$$\frac{\mathsf{E}[\tilde{x}|\tilde{s} + b\tilde{z}] - R_f p(\tilde{s}, \tilde{z})}{\sigma_U^2} \left(\sum_{\text{uninformed } h} \frac{1}{\alpha_h} \right) = \frac{\mathsf{E}[\tilde{x}|\tilde{s} + b\tilde{z}] - R_f p(\tilde{s}, \tilde{z})}{\alpha_U \sigma_U^2}.$$

The market clearing condition is

$$\frac{\mathsf{E}[\tilde{x}|\tilde{s}] - R_f p(\tilde{s}, \tilde{z})}{\alpha_I \sigma_I^2} + \frac{\mathsf{E}[\tilde{x}|\tilde{s} + b\tilde{z}] - R_f p(\tilde{s}, \tilde{z})}{\alpha_U \sigma_U^2} = \tilde{z}.$$

To express the solution of the market clearing condition in a simple form, define the risk tolerances $\tau_I = 1/\alpha_I$ and $\tau_U = 1/\alpha_U$ and the precisions $\phi_I = 1/\sigma_I^2$ and $\phi_U = 1/\sigma_U^2$. The market clearing condition can be written as

$$\tau_I \phi_I \mathsf{E}[\tilde{x}|\tilde{s}] + \tau_U \phi_U \mathsf{E}[\tilde{x}|\tilde{s} + b\tilde{z}] - \tilde{z} = (\tau_I \phi_I + \tau_U \phi_U) R_f p(\tilde{s}, \tilde{z}),$$

implying

$$\left(\frac{\tau_I \phi_I}{\tau_I \phi_I + \tau_U \phi_U} \right) \frac{\mathsf{E}[\tilde{x}|\tilde{s}]}{R_f} + \left(\frac{\tau_U \phi_U}{\tau_I \phi_I + \tau_U \phi_U} \right) \frac{\mathsf{E}[\tilde{x}|\tilde{s} + b\tilde{z}]}{R_f}$$

$$- \frac{\tilde{z}}{(\tau_I \phi_I + \tau_U \phi_U) R_f} = p(\tilde{s}, \tilde{z}). \quad (22.20)$$

To solve for b, let μ_x denote the unconditional mean of $\tilde{x}$, μ_s the unconditional mean of $\tilde{s}$, and μ_z the unconditional mean of $\tilde{z}$. The normal-normal updating rule (22.2) produces

$$E[\tilde{x}|\tilde{s}] = \mu_x + \beta(\tilde{s} - \mu_s),$$

$$E[\tilde{x}|\tilde{s} + b\tilde{z}] = \mu_x + \kappa(\tilde{s} - \mu_s + b\tilde{z} - b\mu_z),$$

where

$$\beta = \frac{\operatorname{cov}(\tilde{x}, \tilde{s})}{\operatorname{var}(\tilde{s})},$$

$$\kappa = \frac{\operatorname{cov}(\tilde{x}, \tilde{s})}{\operatorname{var}(\tilde{s}) + b^2 \operatorname{var}(\tilde{z})}.$$

Substitute these into (22.20) to obtain

$$p(\tilde{s},\tilde{z}) = \left(\frac{\tau_I \phi_I}{\tau_I \phi_I + \tau_U \phi_U}\right) \frac{\mu_x + \beta(\tilde{s} - \mu_s)}{R_f}$$

$$+ \left(\frac{\tau_U \phi_U}{\tau_I \phi_I + \tau_U \phi_U}\right) \frac{\mu_x + \kappa(\tilde{s} - \mu_s + b\tilde{z} - b\mu_z)}{R_f} - \frac{\tilde{z}}{(\tau_I \phi_I + \tau_U \phi_U) R_f}.$$

This equals $a_0 + a_1 \tilde{s} + a_2 \tilde{z}$ if and only if

$$a_0 = \frac{\mu_x}{R_f} - \frac{\tau_I \phi_I \beta + \tau_U \phi_U \kappa}{(\tau_I \phi_I + \tau_U \phi_U) R_f} \mu_s - \frac{\tau_U \phi_U \kappa b}{(\tau_I \phi_I + \tau_U \phi_U) R_f} \mu_z,$$

$$a_1 = \frac{\tau_I \phi_I \beta + \tau_U \phi_U \kappa}{(\tau_I \phi_I + \tau_U \phi_U) R_f},$$

$$a_2 = \frac{\tau_U \phi_U \kappa b - 1}{(\tau_I \phi_I + \tau_U \phi_U) R_f}.$$

Note that $a_1 \neq 0$. The last two equations imply

$$b \stackrel{\text{def}}{=} \frac{a_2}{a_1} = -\frac{1}{\tau_I \phi_I \beta} = -\frac{\alpha_I \sigma_I^2}{\beta}.$$

To obtain explicit formulas for a_0, a_1, and a_2, substitute this formula for b into κ and

$$\sigma_U^2 = \operatorname{var}(x) - \frac{\operatorname{cov}(\tilde{x},\tilde{s})^2}{\operatorname{var}(s) + b^2 \operatorname{var}(\tilde{z})}.$$

For this last fact, see (22.5).

Notice that observing $\tilde{s} + b\tilde{z}$ is equivalent to observing

$$\mu_x - \beta \mu_s + \beta(\tilde{s} + b\tilde{z}) = \mu(\tilde{s}) - \alpha_I \sigma_I^2 \tilde{z}.$$

Therefore, the information revealed by prices is $\mu(\tilde{s})$ perturbed by noise as stated above.

22.5 HELLWIG MODEL

In some circumstances, rational expectations equilibria (whether fully or partially revealing) suffer from what Hellwig (1980) terms "schizophrenia." If the equilibrium price reveals something that is observed by only a single investor, and the investor understands that the price at least partially reveals her information (that is, if the investor has rational expectations), then the investor should also understand that her trades must have affected the price. To be a price taker when formulating demands, as assumed in competitive models, and to simultaneously recognize the dependence of the price on one's private information is "schizophrenic."

The schizophrenia issue does not arise when there are multiple investors with identical information. An investor in such an economy can reasonably assume that the trades of others with the same information affect the price but that her own trades have negligible influence. The model in the previous section is a model of that type.

The schizophrenia issue also does not arise when the price is independent of an investor's signal. For the equilibrium price to be independent of each investor's signal, each investor's signal must be irrelevant for forecasting, given the signals of others. This can be true only if there is a large number (more precisely, an infinite number) of investors. This section presents such a model, due to Hellwig (1980).

Model

Consider the model of the previous section, but suppose all investors have the same absolute risk aversion α and change the signal structure as follows. Suppose each investor h observes $\tilde{s}_h = \tilde{x} + \tilde{\varepsilon}_h$, where the $\tilde{\varepsilon}_h$ are IID zero-mean normal random variables that are independent of $\tilde{x}$. To make clear that the variance of $\tilde{\varepsilon}_h$ is the same for each h, denote the variance by $\text{var}(\tilde{\varepsilon})$. We are going to consider the limit as $H \to \infty$. Assume the supply of the asset is $\tilde{z} = H\tilde{y}$, where $\tilde{y}$ is normally distributed and independent of $\tilde{x}$ and the $\tilde{\varepsilon}_h$. The random variable $\tilde{y}$ is the supply per capita ($\tilde{z}/H$), and it will be held fixed as H is increased. By the strong law of large numbers, $\frac{1}{H}\sum_{h=1}^{H} \tilde{s}_h \to \tilde{x}$ as $H \to \infty$. We are going to work in the limit economy, taking $H = \infty$, so $\tilde{x}$ would be known if we had access to all of the signals observed by investors.[4]

---

[4]. For a more formal model, we could take the set of investors to be a continuum, subject to the issues discussed in Section 11.6 regarding the law of large numbers for a continuum of IID random variables, or we could take $H = \infty$ with the size of any set of investors being defined by a purely finitely additive measure, as is also discussed in Section 11.6.

Revelation

We show below that there is a partially revealing equilibrium in this limit economy in which the equilibrium price $p(\tilde{x},\tilde{y})$ reveals

$$\tilde{x} - \alpha \operatorname{var}(\tilde{\varepsilon})\tilde{y}.$$

Because $\tilde{x}$ can be computed from the signals of any infinite subset of investors—in particular, from the set excluding investor h for any h—no investor's private information can be seen in $p(\tilde{x},\tilde{y})$. Thus, the behavior of investors is not schizophrenic. On the other hand, each investor's private information $\tilde{s}_h$ is useful in equilibrium, because the equilibrium price reveals neither $\tilde{x}$ nor $\tilde{s}_h$. Thus, the Grossman-Stiglitz paradox is avoided. Also, each investor's equilibrium demand depends on her private signal $\tilde{s}_h$, so the Diamond-Verrecchia paradox is avoided.

Notice that the equilibrium price reveals more about $\tilde{x}$ when risk aversion α is smaller or when individual signals are more precise ($\operatorname{var}(\tilde{\varepsilon})$ is smaller). This is similar to the results of the previous section.

Risk Premium

The equilibrium price is

$$p(\tilde{x},\tilde{y}) = \frac{1}{R_f} \lim_{H \to \infty} \frac{1}{H} \sum_{h=1}^{H} \mathsf{E}[\tilde{x} \mid \tilde{x} + b\tilde{y}, \tilde{s}_h] - \frac{\alpha \sigma^2 \tilde{y}}{R_f}, \qquad (22.21)$$

where σ^2 denotes the variance of $\tilde{x}$ conditional on $\tilde{x} + b\tilde{y}$ and $\tilde{s}_h$ and where $b = -\alpha \operatorname{var}(\tilde{\varepsilon})$. The first term on the right-hand side of (22.21) is the average conditional expectation of the discounted asset value, conditional on the information obtained in equilibrium. The last term, $-\alpha \sigma^2 \tilde{y}/R_f$, is a risk premium term, depending on risk aversion, the conditional risk, and the supply of the asset, also as in the previous section.

Continuum of Investors

This model can be solved when investors differ with regard to risk aversion and signal quality. It is easiest to express such a model in a continuum of investors framework, with investors indexed by $h \in [0,1]$. The equilibrium price in the model reveals

$$\tilde{x} - \tilde{y} \bigg/ \int_0^1 \frac{1}{\alpha_h \operatorname{var}(\tilde{\varepsilon}_h)}\, dh\ .$$

See Exercise 22.2.

We will solve for an equilibrium price of the form $p(\tilde{x},\tilde{y}) = a_0 + a_1\tilde{x} + a_2\tilde{y}$ for constants a_0, $a_1 \neq 0$, and a_2. In such an equilibrium, each investor can calculate $\tilde{x} + b\tilde{y}$, where $b = a_2/a_1$, so she has that information in addition to her private signal $\tilde{s}_h$.

The number of shares of the risky asset demanded by investor h is

$$\frac{\mathsf{E}[\tilde{x}|\tilde{x}+b\tilde{y},\tilde{s}_h] - R_f p(\tilde{x},\tilde{y})}{\alpha \sigma^2},$$

where σ^2 denotes the variance of $\tilde{x}$ conditional on $\tilde{x} + b\tilde{y}$ and $\tilde{s}_h$. The market clearing condition, in per capita terms, is

$$\lim_{H \to \infty} \frac{1}{H} \sum_{h=1}^{H} \frac{\mathsf{E}[\tilde{x}|\tilde{x}+b\tilde{y},\tilde{s}_h] - R_f p(\tilde{x},\tilde{y})}{\alpha \sigma^2} = \bar{y},$$

which we can rearrange as (22.21).

Formulas (22.13), (22.14′), and (22.15′) for sequential conditional expectations, conditioning first on $\tilde{x} + b\tilde{y}$ and then on $\tilde{s}_h$, imply

$$\mathsf{E}[\tilde{x}|\tilde{x}+b\tilde{y}] = \bar{x} + \beta(\tilde{x} - \bar{x} + b\tilde{y} - b\bar{y}) \tag{22.22a}$$

and

$$\mathsf{E}[\tilde{x}|\tilde{x}+b\tilde{y},\tilde{s}_h] = \mathsf{E}[\tilde{x}|\tilde{x}+b\tilde{y}] + \kappa\left(\tilde{s}_h - \mathsf{E}[\tilde{x}|\tilde{x}+b\tilde{y}]\right), \tag{22.22b}$$

where

$$\frac{1}{\beta} = 1 + \frac{b^2 \operatorname{var}(\tilde{y})}{\operatorname{var}(\tilde{x})}, \tag{22.22c}$$

$$\frac{1}{\kappa} = 1 + \frac{\operatorname{var}(\tilde{\varepsilon})}{\operatorname{var}(\tilde{x})} + \frac{\operatorname{var}(\tilde{\varepsilon})}{b^2 \operatorname{var}(\tilde{y})}, \tag{22.22d}$$

$$\sigma^2 = \kappa \operatorname{var}(\tilde{\varepsilon}). \tag{22.22e}$$

By the strong law of large numbers and (22.22b),

$$\lim_{H \to \infty} \frac{1}{H} \sum_{h=1}^{H} \mathsf{E}[\tilde{x}|\tilde{x}+b\tilde{y},\tilde{s}_h]$$

$$= \mathsf{E}[\tilde{x}|\tilde{x}+b\tilde{y}] + \kappa\left(\tilde{x} - \mathsf{E}[\tilde{x}|\tilde{x}+b\tilde{y}]\right) = \kappa\tilde{x} + (1-\kappa)\mathsf{E}[\tilde{x}|\tilde{x}+b\tilde{y}].$$

Thus, the market clearing condition (22.21) is equivalent to

$$p(\tilde{x},\tilde{y}) = \frac{\kappa\tilde{x} + (1-\kappa)[\bar{x} + \beta(\tilde{x} - \bar{x} + b\tilde{y} - b\bar{y})] - \alpha\sigma^2\bar{y}}{R_f}. \tag{22.21′}$$

This equals $a_0 + a_1\tilde{x} + a_2\tilde{y}$ if and only if

$$a_0 = \frac{(1-\kappa)[(1-\beta)\bar{x} - \beta b \bar{y}]}{R_f},$$

$$a_1 = \frac{\kappa + (1-\kappa)\beta}{R_f},$$

$$a_2 = \frac{(1-\kappa)\beta b - \alpha\sigma^2}{R_f}.$$

Note that $a_1 \neq 0$. The last two equations imply

$$b \stackrel{\text{def}}{=} \frac{a_2}{a_1} = -\frac{\alpha\sigma^2}{\kappa},$$

and now (22.22e) implies $b = -\alpha \operatorname{var}(\tilde{\varepsilon})$. Thus, the equilibrium price reveals $\tilde{x} - \alpha \operatorname{var}(\tilde{\varepsilon})\tilde{y}$ as claimed.

22.6 NOTES AND REFERENCES

For the history of the rational expectations hypothesis, see Grossman (1981). The weak, semi-strong, and strong forms of the efficient markets hypothesis appear in Fama (1970), who attributes the weak- and strong-form terminology to Harry Roberts.

The effect of short sales constraints on prices is very different when differences in beliefs are due to asymmetric information rather than to heterogeneous priors. In rational expectations equilibria with homogeneous priors, investors understand that short sales constraints preclude the expression of negative opinions, and they account for the possibility that such unexpressed opinions may exist when they attempt to infer the asset value from the price. The consequence of short sales constraints in such a setting is that prices may be less informative but are not biased. This is demonstrated by Diamond and Verrecchia (1987) using a variation of the Glosten-Milgrom model (Section 24.1).

The no-trade theorem is due to Milgrom and Stokey (1982) and Tirole (1982). The proof given in the text follows Tirole (1982). Milgrom and Stokey consider more general economic mechanisms (not just rational expectations equilibria) and show that there is no trade if the initial allocation is Pareto optimal and if investors have "concordant beliefs." Investors have concordant beliefs if they perceive the same conditional distribution of signals given payoffs. They could have different beliefs regarding the marginal distribution of payoffs, so concordant beliefs is a weaker assumption than common priors. A result closely related to the no-trade theorem is the fact that individuals with common priors cannot "agree to disagree" (Aumann, 1976; Rubinstein and Wolinsky, 1990).

The conditional projections discussed in Section 22.2 are frequently applied in dynamic models. Suppose that $\tilde{x}, \tilde{s}_1, \ldots, \tilde{s}_T$ are joint normal, and the information at date t is $(\tilde{s}_1, \ldots, \tilde{s}_t)$. Then,

$$\mathsf{E}_t[\tilde{x}] = \mathsf{E}_{t-1}[\tilde{x}] + \frac{\mathrm{cov}_{t-1}(\tilde{x}, \tilde{s}_t)}{\mathrm{var}_{t-1}(\tilde{s}_t)} \left(\tilde{s}_t - \mathsf{E}_{t-1}[\tilde{s}_t] \right). \qquad (22.23)$$

We call $\tilde{s}_t - \mathsf{E}_{t-1}[\tilde{s}_t]$ an innovation. Formula (22.23) is the discrete-time Kalman filtering formula, and it states that the conditional mean of $\tilde{x}$ is revised in proportion to the innovation. Chapter 23 presents the Kalman filter in continuous time.

The example of a fully revealing equilibrium in Section 22.3 is from Grossman (1976). Grossman (1981) presents a more general treatment of fully revealing equilibria. The Grossman-Stiglitz paradox is named for Grossman (1976) and Grossman and Stiglitz (1976).

The model of a partially revealing equilibrium in Section 22.4 is due to Grossman (1976) and Grossman and Stiglitz (1976). For a continuous-time version of that model, see Wang (1993). Diamond and Verrecchia (1981) solve a variation of the model in Section 22.4 in which each investor observes a private signal $\tilde{s}_h$. They also discuss the issue that Section 22.3 calls the Diamond-Verrecchia paradox. The model of a large number of investors is from Hellwig (1980). Hellwig also presents convergence results, as the number of investors converges to infinity. Breon-Drish (2015) solves the Grossman-Stiglitz model—and an extension in which there are multiple types of investors who receive different signals—without assuming that the asset value is normally distributed. Instead, he assumes that the conditional distribution of the payoff given signals has an exponential distribution. This includes, for example, the case in which the asset value has a Bernoulli (two-point) distribution.

The Grossman-Stiglitz and Hellwig models are a bit unsatisfactory, because they assume a single risky asset and risk-averse investors. We do not really want to assume the single risky asset is the aggregate market, because there is unlikely to be asymmetric information about the market as a whole. On the other hand, we cannot easily regard the models as being about a single asset in a multiasset market, because demands by risk-averse investors depend on correlations between assets, which are not modeled. One possibility is to assume the value of the asset being modeled is independent of all other assets, so demands in the CARA-normal world depend only on the mean and variance of the risky asset and the risk-free return (Section 2.4). In this setting, it is natural to assume that the risk-free return is given exogenously (as in Sections 22.4 and 22.5), because demands for an individual asset should have a negligible effect on the risk-free return. Admati (1985) develops a version of the Hellwig model with multiple assets.

The reason we want to assume the risk-free return is exogenously given is that the inference problem is complicated when the risk-free return depends on signals and asset supplies. For example, $\log R_f$ in (22.16a) is quadratic rather than affine in $\bar{\theta}$ and $\bar{s}$. One justification that is commonly given for taking the risk-free return to be exogenous is that we can take the risk-free asset to be the numeraire. However, when the risk-free asset is the numeraire, the price of the date-0 consumption good in units of the risk-free asset should reveal exactly the same information that would be revealed by the price of the risk-free asset if the consumption good were the numeraire. A different justification that could be given is that there exists a risk-free production technology that is perfectly elastic. However, investors in aggregate short the risk-free asset in these models with positive probability (due to linearity of demands in normally distributed signals), so this assumption would require that it be possible to run the production technology in reverse, borrowing in aggregate from the future.

Grundy and McNichols (1989), Brown and Jennings (1989), and Brennan and Cao (1996) study dynamic (finite-horizon, discrete-time) versions of the Hellwig model. In those models, investors trade at each date even if there is no additional information other than the asset price, because of updating of expectations from the asset price. This is another possible explanation (in addition to heterogeneous priors) of the large volume of trade observed in actual markets. Grundy and McNichols (1989) show that there may be an equilibrium in which prices in a second round of trade fully reveal the asset value. However, equilibrium prices do not depend on demands in that equilibrium (that is, the equilibrium is paradoxical as discussed at the end of Section 22.3). There is also an equilibrium in which prices at each date reveal additional information about the asset value but are not fully revealing. In such an equilibrium, the current price at any date is not a sufficient statistic for predicting the asset value, so investors condition on the history of prices. This could be interpreted as technical analysis (Brown and Jennings, 1989). Brennan and Cao (1996) show that better-informed investors act as contrarians, selling when the asset price rises, and lesser-informed investors act as momentum traders, buying when the asset price rises. Wang (1993) obtains the same result, for some parameter values. Brennan and Cao (1996) also show that investors would not trade after date 0 (assuming there are no liquidity trades after date 0) if there were a derivative asset the payoff of which is quadratic in the underlying asset value. This is because such an asset is sufficient for implementing the Pareto-optimal sharing rules, similar to Exercise 21.1.

Whether information asymmetry affects a firm's cost of capital (the expected return required by shareholders) is a topic that has received considerable attention, particularly in the accounting literature. Easley and O'Hara (2004) observe that making some private information public will reduce the risk premium of a stock. They conclude that information asymmetry matters for the cost of capital of a firm. However, Lambert, Leuz, and Verrecchia (2011) point out that the shift in information described by Easley and O'Hara (2004) has the effect of increasing the average precision of investors' information. Lambert et al. also point out that, given the average precision, the distribution of information across investors is irrelevant for the risk premium. Thus, it is the average amount of information, not the asymmetry of information, that affects the cost of capital.

EXERCISES

22.1. In the economy of Section 22.4, assume the uninformed investors are risk neutral. Find a fully revealing equilibrium, a partially revealing equilibria in which the price reveals $\tilde{s} + b\tilde{z}$ for any b, and a completely unrevealing equilibrium (an equilibrium in which the price is constant rather than depending on $\tilde{s}$ and/or $\tilde{z}$).

22.2. Consider the model of Section 22.5, but assume there is a continuum of investors indexed by $h \in [0,1]$ with possibly differing risk-aversion coefficients α_h and possibly differing error variances $\text{var}(\tilde{\varepsilon}_h)$. Suppose, for some b, that each investor observes $\tilde{x} + b\tilde{y}$ in addition to her private signal $\tilde{s}_h$. The market clearing condition is

$$\int_0^1 \theta_h(\tilde{x} + b\tilde{y}, \tilde{s}_h)\, dh = \tilde{y},$$

where θ_h is the number of shares demanded by investor h. Let σ_h^2 denote the variance of $\tilde{x}$ conditional on $\tilde{x} + b\tilde{y}$ and $\tilde{s}_h$. Set $\phi_h = 1/\sigma_h^2$. Define

$$\tau = \int_0^1 \tau_h\, dh \quad \text{and} \quad \phi = \frac{1}{\tau}\int_0^1 \tau_h \phi_h\, dh,$$

where τ_h is the risk tolerance of investor h.

(a) Show that the equilibrium price is a discounted weighted average of the conditional expectations of $\tilde{x}$ minus a risk premium term, where the weight on investor h is $\tau_h \phi_h/(\tau \phi)$.

(b) Define

$$\beta = \frac{\text{var}(\tilde{x})}{\text{var}(\tilde{x}) + b^2 \text{var}(\tilde{y})} \quad \text{and} \quad \kappa_h = \frac{(1-\beta)\text{var}(\tilde{x})}{(1-\beta)\text{var}(\tilde{x}) + \text{var}(\tilde{\varepsilon}_h)}.$$

Show that

$$\tau_h \phi_h \kappa_h = \frac{1}{\alpha_h \text{var}(\tilde{\varepsilon}_h)}.$$

(c) Assume the strong law of large numbers holds in the sense that

$$\int_0^1 \tau_h \phi_h \kappa_h \tilde{\varepsilon}_h \, dh = 0.$$

Define

$$\kappa = \frac{1}{\tau \phi} \int_0^1 \tau_h \phi_h \kappa_h \, dh.$$

Show that the equilibrium price equals $a_0 + a_1(\tilde{x} + b\tilde{y})$ if and only if

$$a_0 = \frac{(1-\kappa)[(1-\beta)\bar{x} - \beta b \bar{y}]}{R_f},$$

$$a_1 = \frac{(1-\kappa)\beta + \kappa}{R_f},$$

$$b = -1 \bigg/ \int_0^1 \frac{1}{\alpha_h \operatorname{var}(\tilde{\varepsilon}_h)} \, dh.$$

23

Learning

CONTENTS

23.1 Estimating an Unknown Drift   592
23.2 Portfolio Choice with an Unknown Expected Return   594
23.3 More Filtering Theory   597
23.4 Learning Expected Consumption Growth   603
23.5 A Regime-Switching Model   605
23.6 Notes and References   608

In dynamic models, we always assume investors learn over time—their information at time t is greater than at $t-1$ and so forth. However, we have not previously examined the learning process in detail. In this chapter, we study models in which we can derive specific formulas for beliefs at time u (posterior probabilities) given beliefs at time $t < u$ (prior probabilities) and given information that arrives between t and u. This builds on the discussion of normal-normal updating in Chapter 22, but we will go beyond the case of normal distributions in this chapter.

We will study learning in several different contexts:

- Portfolio choice with learning about features of the return distribution,
- Equilibrium with learning about mean consumption growth,
- Equilibrium with learning about mean consumption growth in a regime-switching model.

This is only an illustrative sample of models of learning in finance; however, each example illustrates interesting consequences of learning for portfolio choice and asset pricing. We begin the chapter with a relatively simple example to demonstrate how normal-normal updating works in continuous time. A fairly general treatment of learning (filtering) theory is given in Section 23.3.

23.1 ESTIMATING AN UNKNOWN DRIFT

Model

A simple example of learning is observing a process[1]

$$dY_t = a(t)\tilde{x}\,dt + dB_t, \qquad (23.1)$$

where $\tilde{x}$ is normally distributed (and unobserved), B is a standard Brownian motion that is independent of $\tilde{x}$, and $a(\cdot)$ is a nonrandom function of time (we use the functional notation $a(t)$ rather than a time subscript to emphasize the nonrandomness). In this example, we learn about $\tilde{x}$ by observing Y. The same information is transmitted by the process H defined by

$$dH_t = \frac{1}{a(t)}dY_t = \tilde{x}\,dt + \frac{1}{a(t)}dB_t. \qquad (23.2)$$

The signal dH is truth plus noise, in analogy with the discrete signals discussed in Section 22.2.

Let $\hat{x}_t$ denote the conditional expectation of $\tilde{x}$ given information at time t (the history of Y through time t). Let $\Sigma(t)$ denote the conditional variance of $\tilde{x}$ given information at time t. As will be made clear, Σ is a nonrandom function of time. The initial conditions are given by the prior distribution of $\tilde{x}$ as $\hat{x}_0 = E[\tilde{x}]$ and $\Sigma(0) = \text{var}(\tilde{x})$.

Conditional Variance

The formulas for sequential projections developed in discrete time in Section 22.2 for truth-plus-noise signals apply straightforwardly to this model. In

---

[1]. This includes the seemingly more general model in which

$$dY_t = b(t)\tilde{x}\,dt + c(t)dB_t,$$

because the same information is carried by the process

$$\frac{1}{c(t)}dY_t = \frac{b(t)}{c(t)}\tilde{x}\,dt + dB_t.$$

In other words, the results for the more general model are obtained by substituting $a = b/c$ in the results that follow.

Learning

analogy with (22.15), the precision of the estimate of $\tilde{x}$—meaning the reciprocal of $\Sigma(t)$—increases by the precision of the signal noise term each time a signal is observed. Here, the variance of the signal noise term is

$$\left(\frac{1}{a(t)} dB_t\right)^2 = \frac{1}{a(t)^2} dt,$$

so the precision is $a(t)^2 dt$, and we have

$$\frac{1}{\Sigma(t)} = \frac{1}{\Sigma(0)} + \int_0^t a(s)^2 ds. \qquad (23.3)$$

Filtering Equation

As in (22.13), the change in the conditional expectation of $\tilde{x}$ is proportional to the innovation in the signal. Here, the innovation in the signal (the unpredicted part) is $dH_t - \hat{x}_t dt$, and we have

$$d\hat{x}_t = \beta(t)[dH_t - \hat{x}_t dt], \qquad (23.4)$$

where, in analogy with (22.15'),

$$\beta(t) \frac{1}{a(t)^2} = \Sigma(t) \quad \Rightarrow \quad \beta(t) = a(t)^2 \Sigma(t). \qquad (23.5)$$

Innovation Process

For a more general treatment, it is convenient to work with a standard Brownian motion rather than with the "truth plus noise" H. So, rather than dividing dY by a to construct H above, we will simply subtract the conditional mean of dY to compute the innovation in Y. This produces

$$dZ_t = dY_t - a(t)\hat{x}_t dt = a(t)(\tilde{x} - \hat{x}_t) dt + dB_t. \qquad (23.6)$$

The process Z is a standard Brownian motion relative to the information obtained by observing Y. It is called the innovation process. We can rewrite the filtering equation (23.4) as

$$d\hat{x}_t = a(t)\Sigma(t) dZ_t. \qquad (23.7)$$

The factor $a\Sigma$ is the conditional covariance of $\tilde{x}$ with the drift $a(\tilde{x} - \hat{x})$ on the right-hand side of (23.6). It can also be called the beta (= covariance/variance) of $\tilde{x}$ with respect to dZ, because Z has a unit standard deviation.

23.2 PORTFOLIO CHOICE WITH AN UNKNOWN EXPECTED RETURN

We will apply the theory of Section 23.1 to describe the optimal portfolio of a CRRA investor when there is a single risky asset with an unknown expected return.[2] Two interesting features of this model are (i) the investor trades as a momentum trader, increasing the portfolio weight on the risky asset when it goes up and decreasing the weight when it falls, and (ii) with reasonable risk aversion (greater than 1) the investor holds less of the risky asset on average than if the expected return were known. The reason for momentum trading is that the investor revises her estimate of the expected return upward, causing the optimal portfolio weight to increase, when there are high returns. The reason for the lower average portfolio weight is that the asset return is positively correlated with the estimated expected return, and the investor wants to hedge against drops in the estimated expected return.

In addition to an unknown expected return, we could also study an unknown risk. However, risks can be estimated much more precisely than expected returns, so it is reasonable to assume risks are known. In fact, with continuous observation of returns, the risk σ in (23.8) below can be calculated exactly from

$$\sigma^2 = \frac{1}{t} \int_0^t \left(\frac{dS_u}{S_u} \right)^2 du$$

for any $t > 0$.[3]

Model

Suppose the dividend-reinvested price S of the risky asset satisfies

$$\frac{dS}{S} = \tilde{\mu} \, dt + \sigma \, dB \qquad (23.8)$$

---

2. The phrase "unknown expected return" requires some clarification. If we consider subjective expectations, then the expected return is always known—it is whatever the investor thinks the return will be on average. The phrase "unknown expected return" is meaningful only if we consider the expected return in an objective sense. If this objective expected return is unknown, then the investor's best estimate of it—calculated as described in Section 23.1—is her subjective expected return.

3. Even if σ in (23.8) were a continuous stochastic process, we could calculate it from a continuous record of returns as

$$\sigma_t^2 = \frac{d}{dt} \int_0^t \left(\frac{dS_u}{S_u} \right)^2 du.$$

This reflects the fact that quadratic variation is a path property; thus, it can be calculated from paths.

Learning

for a random variable $\tilde{\mu}$, constant σ, and standard Brownian motion B. Conditional on $\tilde{\mu}$, S is a geometric Brownian motion. However, suppose the investor does not know $\tilde{\mu}$. Suppose her prior distribution for $\tilde{\mu}$ is normal, and she learns about $\tilde{\mu}$ by observing S.

Define

$$Y_t = \frac{1}{\sigma} \log S_t + \frac{\sigma t}{2} = \frac{1}{\sigma} \left[\log S_0 + \left(\tilde{\mu} - \frac{1}{2}\sigma^2 \right) t + \sigma B_t \right] + \frac{\sigma t}{2},$$

so

$$dY_t = \frac{\tilde{\mu}}{\sigma} dt + dB_t. \qquad (23.9)$$

The process Y provides the same information as S, and it is in the standard form (23.1) of an observation process.

Innovation Process

Let $\hat{\mu}_t$ denote the conditional expectation of $\tilde{\mu}$ given the history of S through date t. Then, as in (23.6), the innovation process is

$$dZ_t = dY_t - \frac{\hat{\mu}_t}{\sigma} dt = \frac{\tilde{\mu} - \hat{\mu}_t}{\sigma} dt + dB_t. \qquad (23.10)$$

Substitute from this equation for dB in (23.8) to obtain

$$\frac{dS}{S} = \hat{\mu}_t \, dt + \sigma \, dZ. \qquad (23.11)$$

This expresses the asset return in terms of the subjective expected return $\hat{\mu}_t$ and in terms of the process Z, which is a standard Brownian motion given the investor's information. This is the same model that we have studied before, except that the learning model here gives us particular dynamics for $\hat{\mu}$. Equation (23.11) reflects what is sometimes called the separation principle: We can first estimate expected returns and risks and then solve for the optimal portfolio in the standard way (albeit with perhaps complicated state processes).

Filtering Equation and Conditional Variance

In our current model, the filtering equation (23.7) is

$$d\hat{\mu}_t = \frac{\Sigma(t)}{\sigma} dZ_t, \qquad (23.12)$$

where, from (23.3), the conditional variance $\Sigma(t)$ of $\tilde{\mu}$ satisfies

$$\frac{1}{\Sigma(t)} = \frac{1}{\Sigma(0)} + \frac{t}{\sigma^2} \quad \Rightarrow \quad \Sigma(t) = \frac{\sigma^2 \Sigma(0)}{\sigma^2 + t\Sigma(0)}. \tag{23.13}$$

From (23.12) we see that changes in the conditional expectation $\hat{\mu}$ are larger when there is more uncertainty about $\tilde{\mu}$ (that is, when Σ is larger) and are smaller when returns are noisier (that is, when σ is larger). From (23.13) we see that learning is also slower when returns are noisier. However, learning eventually becomes nearly complete, in the sense that $\Sigma(t) \to 0$ as $t \to \infty$. This is a perhaps unrealistic feature of the model. It reflects the fact that, given enough observations, we can eventually learn anything that does not change. We can obtain a steady state $\Sigma \neq 0$ by allowing $\tilde{\mu}$ to evolve over time (as in Section 23.4). Finally, notice that, due to (23.11), we can also write the filtering equation as

$$d\hat{\mu}_t = \frac{\Sigma(t)}{\sigma^2}\left(\frac{dS_t}{S_t} - \hat{\mu}_t\, dt\right). \tag{23.14}$$

This shows that the estimate of the objective expected return increases (decreases) whenever the realized return dS/S is greater (less) than the current subjective expected return $\hat{\mu}_t\, dt$.

Value Function

Suppose the investor has constant relative risk aversion ρ, no labor income, and an infinite horizon. Assume there is a constant risk-free rate r. Let C denote the investor's consumption, W the investor's wealth, and $J(w, \hat{\mu}, \Sigma)$ the investor's value function as in Chapter 14. The state variables evolve as (use the dynamics (23.11) of the asset price in terms of the innovation process to derive the dynamics of W)

$$\frac{dW}{W} = \left[r + \pi(\hat{\mu} - r) - \frac{C}{W}\right] dt + \pi\sigma\, dZ, \tag{23.15}$$

$$d\hat{\mu}_t = \frac{\Sigma}{\sigma}\, dZ_t, \tag{23.16}$$

$$d\Sigma = -\frac{\Sigma^2}{\sigma^2}\, dt. \tag{23.17}$$

From Section 14.5, we know that the value function is of the form

$$J(w, \hat{\mu}, \Sigma) = \frac{1}{1-\rho} w^{1-\rho} f(\hat{\mu}, \Sigma) \tag{23.18}$$

for a function $f > 0$ that is determined by the HJB equation.

Myopic Demand

As discussed in Section 14.5, the optimal portfolio consists of the myopic demand plus the hedging demand. The myopic demand is, as always for a CRRA investor,

$$\frac{\hat{\mu} - r}{\rho \sigma^2}.$$

It does not depend on Σ, and it increases when $\hat{\mu}$ increases. Because $\hat{\mu}$ increases when there are high returns, the myopic demand is a momentum strategy—the demand rises following good returns and falls after bad returns.

Hedging Demand

There is no hedging demand for Σ, because it evolves nonrandomly. From (14.24′), the hedging demand for $\hat{\mu}$ is[4]

$$\frac{\Sigma}{\rho \sigma^2} \cdot \frac{\partial \log f(\hat{\mu}, \Sigma)}{\partial \hat{\mu}}. \tag{23.19}$$

The hedging demand has opposite signs in the cases $\rho < 1$ and $\rho > 1$, as discussed in Section 14.5. Assume $\rho > 1$. Then, the investor hedges against adverse changes in the investment opportunity set. When $\hat{\mu} > r$, a drop in the estimated expected return $\hat{\mu}$ is an adverse change in the investment opportunity set. The investor hedges against a drop in $\hat{\mu}$ by holding less of the risky asset, the return of which is positively correlated with changes in $\hat{\mu}$.

23.3 MORE FILTERING THEORY

In this section, we explain how the filtering theory of Section 23.1 extends to a more general model. The extension—even to a quite general model—is straightforward. The difficulty comes in implementing the general model, because usually it is not possible to directly calculate the formulas we will give. We consider two special cases for which calculations are possible: the Kalman filter and a Markov chain with hidden states.

---

4. The variables in (14.24′) are as follows: $-J_w/(w J_{ww})$ is the relative risk aversion of the value function and is here equal to ρ, Σ is the covariance matrix of the asset returns and is here equal to σ^2, J_{wx_j}/J_w is the derivative of $\log J_w$ with respect to the state variable and is here equal to $\partial \log f/\partial \hat{\mu}$, and σv_j is the covariance of dS/S with the change in the state variable and is here equal to Σ.

General Filtering Theory

We consider estimating a stochastic process X rather than a random variable $\tilde{x}$. The initial value of X is an unobserved random variable X_0, and

$$dX_t = C_t\, dt + dM_t \tag{23.20}$$

for a stochastic process C and a martingale M. We do not require that M be a Brownian motion or even an Itô process. The martingale M can even have jumps. Some minor regularity conditions that we need regarding C, M, and the process A in (23.21) below are stated at the end of this subsection. The example of Section 23.1 fits within this framework as $X_0 = \tilde{x}$ and $C = M = 0$.

Assume that investors observe Y given by

$$dY_t = A_t\, dt + dB_t \tag{23.21}$$

for a stochastic process A and a standard Brownian motion B. Neither A nor B is directly observable. Assume B is independent of X (this assumption is relaxed below). On the other hand, assume A depends in some way on X. If it did not, then Y would be useless for estimation. We consider specific examples below. Following common convention, use a hat ˆ to denote all conditional expectations with respect to the information generated by Y.

The innovation process is

$$dZ_t \stackrel{\text{def}}{=} dY_t - \hat{A}_t\, dt = (A_t - \hat{A}_t)\, dt + dB_t, \tag{23.22}$$

in parallel with (23.6). The conditional expected value of X given the information in Y is $\hat{X}$ defined by $\hat{X}_0 = \mathsf{E}[X_0]$ and

$$d\hat{X}_t = \hat{C}_t\, dt + \beta_t\, dZ_t. \tag{23.23}$$

The term $\hat{C}_t\, dt$ is the expected value of the drift in (23.20) and hence is the expected change in X, given the information in Y. The term $\beta_t\, dZ_t$ is an update in the estimate of X based on the information in the innovation process Z (equivalently, the information in Y). The formula for $\hat{\beta}$ is

$$\beta_t = \widehat{(XA)}_t - \hat{X}_t \hat{A}_t. \tag{23.24}$$

This is the conditional covariance of X with A, and it can be called the beta of X with respect to dZ, in parallel with (23.5). As mentioned above, the difficulty in implementing the model is the need to calculate $\hat{A}$, $\hat{C}$, and, especially, β.

The filtering equation (23.23) is a special case of equation (4.12) in Fujisaka, Kallianpur, and Kunita (1972). They do not assume that X and B are independent. Under assumptions specified at the end of this paragraph, they show that the sharp bracket process of M and B (with respect to the information generated by X and B) exists and is absolutely continuous with respect to time—see Section 15.4 for an explanation of the sharp bracket process. This means that it has the form $\langle M, B \rangle_t = \int_0^t \gamma_s \, ds$ for some stochastic process γ. The more general version of the filtering equation (23.23)—stated as equation (4.12) in Fujisaka, Kallianpur, and Kunita—is

$$d\hat{X}_t = \hat{C}_t \, dt + \left(\beta_t + \hat{\gamma}_t\right) dZ_t. \qquad (23.25)$$

The assumptions they make are as follows: A and C are measurable in (t, ω); C is adapted to X and B; A is adapted to X and Y; $\mathsf{E}\int_0^T \mathsf{E}[A_t^2] \, dt < \infty$ for all T; $\mathsf{E}[X_t^2] < \infty$ for all t; $\mathsf{E}\int_0^T \mathsf{E}[A_t^2 X_t^2] \, dt < \infty$ for all T; the σ-field generated by A_s and B_s for $0 \le s \le t$ is independent of the σ-field generated by $B_v - B_u$ for $t < u < v$ for all t.

Kalman Filter

Here, we stay within the realm of normal distributions. We generalize Section 23.1 by estimating a stochastic process X that has normal increments. Assume the process X to be estimated satisfies

$$dX_t = c(t) X_t \, dt + dM_t, \qquad (23.26)$$

and the observation process Y satisfies

$$dY_t = a(t) X_t \, dt + dB_t, \qquad (23.27)$$

where a and c are nonrandom functions of time, X_0 is normally distributed, and M and B are standard Brownian motions with correlation $\rho(t) \ne \pm 1$.

As in (23.22), the innovation process is defined by subtracting the drift from the observation process, producing

$$dZ_t = dY_t - a(t)\hat{X}_t \, dt = a(t)(X_t - \hat{X}_t) \, dt + dB_t, \qquad (23.28)$$

and, from (23.25), the filtering equation is

$$d\hat{X}_t = c(t)\hat{X}_t \, dt + [\beta(t) + \rho(t)] \, dZ_t, \qquad (23.29)$$

where, as in (23.24),

$$\beta(t) = a(t)\widehat{X^2}_t - a(t)\hat{X}_t^2 = a(t)\Sigma(t). \qquad (23.30)$$

The conditional variance $\Sigma(t)$ satisfies the ODE

$$\frac{d\Sigma(t)}{dt} = 2c(t)\Sigma(t) - [a(t)\Sigma(t) + \rho(t)]^2 + 1 \qquad (23.31)$$

with initial condition $\Sigma(0) = \mathrm{var}(X_0)$. We explain how the ODE is derived in the end-of-chapter notes.

We can solve the ODE (23.31) and hence obtain the filtering equation (23.29) in closed form when a, c, and ρ are constants. Assume they are constants, and let ϕ denote the positive root of the quadratic equation

$$a^2\phi^2 + 2(a\rho - c)\phi + \rho^2 - 1 = 0. \tag{23.32}$$

If $\mathrm{var}(X_0) = \phi$, then the ODE (23.31) and initial condition $\Sigma(0) = \mathrm{var}(X_0)$ are satisfied by $\Sigma(t) = \phi$ for all t, because (23.31) implies that $d\Sigma(t)/dt = 0$ when $\Sigma(t) = \phi$ satisfies the quadratic equation (23.32). The parameter ϕ is called the steady-state conditional variance. At that level of the conditional variance, learning from the observation process is exactly offset by unobserved changes in X so that the amount of uncertainty about X remains constant. Set

$$\lambda = 2\sqrt{(a\rho - c)^2 + (1 - \rho^2)a^2},$$

$$\phi = \frac{c - a\rho + \lambda/2}{a^2},$$

$$\theta = \frac{a\rho - c + \lambda/2}{a^2},$$

$$\kappa = \frac{\mathrm{var}(X_0) + \theta}{\phi - \mathrm{var}(X_0)}.$$

Then, ϕ is the positive root of the quadratic equation (23.32) and hence is the steady-state conditional variance. If $\mathrm{var}(X_0) \neq \phi$, then straightforward algebra shows that the solution of the ODE (23.31) and initial condition $\Sigma(0) = \mathrm{var}(X_0)$—and hence the conditional variance—is[5]

$$\Sigma(t) = \frac{\kappa\phi e^{\lambda t} - \theta}{\kappa e^{\lambda t} + 1}. \tag{23.33}$$

As $t \to \infty$, $\Sigma(t) \to \phi$, regardless of the initial value $\mathrm{var}(X_0)$. Thus, the conditional variance converges to its steady-state value as time passes.

There is a multivariate version of the filtering equation and ODE. We can take X, Y, M, and B to be vector processes and take $a(t)$ and $c(t)$ to be matrices. If a and c and the correlation matrix of M and B are constant, then the ODE (23.31) is a system of ODEs of the Riccati type, but there is in general no explicit solution in the multivariate case.

---

5. The formulas for the parameters κ, λ, ϕ, and θ can be derived by substituting (23.33) into the ODE (23.31), matching the coefficients of $e^{2\lambda t}$ on both sides, matching the coefficients of $e^{\lambda t}$ on both sides, matching the constants on both sides, and using the initial condition.

Two-State Markov Chain

This subsection applies the general theory (23.20)–(23.24) to the problem of estimating the state of an unobserved Markov chain. An important feature of this model is that beliefs are most sensitive to new information when there is the most uncertainty about which state the Markov process is in.

A Markov chain is a Markov process X that takes only finitely many values. A two-state chain takes only two values. We denote them by x_0 and x_1. We will say that X is in state 0 when $X_t = x_0$ and in state 1 when $X_t = x_1$. As in Section 15.4, we use the standard notation X_{t-} to denote $\lim_{s \uparrow t} X_s$, meaning the value of the Markov process X just before any possible transition between states at time t. If a transition occurs, then $X_t \neq X_{t-}$. The process X is a jump process, jumping between states.

The Markov property implies that the probability of transiting from state i to state j in a time period (t, u), conditional on being in state i at time t, is independent of how much time the process spent in state i before time t. This memoryless property of the transition time is possessed only by the exponential distribution. The density function of the length of time spent in state i before a transition to state j is $f_i(t) = \lambda_i e^{-\lambda_i t}$ for $t \in [0, \infty)$ and some constant $\lambda_i > 0$. Let F_i denote the distribution function corresponding to the density f_i. The hazard rate is

$$\frac{f_i(t)}{1 - F_i(t)} = \lambda_i.$$

It is common to say that the probability of a transition from state i to state j in an instant dt is $\lambda_i \, dt$.

It is convenient to first construct a Markov chain ξ with states 0 and 1 and then build X from ξ. Let N_0 and N_1 be independent Poisson processes (Section 15.4) with parameters λ_0 and λ_1. Define

$$d\xi_t = (1 - \xi_{t-}) dN_{0t} - \xi_{t-} dN_{1t}. \tag{23.34}$$

Equation (23.34) means that, if we are in state 0 prior to time t ($\xi_{t-} = 0$), then we transition to state 1 at t ($d\xi_t = 1$) if there is an arrival at t for the Poisson process N_0 ($dN_{0t} = 1$). On the other hand, if we are in state 1 prior to time t ($\xi_{t-} = 1$), then we transition to state 0 at t ($d\xi_t = -1$) if there is an arrival at t for the Poisson process N_1 ($dN_{1t} = 1$). Thus, the times between jumps for ξ are exponentially distributed, with parameters λ_0 in state 0 and λ_1 in state 1. Define $X_t = (1 - \xi_t)x_0 + \xi_t x_1$. Then, X_t is a two-state Markov chain with states x_0 and x_1, and the transition probability from state i to state j is λ_i, as desired. Take the conditional expectation of the definition $X_t = (1 - \xi_t)x_0 + \xi_t x_1$ to obtain

$$\hat{X}_t = (1 - \hat{\xi}_t)x_0 + \hat{\xi}_t x_1. \tag{23.35}$$

This shows that X can be estimated by estimating ξ. Because $\hat{\xi}_t$ is the conditional expectation of the $\{0,1\}$-valued random variable ξ_t, it is the conditional probability that $\xi_t = 1$; in other words, it is the conditional probability that we are in state 1.

As discussed in Section 15.4, the compensated Poisson processes M_0 and M_1 defined by $M_{0t} = N_{0t} - \lambda_0 t$ and $M_{1t} = N_{1t} - \lambda_1 t$ are martingales. We can rewrite (23.34) as[6]

$$d\xi_t = [(1-\xi_t)\lambda_0 - \xi_t \lambda_1] \, dt + (1-\xi_{t-}) dM_{0t} - \xi_{t-} dM_{1t}. \qquad (23.36)$$

Equation (23.36) fits the general model (23.20)—with ξ now playing the role of X—with

$$C_t = (1-\xi_t)\lambda_0 - \xi_t \lambda_1,$$

$$M_t = \int_0^t \{(1-\xi_{s-}) dM_{0s} - \xi_{s-} dM_{1s}\}.$$

Assume the observation process Y satisfies

$$dY_t = [(1-\xi_t)a_0 + \xi_t a_1] \, dt + dB_t \qquad (23.37)$$

for constants a_0 and a_1 and a standard Brownian motion B that is independent of the Poisson processes N_0 and N_1. The process Y has drift a_0 when $\xi_t = 0$ and drift a_1 when $\xi_t = 1$. Because Y has different drifts in the two states, observing Y provides information about which state we are in.

Subtracting the expectation of the drift from dY produces the innovation process

$$dZ_t = dY_t - \left[(1-\hat{\xi}_t)a_0 + \hat{\xi}_t a_1\right] dt$$

$$= (a_1 - a_0)(\xi_t - \hat{\xi}_t) \, dt + dB_t. \qquad (23.38)$$

As in (23.23), the filtering equation is

$$d\hat{\xi}_t = \hat{C}_t \, dt + \beta_t \, dZ_t, \qquad (23.39)$$

where, as in (23.24), β_t is the conditional covariance between the drift of Z and ξ_t; namely,

$$\beta_t = (a_1 - a_0)\Sigma_t, \qquad (23.40)$$

---

6. To achieve a simpler notation, we substituted ξ_t for ξ_{t-} in the drift term of (23.36). Because the number of jumps of a Poisson process is countable, the Lebesgue integrals of ξ_t and ξ_{t-} are equal, so the substitution is possible in the drift. The substitution is not possible in the dM_i terms, because those changes are sometimes discrete (equal to 1) rather than infinitesimal.

with Σ_t being the conditional variance $\widehat{\xi_t^2} - \hat{\xi}_t^2$. Because ξ only takes the values 0 and 1, $\xi^2 = \xi$. Thus, $\widehat{\xi_t^2} = \hat{\xi}_t$, and

$$\Sigma_t = \hat{\xi}_t(1 - \hat{\xi}_t).$$

Collecting these formulas, we see that $\hat{\xi}$—the conditional probability of being in state 1—is a Markov process with dynamics

$$d\hat{\xi}_t = \left[(1 - \hat{\xi}_t)\lambda_0 - \hat{\xi}_t\lambda_1\right] dt + (a_1 - a_0)\hat{\xi}_t(1 - \hat{\xi}_t) dZ_t. \quad (23.41)$$

It is useful to note that the instantaneous standard deviation of $\hat{\xi}_t$ is proportional to $\hat{\xi}_t(1 - \hat{\xi}_t)$, which is maximized over $\hat{\xi}_t \in [0, 1]$ at $\hat{\xi}_t = 1/2$, that is, when the states are regarded as equally likely. This is the situation in which investors have the most uncertainty about which state they are in. When uncertainty about the state is highest, beliefs are revised the most in response to new information and hence have the highest standard deviation.

23.4 LEARNING EXPECTED CONSUMPTION GROWTH

This section analyzes a representative investor economy in which the average growth rate of consumption is estimated via the Kalman filter. Beliefs about expected consumption growth are revised as consumption is observed. This makes consumption risk more important than in models studied earlier in this book. Consumption shocks are especially important in this model because they affect not only current consumption but also forecasts of future consumption growth. As a consequence of revisions in beliefs, prices are more sensitive to consumption shocks. Hence, the volatility and risk premium of the market portfolio can be higher than in models without learning, helping to fit the empirical equity premium. This effect of learning on volatility occurs in other circumstances also. For example, quarterly earnings announcements can have a large effect on stock prices because analysts and investors use current earnings growth to forecast future earnings growth.

Model

Assume

$$\frac{dC}{C} = \mu \, dt + \sigma \, dB_1, \quad (23.42a)$$

$$d\mu = \kappa(\theta - \mu) \, dt + \gamma \, dB_2, \quad (23.42b)$$

for constants $\sigma, \kappa, \theta,$ and γ and Brownian motions B_1 and B_2. Thus, the expected consumption growth rate μ is an Ornstein-Uhlenbeck process. Assume the two Brownian motions have a constant correlation ζ. Assume C is observed but μ is not. Assume μ_0 is regarded as normally distributed.

Filter

The innovation process is[7]

$$dZ = \frac{\mu - \hat{\mu}}{\sigma} dt + dB_1. \tag{23.43}$$

Substitute for dB_1 in (23.42a) to obtain

$$\frac{dC}{C} = \hat{\mu}\, dt + \sigma\, dZ. \tag{23.44a}$$

This expresses consumption growth in terms of the subjective expected growth rate $\hat{\mu}$. The filtering equation is

$$d\hat{\mu} = \kappa(\theta - \hat{\mu})\, dt + \gamma\beta\, dZ. \tag{23.44b}$$

Furthermore,

$$\beta = \frac{\Sigma}{\gamma\sigma} + \zeta, \tag{23.45}$$

where Σ is the conditional variance of μ. The differences between (23.42) and (23.44) are (i) the instantaneous standard deviation of the expected growth rate is scaled by β in (23.44), and (ii) consumption growth and the expected growth rate are perfectly correlated in (23.44). The reason for the perfect correlation in (23.44) is that updates in the expected growth rate in (23.44) are based on realized consumption growth. In the steady state, $d\Sigma(t)/dt = 0$, and (Exercise 23.1)

$$\gamma\beta = \sqrt{\kappa^2\sigma^2 + 2\kappa\sigma\gamma\zeta + \gamma^2} - \kappa\sigma. \tag{23.46}$$

Risk Premium in Steady State

Assume the representative investor has constant relative risk aversion ρ. In the steady state, the market price-dividend ratio is a strictly increasing function $f(\hat{\mu})$, and the risk premium of the market portfolio is (Exercise 15.1)

---

[7]. We leave it as an exercise to rewrite the model in terms of an unobserved process X and observation process Y with unit standard deviations and to derive the facts stated here from the results of Section 23.3.

$$\rho\sigma^2 + \frac{f'(\hat{\mu})}{f(\hat{\mu})}\rho\sigma\gamma\beta. \qquad (23.47)$$

If investors knew the expected consumption growth rate μ, then the market price-dividend ratio would be a function $g(\mu)$, and the risk premium of the market portfolio would be

$$\rho\sigma^2 + \frac{g'(\mu)}{g(\mu)}\rho\sigma\gamma\zeta. \qquad (23.48)$$

We can calculate (23.47) and (23.48) by solving ODEs for f and g (Exercise 15.4). If $\zeta = 0$ (consumption and expected consumption growth are locally uncorrelated), then the market risk premium is $\rho\sigma^2$ when expected consumption growth is known, but (23.47) shows that the market risk premium is larger than $\rho\sigma^2$ when investors learn about expected consumption growth from realized consumption. More generally, the fact that $\beta > \zeta$ suggests that the risk premium is higher when investors have to learn μ by observing consumption.

23.5 A REGIME-SWITCHING MODEL

In the previous section, the mean consumption growth rate follows an Ornstein-Uhlenbeck process. This section considers a similar economy but with the mean consumption growth rate following a two-state Markov chain. The economy switches between a high-growth and a low-growth regime. The representative investor estimates which regime the economy is in by observing consumption. This produces stochastic volatility. Volatility tends to be higher when the representative investor is less certain about which regime the economy is in.

Assume aggregate consumption evolves as

$$\frac{dC_t}{C_t} = \mu_t\, dt + \sigma\, dB_t, \qquad (23.49)$$

where σ is a constant and B is a Brownian motion. Assume μ follows a two-state Markov chain with states m_0 and m_1 and transition parameters λ_0 and λ_1. Assume the Markov chain is independent of the Brownian motion B. Assume there is a representative investor with constant relative risk aversion ρ who estimates μ_t by observing aggregate consumption. As in the discussion of the two-state Markov chain in Section 23.3, let ξ denote the state indicator process, taking the value 0 when $\mu_t = m_0$ and the value 1 when $\mu_t = m_1$.

Observation Process

Define
$$Y_t = \frac{1}{\sigma}\log C_t + \frac{\sigma}{2}t,$$

so
$$dY_t = \frac{\mu_t}{\sigma}dt + dB_t$$
$$= \frac{(1-\xi_t)\mu_0 + \xi_t\mu_1}{\sigma}dt + dB_t.$$

The process Y provides the same information as C, and it is in the standard form of an observation process.

Filtering

The innovation process is
$$dZ_t = dY_t - \frac{(1-\hat{\xi}_t)\mu_0 + \hat{\xi}_t\mu_1}{\sigma}dt = \frac{(\mu_1 - \mu_0)(\xi_t - \hat{\xi}_t)}{\sigma}dt + dB_t.$$

The filtering equation is
$$\hat{\mu}_t = (1-\hat{\xi}_t)m_0 + \hat{\xi}_t m_1,$$

where
$$d\hat{\xi}_t = [(1-\hat{\xi}_t)\lambda_0 - \hat{\xi}_t\lambda_1]dt + \frac{m_1 - m_0}{\sigma}\hat{\xi}_t(1-\hat{\xi}_t)dZ_t. \quad (23.50)$$

Market Price-Dividend Ratio

Let P_t denote the price of the market portfolio (the claim to aggregate consumption). Using the representative investor's MRS as an SDF process as usual, we can calculate the price of the market portfolio conditional on the history of consumption and conditional on being in state $i \in \{0, 1\}$ at any date t as

$$E\left[\int_t^\infty e^{-\delta(u-t)}\left(\frac{C_u}{C_t}\right)^{-\rho}C_u\,du\,\bigg|\,(C_s)_{s\leq t}, \xi_t = i\right]$$
$$= C_t E\left[\int_t^\infty e^{-\delta(u-t)}\left(\frac{C_u}{C_t}\right)^{1-\rho}du\,\bigg|\,(C_s)_{s\leq t}, \xi_t = i\right]. \quad (23.51)$$

It follows from the specification (23.49) and Itô's formula that

$$\frac{C_u}{C_t} = e^{\int_t^u (\mu_s - \sigma^2/2)\,ds + \sigma(B_u - B_t)}.$$

This depends on the path of μ between t and u and on the innovation $B_u - B_t$. The distribution of those variables conditional on information at time $t < u$ depends entirely on the state at time t because of the Markovian hypothesis. Therefore, the conditional expectation in (23.51) does not depend on the past $(C_s)_{s \leq t}$, and we can define

$$\pi_i = E\left[\int_t^\infty e^{-\delta(u-t)} \left(\frac{C_u}{C_t}\right)^{1-\rho} du \,\bigg|\, (C_s)_{s \leq t}, \xi_t = i\right]. \quad (23.52)$$

In our model, the representative investor does not know the state at time t. So, the market price-dividend ratio is

$$\frac{P_t}{C_t} = E\left[\int_t^\infty e^{-\delta(u-t)} \left(\frac{C_u}{C_t}\right)^{1-\rho} du \,\bigg|\, (C_s)_{s \leq t}\right].$$

Use iterated expectations—conditioning first on the state—and (23.52) to compute the price-dividend ratio as

$$\frac{P_t}{C_t} = \pi_0 \cdot \text{prob}\left(\xi_t = 0 \mid (C_s)_{s \leq t}\right) + \pi_1 \cdot \text{prob}\left(\xi_t = 1 \mid (C_s)_{s \leq t}\right)$$

$$= (1 - \hat{\xi}_t)\pi_0 + \hat{\xi}_t \pi_1. \quad (23.53)$$

Volatility of the Market Return

From (23.50) and (23.53), the differential of the price-dividend ratio is

$$(\pi_1 - \pi_0) d\hat{\xi}_t = (\pi_1 - \pi_0)[(1 - \hat{\xi}_t)\lambda_0 - \hat{\xi}_t \lambda_1]\,dt$$

$$+ \frac{(\pi_1 - \pi_0)(\mu_1 - \mu_0)}{\sigma} \hat{\xi}_t(1 - \hat{\xi}_t)\,dZ_t. \quad (23.54)$$

It follows that the volatility of the market return $(dP + C\,dt)/P$ is[8]

$$\sigma + \frac{(\pi_1 - \pi_0)(\mu_1 - \mu_0)}{\sigma} \cdot \frac{\hat{\xi}_t(1 - \hat{\xi}_t)}{(1 - \hat{\xi}_t)\pi_0 + \hat{\xi}_t \pi_1} > \sigma. \quad (23.55)$$

8. To make this calculation, use the fact that $P = C \times (P/C)$, so

$$\frac{dP}{P} = \frac{dC}{C} + \frac{d(P/C)}{P/C} + \left(\frac{d(P/C)}{P/C}\right)\left(\frac{dC}{C}\right).$$

The inequality in (23.55) is due to the fact that the market price-dividend ratio conditioning on the state is higher in the state with higher consumption growth (so $\pi_1 - \pi_0$ has the same sign as $\mu_1 - \mu_0$). The numerator $\hat{\xi}_t(1 - \hat{\xi}_t)$ in the second factor in (23.55) is largest when $\hat{\xi}_t = 1/2$. Thus, market volatility tends to be higher when investors are less certain about the state. The denominator in the same factor is smallest when investors are confident they are in the worse state, so the peak of market volatility occurs when the conditional probability of the worse state exceeds $1/2$. This is consistent with the empirical relation known as the leverage effect (volatility tends to rise following low returns).

23.6 NOTES AND REFERENCES

Standard references for filtering theory are Kallianpur (1980) and Liptser and Shiryaev (2000). The text by Xiong (2008) is also worth consulting. Early applications of filtering theory to finance include Detemple (1986, 1991), Dothan and Feldman (1986), and Gennotte (1986). Among other things, these papers show that many results in finance that assume known parameters also apply when the parameters are unknown if we replace the unknown parameters by the best estimates of them (for example, replacing an unknown objective expected return with its expected value, which can be interpreted as the subjective expected return).

The analysis in Section 23.2 of portfolio choice when the expected return is unknown is based on Brennan (1998). An important extension is provided by Xia (2001), who models the expected return as being linear in a vector of state variables, with the linear coefficients being unknown. In Xia's model, investors learn about predictability of the market risk premium. Lakner (1998) provides a general characterization of optimal portfolios for maximizing the expected utility of terminal wealth when the drift processes of n risky assets are unobserved, including solutions when the drifts form a vector Ornstein-Uhlenbeck process. Brendle (2006) provides an analytic solution for the portfolio choice problem in Exercise 23.4 for a CRRA investor who maximizes the expected utility of terminal wealth.

The model in Section 23.4 is a simplified version of the model in Brennan and Xia (2001). Brennan and Xia distinguish between aggregate dividends and aggregate consumption. They allow for learning about the expected dividend growth rate from both realized dividends and realized consumption. They show that the model can explain the equity premium, risk-free rate, and excess volatility puzzles. Related work includes Timmermann (1993), who analyzes a discrete-time model with a constant expected consumption growth rate and shows that it can explain the excess volatility puzzle. As discussed in Section 23.4, consumption risk is more important in these models (hence, volatility of the market return is higher) because consumption shocks signal expected future consumption growth.

Versions of the regime-switching model in Section 23.5 are studied by David (1997, 2008a,b), Veronesi (1999, 2000), and David and Veronesi (2013, 2014). The model generates a number of phenomena that do not appear in the model with normal distributions (the Kalman filter). For example, it produces a time-varying correlation between the market return and its volatility. When investors believe the low-growth state is more likely, a positive consumption shock generates a positive return and an increase in uncertainty (because investors become less certain about the state), but when the high-growth state is perceived as more likely, a positive shock generates a positive return and a decrease in uncertainty. Veronesi (2000) analyzes an n-state model with transition probabilities as described in Exercise 23.5.

Learning is frequently a component of models of heterogeneous beliefs. Detemple and Murthy (1994, 1997), Zapatero (1998), Basak (2000), Basak and Croitoru (2000, 2006), and Buraschi and Jiltsov (2006) all study general equilibrium with two classes of agents who learn via filtering theory and have heterogeneous priors. Scheinkman and Xiong (2003) and Dumas, Kurshev, and Uppal (2009) study models in which investors disagree about the interpretation of signals—they learn differently. When an investor reacts too strongly to a signal, it can be regarded as due to overconfidence in the informativeness of the signal. Scheinkman and Xiong show that speculative bubbles as discussed in Section 21.5 arise due to overconfidence. Dumas et al. analyze the optimal response of rational investors (who do not overreact) to the presence of other investors who are overconfident.

Ordinarily, riskier assets should be worth less than safer assets, ceteris paribus, but there are exceptions if the risk involves uncertainty about a parameter that affects returns over multiple periods. For a simple example, suppose an investment of \$1 will grow to $1+g$ dollars in one period and then to $(1+g)^2$ dollars after two periods, at which time the investment can be sold and the value realized. Because the function $g \mapsto (1+g)^2$ is convex, Jensen's inequality implies that $\mathsf{E}[(1+\tilde{g})^2] > (1+\bar{g})^2$, where $\bar{g}$ denotes the mean of a random $\tilde{g}$. Thus, a risk-neutral investor will value an investment with a risky growth rate $\tilde{g}$ higher than an investment with a known growth rate $\bar{g}$, and the same is true for risk-averse investors who are not too risk averse. This issue is explored by Grinblatt and Linnainmaa (2011). The same phenomenon is studied by Pastor and Veronesi (2003) in a model in which investors learn over time about the growth rate of firm cash flows. They show that young firms (about which investors know less and which have longer to grow before becoming "mature") should have higher market-to-book ratios, a relationship that is present in the data.

The reason that explicit calculations are possible in the Kalman filter and for the two-state Markov chain (but not in general) is as follows: The filtering equation to calculate the conditional mean of a process X involves a conditional covariance, as shown in (23.23) and (23.24). We can derive a filtering equation for the conditional covariance of A and X in (23.24) by applying the filtering equation to compute the conditional mean of AX. However, this filtering equation for a second moment involves a conditional third moment. In general, this process does not terminate—computing each moment requires knowing the next-higher (unknown) moment. However, the third central moment of a normal distribution is known to be zero. Thus, with normal distributions, the process terminates once we derive the filtering equation for the conditional variance. Likewise, it is simple to apply the filtering theory to calculate the conditional expectation of a $\{0,1\}$–valued process like the state indicator ξ_t in Section 23.3, because the square of such a process equals itself. Hence, again, the sequence of successively higher moments terminates.

To illustrate this phenomenon, we derive the ODE (23.31) for the Kalman filter. Take a and c to be constant and $\rho = 0$ for simplicity. The same reasoning applies in general, and the same reasoning produces the system of ODEs in the multivariate case. To compute the conditional variance, apply the general theory to the problem of estimating X_t^2. We have

$$dX_t^2 = 2X_t \, dX_t + (dX_t)^2 = 2cX_t^2 \, dt + 2X_t \, dM_t + dt,$$

which fits in the general model (with X_t^2 now being the process to be estimated) with $C_t = 2cX_t^2 + 1$ and with dM redefined as $2X_t \, dM_t$. Recall that the observation process in this model satisfies

$$dY_t = aX_t \, dt + dB_t,$$

which fits in the general model as $A_t = aX_t$. Thus,

$$d\widehat{X^2}_t = [2c\widehat{X^2}_t + 1] \, dt + \zeta_t \, dZ_t,$$

where the conditional covariance (beta) is ζ_t defined by

$$\zeta_t = a\widehat{X^3}_t - a\widehat{X^2}_t \hat{X}_t.$$

The conditional variance of X_t evolves as

$$d\Sigma(t) = d\widehat{X^2}_t - d\hat{X}_t^2$$
$$= d\widehat{X^2}_t - 2\hat{X}_t\, d\hat{X}_t - (d\hat{X}_t)^2$$
$$= [2c\widehat{X^2}_t + 1]\,dt + \zeta_t\,dZ_t - 2\hat{X}_t[c\hat{X}_t\,dt + \beta(t)\,dZ_t] - \beta(t)^2\,dt$$
$$= [2c\Sigma(t) - a^2\Sigma(t)^2 + 1]\,dt + [a\widehat{X^3}_t - a\widehat{X^2}_t\hat{X}_t - 2a\hat{X}_t\Sigma(t)]\,dZ_t$$
$$= [2c\Sigma(t) - a^2\Sigma(t)^2 + 1]\,dt.$$

This is the ODE (23.31). The first equality in the above chain is the definition of the conditional variance, the second follows from Itô's formula, the third is obtained by substituting, the fourth is obtained by substituting, and the last follows from the fact that X_t is conditionally normally distributed given information at t—so its third central moment is zero, which implies that its third noncentral moment is $\widehat{X^3}_t = \hat{X}_t\widehat{X^2}_t + 2\hat{X}_t\Sigma(t)$.

EXERCISES

23.1. In the model of Section 23.4, define

$$X = \frac{\mu - \theta}{\gamma},$$

$$Y = \left(\frac{\sigma}{2} - \frac{\theta}{\sigma}\right)t + \frac{\log C}{\sigma}.$$

Write down the innovation process, filtering equation, and ODE for the conditional variance of X. Explain why these imply (23.43)–(23.46).

23.2. In the model of Section 23.4, assume $\zeta = 0$ (that is, C and μ are locally uncorrelated).

(a) Show that $\beta < 1$. This implies that $\hat{\mu}$ has a lower standard deviation than does μ. Intuitively, why is this the case?
(b) Show that β is a decreasing function of $\kappa\sigma$. Intuitively, why should the standard deviation of $\hat{\mu}$ be smaller when κ is larger or σ is larger?
(c) Show that $\lim_{\kappa\sigma \to \infty} \beta = 0$.

23.3. For the model of Section 23.4, derive the ODE that the market price-dividend ratio satisfies.

23.4. Assume there is a single risky asset with dividend-reinvested price S satisfying
$$\frac{dS}{S} = \mu \, dt + \sigma \, dB_1,$$
where
$$d\mu = \kappa(\theta - \mu) \, dt + \gamma \, dB_2$$
with σ, κ, θ, and γ being constants and where B_1 and B_2 are Brownian motions with constant correlation ρ. Assume the risk-free rate is constant. Assume an investor observes S but does not observe μ. Assume μ_0 is regarded as normally distributed.

(a) Adapt the analysis of Section 23.4 to write dS/S and $d\mu$ in terms of the innovation process.

(b) Refer to Exercise 15.3 and discuss how the unobservability of μ affects the optimal portfolio of a CRRA investor.

23.5. Consider the model of Section 23.5 but assume there are n possible states. Label them as $\{1, \ldots, n\}$. Let N_i be independent Poisson processes with parameters λ_i, for $i = 1, \ldots, n$. Assume the state X_t evolves as
$$dX_t = \sum_{i=1}^{n}(i - X_{t-}) \, dN_{it}.$$

(a) Suppose the economy is in state j just before time t ($X_{t-} = j$). What is the probability of transiting to state $i \neq j$ in an instant dt? Does it depend on j?

(b) Let π_{it} denote the conditional probability that the economy is in state i at time t, for $i = 1, \ldots, n$. Define $X_{it} = 1_{\{X_t = i\}}$. Note that π_{it} is also the conditional probability that the two-state Markov chain X_{it} is in state 1. Write down the dynamics (filtering equation) of π_{it}.

(c) Write down a formula for the market price-dividend ratio in terms of the probabilities π_{it}.

23.6. Verify that $\Sigma(t)$ defined in (23.33) satisfies the ODE (23.31) with initial condition $\Sigma(0) = \text{var}(X_0)$.

24

Information, Strategic Trading, and Liquidity

CONTENTS

24.1 Glosten-Milgrom Model	614
24.2 Kyle Model	616
24.3 Glosten Model of Limit Order Markets	620
24.4 Auctions	624
24.5 Continuous-Time Kyle Model	632
24.6 Notes and References	642

Traders measure their transaction costs in four buckets: commissions, bid-ask spreads, price impacts, and opportunity costs. Commissions are payments to securities brokers for executing trades. The bid-ask spread is the difference between the ask quote and the bid quote, the former being the price at which a trader can buy with a market order (an order to trade that requests execution at the market price) and the latter being the price at which a trader can sell with a market order. For traders who want to trade large quantities, price impacts and opportunity costs are the largest costs. Price impact is how much the market has moved between the time a decision to trade is made and a trade is executed. Opportunity costs are lost profits on trades that were never executed because the market moved too far and the trader decided to cancel her remaining plans to trade.

Traders who are concerned with the price impacts of their trades are called strategic. They are not price takers. Studying strategic traders takes us outside the competitive paradigm that has been employed in all previous parts of this book. Large investors in actual markets are well aware that their trades move prices, and they exert a great deal of effort in attempting to minimize adverse price impacts. So, strategic trading is an important topic.

The magnitudes of bid-ask spreads, price impacts, and opportunity costs depend on the liquidity of the security being traded. Or, rather, they are the same thing as liquidity. If a security is very liquid, then it can be traded at low cost. If it is illiquid, costs are high. This is effectively a definition of liquidity.[1] One cause of illiquidity is asymmetric information. Asymmetric information gives rise to adverse selection. If you offer to trade at a given price, your offer is more likely to be accepted (selected) if other investors have information that the price is favorable to them and consequently unfavorable (adverse) to you.

This chapter is about asymmetric information models of liquidity. It consists of elaborations of the explanation of liquidity given by Treynor (1971)—published under the pseudonym Walter Bagehot—who observes that "the liquidity of a market ... is inversely related to the average rate of flow of new information ... and directly related to the volume of liquidity-motivated transactions." As elsewhere in this part of the book, we only try to describe some basic models. These models are part of the branch of finance called market microstructure. However, many issues in market microstructure (prominently including high-frequency trading, dark pools, make-or-take fees, over-the-counter markets, etc.) are entirely omitted from this chapter.

Section 24.4 on auctions differs a bit from the rest of the chapter. In the models in the rest of the chapter, prices are set by uninformed traders (market makers) who face adverse selection from informed traders. Section 24.4 considers a trade in which the price is set by informed individuals bidding against one another. Competition by informed traders who have common values produces the winner's curse, which is a phenomenon with which any student of markets should be familiar. Furthermore, auctions are widely used to sell new issues of securities (for example, government bonds) and are an important prototype for other models in which price-quoting agents are informed and compete (for example, dealer markets).

24.1 GLOSTEN-MILGROM MODEL

Glosten and Milgrom (1985) study a model in which risk-neutral competitive market makers set bid and ask quotes for transacting a single unit of an asset with a trader who submits a market order. The market-order trader may be informed or she may be trading for other reasons. If trading for other reasons—for example, selling the asset because she is in need of cash for some other purpose—then the

---

[1]. Liquidity has been described as being similar to pornography in the sense that, as famously stated by U.S. Supreme Court Justice Potter Stewart, it is hard to define but we know it when we see it (O'Hara, 1995).

trader is commonly called a liquidity trader, consistent with Treynor's reference to liquidity-motivated transactions.

In the simplest setting, the asset has only two possible values, $L < H$, and the informed trader knows the value and buys if the value is H and sells if the value is L, and the liquidity trader buys or sells with equal probabilities regardless of the quotes (her demands are price inelastic). Adopt the following notation:

$p = \text{prob}(H),$

$\mu = \text{prob}(\text{Informed Order}),$

$\pi_b = \text{prob}(\text{Buy Order}) = p\mu + (1-\mu)/2,$

$\pi_s = \text{prob}(\text{Sell Order}) = (1-p)\mu + (1-\mu)/2,$

$p_b = \text{prob}(H \mid \text{Buy Order})$
$= \dfrac{\text{prob}(\text{Buy Order} \mid H) \cdot \text{prob}(H)}{\text{prob}(\text{Buy Order})} = \dfrac{[\mu + (1-\mu)/2]p}{\pi_b},$

$p_s = \text{prob}(H \mid \text{Sell Order}) = \dfrac{\text{prob}(\text{Sell Order} \mid H) \cdot \text{prob}(H)}{\text{prob}(\text{Sell Order})} = \dfrac{(1-\mu)p/2}{\pi_s}.$

Note that beliefs are revised in the directions of orders in the sense that the probability of H increases with a buy order and the probability of L increases with a sell order—that is, $p_b > p > p_s$. The expected value of the asset conditional on a buy order is

$$A \stackrel{\text{def}}{=} \mathsf{E}[\tilde{v} \mid \text{Buy Order}] = p_b H + (1 - p_b)L. \qquad (24.1)$$

The expected value conditional on a sell order is

$$B \stackrel{\text{def}}{=} \mathsf{E}[\tilde{v} \mid \text{Sell Order}] = p_s H + (1 - p_s)L. \qquad (24.2)$$

Competition between risk-neutral market makers forces the quotes to the expected value of the asset conditional on the order. Thus, (24.1) is the ask price (the price that dealers ask in exchange for selling the asset) and (24.2) is the bid price (the price that dealers bid for the asset). The bid and the ask are regret-free quotes, in the sense that they anticipate the adverse selection of informed market order traders. Each quote already incorporates the information provided by a market order that hits it.

The difference between the bid and the ask is the bid-ask spread $A - B = (p_b - p_s)(H - L)$. The spread is larger in this model when the high and low values H and L are further apart and when orders are more informative in the sense that the posterior probabilities p_b and p_s are further apart. Thus, the market is less liquid when private information is larger and/or orders are more informative. Notice that

$$\pi_b A + \pi_s B = pH + (1-p)L. \qquad (24.3)$$

The left-hand side is the expected transaction price (an expectation of a conditional expectation of the asset value), and the right-hand side is the unconditional expectation of the asset value. Thus, (24.3) is an example of the law of iterated expectations.

In a dynamic version of the model, transaction prices form a martingale, because the transaction price (ask or bid) is always the conditional expectation of the asset value, given the information at the time, and a sequence of conditional expectations is always a martingale. Because transaction prices form a martingale, changes in transaction prices are a martingale difference series and hence are uncorrelated.[2] Changes in transactions prices are correlated in some other models of the bid-ask spread. See the end-of-chapter notes.

24.2 KYLE MODEL

In the Kyle model, there is a strategic informed trader. Illiquidity in the model is measured by what is called Kyle's lambda. Kyle's lambda depends in a very simple way on model parameters that measure private information and liquidity trading. The market is more liquid when there is less private information and/or more liquidity trading.

Model

There are two dates, 0 and 1. The asset is traded with asymmetric information at date 0, and the asset value $\tilde{v}$ is realized at date 1. Assume $\tilde{v}$ is normally distributed and observed by a single risk-neutral informed trader prior to trade at date 0.[3] After observing $\tilde{v}$, the informed trader submits a market order $\tilde{x}$. There are also uninformed (liquidity) trades represented by a random variable $\tilde{z}$ that is normally

---

2. Given dates $s < t < u$ and a martingale M,

$$\text{cov}(M_u - M_t, M_t - M_s) = \mathsf{E}[(M_u - M_t)(M_t - M_s)] = \mathsf{E}[(M_t - M_s)\mathsf{E}_t[M_u - M_t]] = 0.$$

3. Alternatively, we can assume the informed trader observes a signal $\tilde{s}$ that is joint normally distributed with $\tilde{v}$. Because of risk neutrality, the residual risk $\tilde{v} - \mathsf{E}[\tilde{v} \mid \tilde{s}]$ is irrelevant for prices, and all of the results of this section hold if we substitute $\mathsf{E}[\tilde{v} \mid \tilde{s}]$ for $\tilde{v}$ and substitute the variance of $\mathsf{E}[\tilde{v} \mid \tilde{s}]$ for the variance of $\tilde{v}$.

distributed and independent of $\tilde{v}$. No generality is lost in assuming the mean of $\tilde{z}$ is zero.[4]

Market makers observe $\tilde{y} \stackrel{\text{def}}{=} \tilde{x} + \tilde{z}$. This is the aggregate order. If $\tilde{x}$ and $\tilde{z}$ have opposite signs (one is buying and the other selling), then the interpretation is that they trade with each other and only the residual is seen by market makers. However, all trades take place at the same price, which is the price set by market makers after observing $\tilde{x} + \tilde{z}$. Use a risk-free asset as the numeraire, and assume market makers are risk neutral and compete in a Bertrand fashion to fill the aggregate order. Denote the equilibrium price by $p(\tilde{y})$.

Equilibrium

An equilibrium of this model is an informed order $\tilde{x}$ depending on $\tilde{v}$ and a price function p satisfying

$$p(\tilde{y}) = \mathsf{E}[\tilde{v} \mid \tilde{y}], \qquad (24.4\text{a})$$

$$\tilde{x} \in \text{argmax}_x \, x\bigl(\tilde{v} - \mathsf{E}[p(x + \tilde{z})]\bigr). \qquad (24.4\text{b})$$

The first condition states that the price equals the expected asset value conditional on the information in the aggregate order. The second condition states that the informed trader maximizes her conditional expected gain from trade, understanding that her order affects the price.

An equilibrium is said to be linear if there are constants δ, λ, α, and β such that $p(y) = \delta + \lambda y$ and $\tilde{x} = \alpha + \beta \tilde{v}$. Denote the standard deviations of $\tilde{v}, \tilde{z}$, and $\tilde{y}$ by σ_v, σ_z, and σ_y respectively. Denote $\mathsf{E}[\tilde{v}]$ by $\bar{v}$. There is a unique linear equilibrium given by

$$\delta = \bar{v}, \qquad (24.5\text{a})$$

$$\lambda = \frac{\sigma_v}{2\sigma_z}, \qquad (24.5\text{b})$$

$$\alpha = -\delta\beta, \qquad (24.5\text{c})$$

$$\beta = \frac{1}{2\lambda}. \qquad (24.5\text{d})$$

We verify that this is the unique linear equilibrium at the end of the section.

---

4. If the mean $\bar{z}$ of $\tilde{z}$ is nonzero, then all of the results in this section hold when we replace $\tilde{z}$ by $\tilde{z} - \bar{z}$. The irrelevance of the mean of $\tilde{z}$ is a consequence of market makers being risk neutral. This is in contrast to models studied in Chapter 22, in which all investors are risk averse and the mean of liquidity trades affects prices, because it affects the amount of risk other investors must bear in equilibrium.

Kyle's Lambda

The parameter λ is universally denoted by this symbol. In fact, it is universally known as Kyle's lambda. It measures the impact on the equilibrium price of a unit order. Its reciprocal is the size of the trade that can be made with a unit impact on the price. A market in which large trades can be made with only a small price impact is called a deep (or liquid) market, so $1/\lambda$ measures the depth of the market: If $1/\lambda$ is larger, then the market is deeper. Note that a market is deeper if there is less private information in the sense of σ_v being smaller or if there is more liquidity trading in the sense of σ_z being larger. Thus, the formula for lambda is consistent with Treynor's description of liquidity quoted in the introduction to the chapter.

Information Revelation

The variance of $\tilde{v}$ measures the ex ante informational advantage of the informed trader. For example, if σ_v is large, then it will frequently be the case that the informed trader has an important informational advantage, in the sense that her estimate $\tilde{v}$ of the asset value is quite far from the value $\bar{v}$ perceived ex ante by market makers. The information revealed to market makers by the order flow $\tilde{y}$ in the linear equilibrium is very simply described: The variance of $\tilde{v}$ conditional on $\tilde{y}$ is half of the unconditional variance. Moreover, the equilibrium price, because it is affine in $\tilde{y}$, reveals the same information. Thus, the market at large learns half of the private information of the informed trader. This is verified below.

Value of Private Information

Notice that the equilibrium strategy of the informed trader is $\tilde{x} = \beta(\tilde{v} - \bar{v})$. The unconditional expected gain of the informed trader is

$$E\left[\tilde{x}[\tilde{v} - p(\tilde{x} + \tilde{z})]\right] = \beta E\left[(\tilde{v} - \bar{v})[\tilde{v} - \delta - \lambda\beta(\tilde{v} - \bar{v}) - \lambda\tilde{z}]\right]$$
$$= \beta(1 - \lambda\beta)\sigma_v^2$$
$$= \frac{\sigma_v \sigma_z}{2}.$$

Thus, the informed trader's expected profit is higher when she has more private information or when there is more liquidity trading. The expected gain of market makers is zero, because the price at which they trade is the conditional expected value of the asset. Thus, the expected profits of the informed traders are expected losses for the liquidity traders. In fact, the expected gain of liquidity traders is

$$\mathsf{E}\left[\tilde{z}[\tilde{v} - p(\tilde{x} + \tilde{z})]\right] = -\lambda\sigma_z^2 = -\frac{\sigma_v \sigma_z}{2}.$$

The liquidity traders are presumably willing to accept these losses, due to their unmodeled motives for trading.

We want to establish that (24.4) is the unique linear equilibrium, and we want to verify the statements made above about information revelation. Suppose the informed trade is $\tilde{x} = \alpha + \beta\tilde{v}$ for some α and β. Then,

$$\mathsf{E}[\tilde{v}\,|\,\tilde{x} + \tilde{z}] = \mathsf{E}[\tilde{v}\,|\,\beta\tilde{v} + \tilde{z}] = \bar{v} + \frac{\mathrm{cov}(\tilde{v}, \beta\tilde{v} + \tilde{z})}{\mathrm{var}(\beta\tilde{v} + \tilde{z})}(\beta\tilde{v} - \beta\bar{v} + \tilde{z})$$

$$= \bar{v} - \frac{\beta\sigma_v^2}{\beta^2\sigma_v^2 + \sigma_z^2}(\alpha + \beta\bar{v})$$

$$+ \frac{\beta\sigma_v^2}{\beta^2\sigma_v^2 + \sigma_z^2}(\alpha + \beta\tilde{v} + \tilde{z}).$$

Thus, in a linear equilibrium, we must have

$$\lambda = \frac{\beta\sigma_v^2}{\beta^2\sigma_v^2 + \sigma_z^2}, \tag{24.6a}$$

$$\delta = \bar{v} - \lambda(\alpha + \beta\bar{v}). \tag{24.6b}$$

On the other hand, if $p(y) = \delta + \lambda y$ for any δ and λ, then the informed trader's optimization problem is to maximize

$$\tilde{x}\tilde{v} - \tilde{x}\left[\delta + \lambda\tilde{x} + \lambda\mathsf{E}[\tilde{z}]\right] = (\tilde{v} - \delta)\tilde{x} - \lambda\tilde{x}^2.$$

There is a solution to this problem only if $\lambda > 0$, and, in that case, the solution is

$$\tilde{x} = \frac{\tilde{v} - \delta}{2\lambda}.$$

Thus, in a linear equilibrium, we must have

$$\lambda > 0, \tag{24.6c}$$

$$\alpha = \frac{-\delta}{2\lambda}, \tag{24.6d}$$

$$\beta = \frac{1}{2\lambda}. \tag{24.6e}$$

We will show that the unique solution of the system (24.6) is (24.5).

Substitute (24.6a) into (24.6e) to obtain

$$\beta = \frac{1}{2}\left(\frac{\beta^2\sigma_v^2+\sigma_z^2}{\beta\sigma_v^2}\right),$$

so $\beta^2 = \sigma_z^2/\sigma_v^2$. From $\lambda > 0$ and (24.6e), we have $\beta > 0$. Hence, $\beta = \sigma_z/\sigma_v$ as claimed in (24.5d). Now, substitute β into (24.6e) to obtain the formula claimed for λ, and substitute β and λ into (24.6b) and (24.6d) to obtain the formulas claimed for α and δ.

The remaining task is to compute the variance of $\tilde{v}$ conditional on $\tilde{y}$. According to (22.5), the conditional variance is $[1 - \text{corr}(\tilde{v},\tilde{y})^2]\sigma_v^2$. The correlation is

$$\frac{\sigma_v^2/(2\lambda)}{\sigma_v\sigma_y} = \frac{\sigma_z}{\sigma_y} = \frac{\sigma_z}{\sqrt{\sigma_v^2/(4\lambda^2)+\sigma_z^2}} = \frac{1}{\sqrt{2}}.$$

Therefore, the conditional variance is $\sigma_v^2/2$.

24.3 GLOSTEN MODEL OF LIMIT ORDER MARKETS

Most modern exchanges are organized as limit order markets. A buy limit order is an order to buy at a price not exceeding the limit price specified in the order, and a sell limit order is an order to sell at a price not less than the limit price specified in the order. The collection of limit orders residing at an exchange is called the exchange's book of limit orders. Exchanges respect price priority—a bid to buy at a higher price executes before a bid to buy at a lower price, and an offer to sell at a lower price executes before an offer to sell at a higher price. Exchanges also respect time priority—among limit orders at the same price, the first to arrive executes first.[5]

If an order arrives that can execute immediately against another order or orders already in the book, then it is called a marketable limit order. The execution price is the limit price of the order that is already in the book. To illustrate the concept, Table 24-1 presents a simple example of a possible book in a stock. This stock is trading at $49.96 bid, $49.97 ask.

---

5. Exceptions to time priority may be made based on order display. If an exchange allows hidden orders or iceberg orders (where only part of the order is made visible), then it might stipulate that orders that are publicly displayed are executed before hidden or iceberg orders, regardless of the time of arrival.

Table 24-1. An Example of a Limit Order Book.

Price	Orders to Buy	Orders to Sell
$50.00		2,000
$49.99		2,500
$49.97		500
$49.96	1,000	
$49.95	500	
$49.93	2,000	

If a limit order to buy 1,500 shares at $49.99 arrives, then it will execute against the shares offered at $49.97 and against shares offered at $49.99, with 500 shares trading at $49.97 and 1,000 shares trading at $49.99. After execution of the order, the new book will be as shown in Table 24-2.

Table 24-2. The Limit Order Book after Execution of a Marketable Buy Order.

Price	Orders to Buy	Orders to Sell
$50.00		2,000
$49.99		1,500
$49.96	1,000	
$49.95	500	
$49.93	2,000	

The stock is now trading at $49.96 bid, $49.99 ask. The now larger gap between the bid and ask price may attract new orders to the market.

Glosten (1994) analyzes a limit order market in which there is perfect competition among uninformed market makers in posting buy and sell limit orders and then a marketable limit order arrives that may or may not be informed. This is similar to the Glosten-Milgrom model except that there is only one possible order size in the Glosten-Milgrom model, so only the limit orders at the best quotes (the bid and ask) are relevant. In contrast, the Glosten (1994) model allows for multiple order sizes and solves for the entire book of limit orders.

A trader who submits a marketable order into the limit order book faces nonlinear pricing. Consider a buy order. As in the simple example above, the order will "walk up the book," executing at successively higher prices. The cost of the 1,500 share buy order in the example is

$$500 \times 46.97 + 1{,}000 \times 49.99 = \int_0^{1500} P(x)\, dx,$$

where

$$P(x) = \begin{cases} 50.00 & \text{for } 3{,}000 < x \leq 5{,}000, \\ 49.99 & \text{for } 500 < x \leq 3{,}000, \\ 49.97 & \text{for } 0 < x \leq 500. \end{cases}$$

We say that $P(x)$ is the limit price at depth x. In this example, the cost of the order is the amount of money transferred from the trader who submitted the marketable order to the traders whose limit orders were already in the book. The transfer function for market buy orders is defined in general as

$$T(q) = \int_0^q P(x)\, dx.$$

The same definition of the transfer function applies for marketable sell orders, which we regard as negative in sign. The monetary transfer from the trader who submits the marketable order is also negative, representing a payment to the trader. Consider a sell order for 2,000 shares in the example above. We can calculate the associated transfer as

$$T(-2000) = \int_0^{-2000} P(x)\, dx = -\int_{-2000}^0 P(x)\, dx$$
$$= -(500 \times 49.93 + 500 \times 49.95$$
$$+ 1{,}000 \times 49.96),$$

where we extend the definition of P as

$$P(x) = \begin{cases} 49.96 & \text{for } -1{,}000 \leq x < 0, \\ 49.95 & \text{for } -1{,}500 \leq x < -1{,}000, \\ 49.93 & \text{for } -3{,}500 \leq x < -1{,}500. \end{cases}$$

For simplicity, assume the marketable order is actually a market order (its limit price is $+\infty$ if it is a buy order and $-\infty$ if it is a sell order, so the limit price is never constraining). Let $\tilde{x}$ denote the size of the market order, with $\tilde{x} < 0$ denoting a sell order as before, and let $\tilde{v}$ denote the value of the asset. The equilibrium condition in the Glosten model is that, for each $x > 0$,

$$P(x) = E[\tilde{v} \mid \tilde{x} \geq x].$$

The conditional expectation here is called an upper-tail expectation. The reason for taking the upper-tail expectation is that the limit sell order at depth x will execute against any market buy order that is of size x or larger. For $x < 0$, the equilibrium condition is that

$$P(x) = \mathsf{E}[\tilde{v} \mid \tilde{x} \leq x].$$

This is motivated by the fact that a limit buy order at depth x will execute against any market sell order of size $|x|$ or larger.

An important feature of the Glosten model is that there is a small trade spread. The inside quotes (best bid and ask) are

$$\text{BID} = \lim_{x \uparrow 0} P(x) = \mathsf{E}[\tilde{v} \mid \tilde{x} < 0], \qquad \text{ASK} = \lim_{x \downarrow 0} P(x) = \mathsf{E}[\tilde{v} \mid \tilde{x} > 0].$$

If there is any information at all in the market order $\tilde{x}$, then the bid is less than the ask, so there is a discontinuity in $P(\cdot)$ at 0. The small trade spread ASK − BID *does not* stem from small trades being informative. Instead, it stems from the fact that the inside limit orders execute against market orders of any size, including larger orders that are informative. This is another example of adverse selection.

As an example, suppose the market order comes from an investor with constant absolute risk aversion α who has a random endowment $\tilde{w}$ of the asset—known to the investor but not to the market—and a private signal $\tilde{s}$ about the asset value $\tilde{v}$. Define $\tilde{z} = \mathsf{E}[\tilde{v} \mid \tilde{s}]$, and set $\tilde{\varepsilon} = \tilde{v} - \tilde{z}$. Because the investor knows $\tilde{w}$ and $\tilde{s}$, the risk she faces is $\tilde{\varepsilon}$. Assume $\tilde{\varepsilon}$ is normally distributed. Then, as always in a CARA-normal model, the investor will maximize the mean-variance certainty equivalent. Given an order x, the certainty equivalent is

$$(x+\tilde{w})\tilde{z} - \frac{1}{2}\alpha(x+\tilde{w})^2 \operatorname{var}(\tilde{\varepsilon}) - T(x). \tag{24.7}$$

The definition of the transfer function T implies that $T'(x) = P(x)$. Consequently, the first-order condition for the optimal order $\tilde{x}$ is

$$\tilde{z} - \alpha(x+\tilde{w})\operatorname{var}(\tilde{\varepsilon}) = P(\tilde{x}).$$

The left-hand side here is the marginal benefit of buying a little more (or selling a little less) of the asset, and $P(\tilde{x})$ is the marginal cost. Define

$$\tilde{\theta} = \tilde{z} - \alpha \operatorname{var}(\tilde{\varepsilon})\tilde{w}.$$

We can rewrite the first-order condition as

$$\tilde{\theta} = P(\tilde{x}) + \alpha \operatorname{var}(\tilde{\varepsilon})\tilde{x}. \tag{24.8}$$

The first-order condition (24.8) depends on the investor's information only via $\tilde{\theta}$, so different investors with the same $\tilde{\theta}$ make the same choices. We call $\tilde{\theta}$ the investor's type. To compute the tail expectations, note that, for any x,

$$\tilde{x} \geq x \quad \Leftrightarrow \quad P(\tilde{x}) + \alpha \operatorname{var}(\tilde{\varepsilon})\tilde{x} \geq P(x) + \alpha \operatorname{var}(\tilde{\varepsilon})x$$
$$\Leftrightarrow \quad \tilde{\theta} \geq P(x) + \alpha \operatorname{var}(\tilde{\varepsilon})x.$$

Thus, the equilibrium condition is that

$$P(x) = \begin{cases} \mathsf{E}[\tilde{v} \mid \tilde{\theta} \geq P(x) + \alpha \operatorname{var}(\tilde{\varepsilon})x] & \text{if } x > 0, \\ \mathsf{E}[\tilde{v} \mid \tilde{\theta} \leq P(x) + \alpha \operatorname{var}(\tilde{\varepsilon})x] & \text{if } x < 0. \end{cases} \quad (24.9)$$

This is an equation that is to be solved in $P(x)$ for each $x \neq 0$. There are cases in which a solution does not exist, but a solution exists if $\tilde{v}$, $\tilde{s}$, and $\tilde{w}$ are normally distributed. An illustration is provided in Figure 24.1.

24.4 AUCTIONS

Many securities transactions are organized as auctions. In fact, any market in which a request to trade is made and other traders compete to take the opposite side, such as the Kyle model, can be considered an auction. Those who compete to fill the order (for example, the market makers in the Kyle model) are bidders in an auction. Auctions are also used by the Treasury (of the U.S. and many other countries) to sell bonds and bills, are used to sell entire companies when there are competing bidders in M&A transactions, have been used to sell shares in initial public offerings and seasoned offerings, and are used in many other financial settings. Even when a market is not explicitly organized as an auction, the economic theory of auctions is likely to provide insight into the working of the market and perhaps also into why the market is organized as it is. This section describes a few important issues in auctions, including the winner's curse and the linkage principle.

Assume there is a seller offering one unit of an asset for sale and there are multiple potential buyers. If the buyers all have the same information and the same value for the asset, and the seller asks them to bid for the asset, then the game is a Bertrand pricing game, and the equilibrium is for all buyers to bid their value. This is the assumption made in the Kyle model. Here, we consider the case in which the bidders have different information and possibly different values.

Model

Assume the bidders are risk neutral and each bidder $i = 1, \ldots, n$ observes a signal $\tilde{s}_i$. Assume for convenience that the signals are continuously distributed and there is a minimum possible signal $\underline{s}$; that is, assume the support of each $\tilde{s}_i$ is the interval $[\underline{s}, \infty)$, or, for some $\bar{s}$, assume the support of each $\tilde{s}_i$ is the interval $[\underline{s}, \bar{s}]$.

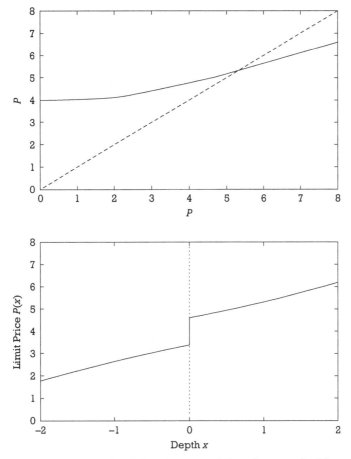

Figure 24.1 Example of the Glosten model. In this example, $\tilde{w}$ has a zero mean and is independent of $\tilde{v}$ and $\tilde{s}$. Also, $\mathsf{E}[\tilde{v}] = 4$, $\text{stdev}(\tilde{\theta}) = 1$, $\text{cov}(\tilde{v}, \tilde{\theta})/\text{var}(\tilde{\theta}) = 0.5$, and $\alpha \, \text{var}(\tilde{\varepsilon}) = 1$. The left panel illustrates the solution of (24.9) for $P(x)$ when $x = 1$. The dashed line is the 45° line, and the solid line is the upper-tail expectation of $\tilde{v}$ on the right-hand side of (24.9). The solution is $P(x) = 5.34$. The right panel plots the limit order book; that is, it plots the solution $P(x)$ of (24.9). The inside quotes are BID = 3.39 and ASK = 4.61.

Let $\tilde{x}_i$ denote the value of the object to bidder i in monetary units, and set $\tilde{v}_i = \mathsf{E}[\tilde{x}_i \mid \tilde{s}_1, \ldots \tilde{s}_n]$. Because of risk neutrality, bidders would act the same if their true values were $\tilde{v}_i$ rather than $\tilde{x}_i$, so we henceforth call $\tilde{v}_i$ the value of bidder i. The definition as a conditional expectation implies that $\tilde{v}_i = g_i(\tilde{s}_1, \ldots \tilde{s}_n)$ for some function g_i. Assume the bidders are symmetric in the sense that

- the joint distribution of $(\tilde{s}_1, \ldots \tilde{s}_n)$ is invariant with respect to a permutation of the indices,
- $g_i(s_1, \ldots, s_n) = g(s_i, s_{-i})$ for some function g, where s_{-i} denotes the list of signals of bidders $j \neq i$,
- g is a symmetric function of the components of s_{-i}.

Assume further that g is a nondecreasing function in each of its arguments. Finally, assume that the signals are affiliated.[6]

Each bidder chooses a bid $\tilde{b}_i$ depending on her signal. The allocation of the asset and the monetary transfers between the buyers and seller are functions of the vector of bids $(b_1, \ldots, b_n)$. In a symmetric equilibrium, there is a function $b(\cdot)$ such that each bidder i bids $\tilde{b}_i = b(\tilde{s}_i)$. Given that the asset value is monotonically related to signals, the function b will be monotone.

Common Values and Independent Private Values

There are two important special cases of the auction model. In the common values model, all bidders have the same value. This is the special case of the model in which g is a symmetric function of all of its arguments, so the value depends on the collection of signals but not on who gets which signal. The second important special case is the independent private values (IPV) model, which assumes that each bidder's value depends only on her own signal—that is, the function g depends only on s_i—and the signals are IID. In that case, we can define the signal to be the value (taking $g(s) = s$). In a financial setting, the possibility of reselling an asset after it is auctioned creates a common value aspect for the bidders. Of the two special cases, the common values model is usually more appropriate for financial auctions.

---

6. Affiliation is a relationship among random variables that is in the spirit of positive correlation. It means that if some subset of the random variables are all large, then it is more likely that the others are also large. Suppose the signals have a joint density function f. Given two signal vectors $s = (s_1, \ldots, s_n)$ and $s' = (s'_1, \ldots, s'_n)$, let $s \vee s'$ denote the vector with elements $\max(s_i, s'_i)$, and let $s \wedge s'$ denote the vector with elements $\min(s_i, s'_i)$. The signals are affiliated if

$$(\forall s, s') \quad f(s \vee s')f(s \wedge s') \geq f(s)f(s').$$

Roughly speaking, this means that it is more likely that the signals are all large or all small than that they are mixed. The property of affiliation is used in this section in the discussion of the winner's curse and the discussion of revenue rankings.

Auction Formats

Auctions are organized in a variety of ways. Common formats include ascending price, descending price, first-price sealed bid, and second-price sealed bid. In an ascending price auction, the price rises continuously and bidders drop out successively until only one bidder remains. The winning bidder pays the price at which the second-highest bidder dropped out. In a descending price auction, the price starts high and falls until some bidder accepts it. In sealed bid auctions, bidders submit bids without observing other bidders' actions. The highest bidder wins the asset and pays her bid in a first-price auction or pays the second-highest bid in a second-price auction. Each of these auction formats is called a standard auction, meaning an auction in which the asset is awarded to the highest bidder, and only the highest bidder makes a payment to the seller.

The descending price auction is used in Amsterdam to sell flowers and is called a Dutch auction. In a descending price auction, each bidder chooses a price at which to step in and claim the asset. When a bidder does so, other bidders have not stepped in, so nothing has been learned about the choices of other bidders. Consequently, a descending price auction is strategically equivalent to a first-price sealed bid auction.

The ascending price auction is a simple model of auctions such as art auctions, though it omits some details (the choice of a bidder to bid only slightly higher or much higher than the outstanding high bid, for example). In an ascending price auction, each bidder chooses a price at which to drop out. An ascending price auction is strategically equivalent to a second-price sealed bid auction if bidders in the ascending price auction do not learn anything from the prices at which other bidders drop out. This is the case when there are only two bidders (because the auction ends when one drops out) and in the IPV model (because then the exit prices of other bidders are irrelevant for each bidder's value).

Dominant Strategy Equilibria in the IPV Model

In the IPV model, it is a dominant strategy in a second-price sealed bid auction for each bidder to bid her own value and a dominant strategy in an ascending price auction to drop out when the price reaches her value. If a bidder were to bid higher (or to drop out later), then she would change the outcome (winning when she would otherwise have lost) only when the price she has to pay exceeds her value. If a bidder were to bid lower (or drop out earlier), then she would change the outcome (losing when she would otherwise have won) only when her value exceeds the price she would have paid. Thus, neither deviation is ever profitable.

Winner's Curse with Common Values

In a standard auction that has a common value aspect, a bidder evaluates different possible bids by computing her expected gain (expected value minus expected payment) conditional on her signal and conditional on winning. Conditioning on winning is important, because there is important information in the fact of winning. A bidder who wins with a bid x knows that the other bidders' signals lie below $b^{-1}(x)$. This information about an upper bound on others' signals causes the bidder (except in the IPV model) to revise her estimate of the asset value downward. This phenomenon is called the winner's curse. It is both good news and bad news for a bidder to win in an auction. The good news is of course that she won; the bad news is that the asset is probably worth less than she thought. A rational bidder recognizes that this "bad news" aspect of winning is inevitable, and she takes it into account before bidding when calculating her expected gain and in determining her optimal bid.

Equilibrium of an Auction

Consider a standard auction, and let $\pi(b_1, \ldots, b_n)$ denote the amount paid by the winning bidder. Assume π is symmetric in its arguments, meaning that the payment depends on bids but not on bidder names. We want to characterize a symmetric equilibrium bidding strategy $b(\cdot)$. Define

$$h(s_1, \ldots, s_n) = \pi(b(s_1), \ldots, b(s_n))$$

so that h is the equilibrium amount paid by the winning bidder as a function of signals. Define

$$T(s,z) = \mathsf{E}[h(\tilde{s}_1, \ldots, \tilde{s}_{i-1}, z, \tilde{s}_{i+1}, \ldots, \tilde{s}_n) \mid \tilde{s}_i = s, \max_{j \neq i} \tilde{s}_j < z].$$

Thus, $T(s,z)$ denotes the expected payment—conditional on winning—by a bidder with signal s who submits a bid as if she had signal z. Similarly, denote the expected asset value conditional on winning when the bidder's signal is s and she acts as if it were z by

$$V(s,z) = \mathsf{E}[g(\tilde{s}_i, \tilde{s}_{-i}) \mid \tilde{s}_i = s, \max_{j \neq i} \tilde{s}_j < z].$$

Denote the conditional probability of winning in this circumstance by

$$P(s,z) = \mathrm{prob}(\max_{j \neq i} \tilde{s}_j < z \mid \tilde{s}_i = s).$$

The bidder's expected gain with a signal of s and a bid of $b(z)$ is

$$[V(s,z) - T(s,z)]P(s,z). \qquad (24.10)$$

In equilibrium, this function of z must be maximized at $z = s$, for each s. The first-order condition at $z = s$ is

$$[V_2(s,s) - T_2(s,s)]P(s,s) + [V(s,s) - T(s,s)]P_2(s,s) = 0. \qquad (24.11)$$

Another necessary condition for equilibrium is that bids never exceed expected values conditional on winning. This means that $b(s) \leq V(s,s)$. In particular, $b(\underline{s}) \leq V(\underline{s},\underline{s})$.

Example

Consider two bidders who have a common value $\tilde{v} = \tilde{s}_1 + \tilde{s}_2$. Assume $\tilde{s}_1$ and $\tilde{s}_2$ are independently uniformly distributed on $[0,1]$. The equilibrium of a first-price auction is calculated as follows: Let $b(\cdot)$ be a candidate strictly increasing equilibrium bid function. The probability of winning conditional on a signal of s and a bid of $b(z)$ is the probability that the other bidder's signal lies below z, which is z. Thus, $P(s,z) = z$. The payment conditional on winning with a bid of b is b. Thus, $T(s,z) = b(z)$. The expected value conditional on winning with a signal of s and a bid of z is

$$V(s,z) = s + \mathsf{E}[\tilde{s}_2 \mid \tilde{s}_1 = s, \tilde{s}_2 < z] = s + \mathsf{E}[\tilde{s}_2 \mid \tilde{s}_2 < z] = s + \frac{z}{2}.$$

Thus, the first-order condition (24.11) is

$$\left[\frac{1}{2} - b'(s)\right]s + \left[\frac{3s}{2} - b(s)\right] = 0. \qquad (24.12)$$

The condition $b(\underline{s}) \leq V(\underline{s},\underline{s})$ means that $b(0) \leq 0$. The solution of the ODE (24.12) with initial condition $b(0) = 0$ is the equilibrium. The solution is $b(s) = s$.

Linkage Principle and Revenue Rankings

The linkage principle states that expected revenue is higher in a standard auction if the expected payment conditional on winning is linked to signals in addition to being linked to bids. Consider two different standard auctions. They differ in having different payment functions T. Let T^A and T^B denote the payment functions in two auctions A and B. The linkage principle states that if $T_1^A(s,s) \geq T_1^B(s,s)$ for all s, where the subscript 1 denotes the partial derivative with respect to the first argument, then the expected revenue in auction A is at least as great as the expected revenue in auction B. The linkage principle is proven below. The linkage principle means that higher average revenue is obtained if the payment is linked to a bidder's signal (the first argument of T), rather than just to her action (the second argument of T). An auction can link the payment to bidder i's signal by linking the payment to the bids of bidders $j \neq i$, which depend on the signals of bidders $j \neq i$, which are typically correlated with the signal of bidder i. Linkage can also be created by tying the payment price to the future realized value of the asset (for example, through royalties on production if the asset is an oil lease) or by tying the payment to information possessed by the seller via disclosure of the seller's information prior to bidding—one implication of the linkage principle is that a seller should commit to truthfully disclosing any information she possesses prior to soliciting bids, if such a commitment is feasible.

The linkage principle implies that the expected revenue from a second-price auction is at least as large as that from a first-price auction. This may seem somewhat counterintuitive, because, for a given vector of bids, the payment is obviously higher in the first-price auction. However, the equilibrium bids are higher in the second-price auction. To deduce the result from the linkage principle, observe that, in a first-price auction, we have $T(s,z) = b(z)$, so $T_1(s,s) = 0$. In a second-price auction,

$$T(s,z) = \mathsf{E}\left[b(\max_{j \neq i} \tilde{s}_i) \,\Big|\, \tilde{s}_i = s, \max_{j \neq i} \tilde{s}_i < z\right].$$

Because the signals are affiliated, $T(s,z)$ is a nondecreasing function of s, implying $T_1(s,s) \geq 0$. Therefore, the linkage principle together with affiliation implies that the expected revenue from a second-price auction is at least as large as that from a first-price auction.

Another implication of the linkage principle is that the expected revenue from a second-price auction is the same as the expected revenue from a first-price auction if bidders' signals are independent. In fact, every standard auction produces the same expected revenue when signals are independent. This is true even if values are common. This is called the revenue equivalence theorem. It follows from the linkage principle and the fact that $T(s,z)$ does not depend on s in a standard auction when signals are independent; thus, $T_1(s,s) = 0$ for every standard auction when signals are independent.

Assume $T_1^A(s,s) \geq T_1^B(s,s)$ for all s. We will show that this implies

$$T^A(s,s) - T^B(s,s) \geq T^A(\underline{s},\underline{s}) - T^B(\underline{s},\underline{s}) = 0 \qquad (24.13)$$

for each s. Because of symmetry, the expected revenue in the auction is the expected payment by any buyer conditional on winning; that is, the expected revenue is $\mathsf{E}[T(\tilde{s}_i, \tilde{s}_i)]$ for any i. Therefore, (24.13) implies that the expected revenue from auction A is at least as high as that from auction B.

We begin by establishing the equality in (24.13). Consider either of the auctions and drop the superscript. A bidder who bids as if she had the minimum possible signal has a zero probability of winning, so $P(s, \underline{s}) = 0$ for all s. Therefore, the value of the objective function (24.10) is zero at $z = s = \underline{s}$. For truthful bidding to be optimal for a bidder with signal $\underline{s}$, we must consequently have

$$[V(\underline{s}, \underline{s} + \varepsilon) - T(\underline{s}, \underline{s} + \varepsilon)]P(\underline{s}, \underline{s} + \varepsilon) \leq 0$$

for $\varepsilon > 0$. Thus,

$$V(\underline{s}, \underline{s} + \varepsilon) \leq T(\underline{s}, \underline{s} + \varepsilon),$$

and taking the limit as $\varepsilon \to 0$ yields

$$V(\underline{s}, \underline{s}) \leq T(\underline{s}, \underline{s}).$$

On the other hand, for truthful bidding to be optimal for a bidder with signal $\underline{s} + \varepsilon$, we must have

$$[V(\underline{s} + \varepsilon, \underline{s} + \varepsilon) - T(\underline{s} + \varepsilon, \underline{s} + \varepsilon)]P(\underline{s} + \varepsilon, \underline{s} + \varepsilon)$$
$$\geq [V(\underline{s} + \varepsilon, \underline{s}) - T(\underline{s} + \varepsilon, \underline{s})]P(\underline{s} + \varepsilon, \underline{s}) = 0,$$

the equality following again from the fact that $P(s, \underline{s}) = 0$ for all s. Thus,

$$V(\underline{s} + \varepsilon, \underline{s} + \varepsilon) \geq T(\underline{s} + \varepsilon, \underline{s} + \varepsilon),$$

and taking the limit as $\varepsilon \to 0$ yields

$$V(\underline{s}, \underline{s}) \geq T(\underline{s}, \underline{s}).$$

Therefore,

$$T(\underline{s}, \underline{s}) = V(\underline{s}, \underline{s}).$$

The function V is the same for both auctions, so this equality implies the equality in (24.13).

Set $\Delta(s) = T^A(s, s) - T^B(s, s)$. Then, using subscripts to denote partial derivatives,

$$\Delta'(s) = T_1^A(s, s) - T_1^B(s, s) + T_2^A(s, s) - T_2^B(s, s). \tag{24.14}$$

The first-order condition for $z = s$ to maximize (24.10) is

$$[V_2(s, s) - T_2(s, s)]P(s, s) + [V(s, s) - T(s, s)]P_2(s, s) = 0. \tag{24.15}$$

This equation must hold for all s in both auctions A and B. The functions V and P are the same in the two auctions, so subtracting (24.15) for auction A from (24.15) for auction B produces

$$\left[T_2^A(s,s) - T_2^B(s,s)\right]P(s,s) + \left[T^A(s,s) - T^B(s,s)\right]P_2(s,s) = 0.$$

Therefore,

$$T_2^A(s,s) - T_2^B(s,s) = -\left[T^A(s,s) - T^B(s,s)\right]\frac{P_2(s,s)}{P(s,s)} = -\Delta(s)\frac{P_2(s,s)}{P(s,s)}.$$

Substitute this into (24.14) to obtain

$$\Delta'(s) = T_1^A(s,s) - T_1^B(s,s) - \Delta(s)\frac{P_2(s,s)}{P(s,s)}$$

$$\geq -\Delta(s)\frac{P_2(s,s)}{P(s,s)}, \qquad (24.16)$$

using the assumption $T_1^A(s,s) \geq T_1^B(s,s)$ for the inequality. Note that $P_2(s,s)$ is the marginal effect on the probability of winning from increasing the bid and is therefore positive. Thus, the inequality (24.16) shows that if $\Delta(s) < 0$ for any s, then $\Delta'(s) > 0$.

We have shown that Δ is a function that starts at 0 at $s = \underline{s}$ and which has the property that its derivative is positive whenever its value is negative. Such a function can never be negative. Therefore, $\Delta(s) \geq 0$ for all s, which is the inequality in (24.13). [To see that Δ can never be negative, suppose to the contrary that it is negative at some s. Denote this value of s by s_2. Define $s_1 = \max\{s \leq s_2 \mid \Delta(s) \geq 0\}$. By continuity, $\Delta(s_1) = 0$. Furthermore, $\Delta(s) < 0$ for s between s_1 and s_2. Consequently, $\Delta'(s) > 0$ for s between s_1 and s_2. Therefore,

$$\Delta(s_2) = \Delta(s_1) + \int_{s_1}^{s_2} \Delta'(s)\,ds > 0,$$

which is a contradiction.]

24.5 CONTINUOUS-TIME KYLE MODEL

A trader who recognizes that her trades affect the market price will typically want to execute a trade in small pieces.[7] The single-period Kyle model described in Section 24.2 gives the informed trader only one opportunity to trade, which is an unnatural constraint. Relaxing this constraint enables us to examine how the

---

7. For example, if the price is $p_0 + \lambda x$ for a constant λ, where x denotes the cumulative amount purchased, then the cost of buying a quantity A in a single trade is $A(p_0 + \lambda A) = p_0 A + \lambda A^2$,

dynamic optimization of the informed trader affects the evolution of liquidity and the informativeness of prices over time. The continuous-time model affords the informed trader the maximum flexibility in timing her trades. It is also more tractable than a discrete-time dynamic model.

Model

An asset with a terminal value of $\tilde{v}$ is traded over a finite time interval. Use the time interval as the unit in which time is measured, so the interval is $[0,1]$. Suppose $\tilde{v}$ is normally distributed, and the single informed trader observes $\tilde{v}$ at date 0. Denote the mean and standard deviation of $\tilde{v}$ by $\bar{v}$ and σ_v respectively.

Let Z_t denote the number of shares held by liquidity traders at date $t \in [0,1]$, and take $Z_0 = 0$. Assume $dZ = \sigma_z \, dB$, where B is a Brownian motion and σ_z is a constant.[8] The cumulative liquidity trade during the period $[0,1]$ is Z_1, which is normally distributed with mean zero and variance σ_z^2. To compare the continuous-time model to the single-period model, we will interpret Z_1 as corresponding to $\tilde{z}$ in the single-period model. Let X_t denote the number of shares the informed trader purchases by date t.

Market makers observe the stochastic process Y defined by $Y_t = X_t + Z_t$ and set the price P_t to be the expected value of $\tilde{v}$, conditional on the information in Y through date t. The interpretation of this model is that market makers see at each instant the aggregate order $dX_t + dZ_t$ and revise the price based on the information in the order.

Similar to the construction of fully and partially revealing equilibria in Chapter 22, we can assume the informed trader observes Z and then justify the assumption by showing that the equilibrium price reveals Y to the informed trader and hence, given knowledge of X, reveals Z. So, require X to be adapted to $\tilde{v}$ and Z. In fact, assume $dX_t = \theta_t \, dt$ for some stochastic process θ adapted

---

whereas the cost of buying the same quantity via a series of infinitesimal trades is

$$\int_0^A (p_0 + \lambda x) \, dx = p_0 A + \frac{1}{2}\lambda A^2.$$

This distinction between a single purchase and a series of infinitesimal purchases is the distinction between a monopsonist and a perfectly discriminating monopsonist.

8. Choosing the unit of time so that the trading period is $[0,1]$ has an effect on σ_z. Suppose instead that we measure time in years, the standard deviation per year of liquidity trading is σ_z^*, and trading is over a time interval $[0,T]$. When we rescale time so that a unit of time is T years, making the trading interval $[0,1]$, the standard deviation of liquidity trading per unit of time becomes $\sigma_z = \sqrt{T}\sigma_z^*$. We can rewrite all of the formulas in this section in terms of the number of years T and the standard deviation σ_z^* of liquidity trading per year by substituting $\sigma_z = \sqrt{T}\sigma_z^*$.

to $\tilde{v}$ and Z. We discuss this assumption at the end of the section. Given these assumptions, the profit of the informed trader is the mispricing of the asset times the number of shares purchased, added up over time; that is, the profit of the informed trader is

$$\int_0^1 (\tilde{v} - P_t)\theta_t \, dt. \tag{24.17}$$

We need to rule out doubling strategies, similar to the issue discussed in Section 12.2. It suffices to require that θ be such that

$$E \int_0^1 P_t^2 \, dt < \infty. \tag{24.18}$$

The role of this condition is explained below.

Equilibrium

An equilibrium in this model is defined in very much the same way as in the single-period Kyle model. The price at each time must equal the conditional expected value of the asset, given the information in orders, and the informed trader's strategy must maximize her expected profit. The informed trader optimizes with the understanding that her trades affect prices, taking as given the manner in which the price at each time t depends on the history of Y through time t. It is shown below that there is an equilibrium in which

$$P_0 = \bar{v}, \quad dP_t = \lambda \, dY_t, \quad \text{and} \quad \theta_t = \frac{\tilde{v} - P_t}{(1-t)\lambda}, \quad \text{where} \quad \lambda = \frac{\sigma_v}{\sigma_z}. \tag{24.19}$$

In this equilibrium, market depth $1/\lambda$ is constant and only half the depth in the single-period model. Other important properties of the model are

(a) All of the private information is eventually incorporated into the price (the price converges to $\tilde{v}$ by date 1).
(b) The conditional variance of $\tilde{v}$ at date t given the market maker's information is $(1-t)\sigma_v^2$. Thus, information is transmitted at a constant rate.
(c) The equilibrium price process is a Brownian motion with zero drift and standard deviation σ_v, given the market makers' information. The standard deviation does not depend on the level of liquidity trading σ_z.
(d) The expected profit of the informed trader is $\sigma_v \sigma_z$. Thus, the informed trader's expected profit is twice what it is when she is constrained to trade only once. This implies that the expected losses of liquidity traders are also twice what they are in the single-period model.

Filtering Method

There have been numerous extensions of the dynamic Kyle model. The extensions rely on one of two alternative proof techniques. One method is to guess the forms of the price adjustment rule and informed trading strategy, use filtering theory to compute $\hat{v}_t = \mathsf{E}[\tilde{v} \mid (Y_s)_{s \le t}]$, and then solve the equation $P_t = \hat{v}_t$ to find the unknown coefficients in the price adjustment rule and informed trading strategy. We already have explicit formulas for the coefficients in (24.19), but we will use filtering theory to show that $\hat{v}_t = \mathsf{E}[\tilde{v} \mid (Y_s)_{s \le t}]$.

The conjectured equilibrium (24.19) implies that

$$dP_t = \lambda \, dY_t, \tag{24.20a}$$

$$dY_t = \frac{\tilde{v} - P_t}{(1-t)\lambda} \, dt + \sigma_z \, dB_t. \tag{24.20b}$$

The first step is to define the observation and innovation processes. Set $Y_0^* = 0$ and

$$dY_t^* = \frac{1}{\sigma_z} \left(dY_t + \frac{P_t}{(1-t)\lambda} \, dt \right)$$

$$= \frac{\tilde{v}}{(1-t)\sigma_v} \, dt + dB_t.$$

Because Y and hence P are observable to market makers, Y^* is also observable. The process Y^* is an observation process as defined in Section 23.1. The corresponding innovation process is given by

$$dZ_t^* = \frac{\tilde{v} - \hat{v}_t}{(1-t)\sigma_v} \, dt + dB_t,$$

where $\hat{v}_t$ denotes $\mathsf{E}[\tilde{v} \mid (Y_s)_{s \le t}]$. The filtering equation (23.7) is

$$d\hat{v}_t = \frac{1}{(1-t)\sigma_v} \Sigma(t) \, dZ_t^*, \tag{24.21}$$

where $\Sigma(t)$ is the conditional variance of $\tilde{v}$ and is given in (23.3) as

$$\frac{1}{\Sigma(t)} = \frac{1}{\sigma_v^2} + \int_0^t \frac{1}{(1-s)^2 \sigma_v^2} \, ds = \frac{1}{\sigma_v^2} + \frac{t}{(1-t)\sigma_v^2} = \frac{1}{(1-t)\sigma_v^2}. \tag{24.22}$$

Therefore, $\Sigma(t) = (1-t)\sigma_v^2$, and the filtering equation is

$$d\hat{v}_t = \sigma_v \, dZ_t^* = \frac{\tilde{v} - \hat{v}}{1-t} \, dt + \sigma_v \, dB_t.$$

Notice that (24.20) implies

$$dP_t = \frac{\tilde{v} - P_t}{1-t} dt + \sigma_v dB_t. \quad (24.23)$$

Thus, $\hat{v}$ and P satisfy the same stochastic differential equation (SDE). Because $P_0 = \hat{v}_0$, the uniqueness of solutions to SDEs (satisfying Lipschitz conditions) implies $P = \hat{v}$. Thus, we have shown that $P_t = \mathsf{E}[\tilde{v} \mid (Y_s)_{s \leq t}]$, as desired.

One other fact will be useful later. Notice that $\Sigma(t) \to 0$ as $t \to 1$. Thus, $P_t \to \tilde{v}$, as claimed in the previous subsection.

Brownian Bridge

An alternative to filtering theory is to guess that the price at each date t depends only on cumulative orders Y_t rather than on the history of orders—that is, $P_t = f(t, Y_t)$ for some function f. Condition (a) above is a necessary condition for equilibrium, because, if the price differs from $\tilde{v}$ at $t = 1$, then the informed trader had profitable trades remaining that she should have taken before $t = 1$ (buying if $P_1 < \tilde{v}$ and selling if $P_1 > \tilde{v}$). Thus, we guess that $P_t = f(t, Y_t) \to \tilde{v}$ as $t \to 1$, suggesting that Y_t converges to some function of $\tilde{v}$—that is, $Y_t \to f(1, \cdot)^{-1}(\tilde{v})$. This suggests that market makers can estimate $\tilde{v}$ by estimating where Y will end up at date 1.

In the basic dynamic Kyle model, Y is a Brownian motion relative to market makers' information and a Brownian bridge relative to the informed trader's information. We will explain and demonstrate that fact here. From (24.20), we have $P_t = \bar{v} + \lambda Y_t$, so

$$dY_t = \frac{\tilde{v} - P_t}{(1-t)\lambda} dt + \sigma_z dB_t$$

$$= \frac{(\tilde{v} - \bar{v})/\lambda - Y_t}{1-t} dt + \sigma_z dB_t.$$

Now, define $Y^* = Y/\sigma_z$, so we have

$$dY_t^* = \frac{(\tilde{v} - \bar{v})/\sigma_v - Y_t^*}{1-t} dt + dB_t. \quad (24.24)$$

The process Y^* is called a Brownian bridge (see, for example, Karatzas and Shreve, 2004). It satisfies $Y_t^* \to (\tilde{v} - \bar{v})/\sigma_v$ with probability 1 as $t \to 1$. This is equivalent to $Y_t \to (\tilde{v} - \bar{v})/\lambda$, which is equivalent to $P_t \to \tilde{v}$.

The distribution of a Brownian bridge is that of a Brownian motion conditional on knowledge of the terminal value, in this case $(\tilde{v} - \bar{v})/\sigma_v$. If the terminal value is unknown and regarded at date 0 as a standard normal, as is the case for the market makers here due to the fact that $(\tilde{v}-\bar{v})/\sigma_v$ is a standard normal, then the distribution is that of a Brownian motion. Thus, Y^* is a standard Brownian motion and Y is a Brownian motion with standard deviation σ_z relative to market makers' information.

Market makers estimate $\tilde{v}$ by forecasting where Y will end up, knowing that Y ends at

$$Y_1 = \sigma_z Y_1^* = \frac{\sigma_z(\tilde{v}-\bar{v})}{\sigma_v} \quad \Leftrightarrow \quad \tilde{v} = \bar{v} + \frac{\sigma_v Y_1}{\sigma_z}$$

and regarding Y as a Brownian motion with zero drift. Because it is a Brownian motion with zero drift, the best estimate of where it will end is its current value, so the market makers' best estimate of $\tilde{v}$ is $\bar{v} + \sigma_v Y_t/\sigma_z = \bar{v} + \lambda Y_t$. This again verifies that the pricing rule in (24.19) is an equilibrium pricing rule.

Optimality of the Informed Trading Strategy

Having shown by either the filtering or Brownian bridge argument that $P_t = \hat{v}_t$, it remains to show that the informed trader's strategy is optimal. This is fairly simple, because it turns out that any strategy is optimal if it implies $P_t \to \tilde{v}$ as $t \to 1$. The economic explanation of that fact is that, in continuous time, the informed trader can move continuously along the residual supply curve $\Delta P = \lambda \Delta Y$ defined by the market makers' pricing rule. She can buy shares and then resell them at the same price that she bought on average, because liquidity trades do not change the price on average (the effects of liquidity trades on prices is a martingale component $\lambda\, dZ$). So, it does not matter when she buys shares or sells shares or how many she trades prior to $t = 1$. The only requirement for optimality is that the informed trader should eventually exploit all mispricing, buying or selling enough to ensure that there is no gap between the final price and the asset value $\tilde{v}$.

We verify the statements about the informed trader's optimal strategies by using dynamic programming. Fix a realization v of $\tilde{v}$. We regard θ as the control chosen by the informed trader, P as the state variable, and $(v - P_t)\theta$ as the instantaneous utility of the trader. The state variable evolves as $dP = \lambda \theta\, dt + \sigma_v\, dB$. The value function is

$$J(t,p,v) = \max\ \mathsf{E}\left[\int_t^1 (v - P_u)\theta_u\, du \mid P_t = p\right].$$

Keeping in mind that v is regarded as fixed (known to the informed trader), the HJB equation is

$$0 = \max_\theta \left\{(v-p)\theta + J_t + \lambda\theta J_p + \frac{1}{2}\sigma_v^2 J_{pp}\right\}.$$

Note that the objective function in the HJB equation is linear in θ, so the maximum can be zero only if the coefficient of θ is zero. In this case, the HJB equation states that the remaining terms add to zero. Thus, the HJB equation is equivalent to the pair of equations

$$J_p = \frac{p - v}{\lambda}, \tag{24.25a}$$

$$0 = J_t + \frac{1}{2}\sigma_v^2 J_{pp}. \tag{24.25b}$$

Because the coefficient of θ is 0 in the HJB equation, we can guess that any θ is locally optimal. The second equation implies that J will have zero drift (and consequently be a local martingale) if we evaluate it at $(t, \bar{v} + \lambda Z_t)$ instead of at (t, P_t). Note that $\bar{v} + \lambda Z_t$ would be the price process if $\theta = 0$. Thus, it appears that $\theta = 0$ is optimal and

$$J(t, p, v) = \mathsf{E}\left[J\big(1, p + \lambda(Z_1 - Z_t), v\big) \mid Z_t\right].$$

If so, we can compute the value function from the function $p \mapsto J(1, p, v)$. To guess this function, consider trading at time 1 (or very close to time 1) to move the price from p to v. Given the linear relation $\Delta p = \lambda \Delta Y$, the gain from doing so is $(v - p)^2/(2\lambda)$. Thus, we guess that $J(1, p, v) = (v - p)^2/(2\lambda)$ and

$$J(t, p, v) = \mathsf{E}\left[\frac{(v - p - \lambda(Z_1 - Z_t))^2}{2\lambda} \,\bigg|\, Z_t\right] = \frac{(v - p)^2 + (1 - t)\sigma_v^2}{2\lambda}. \tag{24.26}$$

This function J satisfies the HJB equation (24.25).

We prove a verification theorem using the value function guessed above. For any trading strategy θ,

$$J(1, P_1, v) = J(0, P_0, v) + \int_0^1 \left\{J_t\, dt + J_p\, dP + \frac{1}{2}J_{pp}(dP)^2\right\}$$

$$= J(0, P_0, v) + \int_0^1 \left\{-\frac{1}{2}\sigma_v^2 J_{pp}\, dt + \frac{P - v}{\lambda}\, dP + \frac{1}{2}J_{pp}(dP)^2\right\}$$

$$= J(0, P_0, v) - \int_0^1 \frac{v - P_t}{\lambda}\, dP_t$$

$$= J(0, P_0, v) - \int_0^1 (v - P_t)\, dY_t.$$

We used Itô's formula for the first equality, the HJB equation (24.25) for the second equality, $(dP)^2 = \lambda^2 (dY)^2 = \lambda^2 \sigma_z^2\, dt = \sigma_v^2\, dt$ for the third equality, and $dP = \lambda\, dY$ for the fourth. Using $dY = \theta\, dt + \sigma_z\, dB$, we can rearrange this as

$$\int_0^1 (v - P_t)\theta_t\, dt = J(0, P_0, v) - J(1, P_1, v) - \sigma_z \int_0^1 (v - P_t)\, dB_t.$$

The left-hand side is the profit of the informed trader, and the right-hand side is bounded above by

$$J(0, P_0, v) - \sigma_z \int_0^1 (v - P_t)\, dB_t, \qquad (24.27)$$

due to the nonnegativity of $J(1, p \mid v)$.

The no-doubling-strategies condition (24.18) implies that the stochastic integral in (24.27) has a zero expectation. Therefore,

$$\mathsf{E} \int_0^1 (v - P_t)\theta_t\, dt \leq J(0, P_0, v) \qquad (24.28)$$

with equality if and only if $J(1, P_1, v) = 0$, which is equivalent to $P_1 = v$. Thus, $J(0, P_0, v)$ is an upper bound on the informed trader's expected profit, conditional on $\tilde{v} = v$, and the upper bound is realized—and the corresponding strategy is consequently optimal—if and only if $P_1 = v$. We have already shown that the trading strategy in (24.19) implies $P_t \to \tilde{v}$ with probability 1. It follows that the strategy is optimal, provided that it is feasible. To be feasible, it must satisfy the "no doubling strategies" condition (24.18). As discussed previously, the strategy implies that Y is a Brownian motion with standard deviation σ_z relative to market makers' information. Consequently, it implies that $P_t = \bar{v} + \lambda Y_t$ is a Brownian motion with standard deviation σ_v relative to market makers' information. Thus,

$$\mathsf{E}\int_0^1 P_t^2\, dt = \int_0^1 \mathsf{E}[P_t^2]\, dt = \int_0^1 [\bar{v}^2 + \mathrm{var}(P_t)]\, dt = \bar{v}^2 + \int_0^1 \sigma_v^2 t\, dt < \infty.$$

Why the Informed Trader Trades Slowly

To this point, we have only considered informed trading strategies $dX_t = \theta_t\, dt$ for stochastic processes θ. This is a departure from the standard model of portfolio choice. For example, consider the model of Chapter 14 with a single risky asset. Suppose the dividend-reinvested price S is a geometric Brownian motion with expected return μ and volatility σ. Assume there is a constant risk-free rate r and consider an investor with an infinite horizon and CRRA utility with risk aversion ρ. The investor's optimal portfolio is

$$\pi = \frac{\mu - r}{\rho \sigma^2}.$$

This is the fraction of wealth invested in the risky asset. The number of shares held is $X_t = \pi W_t / P_t$, which has dynamics

$$\frac{dX}{X} = \frac{dW}{W} - \frac{dP}{P} - \left(\frac{dW}{W}\right)\left(\frac{dP}{P}\right) + \left(\frac{dP}{P}\right)^2.$$

Assume the asset pays dividends continuously, so $S_t = e^{\int_0^t \delta_u \, du} P_t$. Then, the stochastic part of dP/P is the same as that of dS/S, which equals $\sigma \, dB$. This implies that the stochastic part of dX/X is $(\pi - 1)\sigma \, dB$. The presence of this stochastic part reflects the investor's instantaneous rebalancing. If $\pi < 1$, then the investor sells shares whenever the stock price rises and buys shares whenever it falls. If $\pi > 1$, then she makes the reverse trades.[9] This instantaneous rebalancing causes the number of shares X to have infinite variation.

In contrast, in the Kyle model, the investor's change in the number of shares she owns is of order dt. As usual, we think of dt as being of smaller order than dB, which means that the investor trades more slowly in the Kyle model than in the standard model. There are two differences between the Kyle model and the standard model of portfolio choice that contribute to this difference in the speed of trade. First, the investor is risk neutral in the Kyle model, so she does not rebalance in response to price changes to manage the risk of her portfolio. Second, the investor in the Kyle model recognizes that her trades affect prices. It is costly for the informed trader in the Kyle model to trade excessively.

To see the costs of trading excessively, we need to modify the model to accommodate trading that is of order $\sqrt{dt}$. To obtain the correct formula for the informed trader's profits, we return to the intertemporal budget equation in the standard model of portfolio choice. With no interim consumption, no interim dividends, and a risk-free rate of 0, the intertemporal budget equation is simply $dW = X \, dP$, which says that the change in wealth is the number of shares owned multiplied by the change in price. The investor's total profit is

$$W_1 - W_0 = \int_0^1 X_t \, dP_t.$$

From Itô's formula (integration by parts), we have

$$d(XP) = X \, dP + P \, dX + (dX)(dP).$$

Thus,

$$X \, dP = d(XP) - P \, dX - (dX)(dP),$$

---

9. An investor whose optimal allocation π is less than 1 trades as a contrarian, and an investor whose optimal allocation π is greater than 1 trades as a momentum trader. If $\pi < 1$, then the investor becomes overexposed to the risky asset when it rises, because the fraction of her wealth in the risky asset rises above π when the asset price rises. Thus, she sells shares when the stock price rises. Symmetrically, she buys shares when it falls. On the other hand, if $\pi > 1$, then the investor is levered—she has borrowed to buy the risky asset. In this case, her wealth changes by a greater percentage than the asset price changes when the asset price rises or falls. When the asset rises, she becomes underexposed, so she buys more. Symmetrically, she sells shares when the asset price falls.

and the investor's profit is

$$W_1 - W_0 = X_1 P_1 - X_0 P_0 - \int_0^1 P_t \, dX_t - \int_0^1 (dX_t)(dP_t).$$

Substitute $X_1 P_1 = P_1(X_1 - X_0) + P_1 X_0$ to obtain

$$W_1 - W_0 = (P_1 - P_0)X_0 + \int_0^1 (P_1 - P_t) \, dX_t - \int_0^1 (dX_t)(dP_t). \quad (24.29)$$

This formula is valid in both the Kyle model and in the standard portfolio choice model (with no interim consumption, no interim dividends, and a risk-free rate of 0).

There are three differences between (24.17) and (24.29), two of which are immaterial. The term $(P_1 - P_0)X_0$ in (24.29) is irrelevant, because it is independent of the investor's actions and the investor is risk neutral. We have $P_1 - P_t$ in (24.29), but $\tilde{v} - P_t$ appears in (24.17); however, we have already seen that $P_1 = \tilde{v}$ in equilibrium.[10] The important difference between (24.17) and (24.29) is the last term in (24.29), which is the integral of trades multiplied by price changes. We can combine the two integrals in (24.29) as

$$\int_0^1 (P_1 - P_t - dP_t) \, dX_t.$$

This is the sum of the mispricing multiplied by shares purchased as in (24.17), but here we recognize that the price at which the asset is traded is $P_t + dP_t$. This is natural, because $P_t + dP_t$ is the price after market makers respond to the market order dY_t. The difference between trading at P_t and trading at $P_t + dP_t$ is negligible if the investor trades slowly, because $(dP)(dt) = 0$. But, it is nonneglible if the investor's trades are of order $\sqrt{dt}$.

As noted before, the formula (24.29) is valid in both the Kyle model and in the standard portfolio choice model. However, it leads to slower trading in the Kyle model than in the standard model. The reason is that in the Kyle model the investor considers the effect of her trades on the price. In the Kyle model,

$$(dP)(dX) = \lambda(dY)(dX) = \lambda(dX)^2 + \lambda(dX)(dZ).$$

There is no term in the standard model that is analogous to $\lambda(dX)^2$, because the investor is a price taker in the standard model.

---

10. Furthermore, even out of equilibrium, it is correct to replace P_1 with $\tilde{v}$ in (24.29), because if we do not have $P_t \to \tilde{v}$, then the price will jump to $\tilde{v}$ at the end of the model.

Liquidity Trader Losses

The gain (loss) of liquidity traders is also calculated according to the formula (24.29), but with Z replacing X. Taking $Z_0 = 0$ and using the fact that

$$\mathsf{E}\int_0^1 (P_1 - P_t)\,\mathrm{d}Z_t = 0,$$

we see that expected liquidity trader losses are

$$\mathsf{E}\int_0^1 (\mathrm{d}Z_t)(\mathrm{d}P_t) = \lambda \mathsf{E}\int_0^1 (\mathrm{d}Z_t)^2 = \lambda \sigma_z^2 = \sigma_v \sigma_z.$$

This matches the expected gain of the informed trader, which, from (24.26), is

$$\mathsf{E}[J(0, \bar{v} \mid \tilde{v})] = \frac{2\sigma_v^2}{2\lambda} = \sigma_v \sigma_z.$$

In this model, as in the Glosten-Milgrom model and the single-period Kyle model, market makers lose to informed traders and recoup their losses from liquidity traders, who trade at unfavorable prices because market makers surmise their trades may be informed whereas in reality they are not. The unfavorable prices are captured by the bid-ask spread in the Glosten-Milgrom model, by the product of orders with price changes $\mathsf{E}[\tilde{z} \cdot \Delta P] = \lambda \mathsf{E}[\tilde{z}^2]$ in the single-period Kyle model, and in the sum of orders multiplied by price changes $(\mathrm{d}Z)(\mathrm{d}P)$ in the continuous-time Kyle model.

24.6 NOTES AND REFERENCES

In the models of this chapter, liquidity affects the asset price only indirectly through its effect on optimal trading and the information conveyed by trades. It is generally accepted that liquidity also affects asset prices through its effect on the return required by the marginal investor.[11] The magnitude of this effect is debated. In a model in which illiquidity is manifested in a proportional transactions cost, Constantinides (1986) shows that illiquidity has a first-order effect on volume but only a second-order effect on utility, due to investors optimally reacting to illiquidity by trading less frequently. This implies that the effect of illiquidity on asset prices should be second order. Constantinides assumes IID returns and infinitely

---

11. In most of the models of this chapter, the marginal investor is a risk-neutral market maker who is not restricted by margin requirements and hence can bear any or all of the risk of an asset. In reality, market makers attempt to maintain low inventories, and the asset must be held primarily by other investors, who are adversely affected by illiquidity.

lived investors. Jang, Koo, Liu, and Lowenstein (2007) show that if returns are not IID, then investors want to trade more to adjust to time-varying risks and expected returns; consequently, the effects of illiquidity on utility and asset prices are larger. Amihud and Mendelson (1986) analyze a model in which investors have random finite horizons at which they must liquidate their positions. They also conclude that the effect of transaction costs on asset prices is significant.

There are several possible explanations for the bid-ask spread. The Glosten-Milgrom model derives the spread from adverse selection considerations. Another possible explanation is that market makers are risk averse and must earn a risk premium to compensate for the risk of holding inventory. A third is that there are fixed costs of securities dealing that market makers cover in equilibrium from bid-ask spreads. A fourth is that market makers have monopoly power. Explanations other than adverse selection typically imply that transaction price changes are serially correlated. To explain this phenomenon, suppose the bid and ask prices are fixed at B and A respectively and each market order is equally likely to be a buy or a sell order. Then, there are eight equally likely configurations for a sequence of three transactions: (buy, buy, buy), (buy, buy, sell), ... Three transactions produce two price changes, each of which has an unconditional mean of zero. In this model, there cannot be any continuations of price changes. For example a positive price change from B to A must be followed by a zero change (another A) or a negative change (from A to B). When a reversal occurs, the product of price changes is $-(A - B)^2$, and reversals occur with probability 1/4 (ABA and BAB are the two of the eight possible sequences of prices for which there are reversals). Thus, the covariance between the pair of price changes is $-(A - B)^2/4$. Roll (1984) uses this fact to estimate the bid-ask spread from transaction prices. The martingale property in the Glosten-Milgrom model implies that price changes have zero serial correlation. Inventory control models can imply positive serial correlation (Exercise 24.1).

Glosten and Milgrom (1985) analyze a version of the model presented in Section 24.1 in which the asset value has a general distribution (not just H and L), informed traders may be only partially informed about the asset value and may learn more over time, the probability μ of an informed trade may be a general stochastic process, and the demands of uninformed traders may be price elastic. One observation they make is that the market can break down if there is too much private information and uninformed demands are too elastic. In such cases, no matter how wide the bid-ask spread is set, market makers cannot break even on average, because uninformed traders drop out when the spread widens, exacerbating the adverse selection problem. A major result of Glosten and Milgrom is that, if the market does not break down and the horizon T is sufficiently far away, then the beliefs of market makers and informed traders must converge over time.

There have been many extensions and applications of the Glosten-Milgrom model. Two notable extensions are the study of short sales constraints by Diamond and Verrecchia (1987) and the PIN (Probability of INformed trading) model of Easley, Kiefer, O'Hara, and Paperman (1996). Diamond and Verrecchia (1987) modify the Glosten-Milgrom model by assuming there are some traders who cannot sell short or who find it costly to sell short. When the asset value is low, informed traders who face short sales restrictions may choose not to trade. Diamond and Verrecchia show that this slows the incorporation of negative news into the market, but it does not bias prices. The reason there is no bias is that market makers update their beliefs with the recognition that traders with negative news may have chosen not to trade. This contrasts with heterogeneous prior models in which short sales restrictions cause prices to be biased upward (Section 21.4).

Easley, Kiefer, O'Hara, and Paperman (1996) work in continuous time and assume trades arrive as Poisson processes. At the beginning of the model, there may be an information event, in which case there is an informed trader who knows the asset value in $\{L, H\}$, or there may be no information event, in which case there are only uninformed trades. Thus, the number of buys and the number of sells are each drawn from a mixture distribution, mixing over whether there was an information event. They suggest applying the model empirically by assuming each day is a new iteration of the model and using a sample of daily buys and sells to fit the model parameters. They define PIN to be the expected rate of informed trades (the probability of an information event multiplied by the arrival rate of informed trades) divided by the expected rate of total trades. This model has been widely applied. However, Venter and de Jongh (2006) and Duarte and Young (2009) argue that the unconditional distribution of trades implied by the model does not fit the empirical distribution of trades. There are also some questions about whether empirical estimates of PIN actually measure the probability of informed trading (Aktas, de Bodt, Declerck, and Van Oppens, 2007; Akay, Cyree, Griffiths, and Winters, 2012).

Two notable extensions of the single-period Kyle model are to multiple assets (Caballé and Krishnan, 1994) and to a risk-averse informed trader (Subrahmanyam, 1991, and Exercise 24.2). There have been numerous other extensions and applications of the single-period Kyle model. A different model by Kyle has also been applied many times, though certainly not as many times as the 1985 model has been applied. Kyle (1989) analyzes a model in which informed and uninformed strategic investors with CARA utility submit demand curves for an asset. There are also normally distributed price-inelastic demands from liquidity traders. The equilibrium price is determined by market clearing. Thus, the role of the competitive market makers in the Kyle (1985) model is played by the uninformed strategic traders in Kyle (1989). Because traders submit demand curves (multiple orders at different prices), this is sometimes mistakenly called a limit order model. However, execution in a limit order model involves nonlinear pricing (walking up the book), and in the Kyle (1989) model all orders are executed at the market clearing price. Solving for equilibria in demand curves is difficult in general, because the space of functions is an infinite-dimensional space. However, the model has the property that each trader can compute her optimum one point at a time, solving for the optimum for each possible supply curve the other traders' actions might present her and then compiling these various optima into an optimal demand curve. The same phenomenon appears in Klemperer and Meyer (1989), who describe the curves as passing through ex post optimal points.

The Glosten model assumes an infinite number of risk-neutral uninformed limit order traders. Berhnardt and Hughson (1997) show that, if there is only a finite number of limit order traders, then they will make positive expected profits in equilibrium. Thus, competition in limit orders is different from Bertrand competition in prices. Biais, Martimort, and Rochet (2000) derive a differential equation that may define an equilibrium in a limit order market with a finite number of limit order traders. Back and Baruch (2013) and Biais, Martimort, and Rochet (2013) provide sufficient conditions for the differential equation to define an equilibrium, correcting a misstatement in Biais, Martimort, and Rochet (2000). Dynamic models of limit order markets are developed by Goettler, Parlour, and Rajan (2005, 2009) and Rosu (2009). In these models, as in many models in market microstructure theory, each market order trader wants to buy only a single unit of the asset.

Back and Baruch (2007) show that there should be a small-trade spread as in the Glosten model even in markets that are not organized as limit order markets. The reason is that, as discussed in footnote 7, traders who want to trade large quantities (including informed traders) should split their orders to minimize price impact costs. This is the phenomenon that occurs in the dynamic Kyle model. With order splitting, market makers cannot condition on the full size of a trade when they set the price for a piece of a trade. Thus, they should compute expected values as tail expectations exactly as in the Glosten model.

There is a large literature on auctions, originating with the seminal paper of Vickrey (1961). The revenue rankings described in Section 24.4 for models with common value aspects are due to Milgrom and Weber (1982). The discussion of the linkage principle and its proof in Section 24.4 are based on Milgrom (1989). For the definition and properties of affiliated random variables, see Milgrom and Weber (1982) or Krishna (2009). A topic not discussed in this chapter is the design of an optimal auction. The seminal paper on that topic is Myerson (1981). For surveys of auction theory, see Klemperer (1999, 2004) and Krishna (2009). The "half of her value" result in Part (a) of Exercise 24.3 is an example of a more general result: In first-price auctions with independent private values, it is an equilibrium to bid the expected value of the maximum of the other bidders' values conditional on the maximum being less than the bidder's value (Krishna, 2009, Proposition 2.2). The "twice her signal" result in Exercise 24.4 is also an example of a more general result: In second-price auctions with common values and symmetric signals, it is an equilibrium for each bidder to bid what her value would be if all other bidders had the same signal as her (Krishna, 2009, Proposition 6.1).

Section 24.4 describes the sale of a single unit of an asset. Financial auctions like Treasury auctions are typically for many units, and bidders can submit multiple bids for different quantities at different prices. These are called divisible-good auctions. The two main divisible-good auction formats are uniform price (all winning bidders pay the lowest winning price) and discriminatory (all winning bidders pay the price they bid). There is some analogy between uniform-price divisible-good auctions and second-price single-good auctions and between discriminatory divisible-good auctions and first-price single-unit auctions. However, the revenue ranking theorem for single-unit auctions does not extend to divisible-good auctions. In fact, Wilson (1979) shows that uniform-price auctions can have equilibria that are very bad for the seller. In these equilibria, bidders bid steep demand curves that create a large discrepancy between price and marginal cost for other bidders, encouraging all bidders to bid low for the quantity they expect to win. Correcting a misstatement in Wilson (1979), Back and Zender (1993) show that discriminatory auctions can be strictly superior to uniform-price auctions.

Kyle (1985) solves the discrete-time and continuous-time versions of his model and proves convergence of the equilibria as the length of the time periods in the discrete-time model goes to zero. Thus, the discrete-time model with small time periods and the continuous-time model have equilibria that are approximately the same. There are a few variations of the model in which this is not true. In each of the three cases listed below, the discrete-time model has an approximate strong-form efficiency property when the period length is small. In the limit, there is exact strong-form efficiency, but the strong-form efficient outcome is not an equilibrium of the continuous-time model.

- Holden and Subrahmanyam (1992) consider a model with multiple informed traders who have identical information. When the time periods are short, the competition between the traders is so aggressive that information is revealed almost immediately—the market is approximately strong-form efficient. In the continuous-time version of the model, there is no equilibrium (Back, Cao, and Willard, 2000). The extremely aggressive competition is a consequence of the informed traders having identical information. When they have heterogeneous (correlated but not identical) information, the discrete-time equilibria converge to a continuous-time equilibrium—see Foster and Viswanathan (1996) for the discrete-time model and Back, Cao, and Willard (2000) for continuous time.

- When (i) the asset value evolves over time, (ii) the informed trader learns about the asset value over time, and (iii) the announcement date is random, Caldentey and Stacchetti (2010) show that the discrete-time equilibria converge to something that is not a continuous-time equilibrium. In the limit, there is a finite date T such that, if the announcement does not occur prior to T, then at T and after the market knows everything the informed trader knows. The new information the trader learns after T is communicated to the market instantly. Thus, the market is strong-form efficient after T. This is approximately true in discrete time when the period length is small; however, the continuous-time limit is not an equilibrium of the continuous-time version of the model. Related work includes Back and Pedersen (1998), who derive an equilibrium in a continuous-time model with the informed trader learning about the asset value over time (and in which the asset value can evolve over time) but in which the announcement date is fixed. Their existence result depends on there being a sufficiently high amount of information asymmetry at the beginning of the model. The amount of information asymmetry evidently falls below the required level when the announcement date is random and the announcement does not occur soon enough.

- Chau and Vayanos (2008) study a model similar to that of Caldentey and Stacchetti (2010) but with no announcement date and with an asset that pays dividends each period. In the Chau-Vayanos model, market makers observe a public signal that reveals part of the informed trader's information. Chau and Vayanos analyze the steady state of their model. They show that, as the period length becomes small, the steady states converge to strong-form efficiency. However, informed profits do not converge to zero.

Back and Baruch (2004) analyze a hybrid of the Glosten-Milgrom and dynamic Kyle models. In their model, liquidity trades arrive as Poisson processes and there is a fixed order size. Market makers quote bid and ask prices, as in the Glosten-Milgrom model. Their model also has a random announcement date and a binary (high or low) distribution for the asset value. They show that, as the order size goes to zero and the arrival rates of liquidity trades go to infinity, equilibria converge to the equilibrium of a continuous-time Kyle model. One feature of their model is that, under some circumstances, the informed trader trades with positive probability in the opposite direction of her information (buying when the asset value is low or selling when it is high). This feature of "bluffing" also occurs in a dynamic Kyle model when there is a requirement that trades be disclosed ex post (Huddart, Hughes, and Levine, 2001).

The solution of the continuous-time Kyle model in Section 24.5 using the Brownian bridge argument is based on Back (1992), who solves the model for general continuous—not necessarily normal—asset value distributions (Exercise 24.5). Other extensions of the continuous-time Kyle model include trading an underlying asset and a derivative security (Back, 1993; Back and Crotty, 2015); risk aversion on the part of the informed trader (Baruch, 2002); atoms in the distribution of $\tilde{v}$ (Back and Baruch, 2004; Çetin and Xing, 2013; Back and Crotty, 2015; Back, Crotty, and Li, 2015); private information about the default time of a bond (Campi and Çetin, 2007; Campi, Çetin, and Danilova, 2013); a stochastic process for the standard deviation of liquidity trades (Collin-Dufresne and Fos, 2014); an asset value that can be influenced by the large trader with costly effort (Back, Collin-Dufresne, Fos, Li, and Ljungqvist, 2016).

Large traders (who anticipate how their trades will affect prices) have also been studied in models with symmetric information. Bertsimas and Lo (1998) study how a large trader should split her orders over time to minimize the expected cost of executing a predetermined trade within a prespecified amount of time. They describe the optimum as splitting trades evenly over time plus an adjustment for changing market circumstances. Almgren and Chriss (2000) and Huberman and Stanzl (2005) study how an investor with mean-variance preferences should do the same thing. A risk-averse investor trades more quickly to minimize price risk. The price impact functions in these models are exogenously specified. They include permanent and transitory effects. Huberman and Stanzl (2004) show that the permanent effect must be linear in order to avoid arbitrage opportunities for the large trader. Obizhaeva and Wang (2013) analyze how to minimize expected execution costs when trading into a limit order book when the evolution of the book depends on trades. Brunnermeier and Pedersen (2005), Carlin, Lobo, and Viswanathan (2007), and Teguia (2015) study equilibrium trading by large traders when it becomes known that another trader needs to make a large trade (is distressed). In equilibrium, other traders try to front-run the distressed trader (predatory trading) and then buy to cover their positions after the distressed trader's sale has depressed prices.

Vayanos (2001) studies a model similar to the dynamic Kyle model, except that the large trader's private information is about her endowment shocks rather than about the fundamental value of the asset. This captures the fact that private information of a large trader is frequently about her own trade intentions, which have price consequences when the market is less than perfectly liquid. The illiquidity in the Vayanos (2001) model arises not from adverse selection but because market makers are risk averse. Consequently, the asset's risk premium depends on the inventory that the market makers hold and hence depends on the trades of the large trader. Choi, Larsen, and Seppi (2015) study a dynamic Kyle model in which there are two strategic traders. One has long-lived private information as in the standard Kyle model and the other is constrained to trade a prespecified number of shares known only to her. Thus, Choi, Larsen, and Seppi solve the Kyle model with (somewhat) endogenous liquidity trades, and they also solve the optimal execution problem with endogenous price impacts.

EXERCISES

24.1. Suppose there is a representative market maker with constant absolute risk aversion α, and competition forces the bid and ask to the prices that make the market maker indifferent about trade. Suppose there is no information in market orders, which are of a unit size. Assume that the future value of a unit of the asset is normally distributed with mean μ and variance σ^2.

(a) Let θ denote the number of shares owned by the market maker before trade. Compute the ask and bid prices. Show that, even though the bid and ask depend on θ, the bid-ask spread does not.
(b) Assume buys and sells are equally likely. Consider all possible sequences of three transactions. Compute the transaction price for each transaction in each sequence, and show that the transaction price changes are positively serially correlated.

24.2. In the single-period Kyle model, assume the informed investor has CARA utility. There is a linear equilibrium. Derive an expression for λ as a root of a fifth-order polynomial.

24.3. Assume there are two buyers in an auction who have independent private values. Assume the value of each buyer is uniformly distributed on $[0, 1]$. Each buyer knows her own value but does not know the value of the other buyer.

(a) Assume the auction is conducted as a first-price auction. Show that it is an equilibrium for each buyer to bid one-half of her value. Compute the expected revenue for the seller.
(b) Assume the auction is conducted as a second-price auction. Show that it is an equilibrium for each buyer to bid her signal. Compute the expected revenue for the seller.

24.4. Assume there are two buyers in an auction who have a common value. Assume the buyers receive signals that are independently uniformly distributed on $[0, 1]$, and assume the value is the sum of the signals. Each buyer knows her own signal but does not know the signal of the other buyer. Assume the auction is conducted as a second-price auction. Show that it is an equilibrium for each buyer to bid twice her signal. Compute the expected revenue for the seller and compare it to the expected revenue from the first-price auction solved as an example in Section 24.4.

24.5. In the continuous-time Kyle model, assume $\log \tilde{v}$ is normally distributed instead of $\tilde{v}$ being normally distributed. Denote the mean of $\log \tilde{v}$ by μ and the variance of $\log \tilde{v}$ by σ^2. Set $\lambda = \sigma_v / \sigma_z$. Show that the strategies

$$P_0 = e^{\mu + \frac{1}{2}\sigma_v^2}$$

$$dP_t = \lambda P_t \, dY_t$$

$$dX_t = \frac{(\log \tilde{v} - \mu)/\lambda - Y_t}{1 - t} \, dt$$

form an equilibrium by showing the following:

(a) Define $W_t = Y_t / \sigma_z$. Show that, conditional on $\tilde{v}$, W is a Brownian bridge on $[0, 1]$ with terminal value $(\log \tilde{v} - \mu)/\sigma_v$. Use this fact to show that P satisfies $P_1 = \tilde{v}$ and is a martingale relative to the market makers' information.

(b) For $v > 0$ and $p > 0$, define

$$J(t, p) = \frac{p - v + v(\log v - \log p)}{\lambda} + \frac{1}{2}\sigma_v \sigma_z (1 - t) v.$$

Prove the verification theorem. (The intuition for this J is the same as that described in Section 24.5—take $\theta = 0$ and then trade at the end to the point that $p = v$.)

25

Alternative Preferences

CONTENTS

25.1 Experimental Paradoxes 652
25.2 Betweenness Preferences 658
25.3 Rank-Dependent Preference 663
25.4 First-Order Risk Aversion 665
25.5 Ambiguity Aversion 666
25.6 Notes and References 673

There is considerable experimental evidence that individuals make choices that depart systematically from the predictions of expected utility theory. This chapter reviews some of the evidence and some of the models of decision making that have been developed to accommodate the evidence. The chapter is necessarily a brief introduction to this large literature. Applications of the models in finance constitute part of what is called behavioral finance. The first section describes the Allais and Ellsberg paradoxes and some of the experimental evidence of Kahneman and Tversky (1979) about loss aversion and prospect theory.

Most of the discussion in this chapter is atemporal, in the sense that there is no definite amount of time elapsing between a choice and its consequence. Parts of the chapter consider a single-period portfolio choice problem. While some of the theory discussed in this chapter pertains to general outcomes, the theory will be presented for outcomes that are monetary (or in units of the consumption good). Thus, we will assume outcomes are real numbers, and more is preferred to less.

Much of the discussion in this chapter concerns preferences over gambles that define gains or losses. This is a departure from most of the book, which considers preferences over consumption or wealth. Under conventional assumptions, these are equivalent concepts. Suppose we are given an initial (or reference) wealth w_0, which could itself be random, and a preference relation $\succeq$ over random terminal wealth $\tilde{w}$. Denoting the gain or loss by $\tilde{x} = \tilde{w} - w_0$, the preference relation $\succeq$ over terminal wealth is equivalent to a preference relation $\succeq^*$ over gains and losses defined as

$$\tilde{x} \succeq^* \tilde{x}' \quad \Leftrightarrow \quad \tilde{x} + w_0 \succeq \tilde{x}' + w_0.$$

Thus, under conventional assumptions, it is simply a matter of convenience whether we discuss preferences over gains/losses or preferences over terminal wealth. In most of the chapter, we will move between the two without comment. The qualification "under conventional assumptions" is motivated by the prospect theory of Kahneman and Tversky (1979), who present evidence that the framing of a decision problem can affect the decomposition of terminal wealth $\tilde{w}$ into reference wealth w_0 and the gain/loss $\tilde{x}$. In their theory, preferences over gains/losses are the fundamental objects, and preferences over terminal wealth are induced by preferences over gains/losses in conjunction with a coding process that determines w_0.

25.1 EXPERIMENTAL PARADOXES

We first discuss evidence against objective expected utility as axiomatized by von Neumann and Morgenstern (1947). We then discuss evidence against subjective expected utility as axiomatized by Savage (1954). Finally, we discuss evidence of Kahneman and Tversky (1979) that framing matters and individuals exhibit loss aversion.

THE ALLAIS PARADOX AND THE INDEPENDENCE AXIOM

In the theory of von Neumann and Morgenstern (1947), a person compares different gambles with known probabilities of outcomes. The defining characteristic of expected utility is that it is a utility function over gambles that is linear in probabilities. Specifically, given a set of possible outcomes $\{x_1, \ldots, x_n\}$, the expected utility of a gamble is

$$\sum_{i=1}^{n} p_i u(x_i)$$

for some utility function u, where the p_i denote the probabilities. Thus, the utilities $u(x_i)$ are the linear coefficients on the probabilities p_i. Evidence that preferences are not linear in probabilities is evidence against the expected utility hypothesis.

Consider the following pairs of gambles:

$$A : 100\% \text{ chance of } \$1{,}000{,}000 \quad \text{versus} \quad B : \begin{cases} 10\% \text{ chance of } \$5{,}000{,}000 \\ 89\% \text{ chance of } \$1{,}000{,}000 \\ 1\% \text{ chance of } \$0 \end{cases}$$

$$C : \begin{cases} 11\% \text{ chance of } \$1{,}000{,}000 \\ 89\% \text{ chance of } \$0 \end{cases} \quad \text{versus} \quad D : \begin{cases} 10\% \text{ chance of } \$5{,}000{,}000 \\ 90\% \text{ chance of } \$0 \end{cases}$$

Various researchers, beginning with Allais (1953), have found a propensity for people to prefer A to B and D to C. Apparently, when comparing B to A, the 1% chance of getting 0 seems large compared to the 10% chance of a larger gain. However, when comparing D to C, the extra 1% chance of getting 0 seems small compared to the 10% chance of a larger gain. These preferences are inconsistent with expected utility maximization, and this is known as the Allais paradox.

To see that the preferences are inconsistent with expected utility maximization, let $x_1 = 0$, $x_2 = 1{,}000{,}000$, and $x_3 = 5{,}000{,}000$. If the preferences were consistent with expected utility maximization, we would have[1]

$$A \succ B \implies u(x_2) > 0.01 u(x_1) + 0.89 u(x_2) + 0.10 u(x_3),$$
$$D \succ C \implies 0.90 u(x_1) + 0.10 u(x_3) > 0.89 u(x_1) + 0.11 u(x_2).$$

However, adding $0.89[u(x_2) - u(x_1)]$ to both sides of the bottom inequality gives the reverse of the top inequality. Thus, there is no utility function u for which these preferences are consistent with expected utility maximization.

The preferences $A \succ B$ and $D \succ C$ violate the independence axiom used to derive objective expected utility (Herstein and Milnor, 1953). The independence axiom says that preferences regarding two gambles should be independent of whether they are mixed in the same way with a third gamble. For $0 < \alpha < 1$ and two gambles $P = (p_1, \ldots, p_n)$ and $Q = (q_1, \ldots, q_n)$ on an outcome space $\{x_1, \ldots, x_n\}$, the mixture $\alpha P + (1 - \alpha) Q$ is the gamble that assigns probability $\alpha p_i + (1 - \alpha) q_i$ to outcome i, for $i = 1, \ldots, n$. A mixture is a compound gamble (gamble over gambles): with probability α we get the gamble P and with probability $1 - \alpha$ we get the gamble Q.

The gambles in the Allais paradox can be represented as

$$A = \alpha P + (1 - \alpha) Q \quad \text{versus} \quad B = \alpha P^* + (1 - \alpha) Q$$
$$C = \alpha P + (1 - \alpha) Q^* \quad \text{versus} \quad D = \alpha P^* + (1 - \alpha) Q^*,$$

---

1. We use the standard notation $\succ$ for strict preference, $\sim$ for indifference, and $\succeq$ for weak preference (strict preference or indifference).

where $\alpha = 0.11$, $P = Q = (0, 1, 0)$, $P^* = (0.01/0.11, 0, 0.10/0.11)$, and $Q^* = (1, 0, 0)$. According to the independence axiom, preferences regarding A and B should depend on preferences regarding the gambles P and P^* that occur with probability α, because with probability $1 - \alpha$ we get the same gamble Q in both cases. Thus, the preference $A \succ B$ should imply $P \succ P^*$. However, C and D are also mixtures of P and P^* with a gamble Q^*, so the preference $P \succ P^*$ should imply $C \succ D$, contrary to the preferences expressed by the experimental subjects.

The Allais paradox is a special case of a more general phenomenon that researchers have observed, which is known as the common consequence effect. Suppose P is a gamble with a sure outcome, as in the Allais paradox (\$1,000,000 for sure). Thus, the choice between P and P^* is a choice between a sure outcome and a risky gamble. The common consequence effect is that people seem to be more risk averse in comparing P and P^* when the gamble with which they are mixed is attractive (like Q in the Allais paradox, which is \$1,000,000 for sure). They are less risk averse in comparing P and P^* when both are mixed with a less desirable gamble (like Q^* in the Allais paradox, which is 0 for sure). This produces the preference for P over P^* when A and B are compared but preference for P^* over P when C and D are compared.

The fact that the preferences $A \succ B$ and $D \succ C$ are nonlinear in probabilities can be seen in Figure 25.1. The triangular region (simplex) is the set of probability distributions over the outcomes $(0, 1,000,000, 5,000,000)$ represented as $\{(p_1, p_3) \mid p_1 \geq 0, p_3 \geq 0, p_1 + p_3 \leq 1\}$. Linearity in probabilities means there are some u_i such that

$$U(P) = u_1 p_1 + u_2 p_2 + u_3 p_3$$
$$= u_1 p_1 + u_2 (1 - p_1 - p_3) + u_3 p_3$$
$$= u_2 + (u_1 - u_2) p_1 + (u_3 - u_2) p_3.$$

Thus, a linear utility function is affine in (p_1, p_3). The indifference curves are the parallel lines: $\{(p_1, p_3) \mid b_1 p_1 + b_3 p_3 = \text{constant}\}$, where $b_i = u_i - u_2$. The direction of increasing utility is up (higher p_3) and to the left (lower p_1). As the figure illustrates, the preferences $A \succ B$ and $D \succ C$ imply that at least some of the indifference curves must "fan out" from the origin rather than being parallel.[2]

---

2. The slope of an indifference curve in the probability simplex reflects risk aversion. Moving up and to the right in the simplex creates a riskier gamble (higher probabilities of the extreme outcomes x_1 and x_3), so a steeper slope means that the probability of the best outcome must be increased more in order to maintain indifference when risk is increased. Thus, "fanning out" implies greater risk aversion when comparing more attractive gambles (gambles in the left portion of the simplex).

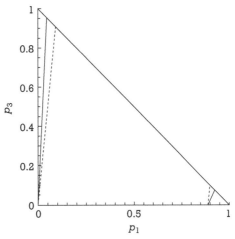

Figure 25.1 Nonlinearity of utility in the Allais paradox. This figure illustrates the Allais paradox in the probability simplex $\{(p1, p3) \mid p1 \geq 0, p3 \geq 0, p1 + p3 \leq 1\}$. The dotted lines are parallel, the one on the left passing through gambles A and B—that is, (0, 0) and (0.01, 0.10)—and the one on the right passing through gambles C and D—that is, (0.89, 0) and (0.90, 0.10). The solid lines are indifference curves consistent with strict preference for A over B and D over C. The indifference curve containing A must lie above and to the left of the dotted line connecting A and B, because $A \succ B$. The indifference curve containing C must lie below and to the right of the dotted line connecting C and D, because $C \prec D$. Therefore, the two indifference curves cannot be parallel. Hence, the preferences are not linear in probabilities.

The Ellsberg Paradox and the Sure Thing Principle

In the Savage (1954) theory of decision making, subjective probabilities are deduced from preferences. Consider two mutually exclusive events A and B. Suppose we offer someone a choice between two gambles: a gamble that pays $1 if A occurs and 0 if B occurs, or a gamble that pays $1 if B occurs and 0 if A occurs. If the person chooses the first gamble, then we can infer that she regards A as more likely than B. This is a simple example of deriving a probability distribution such that expected utility represents preferences.

Savage shows that preferences for gambles over events can be represented by expected utility whenever the preferences satisfy certain axioms. The most important axiom is the sure thing principle. For any event A, let A^c denote the complement of A, and, as usual, let 1_A denote the random variable that is 1 on A and 0 on A^c. The sure thing principle states that, for any four gambles $\tilde{x}, \tilde{y}, \tilde{w}$, and $\tilde{z}$ and any event A,

$$\tilde{x}1_A + \tilde{w}1_{A^c} \succeq \tilde{y}1_A + \tilde{w}1_{A^c} \quad \Rightarrow \quad \tilde{x}1_A + \tilde{z}1_{A^c} \succeq \tilde{y}1_A + \tilde{z}1_{A^c}. \quad (25.1)$$

This is quite similar to the independence axiom. The difference is that here we are mixing over events instead of probabilities. Preference for the gamble $\tilde{x}1_A + \tilde{w}1_{A^c}$ over the gamble $\tilde{y}1_A + \tilde{w}1_{A^c}$ indicates that $\tilde{x}$ is preferred to $\tilde{y}$ conditional on A occurring. The sure thing principle states that this preference for $\tilde{x}$ conditional on A occurring should imply that $\tilde{x}1_A + \tilde{z}1_{A^c}$ is preferred to $\tilde{y}1_A + \tilde{z}1_{A^c}$. The sure thing principle seems reasonable. Indeed, Savage states, "Except possibly for the assumption of simple ordering, I know of no other extralogical principle governing decisions that finds such ready acceptance."

The sure thing principle is contradicted by the following example, due to Ellsberg (1961). Consider an urn in which there are 30 red balls and 60 balls that are either black or yellow. The relative proportion of black and yellow balls is unknown. Consider a gamble that pays a certain amount of money if a ball of a particular color is drawn. The set of states of the world is the set $\{R, B, Y\}$ of colors that can be drawn. Call the amount of money one unit, so, for example, the gamble that pays if red is drawn is $1_{\{R\}}$. The probability of winning when betting on the red ball is $1/3$, but the odds when betting on either black or yellow are unknown. It is common for people to state a strict preference for gambling on the red ball versus either the black or yellow, which we denote by $1_{\{R\}} \succ 1_{\{B\}}$ and $1_{\{R\}} \succ 1_{\{Y\}}$. Thus, people seem to prefer known odds.

Now, consider a gamble that pays the unit of money if either red or black is chosen ($1_{\{R,B\}}$) and a gamble that pays if either yellow or black is chosen ($1_{\{Y,B\}}$). The probability of winning with the latter gamble is $2/3$, but the odds for the former are unknown. It is common for people to be consistent in preferring known odds and to prefer the yellow and black gamble here ($1_{\{Y,B\}} \succ 1_{\{R,B\}}$).

The preferences $1_{\{R\}} \succ 1_{\{Y\}}$ and $1_{\{Y,B\}} \succ 1_{\{R,B\}}$ are inconsistent with the sure thing principle and inconsistent with expected utility maximization. First, consider expected utility maximization and denote a person's subjective probabilities by P. The expected utility of 1_A for any event A is $u(1)P(A)$. Therefore, expected utility maximization and the additivity of probabilities imply that

$$1_{\{R\}} \succ 1_{\{Y\}} \Leftrightarrow P(\{R\}) > P(\{Y\}) \Leftrightarrow P(\{R,B\}) > P(\{Y,B\}) \Leftrightarrow 1_{\{R,B\}} \succ 1_{\{Y,B\}}. \quad (25.2)$$

Thus, expected utility maximization (and additivity of probabilities) is inconsistent with the preferences $1_{\{R\}} \succ 1_{\{Y\}}$ and $1_{\{Y,B\}} \succ 1_{\{R,B\}}$ expressed in the Ellsberg example.

To see that the preferences contradict the sure thing principle, set $A = \{R, Y\}$, $\tilde{x} = 1_{\{R\}}, \tilde{y} = 1_{\{Y\}}, \tilde{w} = 0$, and $\tilde{z} = 1$. Then,

$$\tilde{x}1_A + \tilde{w}1_{A^c} = 1_{\{R\}} \succ 1_{\{Y\}} = \tilde{y}1_A + \tilde{w}1_{A^c}, \qquad (25.3a)$$

but

$$\tilde{x}1_A + \tilde{z}1_{A^c} = 1_{\{R,B\}} \prec 1_{\{Y,B\}} = \tilde{y}1_A + \tilde{z}1_{A^c}. \qquad (25.3b)$$

Framing, Loss Aversion, and Prospect Theory

Kahneman and Tversky (1979) present evidence that people may sometimes not maximize any utility function over outcomes. Instead, they argue that people look at gains and losses rather than final outcomes. Furthermore, they argue that how people identify gains and losses can depend on how gambles are framed. They report responses of experimental subjects to the following scenarios:

(i) In addition to whatever you own, you have been given $1,000. You are now asked to choose between

$$A: \begin{cases} 50\% \text{ chance of } \$1,000 \\ 50\% \text{ chance of } \$0 \end{cases} \quad \text{versus} \quad B: 100\% \text{ chance of } \$500$$

(ii) In addition to whatever you own, you have been given $2,000. You are now asked to choose between

$$C: \begin{cases} 50\% \text{ chance of } \$0 \\ 50\% \text{ chance of } -\$1,000 \end{cases} \quad \text{versus} \quad D: 100\% \text{ chance of } -\$500.$$

A majority of the subjects chose B in case (i) and C in case (ii). This is inconsistent with any preference relation over terminal wealth gambles. The preference for B in case (i) means a preference for $1,500 with certainty over a 50-50 gamble with outcomes $2,000 and $1,000. Of course, this is consistent with risk aversion. However, the preference for C in case (ii) means that the gamble with outcomes $2,000 and $1,000 is preferred to the certain outcome $1,500. Similar results have been obtained by other experimenters. The difference between (i) and (ii) is obviously that the outcomes are framed as gains relative to a reference wealth level in (i) but as losses relative to a reference wealth level in (ii).

Kahneman and Tversky (1979) suggest that choices are determined by preferences defined over gains and losses. A coding process determines the translation of an outcome into a gain/loss, and this coding process is affected by framing.[3] Hence, preferences over terminal wealth, which are induced by preferences over gains/losses and the coding process, are affected by framing. They suggest further that the utility function over gains/losses is concave over gains (reflecting risk aversion, as in the preference for B over A in the example of the preceding section) and convex over losses (reflecting risk seeking, as in the preference for C over D). In addition, they hypothesize that the utility function is steeper in the realm of losses than in the realm of gains, arguing that aversion to losses is greater than the desire for gains. An example of such a utility function is

$$u(x) = \begin{cases} \frac{1}{1-\rho} x^{1-\rho} & \text{if } x \geq 0, \\ -\frac{\gamma}{1-\rho} (-x)^{1-\rho} & \text{if } x < 0, \end{cases} \qquad (25.4)$$

for $0 < \rho < 1$ and $\gamma > 1$, where x denotes the gain or loss rather than the resulting wealth level. The Kahneman-Tversky theory is called prospect theory.

25.2 BETWEENNESS PREFERENCES

There are two main generalizations of expected utility theory accommodating the Allais paradox that have been applied in the finance literature. One ("betweenness") retains the linearity of indifference curves but does not require them to be parallel. The other ("rank-dependent preferences") allows nonlinear indifference curves. Rank-dependent preferences are discussed in Section 25.3.

The betweenness axiom is that

$$P \succ Q \;\Rightarrow\; P \succ \alpha P + (1-\alpha) Q \succ Q \quad \text{and} \quad P \sim Q \;\Rightarrow\; P \sim \alpha P + (1-\alpha) Q \sim Q$$

for all gambles P and Q and $0 < \alpha < 1$. This implies that indifference curves are linear;[4] however, unlike the stronger independence axiom, it does not imply that indifference curves are parallel. Indifference curves can fan out or fan in or do both in different regions of the simplex in Figure 25.1.

A preference relation over gambles on a finite outcome space satisfies the betweenness axiom (and monotonicity and continuity axioms) if and only if

---

3. The coding that translates outcomes into gains/losses is said to be part of a more general editing process that includes, for example, rounding probabilities.

4. Technically, indifference curves are convex in the mixture space. In the probability simplex of Figure 20.1, indifference curves are line segments.

there is a utility function U over gambles with U taking values between 0 and 1 and a function $u : X \times [0,1] \to \mathbb{R}$ that is strictly monotone in x such that

(a) For all gambles P and Q, $P \succeq Q$ if and only if $U(P) \geq U(Q)$, and
(b) For all gambles P,
$$U(P) = \mathsf{E}^P[u(x, U(P))]. \tag{25.5}$$

In (25.5), the expectation is with respect to the distribution P of the outcome x. The utility $U(P)$ of a gamble P is defined implicitly as the solution of (25.5).

Of course, if the function u depends only on the outcome x and not on $U(P)$, then this is standard expected utility. Along an indifference curve—that is, fixing $U(P)$—the utility function $x \mapsto u(x, U(P))$ represents preferences just as in standard expected utility theory. Consequently, each indifference curve is linear. However, the utility function $x \mapsto u(x, U(P))$ changes when we change indifference curves (because $U(P)$ changes), so the indifference curves need not be parallel.

The utility function U is monotone in the sense that $U(P) \geq U(Q)$ if P first-order stochastically dominates Q. If u is concave in x, then the preferences are risk averse, in the sense of aversion to mean-preserving spreads.

WEIGHTED UTILITY

Weighted utility is a special case of betweenness preferences. Weighted utility $U(P)$ is defined implicitly by

$$v(U(P)) = \frac{\mathsf{E}^P[\lambda(x)v(x)]}{\mathsf{E}^P[\lambda(x)]} \tag{25.6}$$

for a strictly monotone function v and a positive function λ. For each outcome x, let δ_x denote the gamble that produces outcome x with probability 1. From (25.6),

$$v(U(\delta_x)) = \frac{\lambda(x)v(x)}{\lambda(x)} = v(x),$$

which implies $U(\delta_x) = x$. Thus, we have normalized utility so that the utility of a sure outcome is the outcome itself.[5] In fact, this normalization means that the utility of any gamble P is the certainty equivalent of P. To see this, note that the certainty equivalent of a gamble P is the outcome x such that $U(\delta_x) = U(P)$. From $x = U(\delta_x)$, we obtain $U(P) = x$.

---

5. We can equivalently define weighted utility by replacing $v(U(P))$ on the left-hand side of (25.6) by $U(P)$. With this definition, $v(x)$ is the utility of a sure outcome x.

If there is a finite number n of outcomes, then we can write (25.6) as

$$v(U(P)) = \sum_{i=1}^{n} \hat{p}_i v(x_i),$$

where

$$\hat{p}_i = \frac{\lambda(x_i) p_i}{\sum_{i=1}^{n} \lambda(x) p_i}.$$

Note that the $\hat{p}_i$ are nonnegative and sum to 1, so we can regard them as transformed probabilities.[6] For example, if λ is a decreasing function, then the $\hat{p}_i$ overweight low values of x, compared to the objective probabilities p_i. Weighted utility can be written in the general betweenness form (25.5) by defining[7]

$$u(x, a) = a + \lambda(x)[v(x) - v(a)]. \tag{25.7}$$

To see that weighted utility with this definition of u satisfies (25.5), note that

$$\mathsf{E}^P[u(x, U(P))] = U(P) + \mathsf{E}^P[\lambda(x) v(x)] - v(U(P)) \mathsf{E}^P[\lambda(x)] = U(P),$$

the last equality following from (25.6).

Disappointment Aversion

Disappointment aversion is another special case of betweenness. Disappointment-averse utility $U(P)$ is defined implicitly by

$$v(U(P)) = \frac{\mathsf{E}^P[\lambda(x, U(P)) v(x)]}{\mathsf{E}^P[\lambda(x, U(P))]}. \tag{25.8a}$$

This is similar to weighted utility, except that the weighting function λ depends on $U(P)$ here in a specific manner specified in (25.8b) below. As for weighted utility, the property (25.8a) implies that $U(P)$ is the certainty equivalent of P. Thus, the second argument of λ in (25.8a) is the certainty equivalent. The function λ is defined as

$$\lambda(x, a) = \begin{cases} 1 + \beta & \text{if } x < a, \\ 1 & \text{if } x \geq a, \end{cases} \tag{25.8b}$$

---

6. In general, $\lambda(x)/\mathsf{E}^P[\lambda(x)]$ is the Radon-Nikodym derivative $d\hat{P}/dP$ of a probability $\hat{P}$ with respect to P, so $v(U(P)) = \mathsf{E}^{\hat{P}}[v(x)]$.

7. There are many different ways to do this: We can take $u(x, a) = a + f(a) \lambda(x)[v(x) - v(a)]$ for any function f.

for some $\beta > 0$. Thus, these preferences overweight outcomes worse than the certainty equivalent and overweight all such outcomes by the same factor β.

As for weighted utility, disappointment-averse utility can be written in the general betweenness form (25.5) by defining

$$u(x,a) = a + \lambda(x,a)[v(x) - v(a)].$$

Another special case of betweenness is generalized disappointment aversion, in which the conditions $x < a$ and $x \geq a$ in (25.8b) are replaced by $x < \delta a$ and $x \geq \delta a$ for a constant $\delta \leq 1$.

Constant Relative Risk Aversion

CRRA betweenness preferences can be constructed as follows: Suppose there is a utility function U and strictly monotone function g with $g(1) = 0$ such that

$$E^P\left[g\left(\frac{x}{U(P)}\right)\right] = 0 \qquad (25.9)$$

for each gamble P. Then, as is shown below, the preferences satisfy betweenness, and the utility $U(P)$ of any gamble P is its certainty equivalent. The preferences represented by the utility function U exhibit constant relative risk aversion. A general definition of constant relative risk aversion is that scaling the outcomes of a gamble by a positive constant scales the certainty equivalent by the same constant. This is obvious from (25.9), because scaling the numerator and denominator in the ratio in (25.9) by the same number leaves the equality unchanged. This linear homogeneity of the certainty equivalent is equivalent to the following property: The proportion of initial wealth an individual would pay to avoid a gamble proportional to initial wealth is independent of initial wealth (compare to Exercise 1.5). This means that if $w_0(1-\pi)$ is the certainty equivalent of $w_0(1+\varepsilon)$ for any constant $w_0 > 0$ and a random ε, then $\hat{w}_0(1-\pi)$ is the certainty equivalent of $\hat{w}_0(1+\varepsilon)$ for every $\hat{w}_0 > 0$. Clearly, this is the same as the certainty equivalent being linearly homogeneous.

A CRRA weighted utility is obtained by taking

$$v(x) = \frac{1}{1-\rho}x^{1-\rho} \quad \text{and} \quad \lambda(x) = x^\gamma \qquad (25.10)$$

for constants ρ and γ such that $\gamma \leq 0$ and $\rho \leq \gamma + 1$ with at least one of these being a strict inequality. The proof that this is a CRRA utility is given below. The relative risk aversion of CRRA weighted utility should be regarded as $\rho - 2\gamma$

(Exercises 25.3 and 25.4). A CRRA disappointment-averse utility is obtained by taking

$$v(w) = \frac{1}{1-\rho} w^{1-\rho} \qquad (25.11)$$

for any $\rho > 0$. Again, the proof that this is a CRRA utility is given below.

To see that (25.9) implies betweenness, define

$$u(x,a) = a + g\left(\frac{x}{a}\right). \qquad (25.12)$$

Then,

$$E^P[u(x, U(P))] = U(P) + E^P\left[g\left(\frac{x}{U(P)}\right)\right] = U(P)$$

by virtue of (25.9). To see that $U(P)$ is the certainty equivalent of P, note that, for any outcome x, (25.9) implies

$$g\left(\frac{x}{U(\delta_x)}\right) = 0.$$

Hence, the assumption that g is strictly monotone with $g(1) = 0$ implies $x/U(\delta_x) = 1$. This implies that the utility of any gamble is its certainty equivalent, as discussed for weighted utility.

To see that the weighted utility (25.10) has constant relative risk aversion, define

$$g(y) = \lambda(y)\left[y^{1-\rho} - 1\right]. \qquad (25.13)$$

The proof that the restrictions on the parameters ρ and γ imply g is strictly monotone is left for the exercises. We have

$$E^P\left[g\left(\frac{x}{U(P)}\right)\right] = E^P\left[\lambda\left(\frac{x}{U(P)}\right)\left(\frac{x}{U(P)}\right)^{1-\rho}\right] - E^P\left[\lambda\left(\frac{x}{U(P)}\right)\right]$$

$$= U(P)^{-\gamma} \frac{E^P[\lambda(x)v(x)]}{v(U(P))} - U(P)^{-\gamma} E^P[\lambda(x)]$$

$$= 0,$$

using $v(x)/v(U(P)) = (x/U(P))^{1-\rho}$ for the second equality and the definition (25.6) of weighted utility $U(P)$ for the third. Thus, (25.9) holds.

To see that (25.11) is a CRRA disappointment-averse utility, set

$$g(y) = \begin{cases} (1+\beta)(1-\rho)v(y) - 1 - \beta & \text{if } y < 1, \\ (1-\rho)v(y) - 1 & \text{if } y \geq 1. \end{cases}$$

A calculation similar to that given for weighted utility shows that g satisfies (25.9).

Portfolio Choice and SDFs

Consider a single-period portfolio choice problem. Let $\tilde{\mathbf{R}}$ denote the vector of returns, and let ι denote a vector of 1's. The investor's final wealth is

$$\tilde{w} = w_0 R_f + w_0 \pi'(\tilde{\mathbf{R}} - R_f \iota).$$

To emphasize the dependence of the end-of-period wealth on the portfolio, write $\tilde{w}_\pi$ for $\tilde{w}$. To map this to the discussion in this chapter, the probability distribution P of outcomes (final wealth) is determined by the portfolio π. It is convenient to write $U(\pi)$ instead of $U(P)$ for the utility corresponding to a portfolio π. For betweenness preferences, we have

$$U(\pi) = \mathsf{E}[u(\tilde{w}_\pi, U(\pi))].$$

Denote partial derivatives by subscripts. The first-order condition for optimization in π is

$$0 = U'(\pi) = \mathsf{E}\left[u_1(\tilde{w}_\pi, U(\pi)) w_0 (\tilde{\mathbf{R}} - R_f \iota)\right] + \mathsf{E}\left[u_2(\tilde{w}_\pi, U(\pi))\right] U'(\pi)$$
$$= \mathsf{E}\left[u_1(\tilde{w}_\pi, U(\pi))(\tilde{\mathbf{R}} - R_f \iota)\right]. \tag{25.14}$$

This means that marginal utility $u_1(\tilde{w}_\pi, U(\pi))$ is proportional to an SDF. Consequently, many of the asset pricing results obtained from the expected utility hypothesis can be straightforwardly generalized to betweenness preferences. This is due to the fact that, for each fixed utility level (indifference curve), betweenness preferences are the same as expected utility preferences; that is, each indifference curve is linear.

25.3 RANK-DEPENDENT PREFERENCES

Rank-dependent preferences do not satisfy either the independence axiom or the betweenness axiom. Rank-dependent preferences are defined by a strictly monotone function v and a strictly monotone function $f : [0,1] \to [0,1]$ satisfying $f(0) = 0$ and $f(1) = 1$. Given a finite number of possible outcomes $x_1 < x_2 < \cdots < x_n$, the utility of a gamble P is defined as the solution $U(P)$ of

$$v(U(P)) = \sum_{i=1}^n v(x_i) \left[f\left(\sum_{j=i}^n p_j\right) - f\left(\sum_{j=i+1}^n p_j\right) \right]. \tag{25.15}$$

This equation implies $U(\delta_{x_i}) = x_i$, so we are measuring utility in certainty equivalent terms again. If $f(a) = a$ for all a, then this is standard expected utility $\sum_i p_i v(x_i)$. In general, rank-dependent utility is expected utility with respect to transformed probabilities

$$\hat{p}_i = f\left(\sum_{j=i}^{n} p_j\right) - f\left(\sum_{j=i+1}^{n} p_j\right).$$

Note that the $\hat{p}_i$ are nonnegative and sum to $f(1) - f(0) = 1$ (we adopt the usual convention that $\sum_{j=n+1}^{n} p_j = 0$).

The transformed probabilities depend on the cumulative objective probabilities.[8] If we fix P and vary the outcomes, then the transformed probability attached to any outcome depends only on the rank of the outcome within the set of outcomes. This is in contrast to weighted or disappointment-averse utility, where the transformed probability depends on the value of the weighting function λ at the outcome. Which outcomes are underweighted or overweighted, relative to the objective probabilities, depends on the curvature of f. For example, if f is convex with $f(0) = 0$ and $f(1) = 1$, then $f(a) \leq a$ for all a.[9] Consequently, if there are two outcomes with $p_1 = p_2 = 0.5$, then

$$\hat{p}_1 = f(p_1 + p_2) - f(p_2) = f(1) - f(0.5) = 1 - f(0.5) \geq 0.5$$

and $\hat{p}_2 = f(0.5) \leq 0.5$. Hence, the worse outcome is overweighted. There is experimental evidence that f should be assumed concave on $[0, a]$ for some a and convex on $[a, 1]$. See Starmer (2000) for references. This implies that both very low and very high outcomes are overweighted.[10]

CRRA rank-dependent preferences are obtained by taking v to be power or log utility. Scaling outcomes by a constant does not affect any of the $\hat{p}_i$ (because they depend only on ranks, given P), so the certainty equivalent is scaled by the same factor, just as for power or log expected utility.

---

8. We can define rank-dependent utility equivalently as

$$v(U(P)) = \sum_{i=1}^{n} v(x_i) \left[g\left(\sum_{j=1}^{i} p_j\right) - g\left(\sum_{j=1}^{i-1} p_j\right) \right] \quad (25.16)$$

by setting $g(a) = 1 - f(1 - a)$.

9. We have $f(a) = f((1-a) \times 0 + a \times 1) \leq (1-a) \times f(0) + a \times f(1) = a$.

10. For example, if there are four possible outcomes with equal objective probabilities and f is strictly concave on $[0, 1/2]$ and strictly convex on $[1/2, 1]$, then $\hat{p}_1 > \hat{p}_2$ and $\hat{p}_4 > \hat{p}_3$.

25.4 FIRST-ORDER RISK AVERSION

As discussed in Section 1.2, expected utility implies approximate risk neutrality with respect to small gambles. Of course, we can obtain moderate aversion to small gambles by assuming a high degree of risk aversion. However, this implies extreme aversion to large gambles. Rabin (2000) shows the following. Suppose a person whose preferences are represented by concave expected utility will turn down a 50-50 gamble in which she loses $100 or gains $110 (that is, a gamble in which a loss of $100 and a gain of $110 occur with probability 1/2 each) given any initial wealth level. Then, the person will, for any initial wealth level, turn down a 50-50 gamble in which she loses $1,000 or gains any amount of money. This example assumes an aversion to a small gamble for all wealth levels, which may be unreasonable. Perhaps more striking is another fact shown by Rabin (2000). Suppose a person with concave expected utility will turn down a 50-50 gamble in which she loses $100 or gains $105 at any initial wealth level less than $350,000. Then, from an initial wealth of $340,000, the person will turn down a 50-50 gamble in which she loses $4,000 or gains $635,670. For similar examples in the special case of CRRA utility, see Exercise 1.2.

The tight link between aversion to small gambles and aversion to large gambles in the expected utility framework stems from the differentiability of expected utility. See, for example, the proof in Section 1.2 that the risk premium for a small gamble is proportional to the variance of the gamble when preferences are represented by expected utility. This proportionality to the variance is called second-order risk aversion. Like expected utility, weighted utility has second-order risk aversion (Exercises 25.3 and 25.4).

On the other hand, disappointment-averse utility has first-order risk aversion, due to the discontinuity in the weighting function λ at the certainty equivalent. For example, consider an individual with CRRA disappointment-averse utility who has initial wealth w. Let $w - \pi$ be the certainty equivalent of the gamble paying $w + \varepsilon$ or $w - \varepsilon$ with equal probabilities, for a constant ε. From the definition (25.8a) of disappointment-averse utility, we have

$$\frac{1}{1-\rho}(w-\pi)^{1-\rho} = \frac{1}{1-\rho}(U(P))^{1-\rho} = \frac{1}{1-\rho}\frac{0.5(w+\varepsilon)^{1-\rho} + 0.5(w-\varepsilon)^{1-\rho}}{1+0.5\beta}.$$

Solving for π, a straightforward calculation gives

$$\pi'(0) = \frac{\beta}{2+\beta},$$

so the risk premium for a small gamble is approximately $\beta\varepsilon/(2+\beta)$. Therefore, the risk premium is approximately proportional to the standard deviation ε as the gamble becomes small. See Exercise 25.5 for another example.

Rank-dependent preferences also have first-order risk aversion. Consider CRRA rank-dependent utility and the same gamble as in the previous paragraph. Another straightforward calculation shows that

$$\pi'(0) = \hat{p}_1 - \hat{p}_2,$$

where $\hat{p}_1 = f(1) - f(0.5)$ is the transformed probability of the low outcome $w - \varepsilon$ and $\hat{p}_2 = f(0.5)$ is the transformed probability of the high outcome $w + \varepsilon$. The risk premium for a small gamble is therefore approximately $(\hat{p}_1 - \hat{p}_2)\varepsilon$, which is the negative of the mean of the gamble under the transformed probabilities. This reflects the fact that someone with rank-dependent preferences is approximately risk neutral for small gambles but under the transformed rather than the objective probabilities. If f is convex so that the low outcome is overweighted, then the mean of the gamble under the transformed probabilities is negative, and the risk premium is positive.

25.5 AMBIGUITY AVERSION

The Ellsberg experiment (Section 25.1) indicates that people may dislike bets with unknown odds. This is called ambiguity aversion. Ambiguity is also known as Knightian uncertainty (see the end-of-chapter notes). There are two closely related generalizations of Savage's theory that accommodate ambiguity aversion. One generalization replaces the subjective probability with a subjective nonadditive set function. The other replaces it with a set of subjective probabilities (multiple priors) and represents preferences by the worst-case expected utility. The optimal gamble is the one that maximizes this minimum expected utility. We first describe the multiple-priors approach, which is the model that has been more often used in finance. Nonadditive set functions are discussed at the end of the section.

To see how multiple priors resolve the Ellsberg paradox, let $\mathcal{P}$ denote the class of probabilities P that satisfy $P(\{R\}) = 1/3$. Denote expectation with respect to P by E^P. Define the utility U of a gamble to be the minimum expected value of the gamble, where the minimum is taken over $P \in \mathcal{P}$. The interpretation is that the decision maker is unsure of the probabilities of black and yellow balls, knowing only that they sum to $2/3$, and evaluates any gamble according to the worst-case distribution. For example,

$$U(1_{\{R\}}) = \min_{P \in \mathcal{P}} E^P[1_{\{R\}}] = \min_{P \in \mathcal{P}} P(\{R\}) = 1/3.$$

Likewise, $U(1_{\{B\}}) = \min_{P \in \mathcal{P}} P(\{B\}) = 0$, and $U(1_{\{Y\}}) = \min_{P \in \mathcal{P}} P(\{Y\}) = 0$, so the minimum expected utility U is consistent with the preference ordering $1_{\{R\}} \succ 1_{\{B\}}$ and $1_{\{R\}} \succ 1_{\{Y\}}$. Also, $U(1_{\{Y,B\}}) = 2/3$, $U(1_{\{R,Y\}}) = 1/3$, and

$U(1_{\{R,B\}}) = 1/3$. Thus, the minimum expected utility U is also consistent with the preferences $1_{\{Y,B\}} \succ 1_{\{R,Y\}}$ and $1_{\{Y,B\}} \succ 1_{\{R,B\}}$.

NONPARTICIPATION

A simple but important consequence of ambiguity aversion is that investors may decline to participate in security markets. Suppose there is a risk-free asset with return R_f and a risky asset with return $\tilde{R}$. Let $\mathcal{P}$ be a class of probability distributions for the return $\tilde{R}$ such that $\mathsf{E}^{P_1}[\tilde{R}] < R_f < \mathsf{E}^{P_2}[\tilde{R}]$ for some $P_1, P_2 \in \mathcal{P}$. This is a model for an investor who is unsure whether the risk premium is positive. Suppose the utility function u is concave. For a long position $\pi > 0$, concavity implies that

$$\min_{P \in \mathcal{P}} \mathsf{E}^P\left[u\left(w_0 R_f + w_0 \pi (\tilde{R} - R_f)\right)\right] \leq \mathsf{E}^{P_1}\left[u\left(w_0 R_f + w_0 \pi (\tilde{R} - R_f)\right)\right]$$

$$\leq u\left(w_0 R_f + w_0 \pi \mathsf{E}^{P_1}[\tilde{R} - R_f]\right)$$

$$< u(w_0 R_f).$$

Therefore, $\pi = 0$ is preferred to any $\pi > 0$. Likewise, for a short position,

$$\min_{P \in \mathcal{P}} \mathsf{E}^P\left[u\left(w_0 R_f + w_0 \pi (\tilde{R} - R_f)\right)\right] \leq \mathsf{E}^{P_2}\left[u\left(w_0 R_f + w_0 \pi (\tilde{R} - R_f)\right)\right] < u(w_0 R_f).$$

Thus, $\pi = 0$ is optimal. Hence, ambiguity aversion can potentially explain why there are many investors who do not participate in the stock market and why many others participate only selectively, holding undiversified positions. Recall that it is rare for $\pi = 0$ to be optimal in the standard model, because in a single-asset model a long (short) position is optimal if the expected return is greater than (less than) the risk-free return (Section 2.2). Thus, $\pi = 0$ is optimal in the standard model only if the risk premium is exactly zero.

As an example, suppose there is a single risky asset and $\mathcal{P}$ is a class of normal distributions for its return $\tilde{R}$. Suppose there is a risk-free asset and the investor has a CARA utility function. Let $\phi = w_0 \pi$ denote the investment in the risky asset. The investor chooses ϕ to maximize the worst-case certainty equivalent:

$$\min_{P \in \mathcal{P}} \phi \left(\mathsf{E}^P[\tilde{R}] - R_f\right) - \frac{1}{2}\alpha \phi^2 \operatorname{var}^P(\tilde{R}).$$

Suppose $\mathcal{P}$ is the class of all normal distributions for which $\mu_a \leq \mathsf{E}^P[\tilde{R}] \leq \mu_b$ and $\sigma_a^2 \leq \operatorname{var}^P(\tilde{R}) \leq \sigma_b^2$ for constants $\mu_a < \mu_b$ and $\sigma_a < \sigma_b$. Then, the worst-case distribution for a long position is the lowest mean μ_a and the maximum variance σ_b^2, and the worst-case distribution for a short position is the highest mean μ_b and the maximum variance σ_b^2. A long position is optimal if $\mu_a > R_f$, a short position is optimal if $\mu_b < R_f$, and $\phi = 0$ is optimal if $\mu_a < R_f < \mu_b$.

Euler Inequalities

To see how the usual relationship between marginal utility and SDFs is affected by ambiguity aversion, let ϕ^* be an optimal portfolio and let $\mathcal{Q}(\phi^*) \subset \mathcal{P}$ be the class of worst-case distributions for ϕ^*. In the CARA–normal example in the preceding subsection, if $\phi^* \neq 0$ (the investor chooses to go long or short), then $\mathcal{Q}(\phi^*)$ contains only a single distribution P_{ϕ^*} (the lowest mean and highest variance if long and the highest mean and highest variance if short), and the first-order condition must hold relative to that distribution—that is,

$$\mathsf{E}^{P_{\phi^*}}\left[u'(\tilde{w}^*)(\tilde{R}-R_f)\right]=0,$$

where $\tilde{w}^* = w_0 R_f + \phi^*(\tilde{R}-R_f)$. Thus, pricing is as in a single-prior model with prior P_{ϕ^*}. On the other hand, if $\phi^* = 0$, then $\mathcal{Q}(\phi^*) = \mathcal{P}$ and

$$\min_{P \in \mathcal{Q}(\phi^*)} \mathsf{E}^P\left[u'(\tilde{w}^*)(\tilde{R}-R_f)\right] = u'(w_0 R_f)(\mu_a - R_f)$$

$$< 0 < u'(w_0 R_f)(\mu_b - R_f) = \max_{P \in \mathcal{Q}(\phi^*)} \mathsf{E}^P\left[u'(\tilde{w}^*)(\tilde{R}-R_f)\right].$$

Thus, in this example, the asset is not correctly priced by marginal utility relative to all of the worst-case distributions (though it is correctly priced relative to the distributions with $\mathsf{E}^P[\tilde{R}] = R_f$). Whether ϕ^* is zero or not, we have

$$\min_{P \in \mathcal{Q}(\phi^*)} \mathsf{E}^P\left[u'(\tilde{w}^*)(\tilde{R}-R_f)\right] \leq 0 \leq \max_{P \in \mathcal{Q}(\phi^*)} \mathsf{E}^P\left[u'(\tilde{w}^*)(\tilde{R}-R_f)\right]. \quad (25.17)$$

Epstein and Wang (1994) call (25.17) the Euler inequalities.

Updating Multiple Priors

The issue of learning from information when there are multiple priors is complex. The fundamental object is preferences, so the basic question is how preferences are updated when new information is obtained. Consider the Ellsberg experiment, in which the numbers of black and yellow balls are unknown, and suppose an ambiguity-averse person is informed that the ball drawn is either red or yellow. How will the person now evaluate the gambles $1_{\{R\}}$ and $1_{\{Y\}}$? It is impossible to give a general answer to this question, because the violation of the sure thing principle means that conditional preferences—here, conditional on the event $A = \{R, Y\}$—are not uniquely determined. The preferences depend on what would have happened on the complementary event (when a black ball is drawn). If someone with the Ellsberg preferences gets 0 when a black ball is drawn, then she prefers the bet $1_{\{R\}}$ to $1_{\{Y\}}$ conditional on A. But if the same person gets 1 when a black ball is drawn, then she prefers $1_{\{Y\}}$ to $1_{\{R\}}$ conditional on A. See (25.3).

Gilboa and Schmeidler (1993) suggest two different updating rules. One is to preserve the ranking the gambles had when the worst possible outcome is paid on the complement of A. This is called optimistic, because A is good news in this circumstance. The other is to preserve the ranking when the best possible outcome is paid on the complement of A, which is called pessimistic. In the Ellsberg experiment, 0 is the worst outcome and 1 is the best, so the optimistic ranking of the gambles conditional on $A = \{R, Y\}$ is $1_{\{R\}} \succ 1_{\{Y\}}$, as shown in (25.3a). On the other hand, the pessimistic ranking is the reverse, as shown in (25.3b).

The different updating rules for preferences correspond to different updating rules for multiple priors (and also for nonadditive set functions). The optimistic ranking corresponds to applying Bayes' rule to each prior and then evaluating the gambles according to the worst-case conditional distribution. The pessimistic ranking corresponds to using maximum likelihood to update the set of priors.

As an example, consider the Ellsberg experiment, taking $\mathcal{P}$ as before to be the set of probabilities for which $P(\{R\}) = 1/3$. Applying Bayes' rule to each probability produces the set of conditional probabilities

$$P(R \mid R, Y) = \frac{1/3}{1/3 + p},$$

$$P(Y \mid R, Y) = \frac{p}{1/3 + p},$$

where the prior probability p of yellow ranges between 0 and $2/3$. Thus, the conditional distributions assign probability between $1/3$ and 1 to red and the complementary probability to yellow. The implied conditional preferences are the optimistic preferences:

$$U(1_{\{R\}} \mid R, Y) = \min_{P \in \mathcal{P}} P(R \mid R, Y) = \min_{0 \le p \le 2/3} \frac{1/3}{1/3 + p} = 1/3$$

$$> 0 = \min_{0 \le p \le 2/3} \frac{p}{1/3 + p} = \min_{P \in \mathcal{P}} P(Y \mid R, Y) = U(1_{\{Y\}} \mid R, Y).$$

Updating by maximum likelihood in this example implies that the set of priors shrinks to a single conditional distribution. For each prior, the likelihood of observing the event $\{R, Y\}$ is $1/3 + p$, where p is the prior probability of yellow, so the likelihood is maximized at $p = 2/3$. For this prior, the conditional probability of red is $1/3$ and the conditional probability of yellow is $2/3$. The implied conditional preferences are the pessimistic preferences: $1_{\{Y\}} \succ 1_{\{R\}}$.

For another example, consider a risky asset with payoff $\tilde{x}$. Suppose there is no ambiguity about the marginal distribution of $\tilde{x}$. In particular, suppose it is normal with mean μ and variance σ^2. Assume that a signal $\tilde{s} = \tilde{x} + \tilde{\varepsilon}$ is observed. Assume that $\tilde{\varepsilon}$ is known to be normally distributed, independent of $\tilde{x}$, and to have a zero mean, but there is ambiguity about its standard deviation ϕ, with any standard deviation in an interval $[\phi_a, \phi_b]$ being possible. Thus, there is ambiguity about the quality of the signal $\tilde{s}$. Applying Bayes' rule to each prior produces the class of normal conditional distributions for $\tilde{x}$ with mean $\mu + \beta(\tilde{s} - \mu)$ and variance $(1 - \beta)\sigma^2$, where

$$\beta = \frac{\sigma^2}{\sigma^2 + \phi^2}. \tag{25.18}$$

On the other hand, the maximum likelihood estimate is

$$\phi^2 = \max\left(0, (\tilde{s} - \mu)^2 - \sigma^2\right).$$

Again, updating by maximum likelihood shrinks the set of priors to a single conditional distribution. It seems a bit unreasonable that a single realization of the signal $\tilde{s}$ should completely resolve the ambiguity about its distribution. The model that has been used in finance is Bayesian updating of the set of priors.

Dynamic Consistency

One desired characteristic of a model is that preferences be dynamically consistent, meaning that plans formulated at one date for implementation at a later date remain optimal when the later date is reached. To see the implications of ambiguity aversion for dynamic consistency, consider a three-date ($t = 0, 1, 2$) version of the model in the previous subsection. Assume the asset pays $\tilde{x}$ at date 2 and pays no dividends at prior dates. Assume $\tilde{x}$ is known to be normally distributed with mean μ and variance σ^2. Assume the investor consumes only at date 2. Assume a signal $\tilde{s} = \tilde{x} + \tilde{\varepsilon}$ is observed before trade at date 1, with the ambiguity about the signal quality being the same as in the preceding subsection. This means that $\tilde{x}$ is conditionally normal with mean $\mu + \beta(\tilde{s} - \mu)$ and variance $(1 - \beta)\sigma^2$, where β is defined in (25.18). Because ϕ is in the interval $[\phi_a, \phi_b]$, β is in the interval $[\beta_a, \beta_b]$, where

$$\beta_a = \frac{\sigma^2}{\sigma^2 + \phi_b^2} \quad \text{and} \quad \beta_b = \frac{\sigma^2}{\sigma^2 + \phi_a^2}.$$

Assume the risk-free return is R_f in each period. Let θ_t denote the number of shares the investor chooses to hold at date t. The investor's wealth W evolves as

$$W_{t+1} = W_t R_f + \theta_t (P_{t+1} - P_t R_f),$$

where $P_2 = \tilde{x}$ and P_1 depends on $\tilde{s}$. This model is not dynamically consistent. To see the difficulty, note that θ_1 will be chosen to maximize

$$\min_{\beta_a \leq \beta \leq \beta_b} \mathsf{E}^\beta[u(W_2) \mid W_1, \tilde{s}],$$

where E^β denotes expectation with regard to $\tilde{x}$ being normal with mean $\mu + \beta(\tilde{s} - \mu)$ and variance $(1-\beta)\sigma^2$. The worst-case β will in general depend on $\tilde{s}$, and this dependence will in general affect the choice of θ_1. However, when the investor chooses θ_0 and formulates a plan $s \mapsto \theta_1(s)$ at date 0, she does so to maximize $\min_\beta \mathsf{E}^\beta[u(W_2)]$, where now the expectation is over the joint distribution of $\tilde{s}$ and $\tilde{x}$, with $\tilde{s}$ being normal with mean μ and variance σ^2/β.[11] This minimization does not allow β to depend on $\tilde{s}$. In particular, the worst-case β at date 0 need not be the worst-case β at date 1. Hence, the plan that is selected at date 0 will not in general be optimal when date 1 arrives. Furthermore, the date–0 portfolio that is optimal in conjunction with the inconsistent plan will not generally be the same as the portfolio that would be chosen if the actual choice of θ_1 were anticipated correctly. In a nutshell, dynamic inconsistency is a failure of Bellman's principle of optimality: Backward induction produces different decisions than forward planning. This phenomenon is illustrated in Exercise 25.6, which analyzes this model with a risk-neutral representative investor (a model due to Epstein and Schneider, 2008).

We must either give up dynamic consistency or modify the model. To modify the model, we can start with the dynamic programming solution and change the forward-planning model to match it. Thus, in the example, we can take the marginal distribution of $\tilde{s}$ to be normal with mean μ and variance σ^2/β_0 for $\beta_a \leq \beta_0 \leq \beta_b$ and take the conditional distribution of $\tilde{x}$ given $\tilde{s}$ to be normal with mean $\mu + \beta_1(\tilde{s})(\tilde{s} - \mu)$ and variance $(1 - \beta_1(\tilde{s}))\sigma^2$, with the investor believing that all (measurable) functions $\beta_1 : \mathbb{R} \to [\beta_a, \beta_b]$ are possible. In this modified model, when the investor formulates a plan at date 0, she maximizes the worst case over all constants β_0 and functions β_1. The modified model possesses the property of rectangularity defined by Epstein and Schneider (2003), who show that rectangularity implies dynamic consistency when the set of priors is updated by Bayes' rule. The modified model may seem artificial, but a multiple-priors model (like expected utility) is always an "as if" construction—solving a max-min problem produces the same decisions an Ellsberg agent would make, but it is not necessary that the agent literally believe in the multiple priors. In this example, we can take the view that the backward induction with ambiguity represents the investor's preferences. If so, then the modified forward-planning model is simply another way to represent them.

---

11. The variance of $\tilde{s}$ is $\sigma^2 + \phi^2$ and $\beta = \sigma^2/(\sigma^2 + \phi^2)$, so the variance of $\tilde{s}$ is σ^2/β.

Nonadditive Set Functions

To describe the other generalization of Savage's theory, let ϕ be a function of events $A \subset \Omega$ with the properties that $\phi(\emptyset) = 0$, $\phi(\Omega) = 1$, and $\phi(A) \leq \phi(B)$ if $A \subset B$. Additivity is the property that $\phi(A \cup B) = \phi(A) + \phi(B)$ when A and B are disjoint sets. Probabilities are additive (in fact, they are additive for countable unions of disjoint sets). However, here we do not require additivity. A nonadditive set function ϕ consistent with the Ellsberg preferences—in the sense that, for all sets $A_0, A_1 \subset \{R, B, Y\}$, $1_{A_0} \succeq 1_{A_1}$ if and only if $\phi(A_0) \geq \phi(A_1)$—is the set function $\phi(\emptyset) = 0$, $\phi(\{R\}) = 1/3$, $\phi(\{B\}) = \phi(\{Y\}) = 0$, $\phi(\{R,B\}) = \phi(\{R,Y\}) = 1/3$, $\phi(\{Y,B\}) = 2/3$, $\phi(\{R,B,Y\}) = 1$. Additivity fails because $\phi(\{Y,B\}) = 2/3 > 0 = \phi(\{Y\}) + \phi(\{B\})$, and this breaks the chain of implications (with $P = \phi$) in (25.2).

A set function ϕ is said to be ambiguity averse if

$$\phi(A \cup B) + \phi(A \cap B) \geq \phi(A) + \phi(B)$$

for all events A and B.[12] Given ϕ, let $\mathcal{P}$ be the set of additive set functions P satisfying $P(A) \geq \phi(A)$ for all events A. The set $\mathcal{P}$ is called the core of ϕ. Then, for all events A,

$$\phi(A) = \min_{P \in \mathcal{P}} P(A).$$

For example, with ϕ as in the preceding paragraph for the Ellsberg experiment, $\mathcal{P}$ is the class of probabilities P satisfying $P(\{R\}) = 1/3$, as discussed earlier in the multiple-priors resolution of the Ellsberg paradox. The Ellsberg preferences are represented both by the nonadditive set function ϕ and by its core $\mathcal{P}$.

If ϕ is an ambiguity-averse set function and $\mathcal{P}$ is the core of ϕ, then, for any measurable function u and gamble $\tilde{x}$,

$$\mathrm{E}^\phi[u(\tilde{x})] = \min_{P \in \mathcal{P}} \mathrm{E}^P[u(\tilde{x})],$$

where the superscript denotes the set function with respect to which the "expectation" is taken. Thus, a utility function u defines the same preferences over gambles, whether we use it with the nonadditive set function ϕ or with the core $\mathcal{P}$ of ϕ. The "expectation" of a function u with respect to a nonadditive set function ϕ is defined as

$$\mathrm{E}^\phi[u(\tilde{x})] = \int_0^\infty \phi(\{\omega \mid u(\tilde{x}(\omega)) \geq a\}) \, da - \int_{-\infty}^0 \left[1 - \phi(\{\omega \mid u(\tilde{x}(\omega)) \geq a\})\right] da.$$

---

12. If ϕ were additive, we would have $\phi(A \cup B) + \phi(A \cap B) = \phi(A) + \phi(B)$.

This is called the Choquet integral.[13]

25.6 NOTES AND REFERENCES

Another example of framing given by Kahneman and Tversky (1979) concerns the example in Exercise 25.1. They posed the same example to subjects as a two-stage (compound) gamble. In the first stage, there is a 75% chance of the game ending with no payment. With 25% probability, the game continues and the subjects get the choice between A and B. However, the choice must be made before the game begins, so the actual gambles faced are C and D. A majority of subjects chose C over D when the choice was described as this compound gamble, in contrast to the dominant preference for D over C when described as a single gamble.

Other examples of loss aversion given by Kahneman and Tversky (1979) concern the examples in Exercises 25.1 and 25.2. When the gains in those examples (e.g., $3,000) were changed to losses (e.g., −$3,000) the opposite pattern of preferences was obtained (still violating the independence axiom). For example, the preferences $A \succ B$ and $D \succ C$ in Exercise 25.1 indicate that, when gains are probable, the sure outcome is preferred, but when gains are less probable, the riskier gamble (the gamble with the larger possible outcome) is preferred. However, when cast as losses, subjects generally preferred the riskier gamble when losses are probable (preferring to avoid a sure loss) but the safer outcome when losses are less likely.

Tversky and Kahneman (1992) extend prospect theory to incorporate ambiguity aversion and rank-dependent preferences. They recommend a utility function of the form (25.4) with a nonadditive set function ϕ replacing the subjective probability and with a transformation of the "probabilities" $\phi(\{x_i\})$ as in rank-dependent preferences.

Benartzi and Thaler (1995) argue that prospect theory can explain the equity premium puzzle, because loss aversion combined with frequent portfolio evaluation makes people unwilling to hold equities even when the equity premium is large. The importance of frequent portfolio evaluation is that the probability of loss with equities is small over long horizons when the equity premium is substantial; hence, loss aversion would have little effect if gains/losses over long

---

13. To interpret the Choquet integral, consider the example of u being nonnegative and monotone and $\tilde{x}$ being uniformly distributed on $[0, 1]$ relative to a probability ϕ (i.e., $\phi(\{\omega \,|\, \tilde{x}(\omega) \leq x\}) = x$ for $x \in [0, 1]$). If we graph the function u with x on the horizontal axis and $u(x)$ on the vertical, then the Choquet integral is the area between the utility function and 0, computed by integrating over the vertical axis, instead of by integrating over the horizontal axis, as we would normally do.

horizons were the issue. However, Benartzi and Thaler argue that even investors with long horizons are likely to evaluate their portfolios frequently, adjusting the reference wealth level to the current value each time, and suffer disutility each time a loss occurs. Barberis and Huang (2001) and Barberis, Huang, and Santos (2001) develop more formal dynamic models of prospect theory.

Betweenness preferences are often called Chew-Dekel preferences, in recognition of Chew (1983, 1989) and Dekel (1986). The utility representation (25.5) for betweenness preferences is due to Dekel (1986). Weighted utility is axiomatized by Chew (1983). A generalization of weighted utility that also satisfies betweenness, called semiweighted utility, is axiomatized by Chew (1989). Disappointment aversion is axiomatized by Gul (1991). Generalized disappointment aversion is due to Routledge and Zin (2010). They show that it produces countercyclical risk aversion, a large equity premium and a small and slowly varying risk-free rate. Thus, it can explain the major aggregate puzzles.

Chew and Epstein (1989) and Skiadas (1998) axiomatize recursive utility (Section 11.6) with non-expected-utility certainty equivalents, extending Kreps and Porteus (1978). Epstein and Zin (1989) develop EZW utility for betweenness preferences, replacing the certainty equivalent (11.27) with a certainty equivalent defined from (25.5). They also describe portfolio choice with betweenness preferences.

Rank-dependent preferences are axiomatized by Yaari (1987) and Segal (1990). The distinction between first- and second-order risk aversion is made by Segal and Spivak (1990). For surveys of non-expected-utility theory that are more extensive on some dimensions and for additional references, see Machina (1987), Starmer (2000), and Backus, Routledge, and Zin (2005). The last of these focuses on applications to finance and macroeconomics and is the source for Exercises 25.4(c) and 25.5.

It is common to cite Knight (1921) as making a distinction between risk and uncertainty, defining risk as a circumstance in which probabilities can be attached to events and uncertainty as a situation that is so unfamiliar that probabilities cannot be assessed. In the economics and finance literature, "Knightian uncertainty" and "ambiguity" are synonymous. However, LeRoy and Singell (1987) argue that the distinction between risk and uncertainty is made by Keynes (1921) and that Knight intended no such distinction.

Expected utility with respect to a nonadditive set function is axiomatized by Gilboa (1987) and Schmeidler (1989). Max-min utility with multiple priors is axiomatized by Gilboa and Schmeidler (1989). The connection between nonadditive set functions and multiple priors described in Section 25.5 is due to Schmeidler (1986). The result on nonparticipation with ambiguity aversion is due to Dow and Werlang (1992), though they use the nonadditive set function approach instead of multiple priors. The Euler inequalities are derived in a more general setting by Epstein and Wang (1994). Chen and Epstein (2002) study ambiguity aversion in continuous time. They extend stochastic differential utility (Duffie and Epstein, 1992b) to accommodate multiple priors.

The three-date asset pricing model with ambiguity aversion described in Section 25.5 is due to Epstein and Schneider (2008). The primary conclusion of that model is that there is a price discount (an ambiguity premium) even when investors are risk neutral and have no ambiguity about the marginal distribution of the asset payoff—see Exercise 25.6(c) for the precise formula. Thus, ambiguity aversion is another possible explanation of the equity premium puzzle. Illeditsch (2011) extends the Epstein-Schneider model by allowing the representative investor to be risk averse. He shows that the model exhibits portfolio inertia and excess volatility. Furthermore, he shows that the price function $P_1(s)$ has a discontinuity. Thus, a small change in the information $\tilde{s}$ can have a large effect on the price.

EXERCISES

25.1. Consider the following pairs of gambles:

A : 100% chance of \$3,000   versus $B : \begin{cases} 80\% \text{ chance of } \$4,000 \\ 20\% \text{ chance of } \$0 \end{cases}$

$C : \begin{cases} 25\% \text{ chance of } \$3,000 \\ 75\% \text{ chance of } \$0 \end{cases}$ versus $D : \begin{cases} 20\% \text{ chance of } \$4,000 \\ 80\% \text{ chance of } \$0 \end{cases}$

(a) Show that an expected utility maximizer who prefers A to B must also prefer C to D.
(b) Show that the preferences $A \succ B$ and $D \succ C$ violate the independence axiom by showing that $C = \alpha A + (1-\alpha)Q$ and $D = \alpha B + (1-\alpha)Q$ for some $0 < \alpha < 1$ and some gamble Q.
(c) Plot the gambles A, B, C, and D in the probability simplex of Figure 20.1, taking p_1 to be the probability of \$0 and p_3 to be the probability of \$4,000. Show that the line connecting A with B and the line connecting C with D are parallel.

Note: The preferences $A \succ B$ and $D \succ C$ are common. This example is due to Allais (1953) and is a special case of the common ratio effect. See, for example, Starmer (2000).

25.2. Consider the following pairs of gambles:

$$A : \begin{cases} 90\% \text{ chance of } \$3{,}000 \\ 10\% \text{ chance of } \$0 \end{cases} \quad \text{versus} \quad B : \begin{cases} 45\% \text{ chance of } \$6{,}000 \\ 55\% \text{ chance of } \$0 \end{cases}$$

$$C : \begin{cases} 0.2\% \text{ chance of } \$3{,}000 \\ 99.8\% \text{ chance of } \$0 \end{cases} \quad \text{versus} \quad D : \begin{cases} 0.1\% \text{ chance of } \$6{,}000 \\ 99.9\% \text{ chance of } \$0 \end{cases}.$$

Show that an expected utility maximizer who prefers A to B must also prefer C to D. Note: The preferences $A \succ B$ and $D \succ C$ are common. This example is due to Kahneman and Tversky (1979).

25.3. Consider weighted utility. Let $\tilde{\varepsilon}$ have zero mean and unit variance. For a constant σ, denote the certainty equivalent of $w + \sigma\tilde{\varepsilon}$ by $w - \pi(\sigma)$. Assume $\pi(\cdot)$ is twice continuously differentiable. By differentiating

$$v(w - \pi(\sigma))\mathsf{E}[\lambda(w + \sigma\tilde{\varepsilon})] = \mathsf{E}[\lambda(w + \sigma\tilde{\varepsilon})v(w + \sigma\tilde{\varepsilon})],$$

assuming differentiation and expectation can be interchanged, show successively that $\pi'(0) = 0$ and

$$\pi''(0) = -\frac{v''(w)}{v'(w)} - \frac{2\lambda'(w)}{\lambda(w)}.$$

Note: This implies that for CRRA weighted utility and small σ, $\pi(\sigma)/w \approx (\rho - 2\gamma)\operatorname{var}(\sigma\tilde{\varepsilon}/w)/2$.

25.4. Consider CRRA weighted utility.

(a) Show that g in (25.13) is strictly monotone in $y > 0$—so the preferences are monotone with regard to stochastic dominance—if and only if $\gamma \leq 0$ and $\rho \leq \gamma + 1$ with at least one of these being a strict inequality.
(b) Show that g in (25.13) is strictly monotone and concave if and only if $\gamma \leq 0$ and $\gamma \leq \rho \leq \gamma + 1$ with either $\gamma < 0$ or $\rho < \gamma + 1$.
(c) Consider a lognormal gamble: $\tilde{w} = w(1 + \tilde{\varepsilon})$ where $\log(1 + \tilde{\varepsilon})$ is normally distributed with variance σ^2 and mean $-\sigma^2/2$ (implying $\mathsf{E}[\tilde{\varepsilon}] = 0$). Show that the certainty equivalent is $w(1 - \pi)$, where

$$\pi = 1 - e^{-(\rho - 2\gamma)\sigma^2/2}.$$

Note: This implies that $\pi \approx (\rho - 2\gamma)\sigma^2/2$ for small σ. Compare Exercise 1.5.

25.5. Consider CRRA disappointment-averse utility and a random wealth $\tilde{w} = e^{\tilde{z}}$, where $\tilde{z}$ is normally distributed with mean μ and variance σ^2. Let ξ denote the certainty equivalent of $\tilde{w}$, and set $\theta = \log \xi$.

(a) Show that θ satisfies the equation

$$\theta = \mu + \frac{1}{2}(1-\rho)\sigma^2 + \frac{1}{1-\rho}\log\left[\frac{1+\beta N\left(\frac{\theta - \mu - (1-\rho)\sigma^2}{\sigma}\right)}{1+\beta N\left(\frac{\theta - \mu}{\sigma}\right)}\right],$$

where N denotes the standard normal distribution function. Hint: See the calculation of $\mathsf{E}\left[e^{b\tilde{z}}1_{\{\tilde{z}<a\}}\right]$ for a normal random variable $\tilde{z}$ in Section 7.5.

(b) Let $\mu = -\sigma^2/2 + \log w$ for a constant w. (Defining the standard normal $\tilde{x} = (\tilde{z}-\mu)/\sigma$, we then have $\tilde{w} = w(1+\tilde{\varepsilon})$, where $\tilde{\varepsilon} \stackrel{\text{def}}{=} e^{-\sigma^2/2+\sigma\tilde{x}} - 1$ has mean zero.) Define $\pi = (w-\xi)/w$. (Then, $w(1-\pi)$ is the certainty equivalent of $w(1+\tilde{\varepsilon})$.) Show numerically that π/σ^2 appears to increase without bound and π/σ appears to converge to a positive constant as $\sigma \downarrow 0$.

25.6. Consider the portfolio choice model with a single risky asset described in Section 25.5, in which there is no ambiguity about the marginal distribution about the asset payoff but there is ambiguity about the informativeness of a signal. Assume the investor is a representative investor, there is a single unit of the risky asset, and the risk-free asset is in zero net supply. Assume Bayesian updating of the set of priors, and assume the representative investor is risk neutral (but ambiguity averse). Let P_t denote the price of the risky asset at date t, with $P_2 = \tilde{x}$. Take $R_f = 1$. The intertemporal budget equation is

$$W_{t+1} = W_t + \theta_t(P_{t+1} - P_t),$$

where θ_t denote the number of shares of the risky asset chosen at date t. The distribution of $\tilde{x}$ conditional on $\tilde{s}$ is normal with mean $\mu + \beta(\tilde{s}-\mu)$ and variance $(1-\beta)\sigma^2$, and the marginal distribution of $\tilde{s}$ is normal with mean μ and variance σ^2/β, for $\beta_a \leq \beta \leq \beta_b$. Take the backward induction (dynamic programming) approach.

(a) Suppose $\tilde{s} < \mu$. Show that $\theta_1 = 1$ maximizes

$$\min_{\beta} \mathsf{E}^{\beta}[W_2 \mid \tilde{s}]$$

if and only if

$$P_1 = \mu + \beta_b(\tilde{s} - \mu).$$

(b) Suppose $\tilde{s} > \mu$. Show that $\theta_1 = 1$ maximizes

$$\min_{\beta} \mathsf{E}^{\beta}[W_2 \mid \tilde{s}]$$

if and only if
$$P_1 = \mu + \beta_a(\tilde{s} - \mu).$$

(c) Suppose that P_1 depends on $\tilde{s}$ as described in the previous parts. Show that $\theta_0 = 1$ maximizes
$$\min_\beta \mathsf{E}^\beta[W_1]$$
if and only if
$$P_0 = \mu - \frac{(\beta_b - \beta_a)\sigma}{\sqrt{2\pi \beta_a}}.$$

Hint: The function P_1 is concave in $\tilde{s}$. Hence, $\mathsf{E}^\beta[P_1(\tilde{s})] < P_1(\mu)$, and the difference $P_1(\mu) - \mathsf{E}^\beta[P_1(\tilde{s})]$ is maximized at the maximum variance for $\tilde{s}$.

(d) Now take the forward-planning approach. Let P_0 and $P_1(s)$ be as described in the previous parts. The investor chooses θ_0 and a plan $s \mapsto \theta_1(s)$ at date 0 to maximize
$$\min_\beta \mathsf{E}^\beta[W_2] = \min_\beta \mathsf{E}^\beta[W_0 + \theta_0\{P_1(\tilde{s}) - P_0\} + \theta_1(s)\{\tilde{x} - P_1(\tilde{s})\}].$$

Show that the investor can achieve unbounded worst-case expected wealth. In particular, choosing $\theta_0 = 1$ and $\theta_1(s) = 1$ for all s is not optimal.

Some Probability and Stochastic Process Theory

CONTENTS

A.1	Random Variables	679
A.2	Probabilities	680
A.3	Distribution Functions and Densities	681
A.4	Expectations	681
A.5	Convergence of Expectations	682
A.6	Interchange of Differentiation and Expectation	683
A.7	Random Vectors	684
A.8	Conditioning	685
A.9	Independence	686
A.10	Equivalent Probability Measures	687
A.11	Filtrations, Martingales, and Stopping Times	688
A.12	Martingales under Equivalent Measures	688
A.13	Local Martingales	689
A.14	The Usual Conditions	690

Shiryayev (1984) is an accessible reference for most of the results stated in this appendix.

A.1 RANDOM VARIABLES

A random variable is a variable that depends on the random state of the world. The set of states of the world is denoted by Ω and a typical state of the world by ω. We denote the real line $(-\infty, \infty)$ by $\mathbb{R}$. Thus, a real-valued random variable is a function $\tilde{x}$ from Ω to $\mathbb{R}$. More precisely, a real-valued random variable is a measurable function from Ω to $\mathbb{R}$, a concept that we will now define.

There is a given class $\mathcal{F}$ of subsets of Ω with the interpretation that any set A of states of the world that can potentially be distinguished is included in $\mathcal{F}$. An element of $\mathcal{F}$ is called an event. The class $\mathcal{F}$ includes Ω and is closed under complements and under countable unions and intersections:

$$A \in \mathcal{F} \quad \Rightarrow \quad \{\omega \mid \omega \notin A\} \in \mathcal{F}, \tag{A.1a}$$

$$A_1, A_2, \ldots \in \mathcal{F} \quad \Rightarrow \quad \bigcup_{i=1}^{\infty} A_i \in \mathcal{F}, \tag{A.1b}$$

$$A_1, A_2, \ldots \in \mathcal{F} \quad \Rightarrow \quad \bigcap_{i=1}^{\infty} A_i \in \mathcal{F}. \tag{A.1c}$$

These properties identify $\mathcal{F}$ as a σ-field.

A function $\tilde{x}$ from Ω to the real line is measurable (with respect to $\mathcal{F}$) if $\{\omega \mid \tilde{x}(\omega) \leq a\} \in \mathcal{F}$ for each real a. The Borel σ-field is the smallest σ-field of subsets of the real line containing the set $(-\infty, a]$ for each real a, and one calls $A \subset \mathbb{R}$ a Borel set if A belongs to the Borel σ-field. An equivalent definition of $\tilde{x}$ being measurable is that $\{\omega \mid \tilde{x}(\omega) \in A\} \in \mathcal{F}$ for each Borel set A.

We also need the concept of a measurable function defined on the real line. A function $g : \mathbb{R} \to \mathbb{R}$ is measurable if the set $\{x \mid g(x) \leq a\}$ is a Borel set for each real a. Every continuous function is measurable. If $g : \mathbb{R} \to \mathbb{R}$ is a measurable function and $\tilde{x}$ is a random variable, then $g(\tilde{x})$ is also a random variable.

Let $\mathcal{G}$ be a sub-σ-field of $\mathcal{F}$, meaning that $\mathcal{G}$ is a σ-field and $\mathcal{G} \subset \mathcal{F}$. A random variable $\tilde{x}$ is measurable with respect to $\mathcal{G}$ if $\{\omega \mid \tilde{x}(\omega) \in A\} \in \mathcal{G}$ for every Borel $A \subset \mathbb{R}$. The interpretation of measurability with respect to $\mathcal{G}$ is that if one has the information $\mathcal{G}$, then one knows the realization of $\tilde{x}$.

For a random variable $\tilde{x}$, let $\mathcal{G}$ denote the collection of sets $\{\omega \mid \tilde{x}(\omega) \in A\}$ where A ranges over the Borel subsets of $\mathbb{R}$. Then, $\mathcal{G}$ is a sub-σ-field of $\mathcal{F}$ and is called the σ-field generated by $\tilde{x}$. The interpretation is that the information $\mathcal{G}$ is exactly that obtained by observing $\tilde{x}$. Any other random variable $\tilde{y}$ is measurable with respect to $\mathcal{G}$ if and only if it is a function of $\tilde{x}$, meaning $\tilde{y} = g(\tilde{x})$ for a measurable $g : \mathbb{R} \to \mathbb{R}$.

A.2 PROBABILITIES

The probability of each event $A \in \mathcal{F}$ is denoted by $\mathbb{P}(A)$. Of course, $0 \leq \mathbb{P}(A) \leq 1$ and $\mathbb{P}(\Omega) = 1$. The set-function $\mathbb{P}$ is also assumed to have the property that if $A_1, A_2, \ldots$ are disjoint events (meaning that $A_i \cap A_j = \emptyset$ for each $i \neq j$), then

Some Probability and Stochastic Process Theory

$$\mathbb{P}\left(\bigcup_{i=1}^{\infty} A_i\right) = \sum_{i=1}^{\infty} \mathbb{P}(A_i).$$

These properties identify $\mathbb{P}$ as a probability measure.

If $\tilde{x}$ is a real-valued random variable and A is a Borel set, then we define

$$\text{prob}(\tilde{x} \in A) = \mathbb{P}(\{\omega \mid \tilde{x}(\omega) \in A\}).$$

The set function $A \mapsto \text{prob}(\tilde{x} \in A)$ is a probability measure on the Borel σ-field of the real line; it is the measure "induced" by the random variable $\tilde{x}$. We call this set function the distribution of $\tilde{x}$.

A.3 DISTRIBUTION FUNCTIONS AND DENSITIES

The cumulative distribution function of a random variable $\tilde{x}$ is the function F defined by

$$F(b) = \text{prob}(\tilde{x} \leq b)$$

for each real b. A function f is a density function of $\tilde{x}$ if

$$F(b) = \int_{-\infty}^{b} f(a) \, da$$

for each b. A density function is not uniquely defined; however, if f_1 and f_2 are density functions of the same random variable $\tilde{x}$, then $\{a \mid f_1(a) \neq f_2(a)\}$ has "zero Lebesgue measure." In most applications, the cumulative distribution function is continuously differentiable, and there is a unique continuous density function, namely $f(a) = F'(a)$.

A.4 EXPECTATIONS

A real-valued random variable $\tilde{x}$ is simple if there is a finite set of real numbers $A = \{a_1, \ldots, a_\ell\}$ such that $\tilde{x}(\omega) \in A$ for each $\omega \in \Omega$. The expectation of this simple $\tilde{x}$ is defined as

$$\mathsf{E}[\tilde{x}] = \sum_{i=1}^{\ell} a_i \, \text{prob}(\tilde{x} = a_i).$$

If $\tilde{x}$ is a nonnegative random variable, then there exists a sequence of nonnegative simple random variables $\tilde{x}_n$ converging upward to $\tilde{x}$ in the sense that $\tilde{x}_j(\omega) \leq \tilde{x}_k(\omega)$ for each ω and each $j < k$ and $\lim_{n \to \infty} \tilde{x}_n(\omega) = \tilde{x}(\omega)$ for each ω.

The expectation of a nonnegative random variable $\tilde{x}$ is defined as

$$E[\tilde{x}] = \lim_{n \to \infty} E[\tilde{x}_n].$$

This limit is possibly equal to infinity. The limit is independent of the approximating sequence: For any sequence of nonnegative simple random variables converging upward to $\tilde{x}$ as just described, the limit is the same.

For any random variable $\tilde{x}$, define

$$\tilde{x}^+(\omega) = \max(0, \tilde{x}(\omega)) \quad \text{and} \quad \tilde{x}^-(\omega) = \max(0, -\tilde{x}(\omega)),$$

so that $\tilde{x} = \tilde{x}^+ - \tilde{x}^-$ with both $\tilde{x}^+$ and $\tilde{x}^-$ being nonnegative. If at least one of the random variables $\tilde{x}^+$ and $\tilde{x}^-$ has a finite expectation, then we say that the expectation of $\tilde{x}$ exists, and it is defined as

$$E[\tilde{x}] = E[\tilde{x}^+] - E[\tilde{x}^-].$$

If a random variable $\tilde{x}$ has a density function f and if $g : \mathbb{R} \to \mathbb{R}$ is a measurable function, then

$$E[g(\tilde{x})] = \int_{-\infty}^{\infty} g(a) f(a)\, da,$$

in the sense that if either of these exists, then the other does also and they are equal.

The expectation operator is linear, meaning:

(i) If the expectation of $\tilde{x}$ exists and a is a real number, then the expectation of $a\tilde{x}$ also exists and

$$E[a\tilde{x}] = a E[\tilde{x}].$$

(ii) If the expectation of $\tilde{x}_1$ and $\tilde{x}_2$ both exist and are finite, then the expectation of $\tilde{x}_1 + \tilde{x}_2$ exists and is finite and

$$E[\tilde{x}_1 + \tilde{x}_2] = E[\tilde{x}_1] + E[\tilde{x}_2].$$

A.5 CONVERGENCE OF EXPECTATIONS

Given a random variable $\tilde{x}$ and a sequence of random variables $\tilde{x}_n$ with the property that $\lim_{n \to \infty} \tilde{x}_n(\omega) = \tilde{x}(\omega)$, either of the following is a sufficient condition for $\lim_{n \to \infty} E[\tilde{x}_n] = E[\tilde{x}]$:

Monotone Convergence: Either (i) $\tilde{x}_j(\omega) \leq \tilde{x}_k(\omega)$ for each ω and each $j < k$ and $E[\tilde{x}_1(\omega)] > -\infty$, or (ii) $\tilde{x}_j(\omega) \geq \tilde{x}_k(\omega)$ for each ω and each $j < k$ and $E[\tilde{x}_1(\omega)] < \infty$.

Dominated Convergence: There exists a random variable $\tilde{y}$ such that $\mathsf{E}[\tilde{y}]$ is finite and $|\tilde{x}_n(\omega)| \leq \tilde{y}(\omega)$ for each ω.

Note that a special case of the first is when the random variables are nonnegative and converge upward to $\tilde{x}$ or nonpositive and converge downward to $\tilde{x}$, and a special case of the second is when the random variables are all bounded by some constant. There is a more general sufficient condition for the convergence of expectations, which is that the sequence $\tilde{x}_1, \tilde{x}_2, \ldots$ be "uniformly integrable," but the above conditions are adequate for most applications.

A related result that is sometimes useful is Fatou's lemma, which can be expressed as: Given a random variable $\tilde{x}$ and a nonnegative sequence of random variables $\tilde{x}_n$ with the property that $\lim_{n \to \infty} \tilde{x}_n(\omega) = \tilde{x}(\omega)$ for each ω and $\lim_{n \to \infty} \mathsf{E}[\tilde{x}_n] = a$ for some a, it must be that $\mathsf{E}[\tilde{x}] \leq a$.

A.6 INTERCHANGE OF DIFFERENTIATION AND EXPECTATION

Consider a random variable $f(\theta, \omega)$ depending on a real-valued parameter θ. Assume f is differentiable in θ for each ω. Frequently, we want to interchange differentiation and expectation as:

$$\frac{d}{d\theta} \mathsf{E}[f(\theta, \omega)] = \mathsf{E}\left[\frac{d}{d\theta} f(\theta, \omega)\right].$$

This is certainly possible if there are only finitely many states of the world, because differentiation is a linear operation and hence interchanges with a sum. If there are infinitely many states of the world, the results stated in the preceding subsection can be applied to justify the interchange, in most circumstances commonly encountered. The interchange means

$$\lim_{\Delta \to 0} \mathsf{E}\left[\frac{f(\theta + \Delta, \omega) - f(\theta, \omega)}{\Delta}\right] = \mathsf{E}\left[\lim_{\Delta \to 0} \frac{f(\theta + \Delta, \omega) - f(\theta, \omega)}{\Delta}\right].$$

By the mean value theorem, for each θ, there exists $\xi(\Delta, \omega)$ between θ and $\theta + \Delta$ such that

$$\frac{f(\theta + \Delta, \omega) - f(\theta, \omega)}{\Delta} = f'(\xi(\Delta, \omega), \omega),$$

where the prime denotes the derivative in the first argument. Hence, the interchange is possible if

$$\lim_{\Delta \to 0} \mathsf{E}[f'(\xi(\Delta, \omega), \omega)] = \mathsf{E}\left[\lim_{\Delta \to 0} f'(\xi(\Delta, \omega), \omega)\right].$$

Thus, it is possible if, for some $\varepsilon > 0$, the collection of random variables $\{f'(\theta + \Delta, \omega) \,|\, |\Delta| < \varepsilon\}$ is dominated by a random variable having finite mean, or, more generally, if the collection is uniformly integrable.

A.7 RANDOM VECTORS

For any positive integer k, $\mathbb{R}^k$ denotes the set of k–dimensional vectors (= the set of ordered k–tuples of real numbers = the k–fold copy of $\mathbb{R}$). We are not doing matrix algebra in this appendix, so the row or column orientation of a vector is immaterial.

The Borel σ–field of $\mathbb{R}^k$ is the smallest σ–field containing all of the sets

$$A_1 \times \cdots \times A_k = \{(a_1, \ldots, a_k) \,|\, a_1 \in A_1, \ldots a_k \in A_k\},$$

where the A_i are Borel subsets of $\mathbb{R}$. A function $g : \mathbb{R}^k \to \mathbb{R}$ is measurable if $\{x \,|\, g(x) \leq a\}$ is a Borel set in $\mathbb{R}^k$ for each real a.

If $\tilde{x}_1, \ldots, \tilde{x}_k$ are random variables, then we say that $\tilde{x} = (\tilde{x}_1, \ldots, \tilde{x}_k)$ is a random vector. If $\tilde{x}$ is a random vector, then $\{\omega \,|\, \tilde{x}(\omega) \in A\} \in \mathcal{F}$ for each Borel subset A of $\mathbb{R}^k$. We define

$$\text{prob}(\tilde{x} \in A) = \mathbb{P}(\{\omega \,|\, \tilde{x}(\omega) \in A\}).$$

The set function $A \mapsto \text{prob}(\tilde{x} \in A)$ is a probability measure on the Borel σ–field of $\mathbb{R}^k$ and is called the distribution of the random vector $\tilde{x}$. If $A = A_1 \times \cdots \times A_k$, then we also write

$$\text{prob}(\tilde{x}_1 \in A_1, \ldots, \tilde{x}_k \in A_k)$$

for $\text{prob}(\tilde{x} \in A)$, that is, for the probability that $\tilde{x}_i \in A_i$ for each i.

The cumulative distribution function of the random vector $\tilde{x}$ is the function $F : \mathbb{R}^k \to \mathbb{R}$ defined by

$$F(b_1, \ldots, b_k) = \text{prob}(\tilde{x}_1 \leq b_1, \ldots, \tilde{x}_k \leq b_k).$$

The random vector $\tilde{x}$ has a density function f if

$$F(b_1, \ldots, b_k) = \int_{-\infty}^{b_1} \cdots \int_{-\infty}^{b_k} f(a_1, \ldots, a_k) \, da_1 \cdots da_k$$

for each $(b_1, \ldots, b_k) \in \mathbb{R}^k$. The function f is also called the joint density function of the random variables $\tilde{x}_1, \ldots, \tilde{x}_k$.

If the random variables $\tilde{x}_1, \ldots \tilde{x}_k$ have a joint density function, then each of the random variables $\tilde{x}_i$ has a density function, and the term marginal density function is a synonym for density function of one of the random variables $\tilde{x}_i$ (or, more generally, a subvector of $\tilde{x}$) as defined before. The marginal density function of $\tilde{x}_i$ is obtained by "integrating over" the other random variables. Without loss of generality, consider $i = 1$. The density function of $\tilde{x}_1$ is f_1 defined by

$$f_1(a_1) = \int_{-\infty}^{\infty} \cdots \int_{-\infty}^{\infty} f(a_1, \ldots, a_k) \, da_2 \cdots da_k.$$

If $\tilde{x} = (\tilde{x}_1, \ldots, \tilde{x}_k)$ is a random vector having a density function f, and $g : \mathbb{R}^k \to \mathbb{R}$ is a measurable function such that $E[g(\tilde{x})]$ exists, then

$$E[g(\tilde{x})] = \int_{-\infty}^{\infty} \cdots \int_{-\infty}^{\infty} g(a_1, \ldots, a_k) f(a_1, \ldots, a_k) \, da_1 \cdots da_k.$$

A.8 CONDITIONING

Let $\tilde{x}$ be a nonnegative random variable, and let $\mathcal{G}$ be a sub-σ-field of $\mathcal{F}$. The expectation of $\tilde{x}$ conditional on $\mathcal{G}$ is defined to be the random variable $\tilde{z}$ that is measurable with respect to $\mathcal{G}$ and satisfies

$$(\forall G \in \mathcal{G}) \quad E[1_G \tilde{x}] = E[1_G \tilde{z}], \tag{A.2}$$

where 1_G denotes the random variable that equals 1 when $\omega \in G$ and 0 otherwise (it is called the indicator function of G). This random variable $\tilde{z}$ exists (though possibly equal to ∞) and is unique up to null sets, meaning that if $\tilde{z}$ and $\tilde{z}'$ are both measurable with respect to $\mathcal{G}$ and satisfy (A.2), then $\mathbb{P}(\{\omega \mid \tilde{z}(\omega) \neq \tilde{z}'(\omega)\}) = 0$. We denote the random variable $\tilde{z}$ by $E[\tilde{x} \mid \mathcal{G}]$. For a general random variable $\tilde{x}$, we define

$$E[\tilde{x} \mid \mathcal{G}] = E[\tilde{x}^+ \mid \mathcal{G}] - E[\tilde{x}^- \mid \mathcal{G}],$$

provided that, with probability 1, at least one is finite. If $\mathcal{G}$ is the σ-field generated by a random variable $\tilde{y}$, then we write $E[\tilde{x} \mid \tilde{y}]$ for $E[\tilde{x} \mid \mathcal{G}]$.

Conditional expectations have the following properties:

Iterated Expectations: $E[E[\tilde{x} \mid \mathcal{G}]] = E[\tilde{x}]$.
Conditional Linearity: If $\tilde{y}$ is measurable with respect to $\mathcal{G}$, then $E[\tilde{x}\tilde{y} \mid \mathcal{G}] = \tilde{y} E[\tilde{x} \mid \mathcal{G}]$.

These two properties are actually equivalent to the definition (A.2). To see this, note that the right-hand side of (A.2) is

$$E[1_G E[\tilde{x} \mid \mathcal{G}]] = E[E[1_G \tilde{x} \mid \mathcal{G}]] = E[1_G \tilde{x}],$$

using conditional linearity for the first equality and iterated expectations for the second. A special case of conditional linearity is

$$E[g(\tilde{y})\tilde{x} \mid \tilde{y}] = g(\tilde{y}) E[\tilde{x} \mid \tilde{y}]$$

for each measurable g.

Suppose $\tilde{x}$ and $\tilde{y}$ are random variables with a joint density function f. Then, we can compute $\mathsf{E}[\tilde{x}\,|\,\tilde{y}]$ as follows: The density function of $\tilde{x}$ conditional on $\tilde{y} = b$ is denoted by $f_{x|y}(\cdot\,|\,b)$ and defined, for any real a, as

$$f_{x|y}(a\,|\,b) = \frac{f(a,b)}{f_y(b)},$$

where f_y denotes the marginal density of $\tilde{y}$, and where we set $f_{x|y}(a\,|\,b) = 0$ if $f_y(b) = 0$. The expectation of $\tilde{x}$ conditional on $\tilde{y} = b$ is defined as

$$\mathsf{E}[\tilde{x}\,|\,\tilde{y} = b] = \int_{-\infty}^{\infty} a f_{x|y}(a\,|\,b)\,da,$$

provided the integral exists. The expectation of $\tilde{x}$ conditional on $\tilde{y}$ is the random variable

$$\int_{-\infty}^{\infty} a f_{x|y}(a\,|\,\tilde{y})\,da.$$

To see this, recall that each G in the σ-field generated by $\tilde{y}$ is of the form $\{\omega\,|\,\tilde{y}(\omega) \in A\}$ for some Borel A. We want to verify (A.2), and we have

$$\mathsf{E}\left[1_G\left(\int_{-\infty}^{\infty} a f_{x|y}(a\,|\,\tilde{y})\,da\right)\right] = \int_{-\infty}^{\infty} 1_A(b) \left(\int_{-\infty}^{\infty} a f_{x|y}(a\,|\,b)\,da\right) f_y(b)\,db$$

$$= \int_{-\infty}^{\infty}\int_{-\infty}^{\infty} 1_A(b) a f(a,b)\,da\,db$$

$$= \mathsf{E}[1_G \tilde{x}],$$

where $1_A(b) = 1$ if $b \in A$ and $1_A(b) = 0$ otherwise.

A.9 INDEPENDENCE

We say that random variables $\tilde{x}_1, \ldots, \tilde{x}_k$ are independent if the cumulative distribution function F of the random vector $\tilde{x} = (\tilde{x}_1, \ldots, \tilde{x}_k)$ satisfies

$$F(a_1, \ldots, a_k) = \prod_{i=1}^{k} F_i(a_i),$$

for each $(a_1, \ldots, a_k) \in \mathbb{R}^k$, where F_i denotes the cumulative distribution function of $\tilde{x}_i$. This implies the seemingly stronger property that

$$\mathrm{prob}(\tilde{x}_1 \in A_1, \ldots, \tilde{x}_k \in A_k) = \prod_{i=1}^{k} \mathrm{prob}(\tilde{x}_i \in A_i) \qquad (\text{A.3})$$

for all Borel sets $A_1, \ldots, A_k \subset \mathbb{R}$. It also implies, if the expectations $\mathsf{E}[\tilde{x}_i]$ exist,

$$\mathsf{E}\left[\prod_{i=1}^{k} \tilde{x}_i\right] = \prod_{i=1}^{k} \mathsf{E}[\tilde{x}_i].$$

Furthermore, if the random variables $\tilde{x}_1, \ldots, \tilde{x}_k$ are independent and have a joint density function f, then

$$f(a_1, \ldots, a_k) = \prod_{i=1}^{k} f_i(a_i)$$

for each $(a_1, \ldots, a_k) \in \mathbb{R}^k$, where f_i denotes the density function of $\tilde{x}_i$. Finally, if $\tilde{x}$ and $\tilde{y}$ are independent, then $\mathsf{E}[\tilde{x}|\tilde{y}] = \mathsf{E}[\tilde{x}]$.

A.10 EQUIVALENT PROBABILITY MEASURES

A probability measure $\mathbb{Q}$ on $(\Omega, \mathcal{F})$ is said to be absolutely continuous with respect to a probability measure $\mathbb{P}$ on $(\Omega, \mathcal{F})$ if $\mathbb{Q}(A) = 0$ for all $A \in \mathcal{F}$ such that $\mathbb{P}(A) = 0$. The two measures are said to be equivalent if $\mathbb{Q}(A) = 0 \Leftrightarrow \mathbb{P}(A) = 0$.

Let $\mathsf{E}^\mathbb{P}$ denote the expectation operator with respect to $\mathbb{P}$. The measure $\mathbb{Q}$ is absolutely continuous with respect to $\mathbb{P}$ if and only if there exists a random variable $\tilde{z}$ with $\mathbb{P}\{\omega \,|\, \tilde{z}(\omega) \geq 0\} = 1$ and $\mathsf{E}^\mathbb{P}[\tilde{z}] = 1$ such that

$$\mathbb{Q}(A) = \mathsf{E}^\mathbb{P}[\tilde{z} 1_A]$$

for every $A \in \mathcal{F}$. The random variable $\tilde{z}$ is called the Radon-Nikodym derivative of $\mathbb{Q}$ with respect to $\mathbb{P}$, and we write $\tilde{z} = d\mathbb{Q}/d\mathbb{P}$. Letting $\mathsf{E}^\mathbb{Q}$ denote the expectation operator with respect to $\mathbb{Q}$, we have

$$\mathsf{E}^\mathbb{Q}[\tilde{x}] = \mathsf{E}^\mathbb{P}[\tilde{z}\tilde{x}]$$

for every random variable $\tilde{x}$ in the sense that, if one expectation exists, then the other does also and they are equal. The measures $\mathbb{P}$ and $\mathbb{Q}$ are equivalent if and only if $\tilde{z} = d\mathbb{Q}/d\mathbb{P}$ satisfies $\mathbb{P}\{\omega \,|\, \tilde{z}(\omega) > 0\} = 1$.

A risk-neutral probability $\mathbb{Q}$ is an example of a probability measure equivalent to $\mathbb{P}$, and we have

$$\frac{d\mathbb{Q}}{d\mathbb{P}} = \frac{\tilde{m}}{\mathsf{E}^\mathbb{P}[\tilde{m}]},$$

where $\tilde{m}$ is the strictly positive SDF defining $\mathbb{Q}$.

A.11 FILTRATIONS, MARTINGALES, AND STOPPING TIMES

Let $\mathcal{T}$ be a subset of the nonnegative real numbers, for example $\mathcal{T} = [0, T]$, or $\mathcal{T} = \{0, 1, 2, \ldots\}$. A filtration with index set $\mathcal{T}$ is a collection of σ-fields $\{\mathcal{F}_t \mid t \in \mathcal{T}\}$ such that for all $s, t \in \mathcal{T}$ with $s \leq t$ we have $\mathcal{F}_s \subset \mathcal{F}_t$. We can interpret $\mathcal{F}_t$ as the information available at time t, and the condition $\mathcal{F}_s \subset \mathcal{F}_t$ means that there is at least as much information available at t as is available at s. In other words, nothing is ever forgotten.

A collection of random variables $\{X_t \mid t \in \mathcal{T}\}$ is said to be adapted to the filtration if X_t is $\mathcal{F}_t$-measurable for each t. The collection of random variables is said to be a martingale if (i) it is adapted to the filtration, (ii) $\mathsf{E}[|X_t|] < \infty$ for each $t \in \mathcal{T}$, and (iii) $X_s = \mathsf{E}[X_t \mid \mathcal{F}_s]$ for all $s, t \in \mathcal{T}$ with $s \leq t$. If (i) and (ii) hold and (iii) is replaced by $X_s \geq \mathsf{E}[X_t \mid \mathcal{F}_s]$, then the stochastic process X is said to be a supermartingale. If (iii) is replaced by $X_s \leq \mathsf{E}[X_t \mid \mathcal{F}_s]$, then X is said to be a submartingale. A sufficient condition for a supermartingale or submartingale X to be a martingale on a finite time horizon $[0, T]$ or $\{0, 1, \ldots, T\}$ is that $X_0 = \mathsf{E}[X_T]$.

A nonnegative random variable τ is said to be a stopping time of the filtration $\{\mathcal{F}_t \mid t \in \mathcal{T}\}$ if $\{\omega \mid \tau(\omega) \leq t\} \in \mathcal{F}_t$ for each $t \in \mathcal{T}$. Given a stopping time τ, we denote by $\mathcal{F}_\tau$ the σ-field of events A such that $A \cap \{\omega \mid \tau \leq t\} \in \mathcal{F}_t$ for each $t \in \mathcal{T}$. $\mathcal{F}_\tau$ is interpreted as the set of events that are known by the random time τ. Given stopping times $\tau_1 \leq \tau_2$ and a martingale X (or supermartingale or submartingale) we frequently want to know that $X_{\tau_1} = \mathsf{E}[X_{\tau_2} \mid \mathcal{F}_{\tau_1}]$ (or $X_{\tau_1} \geq \mathsf{E}[X_{\tau_2} \mid \mathcal{F}_{\tau_1}]$ or $X_{\tau_1} \leq \mathsf{E}[X_{\tau_2} \mid \mathcal{F}_{\tau_1}]$). Sufficient conditions are that τ_1 and τ_2 are bounded and either $\mathcal{T} = \{0, 1, \ldots\}$ or X has right-continuous paths. See, for example, Shiryayev (1984, VII.2) for the discrete-time case and Dellacherie and Meyer (1982, VI.10) for the continuous-time case. There are other sets of sufficient conditions. Results of this type are generically called the stopping theorem.

A.12 MARTINGALES UNDER EQUIVALENT MEASURES

Let $\mathcal{T}$ be a subset of the nonnegative real numbers and consider a filtration $\{\mathcal{F}_t \mid t \in \mathcal{T}\}$ and an adapted collection of random variables $\{X_t \mid t \in \mathcal{T}\}$. Let $\mathbb{P}$ and $\mathbb{Q}$ be equivalent probability measures. Let ξ denote the Radon-Nikodym derivative $\mathrm{d}\mathbb{Q}/\mathrm{d}\mathbb{P}$, and define $\xi_t = \mathsf{E}^\mathbb{P}[\xi \mid \mathcal{F}_t]$ for $t \in \mathcal{T}$. We will show the following: The collection $\{X_t \mid t \in \mathcal{T}\}$ is a martingale with respect to $\mathbb{Q}$ if and only if the collection $\{\xi_t X_t \mid t \in \mathcal{T}\}$ is a martingale with respect to $\mathbb{P}$.

We have $\mathsf{E}^\mathbb{Q}[|X_t|] = \mathsf{E}^\mathbb{P}[\xi |X_t|] = \mathsf{E}^\mathbb{P}[\xi_t |X_t|]$, using iterated expectations to obtain the second equality. Hence, $\mathsf{E}^\mathbb{Q}[|X_t|] < \infty$ if and only if $\mathsf{E}^\mathbb{P}[|\xi_t X_t|] < \infty$.

Given adaptedness, the condition $E^{\mathbb{Q}}[X_t | \mathcal{F}_s] = X_s$ means that

$$E^{\mathbb{Q}}[(X_t - X_s)1_A] = 0$$

for each $A \in \mathcal{F}_s$. Likewise, the condition $E^{\mathbb{P}}[\xi_t X_t | \mathcal{F}_s] = \xi_s X_s$ means that

$$E^{\mathbb{P}}[(\xi_t X_t - \xi_s X_s)1_A] = 0$$

for each $A \in \mathcal{F}_s$. Hence, it suffices to show that

$$E^{\mathbb{Q}}[(X_t - X_s)1_A] = E^{\mathbb{P}}[(\xi_t X_t - \xi_s X_s)1_A].$$

The left-hand side equals $E^{\mathbb{P}}[\xi(X_t - X_s)1_A]$. We can compute this by iterated expectations as

$$E^{\mathbb{P}}[X_t 1_A E^{\mathbb{P}}[\xi | \mathcal{F}_t]] - E^{\mathbb{P}}[X_s 1_A E^{\mathbb{P}}[\xi | \mathcal{F}_s]] = E^{\mathbb{P}}[(\xi_t X_t - \xi_s X_s)1_A].$$

A.13 LOCAL MARTINGALES

Let $\mathcal{T} = [0, \infty)$. Consider a filtration $\{\mathcal{F}_t | t \in \mathcal{T}\}$ and an adapted collection of random variables $\{X_t | t \in \mathcal{T}\}$. Assume X_0 is a constant and X has right-continuous paths. X is a local martingale if there exists an increasing sequence of stopping times τ_n such that $\lim_{n \to \infty} \tau_n(\omega) = \infty$ with probability 1 and such that for each n the collection $\{X_t^n | t \in \mathcal{T}\}$ is a martingale, where X_t^n is defined as

$$X_t^n = \begin{cases} X_t & \text{if } t \leq \tau_n, \\ X_{\tau_n} & \text{if } t > \tau_n. \end{cases}$$

If X is a nonnegative local martingale, then it is a supermartingale. This follows from:

$$X_s = \lim_{n \to \infty} X_s^n = \lim_{n \to \infty} E[X_t^n | \mathcal{F}_s] \geq E[\lim_{n \to \infty} X_t^n | \mathcal{F}_s] = E[X_t | \mathcal{F}_s],$$

using the martingale property of X^n, using the fact that $X_u^n \to X_u$ for each u (due to $\tau_n \to \infty$), and using a conditional version of Fatou's lemma for the inequality. More generally, if X is a local martingale, Y is a martingale with right-continuous paths, and $X \geq -Y$, then X is a supermartingale. This follows from the fact that $X + Y$ is a nonnegative local martingale (for the local martingale property, we need to apply the stopping theorem to Y), hence a supermartingale, and from the fact that the difference $(X + Y) - Y$ of a supermartingale and a martingale is a supermartingale.

A.14 THE USUAL CONDITIONS

Certain results in probability theory depend on the probability space being "complete." For example, the existence of an optimum to a problem of the form:

$$\text{choose } \tilde{x} \text{ to maximize } E[f(\omega, \tilde{x}(\omega))]$$

requires the existence of a measurable selection

$$\tilde{x}(\omega) \in \operatorname{argmax} f(\omega, \cdot).$$

This issue arises in dynamic programming, and the existence of a (measurable) optimum may depend on the probability space being complete (see Bertsekas and Shreve (1978)). A probability space—that is, a triple $(\Omega, \mathcal{F}, \mathbb{P})$—is said to be complete if $A \in \mathcal{F}$ whenever $B_1 \subset A \subset B_2$ for some $B_1, B_2 \in \mathcal{F}$ with $\mathbb{P}\{\omega \mid \omega \in B_2, \omega \notin B_1\} = 0$. If the probability space is not complete, it can easily be completed by simply including all sets A of the above form. Because null events (sets $B \in \mathcal{F}$ with $\mathbb{P}(B) = 0$) are generally irrelevant, and because A in the above definition would be equal to B_1 or B_2 up to a null event if it were in $\mathcal{F}$, there is no practical loss in adding sets A of this form to $\mathcal{F}$ and making the probability space complete. For example, the Lebesgue σ-field on the real line is the completion of the Borel σ-field defined in Appendix A.1.

Some results in stochastic process theory likewise depend on the probability space being complete and related conditions. In the continuous-time ($\mathcal{T} = [0, T]$ or $\mathcal{T} = [0, \infty)$) models in this book, we can use the following structure. Let B be a vector of Brownian motions, and let $\{\mathcal{G}_t \mid t \in \mathcal{T}\}$ be the filtration it generates, meaning that $\mathcal{G}_t$ is the smallest σ-field with respect to which B_s is measurable for all $s \leq t$. Let $\mathcal{G}$ denote the smallest σ-field containing each of the $\mathcal{G}_t$. Let $\mathcal{F}$ be the completion of $\mathcal{G}$, and let $\mathcal{F}_t$ be the smallest σ-field containing $\mathcal{G}_t$ and the null sets in $\mathcal{F}$. We take $\mathcal{F}_t$ to be the information available at date t. The filtration $\{\mathcal{F}_t \mid t \in \mathcal{T}\}$ satisfies the following "usual conditions:"

(a) $\mathcal{F}$ is complete.
(b) For each t, $\mathcal{F}_t$ contains all the null events in $\mathcal{F}$.
(c) For each s, if $A \in \mathcal{F}_t$ for all $t > s$, then $A \in \mathcal{F}_s$.

Moreover, B is a Brownian motion with respect to the filtration $\{\mathcal{F}_t \mid t \in \mathcal{T}\}$; see Karatzas and Shreve (2004, 2.7.A). The filtration $\{\mathcal{F}_t \mid t \in \mathcal{T}\}$ is called the augmented filtration generated by the Brownian motion B.

BIBLIOGRAPHY

Abel, Andrew B., 1985, A stochastic model of investment, marginal q, and the market value of the firm, *International Economic Review* 26, 305–322.

——— , 1990, Asset prices under habit formation and catching up with the Joneses, *American Economic Review* 80, 38–42.

——— , and Janice C. Eberly, 1996, Optimal investment with costly reversibility, *Review of Economic Studies* 63, 581–593.

Admati, Anat R., 1985, A noisy rational expectations equilibrium for multi-asset securities markets, *Econometrica* 53, 629–658.

Aguerrevere, Felipe L., 2009, Real options, product market competition, and asset returns, *Journal of Finance* 64, 957–983.

Ahn, Dong-Hyun, Robert F. Dittmar, and A. Ronald Gallant, 2002, Quadratic term structure models: Theory and evidence, *Review of Financial Studies* 15, 243–288.

Akay, Ozgur, Ken B. Cyree, Mark D. Griffiths, and Drew B. Winters, 2012, What does PIN identify? Evidence from the T-bill market, *Journal of Financial Markets* 15, 29–46.

Aktas, Nihat, Eric de Bodt, Fany Declerck, and Hervé Van Oppens, 2007, The PIN anomaly around M&A announcements, *Journal of Financial Markets* 10, 160–191.

Allais, M., 1953, Le comportement de l'homme rationnel devant le risque: Critique des postulats et axiomes de l'ecole Americane, *Econometrica* 21, 503–546.

Almgren, Robert, and Neil Chriss, 2000, Optimal execution of portfolio transactions, *Journal of Risk* 3, 5–39.

Amihud, Yakov, and Haim Mendelson, 1986, Asset pricing and the bid-ask spread, *Journal of Financial Economics* 17, 223–249.

Anderson, Evan W., Eric Ghysels, and Jennifer L. Juergens, 2005, Do heterogeneous beliefs matter for asset pricing?, *Review of Financial Studies* 18, 875–924.

Anderson, Theodore W., 2003, *An Introduction to Multivariate Statistical Analysis* (Wiley).

Arditti, Fred D., 1967, Risk and the required return on equity, *Journal of Finance* 22, 19–36.

Arrow, Kenneth J., 1951, An extension of the basic theorems of classical welfare economics, in J. Neyman, ed.: *Proceedings of the 2nd Berkeley Symposium on Mathematical Statistics and Probability* (University of California Press: Berkeley).

——— , 1953, The role of securities in the optimal allocation of risk-bearing, *Économétric*. Translation in *Review of Economic Studies* 31, 1964, 91–96.

———, 1965, The theory of risk aversion, in *Aspects of the Theory of Risk Bearing* (Yrjö Jahnssonin Säätiö: Helsinki).

———, 1971, An exposition of the theory of choice under uncertainty, in C. B. McGuire, and R. Radner, ed.: *Decision and Organization* (North Holland: Amsterdam).

———, and Gerard Debreu, 1954, Existence of an equilibrium for a competitive economy, *Econometrica* 22, 265–290.

Asness, Clifford S., Tobias J. Moskowitz, and Lasse Heje Pedersen, 2013, Value and momentum everywhere, *Journal of Finance* 68, 929–985.

Aumann, Robert J., 1976, Agreeing to disagree, *The Annals of Statistics* 4, 1236–1239.

Bachelier, M. L., 1900, *Théorie de la Spéculation* (Gauthier-Villars).

Back, Kerry, 1991, Asset pricing for general processes, *Journal of Mathematical Economics* 20, 371–395.

———, 1992, Insider trading in continuous time, *Review of Financial Studies* 5, 387–409.

———, 1993, Asymmetric information and options, *Review of Financial Studies* 6, 435–472.

———, 2005, *A Course in Derivative Securities: Introduction to Theory and Computation* (Springer: Berlin).

———, 2010, Martingale pricing, *Annual Review of Financial Economics* 2, 235–250.

———, and Shmuel Baruch, 2004, Information in securities markets: Kyle meets Glosten and Milgrom, *Econometrica* 72, 433–465.

———, 2007, Working orders in limit order markets and floor exchanges, *Journal of Finance* 62, 1589–1621.

———, 2013, Strategic liquidity provision in limit order markets, *Econometrica* 81, 363–392.

Back, Kerry, Pierre Collin-Dufresne, Vyacheslav Fos, Tao Li, and Alexander Ljungqvist, 2016, Activism, strategic trading, and liquidity, Working Paper.

Back, Kerry, C. Henry Cao, and Gregory A. Willard, 2000, Imperfect competition among informed traders, *Journal of Finance* 55, 2117–2155.

Back, Kerry, and Kevin Crotty, 2015, The informational role of stock and bond volume, *Review of Financial Studies* 28, 1381–1427.

———, and Tao Li, 2015, Estimating information asymmetry in securities markets, Working Paper.

Back, Kerry, Ruomeng Liu, and Alberto Teguia, 2015, Increasing risk aversion, habits, and life-cycle investing, Working Paper.

Back, Kerry, and Dirk Paulsen, 2009, Open loop equilibria and perfect competition in option exercise games, *Review of Financial Studies* 22, 4531–4552.

Back, Kerry, and Hal Pedersen, 1998, Long-lived information and intraday patterns, *Journal of Financial Markets* 1, 385–402.

Back, Kerry, and Jaime P. Zender, 1993, Auctions of divisible goods: On the rationale for the treasury experiment, *Review of Financial Studies* 6, 733–764.

Backus, David, Mikhail Chernov, and Ian Martin, 2011, Disasters implied by equity index options, *Journal of Finance* 66, 1969–2012.

Backus, David K., Bryan R. Routledge, and Stanley E. Zin, 2005, Exotic preferences for macroeconomists, in Mark Gertler, and Kenneth Rogoff, ed.: *NBER Macroeconomics Annual 2004* (MIT Press: Cambridge, MA).

Baldursson, Fridrik M., 1998, Irreversible investment under uncertainty in oligopoly, *Journal of Economic Dynamics and Control* 22, 627–644.

Banerjee, Snehal, Ron Kaniel, and Ilan Kremer, 2009, Price drift as an outcome of differences in higher order beliefs, *Review of Financial Studies* 22, 3707–3734.

Bank, Peter, 2005, Optimal control under a dynamic fuel constraint, *SIAM* 44, 1529–1541.

Bansal, Ravi, Dana Kiku, and Amir Yaron, 2012, An empirical evaluation of the long-run risks model for asset prices, *Critical Finance Review* 1, 183–221.

Bansal, Ravi, and Amir Yaron, 2004, Risks for the long run: A potential resolution of asset pricing puzzles, *Journal of Finance* 59, 1481–1509.

Banz, Rolf W., 1981, The relationship between return and market value of common stocks, *Journal of Financial Economics* 9, 3–18.

Barber, Brad M., and Terrance Odean, 2000, Trading is hazardous to your wealth: The common stock investment performance of individual investors, *Journal of Finance* 55, 773–806.

Barberis, Nicholas, and Ming Huang, 2001, Mental accounting, loss aversion, and individual stock returns, *Journal of Finance* 56, 1247–1292.

———, and Tano Santos, 2001, Prospect theory and asset prices, *Quarterly Journal of Economics* 116, 1–53.

Barro, Robert J., 2006, Rare disasters and asset markets in the twentieth century, *Quarterly Journal of Economics* 121, 823–866.

———, 2009, Rare disasters, asset prices, and welfare costs, *American Economic Review* 99, 243–264.

Baruch, Shmuel, 2002, Insider trading and risk aversion, *Journal of Financial Markets* 5, 451–464.

Basak, Suleyman, 2000, A model of dynamic equilibrium asset pricing with heterogeneous beliefs and extraneous risk, *Journal of Economic Dynamics and Control* 24, 63–95.

———, and Benjamin Croitoru, 2000, Equilibrium mispricing in a capital market with portfolio constraints, *Review of Financial Studies* 13, 715–748.

———, 2006, On the role of arbitrageurs in rational markets, *Journal of Financial Economics* 81, 143–173.

Basu, Sanjoy, 1983, The relationship between earnings yield, market value and return for NYSE common stocks: Further evidence, *Journal of Financial Economics* 12, 129–156.

Beeler, Jason, and John Y. Campbell, 2012, The long-run risks model and aggregate asset prices: An empirical assessment, *Critical Finance Review* 1, 141–182.

Beja, Avraham, 1971, The structure of the cost of capital under uncertainty, *Review of Economic Studies* 38, 359–368.

Bekaert, Geert, and Robert J. Hodrick, 2001, Expectations hypotheses tests, *Journal of Finance* 56, 1357–1394.

———, and David A. Marshall, 1997, On biases in tests of the expectations hypotheses of the term structure of interest rates, *Journal of Financial Economics* 44, 309–348.

Benartzi, Shlomo, and Richard H. Thaler, 1995, Myopic loss aversion and the equity premium puzzle, *Quarterly Journal of Economics* 110, 73–92.

Bensoussan, A., 1984, On the theory of option pricing, *Acta Applicandae Mathematicae* 2, 139–158.

Berhnardt, Dan, and Eric Hughson, 1997, Splitting orders, *Review of Financial Studies* 10, 69–101.

Berk, Jonathan B., 1995, A critique of size-related anomalies, *Review of Financial Studies* 8, 275–286.

———, 1997, Necessary conditions for the CAPM, *Journal of Economic Theory* 73, 245–257.

———, Richard C. Green, and Vasant Naik, 1999, Optimal investment, growth options, and security returns, *Journal of Finance* 54, 1553–1607.

Bertsekas, Dimitri P., and Steven E. Shreve, 1978, *Stochastic Optimal Control: The Discrete-Time Case* (Academic Press: Orlando, FL).

Bertsimas, Dimitris, and Andrew W. Lo, 1998, Optimal control of execution costs, *Journal of Financial Markets* 1, 1–50.

Biais, Bruno, David Martimort, and Jean-Charles Rochet, 2000, Competing mechanisms in a common value environment, *Econometrica* 68, 799–837.

———, 2013, Corrigendum to "Competing mechanisms in a common value environment", *Econometrica* 81, 393–406.

Black, Fischer, 1972, Capital market equilibrium with restricted borrowing, *Journal of Business* 45, 444–445.

———, 1976, The pricing of commodity contracts, *Journal of Financial Economics* 3, 167–179.

———, and Myron Scholes, 1973, The pricing of options and corporate liabilities, *Journal of Political Economy* 81, 637–654.

Bodie, Zvi, Jerome Detemple, Susanne Otruba, and Stephan Walter, 2004, Optimal consumption-portfolio choices and retirement planning, *Journal of Economic Dynamics and Control* 28, 1115–1148.

Borovička, Jaroslav, Lars Peter Hansen, and José Scheinkman, forthcoming, Misspecified recovery, *Journal of Finance*.

Brandt, Michael W., Amit Goyal, Pedro Santa-Clara, and Jonathan R. Stroud, 2005, A simulation approach to dynamic portfolio choice with an application to learning about return predictability, *Review of Financial Studies* 18, 831–873.

Breeden, Douglas T., 1979, An intertemporal asset pricing model with stochastic consumption and investment opportunities, *Journal of Financial Economics* 7, 265–296.

———, 1986, Consumption, production, inflation and interest rate: A synthesis, *Journal of Financial Economics* 16, 3–39.

———, and Robert H. Litzenberger, 1978, Prices of state-contingent claims implicit in option prices, *Journal of Business* 51, 621–651.

Brendle, Simon, 2006, Portfolio selection under incomplete information, *Stochastic Processes and their Applications* 116, 701–723.

Brennan, Michael J., 1971, Capital market equilibrium with divergent borrowing and lending rates, *Journal of Financial and Quantitative Analysis* 6, 1197–1205.

———, 1998, The role of learning in dynamic portfolio decisions, *European Finance Review* 1, 295–306.

———, and H. Henry Cao, 1996, Information, trade, and derivative securities, *Review of Financial Studies* 9, 163–208.

Brennan, Michael J., and Eduardo S. Schwartz, 1985, Evaluating natural resource investments, *Journal of Business* 58, 135–157.

Brennan, Michael J., and Yihong Xia, 2001, Stock price volatility and equity premium, *Journal of Mathematical Economics* 47, 249–283.

Breon-Drish, Bradyn, 2015, On existence and uniqueness of equilibrium in a class of noisy rational expectations models, *Review of Economic Studies* 82, 868–921.

Broadie, Mark, Mikhail Chernov, and Michael Johannes, 2009, Understanding index option returns, *Review of Financial Studies* 22, 4493–4529.

Broadie, Mark, Mikhail Chernov, and Suresh Sundaresan, 2007, Optimal debt and equity values in the presence of Chapter 7 and Chapter 11, *Journal of Finance* 62, 1341–1377.

Brocas, Isabelle, Juan D. Carrillo, Aleksandar Giga, and Fernando Zapatero, 2015, Risk aversion in a dynamic asset allocation experiment, Working Paper.

Brockett, Patrick L., and Yehuda Kahane, 1992, Risk, return, skewness and preference, *Management Science* 38, 851–866.

Brown, David P., and Robert H. Jennings, 1989, On technical analysis, *Review of Financial Studies* 2, 527–551.

Brunnermeier, Markus K., and Lasse Heje Pedersen, 2005, Predatory trading, *Journal of Finance* 60, 1825–1863.

Buraschi, Andrea, and Alexei Jiltsov, 2006, Model uncertainty and option markets with heterogeneous agents, *Journal of Finance* 61, 2841–2897.

Bustamante, M. Cecilia, 2012, The dynamics of going public, *Review of Finance* 16, 577–618.

Caballé, Jordi, and Murugappa Krishnan, 1994, Imperfect competition in a multi-security market with risk neutrality, *Econometrica* 62, 695–704.

Caldentey, Ren'e, and Ennio Stacchetti, 2010, Insider trading with a random deadline, *Econometrica* 78, 245–283.

Campbell, John Y., 1986, A defense of traditional hypotheses about the term structure of interest rates, *Journal of Finance* 41, 183–193.

———, 1987, Stock returns and the term structure, *Journal of Financial Economics* 18, 373–399.

———, 2003, Consumption-based asset pricing, in George Constantinides, Milton Harris, and René M. Stulz, ed.: *Handbook of the Economics of Finance: Vol. 1B, Financial Markets and Asset Pricing* (Elsevier: Amsterdam).

———, and John H. Cochrane, 1999, By force of habit: A consumption-based explanation of aggregate stock market behavior, *Journal of Political Economy* 107, 205–251.

Campbell, John Y., and Robert J. Shiller, 1988, Stock prices, earnings, and expected dividends, *Journal of Finance* 43, 661–676.

———, 1991, Yield spread and interest rate movements: A bird's eye view, *Review of Economic Studies* 58, 495–514.

Campi, Luciano, and Umut Çetin, 2007, Insider trading in an equilibrium model with default: A passage from reduced-form to structural modeling, *Finance & Stochastics* 11, 591–602.

———, and Albina Danilova, 2013, Equilibrium model with default and dynamic insider information, *Finance & Stochastics* 17, 565–585.

Cao, H. Henry, and Hui Ou-Yang, 2005, Bubbles and panics in a frictionless market with heterogeneous expectations, Duke University.

———, 2009, Differences of opinion of public information and speculative trading in stocks and options, *Review of Financial Studies* 22, 299–335.

Carhart, Mark M., 1997, On persistence in mutual fund performance, *Journal of Finance* 52, 57–82.

Carlin, Bruce Ian, Miguel Sousa Lobo, and S. Viswanathan, 2007, Episodic liqudiity crises: Cooperative and predatory trading, *Journal of Finance* 62, 2235–2274.

Carlson, Murray, Adlai Fisher, and Ron Giammarino, 2004, Corporate investment and asset price dynamics: Implications for the cross-section of returns, *Journal of Finance* 59, 2577–2603.

———, 2006, Corporate investment and asset price dynamics: Implications for SEO event studies and long-run performance, *Journal of Finance* 61, 1009–1034.

Cass, David, and Joseph E. Stiglitz, 1970, The structure of investor preferences and asset returns, and separability in portfolio allocation: A contribution to the pure theory of mutual funds, *Journal of Economic Theory* 2, 122–160.

Çetin, Umut, and Hao Xing, 2013, Point process bridges and weak convergence of insider trading models, *Electronic Journal of Probability* 18, 1–24.

Chamberlain, Gary, 1983a, A characterization of the distributions that imply mean-variance utility functions, *Journal of Economic Theory* 29, 185–201.

———, 1983b, Funds, factors, and diversification in arbitrage pricing models, *Econometrica* 51, 1305–1324.

———, 1988, Asset pricing in multiperiod securities markets, *Econometrica* 56, 1283–1300.

———, and Michael Rothschild, 1983, Arbitrage, factor structure, and mean-variance analysis on large asset markets, *Econometrica* 51, 1281–1304.

Chan, Yeung Lewis, and Leonid Kogan, 2002, Catching up with the Joneses: Heterogeneous preferences and the dynamics of asset prices, *Journal of Political Economy* 110, 1255–1285.

Chapman, David A., 1998, Habit formation and aggregate consumption, *Econometrica* 66, 1223–1230.

———, 2002, Does intrinsic habit formation actually resolve the equity premium puzzle?, *Review of Economic Dynamics* 5, 618–645.

Chau, Minh, and Dimitri Vayanos, 2008, Strong-form efficiency with monopolistic insiders, *Review of Financial Studies* 21, 2275–2306.

Chen, Joseph, Harrison Hong, and Jeremy C. Stein, 2002, Breadth of ownership and stock returns, *Journal of Financial Economics* 66, 171–205.

Chen, Nan, and S. G. Kou, 2009, Credit spreads, optimal capital structure, and implied volatility with endogenous default and jump risk, *Mathematical Finance* 19, 343–378.

Chen, Zengjing, and Larry Epstein, 2002, Ambiguity, risk, and asset returns in continuous time, *Econometrica* 70, 1403–1443.

Cheridito, Patrick, Damir Filipović, and Robert L. Kimmel, 2007, Market price of risk specifications for affine models: Theory and evidence, *Journal of Financial Economics* 83, 123–170.

Chew, Soo Hong, 1983, A generalization of the quasilinear mean with applications to the measurement of income inequality and decision theory resolving the Allais paradox, *Econometrica* 51, 1065–1092.

Chew, S. H., 1989, Axiomatic utility theories with the betweenness property, *Annals of Operation Research* 19, 273–298.

———, and Larry Epstein, 1989, The structure of preferences and attitudess towards the timing of the resolution of uncertainty, *International Economic Review* 30, 103–117.

Choi, Jin Hyuk, Kasper Larsen, and Duane J. Seppi, 2015, Information and trading targets in a dynamic market equilibrium, Working Paper.

Chu, K'ai-Ching, 1973, Estimation and decision for linear systems with elliptical random processes, *IEEE Transactions on Automatic Control* 18, 499–505.

Cochrane, John H., 1991, Production-based asset pricing and the link between stock returns and economic fluctuation, *Journal of Finance* 46, 209–237.

———, 2001, *Asset Pricing* (Princeton University Press: Princeton, NJ).

———, 2011, Presidential address: Discount rates, *Journal of Finance* 66, 1047–1108.

———, 2014, A mean-variance benchmark for intertemporal portfolio theory, *Journal of Finance* 69, 1–49.

———, Francis A. Longstaff, and Pedro Santa-Clara, 2008, Two trees, *Review of Financial Studies* 21, 348–385.

Collin-Dufresne, Pierre, and Vyacheslav Fos, 2014, Insider trading, stochastic liquidity and equilibrium prices, Working Paper.

Collin-Dufresne, Pierre, and Robert S. Goldstein, 2002, Do bonds span the fixed income markets? Theory and evidence for unspanned stochastic volatility, *Journal of Finance* 57, 1685–1730.

———, and Christopher S. Jones, 2008, Identification of maximal affine term structure models, *Journal of Finance* 63, 743–795.

Connor, Gregory, 1984, A unified beta pricing theory, *Journal of Economic Theory* 34, 13–31.

Constantinides, George M., 1982, Intertemporal asset pricing with heterogeneous consumers and without demand aggregation, *Journal of Business* 55, 253–267.

———, 1984, Warrant exercise and bond conversion in competitive markets, *Journal of Financial Economics* 13, 371–397.

———, 1986, Capital market equilibrium with transaction costs, *Journal of Political Economy* 94, 842–862.

———, 1990, Habit formation: A resolution of the equity premium puzzle, *Journal of Political Economy* 98, 519–543.

———, 1992, A theory of the nominal term structure of interest rates, *Review of Financial Studies* 5, 531–552.

———, and Darrell Duffie, 1996, Asset pricing with heterogeneous consumers, *Journal of Political Economy* 104, 219–240.

Cooper, Ilan, 2006, Asset pricing implications of nonconvex adjustment costs and irreversibility of investment, *Journal of Finance* 61, 139–170.

Coval, Joshua D., and Tobias J. Moskowitz, 1999, Home bias at home: Local equity preference in domestic portfolios, *Journal of Finance* 54, 2045–2073.

Cox, Alexander M. G., and David G. Hobson, 2005, Local martingales, bubbles and option prices, *Finance & Stochastics* 9, 477–492.

Cox, John C., and Chi-Fu Huang, 1989, Optimal consumption and portfolio policies when asset prices follow a diffusion process, *Journal of Economic Theory* 49, 33–83.

Cox, John C., Jonathan E. Ingersoll, and Stephen A. Ross, 1981, A re-examination of traditional hypotheses about the term structure of interest rates, *Journal of Finance* 36, 769–799.

———, 1985, A theory of the term structure of interest rates, *Econometrica* 53, 385–408.

Cox, John C., and Stephen A. Ross, 1976a, A survey of some new results in financial option pricing theory, *Journal of Finance* 31, 383–402.

———, 1976b, The valuation of options for alternative stochastic processes, *Journal of Financial Economics* 3, 145–166.

Cvitanic, Jaksa, and Ioannis Karatzas, 1992, Convex duality in constrained portfolio optimization, *Annals of Applied Probability* 2, 767–818.

Dai, Qiang, and Kenneth J. Singleton, 2000, Specification analysis of affine term structure models, *Journal of Finance* 55, 1943–1978.

Dalang, R. C., Andrew Morton, and Walter Willinger, 1990, Equivalent martingale measures and no-arbitrage in a stochastic securities market model, *Stochastics and Stochastics Reports* 29, 185–201.

David, Alexander, 1997, Fluctuating confidence in stock markets: Implications for returns and volatility, *Journal of Financial and Quantitative Analysis* 32, 427–462.

———, 2008a, Heterogeneous beliefs, speculation, and the equity premium, *Journal of Finance* 63, 41–83.

———, 2008b, Inflation uncertainty, asset valuations, and the credit spreads puzzle, *Review of Financial Studies* 21, 2487–2534.

———, and Pietro Veronesi, 2013, What ties return volatilities to price valuations and fundamentals?, *Journal of Political Economy* 121, 682–746.

———, 2014, Investors' and central banks' uncertainty embedded in index options, *Review of Financial Studies* 27, 1661–1716.

Davis, James L., Eugene F. Fama, and Kenneth R. French, 2000, Characteristics, covariances, and average returns: 1929–1997, *Journal of Finance* 55, 389–406.

Davis, M. H. A., and A. R. Norman, 1990, Portfolio selection with transaction costs, *Mathematics of Operations Research* 15, 676–713.

Debreu, Gerard, 1954, Valuation equilibrium and pareto optimum, *Proceedings of the National Academy of Sciences of the U.S.A.* 40, 588–592.

Dekel, Eddie, 1986, An axiomatic characterization of preferences under uncertainty: Weakening the independence axiom, *Journal of Economic Theory* 40, 304–318.

Delbaen, Freddy, and Walter Schachermayer, 1994, A general version of the fundamental theorem of asset pricing, *Mathematische Annalen* 300, 463–520.

———, 2006, *The Mathematics of Arbitrage* (Springer: Berlin).

Dellacherie, Claude, and Paul-André Meyer, 1982, *Probabilities and Potential B: Theory of Martingales* (North Holland).

DeMarzo, Peter, Ron Kaniel, and Ilan Kremer, 2004, Diversification as a public good: Community effects in portfolio choice, *Journal of Finance* 59, 1877–1715.

DeMarzo, Peter, and Costis Skiadas, 1998, Aggregation, determinacy, and informational efficiency for a class of economies with asymmetric information, *Journal of Economic Theory* 80, 123–152.

DeMarzo, Peter M., Ron Kaniel, and Ilan Kremer, 2008, Relative wealth concerns and financial bubbles, *Review of Financial Studies* 21, 19–50.

Derman, Emanuel, and Iraj Kani, 1998, Stochastic implied trees: Arbitrage pricing with stochastic term and strike structure of volatility, *International Journal of Theoretical and Applied Finance* 1, 61–110.

Detemple, Jerome, 1986, Asset pricing in a production economy with incomplete information, *Journal of Finance* 41, 383–391.

———, 1991, Further results on asset pricing with incomplete information, *Journal of Economic Dynamics and Control* 15, 425–453.

———, and Shashidhar Murthy, 1994, Intertemporal asset pricing with heterogeneous beliefs, *Journal of Economic Theory* 62, 294–320.

———, 1997, Equilibrium asset prices and no-arbitrage with portfolio constraints, *Review of Financial Studies* 10, 1133–1174.

Detemple, Jerome, and Larry Selden, 1991, A general equilibrium analysis of option and stock market interactions, *International Economic Review* 32, 279–303.

Detemple, Jerome, and Fernando Zapatero, 1991, Asset prices in an exchange economy with habit formation, *Econometrica* 59, 1633–1657.

Diamond, Douglas W., and Robert E. Verrecchia, 1981, Information aggregation in a noisy rational expectations economy, *Journal of Financial Economics* 9, 221–235.

———, 1987, Constraints on short-selling and asset price adjustment to private information, *Journal of Financial Economics* 18, 277–311.

Dittmar, Robert F., 2002, Nonlinear pricing kernels, kurtosis preference, and evidence from the cross section of equity returns, *Journal of Finance* 57, 369–403.

Dixit, Avinash K., and Robert S. Pindyck, 1994, *Investment under Uncertainty* (Princeton University Press: Princeton, NJ).

Dothan, Michael U., and David Feldman, 1986, Equilibrium interest rates and multi-period bonds in a partially observable economy, *Journal of Finance* 41, 369–382.

Dow, James, and Sergio Ribeiro da Costa Werlang, 1992, Uncertainty aversion, risk aversion, and the optimal choice of portfolio, *Econometrica* 60, 197–204.

Dreze, Jacques H., 1970, Market allocation under uncertainty, *European Economic Review* 71, 133–165.

Duan, Jin-Chuan, 1995, The GARCH option pricing model, *Mathematical Finance* pp. 16–32.

Duarte, Jefferson, and Lance Young, 2009, Why is PIN priced?, *Journal of Financial Economics* 91, 119–138.

Duffee, Gregory, 2002, Term premia and interest rate forecasts in affine models, *Journal of Finance* 57, 405–443.

Duffie, Darrell, 2001, *Dynamic Asset Pricing Theory* (Princeton University Press: Princeton, NJ) 3rd edn.

———, and Larry Epstein, 1992a, Asset pricing with stochastic differential utility, *Review of Financial Studies* 5, 411–436.

Duffie, Darrell, and Larry G. Epstein, 1992b, Stochastic differential utility, *Econometrica* 60, 353–394.

Duffie, Darrell, Damir Filipović, and Walter Schachermayer, 2003, Affine processes and applications in finance, *Annals of Applied Probability* 13, 984–1053.

Duffie, Darrell, and Chi-fu Huang, 1985, Implementing Arrow-Debreu equilibria by continuous trading of few long-lived securities, *Econometrica* 53, 1337–1356.

Duffie, Darrell, and Rui Kan, 1996, A yield-factor model of interest rates, *Mathematical Finance* 6, 379–406.

Duffie, Darrell, and David Lando, 2001, Term structure of credit spreads with incomplete accounting information, *Econometrica* 69, 633–664.

Duffie, Darrell, Jun Pan, and Kenneth J. Singleton, 2000, Transform analysis and asset pricing for affine jump-diffusions, *Econometrica* 68, 1343–1376.

Dumas, Bernard, 1992, Dynamic equilibrium and the real exchange rate in a spatially separated world, *Review of Financial Studies* 5, 153–180.

———, Alexander Kurshev, and Raman Uppal, 2009, Equilibrium portfolio strategies in the presence of sentiment risk and excess volatility, *Journal of Finance* 64, 579–629.

Dybvig, Philip H., 1983, An explicit bound on individual assets' deviation from APT pricing in a finite economy, *Journal of Financial Economics* 12, 483–496.

———, 1984, Short sales restrictions and kinks on the mean variance frontier, *Journal of Finance* 39, 239–244.

———, 1995, Dusenberry's racheting of consumption: Optimal dynamic consumption and investment given intolerance for any decline in standard of living, *Review of Economic Studies* 62, 287–313.

———, and Chi-Fu Huang, 1988, Nonnegative wealth, absence of arbitrage, and feasible consumption plans, *Review of Financial Studies* 1, 377–401.

Dybvig, Philip H., and Jonathan E. Ingersoll, 1982, Mean-variance theory in complete markets, *Journal of Business* 55, 233–251.

Dybvig, Philip H., and Steven A. Lippman, 1983, An alternative characterization of decreasing absolute risk aversion, *Econometrica* 51, 223–224.

Dybvig, Philip H., and L. C. G. Rogers, 1997, Recovery of preferences from observed wealth in a single realizations, *Review of Financial Studies* 10, 151–174.

Dybvig, Philip H., and Stephen A. Ross, 1985a, The analytics of performance measurement using a security market line, *Journal of Finance* 40, 401–416.

———, 1985b, Differential information and performance measurement using a security market line, *Journal of Finance* 40, 383–399.

———, 1985c, Yes, the APT is testable, *Journal of Finance* 40, 1173–1188.

———, 1989, Arbitrage, in J. Eatwell, M. Milgate, and P. Newman, ed.: *The New Palgrave: Finance*. pp. 57–71 (W. W. Norton & Co.: New York).

Easley, David, Nicholas M. Kiefer, Maureen O'Hara, and Joseph B. Paperman, 1996, Liquidity, information, and infrequently traded stocks, *Journal of Finance* 51, 1405–1436.

Easley, David, and Maureen O'Hara, 2004, Information and the cost of capital, *Journal of Finance* 59, 1553–1583.

Ellsberg, Daniel, 1961, Risk, ambiguity, and the Savage axioms, *Quarterly Journal of Economics* 75, 643–669.

Emanuel, David C., 1983, Warrant valuation and exercise strategy, *Journal of Financial Economics* 12, 211–235.

Epstein, Larry, and Martin Schneider, 2003, Recursive multiple-priors, *Journal of Economic Theory* 113, 1–31.

———, 2008, Ambiguity, information quality, and asset pricing, *Journal of Finance* 63, 197–228.

Epstein, Larry, and Tan Wang, 1994, Intertemporal asset pricing under Knightian uncertainty, *Econometrica* 62, 283–322.

Epstein, Larry, and Stanley E. Zin, 1989, Substitution, risk aversion, and the temporal behavior of consumption and asset returns: A theoretical framework, *Econometrica* 57, 937–969.

Epstein, Larry G., 1985, Decreasing risk aversion and mean-variance analysis, *Econometrica* 53, 945–962.

Fama, Eugene F., 1970, Efficient capital markets: A review of theory and empirical work, *Journal of Finance* 25, 383–417.

——, 1984a, Forward and spot exchange rates, *Journal of Monetary Economics* 14, 319–338.

——, 1984b, The information in the term structure, *Journal of Finacial Economics* 13, 509–528.

——, and Kenneth R. French, 1988a, Dividend yields and expected stock returns, *Journal of Financial Economics* 22, 3–25.

——, 1988b, Permanent and temporary compents of stock prices, *Journal of Political Economy* 96, 246–273.

——, 1992, The cross-section of expected stock returns, *Journal of Finance* 47, 427–465.

——, 1993, Common risk factors in the returns on stocks and bonds, *Journal of Financial Economics* 33, 3–56.

——, 1996, Multifactor explanations of asset pricing anomalies, *Journal of Finance* 51, 55–84.

Fama, Eugene F, and Kenneth R French, 2015, A five-factor asset pricing model, *Journal of Financial Economics* 116, 1–22.

Fama, Eugene F., and G. William Schwert, 1977, Asset returns and inflation, *Journal of Financial Economics* 5, 115–146.

Feldman, Mark, and Christian Gilles, 1985, An expository note on individual risk without aggregate uncertainty, *Journal of Economic Theory* 35, 26–32.

Foster, F. Douglas, and S. Viswanathan, 1996, Strategic trading when agents forecast the forecasts of others, *Journal of Finance* 51, 1437–1478.

François, Pascal, and Erwan Morellec, 2004, Capital structure and asset prices: Some effects of bankruptcy procedures, *Journal of Business* 77, 387–411.

French, Kenneth R., and James M. Poterba, 1991, Investor diversification and international equity markets, *American Economic Review* 81, 222–226.

Fudenberg, Drew, and Jean Tirole, 1992, *Game Theory* (MIT Press: Cambridge, MA).

Fujisaka, Masotoshi, Gopinath Kallianpur, and Hiroshi Kunita, 1972, Stochastic differential equations for the non linear filtering problem, *Osaka Journal of Mathematics* 9, 19–40.

Gallmeyer, Michael, and Burton Hollifield, 2008, An examination of heterogeneous beliefs with a short-sale constraint in a dynamic economy, *Review of Finance* 12, 323–364.

Geman, Hélyette, Nicole El Karoui, and Jean-Charles Rochet, 1995, Changes of numeraire, changes of probability measure and option pricing, *Journal of Applied Probability* 32, 443–458.

Gennotte, Gerard, 1986, Optimal portfolio choice under incomplete information, *Journal of Finance* 41, 733–746.

Gilboa, Itzhak, 1987, Expected utility with purely subjective non-additive probabilities, *Journal of Mathematical Economics* 16, 65–88.

——, and David Schmeidler, 1989, Maxmin expected utility with non-unique prior, *Journal of Mathematical Economics* 18, 141–153.

——, 1993, Updating ambiguous beliefs, *Journal of Economic Theory* 59, 33–49.

Gilles, Christian, and Stephen F. LeRoy, 1991, On the arbitrage pricing theory, *Economic Theory* 1, 213–229.

Glosten, Lawrence R., 1994, Is the electronic open limit order book inevitable?, *Journal of Finance* 49, 1127–1161.

———, and Paul R. Milgrom, 1985, Bid, ask and transaction prices in a specialist market with heterogeneously informed traders, *Journal of Financial Economics* 14, 71–100.

Goettler, Ronald L., Christine A. Parlour, and Uday Rajan, 2005, Equilibrium in a dynamic limit order market, *Journal of Finance* 60, 2149–2192.

———, 2009, Informed traders and limit order markets, *Journal of Financial Economics* 93, 67–87.

Goldstein, Robert, Nengjiu Ju, and Hayne Leland, 2001, An EBIT-based model of dynamic capital structure, *Journal of Business* 74, 483–512.

Gomes, Francisco, and Alexander Michaelides, 2003, Portfolio choice with internal habit formation: A life-cycle model with uninsurable labor income risk, *Review of Economic Dynamics* 6, 729–766.

Gomes, Joao, Leonid Kogan, and Lu Zhang, 2003, Equilibrium cross section of returns, *Journal of Political Economy* 111, 693–732.

Gordon, Myron J., 1962, *The Investment, Financing, and Valuation of the Corporation* (Irwin: Homewood, IL).

Gorman, W. M., 1953, Community preference fields, *Econometrica* 21, 63–80.

Green, Edward J., 1994, Individual-level randomness in a nonatomic population, University of Minnesota.

Grenadier, Steven R., 2002, Option exercise games: An application to the equilibrium investment strategies of firms, *Review of Financial Studies* 15, 691–721.

———, and Andrey Malenko, 2011, Real options signaling games with applications to corporate finance, *Review of Financial Studies* 24, 3993–4036.

Grinblatt, Mark, and Juhani T. Linnainmaa, 2011, Jensen's inequality, parameter uncertainty, and multi-period investment, *Review of Asset Pricing Studies* 1, 1–34.

Grinblatt, Mark, and Sheridan Titman, 1983, Factor pricing in a finite economy, *Journal of Financial Economics* 12, 497–507.

Grossman, Sanford J., 1976, On the efficiency of competitive stock markets where trades have diverse information, *Journal of Finance* 31, 573–585.

———, 1981, An introduction to the theory of rational expectation under asymmetric information, *Review of Economic Studies* 48, 541–559.

———, 1988, An analysis of the implications for stock and futures price volatility of program trading and dynamic hedging strategies, *Journal of Business* 61, 275–298.

———, and Robert J. Shiller, 1982, Consumption correlatedness and risk measurement in economies with non-traded assets and heterogeneous information, *Journal of Financial Economics* 10, 195–210.

Grossman, Sanford J., and Joseph E. Stiglitz, 1976, Information and competitive price systems, *American Economic Review* 66, 246–253.

Grundy, Bruce D., and Maureen McNichols, 1989, Trade and the revelation of information through prices and direct disclosure, *Review of Financial Studies* 2, 495–526.

Gul, Faruk, 1991, A theory of disappointment aversion, *Econometrica* 59, 667–686.

Haas, Markus, 2007, Do investors dislike kurtosis?, *Economics Bulletin* 2, 1–9.

Hackbarth, Dirk, Christopher A. Hennessy, and Hayne E. Leland, 2007, Can the trade-off theory explain debt structure?, *Review of Financial Studies* 20, 1389–1428.

Hadar, Josef, and William R. Russell, 1969, Rules for ordering uncertain prospects, *American Economic Review* 59, 25–34.

Hakansson, Nils H., 1970, Optimal investment and consumption strategies under risk for a class of utility functions, *Econometrica* 38, 587–607.

Hansen, Lars Peter, and Ravi Jagannathan, 1991, Implications of security market data for models of dynamic economics, *Journal of Political Economy* 99, 225–262.

Hansen, Lars Peter, and Scott F. Richard, 1987, The role of conditioning information in deducing testable restrictions implied by dynamic asset pricing models, *Econometrica* 55, 587–613.

Hansen, Lars Peter, and Kenneth J. Singleton, 1982, Generalized instrument variables estimation of nonlinear rational expectations models, *Econometrica* 50, 1269–1286.

———, 1983, Stochastic consumption, risk aversion, and the temporal behavior of asset returns, *Journal of Political Economy* 91, 249–265.

———, 1984, Errata: Generalized instrument variables estimation of nonlinear rational expectations models, *Econometrica* 52, 267–268.

Harris, Milton, and Arthur Raviv, 1993, Differences of opinion make a horse race, *Review of Financial Studies* 6, 473–506.

Harrison, J. Michael, 1985, *Brownian Motion and Stochastic Flow Systems* (Wiley: New York).

———, 2013, *Brownian Models of Performance and Control* (Cambridge University Press: New York).

———, and David M. Kreps, 1978, Speculative investor behavior in a stock market with heterogeneous expectations, *Quarterly Journal of Economics* 92, 323–336.

———, 1979, Martingales and arbitrage in multiperiod securities markets, *Journal of Economic Theory* 20, 381–408.

Harrison, J. Michael, and Stanley R. Pliska, 1981, Martingales and stochastic integrals in the theory of continuous trading, *Stochastic Processes and their Applications* 11, 215–260.

Harvey, Campbell R., Yan Liu, and Heqing Zhu, 2014, …and the cross-section of expected returns, Duke University.

Havránek, Tomáš, 2015, Measuring intertemporal substitution: The importance of method choices and selective reporting, *Journal of the European Economic Association*.

Hayashi, Fumio, 1982, Tobin's marginal q and average q: A neoclassical interpretation, *Econometrica* 50, 213–224.

He, Hua, and Neil D. Pearson, 1991a, Consumption and portfolio policies with incomplete markets and short-sale constraints: The finite-dimensional case, *Mathematical Finance* 1, 1–10.

———, 1991b, Consumption and portfolio policies with incomplete markets and short-sale constraints: The infinite-dimensional case, *Journal of Economic Theory* 54, 259–304.

He, Zhiguo, and Konstantin Milbradt, 2014, Endogenous liquidity and defaultable bonds, *Econometrica* 82, 1443–1509.

———, 2015, Dynamic debt maturity, Working Paper.

He, Zhiguo, and Wei Xiong, 2012, Rollover risk and credit risk, *Journal of Finance* 67, 391–429.

Heath, David, Robert Jarrow, and Andrew Morton, 1992, Bond pricing and the term structure of interest rates: A new methodology for contingent claim valuation, *Econometrica* 60, 77–105.

Heaton, J. B., and Deborah J. Lucas, 1996, Evaluating the effects of incomplete markets on risk sharing and asset pricing, *Journal of Political Economy* 104, 443–487.

Hellwig, Martin F., 1980, On the aggregation of information in competitive markets, *Journal of Economic Theory* 22, 477–498.

Herstein, I. N., and John Milnor, 1953, An axiomatic approach to measurable utility, *Econometrica* 21, 291–297.

Heston, Steven L., 1993, A closed-form solution for options with stochastic volatility with applications to bond and currency options, *Review of Financial Studies* 6, 327–343.

———, Mark Lowenstein, and Gregory A. Willard, 2007, Options and bubbles, *Review of Financial Studies* 20, 359–390.

Heston, Steven L., and Saikat Nandi, 2000, A closed-form GARCH option valuation model, *Review of Financial Studies* 13, 585–625.

Hinderer, K., 1970, *Foundations of Non-Stationary Dynamic Programming with Discrete Time Parameter* (Springer: Berlin).

Hindy, Ayman, and Chi-Fu Huang, 1993, Optimal consumption and portfolio rules with duality and local substitution, *Econometrica* 61, 85–121.

Ho, Thomas S. Y., and Sang-Bin Lee, 1986, Term structure movements and pricing interest rate contingent claims, *Journal of Finance* 41, 1011–1029.

Hobson, David G., and L. C. G. Rogers, 1998, Complete models with stochastic volatility, *Mathematical Finance* 7, 27–48.

Holden, Craig W., and Avanidhar Subrahmanyam, 1992, Long-lived private information and imperfect competition, *Journal of Finance* 47, 247–270.

Hong, Harrison, and Jeremy C. Stein, 2003, Differences of opinion, short-sales constraints, and market crashes, *Review of Financial Studies* 16, 487–525.

———, 2007, Disagreement and the stock market, *Journal of Economic Perspectives* 21, 109–128.

Hou, Kewei, Chen Xue, and Lu Zhang, 2015, Digesting anomalies: An investment approach, *Review of Financial Studies* 28, 650–705.

Huang, Jing-Zhi, and Ming Huang, 2012, How much of the corporate-treasury yield spread is due to credit risk?, *Review of Asset Pricing Studies* 2, 153–202.

Huberman, Gur, and Werners Stanzl, 2004, Price manipulation and quasi-arbitrage, *Econometrica* 72, 1247–1275.

———, 2005, Optimal liquidity trading, *Review of Finance* 9, 165–200.

Huddart, Steven, John S. Hughes, and Carolyn B. Levine, 2001, Public disclosure and dissimulation of insider trades, *Econometrica* 69, 665–681.

Hull, John, and Alan White, 1987, The pricing of options on assets with stochastic volatilities, *Journal of Finance* 42, 281–300.

———, 1990, Pricing interest-rate derivative securities, *Review of Financial Studies* 3, 573–592.

Illeditsch, Philipp Karl, 2011, Ambiguous information, portfolio inertia, and excess volatility, *Journal of Finance* 66, 2213–2247.

Jackwerth, Jens Carsten, 2000, Recovering risk aversion from option prices and realized returns, *Review of Financial Studies* 13, 433–451.

———, and Mark Rubinstein, 1996, Recovering probability distributions from option prices, *Journal of Finance* 51, 1611–1631.

Jagannathan, Ravi, and Tongshu Ma, 2003, Risk reduction in large portfolios: Why imposing the wrong constraint helps, *Journal of Finance* 58, 1651–1683.

Jagannathan, Ravi, and Zhenyu Wang, 1996, The conditional CAPM and the cross-section of expected returns, *Journal of Finance* 51, 3–53.

Jamshidian, Farshid, 1989, An exact bond option formula, *Journal of Finance* 44, 205–209.

Jang, Bong-Gyu, Hyeng Keun Koo, Hong Liu, and Mark Lowenstein, 2007, Liquidity premia and transaction costs, *Journal of Finance* 62, 2329–2356.

Jegadeesh, Narasimhan, and Sheridan Titman, 1993, Returns to buying winners and selling losers: Implications for stock market efficiency, *Journal of Finance* 48, 65–91.

Jensen, Michael C., 1969, Risk, the pricing of capital assets, and the evaluation of investment portfolio, *Journal of Business* 42, 167–247.

Jouini, Elyes, and Clotilde Napp, 2006, Heterogeneous beliefs and asset pricing in discrete time: An analysis of pessimism and doubt, *Journal of Economic Dynamics and Control* 30, 1233–1260.

Judd, Kenneth L., 1985, The law of large numbers with a continuum of iid random variables, *Journal of Economic Theory* 35, 19–25.

Julliard, Christian, and Anisha Ghosh, 2012, Can rare events explain the equity premium puzzle?, *Review of Financial Studies* 25, 3037–3076.

Kahneman, Daniel, and Amos Tversky, 1979, Prospect theory: An analysis of decision under risk, *Econometrica* 47, 263–292.

Kallianpur, Gopinath, 1980, *Stochastic Filtering Theory* (Springer-Verlag: New York).

Kandel, Eugene, and Neil D. Pearson, 1995, Different interpretation of public signals and trade in speculative markets, *Journal of Political Economy* 103, 831–872.

Kandel, Shmuel, and Robert F. Stambaugh, 1991, Asset returns and intertemporal preferences, *Journal of Monetary Economics* 27, 39–71.

Karatzas, Ioannis, 1988, On the pricing of American options, *Applied Mathematics and Optimization* 17, 37–60.

———, John P. Lehoczky, and Steven E. Shreve, 1987, Optimal portfolio and consumption decisions for a "small investor" on a finite horizon, *SIAM Journal on Control and Optimization* 25, 1557–1586.

———, and Gan-Lin Xu, 1991, Martingale and duality methods for utility maximization in an incomplete market, *SIAM Journal on Control and Optimization* 29, 702–730.

Karatzas, Ioannis, and Steven E. Shreve, 1984, Connections between optimal stopping and singular stochastic control i. monotone follower problems, *SIAM Journal on Control and Optimization* 22, 856–877.

———, 1998, *Methods of Mathematical Finance* (Springer: New York).

———, 2004, *Brownian Motion and Stochastic Calculus* (Springer: New York) 8th edn.

Keynes, John M., 1921, *A Treatise on Probability* (Macmillan: London).

Kimball, Miles S., 1990, Precautionary saving in the small and in the large, *Econometrica* 58, 53–73.

———, 1993, Standard risk aversion, *Econometrica* 61, 589–611.

Klemperer, Paul, 1999, Auction theory: A guide to the literature, *Journal of Economic Surveys* 13, 227–286.

———, 2004, *Auctions: Theory and Practice* (Princeton University Press: Princeton, NJ).

Klemperer, Paul D., and Margaret A. Meyer, 1989, Supply function equilibria in oligopoly under uncertainty, *Econometrica* 57, 1243–1277.

Knight, Frank H., 1921, *Risk, Uncertainty and Profit* (Houghton Mifflin: Boston).

Kocherlakota, Narayana R., 1996, The equity premium: It's still a puzzle, *Journal of Economic Literature* 34, 42–71.

Kogan, Leonid, 2004, Asset prices and real investment, *Journal of Financial Economics* 73, 411–431.

———, and Dimitris Papanikolaou, 2012, Economic activity of firms and asset prices, *Annual Review of Financial Economics* 4, 361–386.

Kraus, Alan, and Robert H. Litzenberger, 1976, Skewness preference and the valuation of risky assets, *Journal of Finance* 31, 1085–1100.

Kreps, David M., 1981, Arbitrage and equilibrium in economies with infinitely many commodities, *Journal of Mathematical Economics* 8, 15–35.

———, and Evan L. Porteus, 1978, Temporal resolution of uncertainty and dynamic choice theory, *Econometrica* 46, 185–200.

Krishna, Vijay, 2009, *Auction Theory* (Academic Press: Burlington, MA) 2nd edn.

Kyle, Albert S., 1985, Continuous auctions and insider trading, *Econometrica* 53, 1315–1336.

———, 1989, Informed speculation with imperfect competition, *Review of Economic Studies* 56, 317–355.

Lakner, Peter, 1998, Optimal trading strategy for an investor: The case of partial information, *Stochastic Processes and their Applications* 76, 77–97.

Lambert, Richard A., Christian Leuz, and Robert E. Verrecchia, 2011, Information asymmetry, information precision, and the cost of capital, *Review of Finance* 16, 1–29.

Lando, David, 2004, *Credit Risk Modeling: Theory and Applications* (Princeton University Press: Princeton, NJ).

Langetieg, Terence C., 1980, A multivariate model of the term structure, *Journal of Finance* 35, 71–97.

Leahy, John V., 1993, Investment in competitive equilibrium: The optimality of myopic behavior, *Quarterly Journal of Economics* 108, 1105–1133.

Ledoit, Olivier, and Michael Wolf, 2004, Honey, I shrunk the sample covariance matrix, *Journal of Portfolio Management* 4, 110–119.

Leippold, Markus, and Liuren Wu, 2002, Asset pricing under the quadratic class, *Journal of Financial and Quantitative Analysis* 37, 271–295.

Leland, Hayne E., 1994, Corporate debt value, bond covenants, and optimal capital structure, *Journal of Finance* 49, 1213–1252.

———, 1998, Agency costs, risk management, and capital structure, *Journal of Finance* 53, 1213–1243.

———, 2004, Predictions of default probabilities in structural models of debt, *Journal of Investment Management* 2, 5–20.

———, and Klaus Bjerre Toft, 1996, Optimal capital structure, endogenous bankruptcy, and the term structure of credit spreads, *Journal of Finance* 51, 987–1019.

LeRoy, Stephen F., 1996, Stock price volatility, in G. S. Maddala, and C. R. Rao, ed.: *Handbook of Statistics: Vol. 14, Statistical Methods in Finance* (Elsevier: Amsterdam).

———, 2004, Rational exuberance, *Journal of Economic Literature* 42, 783–804.

———, and Richard D. Porter, 1981, The present-value relation: Tests based on implied variance bounds, *Econometrica* 49, 555–574.

LeRoy, Stephen F., and Larry D. Singell, 1987, Knight on risk and uncertainty, *Journal of Political Economy* 95, 394–406.

Leshno, Moshe, Haim Levy, and Yishay Spector, 1997, A comment on Rothschild and Stiglitz's "Increasing risk: I. A definition", *Journal of Economic Theory* 77, 223–228.

Levendorskiĭ, Sergei, 2004, Consistency conditions for affine term structure models, *Stochastic Processes and their Applications* 109, 225–261.

Levental, Shlomo, and Antolii V. Skorohod, 1995, Necessary and sufficient condition for absence of arbitrage with tame portfolios, *Annals of Applied Probability* 5, 906–925.

Lewellen, Jonathan, Stefan Nagel, and Jay Shanken, 2007, A skeptical appraisal of asset pricing tests, Stanford University.

Li, Erica X. N., Dmitry Livdan, and Lu Zhang, 2009, Anomalies, *Review of Financial Studies* 22, 4301–4334.

Lintner, John, 1969, The aggregation of investor's diverse judgments and preferences in purely competitive security markets, *Journal of Financial and Quantitative Analysis* 4, 347–400.

Liptser, Robert S., and Albert N. Shiryaev, 2000, *Statistics of Random Processes*. vol. II (Springer: Berlin) 2nd edn.

Liu, Jun, 2007, Portfolio selection in stochastic environments, *Review of Financial Studies* 20, 1–39.

Livdan, Dmitry, Horacio Sapriza, and Lu Zhang, 2009, Financially constrained stock returns, *Journal of Finance* 64, 1827–1862.

Longstaff, Francis A., and Eduardo S. Schwartz, 1992, Interest rate volatility and the term structure: A two-factor general equilibrium model, *Journal of Finance* 47, 1259–1282.

Lowenstein, Mark, and Gregory A. Willard, 2000, Rational equilibrium asset-pricing bubbles in continuous trading models, *Journal of Economic Theory* 91, 17–58.

Lucas, Robert E., 1978, Asset prices in an exchange economy, *Econometrica* 46, 1429–1445.

Luttmer, Erzo G. J., 1996, Asset pricing in economies with frictions, *Econometrica* 64, 1439–1467.

Machina, Mark J., 1982, A stronger characterization of declining risk aversion, *Econometrica* 50, 1069–1080.

———, 1987, Choice under uncertainty: Problems solved and unsolved, *Journal of Economic Perspectives* 1, 121–154.

———, and John W. Pratt, 1997, Increasing risk: Some direct constructions, *Journal of Risk and Uncertainty* 14, 103–127.

Madan, Dilip B., 1988, Risk measurement in semimartingale models with multiple consumption goods, *Journal of Economic Theory* 44, 398–412.

Maginn, John L., Donald L. Tuttle, Dennis W. McLeavey, and Jerald E. Pinto, 2007, *Managing Investment Portfolios: A Dynamic Approach* (John Wiley & Sons: Hoboken, NJ).

Margrabe, William, 1978, The value of an option to exchange one asset for another, *Journal of Finance* 33, 177–186.

Markowitz, Harry M., 1952, Portfolio selection, *Journal of Finance* 7, 77–91.

———, 1959, *Portfolio Selection: Efficient Diversification* (John Wiley & Sons: New York).
Martin, Ian, 2013a, Consumption-based asset pricing with higher cumulants, *Review of Economic Studies* 80, 745–773.
———, 2013b, The Lucas orchard, *Econometrica* 81, 55–111.
———, 2015, What is the expected return on the market?, Working Paper.
McDonald, Robert, and Daniel Siegel, 1986, The value of waiting to invest, *Quarterly Journal of Economics* 101, 707–728.
McKean, Jr., Henry P., 1965, Appendix: A free boundary problem for the heat equation arising from a problem in mathematical economics, *Industrial Management Review* 6, 32–39.
Mehra, Rajnish, and Edward C. Prescott, 1985, The equity premium: A puzzle, *Journal of Monetary Economics* 15, 145–161.
———, 1988, The equity risk premium: A solution?, *Journal of Monetary Economics* 22, 133–136.
———, 2003, The equity premium in retrospect, in George Constantinides, Milton Harris, and René M. Stulz, ed.: *Handbook of the Economics of Finance: Vol. 1B, Financial Markets and Asset Pricing* (Elsevier: Amsterdam).
Merton, Robert C., 1969, Lifetime portfolio selection under uncertainty: The continuous-time case, *Review of Economics and Statistics* 51, 247–257.
———, 1973a, An intertemporal capital asset pricing model, *Econometrica* 41, 867–887.
———, 1973b, Theory of rational option pricing, *Bell Journal of Economics* 4, 141–183.
———, 1974, On the pricing of corporate debt: The risk structure of interest rates, *Journal of Finance* 29, 449–470.
———, 1987, A simple model of capital market equilibrium with incomplete information, *Journal of Finance* 42, 483–510.
Milgrom, Paul, and Nancy Stokey, 1982, Information, trade and common knowledge, *Journal of Economic Theory* 26, 17–27.
Milgrom, Paul R., 1989, Auction theory, in Truman Fassett Bewley, ed.: *Advances in Economic Theory: Fifth World Congress* (Cambridge University Press: Cambridge, UK).
———, and Robert J. Weber, 1982, A theory of auctions and competitive bidding, *Econometrica* 50, 1089–1122.
Miller, Edward M., 1977, Risk, uncertainty, and divergence of opinion, *Journal of Finance* 32, 1151–1168.
Morellec, Erwan, Boris Nikolov, and Norman Schürhoff, 2012, Corporate governance and capital structure dynamics, *Journal of Finance* 67, 803–848.
Morellec, Erwan, and Norman Schürhoff, 2011, Corporate investment and financing under asymmetric information, *Journal of Financial Economics* 99, 262–288.
Mossin, Jan, 1966, Equilibrium in a capital asset market, *Econometrica* 34, 768–783.
———, 1968, Optimal multiperiod portfolio policies, *Journal of Business* 41, 215–229.
Munk, Claus, 2008, Portfolio and consumption choice with stochastic investment opportunities and habit formation in preferences, *Journal of Economic Dynamics and Control* 32, 3560–3589.
Myerson, Roger B., 1981, Optimal auction design, *Mathematics of Operations Research* 6, 58–73.

Myneni, Ravi, 1992, The pricing of the American option, *Annals of Applied Probability* 2, 1–23.
Nielsen, Lars Tyge, and Maria Vassalou, 2006, The instantaneous capital market line, *Economic Theory* 28, 651–664.
Obizhaeva, Anna, and Jiang Wang, 2013, Optimal trading strategy and supply/demand dynamics, *Journal of Financial Markets* 16, 1–32.
O'Hara, Maureen, 1995, *Market Microstructure Theory* (Blackwell: Cambridge).
Øksendal, Bernt, 2003, *Stochastic Differential Equations: An Introduction with Applications* (Springer: Berlin) 6th edn.
——— , and Agnès Sulem, 2007, *Applied Stochastic Control of Jump Diffusions* (Springer: Berlin).
Owen, Joel, and Ramon Rabinovitch, 1983, On the class of elliptical distributions and their applications to the theory of portfolio choice, *Journal of Finance* 38, 745–752.
Pastor, Lubos, and Pietro Veronesi, 2003, Stock valuation and learning about profitability, *Journal of Finance* 58, 1749–1789.
Pemberton, Malcolm, and Nicholas Rau, 2011, *Mathematics for Economists: An Introductory Textbook* (University of Toronto Press).
Phelps, Edmund S., and Robert A. Pollak, 1986, On second-best national saving and game-equilibrium growth, *Review of Economic Studies* 35, 201–208.
Piazzesi, Monika, 2006, Affine term structure models, in Yacine Aït-Sahalia, and Lars Hansen, ed.: *Handbook of Financial Econometrics* (Elsevier: Amsterdam).
Polkovnichenko, Valery, 2007, Life-cycle portfolio choice with additive habit formation preferences and uninsurable labor income risk, *Review of Financial Studies* 20, 83–124.
Pratt, John W., 1964, Risk aversion in the small and in the large, *Econometrica* 32, 122–136.
——— , 1976, Erratum: Risk aversion in the small and in the large, *Econometrica* 44, 420–420.
——— , and Richard J. Zeckhauser, 1987, Proper risk aversion, *Econometrica* 55, 143–154.
Protter, Philip, 1990, *Stochastic Integration and Differential Equations* (Springer: Berlin).
Quirk, James P., and Rubin Saposnik, 1962, Admissibility and measurable utility functions, *Review of Economic Studies* 29, 140–146.
Rabin, Matthew, 2000, Risk aversion and expected-utility theory: A calibration theorem, *Econometrica* 68, 1281–1292.
Radner, Roy, 1972, Existence of equilibrium of plans, prices, and price expectations in a sequence of markets, *Econometrica* 40, 289–303.
Ramsey, Frank P., 1931, Truth and probability, in R. B. Braithwaite, ed.: *The Foundations of Mathematics and other Logical Essays* (Harcourt, Brace and Company: New York).
Reisman, Haim, 1988, A general approach to the arbitrage pricing theory (APT), *Econometrica* 56, 473–476.
——— , 1992, Reference variables, factor structure, and the approximate multibeta representation, *Journal of Finance* 47, 1303–1314.
Revuz, Daniel, and Marc Yor, 1991, *Continuous Martingales and Brownian Motion* (Springer-Verlag: Berlin).
Rietz, Thomas A., 1988, The equity risk premium: A solution, *Journal of Monetary Economics* 22, 117–131.

Rockafellar, R. Tyrrell, 1970, *Convex Analysis* (Princeton University Press: Princeton, NJ).

Rogers, L. C. G., 1994, Equivalent martingale measures and no-arbitrage, *Stochastics and Stochastics Reports* 51, 41–49.

———, 1997, The potential approach to the term structure of interest rates and foreign exchange rates, *Mathematical Finance* 2, 157–164.

Roll, Richard, 1977, A critique of the asset pricing theory's tests, *Journal of Financial Economics* 4, 129–176.

———, 1984, A simple implicit measure of the effective bid-ask spread in an efficient market, *Journal of Finance* 39, 1127–1139.

Ross, Steve, 2015, The recovery theorem, *Journal of Finance* 70, 615–648.

Ross, Stephen A., 1976a, The arbitrage theory of capital asset pricing, *Journal of Economic Theory* 13, 341–360.

———, 1976b, Options and efficiency, *Quarterly Journal of Economics* 90, 75–89.

———, 1977, The capital asset pricing model (CAPM), short-sale restrictions and related issues, *Journal of Finance* 32, 177–183.

———, 1978a, Mutual fund separation in financial theory—the separating distributions, *Journal of Economic Theory* 17, 254–286.

———, 1978b, A simple approach to the valuation of risky steams, *Journal of Business* 51, 453–475.

———, 1981, Some stronger measure of risk aversion in the small and the large with applications, *Econometrica* 49, 621–638.

Rosu, Ioanid, 2009, A dynamic model of the limit order book, *Review of Financial Studies* 22, 4601–4641.

Rothschild, Michael, and Joseph E. Stiglitz, 1970, Increasing risk: I. A definition, *Journal of Economic Theory* 2, 225–243.

Routledge, Bryan R., and Stanley E. Zin, 2010, Generalized disappointment aversion and asset prices, *Journal of Finance* 65, 1303–1332.

Rubinstein, Ariel, 2003, 'Economics and psychology'? the case of hyperbolic discounting, *International Economic Review* 44, 1207–1216.

———, and Asher Wolinsky, 1990, On the logic of "Agreeing to disagree" type results, *Journal of Economic Theory* 51, 184–193.

Rubinstein, Mark, 1974, An aggregation theorem for securities markets, *Journal of Financial Economics* 1, 225–244.

———, 1976, The valuation of uncertain income streams and the pricing of options, *Bell Journal of Economics* 7, 407–425.

———, 1994, Implied binomial trees, *Journal of Finance* 49, 771–818.

———, 2006, *A History of the Theory of Investments* (John Wiley & Sons: Hoboken, NJ).

Rubinstein, Mark E., 1973, The fundamental theorem of parameter-preference security valuation, *Journal of Financial and Quantitative Analysis* 8, 61–69.

Samuelson, Paul A., 1969, Lifetime portfolio selection by dynamic stochastic programming, *Review of Economics and Statistics* 51, 239–246.

Santos, Tano, and Pietro Veronesi, 2010, Habit formation, the cross section of stock returns and the cash-flow risk puzzle, *Journal of Financial Economics* 98, 385–413.

Savage, Leonard J., 1954, *The Foundations of Statistics* (John Wiley & Sons: New York).

Schachermayer, W., M. Sirbu, and E. Taflin, 2009, In which financial markets do mutual fund theorems hold true?, *Finance & Stochastics* 13, 49–77.

Scheinkman, Jose A., and Wei Xiong, 2003, Overconfidence and speculative bubbles, *Journal of Political Economy* 111, 1183–1219.

Schmeidler, David, 1986, Integral representation without additivity, *Proceedings of the American Mathematical Society* 97, 253–261.

———, 1989, Subjective probability and expected utility without additivity, *Econometrica* 57, 571–587.

Schroder, Mark, 1999, Changes of numeraire for pricing futures, forwards, and options, *Review of Financial Studies* 12, 1143–1163.

———, and Costis Skiadas, 1999, Optimal consumption and portfolio selection with stochastic differential utility, *Journal of Economic Theory* 89, 68–126.

———, 2002, An isomorphism between asset pricing models with and without linear habit formation, *Review of Financial Studies* 15, 1189–1221.

Segal, Uzi, 1990, Two-stage lotteries without the independence axiom, *Econometrica* 58, 349–377.

———, and Avia Spivak, 1990, First order versus second order risk aversion, *Journal of Economic Theory* 51, 111–125.

Shanken, Jay, 1982, The arbitrage pricing theory: Is it testable?, *Journal of Finance* 37, 1129–1140.

———, 1992, The current state of the arbitrage pricing theory, *Journal of Finance* 47, 1569–1574.

Sharpe, William F., 1963, A simplified model for portfolio analysis, *Management Science* 9, 499–510.

———, 1964, Capital asset prices: A theory of market equilibrium under conditions of risk, *Journal of Finance* 19, 425–442.

Shefrin, Hersh, and Meir Statman, 1985, The disposition to sell winners too early and ride losers too long: Theory and evidence, *Journal of Finance* 40, 777–790.

Shiller, Robert J., 1981, Do stock prices move too much to be justified by subsequent changes in dividends?, *American Economic Review* 71, 421–436.

Shiryayev, A. N., 1984, *Probability* (Springer: New York) translated by R. P. Boas.

Shreve, Steven E., 2004, *Stochastic Calculus for Finance II: Continuous-Time Models* (Springer: New York).

Singleton, Kenneth J., 2006, *Empirical Dynamic Asset Pricing: Model Specifiication and Econometric Assessment* (Princeton University Press: Princeton, NJ).

Skiadas, Costis, 1998, Recursive utility and preferences for information, *Economic Theory* 12, 293–312.

Smith, James E., and Robert F. Nau, 1995, Valuing risky projects: Option pricing theory and decision analysis, *Management Science* 41, 795–816.

Spatt, Chester S., and Frederic P. Sterbenz, 1988, Warrant exercise, dividends, and reinvestment policy, *Journal of Finance* 43, 493–506.

Starmer, Chris, 2000, Developments in non-expected utility theory: The hunt for a descriptive theory of choice, *Journal of Economic Literature* 38, 332–382.

Steg, Jan-Henrik, 2012, Irreversible investment in oligopoly, *Finance & Stochastics* 16, 207–24.

Stein, Charles, 1973, Estimation of the mean of a multivariate normal distribution, *Proceedings of the Prague Symposium on Asymptotic Statistics*.

Stokey, Nancy L., 2009, *The Economics of Inaction: Stochastic Control with Fixed Costs* (Princeton University Press: Princeton, NJ).

Stoll, Hans, 1978, The supply of dealer services in securities markets, *Journal of Finance* 33, 1133–1151.

Strassen, V., 1965, The existence of probability measures with given marginals, *Annals of Mathematical Statistics* 36, 423–439.

Strebulaev, Ilya A., 2007, Do tests of capital structure theory mean what they say?, *Journal of Finance* 62, 1747–1787.

———, and Toni M. Whited, 2011, Dynamic models and structural estimation in corporate finance, *Foundations and Trends in Finance* 6, 1–163.

———, 2013, Dynamic corporate finance is useful: A comment on Welch, *Critical Finance Review* 2, 173–191.

Strotz, Robert H., 1956, Myopia and inconsistency in dynamic utility maximization, *Review of Economic Studies* 23, 165–180.

Subrahmanyam, Avanidhar, 1991, Risk aversion, market liquidity, and price efficiency, *Review of Financial Studies* 4, 417–441.

Sun, Yeneng, 2006, The exact law of large numbers via Fubini extension and characterization of insurable risks, *Journal of Economic Theory* 126, 31–69.

Teguia, Alberto, 2015, Asymmetric information and liquidity provision, Working Paper.

Telmer, Chris I., 1993, Asset-pricing puzzles and incomplete markets, *Journal of Finance* 48, 1803–1832.

Timmermann, Allan G., 1993, How learning in financial markets generates excess volatility and predictability in stock prices, *Quarterly Journal of Economics* 108, 1135–1145.

Tirole, Jean, 1982, On the possibility of speculation under rational expectations, *Econometrica* 50, 1163–1182.

Tobin, James J., 1969, A general equilibrium approach to monetary theory, *Journal of Money, Credit, and Banking* 1, 15–29.

Treynor, Jack, 1971, The only game in town, *Financial Analysts Journal* 22, 12–14 (pseud.).

Treynor, Jack L., 1999, Toward a theory of market value of risky assets, in Robert A. Korajczyk, ed.: *Asset Pricing and Portfolio Performance* (Risk Publications: London).

Trigeorgis, Lenos, 1996, *Real Options: Managerial Flexibility and Strategy in Resource Allocation* (MIT Press: Boston, MA).

Tversky, Amos, and Daniel Kahneman, 1992, Advances in prospect theory: Cumulative representation of uncertainty, *Journal of Risk and Uncertainty* 5, 297–323.

van Moerbeke, Pierre, 1976, On optimal stopping and free boundary problems, *Archive for Rational Mechanics and Analysis* 60, 101–148.

Vasicek, Oldrich, 1977, An equilibrium characterization of the term structure, *Journal of Financial Economics* 5, 177–188.

Vayanos, Dimitri, 2001, Strategic trading in a dynamic noisy market, *Journal of Finance* 56, 131–171.

Veblen, Theodore B., 1899, *The Theory of the Leisure Class: An Economic Study of Institutions* (Penguin: New York).

Venter, Johannes Hendrik, and David de Jongh, 2006, Extending the EKOP model to estimate the probability of informed trading, *Studies in Economics and Econometrics* 30, 25–39.

Veronesi, Pietro, 1999, Stock market overreaction to bad news in good times: A rational expectations equilibrium model, *Review of Financial Studies* 12, 975–1007.

———, 2000, How does information quality affect stock returns?, *Journal of Finance* 55, 807–837.

Vickrey, William, 1961, Counterspeculation, auctions, and competitive sealed tenders, *Journal of Finance* 16, 8–37.

von Neumann, John, and Oskar Morgenstern, 1947, *Theory of Games and Economic Behavior* (Princeton University Press: Princeton, NJ) 2nd edn.

Wachter, Jessica, 2002, Portfolio choice and consumption decisions under mean-reverting returns: An exact solution for complete markets, *Journal of Financial and Quantitative Analysis* 37, 63–91.

Wang, Jiang, 1993, A model of intertemporal asset prices under asymmetric information, *Review of Economic Studies* 60, 249–282.

———, 1996, The term structure of interest rates in a pure exchange economy with heterogeneous investors, *Journal of Financial Economics* 41, 75–110.

Weil, Philippe, 1989, The equity premium puzzle and the risk-free rate puzzle, *Journal of Monetary Economics* 24, 401–421.

———, 1990, Nonexpected utility in macroeconomics, *Quarterly Journal of Economics* 105, 29–42.

Welch, Ivo, 2013, A critique of recent quantitative and deep-structure modeling in capital structure research and beyond, *Critical Finance Review* 2, 131–172.

Wilson, Robert, 1969, The theory of syndicates, *Econometrica* 36, 119–132.

———, 1979, Auctions of shares, *Quarterly Journal of Economics* 93, 675–689.

Xia, Yihong, 2001, Learning about predictability: The effects of parameter uncertainty on dynamic asset allocation, *Journal of Finance* 56, 205–246.

Xiong, Jie, 2008, *An Introduction to Stochastic Filtering Theory* (Oxford University Press: Oxford).

Yaari, Menahem E., 1987, The dual theory of choice under risk, *Econometrica* 55, 95–115.

Zapatero, Fernanco, 1998, Effects of financial innovations on market volatility when beliefs are heterogeneous, *Journal of Economic Dynamics and Control* 22, 597–626.

Zhang, Lu, 2005, The value premium, *Journal of Finance* 60, 67–103.

INDEX

abandonment option, 517–18
absence of wealth effects, 12, 39–40
action and inaction regions, 532–33, 533f
adapted process, 377n4
adjustment costs, 514–15
affine models, 461–63
affine sharing rules, 90–91
agency problem, 498–99
aggregate absolute risk aversion, 7
aggregate puzzles, xx–xxi
aggregate risk tolerance, 7
Allais paradox, 652–54, 658f
alphas, 129
ambiguity aversion, 283, 666–72
ambiguity premium, 675
American options, 406, 418–23
arbitrage, 192
arbitrage-free price, 419–20
arbitrage opportunity, 57, 192, 419–20
Arbitrage Pricing Theory, 147–50
arbitrary dividends, 441–43
Arrow-Debreu economy, 85
Arrow-Pratt measures of risk aversion, 20
Arrow security, 53
artificial economy, 577
asset risk premia, 375
assets in place, 526–27, 543–44
asset span, 62–63, 69–70
at-the-money option, 444
auctions, 624–31

autarkic equilibrium, 75, 281
average q, 538–40

backward induction, 210–11, 671
Bayes' rule, 573
beginning-of-period consumption, 47, 93–95
Bellman Equation, 211, 213–14, 224–26, 228–29, 271
bequest, 205, 342
Berk-Green-Naik model, 541–46
Bessel process, 449
beta pricing, 286
betweenness preferences, 657–63
binary option, 409
Black-Scholes-Merton formula, 409–13, 424, 437
bliss level, 86–87
bond yields, 284
book-to-market, 546
bounded dynamic programming, 389
Brownian bridge, 636–37, 648
Brownian motion, 290–92, 320, 536–37, 537f
bubbles, 195–97, 337–38, 564
butterfly spread, 95

call option, 403–6, 420f
calls are better alive than dead, 418
Campbell-Cochrane model, 263–66

Capital Asset Pricing Model, 127–35, 168–69, 234–46
capital market line, 107
capital stock process, 525–26
catching up with the Joneses preferences, 261
Cauchy-Schwartz inequality, 74, 150
cautiousness parameter, 79
certainty equivalent, 8–11, 8f, 39, 268–69
CES aggregator, 286
changing numeraires, 407
changing probabilities, 407–8
Chew-Dekel preferences, 674
Cholesky decomposition, 141
Choquet integral, 672–73
coefficients of risk aversion, 5–7, 87, 92
common consequence effect, 654
compensated Poisson processes, 375–76
competitive equilibria, 83–84, 247–48
complementary goods, 286, 381
completely affine models, 463–64
complete markets, 59–60, 84–86, 192–95, 333–35, 345–46
compound Poisson processes, 376–77
concave function, 5
concordant beliefs, 586
conditional beta, 236–37
conditional expectation, 18
conditional mean, 573
conditional models, 246–47
conditional projections, 587
conditional variance, 573
conditioning, 685–86
conspicuous consumption, 261
constant absolute risk aversion, 11–13, 38–41, 557–58
constant capital market line, 347–48
constant dividend yield, 423, 494
constant elasticity of substitution, 268–69
constant elasticity of variance, 13–14, 661–62
constant mimicking return, 114–15
constant projections, 65–67
constant relative risk aversion, 13–16, 14f, 216–18, 661–62

constant relative risk aversion utility, 169–71
consumption-based asset pricing, 167–71
consumption-based Capital Asset Pricing Model, 234–40, 245–46
consumption growth, 603–5
continuation region, 421
continuous function, 680
continuously compounded rate of return, 305
continuously compounded risk-free rate, 542–43
continuous martingales, 292
continuous-time Kyle model, 632–42
continuum of investors, 584–86
contrarian trader, 640
convertible bonds, 427
corporate claims, 496–97
correlation process, 306, 309
Coskewness-Cokurtosis pricing model, 171–72
cost of capital, 129
covariance matrix, 36–37
covariances, 110–11, 138–39
covariation process, 306
covered calls, 403–4
Cox-Ingersoll-Ross models, 465
cross-sectional anomalies, xviii–xx
curse of dimensionality, 362

date-state price, 194
deadwieght costs of bankruptcy, 502–3
debt value, 502, 506–8
decision tree, 209f
decreasing absolute prudence, 16–17
decreasing absolute risk aversion, 7–8, 34–35
default and bankruptcy costs, 496
delta hedging, 415–18, 420f
derivative security, 401
Diamond-Verrecchia paradox, 578
differentiation, 683–84
digital options, 414
disappointment aversion, 660–61
discount bond, 402
dispersion of beliefs, 565

distribution functions and densities, 681
diversification, 38
diversified portfolio, 147–49
dividend-reinvested asset prices, 319
dividends, 423–24, 493
dividend yield, 234
doubling strategy, 294–95
drift of the value function, 351
dynamic capital structure, 500
dynamic consistency, 670–71
dynamic markets, 558–61
dynamic programming, 208–19
dynamic programming for portfolio choice, 212–13
dynamic programming under uncertainty, 212
dynamic resolution of uncertainty, 194f

elasticity of intertemporal substitution, 50, 565
elliptical distribution, 123, 133
Ellsberg paradox, 655–56
empirical performance of popular models, 150–52
end-of-period wealth, 4
endogenous default, 494–97
envelope condition, 215–16, 359–60
Epstein-Zin-Weil utility, 268–76
equilibria, 577–78
equilibrium asset pricing, 445
equilibrium of plans, prices, and price expectations, 257
equilibrium reflection point, 540
equity premium, 175, 257, 386
equity premium puzzle, xx, 170
equity value, 504, 508
equivalent martingale measure, 75
equivalent measures, 688–89
equivalent probability measures, 687
Euler equation, 202–5, 343, 668
Euler inequalities, 668
European option, 439
ex ante optima, 490–92
excess return, 53–54, 112, 136–37, 141–42
excess volatility puzzle, xx
exchange options, 437–39

exchange rates, 198
exercise boundary, 421f
exercise price, 426–27
exogenous risks, 571–72
expectation operator, 112
expectations, 681–83
expectations hypothesis, 436–37
expected consumption growth, 603–5
expected returns, 546
expected utility, 221, 665
experimental paradoxes, 652–57
exponentially decaying habits, 380–83
ex post optima, 490–92
extended affine model, 450, 478
external habit, 260–66

factor models, 139–40, 460–61
factor risk premium, 136
factor structure, 156
Fama-French-Carhart model, 150–51
filtering method, 635–36
filtering theory, 597–603
filtrations, 688
financial economics, xvii–xviii
finite horizon, 190–92, 221–23
finite maturity debt, 505–6
first-order condition, 29–32, 531
first-order risk aversion, 665–66
first welfare theorem, 87n4, 91–93
foreign exchange, 198
forward contract, 433–34
forward measures, 432–33
forward options, 440–41
forward price, 436, 439
forward rates, 459–60
forwards and futures, 433–34
framing, 656–57
frontier portfolios, 106–7
frontier returns, 111–19
fundamental ODE, 486–87
fundamental partial differential equation, 367–70, 373–74, 414–15, 460–61
fundamental theorem of asset pricing, 75
futures, 433–37, 439–41
futures contracts, 435
futures option, 440–41

gamma, 417
GARCH option pricing models, 449
Gaussian affine models, 463
Gaussian term structure model, 463–64
general factor models, 135–42
generalized disappointment aversion, 661
geometric average, 284–85
geometric Brownian motion, 303–5
Girsanov's theorem, 371–72
global minimum variance portfolio, 102–3
Glosten-Molgrom model, 614–16, 620–24, 625f
Gomes-Kogan-Zhang model, 548
Gordon growth model, 495
Gorman aggregation, 89–90, 95
Gram-Schmidt orthogonalization, 72–74
graphical analysis, 100–101, 101f
Greeks, 415–18
Grossman-Stiglitz paradox, 578–82
growth condition, 477
growth firm, 548
growth-optimal portfolio, 347
growth options, 526–27, 544–46

Hamilton-Jacobi-Bellman equation, 296
Hansen-Jagannathan bounds, 67–70, 116
harmonic mean, 7
heat equation, 414, 426
Heath-Jarrow-Morton model, 475
hedging demands, 214–15, 353–54
Hellwig model, 583–86
herding, 284
Heston model, 445, 449
Hilbert spaces, 72–74
hitting time, 487–88, 492–93, 501–2
Hölder's inequality, 561
Ho-Lee model, 478
home bias, xxi
homogeneous function, 216
homothetic function, 216
Hull-White model, 478
hysteresis, 530

idiosyncratic income risk, 279
idiosyncratic risks, 146
implied volatility, 443

inaction region, 532–33
inactivity, 522
increase in precision, 575
independence, 686–87
independence axiom, 19, 652–54
independent increments, 375–76
indicator function, 685
indivisible investment project, 515–18
industry investment, 538
industry output price, 536
infinite horizon, 223–26
infinite variation, 640
inflation, 198
informed trading strategy, 637–41
innovation process, 593, 595
instantaneous risk-free rate, 435
intermediate rewards, 211
internal habits, 380–87
intertemporal budget constraint, 185, 322–23
Intertemporal Capital Asset Pricing Model, 218, 234–35, 240–46, 357–60
in-the-money options, 444
intrinsic value, 406
irreversible investment, 535–41, 540–41
issuance costs, 503–4
Itô integral, 292–94
Itô processes, 292–94
Itô's formula, 299–303, 305–7
Itô's lemma, 299

Jensen's inequality, 5, 6f, 18
jump risks, 374–80

Kalman filter, 599–600
Knightian uncertainty, 674
knock-out boundaries, 492
Kyle model, 616–20
Kyle's lambda, 618

latent factor, 462
law of iterated expectations, 18
Law of One Price, 57–58, 72, 192
least favorable fictitious completion, 227
Lebesgue measure, 528
leverage, 404

INDEX

Levy's Theorem, 292
linear equilibrium, 579–80
linear risk tolerance, 11–16, 43–46, 86–93, 165–67
linear span, 73
linkage principle, 629–32
liquidity, 614–15
liquidity trader losses, 642
local martingales, 294–95, 689
local volatility, 444–45
logarithmic utility, 14
lognormal consumption growth, 169–71
log utility, 216, 556–57, 559–60
long-run mean, 475
Longstaff-Schwartz model, 478
loss aversion, 656–57
low volatility consumption, 385–86
Lucas economy, 257

margin account, 435
marginal q, 522, 528–30, 538–40
marginal utility, 54–55
margin requirements, 29
Margrabe's formula, 437, 440
marketed payoff, 56
market makers, 617
market portfolio, 130–33
market price-dividend ratio, 606–7
market return, 130–31
marking to market, 435
Markov chain, 601–3
Markov process, 538
martingale, 688–89
martingale property, 189–90
martingale representation theorem, 298–99
matrix notation, 36
maximum likelihood, 669
maximum Sharpe ratio, 107, 116
mean-preserving spread, 659
mean-standard deviation trade-off, 105–6
mean-variance efficiency, 71, 131–32
mean-variance frontier, 101–4, 102f, 106–11, 109f–110f, 132–33
mean-variance preferences, 41–43
Merton's formula, 424, 437

microeconomics, 3
minimum standard deviation stochastic discount factor, 68
minimum variance return, 114
momentum, xviii
momentum traders, 588, 594
money market account, 318
multifactor CIR models, 466–68
multifactor models, 137–38
multiple priors, 668–70
multiple risky assets, 35–38, 40–41
multivariate signals, 575
myopic demand, 353–54

negative convexity, 417
negative exponential utility, 11
no arbitrage assumptions, 406–7
no-borrowing constraint, 186, 223
noise trader, 578–79
nonadditive habit model, 261
nonadditive set functions, 672–73
non-dividend-paying, 423
nonincreasing risk aversion, 21
nonnegative wealth constraint, 346
nonparticipation, 667
normal-normal updating, 573–76
no-trade theorem, 570–72
Novikov's condition, 332
numeraire, 514

observationally equivalent models, 462
operating cash flows, 523, 525
operating leverage, 549
optimal capital stock process, 527–28, 529f
optimal coupon, 499
optimal default boundary, 497
optimal exercise, 526
optimal exercise boundary, 419–20, 516–17
optimal fraction, 40
optimal portfolio, 39–40, 369–70
option bounds, 405–6
option premium, 403
options, 403–5, 439–41
Ornstein-Uhlenbeck process, 605, 609

orthogonality to excess returns, 53–54
orthogonalizing factors, 141
orthogonal projections, 62, 63f, 65f, 111, 206–8, 325–27
out-of-the money options, 405, 419

parallel wealth expansion paths, 43–44
Pareto, Vilfredo, 80
Pareto optimal, 80–83, 163–65, 572
partial differential equation, 296
payoff diagrams, 404f
performance evaluation, 142–44
perpetual calls, 488–89
perpetual debt, 494–97
perpetual options, 486–92
perpetual puts, 489
pessimistic investors, 562–63
Poisson process, 375–76
Ponzi schemes, 186, 195–98
portfolio choice, 594–97, 663
portfolio choice model, 184–87
portfolio puzzles, xxi
portfolio variance, 36–37
power utility, 216
precautionary premium, 51
precautionary savings, 51
predictability, xx–xxi
predictable representation theorem, 298
priced factor, 127
price of risk process, 463–64
price processes, 538
prices, 55–56
pricing kernel, 53
probabilities, 680–81
probability measure, 61n5, 258
probability simplex, 654n2
production economy, 384–87
project cash flows, 543
projecting factors, 141
proper risk aversion, 22
prospect theory, 656–57
protective puts, 403
prudence, 22
purely finitely additive measure, 285
put-call parity, 403–6
put option, 403

q theory, 518–30
quadratic models, 469
quadratic utility, 15–16, 70–72, 206–8
quadratic variation, 308, 377–78

Radon-Nikodym derivative, 448
random horizons, 186–87
random interest rates, 441–43
random variables, 679–80
random vectors, 684–85
rank-dependent preferences, 663–64
rate of mean reversion, 464
rate of return, 27n1, 318, 472
rational expectations equilibria, 569–71, 583
real options, 524–30
rectangularity, 671
recursive utility, 268, 286
redundant assets, 194–95
reflected geometric Brownian motion, 536–37
relative risk aversion, 238–39
representative investor model, 260, 384–87
representative investor pricing, 343–44
representative investors, 163–65, 249–51, 274
return decomposition, 112–14
revelation, 578–80
revenue rankings, 629–32
rho, 417–18
Riccati equation, 467
Riesz representation, 74
risk aversion, 6f, 17–19, 40, 82
risk-free asset, 116
risk-free rate puzzle, xx
risk-free return proxies, 114–15
risk-neutral distribution, 444, 449
risk-neutral expectation, 519
risk-neutral investors, 199, 580–82, 609
risk-neutral market makers, 615, 642n11
risk-neutral martingale, 426
risk-neutral pricing, 53, 256
risk-neutral probabilities, 61–62, 258, 370–71
risk-neutral valuation, 371

INDEX

risk-neutral variance, 175
risk premia, 9, 55, 110–11
risk premium, 8f, 33–34, 580
rotations, 477
Rubinstein option pricing model, 172–74

SAINTS model, 478
satiation level of wealth, 15
scaling, 499
schizophrenia, 583
second-order risk aversion, 10
self-financing wealth processes, 188–89
separating distributions, 119–23
sequential projections, 575–76
share digital, 409–11
sharing rules, 82–83
sharp brackets, 377–80
shifted power utility, 15–16
short rate, 297
short sale constraints, 562–63, 586
short sale over-pricing, 562–63
single-factor models, 136–37
single-period markets, 555–58
single-period stochastic discount factor, 188
single risky asset, 32–35, 38–39
singular process, 528
size and book-to-market sorted portfolios, 152–54
size and momentum sorted portfolios, 154–55
smiles and smirks, 444
smooth pasting condition, 418–23, 489–90, 491f
social planner, 81–82, 556
solvency constraint, 29
speculative trade, 564–65
spot-forward convergence, 435
spot-forward parity, 434–35
spot-futures convergence, 435
spot price, 434–35
square brackets, 377–80
square-root process, 449
standard risk aversion, 22
state-dependent utility formulation, 554–55

state price density, 53
state prices, 53
state transition equation, 210–11
state variables, 212
static approach in complete markets, 205–6
static budget constraint, 344, 360–61
static capital structure, 498–99
static problem, 347–48
statistical factors, 145–47
Stein's lemma, 134
stochastic differential equation, 477
stochastic differential utility, 286
stochastic discount factor, 52–55, 57–60, 65f, 117–19
stochastic discount factor processes, 187–92, 323–30
stochastic discount factor projections, 64–65
stochastic dominance, 22
stochastic integral, 297–98
stochastic part of Ito process, 295
stochastic process, 183, 690
stochastic volatility, 445–48
stopping times, 688
St. Petersburg paradox, 20
strategic traders, 613
strictly concave, 523
strike price, 403
strong-form efficiency, 569, 646
subjective probability, 666
subsistence level of consumption, 15, 227
subspace, 112
substitute goods, 286
sufficient statistic, 212–13, 588
suicide strategy, 361
supermartingale, 350–51
sure thing principle, 655–56
switching model, 605–8
synthetic forward, 434
systematic risk, 145

tangency portfolio, 108
taxes, 494–95
tax payments, 503–4
Taylor series expansion, 171

technical analysis, 588
terminal wealth, 221
term structure of interest rates, 469–70
term structure of volatilities, 443
theta, 415, 417
time-additive utility, 185
time aggregation issue, 338
time aggregator, 268–69
time-independent derivatives, 492
total variation, 291–92
trading volatility, 404
transversality conditions, 195–98
truth plus noise signals, 574–75
two-fund separation, 44
two-fund spanning, 105, 108–10, 114

unconditional expectation, 246
unconditional mean, 643
unconditional variance, 618
unknown drift, 592–93
unspanned endowments, 134–35
unspanned stochastic volatility, 477
usual conditions, 690
utility function, 4

value function, 210, 210f, 216–18, 349, 596
value matching condition, 421
value premium, 549
valuing cash flows, 487–88, 492–93
vasicek model, 464–65
vega, 417–18
verification theorem, 387–90
viability, 390
volatility, 443, 608
volatility of the market return, 607–8
volatility puzzle, 608
volume, 566, 642

warrant, 426–27
weak correlation, 156
wealth adjusted for inflation, 3
weighted utility, 659–60
well diversified portfolio, 120, 149

yield, 402
yield curve, 474–76

zero beta return, 136, 240n1
zero-coupon bond, 184